Soil Mechanics

2nd edition

Also available from Spon Press

An Introduction to Geotechnical Processes
John Woodward

	Pb: 0–415–28646–8
Spon Press	Hb: 0–415–28645–X

Geotechnical Modelling
David Muir Wood

	Pb: 0–419–23730–5
Spon Press	Hb: 0–415–34304–6

Foundations of Engineering Geology
Tony Waltham

	Pb: 0–415–25450–7
Spon Press	Hb: 0–415–25449–3

Applied Analysis in Geotechnics
Fethi Azizi

	Pb: 0–419–25350–5
Spon Press	Hb: 0–419–25340–8

Information and ordering details

For price availability and ordering visit our website **www.sponpress.com**

Alternatively our books are available from all good bookshops.

Soil Mechanics

Concepts and applications
2nd edition

William Powrie

Spon Press
Taylor & Francis Group

LONDON AND NEW YORK

First published 1997 by E & FN Spon
Reprinted 1997
Reprinted 2002 by Spon Press

Second edition published 2004 by Spon Press
2 Park Square, Milton Park, Abingdon, Oxon, OX14 4RN

Simultaneously published in the USA and Canada
by Spon Press
270 Madison Avenue, New York, NY 10016

Reprinted 2005, 2006

Spon Press is an imprint of the Taylor & Francis Group

Typeset in Sabon by
Newgen Imaging Systems (P) Ltd, Chennai, India
Printed and bound in Great Britain by
TJ International Ltd, Padstow, Cornwall

British Library Cataloguing in Publication Data
A catalogue record for this book is available from the British Library

Library of Congress Cataloging in Publication Data
    Soil mechanics: concepts and applications/William Powrie. – 2nd ed.
      p. cm.
    Includes bibliographical references and index.
    ISBN 0–415–31155–1 (hb: alk. paper)
    ISBN 0–415–31156–X (pb: alk. paper)
    1. Soil mechanics. I. Title.

TA710.P683 2004
624.1′5136–dc22   2004002128

ISBN 10: 0–415–31155–1 (hbk)
ISBN 10: 0–415–31156–X (pbk)

ISBN 13: 978–0–415–31155–7 (hbk)
ISBN 13: 978–0–415–31156–4 (pbk)

Contents

Preface

In preparing the second edition of this book, I have taken the opportunity to

- introduce a limited amount of new material, for example, on shallow foundations
- align the text, where appropriate, with the design philosophy of Eurocode 7 by embracing more closely the concepts of limit state design and, in an ultimate limit state calculation, applying the factor of safety to the soil strength (rather than to the load or some other parameter)
- update material where there have been other changes in the interpretation of knowledge or design guidance
- re-order the material in Chapter 5 (the triaxial test and soil behaviour) to enable a completely traditional interpretation, not including Cam clay, to be followed more easily if required, and
- correct a number of errors and ambiguities regrettably present in the first edition.

The underlying philosophy of the book, however, remains the same.

The main changes are as follows:

- In Chapter 1, the discussion on the concept of effective stress has been expanded slightly (section 1.7); and the filter rules given in section 1.9 have been changed to reflect those recommended in the revised Construction Industry Research and Information Association Report C515 on groundwater control (Preene *et al.*, 2000).
- In Chapter 5, the sections on peak strength and undrained shear strengths (including the associated worked examples) have been placed before the sections leading up to Cam clay. As indicated above, this will enable lecturers should they so wish to offer their students a complete traditional interpretation of triaxial test data, including critical, peak and undrained stengths, without going into the concepts of yield and Cam clay. For lecturers wishing to cover Cam clay more thoroughly, a new section has been added that includes the derivation of the Cam clay yield surface.
- In Chapter 7, the long case study of the Cricklewood retaining wall has been split into four worked examples. A significant omission in the assessment of the long-term stability of the wall in overturning has been corrected, and I have revised both this and the short-term calculation to give a treatment of factor of safety that is more consistent with the idea of a factor of safety on soil strength

(strength mobilization factor) promoted by BS8002, Eurocode 7 and the revised Construction Industry Research and Information Association Report C580 on embedded retaining walls (Gaba *et al.*, 2003). I had considered replacing this case study with a number of simpler worked examples, but the opinion of other teachers of geotechnical engineering subjects was that a single problem containing all parts of the calculation was helpful.

- Chapter 8 contains a new section on shallow foundations subject to combined vertical, horizontal and moment loading, based on the work of Roy Butterfield and his colleagues. I hope that both teachers and students of geotechnical engineering will find this elegant approach both interesting and useful. The general thrust of this chapter has also been changed towards applying the factor of safety to the soil strength, rather than to the load as in the first edition. This is in line with the philosophy of Eurocode 7 (BSI, 1995), but as with the first edition this book is intended as a guide to soil mechanics principles and their application to geotechnical engineering, not codes of practice.
- Chapter 10 has also been revised to align it more closely with the now generally accepted idea of applying the factor of safety for an embedded retaining wall mainly to the soil strength, in accordance with BS8002, Eurocode 7 and CIRIA report C580. I have therefore reduced substantially the discussion of the different definitions of factor of safety that used to be associated with retaining walls.

In addition, I have made numerous small but nonetheless important changes throughout the text in the interests of (I hope) improving clarity and giving references to relevant new developments in soil mechanics and geotechnical engineering. I have also removed the distinction between case studies and worked examples (they are all now worked examples), not least because the grey background to the case studies in the first edition did not make the text any easier to read. A free solutions manual is available at www.sponpress.com/supportmaterial.

Educationally, things have moved on since 1996 when the first edition came out. For students working towards chartered engineer status, a 4-year MEng is a much more common route if not the norm. (At the time of writing, it is quite possible that revised professional registration requirements will cause this to change.) There is also, quite rightly, an increased emphasis on a broader syllabus that encourages the development of students' communications, IT, teamworking and other transferable skills. This changing educational context has influenced the course structure suggested in the preface to the first edition: at Southampton, our soil mechanics and geotechnical engineering syllabus is now as roughly as follows:

Year 1:   Chapter 1, in courses on engineering geology and civil and environmental engineering materials.

Year 2:   Chapter 2 up to and including section 2.11; Chapter 3 up to and including section 3.15; Chapter 4 up to and including parts of section 4.7 (section 4.8 is covered in parallel in the mathematics course on differential equations); Chapter 5 up to and including section 5.13, except for the derivation of the Cam clay model and detailed calculations of state paths using it; Chapter 6 up to and including section 6.5; and an introduction to retaining walls,

	foundations and slopes covering sections 7.1 to 7.6, 8.1 to 8.4, and 8.9.
Year 3:	Retaining walls, foundations and slopes in more detail, that is, the parts of Chapters 7 and 8 not covered in the second year and sections 10.1 to 10.6, 10.8 and 10.9.
Year 4 (option):	Modelling and analysis including numerical predictions of state paths using Cam clay, Chapter 9 and selected sections from Chapter 11.

This material is, of course, used and developed in design and individual projects in all years of the degree programme.

I am grateful to a number of people including Emmanuel Detournay, Andrew Drescher, Susan Gourvenec, Bill Hewlett, Adrian Oram and Toby Roberts for bringing to my attention various errors and ambiguities in the first edition; Alan Bloodworth, Malcolm Bolton, Roy Butterfield, Asim Gaba, Richard Harkness, David Richards and Antonis Zervos for their help with some of the more major revisions I have made; and most especially to Joel Smethurst who commented on drafts of much of the new text, checked the new calculations and drew the new figures.

William Powrie
2 May 2004

Preface to the first edition

My original aims in writing this book were:

- To encourage students of soil mechanics to develop an understanding of fundamental concepts, as opposed to the formula-driven approach which seems to be used by many authors.
- To assist the student to build a framework of basic ideas, which would be robust and adaptable enough to support and accommodate the more complex problems and analytical procedures which confront the practising geotechnical engineer.
- To illustrate, with reference to real case histories, that the sensible application of simple ideas and methods can give perfectly acceptable engineering solutions to many classes of geotechnical problem.
- To avoid the unnecessary use of mathematics.
- To cover the soil mechanics and geotechnical engineering topics usually included in typical BEng-level university courses in civil engineering and related subjects, without the additional material which clutters many existing textbooks.

While these aims have probably not been compromised to any significant extent, reality is such that

- Civil engineers must be numerate, and possess a reasonable degree of mathematical ability: they must be able to do sums.
- Different lecturers will have different views on the content of a core syllabus in the soil mechanics/geotechnical engineering subject area.
- Some material may not be suitable for formal presentation in lectures, but is nonetheless essential background reading.

Furthermore, the current trend towards four-year MEng courses for the most academically gifted undergraduates means that some of the material which has traditionally been taught at MSc level will inevitably find its way into MEng syllabuses. The result of all this is that while perhaps 75–80% of the book is indisputably core material at BEng level, there are some sections which are useful background, some sections which might be covered in some courses but not in others, and one or two sections (e.g. section 4.8) which are almost certainly for reference only.

I had originally thought that I would mark the noncore sections in the text, but eventually decided not to do so. This is because of the subjectivity involved in deciding what should be included in a core syllabus, and the ensuing need to distinguish further between non-core sections which were (a) desirable background reading; (b) for possible future use; and (c) purely for reference. Instead, I have tried to ensure that separable topics and sub-topics are covered in separate sections and subsections, with clear and unambiguous titles and subtitles. This should enable a university lecturer to draw up a personalized reading schedule, appropriate to his or her own course.

The book is based on the undergraduate courses in soil mechanics and geotechnical engineering that I helped to develop at King's and Queen Mary and Westfield Colleges in the University of London, over the period 1985–1994. The material in the first eight chapters would probably comprise a core BEng-level syllabus, covering the subject in sufficient detail for those not specializing in geotechnical engineering. Some of the sections – particularly those at the end of Chapters 3 and 5 – would be considered to be background reading or for possible future use, while section 4.8 is included primarily for reference on a 'need to know' basis. The material in Chapters 9–11 might well be included in 3rd year options for intending geotechnical specialists in BEng courses, or taught at MEng / MSc level.

My suggested course structure would be:

Year 1:   Chapters 1 and 2, except for sections 2.12 and 2.13.
Year 2:   Chapter 3 (up to and including section 3.15); Chapter 4 (up to and including section 4.6, plus one of the case studies from section 4.7); Chapter 5 (up to and including section 5.13); Chapter 6 (up to and including section 6.5); Chapter 7 (up to and including section 7.7); and Chapter 8 (omitting sections 8.7, 8.8 and 8.10).
Year 3 (optional): Selected material from Chapters 9–11, plus sections from Chapters 7 and 8 not already covered in Year 2.

Some of the early chapters are structured around the standard laboratory tests that are used to investigate a particular class of soil behaviour. These are Chapter 2 (the shearbox test; friction), Chapter 4 (the oedometer test, one-dimensional compression and consolidation) and Chapter 5 (the triaxial test, more general aspects of soil behaviour). This approach is not new, but is by no means universal. In my experience, the integration of the material covered in lectures with laboratory work, by means of coursework assignments in which laboratory test results are used in an appropriate geotechnical engineering calculation, is entirely beneficial.

The order in which the material is presented is based on the belief that students need time to assimilate new concepts, and that too many new ideas should not be introduced all at once. This may have led to the division of what might be seen as a single topic (for example, soil strength or retaining walls) between two chapters. Where this occurs, the topic is initially addressed at a fairly basic level, with more detailed or advanced coverage reserved for a later stage.

For example, while some authors deal with both the shearbox test and the triaxial test under the same heading (such as 'laboratory testing of soils' or 'soil strength'), I have covered them separately. This is because the shearbox serves as a relatively straightforward introduction to the behaviour of soils at failure in terms of simple stress and volumetric parameters, and to the concept of a critical state. The triaxial test introduces more general stress states, the difference between isotropic compression and shear, the generation of pore water pressures in undrained tests, and the behaviour of clay soils before and after yield. Similarly, I have endeavoured to establish the basic principles of earth pressures and collapse calculations with reference to relatively simple retaining walls in Chapter 7, before addressing soil/wall friction more rigorously in Chapter 9, and more complex earth retaining structures in Chapter 10. In the course structure suggested above, the shearbox – basic soil behaviour – is covered in the first year, and the triaxial test – more advanced soil behaviour – in the second. Basic retaining walls, slopes and foundations are covered in the second year, and more complex situations and methods of analysis in the third.

The book assumes a knowledge of basic engineering mechanics (equilibrium of forces and moments, elastic and plastic material behaviour, Mohr circles of stress and strain etc.). Also, it is written to be followed in sequence. Where necessary, a qualitative description of an aspect of soil behaviour which has not yet been covered is given in order to allow the development of a fuller understanding of another. For example, the generation of excess pore water pressures during shear is mentioned qualitatively in Chapter 2 in order to explain the need for drained shear box tests on clay soils to be carried out slowly. For the experienced reader, it is hoped that the section and subsection headings are sufficiently descriptive to enable the required information to be extracted with the minimum of effort.

More than 50 worked examples and case studies are included within the text, with further questions at the end of each chapter. Some of these were provided by my colleagues Dr R.H. Bassett, Dr R.N. Taylor, Dr N.W.M. John, Dr M.R. Cooper and Professor J.B. Burland, to whom I would like to express my gratitude. So far as I am aware, the other examples, case studies and questions are original, but I apologize for any that I have inadvertently 'borrowed'.

The book has been influenced by those from whom I have learnt about soil mechanics and geotechnical engineering, including (as teachers, colleagues or both) John Atkinson, Malcolm Bolton, David Muir Wood, Toby Roberts, Andrew Schofield, Neil Taylor and Jim White. I am indebted to them, and also to the undergraduate and postgraduate students at King's College London, Queen Mary and Westfield College London, and Southampton University, from whom I have learnt a great deal about teaching soil mechanics and geotechnical engineering. I am grateful to Richard Harkness for his help with parts of Chapters 2 and 3. My special thanks go to Susan Gourvenec, who read through the penultimate draft of the book and suggested many changes which have (I hope) improved its clarity.

William Powrie
25 September 1996

General symbols

Note: simple dimensions (A for cross-sectional area, D or d for depth or diameter, H for height, R for radius etc.) and symbols used as arbitrary constants are not included. Subscripts are not listed where their meaning is clear (e.g. crit for critical, max for maximum, ult for ultimate). Effective stresses and effective stress parameters are denoted in the text by a prime ($'$).

A	Air content of unsaturated soil (section 1.5); Activity (section 1.11); A soil parameter used in the description of creep (section 5.17)
A_c	Projected area of cone in cone penetration test (section 11.3.2)
A_n	Fourier series coefficient (section 4.8)
A_n	Area of shaft of cone penetrometer
B	$\Delta u/\Delta\sigma_c$ in undrained isotropic loading (Chapter 5)
B_q	Pore pressure ratio in cone penetration test (section 11.3.2)
C	Tunnel cover (depth of crown below ground surface) (section 10.11)
C	Parameter used in analysis of shallow foundations (section 8.5)
C_c	Compression index – slope of one dimensional normal compression line on a graph of e against $\log_{10}\sigma_v'$
C_s	Swelling index – slope of one dimensional unload/reload lines on a graph of e against $\log_{10}\sigma_v'$
C_N	Correction factor applied to SPT blowcount (section 11.3.1)
D	Drag force (section 1.8)
D_{10} etc.	Largest particle size in smallest 10% etc. of particles by mass
E	Young's modulus. Subscripts may be used as follows: h (horizontal); v (vertical); u (undrained)
E_0'	One-dimensional stiffness modulus
E^*	Rate of increase of Young's modulus with depth
E	Horizontal side force in slope stability analysis (section 8.10)
EI	Bending stiffness of a retaining wall
E_r	Young's modulus of raft foundation
E_s	Young's modulus of soil
F	Shear force
F	Prop load (propped retaining wall)
F	Factor of safety. A subscript may be used to indicate how the factor of safety is applied: see section 10.2
F_r	Normalized friction ratio in cone penetration test

F_s Factor of safety applied to soil strength

F_F, F_T Factor of safety or load factor against frictional failure (pull-out) and tensile failure respectively, for a reinforced soil retaining wall

G Shear modulus

G^* Modified shear modulus in the presence of shear/volumetric coupling (section 6.10); Rate of increase of shear modulus with depth (section 10.6)

G_s Relative density ($= \rho_s/\rho_w$) of soil grains (also known as the grain specific gravity)

H Rating on mineral hardness scale (Chapter 1)

H Horizontal load or force

H Overall head drop (e.g. across flownet)

H Slope of Hvorslev surface on a graph of q against p'

H Hydraulic head at the radius of influence in a well pumping test

H Limiting lateral load on a pile (section 8.9)

I_D Density index

I_L Liquidity index

I_P Plasticity index ($= w_{LL} - w_{PL}$)

I_ρ, I_σ Influence factor for settlement and stress respectively (Chapter 6)

J Parameter describing effect of shear/volumetric coupling (section 6.10)

K Intrinsic permeability (Chapter 3)

K Earth pressure coefficient, σ'_h/σ'_v. Subscripts may be used as follows: a (to denote active conditions); p (passive conditions); i (prior to excavation in front of a diaphragm-type retaining wall); 0 (*in situ* stress state in the ground); nc (for a normally consolidated clay); oc (for an overconsolidated clay)

K Elastic bulk modulus (subscript u denotes undrained)

K^* Modified bulk modulus in the presence of shear/volumetric coupling (section 6.10)

K_{ac}, K_{pc} Multipliers applied to τ_u in the calculation of active and passive total pressures respectively (undrained shear strength model)

K_T Total stress earth pressure coefficient, σ_h/σ_v (section 10.7)

L_0 Distance of influence of a dewatering system idealized as a pumped well

LF Tunnel load factor (section 10.11)

M Bending moment. Subscripts may be used as follows: des (to denote the design bending moment); le (retaining wall bending moment calculated from a limit equilibrium analysis); p or ult (fully plastic or ultimate bending moment of beam or retaining wall)

M Moment load

M Mobilization factor on soil strength

M'_0 Constrained or one-dimensional modulus

N Normal force, for example, on rupture surface or soil/structure interface

N SPT blowcount

N_k Cone factor relating q_c and τ_u (section 11.3.2)

N_1 SPT blowcount normalized to a vertical effective stress of $100\,\text{kPa}$ (section 11.3.1)

N_{60} Corrected SPT blowcount, for an energy ratio of 60% (section 11.3.1)

N_1, N_2, N_3	Interblock normal forces, mechanism analysis for shallow foundation (section 9.9)
N_F, N_H	Number of flowtubes and potential drops respectively in a flownet
N_c	Basic bearing capacity factor: undrained shear strength analysis
N_p	Value of v at $\ln p' = 0$ on isotropic normal compression line on a graph of v against $\ln p'$ (Chapter 5)
N_q	Basic bearing capacity factor: frictional soil strength analysis
N_γ	Term in bearing capacity equation to account for self-weight effects
OCR	Overconsolidation ratio
P	Prop load; tensile strength of reinforcement strip (reinforced soil retaining wall)
Q	Ram load in triaxial test (Chapter 5)
Q	Equivalent toe force in simplified stress analysis for unpropped retaining wall
Q_t	Normalized cone resistance in cone penetration test
R	Proportional settlement ρ/ρ_{ult}
R	Resultant force, for example, on rupture surface or soil/structure interface
R	Dimensionless flexibility number $R = m\rho$ (section 10.6)
R	Depth of tunnel axis below ground level (section 10.11)
R_r, R_z	Degree of consolidation due to radial and vertical flow alone, respectively (section 4.9)
R_0	Radius of influence of a pumped well
S, S_r	Saturation ratio
S	Slope of graph; drain spacing (section 4.9); sensitivity (section 5.15)
S	Surface settlement due to tunnelling (section 10.11)
T	Total shear resistance of soil/pile interface (section 2.11)
T	Shear force, for example, on rupture surface or soil/structure interface
T	Surface tension at air/water interface
T	Dimensionless time factor $c_v t/d^2$ in consolidation problems
T	Anchor load (anchored retaining wall)
T	Torque
T_C	Tunnel stability number at collapse (section 10.11)
T_{des}	Design anchor load (anchored retaining wall)
T_r	Dimensionless time factor for radial consolidation $c_{hv} t/r_w^2$
T_v	Torque due to shear stress on vertical surfaces in shear vane test (section 11.3.4)
U	Coefficient of Uniformity $= D_{60}/D_{10}$
U	Average excess pore water pressure (consolidation problems)
U	Force, for example, on rupture surface or soil/structure interface due to pore water pressure
U_e	Pore water suction at air entry
U_r, U_z	Average excess pore water pressure if drainage were by radial flow or vertical flow alone, respectively (section 4.9)
V	Volume (total)
V	Electrical potential difference (voltage) (section 11.4.1)
V	Vertical load or force

V_a	Volume of air voids in soil sample
V_s	Volume of soil solids in soil sample
V_t	Total volume occupied by a soil sample
V_{ti}	Total volume of triaxial test sample as prepared (Chapter 5)
V_{to}	Total volume of triaxial test sample at start of shear test (Chapter 5)
V_v	Volume of voids in soil sample
V_w	Volume of water in soil sample
V_L	Volume loss in tunnelling (section 10.11)
V_{tunnel}	Nominal volume of tunnel (section 10.11)
W	Weight of a block of soil
W	Mass of falling weight used in heavy tamping (section 11.4.6)
W_c	Set of collapse loads for a structure (in plastic analysis)
W_t	Total weight of a soil sample
X	Vertical side force in slope stability analysis (section 8.11)
Z	Coefficient of curvature $= (D_{30})^2/(D_{60} \cdot D_{10})$
Z	Reference depth in a Newmark chart analysis (Chapter 6)
a	Acceleration (section 11.2.2)
a	Area ratio A_n/A_c in cone penetration test (section 11.3.2)
av	Subscript indicating the average value of a parameter
b	Parameter defining the intermediate principal stress (section 5.10)
c	Subscript denoting the initial state
crit	Subscript denoting a critical condition
current	Subscript indicating the current value of a parameter
c'	Intersection with τ-axis of extrapolated straight line joining peak strength states on a graph of τ against σ'
c_{hv}, c_v	Consolidation coefficient – vertical compression due to horizontal flow, and vertical compression due to vertical flow, respectively
d	Equivalent particle size (section 1.8.1)
d	Prefix denoting infinitesimally small increment (e.g. of stress, strain or length)
d	Half depth of oedometer test sample (maximum drainage path length)
d_c, d_c^*	Depth factors (bearing capacity analysis)
d_q, d_γ	Depth factors (bearing capacity analysis)
ds	Subscript indicating parameter measured in direct shear
e	Void ratio
f	Subscript used to denote 'final' conditions at the end of a test
f	Frequency
f_c	Sleeve friction (stress) in cone penetration test (section 11.3.2)
f_t	Corrected sleeve friction (stress) in CPT
f_1	Parameter relating undrained shear strength to SPT blowcount (section 11.3.1)
g	Acceleration due to Earth's gravity ($= 9.81 \, \mathrm{m/s^2}$)
g	Constant used to define Hvorslev surface (section 5.13)
h	Total or excess head, height of sample in shearbox test
h	Subscript: horizontal

h_c	Critical height of backfill in analysis of compaction stresses behind a retaining wall (section 10.7)
h_{crit}	Critical hydraulic head drop across an element of soil at fluidization
h_e	Excess head (consolidation analysis)
h_i	Height of triaxial test sample as prepared (Chapter 5)
h_0	Height of triaxial test sample at start of shear test (Chapter 5)
h_0	Initial depth of block of soil in analysis of settlement due to change in water content (section 1.13); initial height of soil sample in shearbox test; drawdown at a line of ejector wells analysed as a pumped slot (section 4.7.3)
h_w	Head in a pumped or equivalent well
i	Hydraulic gradient. Subscripts x, y or z may be used to indicate the direction
i	Parameter quantifying width of settlement trough due to tunnelling (section 10.11)
i	Subscript denoting an initial state (the pre-excavation state in the case of an *in situ* retaining wall)
i_{crit}	Critical hydraulic gradient across an element of soil at fluidization
i_e	Electrical potential (voltage) gradient (section 11.4.1)
k	Permeability used in Darcy's Law. Subscripts may be used as follows: h (horizontal); v (vertical); x, y or z (x-, y- or z-direction), t (transformed section); i or f (at start or end of a permeability test)
k_e	Electro-osmotic permeability (section 11.4.1)
l	Limit of range of a Fourier series (section 4.8)
l	Length of part of a slip surface (sections 8.11 and 9.9)
m	Soil stiffness parameter (section 10.6); Rate of increase of soil Young's modulus with depth (section 11.2)
m	A soil parameter used in the description of creep (section 5.17)
m_a	Mass of air in soil sample
m_s	Mass of soil solids in soil sample
m_t	Mass of tin or container
m_w	Mass of water in soil sample
max	Subscript indicating the maximum value of a parameter
n	Porosity
n	Overconsolidation ratio based on vertical effective stresses
n_p	Overconsolidation ratio based on average effective stresses
n	Number of 'squares' of Newmark chart covered by a loaded area (Chapter 6)
n	Centrifuge model scale factor (Chapter 11)
p, p'	Average principal total and effective stress, respectively: $p = (\sigma_1 + \sigma_2 + \sigma_3)/3$
p'_0	Maximum previous value of p'; value of p' at tip of current yield locus (Chapter 5)
p_u, p'_u	Lateral load capacity (per metre depth) of a pile (section 8.9)
p	Subscript denoting prototype (section 11.2.2)
p	Cavity pressure in pressuremeter test (section 11.3.3)

p_p	Cavity pressure at onset of plastic behaviour in pressuremeter test (section 11.3.3)
p_L	Extrapolated 'limit pressure' in analysis of the plastic phase of the pressuremeter test (section 11.3.3)
p_b	Passive side earth pressure (section 10.6)
p'_e	Equivalent consolidation pressure: value of p' on isotropic normal compression line at current specific volume (Chapter 5)
ps	Subscript indicating parameter measured in plane strain
q	Deviator stress
q	Volumetric flowrate
q	Surface surcharge or line load
q_c	Measured cone resistance (stress) (section 11.3.2)
q_t	Corrected cone resistance
r	Wall roughness angle
r_e	Radius of an equivalent well used to represent an excavation
r_p	Radius of plastic zone in the soil around a pressuremeter (section 11.3.3)
r_u	Pore pressure ratio, $r_u = u/\gamma z$
r_γ	Reduction factor (bearing capacity analysis)
s	Average total stress $(\sigma_1 + \sigma_3)/2$: locates centre of Mohr circle on σ-axis
s_c, s_c^*	Shape factors (bearing capacity analysis)
s_q, s_γ	Shape factors (bearing capacity analysis)
s	$[(1 + \sin\psi)\sin\phi'_{peak}]/(1 + \sin\phi'_{peak})]$ (section 11.3.3)
s'	Average effective stress $(\sigma'_1 + \sigma'_3)/2$: locates centre of Mohr circle on σ'-axis
t	Time
t	Radius of Mohr circle of stress, $t = (\sigma'_1 - \sigma'_3)/2 = (\sigma_1 - \sigma_3)/2$
t_h, t_m	Parameters used in analysis of shallow foundations (section 8.5)
t_r	Thickness of raft foundation
t_x	Reference point on time axis used to determine consolidation coefficient c_v from oedometer test data
u	Pore water pressure
u	Subscript: undrained
u_e	Excess pore water pressure (consolidation analysis)
u_{ed}	Excess pore water pressure at mid-depth of an oedometer test sample of overall depth $2d$ (Chapter 4)
u_2	Pore water pressure measured in cone penetration test
ult	Subscript denoting the ultimate value of a parameter (e.g. settlement)
v	Specific volume
v	Particle settlement velocity (section 1.8); velocity of relative sliding (section 9.9)
v	Subscript: vertical
v, v_D	Darcy seepage velocity. Subscripts x, y or z may be used to indicate the direction
v_0	Reference velocity for mechanism analysis (section 9.9)
v_κ	Intersection of unload/reload line with $\ln p' = 0$ axis
v_{true}	True average fluid seepage velocity

w	Water content
w_{LL}, w_{PL}	Water content at liquid limit and plastic limit, respectively
w	Weight of a soil element (sections 8.10 and 10.11)
x	Relative horizontal movement in shearbox test
y	Upward movement of shearbox lid
y_c	Outward movement of cavity wall in pressuremeter test (section 11.3.3)
y_{rp}	Outward displacement of soil at the plastic radius r_p in a pressuremeter test (section 11.3.3)
z	Depth coordinate
z_c	Critical layer thickness for compaction of soil behind a retaining wall (section 10.7)
z_0	Depth of tunnel axis below ground level (section 10.11)
z_p	Depth of pivot point below formation level (unpropped embedded retaining wall)
Γ	Value of v at $\ln p' = 0$ on critical state line on a graph of v against $\ln p'$ space (Chapter 5)
Δ	Prefix denoting increment (e.g. of stress, strain or length)
Δ	Angle used in Mohr circle constructions for stress analyses (Chapter 9)
Δ	Multipropped wall flexibility parameter (Chapter 10)
M	Slope of critical state line on a graph of q against p'
Ψ	$v - v_c$ (section 5.20)
$\Delta V_{tc}, \Delta V_{tq}$	Volume change of triaxial test sample during consolidation and shear, respectively (Chapter 5)
$\Delta y, \Delta z$	Width and depth, respectively, of reinforced soil retaining wall facing panel
$\Delta p_{u/r,max}$	Maximum reduction in cavity pressure in a pressuremeter test that can be applied without causing plastic behaviour in unloading (section 11.3.3)
α	Transformation factor for flownet in a soil with anisotropic permeability (Chapter 3)
α	A soil parameter used in the description of creep (section 5.17)
α	Angle of inclination of slip surface to the horizontal (section 8.11)
α	Term applied to one of the two characteristic directions, along which the full strength of the soil is mobilized (Chapter 9)
α	Retained height ratio h/H of a retaining wall (section 10.6)
α	Soil/wall adhesion reduction factor
β	Angle between flowline and the normal to an interface with a soil of different permeability
β	Term applied to one of the two characteristic directions, along which the full strength of the soil is mobilized (Chapter 9)
β	Slope angle
β	Parameter quantifying depth to anchor for an anchored retaining wall (section 10.6)
γ	Engineering shear strain
γ	Unit weight ($= \rho g$)
γ_{dry}	Unit weight of soil at same void ratio but zero water content
γ_f	Unit weight of permeant fluid (section 3.3)
γ_{sat}	Unit weight of soil when saturated

γ_w	Unit weight of water
δ	Soil/wall interface friction angle
δ	Strength mobilized along a discontinuity (Chapter 9)
δ	Prefix denoting increment (e.g. of stress, strain or length)
δ	Displacement
δ_{mob}	Mobilized soil/wall interface friction angle
ε	Direct strain. Subscripts may be used to indicate the direction as follows: h (horizontal); v (vertical); r (radial); θ (circumferential)
ε_c	Cavity strain in pressuremeter test (section 11.3.3)
ε_q	Triaxial shear strain $\varepsilon_q = (2/3)(\varepsilon_v - \varepsilon_h)$
ε_{vol}	Volumetric strain
$\varepsilon_1, \varepsilon_3$	Major and minor principal strains, respectively
ζ	Electro-kinetic or zeta potential
η	Stress ratio q/p'
η_f	Dynamic viscosity
θ	Rotation of stress path on a graph of q against p'
θ	Rotation of principal stress directions; included angle in a fan zone (Chapter 9)
κ	Slope of idealized unload/reload lines on a graph of v against $\ln p'$
κ_0	Slope of idealized unload/reload lines on a graph of v against $\ln \sigma'_v$
λ	Slope of critical state line and of one-dimensional and isotropic normal compression lines on a graph of v against $\ln p'$
λ	Load factor in structural design
λ_0	Slope of one-dimensional normal compression line on a graph of v against $\ln \sigma'_v$
μ	Coefficient of friction
ν_r	Poisson's ratio of raft foundation
ν_s	Poisson's ratio of soil
ν	Poisson's ratio
ν_u	Undrained Poisson's ratio
ρ	Mass density. Subscripts may be used as follows: b (for the overall or bulk density of a soil); s (for the density of the soil grains); w (for the density of water $= 1000 \, kg/m^3$ at $4°C$)
ρ	Settlement
ρ	Wall flexibility H^4/EI; a subscript c may be used to denote a critical value
ρ	Cavity radius in pressuremeter test (section 11.3.3); a subscript 0 may be used to denote the initial value
ρ	Parameter used in analysis of shallow foundations (section 8.5)
σ, σ'	Total and effective stress, respectively. Subscripts may be used to indicate the direction as follows: a (axial, in a triaxial test); h (horizontal); h_0 (horizontal, *in situ*); v (vertical); v_0 (vertical, maximum previous); r (radial); θ (circumferential); n (normal)
σ_c	Cell pressure in a triaxial test
σ_f, σ'_f	Normal total and effective stress (respectively) on a shallow foundation at failure

σ_0, σ_0'	Normal total and effective stress (respectively), acting on either side of a shallow foundation at failure
$\sigma_{xx}, \sigma_{xx}'$	Total and effective stresses on the plane whose normal is in the x direction, acting in the x direction
$\sigma_1, \sigma_2, \sigma_3$	Major, intermediate and minor principal total stress, respectively
$\sigma_1', \sigma_2', \sigma_3'$	Major, intermediate and minor principal effective stress, respectively
σ_T	Tunnel support pressure (section 10.11)
σ_{TC}	Tunnel support pressure required just to prevent collapse (section 10.11)
σ_{uc}	Unconfined compressive strength
τ	Shear stress
τ_c	Shear stress at cavity wall in pressure meter test (section 11.3.3)
τ_u	Undrained shear strength
$\tau_{u,design}$	Design value of undrained shear strength
τ_w	Shear strength mobilized on soil/wall interface
τ_{xy}	Shear stress on the plane whose normal is in the x direction, acting in the y direction
ϕ'	Soil strength or angle of shearing resistance (effective angle of friction)
ϕ'_{crit}	Critical state strength
ϕ'_{design}	Design strength
ϕ'_{mob}	Mobilized strength
ϕ'_{peak}	Peak strength
ϕ'_{tgt}	Slope of best-fit straight line joining peak strength states on a graph of τ against σ'
ϕ_w	True friction angle between soil grain and wall materials
ϕ'_μ	True friction angle of soil grain material
χ	Parameter used in the description of unsaturated soil behaviour (section 5.19)
ψ	Angle of dilation
ω	Angular velocity; angle of retaining wall batter (Chapter 9)
0	Subscript denoting an initial state (at $t = 0$), a value at $x = 0$ or $z = 0$, or the initial *in situ* state in the ground

Chapter 1

Origins and classification of soils

1.1 Introduction: what is soil mechanics?

Soil mechanics may be defined as the study of the engineering behaviour of soils, with reference to the design of civil engineering structures made from or in the earth. Examples of these structures include embankments and cuttings, dams, earth retaining walls, tunnels, basements, sub-surface waste repositories, and the foundations of buildings and bridges. An embankment, cutting or retaining wall often represents a major component, if not the whole, of a civil engineering structure, and is usually (for better or for worse) clearly visible in its finished form (Figure 1.1). Tunnels and basements are generally only visible from inside the structure, while foundations and underground waste repositories – once completed – are not usually visible at all. By definition, foundations form only a part of the structure which they support. Although out of sight, the foundation is nonetheless important: if it is deficient in its design or construction, the entire building may be at risk (Figure 1.2).

Problems in soil mechanics had begun to be identified and addressed analytically by the beginning of the eighteenth century (Heyman, 1972). Despite this, the growth of soil mechanics as a core discipline within civil engineering, taught at universities with almost the same emphasis as structures and hydraulics, has taken place largely within the last fifty years or so. The expansion of the subject during this time has been very rapid, and the term **geotechnical engineering** has been introduced to describe the application of soil mechanics principles to the analysis, design and construction of civil engineering structures which are in some way related to the earth.

The development of geotechnical processes and techniques has been led primarily by innovation in construction practice. The terms **ground engineering** and **geotechnology** are often used to describe the study of geotechnical processes and practical issues, including techniques for which the only available methods of assessment are either qualitative or empirical.

If these somewhat arbitrary definitions are accepted, the various terms cover a spectrum from soil mechanics (at the theoretical end), through geotechnical engineering (which is analytical but applied) to ground engineering and geotechnology, where the methods used in design may be largely empirical. This book is concerned primarily with soil mechanics and its application to geotechnical engineering (although section 11.4, on ground improvement techniques, could probably be classed as ground engineering or geotechnology). It describes the mechanical (e.g. strength and stress–strain)

Figure 1.1   A visible and, at the time of its construction, controversial road cutting (the M3 motorway at Twyford Down, near Winchester, Hampshire, England). (Photograph courtesy of Mott MacDonald.)

behaviour of soils in general terms, and shows how this knowledge may be used in the analysis of geotechnical engineering structures.

The book does not (apart from the very brief overview given in section 1.3) cover engineering geology; nor does it examine the mineralogy, physics, chemistry or materials science of soils. The book takes a macroscopic view, and does not address at the microscopic level the issues which constitute what Mitchell (1993) calls the *why* aspect of soil behaviour. This is not to say that these issues are unimportant. A study of engineering geology, and the geological history of an individual site, will give an invaluable understanding of the structure and characteristics of the soil and rock formations present. It might also lead the engineer to anticipate the presence of potentially troublesome features, such as buried river beds which form preferential groundwater flow paths, and historic landslips which give rise to pre-existing planes of weakness in the ground. At least a basic knowledge of soil mineralogy and soil chemistry is essential for anyone involved in the increasingly important issue of the movement of contaminants (e.g. from landfill sites) through the ground.

These subjects are covered in more detail by Blyth and de Freitas (1984: engineering geology); Marshall *et al.* (1996: soil physics); and Mitchell (1993: mineralogy and soil chemistry). Full references to these works are given at the end of this chapter.

Figure 1.2 A well-known building with an inadequate foundation (Pisa, Italy). (Photograph courtesy of Professor J.B. Burland.)

Objectives

After having worked through this chapter, you should have gained an appreciation of:

- the origin, nature and mineralogy of soils (sections 1.2–1.4)
- the influence of depositional and transport mechanisms and soil mineralogy on soil type, structure and behaviour (sections 1.3 and 1.4)
- the principles and objectives of a site investigation (section 1.14).

You should understand:

- the three-phase nature of soil, including the relationships between the phases and how these are quantified (sections 1.5 and 1.6)
- the need to separate the **total stress** σ into the component carried by the soil skeleton as **effective stress** σ' and the **pore water pressure** u, by means of **Terzaghi's equation**, $\sigma = \sigma' + u$ (section 1.7)
- the importance of soil description, and classification with reference to particle size and index tests (sections 1.8, 1.10 and 1.11).

You should be able to:

- manipulate the phase relationships to obtain expressions for the unit weight of the soil (section 1.6)
- determine water content, unit weight, grain specific gravity, saturation ratio, liquid and plastic limits, and optimum water content from laboratory test data (sections 1.5, 1.6, 1.11 and 1.12)
- calculate the vertical total stress at a given depth in a soil deposit and, given the pore water pressure, the vertical effective stress (section 1.7)
- construct a particle size distribution curve from sieve and sedimentation test data (section 1.8)
- design a granular filter (section 1.9)
- apply the phase relationships to the practical situations of compaction of fill and the settlement of houses founded on clay soils (sections 1.12 and 1.13).

1.2 The structure of the Earth

Robinson (1977) points out that the highest mountain (Everest) has a height of 8.7 km above mean sea level, while the deepest known part of the ocean (the Mariana Trench, off the island of Guam in the Pacific) has a depth of 11.3 km. This gives a total range of 20 km, or about 0.3% of the radius of the Earth (which is approximately 6440 km). If a cross-section through the Earth were represented by a circle 10 cm in diameter, drawn using a reasonably sharp pencil, the variation in the position of the Earth's surface would be contained within the thickness of the pencil line. The depths of soil with which civil engineers are concerned – usually only a few tens of metres – are even smaller in comparison with the radius of the Earth. Even the deepest mines have not penetrated more than 6 km or so below the surface of the Earth.

Although the civil engineer is concerned primarily with the behaviour of the soils and rocks within 50 m or so of the surface of the Earth, an appreciation of the overall structure of the planet provides a useful starting point. In the descriptions which follow, it must be borne in mind that the theories concerning the nature and composition of the Earth beyond a depth of a few kilometres are based mainly on geophysical tests and the interpretation of geological evidence. They cannot be verified by direct visual observation, or even by the recovery and testing of material, and therefore remain, at least to some extent, conjectural.

The Earth consists of a number of roughly concentric zones of differing composition and thickness. It has been possible to identify the three main zones – the **crust**, the **mantle** and the **core** – because of the changes in the resistance to the passage of seismic (earthquake) waves which occur at the interfaces. The interface between the crust and the mantle is known as the **Mohorovicic discontinuity** (sometimes abbreviated to **Moho**), while the interface between the mantle and the core is known as the **Gutenburg discontinuity**. In both cases, the interfaces or discontinuities are named after their discoverers. The crust is approximately 32–48 km thick, and the mantle 2850 km. The Gutenburg discontinuity, which defines the interface between the mantle and the core, is therefore about 2890 km below the Earth's surface.

The crust and the core may each be subdivided into inner and outer layers. The outer crust is composed primarily of **crystalline granitic rock**, with a comparatively thin and discontinuous covering of **sedimentary rocks**[1] (e.g. sandstone, limestone and shale). The rocks forming the outer crust are composed primarily of silica and aluminium, and have a relative density (or specific gravity) generally in the range 2.0–2.7 (i.e. they are 2.0 to 2.7 times denser than water). The outer crust is known as **sial**, from 'si' for silica and 'al' for aluminium. Below the outer crust, there is a layer of denser basaltic rocks, which have a specific gravity of about 2.7–3.0. These denser rocks are composed primarily of silica and magnesium, and the lower crust is known as **sima** ('si' for silica and 'ma' magnesium). The inner crust or sima is continuous, while the outer crust or sial is discontinuous, and appears to be confined to the continental land masses: it is not generally present under the sea. For this reason, the denser sima is known as **oceanic crust**, while the overlying sial is known as **continental crust.**

The mantle consists mainly of the mineral **olivine**, a dense silicate of iron and magnesium, possibly in a fairly fluid or plastic state. The specific gravity of the mantle increases from about 3 at the Mohorovicic discontinuity (approximately 40 km deep) to about 5 at the Gutenburg discontinuity (approximately 2890 km deep).

The core is composed largely of an alloy of nickel and iron, which is sometimes given the acronym **nife** (from 'ni' for nickel and 'fe' for iron). As the core does not transmit the transverse or S-waves which arise from earthquakes, at least part of it must be in liquid form. There is some evidence that the outer core may be liquid, while the inner core (with a radius of 1440 km or so) is solid. The temperature of the core is estimated to be in excess of 2700°C. The specific gravity of the core material varies from 5 to 13 or more.

According to the theory of plate tectonics, the crust is divided into a number of large slabs or plates, which float on the mantle and move relative to each other as a result of convection currents within the mantle. Although individual plates are fairly stable, relative movements at the plate boundaries are responsible for many geological processes. Sideways movements create tear faults and are responsible for earthquakes: an example of this is the San Andreas fault in California. Where the plates tend to move away from each other or diverge, new oceanic crust is formed by the emergence of molten material from the mantle through volcanoes. Where the plates tend to collide or converge, the oceanic plate (sima) is forced down into the mantle where it tends to melt. The continental plate (sial) rides over the oceanic plate, and is crumpled and thickened to form a mountain chain (e.g. the Andes in South America). The geological process of mountain building is known as **orogenesis**.

1.3 The origin of soils

Soil is the term given to the unbonded, granular material which covers much of the surface of the Earth that is not under water. It is worth mentioning here that civil and geotechnical engineers are not usually interested in the properties of the top metre or so of soil – known as topsoil – in which plants grow, but in the underlying layers or strata of rather older geological deposits. The topsoil is not generally suitable for use as an engineering material, as it is too variable in character, too near the surface, too loose and compressible, has too high an organic content and is too susceptible to the effects of plants and animals and to seasonal changes in groundwater level.

Soil consists primarily of solid particles, which may range in size from less than a micron to several millimetres. Because many aspects of the engineering behaviour of

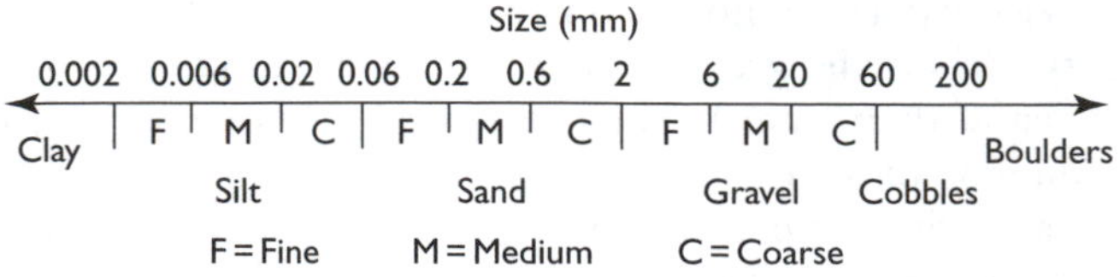

Size (mm)

0.002 0.006 0.02 0.06 0.2   0.6   2   6   20   60   200

Clay | F | M | C | F | M | C | F | M | C | | Boulders

Silt   Sand   Gravel   Cobbles

F = Fine   M = Medium   C = Coarse

Figure 1.3 Classification of soils according to particle size.

soils depend primarily on the typical particle size, civil engineers use this criterion to classify soils as clays, silts, sands or gravels. The system of soil classification according to particle size used in the UK is shown in Figure 1.3. There are other systems in use around the world – particularly in the USA – which differ slightly in detail, but the principle is the same (e.g. Winterkorn and Fang, 1991).

Most soils result from the breakdown of the rocks which form the crust of the Earth, by means of the natural processes of weathering due to the action of the sun, rain, water, snow, ice and frost, and to chemical and biological activity. The rock may be simply broken down into particles. It may also undergo chemical changes which alter its chemical composition or mineralogy. If the soil retains the characteristics of the parent rock and remains at its place of origin, it is known as a **residual soil**. More usually, the weathered particles will be transported by the wind, a river or a glacier to be deposited at some new location. During the transport process, the particles will probably be worn and broken down further, and sorted by size to some extent.

Many soil deposits may be up to 65 million years old. Geotechnical engineers frequently encounter sedimentary rocks, such as chalk, limestone and sandstone, which may be hundreds of millions of years old. The Earth itself is thought to be 4–5000 million years old, and anything which occurred after the end of the last glacial period of the Ice Age (10 000 years ago) is described by geologists as **Recent**. Soils and rocks are classified by geologists according to their age, with reference to a geological timescale divided into four **eras**. The eras are named according to the life-forms which existed at the time. They are:

- **Archaeozoic** (before any form of life, as evidenced by observable fossil remains): more usually known as Pre-Cambrian. This period covers perhaps 3900 million years, from the creation of the Earth up to about 570 million years before the present.
- **Palaeozoic** (ancient forms of life, also known as Primary): 225–570 million years ago.
- **Mesozoic** (intermediate forms of life, also known as Secondary): 65–225 million years ago.
- **Cainozoic** (recent forms of life): commonly but probably artificially subdivided into the Quaternary (0–2 million years ago) and the Tertiary (2–65 million years ago).

The four eras are subdivided into **periods** on the basis of the animal and plant fossils present. The periods are in turn subdivided into rock **series**. During a given **period** of time within an **era**, a **series** of rocks (e.g. shales, sandstones, limestones), containing certain types of fossil, was deposited.

The periods are named in different ways, which may describe the types of rock laid down (cretaceous for chalk, carboniferous for coal); the nature of the fossil content (e.g. holocene, meaning recent); the names of the places where the rocks were first recognized (e.g. Devonian for Devon, Cambrian for Wales); tribal names (Silurian from the Silures and Ordovician from the Ordovices, both ancient Celtic tribes in Wales); or the number of series within the period (e.g. Triassic for three). The names of the eras and periods, together with an indication of the major geological activities, rocks and forms of life, are given in Table 1.1.

In view of the age of most soil deposits, the environment in which a particular soil deposit was laid down is unlikely to be the same as the environment at the same place today. Nonetheless, the transport process and the depositional environment of a particular stratum or layer of soil have a significant influence on its structure and fabric, and probably on its engineering behaviour. They are therefore worthy of some comment.

1.3.1 Transport processes and depositional environments

Water

Small particles settle through water very slowly. They therefore tend to remain in suspension, enabling them to be transported much further by rivers than larger particles. The largest particles are carried – if at all – by being washed along the bed of a river, rather than in suspension. Pebbles, gravels and coarse sands tend to be deposited on the bed of the river along most of its course. As the river changes its course due to the downstream migration of meanders (bends), or erodes a deeper channel in a process known as rejuvenation (following, e.g. a fall in sea level), the coarse material is left behind to form a terrace. Silts and fine particles may also be deposited on either side of the river following a flood, because the floodwater is comparatively still. A soil deposited along the flood plain of a river is known as **alluvium**, or an alluvial deposit.

A river tends to flow more rapidly in its upper reaches than in its lower course. For example, the Amazon has a gradient of about 1 in 70 000 in its lowest reaches, compared with gradients as high as 1 in 100 in many of the upper streams (Robinson, 1977). This means that particles which were carried in suspension in the upper reaches of a river begin to be deposited downstream as the flow velocity falls. At the mouth of the river, sediment builds up on the river bed, and constant dredging is usually required if shipping channels are to remain navigable. Sediment is also carried into the sea and deposited: if it is not removed by the tide, a build-up of sediment known as a **delta** is formed, gradually extending seaward from the coast.

The structure of a typical deltaic deposit is illustrated in Figure 1.4. The **bottomset beds** are made up of the finer particles, which have been carried furthest in suspension beyond the delta slope before settling out. The **foreset beds** are made up of coarser material, which has carried along the river bed before coming to rest on the advancing face of the delta. The **topset beds** are deposited on top of the foreset beds, in much the same way as the alluvial deposits further upstream. Deltaic deposits generally comprise clays and silts, with some sands and organic matter.

Table 1.1 Simplified geological classification of soils and rocks in terms of eras and periods of time (from Robinson, 1977)

Era	Period	Years ago (in millions) (up to)	Geological activity	Chief rocks (with examples from British Isles)	Forms of life
Quaternary	Holocene or Recent	0.01	Period of Great Ice Age	Alluvium (Fenlands) Sands (Culbin) East Anglian boulder clay	*Homo sapiens* emerged about 40 000 years ago Species of man and present day animals appear
	Pleistocene	1.5–2			
Cainozoic or Tertiary	Pliocene	7		Sands and gravels (East Anglia)	
	Miocene	26	Alpine orogeny: Alps, Dinarics, Himalayas, etc.		Mammals dominant creatures Flowering plants Emergence of earliest primates
	Oligocene	38			
	Eocene	54		Clays and sands (Hampshire Basin) Clays, sands, and gravels (London Basin)	
	Palaeocene	65		Thanet sands, Reading beds	
Mesozoic or Secondary	Cretaceous	136	Development of geosynclinal sea of Tethys, which covered southern Europe, the Mediterranean and north Africa, and possibly extended into south-east Asia	Chalk of Downs and Wealden clays and sands	Mammals appear; modern type plants appear
	Jurassic	195		Limestone escarpments of eastern England; sandstones and shales	Great age of reptiles; first primitive birds Reptiles abundant
	Triassic	225		New Red Sandstone and marls of Midlands	
	Permian	280		Magnesian limestone	
Palaeozoic or Primary	Carboniferous	345	Hercynian orogeny; building of mountains now represented by blocks of central Europe	Coal Measures, Millstone Grit and Carboniferous Limestone (Pennines and Mendips)	Land plant life develops
	Devonian	395	Caledonian orogeny; building of ranges now represented by stumps of mountains found in Scotland and Scandinavia	Old Red Sandstone of Cheviots and Exmoor	Amphibians
	Silurian	440		Sandstones, shales, and limestones	First primitive fishes
	Ordovician	500			
	Cambrian	570		Sandstones and slates (central Wales) Slates, shales, and grits (North Wales)	Spineless creatures Marine life clearly distinguishable; invertebrates
Proterozoic Eozoic Azoic	Pre-Cambrian	4500	Three, and perhaps more, phases of mountain building occurred in this remote period which stretched over some 3000 million years	Sandstones in extreme north-west of Scotland	

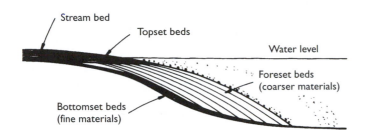

Figure 1.4 Structure of a deltaic deposit (after Robinson, 1977).

Wind

Approximately one-third of the Earth's land surface is classed as arid or semi-arid. Although it is likely that the original weathering processes took place when the climate was more humid than it is now, the primary transport process for soils in desert regions is the wind. Sand dunes gradually migrate in the direction of the wind. Fine particles may be carried for hundreds of kilometres as wind-borne dust. Dust may eventually arrive at a more humid area where it is washed out of the atmosphere by rain. It then settles and accumulates as a non-stratified, lightly cemented material known as **loess**. The cementing is due to the presence of calcium carbonate deposits, from decayed vegetable matter. If the soil becomes saturated with water, the light cementitious bonds are destroyed, and the structure of loess collapses. Extensive deposits of loess are found in north-western China. A soil which has been laid down by the wind is known as an **aeolian deposit**.

Ice

Ice sheets and valley glaciers are particularly efficient at both eroding rock and transporting the resulting debris. Material may be carried along on top of, within, and underneath an ice sheet or glacier as it advances. The effectiveness of ice as a mechanism of transportation does not (unlike water and wind) depend on particle size. It follows that deposits which have been laid down directly by ice action (known as **moraines**) are generally not sorted, and so encompass a large range of particle size. A mound deposited at the end of a glacier is termed a **terminal moraine**, while the sheet deposit below the glacier is known as a **ground moraine** (Figure 1.5). Unsorted glacial moraine is known as **glacial till** or **boulder clay**. The particles found in glacial tills are generally fairly angular, in contrast to the more rounded particles associated with typical water-borne deposits.

Ice and water

Material from on top of or within a melting glacier or ice sheet might be carried away by the meltwater before finally coming to rest. This would result in a degree of sorting according to particle size, with the finer materials being carried further from the end of the glacier. Soils which have been transported, sorted and deposited in this way are described as **fluvio-glacial** materials. The outwash from an ice sheet can cover

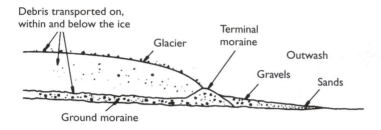

Figure 1.5 Depositional mechanisms associated with glaciers and ice sheets.

a considerable area, forming an extensive **out-wash plain** of fluvio-glacial material (Figure 1.5).

In some cases, the till may be carried by the meltwater into a lake formed by water trapped near the end of the retreating glacier or ice sheet. The larger particles then settle relatively quickly, forming a well-defined layer on the bottom of the lake. The smaller particles settle more slowly, but eventually form an overlying layer of finer material. With the next influx of meltwater, the process is repeated. Eventually, a soil deposit builds up which consists of alternating layers of fine and coarse material, each perhaps only a few millimetres thick (Figure 3.15). This layered or varved structure can have a significant effect on the engineering behaviour of the soil, as discussed in section 3.6.

Material transported by ice, and deposited either directly or sorted and re-laid by outwash streams, is known as **drift**. The principal depositional mechanisms associated with glaciers and ice sheets are summarized in Figure 1.5.

In this section we have discussed the breakdown of rocks into soils. We should note in passing that this is only one-half of the geological cycle. As soils become buried by the deposition of further material on top, they can be converted back into rocks (sedimentary or metamorphic) by the application of increased pressure, and perhaps chemical changes. They might also be converted into igneous rocks, by means of tectonic activity. However, this book is concerned with soils rather than rocks, and a discussion of the formation of rocks is beyond its scope.

1.4   Soil mineralogy

1.4.1   Composition of soils

Soils are composed of **minerals**,[2] which are in turn made up from the elements present in the crust of the Earth. These elements are primarily oxygen (approximately 46.6% by mass), silicon (27.7%), aluminium (8.1%), iron (5.0%), calcium (3.6%), sodium (2.8%), potassium (2.6%) and magnesium (2.1%) (Robinson, 1977; Blyth and de Freitas, 1984). Many of the other elements (such as gold, silver, tin and copper) are rare in a global sense, but are found in concentrated deposits from which they can be extracted economically. The most common elements occur in rocks as oxides, 75% of which are oxides of silicon and aluminium.

Most soils are silicates, which are minerals comprising predominantly silicon and oxygen. The basic unit of a silicate is a group comprising one silicon ion surrounded by four oxygen ions at the corners of a regular tetrahedron: $(SiO_4)^{4-}$. The superscript 4− indicates that the silica tetrahedron has a net negative charge equivalent to four electrons, or valency −4. This is because the silicon ion is Si^{4+}, while the oxygen ion is O^{2-}. In order to become neutrally charged, the silica tetrahedron would need to combine with, for example, two metal ions of valency +2, such as magnesium Mg^{2+}, to give Mg_2SiO_4 (olivine).

The $(SiO_4)^{4-}$ groups may link together in different ways with metal ions and with each other, to form different crystal structures. Although there are many silicate minerals, their properties (such as hardness and stability) depend primarily on their structure.

The $(SiO_4)^{4-}$ tetrahedra may be independent – joined entirely with metal ions, rather than to each other – as in the olivine group of minerals. Alternatively, they may be joined at the corners to form pairs (amermanite: each silica tetrahedron shares one oxygen ion), single chains (pyroxenes: each tetrahedron shares two oxygen ions), double chains or bands (amphiboles: two or three oxygen ions shared, depending on the position of the tetrahedron in the band) or rings (e.g. beryl: two oxygen ions shared). Some of the silicon ions (Si^{4+}) may be replaced by aluminium ions (Al^{3+}), as in augite and hornblende. In this case, the additional negative charge (which arises because of the different valencies of aluminium and silicon) can be balanced by the incorporation of metal ions such as sodium Na^+ and potassium K^+.

Sheet silicates (also known as phyllosilicates or layer-lattice minerals), such as mica, chlorite and the clay minerals, are formed when three of the four oxygen ions are shared with other tetrahedra. Sheet silicates are generally soft and flaky.

The strongest silicate minerals are those in which all four oxygen ions of each $(SiO_4)^{4-}$ tetrahedron are shared with other tetrahedra, resulting in a three-dimensional framework structure.

The arrangements of the silica tetrahedra found in the various silicate minerals are shown diagrammatically in Figure 1.6.

1.4.2 The clay minerals

The clay minerals represent an important sub-group of the sheet silicates or phyllosilicates. In the context of soil mineralogy, the term **clay** is used to denote particular mineralogical properties, in addition to a small particle size. These include a net negative electrical charge, plasticity when mixed with water and a high resistance to weathering. A further distinction between clay and non-clay minerals is that particles of non-clay minerals are generally bulky or rotund, while clay mineral particles are usually flat or platey.

Essentially, the clay minerals can be considered to be made up of basic units or layers comprising two or three alternating sheets of silica, and either brucite [$Mg_3(OH)_6$] or gibbsite [$Al_2(OH)_6$]. Generally, the bonding between the sheets of silica and gibbsite or brucite within each layer is strong, but the bonding between layers may be weak. (Note that the terms **sheet** and **layer** are used quite distinctly: a **layer** of the mineral is made up of two or three **sheets** of silica and gibbsite/brucite.) The most common clay mineral groups are kaolinite, montmorillonite or smectite, and illite. Some clay

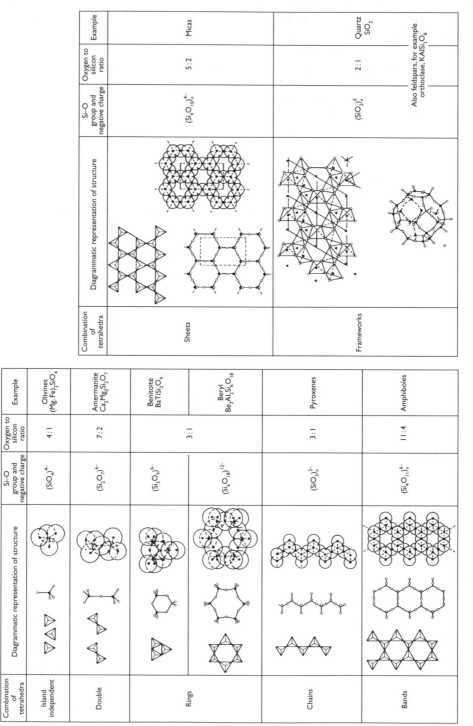

Figure 1.6 Chemistry and structure of silicate minerals. (Redrawn from J.E. Gillott, Clay in Engineering Geology, 1968, pp. 96–97, with kind permission from Elsevier Science – NL, Sara Burgerhartstraat 25, 1055 KV Amsterdam, The Netherlands.)

minerals contain loosely bonded metal ions (cations), which can easily be exchanged for other species (e.g. sodium is readily displaced by calcium), depending on local ion concentrations (e.g. in the pore water). This process is known as **base exchange**.

Kaolinite

Kaolinite has a two-sheet structure, comprising silica and gibbsite. It is the principal component of china clay and results from the destruction of alkali feldspars (section 1.4.3) under acidic conditions. It has few or no exchangeable cations, and the interlayer bonds are reasonably strong. For these reasons, kaolin might be described as the least clay-like of the clay minerals, and it tends to form particles which – for a clay – are relatively large. Particles of well-crystallized kaolin appear as hexagonal plates, with lateral dimensions in the range 0.1–4 μm, and thicknesses of 0.05 to 2 μm. Poorly crystallized kaolinite tends to form platey particles which are smaller and less distinctly hexagonal.

Montmorillonite

The montmorillonite or smectite group of clay minerals have a three-sheet structure comprising a sheet of gibbsite sandwiched between two silica sheets. Montmorillonites (smectites) have a similar basic structure to the non-clay mineral group known as pyrophyllites. The difference is that in smectites there is extensive substitution of silicon (by aluminium) in the silica sheets, and of aluminium (by magnesium, iron, zinc, nickel, lithium and other cations) in the gibbsite sheet. The additional negative charges which result from these substitutions are balanced by exchangeable cations, such as sodium and calcium, located between the layers and on the surfaces of the particles. The interlayer bonds are weak, and layers are easily separated by cleavage or by the adsorption of water. Thus smectite particles are very small (often only one layer or 1 nm thick), and can swell significantly by the adsorption of water. Soils which contain montmorillonites (smectites) exhibit a considerable potential for volume change: because of this characteristic, they are sometimes known as **expansive soils**.

Bentonite is a particular type of montmorillonite which is used extensively in geotechnical engineering. A suspension of 5% bentonite (by mass) in water will form a viscous mud, which is used to support the sides of boreholes and trench excavations, which are later filled with concrete to create deep foundations (known as **piles**: Chapter 8) and certain types of soil retaining wall. It has many other uses, including the sealing of boreholes and the construction of barriers to groundwater flow, known as **cut-off walls** (section 3.3, Example 3.1).

Montmorillonite particles are generally 1–2 μm in length. Particle thicknesses occur in multiples of 1 nm – the thickness of a single silica/gibbsite/silica layer – from 1 nm up to about 1/100 of the particle length.

Illite

Illite also has a three-sheet structure, comprising a sheet of gibbsite sandwiched between two silica sheets. In illite, the layers are separated by potassium ions, whereas in montmorillonite the layers are separated by cations in water. Illites have the same basic structure as the non-clay minerals muscovite mica and pyrophyllite. Muscovite differs from pyrophyllite in that 25% of the silicon positions are taken by aluminium, and the resulting excess negative charges are balanced by potassium ions between the layers. Illite differs from muscovite in that fewer of the Si^{4+} positions are taken by Al^{3+}, so that there is less potassium between the layers. Also, the layers are more randomly stacked, and illite particles are smaller than mica particles. Illite may contain magnesium and iron as well as aluminium in the gibbsite sheet. Iron-rich illite, which has a distinctive green hue, is known as glauconite.

Illites usually occur as small, flaky particles mixed with other clay and non-clay minerals. Illite particles range generally from 0.1 μm to a few micrometre in length, and may be as small as 3 nm thick. Unlike kaolinite and montmorillonite, their occurrence in high-purity deposits is unknown.

Other clay minerals

There are two other groups of clay minerals: vermiculites, which have a similar tendency to swell as montmorillonites; and palygorskites, which are not common, and have a chain (rather than a sheet) structure.

1.4.3 Non-clay minerals

The most abundant non-clay mineral in soils generally is quartz (SiO_2). Quartz is a framework silicate, in which the silica tetrahedra are grouped to form spirals. Small amounts of feldspar and mica are sometimes present, but pyroxenes and amphiboles (single and double chain silicates) are rare. This is very different from the typical composition of igneous rock, the parent material from which many soils were broken down, which might be 60% feldspars, 17% pyroxenes and amphiboles, 12% quartz and 4% micas (Mitchell, 1993).

Quartz is quite hard (rated $H = 7$ on an arbitrary 10-point scale where diamond, the hardest, has $H = 10$ and talc, the softest, has $H = 1$) and resistant to abrasion. It is also chemically and mechanically very stable, as it is already an oxide and has a structure without cleavage planes, along which the material can easily be split. These factors explain its persistence and prevalence in non-clay soils (sands and gravels), which have a comparatively large particle size.

Feldspars also have three-dimensional framework structures, but some of the silicon ions have been replaced by aluminium. The resultant excess negative charge is balanced by the inclusion of cations such as potassium, sodium and calcium. This leads to a more open structure, with lower bond strengths between structural units. Thus feldspars are not as hard as quartz (they will **cleave** or split along weakly bonded planes) and they are more easily broken down. This is why they are not as prevalent in soils generally as they are in igneous rocks. Pyroxenes, amphiboles and olivines are also relatively easily broken down, which is again why they are absent from many soils.

1.4.4 Surface forces

In a solid material, atoms are bonded together in a three-dimensional structure. At the surface of the solid, the structure is interrupted, leaving unbalanced molecular forces.

Table 1.2 Specific surface area of sand and clay particles (data from Mitchell, 1993)

Mineral group	Particle length	Particle thickness	Specific surface area (m^2/g)
Sand	2 mm	2 mm	5×10^{-4}
Sand	1 mm	1 mm	10^{-3}
Kaolinite	0.1–4 μm	0.05–2 μm	10–20
Illite	0.1–4 μm	≥ 3 nm	65–100
Montmorillonite	1–2 μm	1–20 nm	up to 840

Equilibrium across the surface may be restored by the attraction and adsorption of molecules from the adjacent phase (in soils, from the pore water); by cohesion (i.e. sticking together) with another mass of the same material; or by the adjustment of the molecular structure at the surface of the solid.

An unbalanced bond force is significant in comparison with the weight of a molecule, but not in comparison with the weight of a soil particle which is as large as a grain of sand. However, as the particle size is reduced, the ratio of the surface area to the volume or mass of the particle increases dramatically, as indicated in Table 1.2. The total surface area of the particles in 10 g of montmorillonite is equivalent to a football pitch.

It might, therefore, be supposed that surface forces could have a significant influence on the behaviour of clay soils. At low stresses – for example, when clay particles are dispersed in a column of water – this is indeed the case, and many clays behave as **colloids** in these circumstances (i.e. the clay particles are able to remain suspended in water, because the forces which tend to support them are greater than the gravitational force which tends to cause them to settle out). This is partly due to the small size of the clay particles, and partly due to the electrical surface forces which result from the substitution of ions – for example, Al^{3+} for Si^{4+} – within their structure.

In most geotechnical engineering applications, however, the appropriate comparison is between the surface forces and the gravitational force due not just to the mass of a single particle, but to the total mass of soil above the particle in a deposit. This is more or less the same, whether the deposit is a sand or a clay. Thus in soil mechanics and geotechnical engineering, the surface forces between clay particles are not generally significant, and they do not have to be taken into account by means of some special form of analysis. Although surface forces and pore water chemistry might influence the structure of a newly deposited clay, the same laws apply in practical terms to soils made up of clay particles as to soils made up of non-clay particles. Certain effects might be more pronounced in clays than in sands (see, in particular, Chapter 4), but this is due to the difference in particle size, rather than to the influence of surface chemistry.

The strength of an assemblage of soil particles, be they sand or clay, comes primarily from interparticle friction. In some natural deposits the particles may be lightly cemented together, but this is more common in sands than in clays. Although a lump of moist clay can be moulded in the hand (whereas a lump of moist sand would fall apart) this is not due to interparticle or 'cohesive' bonds. If it were, the clay would remain intact if it were immersed in water for a week or so. (Unless the particles are cemented – in which case, the soil will probably be too hard or brittle to mould by hand – a small lump of sand or clay will disintegrate very easily if it is kept immersed in water for long enough.) Clay soils can be moulded in the hand because the spaces or voids between the clay particles are small enough to hold water at a negative pressure, essentially by capillary action. (It may be, however, that the negative pore water pressures in a stiff clay paste which has just been made by mixing clay particles with water result from the tendency of the clay particles to adsorb water, and is therefore due to surface effects.) This negative pore water pressure – or suction – pulls the particles together, giving the soil mass some shear strength. This concept is discussed in section 1.7.

1.4.5 Organic (non-mineral) soils

Some soils (notably peat) do not result from the breakdown of rocks, but from the decay of organic matter. Like topsoil, these soils are not suitable for engineering purposes. Peat is very highly compressible, and will often have a mass density which is only slightly greater than that of water. Unlike topsoil, organic soils may be naturally buried below the surface, and their presence is not necessarily obvious. This can make life difficult for the civil engineer, because it is important that these soils are detected at an early stage of a project, and if necessary, removed. They should not be relied on for anything, except to cause trouble.

Most of the factual content of section 1.4 is taken from the account given by Mitchell (1993), to which book the reader is referred for further details.

1.5   Phase relationships for soils

Soil is made up essentially of solid particles, with spaces or voids in between. The assemblage of particles in contact is usually referred to as the **soil matrix** or the **soil skeleton**. In conventional soil mechanics, it is assumed that the voids are in general occupied partly by water and partly by air. This means that an element of 'soil' (by which we mean the solid particles plus the substance(s) in the voids they enclose) may be a three-phase material, comprising some solid (the soil grains), some liquid (the pore water) and some gas (the pore air). This is illustrated schematically in Figure 1.7. A given mass of soil grains in particulate form occupies a larger volume than it would if it were in a single solid lump, because of the volume taken up by the voids.

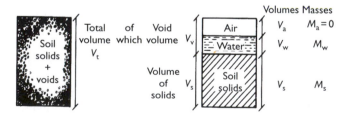

Figure 1.7 Soil as a three-phase material.

Figure 1.7 gives rise to a number of fundamental definitions, known as **phase relationships**, which tell us about the relative volumes of solid, liquid and gas. The phase relationships are important in beginning to characterize the state of the soil. They are:

1. The **void ratio,** which is defined as the ratio of the volume of the voids to the volume of solids (i.e. the soil particles), and is conventionally given the symbol e:

    Void ratio e = volume of voids ÷ volume of solids = V_v/V_s (1.1)

2. The **specific volume,** which is defined as the actual volume occupied by a unit volume of soil solids. It is conventionally given the symbol v:

    Specific volume v = total volume ÷ volume of solids

    $$= (V_s + V_v)/V_s = 1 + e \qquad (1.2)$$

3. The **porosity,** which is defined as the volume of voids per unit total volume, and is given the symbol n:

    Porosity n = volume of voids ÷ total volume

    $$= V_v/(V_s + V_v) = e/(1 + e) = (v - 1)/v \qquad (1.3)$$

4. The **saturation ratio,** which is defined as the ratio of the volume of water to the volume of voids, and is usually given the symbol S or S_r:

    Saturation ratio S_r = volume of water ÷ volume of voids

    $$= V_w/V_v \qquad (1.4)$$

    The saturation ratio must lie in the range $0 \leq S_r \leq 1$. If the soil is dry, $S_r = 0$. If the soil is fully saturated, $S_r = 1$.

    Alternatively, the state of saturation of the soil may be quantified by means of the air content A, which is defined as the ratio of the volume of air to the total volume,

    Air content A = volume of air ÷ total volume

    $$= V_a/(V_s + V_v)$$

    Substituting $V_a = V_v - V_w$, and dividing through the top and the bottom of the definition of A by the volume of voids V_v, it can be shown that

    $$A = (v - 1)(1 - S_r)/v = n(1 - S_r)$$

5. The **water content** (or moisture content), which is defined as the ratio of the mass of water to the mass of soil solids, and is given the symbol w:

    Water content $w = m_w/m_s$ (1.5)

The void ratio, the specific volume and the porosity are all indicators of the efficiency with which the soil particles are packed together. They are not independent: if one is known, the other two may be calculated. The choice of which one to use is largely a matter of personal preference. The specific volume v and the void ratio e

are more commonly used than the porosity n. The specific volume v is often the most mathematically convenient.

Sands normally have specific volumes in the range 1.3–2.0 ($e = 0.3$–1.0). For clays, the specific volume depends on the current stress state and the stress history, and also the mineralogy. The specific volume of a montmorillonite (such as the bentonite mud used as a temporary support for boreholes and trench excavations), in which surface forces are particularly significant, may be as high as 10 at low stresses. In contrast, the maximum specific volume of a kaolinite is unlikely to exceed 3.5. At high stresses, specific volumes as low as 1.3 can be achieved.

The specific volume and the saturation ratio cannot be measured directly. The water content is measured by taking a sample of the soil and weighing it to find its mass: this gives the mass of soil solids m_s plus the mass of water m_w. The soil sample is then dried in an oven at a temperature of 105°C for 24 h, in order to evaporate the water. It is then re-weighed, to determine the mass of the soil particles m_s. The water content of the original sample, m_w/m_s, may then be calculated. A typical calculation is shown in Example 1.1.

Example 1.1 Determination of water content

(i)   mass of container, empty $(m_t) = 21.32\,g$
(ii) mass of container + wet soil sample $(m_t + m_s + m_w) = 83.76\,g$
(iii) mass of container + dry soil sample $(m_t + m_s) = 65.49\,g$

Hence

(iv) mass of soil solids $m_s = $ (iii) − (i) $= 65.49\,g - 21.32\,g = 44.17\,g$
(v)   mass of water $m_w = $ (ii) − (iii) $= 83.76\,g - 65.49\,g = 18.27\,g$
      water content $w = m_w \div m_s = $ (v) $\div$ (iv) $= 18.27\,g \div 44.17\,g = 41.36\%$

In many circumstances in the ground, the voids are full of water. In this state the saturation ratio $S_r = 1$ (because the volume of air $V_a = 0$ and the volume of water V_w is equal to the total void volume V_v), and the soil is described as **saturated** or **fully saturated**. This reduces the number of phases present to two, which simplifies the description of the mechanical behaviour enormously. In this book, except for section 5.19 on partly saturated soils and elsewhere as explicitly stated, it is assumed that the soil is fully saturated.

If the soil is dry, $S_r = 0$ (because the volume of water $V_w = 0$). A dry soil can be treated as a single-phase material, and is usually easier to analyse than a saturated soil. In temperate regions, the natural soils which are of interest to civil engineers are usually below the water table or groundwater level, and are therefore effectively saturated.

Compression of a soil element requires a rearrangement of the soil particles relative to one another, to bring about a reduction in the void ratio of the **soil skeleton**. In principle this could be accompanied by the compression of the soil grains and/or the compression of the pore fluid (gas or liquid). In practice, except perhaps at very high stresses or with extremely dense soils, it is usually found that the soil particles and the water in the pores are incompressible in comparison with the soil skeleton. This means that any change in the overall volume of an element of fully saturated soil must be due to a change in the void ratio alone. Compression or expansion must therefore be accompanied by the flow of water out of or into the soil element. In an unsaturated (or partly saturated) soil, changes in void volume can be accommodated by the compression or expansion of the air. Air cannot reasonably be regarded as incompressible in comparison with the soil matrix. This is one reason why unsaturated or partly saturated soils are much more difficult to analyse than fully saturated soils.

The void ratio, specific volume and porosity are not soil properties or constants, but vary depending on whether the soil particles are densely or loosely packed. It will be seen in section 4.2 that the void ratio of a saturated clay depends on the maximum vertical load to which it has been subjected in the past, and the vertical load which is currently applied – in other words, on its stress history and current stress state. For sands, the void ratio is not uniquely related to the stress state and the stress history. One reason for this is that the initial void ratio of a sandy deposit depends largely on the conditions under which it was laid down (e.g. in air or under water). Subsequent densification of a sand is usually achieved more easily by vibration than by the application of a static stress.

Dense soils, which have low specific volumes, are generally stiffer than loose soils, which have high specific volumes. The initial judgement as to whether a particular specific volume v is 'low' or 'high' depends on where it lies in relation to the maximum and minimum achievable specific volumes v_{max} and v_{min} for the soil in question (Kolbuszewski, 1948). For sandy soils, this can be quantified by means of the **density index** (also known as the relative density), which is given the symbol I_D:

$$I_D = (v_{max} - v)/(v_{max} - v_{min}) = (e_{max} - e)/(e_{max} - e_{min}) \qquad (1.6)$$

If $v = v_{max}$, the sand is as loose as it can be and $I_D = 0$. If $v = v_{min}$, the sand is as dense as it can be and $I_D = 1$.

1.6   Unit weight

The **unit weight** of a soil is defined as the weight of a unit volume, in kN/m^3. It is conventionally given the symbol γ. The unit weight is equal to the overall mass density of the soil – sometimes called the **bulk density**, ρ_b – (in Mg/m^3) multiplied by the gravitational constant $g = 9.81 \, m/s^2$. In soil mechanics, the unit weight is usually used in preference to the mass density. This is because it facilitates the calculation of vertical stresses at depth, which often arise primarily as a result of the weight of overlying soil.

With reference to Figure 1.7, the unit weight γ can be calculated as follows:

$$\gamma = (\text{Total weight}) \div (\text{Total volume})$$

$$= (g \times \text{Total mass}) \div (\text{Total volume})$$

$$= g \times (m_s + m_w + m_a) \div (V_s + V_w + V_a) \qquad (1.7)$$

Now, the mass of air $m_a \approx 0$, and the volume of water V_w plus the volume of air V_a is equal to the volume of voids V_v. Also, $V_v = e \times V_s$, where e is the void ratio (equation (1.1)), and $m_w = w \times m_s$, where w is the water content (equation (1.5)).

Substituting these into equation (1.7),

$$\gamma = [g \times (1 + w)m_s] \div [V_s(1 + e)] = [g \times m_s/V_s] \times [(1 + w)/(1 + e)]$$

But $m_s/V_s = \rho_s$, where ρ_s is the density of the soil grains, and $\rho_s = G_s\rho_w$ where G_s is the density of the soil particles relative to that of water (section 1.6.1) and ρ_w is the density of water ($\rho_w = 1000\,kg/m^3$ at a temperature of 4°C). Also, $1 + e = v$. Thus

$$\gamma = [(gG_s\rho_w) \times (1 + w)]/[v] = [G_s(1 + w)/v]\gamma_w \tag{1.8}$$

where $\gamma_w = g\rho_w$ is the unit weight of water. (Taking $\rho_w = 1000\,kg/m^3$, $\gamma_w = 9.81\,kN/m^3$. γ_w is often taken as $10\,kN/m^3$ in geotechnical engineering calculations.) Alternatively, we could substitute into equation (1.7) as follows:

$$m_a \approx 0$$
$$V_w + V_a = V_v$$
$$V_v = e \times V_s \qquad \text{(from equation (1.1))}$$
$$m_w = \rho_w \times V_w \qquad \text{(mass of water = density of water × volume of water)}$$
$$V_w = S_r \times V_v \qquad \text{(from equation (1.4))} = S_r \times e \times V_s$$
$$m_s = G_s \times \rho_w \times V_s \qquad \text{(mass of soil = density of soil × volume of soil)}$$

giving

$$\gamma = [(g \times \rho_w \times V_s)(G_s + eS_r)] \div [V_s(1 + e)]$$
$$= \{[g\rho_w][G_s + (v - 1)S_r]\}/v$$
$$= \{[G_s + (v - 1)S_r]/v\}\,\gamma_w \tag{1.9}$$

If equations (1.8) and (1.9) are to be compatible,

$$[(G_s\gamma_w) \times (1 + w)] = [(\gamma_w)(G_s + eS_r)]$$

Dividing both sides by $G_s\gamma_w$

$$1 + w = 1 + (eS_r/G_s)$$

giving

$$w = eS_r/G_s$$

or

$$S_r = wG_s/e = wG_s/(v - 1) \tag{1.10}$$

From equation (1.4), $S_r = V_w/V_v$. But $V_w = m_w/\rho_w$, and $V_v = eV_s = em_s/(G_s\rho_w)$. Substituting these into equation (1.4)

$$S_r = V_w/V_v = [m_w/\rho_w] \div [em_s/(G_s\rho_w] = (m_w/m_s)G_s/e = wG_s/e$$

and equation (1.10) is shown to be correct.

If the soil is saturated, substitution of $S_r = 1$ into equation (1.9) gives the **saturated unit weight**,

$$\gamma_{sat} = [(G_s + v - 1)/v]\gamma_w \qquad (1.11)$$

Typically, γ_{sat} will be in the range 16–22 kN/m^3, unless the soil particles are of predominantly organic origin (e.g. peat).

Also, for a saturated soil (e.g. by substitution of $S_r = 1$ into equation (1.10)),

$$v = 1 + wG_s \quad \text{or} \quad e = wG_s$$

If the soil is dry, substitution of $w = 0$ into equation (1.8), or substitution of $S_r = 0$ into equation (1.9), gives the **dry unit weight** γ_{dry} as

$$\gamma_{dry} = [(G_s\gamma_w)/v] \qquad (1.12)$$

At a given specific volume or void ratio, the actual unit weight of a soil will lie somewhere between γ_{dry} (if $S_r = 0$) and γ_{sat} (if $S_r = 1$).

Example 1.2 Calculating the specific volume and the saturation ratio from the water content and the unit weight

A sample of soil has a water content $w = 14.7\%$, and a cube of dimensions 10 cm × 10 cm × 10 cm weighs 18.4 N. The particle relative density $G_s = 2.72$. Calculate the unit weight, the specific volume and the saturation ratio. What would be the unit weight and the water content if the soil had the same specific volume, but was saturated? What would be the unit weight if the soil had a water content of zero?

Solution

The unit weight γ is given by the total weight divided by the total volume of the cube sample:

$$\gamma = 18.4\,\text{N} \div (0.1 \times 0.1 \times 0.1)\,\text{m}^3 = 18\,400\,\text{N/m}^3 = 18.4\,\text{kN/m}^3$$

Rearranging equation (1.8) to determine the specific volume v,

$$v = [G_s \times (1 + w)] \div (\gamma/\gamma_w)$$

$$= [2.72 \times 1.147] \div (18.4/9.81) = 1.663$$

The saturation ratio S_r is calculated using equation (1.10):

$$S_r = wG_s/(v - 1)$$

$$S_r = 0.147 \times 2.72 \div 0.663 = 0.603 \text{ or } 60.3\%$$

If the soil were saturated at the same specific volume, the water content w_{sat} would be given by equation (1.10) with $S_r = 1$:

$$w_{sat} = (v - 1)/G_s = 0.663 \div 2.72 = 0.244 \text{ or } 24.4\%$$

The saturated unit weight γ_{sat} would be given by equation (1.11):

$$\gamma_{sat} = [(G_s + v - 1)/v]\gamma_w$$

$$\gamma_{sat} = [(2.72 + 0.663)/1.663] \times 9.81\,\text{kN/m}^3 = 19.96\,\text{kN/m}^3$$

The unit weight at the same specific volume but zero water content is the dry unit weight γ_{dry}, given by equation (1.12):

$$\gamma_{dry} = [(G_s\gamma_w/v]$$

$$\gamma_{dry} = [(2.72 \times 9.81\,\text{kN/m}^3) \div 1.663] = 16.05\,\text{kN/m}^3$$

Although several of the equations (1.8)–(1.12) have been used in Example 1.2, there is no point in trying to remember them: they are too complex. If you try, you will almost certainly remember them incorrectly. What is important is that you should be able to derive equations (1.8)–(1.12) yourself, starting from the conceptual model of soil as a three-phase material shown in Figure 1.7, and the phase relationships that arise from it.

1.6.1 Measuring the particle relative density G_s

The density of the soil grains relative to water G_s (known as the **particle relative density** or the **grain specific gravity**) is measured using a **Eureka can**, which is a metal container with an overflow device set at a certain level. The can is filled with water until it starts to overflow. After any excess water has drained off, a known mass of dry soil grains is poured gently into the can. The volume of water which overflows from the can due to the immersion of the soil grains is measured: this is equal to the volume of the grains. The density and the relative density of the soil grains can then be calculated.

An alternative procedure involves the use of a 500 ml gas jar or a conical-topped **pycnometer** for coarse soils, or a narrow-necked 50 ml **density bottle** for fine soils. First, the empty container is weighed (m_1). A quantity of dry soil is then placed in the container, and the total mass of the container and the dry soil is determined (m_2). The soil sample is then flooded with de-aired water, and the container is agitated to remove air bubbles. The container is filled to the top with water, and weighed again (m_3). The container is then washed out, filled to the top with water only, and weighed again (m_4). The mass of the dry soil particles is given by ($m_2 - m_1$). The difference between the mass of water required to fill the whole container ($m_4 - m_1$) and the mass of water required to fill the part of the container not occupied by the dry soil grains ($m_3 - m_2$) gives the mass of water displaced by the soil particles, ($m_4 - m_1$) − ($m_3 - m_2$). The relative density of the particles is equal to the mass of the dry soil particles divided by the mass of water they displace:

$$G_s = (m_2 - m_1)/[(m_4 - m_1) - (m_3 - m_2)]$$

The accuracy of the second method depends on filling the container exactly to the top before weighing it to determine the masses m_3 and m_4. The containers used are all designed to facilitate this. The narrow neck of the density bottle will minimize the effect of unavoidable small discrepancies in level. The top of a gas jar is closed off with a ground glass plate, slid into place over a ground glass top-flange, in order to exclude air. A pycnometer has a conical top (like an inverted, cut-off funnel), which again minimizes errors due to small discrepancies in the level to which the vessel is filled.

The relative density of the soil grains depends on their mineralogy. $G_s \approx 2.65$ for quartz (Blyth and de Freitas, 1974), and most common soils have G_s in the range 2.6–2.8.

Example 1.3 Determining the particle relative density, specific volume and unit weight from experimental results

(a) 2 kg of dry sand is poured into a Eureka can, where it displaces 755 cm^3 of water. Calculate the relative density (specific gravity) of the sand particles.

(b) When 2 kg of the same sand is poured into a measuring cylinder of diameter 6 cm, it occupies a total volume of 1200 cm^3. Calculate the specific volume of the sand and its unit weight in this state.

(c) The measuring cylinder is carefully filled with water up to the level of the top of the sand, so that the total volume is still 1200 cm^3. Calculate the unit weight of the saturated sand in this condition.

(d) The side of the measuring cylinder is tapped gently several times, and the level of the sand surface settles to an indicated volume of 1130 cm^3. Calculate the specific volume and the unit weight of the saturated sand in its dense state.

Take the unit weight of water γ_w as 9.81 kN/m^3, and the mass density of water as 1000 kg/m^3.

Solution

(a) The Eureka can experiment tells us that the volume occupied by 2 kg of sand particles is 755 cm^3. This does not change, irrespective of the overall volume occupied: changes in the total volume occur due to changes in the volume of the voids only.

The mass density of the sand particles is given by the mass divided by the volume actually occupied, $\rho_s = 2 \text{ kg} \div (755 \times 10^{-6} \text{ m}^3) = 2649 \text{ kg/m}^3$.

The relative density (specific gravity) of the soil particles G_s is equal to the mass density of the sand particles divided by the mass density of water,

$$G_s = 2649 \text{ kg/m}^3 \div 1000 \text{ kg/m}^3 = 2.65$$

(b) In the loose state, the specific volume v (defined as the total volume occupied by a unit volume of soil particles) is given by

$$v = V_t/V_s = 1200\,\text{cm}^3/755\,\text{cm}^3 = 1.589$$

(The void ratio $e = v - 1 = 0.589$; and the porosity $n = (v-1)/v = 0.371$.)
   The unit weight γ is equal to the total weight divided by the total volume:

$$\gamma = (2\,\text{kg} \times 9.81\,\text{N/kg}) \div (1200 \times 10^{-6}\,\text{m}^3)$$
$$= 16\,350\,\text{N/m}^3 = 16.35\,\text{kN/m}^3$$

(c) The initial volume of voids is $(1200 - 755) = 445\,\text{cm}^3$. On flooding the sand sample without change in volume, the total weight has been increased by the weight of $445\,\text{cm}^3$ of water, that is, $445 \times 10^{-6}\,\text{m}^3 \times 1000\,\text{kg/m}^3 = 0.445\,\text{kg}$. The total weight of the soil is now 2.445 kg.
   The unit weight is now

$$\gamma = W_t/V_t = (2.445\,\text{kg} \times 9.81\,\text{N/kg}) \div (1200 \times 10^{-6}\,\text{m}^3)$$
$$= 19\,989\,\text{N/m}^3 = 19.99\,\text{kN/m}^3$$

(d) When the sand is densified, the level of the water surface remains the same. In effect, the sand settles through the water. The new specific volume v is given by

$$v = V_t/V_s = 1130\,\text{cm}^3/755\,\text{cm}^3 = 1.497$$

(The void ratio $e = v - 1 = 0.497$; and the porosity $n = (v-1)/v = 0.332$.)
   The unit weight γ is equal to the total weight divided by the total volume. The new volume of voids is $(1130 - 755) = 375\,\text{cm}^3$. The total weight of the sand sample, not including the surface water, is now 2 kg plus the weight of $375\,\text{cm}^3$ of water, that is, 2.375 kg.

$$\gamma = W_t/V_t = (2.375\,\text{kg} \times 9.81\,\text{N/kg}) \div (1130 \times 10^{-6}\,\text{m}^3)$$
$$= 20\,618\,\text{N/m}^3 = 20.62\,\text{kN/m}^3$$

1.7   Effective stress

We have already established that a saturated soil comprises two phases: the soil particles and the pore water. The strengths of these two phases, in terms of their ability to withstand shear stress (i.e. a stress that acts parallel to a plane, causing a shearing distortion of the body to which it is applied), are very different. The shear strength of water is zero. The only form of stress that static water can sustain is an isotropic pressure, which is the same in all directions. The soil skeleton, however, can resist shear. It does so partly because of the interlocking of the particles, but mainly because of interparticle friction. The frictional nature of the strength of the soil skeleton means that the higher the normal stress pushing the particles together, the greater the shear stress that can be applied before relative slip between particles starts to occur.

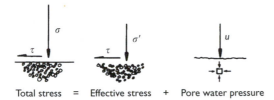

Figure 1.8 The principle of effective stress.

Friction and interlocking are important concepts, to which we will return in Chapter 2. For the present, the main point is that because the strengths of the soil skeleton and the pore water are so different, it is necessary to consider the stresses acting on each phase separately.

As the pore water cannot take shear, all shear stresses must be carried by the soil skeleton. The normal total stress applied to a soil element may be separated quite simply, by means of the **principle of effective stress** (Terzaghi, 1936). The effective stress σ' is the component of normal stress taken by the soil skeleton. It is the effective stress which controls the volume and strength of the soil. For saturated soils, the effective stress may be calculated from the total normal stress σ and the gauge[3] pore water pressure u by Terzaghi's equation (Figure 1.8):

$$\sigma' = \sigma - u \qquad\qquad (1.13)$$

Equation (1.13) is without doubt the most important equation in soil mechanics. There are not many equations in soil mechanics that it is necessary to remember, but this is one of them. Terzaghi is universally regarded as the leading founder of modern soil mechanics: most of the techniques described in this book are underpinned by the concept of effective stress.

The important point about the effective stress as defined by equation (1.13) is that it controls the volumetric behaviour and strength of the soil. It is not intended to represent the intergranular pressure at the grain contact points, although in some texts the terms are treated as synonymous. Mitchell (1993) shows that the intergranular pressure and Terzaghi's effective stress (equation (1.13)) will be similar unless the 'long-range' interparticle forces (i.e. those that do not depend on particle-to-particle contact – van der Waal's and electrostatic attractions, and double layer repulsions) are significantly out of balance. In most soils (as discussed in section 1.4.4), this is not the case.

Skempton (1960: see also Mitchell, 1993) showed that strictly, the effective stress controlling shear strength in soils, concrete and rocks is

$$\sigma' = \sigma - \left(1 - \frac{a_c \cdot \tan \psi}{\tan \phi'} \right) \cdot u$$

where a_c is the ratio of interparticle contact area to the total cross-sectional area, ψ is the intrinsic friction angle of the material from which the solid particles are made and ϕ' is the effective friction angle of the soil. Normally in soils, $a_c \cdot \tan \psi$ is very much smaller than $\tan \phi'$ (mainly because a_c is very small) and Terzaghi's equation (1.13) holds.

Similarly, the effective stress controlling volume change is

$$\sigma' = \sigma - \left(1 - \frac{C_s}{C}\right) \cdot u$$

where C_s is the compressibility of the soil grains and C is the overall compressibility of the soil matrix. Again, normally in soils C_s is very much smaller than C (i.e. the soil particles are comparatively incompressible) and the more rigorous expression reduces to equation (1.13). In rocks and concrete, however, a_c and C_s/C may not be small and Terzaghi's effective stress equation (1.13) will not then apply.

1.7.1 Calculating vertical stresses in the ground

The ability to calculate effective stresses in the ground is central to the application of the principle of effective stress in soil mechanics and geotechnical engineering. Effective stresses (σ') are usually deduced from the total stress (σ) and the pore water pressure (u), using equation (1.13).

Pore water pressures (u), and hence vertical effective stresses (σ_v'), are easy to calculate if the water in the soil pores is stationary and the depth below the soil surface at which the (gauge) pore water pressure is zero is known. (In the field, the natural pore water is termed the **groundwater**. The depth at which the pore water pressure is zero is known as the **groundwater level** or the **water table**, or – in three dimensions – the **piezometric surface**.)

Figure 1.9(a) shows a soil element at a depth z below the ground surface. The water table, which indicates the level at which the pore water pressure u is zero, is at a depth h. To calculate the vertical total stress σ_v acting on the soil element, we need to imagine a column of soil above it as shown in Figure 1.9(b). The cross-sectional area of this column is equal to the cross-sectional area of the element A. The height of the column is z, and we will assume that the unit weight of the soil γ is the same above the water table as below it.

There can be no vertical shear stress acting on the sides of the column. (Symmetry requires that any shear stress which does act, acts in the same direction on the adjacent column. The condition of equilibrium requires that it acts in the opposite direction. The conditions of symmetry and equilibrium can therefore only be satisfied if the shear stress is zero.)

The weight of the column of soil is

$$A(\mathrm{m}^2) \times z(\mathrm{m}) \times \gamma(\mathrm{kN/m}^3)$$

The resultant force (kN) due to the total vertical stress $\sigma_v(\mathrm{kN/m}^2)$ acting on its base is

$$A(\mathrm{m}^2) \times \sigma_v(\mathrm{kN/m}^2)$$

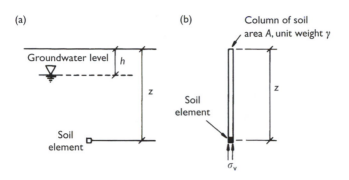

Figure 1.9 Calculation of vertical stress.

If the column of soil is in vertical equilibrium, these are equal and

$$\sigma_v = \gamma z \, (\text{kN/m}^2)$$

In geotechnical engineering, the usual unit of stress is the kiloPascal (kPa). One kilo-Pascal (kPa) is identical to one kiloNewton per square metre (kN/m^2), but the symbol is more convenient to write.

The rate of increase in vertical total stress with depth $d\sigma_v/dz$ is equal to γ, the unit weight of the soil: these stress conditions are sometimes referred to as **geostatic**.

If the groundwater is stationary, the pore water pressure will (by a similar argument) be **hydrostatic** below the water table, giving

$$u = \gamma_w (z - h) \, (\text{kPa})$$

Provided that the groundwater is stationary, and in continuous contact through the pores, the pore water pressure is unaffected by the presence of the soil particles. The pore water pressure at a depth $(z - h)$ below the water table is the same as it would be at a depth $(z - h)$ below the water surface in a lake or a swimming pool.

From equation (1.13), the vertical effective stress σ_v' is

$$\sigma_v' = \sigma_v - u = \gamma z - \gamma_w (z - h) \, (\text{kPa}) \tag{1.14}$$

Equation (1.14) should *not* be committed to memory: it does not represent a general case. The important thing is to understand how to calculate vertical total stresses and pore water pressures, and how to use equation (1.13) to calculate the resulting effective stress.

In a coarse-grained soil (i.e. a sand or a gravel), it is likely that the soil above the water table will not remain saturated. This would lead to a reduction in the unit weight γ of the soil above the water table, which would need to be taken into account in calculating the vertical total stress σ_v at depth z. A fine-grained soil (i.e. a silt or a clay) might remain saturated above the water table by means of capillary action, in which case its unit weight will not be significantly different. This point is discussed more fully in section 3.2.

Surface loads due to embankments and buildings with shallow foundations will impose additional shear and vertical total stresses. Excavation processes will alter pore water pressures and reduce vertical total stresses. The calculation of pore water pressures when the groundwater is not stationary – as will usually be the case in the vicinity of an excavation – is addressed in sections 3.3–3.12. The calculation of horizontal stresses, which are particularly important in the design of retaining walls, is addressed in section 7.5.

Example 1.4 Calculating vertical total and effective stresses

The ground conditions at a particular site are as follows:

Depth below ground level (m)	Description of stratum	Unit weight (kN/m^3)
0–1	Made ground/topsoil	17
1–3	Fine sand	18
3–6	Saturated fine sand	20
Below 6	Saturated stiff clay	19

Pore water pressures are hydrostatic below the groundwater level, which is at a depth of 3 m. Calculate the vertical total stress, the pore water pressure and the vertical effective stress at depths of (a) 1 m, (b) 3 m, (c) 6 m and (d) 10 m. Take the unit weight of water as 10 kN/m^3.

Solution

(a) At a depth of 1 m, the vertical total stress σ_v due to the weight of overlying topsoil is $1\,\text{m} \times 17\,\text{kN/m}^3 = 17\,\text{kPa}$.

   The soil here is above the water table, so assume that the pore water pressure $u = 0$. Hence from equation (1.13), the vertical effective stress $\sigma_v' = \sigma_v - u$, giving $\sigma_v' = 17\,\text{kPa}$.

(b) At a depth of 3 m, the vertical total stress σ_v due to the weight of overlying topsoil and fine sand is $1\,\text{m} \times 17\,\text{kN/m}^3 + 2\,\text{m} \times 18\,\text{kN/m}^3 = 53\,\text{kPa}$.

   The soil here is at the water table level, so the pore water pressure $u = 0$. Using equation (1.13), the vertical effective stress $\sigma_v' = \sigma_v - u$, giving $\sigma_v' = 53\,\text{kPa}$.

(c) At a depth of 6 m, the vertical total stress σ_v due to the weight of overlying topsoil and fine sand is $(1\,\text{m} \times 17\,\text{kN/m}^3) + (2\,\text{m} \times 18\,\text{kN/m}^3) + (3\,\text{m} \times 20\,\text{kN/m}^3) = 113\,\text{kPa}$. The soil here is 3 m below the water table level, so the pore water pressure $u = 3\,\text{m} \times 10\,\text{kN/m}^3 = 30\,\text{kPa}$.

   Using equation (1.13), the vertical effective stress $\sigma_v' = \sigma_v - u = 113\,\text{kPa} - 30\,\text{kPa}$, giving $\sigma_v' = 83\,\text{kPa}$.

(d) At a depth of 10 m, the vertical total stress σ_v due to the weight of overlying topsoil, fine sand and clay is $(1\,\text{m} \times 17\,\text{kN/m}^3) + (2\,\text{m} \times 18\,\text{kN/m}^3) + (3\,\text{m} \times 20\,\text{kN/m}^3) + (4\,\text{m} \times 19\,\text{kN/m}^3) = 189\,\text{kPa}$.

The soil here is 7 m below the water table level, so the pore water pressure $u = 7\,\text{m} \times 10\,\text{kN/m}^3 = 70\,\text{kPa}$.

Using equation (1.13), the vertical effective stress $\sigma'_v = \sigma_v - u = 189\,\text{kPa} - 70\,\text{kPa}$, giving $\sigma'_v = 119\,\text{kPa}$.

1.8 Particle size distributions

We have already mentioned that civil engineers describe and classify soils according to the particle size, rather than according to their age, origin or mineralogy. The principal reason for this is that civil engineers are interested mainly in the mechanical behaviour of soils, which depends primarily on particle size.

We will see in Chapters 3 and 4 that one of the major features which distinguishes a sand from a clay in the mind of the engineer is the ease with which water may flow through the network of voids in between the soil particles. (This property is known as the permeability: it is defined more formally in section 3.3.) As the size of the voids is governed by the size of the smallest particles (because these can fit into the voids between the larger particles), the permeability of an unstructured soil is related approximately to the maximum size of the smallest 10% of particles by mass.

In sands and gravels, water can flow very easily through the voids. The result of this is that a zone of sand or gravel will not be able to sustain a pore water pressure which is very different from that in the surrounding ground, provided it is free to drain. In clays, water can move through the voids only slowly. This means that the pore water pressure in a zone of clay might remain considerably above or below that of the surrounding ground for comparatively long periods of time.

The application of a load (e.g. by the construction of a new building) to a saturated soil will tend to cause a transient (i.e. temporary) increase in pore water pressure, within the loaded area. In a sand, this additional pore water pressure will dissipate very quickly, but in a clay, the process might take years or decades. As the pore pressure dissipates, the effective stresses increase and the soil is compressed. This process is known as consolidation. Clays are generally rather more compressible than sands and gravels, and therefore consolidate more, as well as more slowly. As a result of these two factors acting in combination, an engineer designing the foundations of a building on a clay soil would be concerned about the possibility of large settlements developing over a long period of time. In the design of a building on a sandy soil, the possibility of potentially large, delayed settlements would probably not be a concern.

In principle, both clays and sands consolidate in response to the application of load. In practice, sands consolidate immeasurably quickly, so that there is no point in attempting to carry out an analysis of the process, or even to quantify the parameters which control it. This is one of the reasons why, although the same fundamental rules govern the basic engineering behaviour of most soils, engineers find it useful to try to categorize a soil as a 'clay' or a 'sand'.

When engineers classify a soil as a clay, they do so on the basis of its particle size rather than its mineralogy. It is apparent from section 1.4, however, that there is an approximate relationship between particle size and particle strength and toughness. This means that most clay sized particles are in fact composed of clay minerals.

One common system of soil classification according to particle size was given in Figure 1.3. Natural soils comprise an assortment of many different-sized particles, and should be described with reference to the range and frequency distribution of particle sizes they contain. This is conventionally and conveniently achieved by means of a cumulative frequency curve, which shows the percentage (of the overall mass) of particles that are smaller than a particular size. Particle sizes are plotted to a logarithmic scale on the x-axis, and the percentage of the sample (by mass) that is smaller than a given particle size is plotted on the y-axis. In the UK, it is conventional to plot the particle size so that it increases from left to right. In the USA, the particle size increases from the right of the diagram to the left. Typical particle size distribution curves (a term usually abbreviated to PSD), plotted using the UK convention, are shown in Figure 1.10.

A soil which has a reasonable spread of particle size (represented by a smooth, concave PSD such as curve (a) in Figure 1.10) is conventionally described as **well-graded**. A soil which consists predominantly of a single particle size (whose PSD will have an almost vertical step, as curve (b) in Figure 1.10) is described as **uniform**. A soil which contains small and large particles but few particles of intermediate size (whose PSD will have a horizontal step) is described as **gap-graded**. Uniform and gap-graded soils are sometimes described as **poorly graded**, but this last term is ambiguous and should be avoided. Even the term well-graded presupposes that the soil will need to be compacted: the soil is described as well-graded for this purpose because it contains a wide range of particle sizes, which will pack together well to fill all the voids. A soil suitable for use as a filter would probably be a predominantly single-sized sand or gravel.

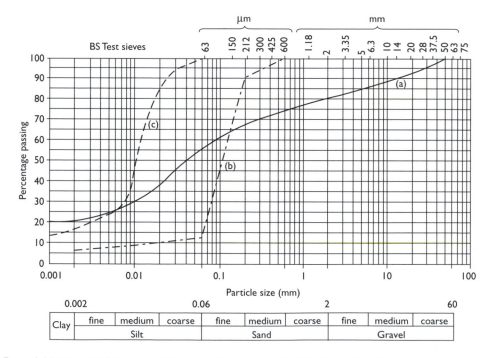

Figure 1.10 Typical (PSD) curves: (a) is probably a glacial till; (b) is Thanet Sand from the London basin; and (c) is an alluvial silt.

Particle size distribution curves for sands, by which we mean the portion of a soil whose particle size is greater than 0.063 mm (63 μm), are obtained by sieving the sample through a stack of sieves. Each sieve has a mesh of a certain single size, so that it will only allow particles smaller than the holes in the mesh to pass through it. The sieves are arranged in order of mesh size, with the largest mesh at the top of the stack and a tray at the bottom to collect the particles that are smaller than the finest mesh, 63 μm. Sieving is usually carried out in a standard way on a sample of standard total dry mass for a standard time, by means of a mechanical shaker with a timing device. By weighing the mass of soil collected on each sieve, and expressing its mass as a percentage of the total, the particle size distribution curve is constructed.

The soil sample should be oven-dried before sieving to determine its dry mass. With many coarse grained soils satisfactory results can be obtained by dry-sieving. Soils containing fine particles, however, must usually be **wet-sieved**. This involves washing the sample through a 63 μm sieve in order to remove the fine particles, which could otherwise stick to each other and to the coarser particles, increasing the apparent particle size. The portion of the sample retained on the 63 μm sieve is then oven-dried and dry-sieved in the usual way.

Particles which are smaller than 63 μm (i.e. anything smaller than fine sand) are too small for size determination by sieving, and so a different technique – **sedimentation** – is used. The velocity v with which a fine particle settles through water decreases with its particle size. This is because the gravitational force, given by the bouyant unit weight multiplied by the volume $(\gamma_s - \gamma_w) \times (4/3)\pi(d/2)^3$, is proportional to the cube of the particle diameter d, while the resistive drag force D is directly proportional to the particle diameter. (From Stokes's Law, $D = 3\pi\eta v d$, where η is the dynamic viscosity of water.) Equating the gravitational and the drag forces, the relationship between the settlement velocity of a spherical particle and its diameter is

$$v = \{(\gamma_s - \gamma_w)/18\eta\}d^2 \tag{1.15}$$

where γ_s is the unit weight of the soil grains and γ_w is the unit weight of water.

Equation (1.15) forms the basis of the sedimentation test. A soil sample containing an assortment of fine particles is shaken up in water to form a fully dispersed suspension, and left to settle out. Samples of the suspension are then taken from a fixed depth z (usually 100 mm) at various times after the start of sedimentation, using a pipette.

A pipette sample taken at a time t from a depth z will contain no particles that are settling at a speed greater than z/t. Substituting $v \le z/t$ into equation (1.15)

$$z/t \ge \{(\gamma_s - \gamma_w)/18\eta\}\,d^2$$

or

$$d \le \{[18\,\eta\,z]/[(\gamma_s - \gamma_w)\mathrm{t}]\}^{1/2} \tag{1.16}$$

Equation (1.16) gives the maximum particle size (expressed as an equivalent particle diameter d) present in a pipette sample taken at time t from a depth z. After the water has evaporated, the mass of solids which remains enables the proportion of particles having an equivalent diameter less than d within the original sample to be calculated. This is equivalent to the percentage of particles by mass which pass through a certain mesh size in a sieve analysis, and is used to construct the particle size distribution curve in exactly the same way.

Full details of the procedures which should be followed in the determination of particle size distributions using sieving and sedimentation will be found in BS1377 (1991). In carrying out standard tests, such as soil grading, description and classification, it is important that standard procedures and methods are followed. This is because the tests are not usually carried out by the people who use the results for engineering design. Indeed, it is extremely unlikely that all of the soil tests carried out in connection with anything but a fairly modest construction project would be carried out by the same person, or even in the same laboratory. Without standardization, the designers would not know whether differences between results were due to differences in the soil, or to differences in the testing procedure. If you carry out soil tests, you should follow the procedures laid down in the appropriate standard (in the UK, BS1377, 1991; in the US, ASTM D2487-1969, 1970). If you need to depart from these procedures for some reason, you should note this and explain why in your report.

Sometimes, for the sake of brevity, it is tempting to attempt to describe a soil by means of a representative particle size, rather than by the entire particle size distribution curve. The term 'representative' has a variety of interpretations, depending on the application the engineer has in mind. For processes in which the pore size rather than the particle size is important, such as groundwater flow and permeability-related problems, the behaviour is controlled by the smallest 10% of the particles. Processes which rely on interparticle contact – such as the generation of frictional strength – might depend on the smallest 25% of the particles. Contacts between larger particles become less frequent, as they are in effect suspended in a matrix of smaller particles. In filtration and clogging processes, which involve the movement or trapping of fine particles, particle sizes which delineate other proportions of the overall mass are significant, as described in section 1.9.

The largest particle size in the smallest 10% of particles is known as the D_{10} particle size. Similarly, the largest particle size in the smallest 25% of the sample is known as the D_{25} particle size. In general, $n\%$ of the soil by mass is smaller than the D_n particle size. The values of D_{10}, D_{25} etc. may be read off from the horizontal axis of the particle size distribution graph (Figure 1.10), at the points on the curve that correspond to 10%, 25% etc. of the sample by mass on the vertical axis.

The information conveyed by quoting a representative particle size such as D_{10} or D_{25} is inevitably somewhat limited. A fuller description is sometimes attempted by quoting the **uniformity coefficient** U, and the **coefficient of curvature**, Z:

$$U = D_{60}/D_{10} \tag{1.17a}$$

$$Z = (D_{30})^2/D_{60}D_{10} \tag{1.17b}$$

U is related to the general shape and slope of the particle size distribution curve. The higher the uniformity coefficient, the larger the range of particle size. According to the Department of Transport (1993: table 6/1), granular materials with a uniformity coefficient U of less than 10 may be regarded as uniformly graded, while granular materials with a uniformity coefficient U of more than 10 may be regarded as well-graded. A well-graded soil generally has a coefficient of curvature Z in the range 1–3. For general application, however, the representative particle sizes D_{10}, D_{25} etc., and the parameters U and Z are not acceptable substitutes for the full grading or particle size distribution curve.

Finally, it should be pointed out that the term 'particle size' as used in soil mechanics refers to a notional dimension. Sieve apertures are square, whereas sand and gravel particles – though generally bulky – are irregular in shape, and often angular rather than rounded. Stokes's Law assumes that soil particles are spherical, whereas clay particles are flat and platey. This does not really matter too much: provided that the standard procedures are followed, any error (which is probably insignificant in most cases) is systematic, and reproduced and accepted by everyone.

The construction of a particle size distribution curve from laboratory sieve and sedimentation test data is illustrated in Example 1.5.

Example 1.5 Constructing a particle size distribution curve from laboratory test data

Table 1.3(a) gives the results of a sieve test on a sample of particulate material of total dry mass 236 g.

Table 1.3(a) Sieve test data

Sieve size (mm)	Mass retained (g)
4.75	5.67
3.35	2.80
2.00	2.57
1.18	5.84
0.6	5.16
0.03	12.49
0.15	17.88
0.063	29.20
Fines (<0.063)	151.75
Total	233.36

Table 1.3(b) gives the results of a sedimentation test carried out on a representative sub-sample of the fine material (<63 μm). The mass of the sub-sample was 12 g, the volume of the pipette was 10 ml, and the overall volume of the suspension was 500 ml.

Table 1.3(b) Sedimentation test data

Equivalent particle diameter (mm)	Mass of sample in 10 ml pipette (g)
<0.02	0.08
<0.006	0.02
<0.002	0.01

Plot the particle size distribution curve, and describe the material.

Solution

The first point to notice is that the total mass given at the bottom of Table 1.3(a) is only 233.36 g, compared with the initial sample mass of 236 g. The missing 2.64 g is probably lodged in the meshes of the sieves, and is known as a sieve loss. For the purpose of calculating the percentage passing each sieve size, the recovered sample mass of 233.36 g should be used. (In effect, this distributes the sieve loss in proportion to the mass retained on each sieve. If the original mass were used, the particle size distribution curve would end at below 100% on the vertical axis.) The mass of material passing each sieve, in grams and as a percentage of the total, is given in Table 1.4(a).

Table 1.4(a) Sieve test data: mass passing each sieve

Sieve size (mm)	Mass retained (g)	Mass passing (g)	Percentage passing, by mass
4.75	5.67	227.69	97.57
3.35	2.80	224.89	96.37
2.00	2.57	222.32	95.27
1.18	5.84	216.48	92.77
0.6	5.16	211.32	90.56
0.03	12.49	198.83	85.20
0.15	17.88	180.95	77.54
0.063	29.20	151.75	65.03
Fines (<0.063)	151.75		
Total	233.36		

The sedimentation test data are already in the required form for the construction of the particle size distribution curve. Each pipette sample gives the mass of particles of less than a certain size, which is effectively the same as the mass passing through an equivalent sieve. It is, however, necessary to multiply the mass of material in each particle size category by the ratio of the volume of the suspension to the volume of the pipette sample (= 500 : 10 ml = 50) in order to determine the mass present in the 12 g sub-sample. Each mass must then be multiplied by the ratio of the mass of fines in the original sample to the mass of the sub-sample (151.75 : 12 g = 12.646), and finally expressed as a percentage of the mass of the entire original sample (233.36 g) for inclusion in the particle size distribution. This is most easily achieved by determining each mass as a percentage of the 12 g sub-sample, and multiplying this by the percentage of fines in the original sample (65.03%), as indicated in Table 1.4(b). (These two procedures give exactly the same result. For a scaled mass (i.e. the mass in the 500 ml suspension, which is 50 times the mass in the 10 ml pipette) of x, the first procedure gives a percentage passing of $x \times (151.75/12) \div 233.36$. The second procedure gives $(x/12) \times (151.75/233.36)$, which is identical.)

The particle size distribution curve, plotted using the data in the last columns of Tables 1.4(a) and 1.4(b), is shown in Figure 1.11.

Table 1.4(b) Sedimentation test data, scaled to account for the suspension : pipette volume ratio

Equivalent particle diameter (mm)	m_1 Mass of sample in 10 ml pipette, (g)	m_2 Mass of sample in 500 ml suspension (g)(= $m_1 \times 50$)	m_3 Percentage of sub-sample (= $m_2/12$)	m_4 Percentage of total sample (= $m_3 \times 0.6503$)
<0.02	0.08	4	33.33	21.67
<0.006	0.02	1	8.33	5.42
<0.002	0.01	0.5	4.17	2.71

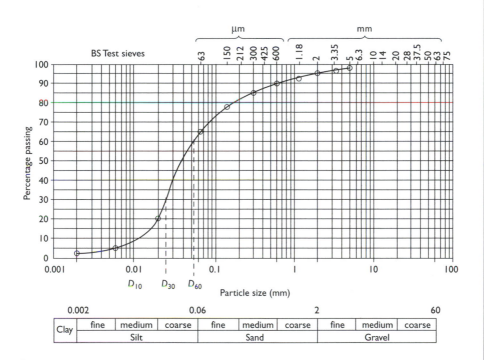

Figure 1.11 Particle size distribution curve for Example 1.5.

From Figure 1.11, the values of D_{10}, D_{30} and D_{60} are:

$D_{10} = 0.010 \, \text{mm}$

$D_{30} = 0.025 \, \text{mm}$

$D_{60} = 0.053 \, \text{mm}$

Hence the uniformity coefficient $U = D_{60}/D_{10}$ (equation (1.17a)) is $0.053/0.010 = 5.3$, and the coefficient of curvature $Z = (D_{30})^2/D_{60}D_{10}$ (equation (1.17b)) is $0.025^2/(0.01 \times 0.053) = 1.18$.

The sample consists mainly of silt-sized particles, with about 35% sand. It could therefore perhaps be just about described as a sandy silt, according to the soil description scheme given later in Table 1.5. In fact, the material is pulverized fuel ash, which is the residue from pulverized coal burned to generate electricity at coal-fired power stations.

1.9 Soil filters

Soil filters are used in the construction of drains and wells, as shown in Figure 1.12. Their function is twofold:

1. to allow water to drain freely out of the natural soil, and in some cases to conduct the outflow away,
2. to support the natural soil, and prevent it from being eroded by the water which drains out of it.

The efficiency with which the filter will perform these functions depends primarily on its particle size distribution, relative to the particle size distribution of the natural soil.

Some migration of fine particles within a limited interface zone between the natural soil and the filter is inevitable during the initial stages of water flow. Indeed, such an

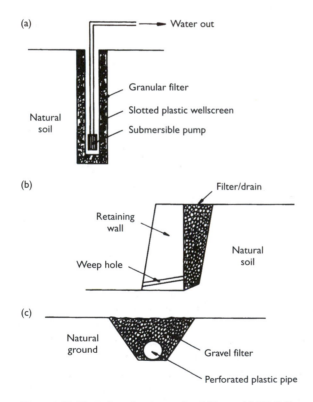

Figure 1.12 Typical applications of soil filters. (a) Well filter; (b) drainage behind a retaining wall; (c) land drainage system.

interface zone normally forms part of the overall filter system. After the initial stage of operation, however, the pore network of the filter must be small enough to prevent the continued loss of fine particles from the natural soil. In experiments carried out by Sherard *et al.* (1984), soil particles smaller than $0.1 \times$ the D_{15} size of the filter were generally found to be able to pass through the filter. Work by Kenney *et al.* (1985) suggested that soil particles of up to $0.2 \times$ the D_{15} size could pass through the filter. The apparent discrepancy between these results may be due to differences in filter type and/or flow pattern, but indicates the need for some conservatism when drawing up empirical rules concerning the relative grading of a filter and the natural soil. In order to prevent the continued migration of fines from the natural soil, Preene *et al.* (2000) suggest for sands and sandy slits with $D_{85s} \geq 0.1$ mm,

$$D_{15f} \leq 5 \times D_{85s} \tag{1.18}$$

where D_{15f} is the D_{15} size of the filter (which is related to the effective aperture size of the filter), and D_{85s} is the D_{85} size of the soil.

The filter must be more permeable than the natural soil, otherwise it would act as a restriction to flow, rather than a preferential drainage path. This is achieved by having

$$D_{15f} > 4 \times D_{15s} \tag{1.19}$$

(Preene *et al.*, 2000) and additionally

$$D_{5f} \geq 63 \, \mu m \tag{1.20}$$

where D_{15s} is the D_{15} size of the soil and D_{15f} is the D_{15} size of the filter etc.

The filter material must be suitably graded, so as to prevent both the migration of fines from the filter and the segregation of the filter material before, during or after installation. In order to meet this performance criterion, Preene *et al.* (2000) suggest that the uniformity coefficient $U (= D_{60f}/D_{10f})$ should be such that

$$(D_{60f}/D_{10f}) < 3 \tag{1.21}$$

although this requirement may be relaxed provided that $U_{filter} < U_{soil}$.

Where a filter is placed in variable ground, it must be able to protect the finest soil present, although in layered soils this may limit the capacity of the well. Preene *et al.* (2000) suggest that equation (1.18) should be applied to the finest soil and equation (1.19) to the coarsest. If the natural soil is gap-graded, the grading limits for the filter should be based only on the finer component of the natural soil. Where the natural soil contains a high proportion of gravel or larger sized particles, the filter might be designed on the basis of the grading curve of the portion of the natural soil finer than 19 mm. If the filter is to be placed against a slotted pipe (e.g. a wellscreen – see section 3.19), the D_{10} size of the filter (D_{10f}) should be of a similar order to the maximum slot width.

In the design of a multi-stage filter, the above rules apply with the term 'natural soil' being taken to refer to the material used in the upstream filter stage.

The application of the above filter rules to the specification of a granular filter grading envelope is illustrated in Example 1.6, which is based on a real case study.

Example 1.6 Specifying a grading envelope for a soil filter

Figure 1.13(a) shows a particle size distribution curve for a soil in which it is intended to install a number of wells of the type shown schematically in Figure 1.12(a). (Figure 1.13(a) is actually the particle size distribution curve of the granular glacial lake deposits shown in Figure 3.15.) Using the filter rules given by Preene *et al.* (2000), construct the envelope within which the particle size distribution curve of the filter material should lie. The well liners will be plastic tubes with 0.5 mm wide slots.

[*University of London 2nd year BEng (Civil Engineering) examination,*
Queen Mary and Westfield College (part question)]

Solution

The grading curve envelope is shown to the right of the particle size distribution curve for the soil in Figure 1.13(b). It has been constructed so as to fulfil the following criteria. (The letters refer to the points indicated on Figure 1.13(b).)

A: $D_{15f} \leq 5 \times D_{85s}$ (to prevent loss of fines from the natural soil; equation (1.18))

B: $D_{15f} > 4 \times D_{15s}$ (to ensure that the filter is sufficiently permeable; equation (1.19))

C: $D_{5f} \geq 63\,\mu m$ (again, to ensure satisfactory filter permeability; equation (1.20))

D: $D_{10f} \approx$ slot width, which is in this case 0.5 mm

E: $(D_{60f}/D_{10f}) < 3$ (equation (1.21))

Curve (b) in Figure 1.13(b) shows a suitable particle size distribution for the filter. It has simply been sketched to lie between the calculated limits. Curve (b) could be moved to the left or to the right, or altered slightly in shape. In practice, any readily available material which met the filter requirements would suffice. In the case shown in Figure 1.13, the material actually used was sand from the on-site concrete batching plant.

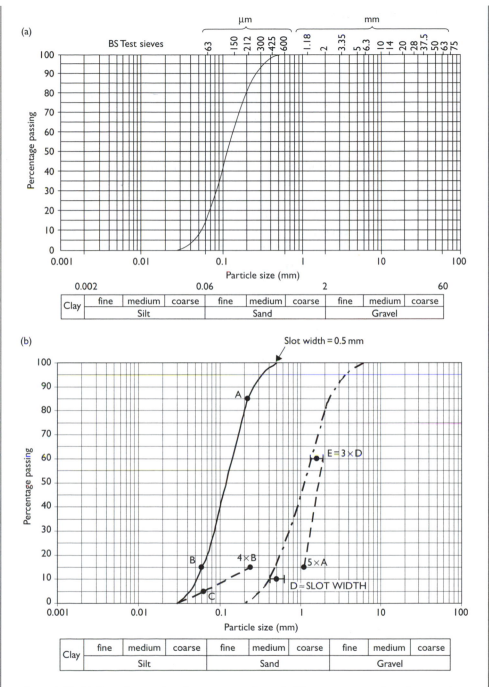

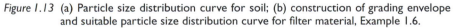

Figure 1.13 (a) Particle size distribution curve for soil; (b) construction of grading envelope and suitable particle size distribution curve for filter material, Example 1.6.

1.10   Soil description

It has already been mentioned that civil engineers are primarily concerned with the mechanical properties of a soil (such as its strength, compressibility and permeability). These properties are investigated using mechanical testing procedures, either in the field or in the laboratory, as described in Chapters 2–5 and 11. Nonetheless, a considerable amount of preliminary information can be conveyed by means of a visual description of the soil as it is found in the field. A systematic procedure for doing this is set out in BS5930, and is reproduced here as Table 1.5.

Table 1.5 shows how the soil can be graded (e.g. as a sand, a silt or a clay) with reference to the appearance and visibility of the particles to the naked eye, and other simple tests and observations (Column 3, headed Visual Identification). Many natural soils contain more than one group of particle sizes: Column 5 (headed Composite soil types) defines the classification system for mixed soils. Column 6 can be used to make an assessment of soil strength, using qualitative terms such as loose or dense, firm or soft. Column 7 is concerned with the structure of the soil, for example, the presence, nature and spacing of laminations or fissures. We will see later that the structure of a soil – which can be destroyed during sampling and transportation back to the laboratory – will often have an important influence on many aspects of its mechanical behaviour. Other aspects of soil description include the colour, which can give an indication of its mineralogy or chemical composition, and – for coarse soils – the particle shape and angularity. The property described as plasticity, which is relevant to clay soils, is discussed in section 1.11.

Table 1.5 demonstrates the need for standard procedures in the description and basic testing of soils. In many cases, a geotechnical designer may never see the soil to which the design calculations relate. To develop a qualitative feeling for the nature of the soil in question, the designer will usually rely entirely on a description prepared by a field or laboratory engineer or technician. In these circumstances, it is vital that when the field engineer describes the soil as a 'firm clay' or a 'slightly silty sand', the design engineer knows exactly what is meant. This can only be achieved if everyone adopts a standard system, in which the terms used to describe soils have a consistent meaning.

The word 'cohesive', which implies a strength which is non-frictional in its nature and is sometimes used to describe clay soils, should be avoided. Although, for the reasons given in section 1.4, surface effects are much more significant in clays than in sands, clays do not normally derive any significant engineering strength from inter-particle bonds or cohesive forces. Clay soils exhibit a property known as plasticity, which might be defined in this context as the ability to be worked and re-moulded in the hand. While surface effects may play some part in this (see section 1.4.4), the main reason that clays can be moulded in the hand is that they can sustain large pore water suctions. These pore water suctions may result in large effective stresses – and hence frictional strength – even if the total stress is zero.

1.11   Index tests and classification of clay soils

A clay soil will only exhibit plasticity between certain limits of water content. If the water content is too low, the soil will be dry and crumbly. If the water content is too high, the soil will behave almost like a liquid, squeezing out from between your fingers as you try to mould it.

Table 1.5 System for the description of soil in the field

Basic soil type		Particle size (mm)	Visual Identification	Particle nature and plasticity	Composite soil types (mixtures of basic soil types)	Compactness/strength		Structure		Interval scales		Colour
						Term	**Field test**	**Term**	**Field Identification**			

Very coarse soils

Basic soil type	Particle size (mm)	Visual Identification	Particle shape	Compactness/strength Term	Field test
Boulders	200	Only seen complete in pits or exposures.	Angular, Subangular, Subrounded, Rounded, Flat, Elongate	Loose	By inspection of voids and particle packing.
Cobbles	60	Often difficult to recover from boreholes		Dense	

Coarse soils (over 65% sand and gravel sizes)

Scale of secondary constituents with coarse soils

Term	% of clay or silt
slightly clayey / slightly silty GRAVEL or SAND	under 5
clayey / silty GRAVEL or SAND	5 to 15
very clayey / very silty GRAVEL or SAND	15 to 35

Sand or gravel an important second constituent of the coarse fraction:
Sandy GRAVEL (Rough / Smooth / Polished texture)
Gravelly SAND

For composite types described as:
clayey: fines are plastic, cohesive;
silty: fines non-plastic or of low plasticity

Basic soil type	Particle size (mm)	Visual Identification	Compactness/strength Term	Field test
Gravels — coarse	20	Easily visible to naked eye; particle shape can be described; grading can be described. Well graded: wide range of grain sizes, well distributed. Poorly graded: not well graded. (May be uniform: size of most particles lies between narrow limits; or gap graded: an intermediate size of particle is markedly under-represented).	Loose	Can be excavated with a spade; 50 mm wooden peg can be easily driven.
Gravels — medium	6		Dense	Requires pick for excavation; 50 mm wooden peg hard to drive.
Gravels — fine	2			
Sands — coarse	0.6	Visible to naked eye; very little or no cohesion when dry; grading can be described. Well graded: wide range of grain sizes, well distributed. Poorly graded: not well graded. (May be uniform: size of most particles lies between narrow limits; or gap graded: an intermediate size of particle is markedly under-represented).	Slightly cemented	Visual examination; pick removes soil in lumps which can be abraded.
Sands — medium	0.2			
Sands — fine	0.06			

Fine soils (over 35% silt and clay sizes)

Scale of secondary constituents with fine soils

Term	% of sand or gravel
Sandy CLAY or SILT	35 to 65
Gravelly — CLAY:SILT	under 35

Examples of composite types (indicating preferred order for description):
- Loose, brown, subangular very sandy, fine to coarse GRAVEL with small pockets of soft grey clay
- Medium dense, light brown, clayey, fine and medium SAND
- Stiff, orange brown, fissured sandy CLAY
- Firm, brown, thinly laminated SILT and CLAY
- Plastic, brown, amorphous PEAT

Basic soil type	Particle size (mm)	Visual Identification	Plasticity	Compactness/strength Term	Field test
Silts — coarse	0.02	Only coarse silt barely visible to naked eye; exhibits little plasticity and marked dilatancy; slightly granular or silky to the touch. Disintegrates in water; lumps dry quickly; possesses cohesion but can be powdered easily between fingers.	Non-plastic or low plasticity	Soft or loose	Easily moulded or crushed in the fingers.
Silts — medium	0.006			Firm or dense	Can be moulded or crushed by strong pressure in the fingers.
Silts — fine	0.002				
Clays		Dry lumps can be broken but not powdered between the fingers; they also disintegrate under water but more slowly that silt; smooth to the touch; exhibits plasticity but no dilatancy; sticks to the fingers and dries slowly; shrinks appreciably on drying usually showing cracks. Intermediate and high plasticity clays show these properties to a moderate and high degree, respectively.	Intermediate plasticity (Lean clay) / High plasticity (Fat clay)	Very soft	Exudes between fingers when squeezed in hand.
				Soft	Moulded by light finger pressure.
				Firm	Can be moulded by strong finger pressure.
				Stiff	Cannot be moulded by fingers. Can be indented by thumb.
				Very stiff	Can be indented by thumb nail.

Organic soils

Basic soil type	Particle size (mm)	Visual Identification	Compactness/strength Term	Field test
Organic clay, silt or sand	Varies	Contains substantial amounts of organic vegetable matter.	Firm	Fibres already compressed together.
			Spongy	Very compressible and open structure.
Peats	Varies	Predominantly plant remains; usually dark brown or black in colour, often with distinctive smell; low bulk density.	Plastic	Can be moulded in hand, and smears fingers.

Structure

Term	Field Identification
Homogeneous	Deposit consists essentially of one type.
Inter-stratified	Alternating layers of varying types or with bands or lenses of other materials. Interval scale for bedding spacing may be used.
Heterogeneous	A mixture of types.
Weathered	Particles may be weakened and may show concentric layering.
Fissured	Breaks into polyhedral fragments along fissures. Interval scale for spacing of discontinuities may be used.
Intact	No fissures.
Homogeneous	Deposit consists essentially of one type.
Inter-stratified	Alternating layers of varying types. Interval scale for thickness of layers may be used.
Weathered	Usually has crumb or columnar structure.
Fibrous	Plant remains recognizable and retain some strength.
Amorphous	Recognizable plant remains absent.

Interval scales

Scale of bedding spacing

Term	Mean spacing (mm)
Very thickly bedded	over 2000
Thickly bedded	2000 to 600
Medium bedded	600 to 200
Thinly bedded	200 to 60
Very thinly bedded	60 to 20
Thickly laminated	20 to 6
Thinly laminated	under 6

Scale of spacing of other discontinuities

Term	Mean spacing (mm)
Very widely spaced	over 2000
Widely spaced	2000 to 600
Medium spaced	600 to 200
Closely spaced	200 to 60
Very closely spaced	60 to 20
Extremely closely spaced	under 20

Colour

Red, Pink, Yellow, Brown, Olive, Green, Blue, White, Grey, Black etc. — Supplemented as necessary with: Light, Dark, Mottled etc. and Pinkish, Reddish, Yellowish, Brownish etc.

The water content below which the clay is brittle and crumbly is known as the **plastic limit**, w_{PL}. The water content above which the clay will behave as a liquid is known as the **liquid limit**, w_{LL}. If the water content is greater than the plastic limit but less than the liquid limit, the soil will behave as a plastic material, and can be moulded in the hand without cracking or running out between the fingers. The range of water content over which the clay behaves in this way is known as the **plasticity index** I_P:

$$I_P = w_{LL} - w_{PL} \tag{1.22}$$

The liquid limit, the plastic limit and the plasticity index are related to both the mineralogy and the amount of clay present in the soil sample. For example, a sample of soil with a high proportion of kaolinite particles might have a similar plasticity index to a different soil with a smaller proportion of illite or smectite particles. The two effects can be separated by means of a parameter known as the activity A:

$$A = I_P/(\text{percentage of the sample by mass with a particle size of} < 2\,\mu\text{m}) \tag{1.23}$$

(Skempton, 1953). For kaolinite, $A \approx 0.5$. For illite, $0.5 \le A \le 1$, and for smectite, $1 \le A \le 7$ (Mitchell, 1993).

The liquid and plastic limit tests are carried out on the part of a soil sample which is smaller than 425 μm in size: this includes silt and some sand. Liquid and plastic limits cannot be determined for non-plastic soils.

The liquid limit test is carried out by mixing up a number of samples of the soil, each with a different water content. A small metal cup containing each sample in turn is placed below a stainless steel cone of mass 80 g and apex angle 30°. The cone is supported with its apex just touching the surface of the soil sample. The cone is then allowed to fall, and the depth to which it has penetrated the sample after five seconds is recorded.

The results are plotted as a graph of cone penetration d (in mm) against the water content of the sample w. The liquid limit is defined (BS1377, 1990) as the water content at which the cone penetration is 20 mm. Provided that the range of penetration is not great, the liquid limit may be determined by drawing a best-fit straight line through the data points (Figure 1.14(a)). We will see in section 5.16 that the relationship between the water content w and the cone penetration d is actually logarithmic, so that a better straight line relationship over a wider range of water content may be obtained by plotting w against $\ln(d)$ (Figure 1.14(b)).

The plastic limit is found by repeatedly rolling and re-rolling a sample of soil from a 6 mm diameter thread to a 3 mm diameter thread. As the test progresses, the soil dries out until eventually it just starts to split and crumble when the diameter of 3 mm is reached. The moisture content at this point defines the plastic limit. This method is not really satisfactory, as it relies too heavily on the judgement of the person carrying out the test. Despite the availability of alternative, less operator-sensitive techniques, such as extrusion (Whyte, 1982), the rolling-out test is still the generally accepted procedure.

The liquid and plastic limit tests are in effect indicators of the shear strength[4] of a clay soil, at two different water contents. The shear strength of the clay at the plastic limit is about 70 times that at the liquid limit (Whyte, 1982). In addition, it will be shown in section 5.16 that the plasticity index is related directly to the compressibility of the soil. The concept of testing clay to determine the water contents at which it undergoes changes in consistency was originally proposed (using different apparatus)

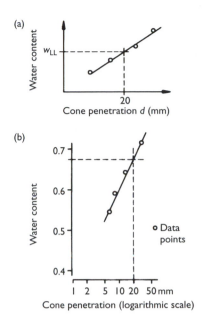

Figure 1.14 Determination of liquid limit from fall cone test data. (a) Interpolation of w_{LL} from a graph of w vs. d over a narrow range; (b) linear relationship between w and $\ln(d)$.

by the Swedish agricultural engineer Albert Atterberg (1911). For this reason, the liquid and plastic limits are sometimes referred to as the **Atterberg limits**.

It was mentioned in section 1.5 that the void ratio of a natural clay soil depends to a large extent on its previous stress history (i.e. the maximum vertical load to which it has been subjected in the past) and its current stress state (i.e. the vertical load to which it is currently subjected). Also, it is not generally possible to densify saturated clay soils by the application of transient loads, for example, by means of a vibrating roller or a pneumatic compactor. (The main reason for this is the time it takes for water to flow from the pores. Although ways of accelerating the process are discussed in later chapters, the only way of densifying a saturated fine-grained soil is to apply a large static load and wait – perhaps years – for the water to drain out of it.)

For these reasons, the concept of a minimum or maximum attainable specific volume or void ratio – which was used to define the density index of a coarse grained soil, equation (1.6) – is inappropriate to clays. However, the liquid and plastic limits for a clay (which, for a saturated soil, may be related to the specific volume by means of equation (1.10) with $S_r = 1$, giving $v = 1 + wG_s$) represent likely limits to the water content or specific volume in practice. As such, they are analogous to the maximum and minimum attainable specific volumes of a coarse soil.

For a fine-grained soil, the nearest analogue to the density index I_D of a coarse soil is the **liquidity index** I_L, which, for a soil of water content w, is defined as

$$I_L = (w - w_{PL})/(w_{LL} - w_{PL}) \tag{1.24}$$

(cf. equation (1.6) for the density index: $I_D = (v_{max} - v)/(v_{max} - v_{min})$).

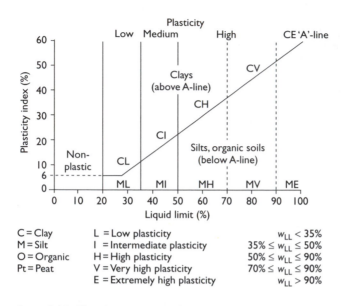

Figure 1.15 Classification system for fine soils, based on plasticity index and liquid limit (modified from BS5930, 1981).

If $w = w_{LL}$, the clay is at the liquid limit and $I_L = 1$. If $w = w_{PL}$, the clay is at the plastic limit and $I_L = 0$. The analogy is inverse, because the density index of a sand increases from 0 to 1 as the soil becomes denser, whereas the liquidity index of a clay increases from 0 to 1 as the clay becomes more liquid (i.e. looser).

Fine-grained soils may be classified as clays or silts of low, intermediate or high plasticity on the basis of their plasticity index and liquid limit, as indicated by the chart shown in Figure 1.15.

1.12 Compaction

When a soil is used as a structural material or a **fill** – for example, in the construction of an embankment, behind a retaining wall to create a raised terrace, or simply to fill in a trench (Figure 1.16) – it is generally compacted into place. This is usually in an attempt to minimize the likelihood of later settlement. When trying to compact unsuitable soils into confined spaces this may be something of a forlorn hope. Figure 1.17 shows the result of inadequate compaction of unsuitable backfill surrounding a soakaway drain in a supermarket car park; and most of us are familiar with the ruts and hollows which develop in a road or pavement within a few weeks of their having been dug up to re-lay a gas or water pipe or a telephone or electricity cable. In less confined spaces, particularly on large civil engineering projects where the client or the developer is prepared to pay for adequate supervision, it is quite possible to achieve satisfactory compaction of the fill material. Indeed, the adequacy of an embankment or an earth dam may depend on it.

Clay soils are generally considered unsuitable as backfill materials for retaining walls and trenches. Apart from the difficulty of expelling the water, clay soils excavated

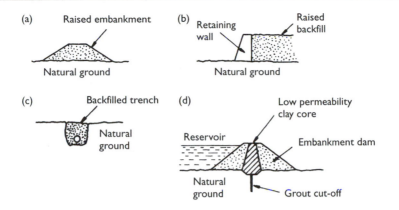

Figure 1.16 Use of soil as a fill material: (a) in an embankment; (b) behind a retaining wall; (c) in a trench; (d) in an earth dam.

Figure 1.17 Surface subsidence following inadequate compaction of backfill around a soakaway.

from the ground tend to be in the form of large clumps or clods. The overall volume occupied by a clay which has formed into clods might be half as much again as the volume occupied *in situ*, due to the extra air voids between the clods. In order to recompact the material, it is necessary to re-mould the clods, which can require the application of unfeasibly high stresses. An additional problem associated with a pre-dominantly clay fill is its potential for swelling in the long term: this is discussed in section 10.7.2. Having said this, clean granular soils are expensive to buy if they are not already available on site, so that many general engineering fills may contain a proportion of clay-sized particles.

In some applications where a low permeability is required, such as the core of an earth dam (Figure 1.16(d)), the use of a clay fill is essential. It is equally essential that the clay core of a dam is placed with an appropriate degree of compaction, and at an appropriate water content, without cracks and large voids which could impair its effectiveness and integrity.

In general, soils used as fills will not be saturated at the time of placement. The water content will depend on factors such as the proportion of clay present; the current climatic conditions; and the length of time for which the soil has been exposed. It is unlikely, however, that the water content will be able to change quickly, and it may therefore be assumed that compaction takes place at constant water content. The reduction in overall volume which is generally the aim of compaction must therefore be achieved by means of a reduction in the volume of the air voids.

The compaction characteristics of a particular soil are traditionally investigated in the laboratory by means of the Proctor compaction test. This involves the compaction of the soil sample (from which particles greater than 20 mm have been removed) into a cylindrical mould of 1 litre capacity and internal diameter 105 mm, by means of a standard rammer of mass 2.5 kg falling freely through a height of 300 mm, or by means of a heavy rammer of mass 4.5 kg falling freely through a height of 450 mm. In the first case, the soil is compacted in three equal layers, each receiving 27 blows of the rammer. In the second case, the soil is compacted in five equal layers, again with 27 blows to each layer. A compaction test is carried out on at least five samples of the same soil, with each sample at a different water content.

Although the efficiency of the packing of the soil grains is fundamentally quantified by the specific volume or the void ratio, the degree of compaction is traditionally assessed with reference to its dry density – that is, the density that the soil would have at the same void ratio but zero water content. From equation (1.12), the dry unit weight is given by

$$\gamma_{dry} = [(G_s\gamma_w)/v]$$

(where G_s is the particle relative density, γ_w is the unit weight of water and v is the specific volume) so that, dividing both sides by the gravitational acceleration g, the dry density ρ_{dry} is

$$\rho_{dry} = [(G_s\rho_w)/v] \tag{1.25}$$

The aim of compaction is to reduce the void ratio and hence the specific volume, v. From equation (1.25), the dry density is proportional to the inverse of the specific volume, so that either parameter will serve as a measure.

From equation (1.8), the actual unit weight of the soil is

$$\gamma = [G_s\gamma_w(1+w)/v]$$

so that (again, dividing through by g), the actual density is

$$\rho = [G_s\rho_w(1+w)/v] \tag{1.26}$$

Dividing equation (1.25) by equation (1.26),

$$\rho_{dry} = \rho/(1+w) \qquad (1.27)$$

This explains the traditional use of ρ_{dry}, rather than the more fundamental specific volume v, as a measure of the success of a compaction process. The actual soil density in the wet state, ρ (equation 1.26) is easily determined from the known mass and volume of soil in the mould after compaction. The water content w has to be measured anyway, as this is the control variable. ρ_{dry} can be calculated directly from ρ and w using equation (1.27), while the determination of v (using equation (1.26)) requires knowledge of G_s.

When the results of the compaction test are plotted as a graph of dry density ρ_{dry} against water content w, curves of the form shown in Figure 1.18(a) should be obtained. These curves show that, as the water content is increased from a low

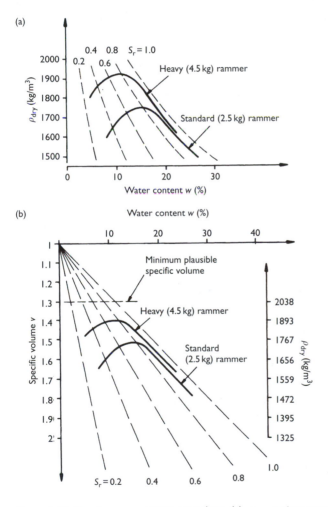

Figure 1.18 Proctor compaction test data: (a) ρ_{dry} against water content; (b) v against water content, with non-linear scale for ρ_{dry}.

value, the dry density increases to a maximum and then starts to decrease again. This is because low water contents imply high suctions, resulting in a soil which is stiff and not readily compactable. At high water contents, a limit to the degree of saturation achievable by compaction is reached. At $S_r \approx 0.9$ any remaining air pockets will be surrounded by water, and virtually impossible to remove by compaction. The specific volume then increases (and the dry density decreases) with increasing water content: from equation (1.10):

$$v = 1 + wG_s/S_r \tag{1.28}$$

The water content at which the compacted soil has the greatest dry density (i.e. the lowest specific volume) is known as the **optimum water content**. This term is potentially misleading. A dense, dry soil may be brittle and prone to cracking, which would not usually be described as the optimum condition for a fill material. Furthermore, a clayey fill which is placed dry might, in the long term, take in water and swell. This would certainly be undesirable if the fill had been placed behind a retaining wall (section 10.7.2).

In Figure 1.18(b), the compaction test data are plotted as specific volume v against water content w. The values of v have been calculated from ρ_{dry} using equation (1.25) with $G_s = 2.65$. Due to the reciprocal relationship between ρ_{dry} and v, the scale for ρ_{dry} in Figure 1.18(b) is non-linear. Lines of constant saturation ratio S_r are also indicated in Figure 1.18(b), calculated using equation (1.28) with $G_s = 2.65$.

The water content of the fill prior to placement and compaction can be regulated on site, in an attempt to control the density of the soil after compaction. As indicated in Figure 1.18, the dry density actually achieved for a given water content will, up to a certain limit, increase with the degree of compactive effort applied. The results of a laboratory compaction test are therefore not directly applicable in the field. However, the range of dry density produced by field compaction plant will probably lie within the limits given by the standard and heavy Proctor tests in the laboratory.

Example 1.7 Analysis of compaction test data

(a)   The results of a heavy (4.5 kg) compaction test carried out on samples of soil A are given in Table 1.6(a).

Table 1.6(a) Compaction test data for Example 1.7

Water content w (%)	8	11	13	15	19
Density ρ (kg/m³)	1945	2090	2120	2080	1990

Determine

(i)   the maximum dry density
(ii)   the optimum water content
(iii)   the saturation ratio at the maximum dry density if the particle relative density $G_s = 2.65$.

(b)   A standard (2.5 kg) compaction test on a sample of soil B gave the following results:

Mass of empty container (g)	14
Mass of container + wet soil (g)	44
Mass of container + dry soil (g)	40
Density in compaction mould (kg/m^3)	1800
Particle relative density G_s	2.54

Which of soil A and soil B would you select for use in a highway embankment, and why?

[University of London 2nd year BEng (Civil Engineering) examination, Queen Mary and Westfield College (part question, slightly modified)]

Solution

(a)   Convert the measured densities to dry densities using equation (1.27), $\rho_{dry} = \rho/(1 + w)$ (Table 1.6(b)). The data are plotted on a graph of ρ_{dry} against w in Figure 1.19. From this figure, the maximum dry density is about 1890 kg/m^3, which occurs at a water content of 11.7%.

Table 1.6(b) Processed compaction test data

Water content w	0.08	0.11	0.13	0.15	0.19
Density ρ (kg/m^3)	1945	2090	2120	2080	1990
Dry density: $\rho_{dry} = \rho/(1 + w)$ (kg/m^3)	1801	1883	1876	1809	1672

The corresponding specific volume is given by rearranging equation (1.25), $v = (G_s\rho_w)/\rho_{dry}$, giving

$$v = (2.65 \times 1000\,\text{kg/m}^3) \div 1890\,\text{kg/m}^3 = 1.402$$

Then, using equation (1.10) to calculate the saturation ratio,

$$S_r = (w \times G_s)/(v - 1) = (0.117 \times 2.65) \div 0.402$$
$$= 0.771 = 77.1\%$$

(These results are quite sensitive to the curve used to interpolate between the data points in Figure 1.19.)

(b)   Assuming that the data given for soil B relate to the optimum water content in the standard test, the water content at the maximum dry density is given by

$$w = (44 - 40) \div (40 - 14) = 0.154 = 15.4\%$$

The maximum dry density is given by equation (1.27), $\rho_{dry} = \rho/(1 + w)$:

$$\rho_{dry} = 1800\,\text{kg/m}^3 \div 1.154 = 1560\,\text{kg/m}^3$$

Using equation (1.25),

$$v = (G_s\rho_w)/\rho_{dry} = (2.54 \times 1000\,\text{kg/m}^3) \div 1560\,\text{kg/m}^3 = 1.628$$

and the saturation ratio

$$S_r = (w \times G_s)/(v - 1) = (0.154 \times 2.54) \div 0.628$$
$$= 0.623 = 62.3\%$$

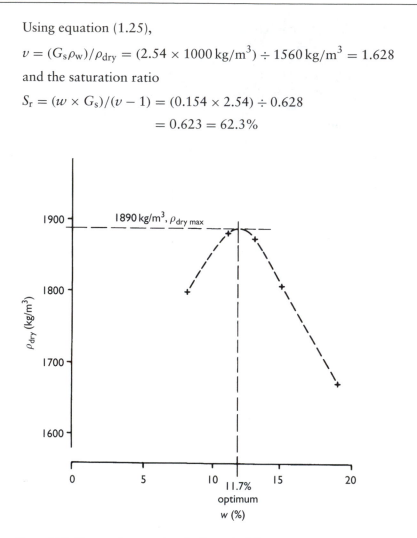

Figure 1.19 Compaction test data for Example 1.7.

Soil A is more suitable for use in a highway embankment, because:

- Soil A has a lower optimum water content, a lower specific volume, a higher maximum density and dry density, and a higher saturation ratio at the optimum water content than soil B. These factors indicate that, as a fill material, soil A will be denser and more stable than soil B.
- The grain specific gravity of soil B is comparatively low. This, together with its low maximum dry density, suggests that soil B may be at least partly of organic origin.

1.13   Houses built on clay

Some of the most densely populated areas of Britain (e.g. London and south-east England) and other parts of the world are built on deposits of clay soils. Many houses in these areas were built with foundations bearing directly onto the clay. As the water content of the clay decreases or increases, the soil will shrink or swell, resulting in the movement of a house whose foundations it supports. This can cause damage to the house, ranging from small hairline cracks (which should be accepted as normal for an older house on a clay soil), to serious structural distress.

The equilibrium water content of the clay will depend on a number of factors, including the imposed vertical effective stress, the availability of water (e.g. from rainfall, soakaways and leaking drains) and the rate at which water is taken from the soil by evaporation, vegetation and natural or engineered drainage. Small changes in water content may occur during the year as a result of normal seasonal variations in rainfall, humidity and temperature. These changes in water content will become more significant during prolonged periods of wet or dry weather. The effects of seasonal or longer term climatic variations will be amplified very considerably by the presence of vegetation: the water demand of some species of tree can be several hundred litres per day in hot weather.

Changes in local vegetation will also have an effect. An increase in the amount of vegetation tends to reduce the equilibrium water content of the clay, leading to shrinkage of the clay and settlement of the building. A decrease in the amount of vegetation tends to increase the equilibrium water content of the clay, leading to swelling of the clay and upward movement or heave of the building. Severe subsidence or heave of the foundation may result in the development of diagonal cracks in plane masonry walls, as shown in Figure 1.20.

In order to confirm that foundation movements are due to changes in the water content of the underlying clay, it is usual to investigate the variation of water content with depth in the soil near to where the building is cracked or otherwise damaged. If – as will often be the case – a localized cause (such as a tree) is suspected, the water content profile close to the house can be compared with that from a location far enough away to be unaffected. (Although tree roots may cause mechanical damage to the underground parts of the structure, the main problem is usually that, in times of comparative drought, the roots will extend further and deeper in an attempt to obtain water from an already moisture-deficient soil.)

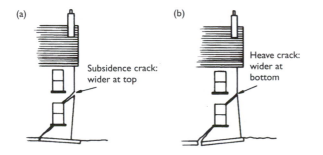

Figure 1.20 Diagonal cracks in walls due to (a) foundation subsidence; (b) foundation heave.

Soil samples from various depths at each location are obtained from boreholes, which are augered using portable, lightweight drilling equipment. The soil samples are sealed on site for transportation to the laboratory, where the water content of each sample is determined by oven drying as described in section 1.5.

The question which then arises is the amount of settlement associated with the measured water content deficiency. If the soil were to remain saturated, and shrink in the vertical direction only, this could be estimated relatively easily from the reduction in water content as follows.

Let the average reduction in water content that occurs over a depth h_0 be Δw from an initial water content of w_0.

The total volume V_t is related to the volume of soil particles V_s by the specific volume v (which is defined as $v = V_t/V_s$, equation (1.2)):

$$V_t = vV_s \tag{1.29}$$

As the volume occupied by the soil grains does not change, any change in total volume ΔV_t must be due entirely to a change in specific volume, Δv. Writing equation (1.29) in difference form

$$\Delta V_t = \Delta v V_s \tag{1.30}$$

Assuming that the soil remains saturated, the specific volume is related to the water content by equation (1.10) with $S_r = 1$:

$$v = 1 + wG_s \tag{1.31}$$

Writing equation (1.31) in difference form

$$\Delta v = \Delta w G_s \tag{1.32}$$

Substituting for Δv from equation (1.32) into equation (1.30)

$$\Delta V_t = \Delta w G_s V_s \tag{1.33}$$

Substituting for v from equation (1.31) into equation (1.29)

$$V_t = (1 + wG_s)V_s$$

For a block of soil of unit area on plan, depth h_0 and water content w_0,

$$V_t = 1 \times h_0 = (1 + w_0G_s)V_s$$

or

$$V_s = h_0/(1 + w_0G_s) \tag{1.34}$$

Substituting for V_s from equation (1.34) into equation (1.33),

$$\Delta V_t = (\Delta w G_s h_0)/(1 + w_0G_s) \tag{1.35}$$

As the block of soil has unit area on plan, the change in total volume ΔV_t is equal to the settlement ρ – assuming that all of the volume change is accommodated by vertical settlement. In reality, however, the clay tends to develop vertical shrinkage cracks as it dries out. This means that some of the volume change is accommodated by lateral shrinkage, so that

$$\Delta V_t = \text{settlement } \rho + \text{lateral shrinkage}$$

or

$$\rho < \Delta V_t$$

Empirical evidence (Driscoll, 1983) suggests that

$$\rho \leq \Delta V_t/3 \quad \text{to} \quad \Delta V_t/4 \tag{1.36}$$

ΔV_t is a reduction in total volume per unit area. It is entirely due to the loss of water from the soil, and is sometimes known as the soil moisture deficit. Another way of looking at the soil moisture deficit is as the depth of rainfall (ignoring losses due to **evapo-transpiration** – the combined effects of evaporation and moisture loss from plant leaves through transpiration) – also a volume per unit area – required to return soil to its original water content.

Foundation problems in domestic buildings on clay soils are often associated with the presence of trees or other vegetation, which provide the means of water content reduction. It is therefore tempting to suppose that the problem can be overcome by removing the offending trees. This is not so: the damage which can be caused to a building whose foundations heave (i.e. move upward) following the removal of trees can be at least as serious as the damage caused by settlement. Ideally, the moisture demand of the tree should be kept as constant as possible, by regular pruning. Additional protection can be provided by the installation of a root barrier, which minimizes the penetration of the tree roots below the house and its foundations.

The typical foundation depth of new houses on clay soils has gradually increased over the last few decades. Step increases in depth have usually followed a period of drought and a corresponding rise in the incidence of settlement damage. It is unlikely, however, that the specification of any reasonable foundation depth will be sufficient to provide absolute protection in all circumstances. Chandler *et al.* (1992) report a case of a house on London Clay which continued to suffer settlement damage even after the foundations had been deepened to 3 m. The continuing problems were shown to be due to water content reductions in the clay at depths of up to 8 m, caused by a row of poplar trees at a distance of 10 m from the house.

Example 1.8 The unpruned cherry tree

In addition to the crack patterns shown in Figure 1.20, cracks may also develop between parts of buildings which have foundations of different depth or type. Many late Victorian and Edwardian (i.e. late nineteenth century and early twentieth century) houses have one- or two-storey bay windows, projecting from the front (and, less frequently, side or rear) elevations (Figure 1.21(a) and (b)). These bay windows were usually built on very shallow foundations: a depth of less than 0.5 m to the underside of the foundation is not uncommon (Figure 1.21(c)). The foundations of the main house will usually be somewhat deeper – especially if the house incorporates a cellar or a basement, in which case the foundation depth could be 2 m or more. Generally, the shallower the foundation, the more susceptible it is to changes in soil water content, which occur primarily near the surface. The difference in foundation depth between the bay and the main house can lead to the development of vertical cracks at the junction between the bay and the main wall, as the bay settles and breaks away.

This type of damage is not necessarily serious. In the particular case shown in Figure 1.21, movement was actually due to the reactivation of an historic

(a)

(b)

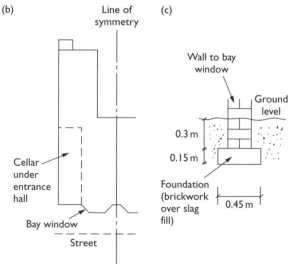

(c)

Figure 1.21 (a) General view, (b) plan and (c) cross-section through foundation of the bay window of a typical late Victorian terraced house on London Clay.

crack. The movement was noticed following a two-year period of particularly dry weather, which happened to coincide with a time during which the local authority decided not to prune the street trees in their care. One of these – a mature cherry tree – was located approximately 4 m from the front corner of the house. Measured profiles of water content with depth, near and remote from the cherry tree, are shown in Figure 1.22. (The remote or control borehole must be far enough away from the tree not to be affected by it, but close enough for the soil profile to be similar. In the present case, the control borehole was located in the back garden of the house.)

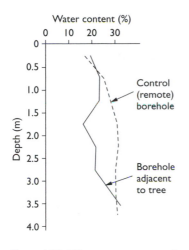

Figure 1.22 Water content as a function of depth, close to and remote from the tree: Example 1.8.

Figure 1.22 suggests that, compared with the control borehole, the clay near the tree has suffered a reduction in water content of up to 10% over a depth range of 0.75 to 3.5 m below the ground surface.

Taking the desiccated depth $h_0 = 2.75$ m, the average initial water content $w_0 = 30\%$, the average change in water content $\Delta w = 6.8\%$, and $G_s = 2.75$, equations (1.35) and (1.36) suggest a maximum settlement in the range 70–90 mm. While this figure is probably on the high side, the calculation does demonstrate that the settlements resulting from the reduction in water content would be sufficient to cause the failure of the bond between the bay and the main front wall of the house, and cracking of the structure.

It has in the past been standard practice to deepen the foundations of a house affected by settlement. This can be carried out by removing the soil from below the foundation and replacing it by concrete, a short length at a time, in a process known as underpinning. Underpinning is expensive, and carries some risk of collapse if it is not carried out very carefully. In the case described above, annual pruning of the cherry tree was resumed, and the crack widths were monitored for a period of 18 months or so. When it was certain that movement had stopped, repairs to the superstructure were carried out. These included the re-establishment of the brickwork bond between the bay window and the main front wall of the house, which had probably never been particularly effective, even when the house was first built.

1.14   Site investigation

One of the first requirements of any civil engineering project is to carry out an investigation of the site, to find out about the ground conditions and the soil types present. This is usually done by making small excavations (up to about 2 m in both diameter and depth) known as **trial pits**, and by drilling boreholes (commonly 150 mm or so in diameter) to depths of up to a few tens of metres, from which samples of soil from various depths can be recovered, inspected and tested. Samples may be tested in the laboratory to investigate characteristics such as the soil strength, compressibility, stress–strain relationship, permeability, particle size distribution, liquid and plastic limits and so on, using the methods described in Chapters 1–5.

The main problem associated with the laboratory testing of soil recovered from trial pits or boreholes is disturbance. Disturbance arises from the sampling process, in transporting the sample to the laboratory, and in preparing the sample for the test. It is very difficult to recover intact samples of non-clay soils, which must usually be tested following recompaction to their estimated *in situ* void ratio. Some sampling techniques will result in the loss of fine particles. This will affect the grading curve, and may give a completely false impression of the likely mechanical behaviour of the soil. For example, the permeability of a soil is sometimes estimated on the basis of the D_{10} particle size (section 3.3). If this procedure is used with a sample from which the fine particles have been lost, the permeability of the soil will be seriously overestimated.

Clay soils can be sampled by means of a sampling tube, which is simply pushed (or, more usually, hammered) into the soil at the current bottom of the borehole. Samples taken in this way are optimistically described as 'undisturbed', but in reality the disturbance caused by sampling can be severe (Figure 3.15). The amount of sample disturbance depends, among other things, on the **area ratio** of the sampler (i.e. the ratio of the wall area to the internal area of the tube). The highest quality samples are taken by thin-walled tubes, pushed into the soil, using an apparatus known as a **piston sampler**.

Conventional, standard size 'U100' sampling tubes have an internal diameter of 100 mm. They are hammered into the soil with a cutting shoe screwed onto the leading end, to facilitate penetration. The internal diameter of the cutting shoe is typically 1 mm less than the internal diameter of the sample tube. This reduces resistance during driving, but allows the sample to expand laterally until it makes contact with the internal walls of the tube. High-quality piston samplers may have an internal diameter of approximately 102 mm and a wall thickness of 2 mm, giving an area ratio of 4%: they are pushed carefully into the soil. Changes in water content between sampling and laboratory testing are minimized by sealing the sample into the sampling tube with wax as soon as it is recovered from the borehole. A detailed investigation into the effects of sample disturbance prior to testing, with reference to a soft estuarine deposit at Bothkennar, Scotland, is presented by Hight *et al.* (1992).

It will be seen in later chapters that the stress–strain behaviour of a soil – particularly if it is a clay – depends on its stress history, its stress state and the changes in stress imposed. It is therefore important that the stress state at the start of a laboratory test, and the changes in stress imposed in the test, should correspond as closely as possible to those anticipated in the field. For example, if a soil sample is taken from a depth of 10 m, the vertical effective stress in the ground at this depth (about 100 kPa, assuming that the groundwater level is near the soil surface) would normally define a sensible starting point for a laboratory test. If the soil in the field is to be subjected to an increase in vertical effective stress of 50 kPa due to the construction of a building, the laboratory test should cover at least the stress range 100–150 kPa.

The effect of soil structure can be very significant: this is discussed with respect to permeability in section 3.6. Unfortunately, it is almost impossible to preserve the structure of a non-clay soil during sampling, while the structure of a clay soil might be severely disrupted.

One way of attempting to overcome the likely disturbance to the soil structure (and, with non-plastic soils, the void ratio) which occurs during sampling is to carry out soil tests in the field or *in situ*. Soil strength and stiffness can be measured *in situ* in a variety of ways: some of the techniques available are described in section 11.3. Soil permeability can be measured *in situ* by means of field pumping tests, as described in section 3.5. An additional advantage of field pumping tests over laboratory investigations is that the average effective permeability of a very large volume of soil may be estimated. One particular parameter which can only be measured directly in the field is the *in situ* horizontal stress, by means of a device known as a pressuremeter (section 11.3.3).

A site investigation report normally consists of a detailed record or log of each borehole and the results of the laboratory and *in situ* tests. Sometimes, an interpretive commentary will also be included. A borehole log should include the following information:

1. Sufficient details to identify the project, the location and the individual borehole.
2. Ground level at the top of the borehole (relative to same identifiable, fixed and reproducible benchmark).
3. A description of the soil types present, using the soil description scheme presented in Table 1.5.
4. The depths and reduced levels of the interfaces between successive soil strata.
5. A record of the depths at which water was first encountered (known as **water strikes**), and standing water levels at the start and end of each day.
6. The depths from which each sample was taken, and the method of sampling.
7. A record of any *in situ* tests carried out.
8. Other details, such as date(s), time(s), personnel, type of drilling rig etc.

A typical borehole log is shown in Figure 1.23.

Boreholes should be deep enough to enable the proposed structure to be designed with confidence. The analytical techniques described in this book give some guidance on this point. For example, the soil below a long foundation is significantly loaded

SITE: LIMEHOUSE LINK ROAD STAGE II								BOREHOLE D2.3

Date: 2-12-87		Hole size 150 mm diameter to 7.50 m			Ground level	3.06 m O.D.	Sheet 1 of 1
							Scale: 1:50

Samples and *In situ* Tests							
Depth (m)	Type	Blowcount or vane shear strength kPa	Water	Legend	O.D. Level (m)	Depth (m)	Descriptions of Strata
							Brick rubble
					2.46	0.60	(MADE GROUND)
0.90–1.40	B1						Clay, gravel and brick rubble
					1.66	1.40	(MADE GROUND)
1.50	D1				1.36	1.70	Firm grey brown and brown mottled silty CLAY
1.70	D2						
			▼				Firm grey mottled silty CLAY
2.60	D3				0.26	2.60	
							Very soft grey silty and sandy CLAY
4.00–4.50	B2		▽		−0.94	4.00	
							Light grey brown medium fine SAND
					−2.84	5.90	
6.00–6.30	B3				−3.24	6.30	Grey brown sandy GRAVEL
6.30	D4						Stiff grey brown fissured silty CLAY
7.50	D5				−4.44	7.50	
							Borehole completed at 7.50 m depth.

KEY

D - Disturbed Sample
B - Bulk Sample
U - Undisturbed Sample
W - Water Sample
V - Vane Shear Test
S - Standard Penetration Test
C - Cone Penetration Test
M - Mackintosh Probe

• S.P.T./C.P.T. Where 0.3 m penetration not achieved, blows given for quoted penetration

N - Blows for 0.3 m in penetration test
▽ Water met
▼ Depth to water on completion
▼ ( ) Depth, hours after completion

REMARKS

(1) Borehole cased to 6.50 m depth.

(2) Water met at 4 m, level rose to 3.80 m, casing at 3.50 m.

(3) Water sealed out by casing at 6.50 m and hole dry on completion.

(4) Depth to water after removing casing 2.30 m.

(5) 50 mm pipe inserted to 7.50 m depth with cover.

Figure 1.23 Typical borehole log. (Reproduced with the permission of the London Docklands Development Corporation, Thames Quay, 191 Marsh Wall, London E14 9TJ.)

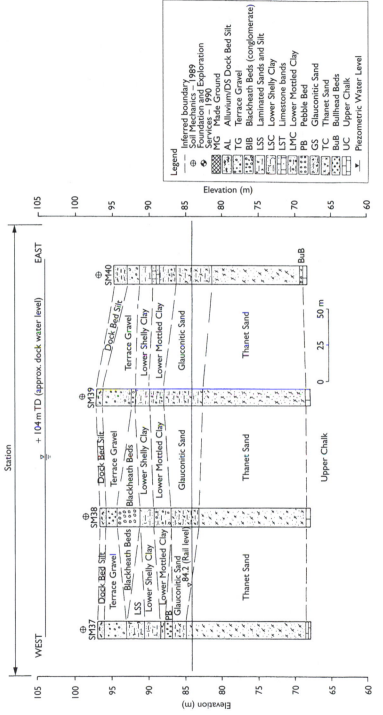

Figure 1.24 A typical geological cross-section, interpolated from borehole data. (Jubilee Line Extension, London Underground Limited.)

to a depth of approximately six times the width of the foundation, while the soil below a square foundation is significantly loaded to a depth of approximately twice the width of the foundation (section 6.3). In a site investigation for an excavation or an underground structure, the boreholes must generally penetrate to at least two or three times the depth of the excavation. This is to allow for the likely depth of an embedded retaining wall (sections 10.1–10.9), and for the estimation of the rate at which groundwater must be pumped from wells during construction, in order to keep the excavation stable and workably dry (sections 3.9–3.19).

Choosing locations of boreholes on plan is more difficult. Geological maps can be helpful, and will sometimes indicate the locations of discontinuities in soil stratification, or provide a warning of the possible presence of features such as buried river beds. In the investigation of a site where there is a significant depth of made ground (i.e. artificially placed material), or which may be contaminated, inspection of the available records and plans relating to its previous use will prove invaluable. In general, the greater the number of boreholes, the smaller the risk of encountering unforeseen ground conditions after the project has commenced, but the higher the cost of the site investigation. Like most things, it is a question of balance; like most things, it is easier to judge whether or not the balance was correct with the benefit of hindsight. It is impossible, however, to eliminate risk entirely.

As an aid to the visualization of the likely sequence of strata below ground, it is usual to plot geological cross-sections along each line of boreholes. The interface levels of the various strata are generally assumed to vary linearly between boreholes. With lenses and other discontinuous layers, which may peter out between boreholes, a degree of informed engineering geological judgement may be required in order to guess their likely extent. A typical geological cross-section is illustrated in Figure 1.24. Three-dimensional geological interpretations can also be made with the aid of an appropriate computer package.

Do not be misled by the fact that the section on site investigation in this book is comparatively short. As mentioned earlier, most of the book is relevant to the site investigation, in that it describes the tests that will be carried out on the samples obtained, or the analyses for which the data are required. However, there are other books (e.g. Clayton *et al.*, 1995) devoted entirely to site investigation, which you would be well advised to read before planning one, and codes of practice (BS5930, 1999) which you would be well advised to follow.

The volume and type of information produced by the site investigation (including the soil tests carried out) must be appropriate to the scale of the project, and relevant to the geotechnical design and analysis procedures that will be employed. There is no point in commissioning a state-of-the-art laboratory testing programme if all the designer wants is a rough indication of the soil strength. Similarly, there is no point in carrying out a sophisticated, state-of-the-art computer analysis on the basis of the results of a few index tests (although it has been done). Again, it is all a question of balance.

Finally, we should bear in mind that the behaviour of a natural soil will depend to some extent on the **soil fabric**, by which we mean the size, shape and arrangement of the solid particles, the organic inclusions and the associated voids. The relevance of soil fabric to site investigation practice is discussed in considerable detail by Rowe (1972).

Key points

- Soil is in general a three-phase material, comprising solid particles with water and air in the voids in between them. The solid particles may be densely or loosely packed, as indicated by the void ratio e or the specific volume v.
- The relative proportions of the three phases are quantified by the void ratio e (voids to solids, by volume), the water content w (water to solids, by mass) and the saturation ratio S_r (water to total voids, by volume).
- Many soils are saturated (i.e. the voids are full of water). This reduces the number of phases present to two, and simplifies analysis considerably.
- In a saturated soil, some of the applied normal stress is carried by the soil skeleton as effective stress, and some by the pore water pressure. It is essential to distinguish between these two components, because of the very different strength characteristics of the solid and liquid phases. The effective stress σ' is calculated from the total stress σ and the pore water pressure u using Terzaghi's equation,

$$\sigma' = \sigma - u$$

All shear stresses must be transmitted through the soil skeleton, as the pore water is incapable of carrying shear.

Questions

Origins and mineralogy of soils

1.1 Describe the main depositional environments and transport processes relevant to soils, and explain their influence on soil fabric and structure.

1.2 Summarize the main effects of soil mineralogy on particle size and soil characteristics.

Phase relationships, unit weight and calculation of effective stresses

1.3 A density bottle test on a sample of dry soil gave the following results:

Mass of 50 ml density bottle empty (g)	25.07
Mass of 50 ml density bottle + 20 g of dry soil particles (g)	45.07
Mass of 50 ml density bottle + 20 g of dry soil particles, with remainder of space in bottle filled with water (g)	87.55
Mass of 50 ml density bottle filled with water only (g)	75.10

Calculate the relative density (specific gravity) of the soil particles. A 1 kg sample of the same soil taken from the ground has a natural water content of 27% and occupies a total volume of 0.52 l. Determine the unit weight, the specific volume and the saturation ratio of the soil in this state. Calculate also the water content and the unit weight that the soil would have if saturated at the same specific volume, and the unit weight at the same specific volume but zero water content.

$$(G_s = 2.65; \ \gamma = 18.865 \text{ kN/m}^3; \ v = 1.75; \ S_r = 0.954;$$

$$w_{sat} = 28.3\%; \ \gamma_{sat} = 19.06 \text{ kN/m}^3; \ \gamma_{dry} = 14.86 \text{ kN/m}^3.)$$

1.4   An office block with an adjacent underground car park is to be built at a site where a 6 m thick layer of saturated clay ($\gamma = 20 \text{ kN/m}^3$) is over-lain by 4 m of sands and gravels ($\gamma = 18 \text{ kN/m}^3$). The water table is at the top of the clay layer, and pore water pressures are hydrostatic below this depth. The foundation for the office block will exert a uniform sur-charge of 90 kPa at the surface of the sands and gravels. The foundation for the car park will exert a surcharge of 40 kPa at the surface of the clay, following removal by excavation of the sands and gravels. Calculate the initial and final vertical total stress, pore water pressure and vertical effective stress, at the mid-depth of the clay layer, (a) beneath the office block and (b) beneath the car park. Take the unit weight of water as 9.81 kN/m^3.

(Initially $\sigma_v = 132$ kPa; $u = 29.4$ kPa; $\sigma_v' = 102.6$ kPa. Finally, beneath the office block $\sigma_v = 222$ kPa; $u = 29.4$ kPa; $\sigma_v' = 192.6$ kPa. Finally, beneath the car park $\sigma_v = 100$ kPa; $u = 29.4$ kPa; $\sigma_v' = 70.6$ kPa.)

1.5   For the measuring cylinder experiment described in Example 1.3, calculate (a) the vertical effective stress at the base of the column of sand in its loose, dry state; (b) the pore water pressure and vertical effective stress at the base of the column in its loose, saturated state; (c) the pore water pressure and vertical effective stress at the base of the column in its dense, saturated state; and (d) the pore water pressure and vertical effective stress at the sand surface in the dense, saturated state. Take the unit weight of water as 9.81 kN/m^3.

((a) $\sigma_v' = 6.94$ kPa; (b) $u = 4.16$ kPa, $\sigma_v' = 4.31$ kPa; (c) $u = 4.16$ kPa, $\sigma_v' = 4.31$ kPa (the weights of water and sand above the base of the column do not change); (d) $u = 0.242$ kPa, $\sigma_v' = 0$ (the water level in the column does not change: as the sand is densified, it settles through the water).)

Particle size analysis and soil filters

1.6   A sieve analysis on a sample of initial total mass 294 g gave the following results:

Sieve size (mm)	6.3	3.3	2.0	1.2	0.6	0.3	0.15	0.063
Mass retained (g)	0	0	30	39	28	28	16	11

A sedimentation test on the 117 g of soil collected in the pan at the base of the sieve stack gave:

Size (μm)	<2	2–6	6–15	15–30	30–63
% of pan sample	0	48	29	14	9

Plot the particle size distribution curve and classify the soil using the system given in Table 1.5. Determine the D_{10} particle size and the uniformity coefficient U, and comment on the grading curve.

($D_{10} = 0.0035$ mm; $D_{60} = 0.52$ mm; $U \approx 150$; soil is approximately 40% silt, 50% sand and 10% fine gravel: this makes it a sandy SILT; soil is poorly graded (almost gap-graded).)

1.7   A sieve analysis on a sample of initial total mass 411 g gave the following results:

Sieve size (mm)	6.3	1.2	0.3	0.063
Mass retained (g)	0	60	126	92

A sedimentation test on the 121 g of soil collected in the pan at the base of the sieve stack gave:

Size (μm)	<2	2–10	10–60
% of pan sample	33	24	43

Plot the particle size distribution curve and classify the soil using the system given in Table 1.5. On the PSD diagram, sketch a suitable curve for a granular filter to be used between this soil and a drainage pipe with 3 mm perforations.

(Soil is approximately 10% clay, 20% silt, 60% sand and 10% fine gravel: this makes it a clayey, very silty SAND. $D_{15s} \approx 0.007$ mm, $D_{85s} \approx 1.2$ mm; filter PSD curve has $D_{5f} \geq 0.063$ mm; $D_{15f} \leq 6$ mm; $D_{10f} \sim 3$ mm and $D_{60f} \sim 9$ mm ($U \sim 3$).)

Index tests and classification

1.8   The following results were obtained from a series of cone penetrometer tests using a standard 80 g, 30° cone:

Mass of tin empty (g)	18.2	19.1	17.7	18.6
Mass of tin + sample wet (g)	51.5	45.5	50.7	43.4
Mass of tin + sample dry (g)	37.8	35.6	39.7	36.3
Cone penetration d (mm)	25.0	14.2	8.5	5.1

Determine the water content w of each sample. Plot a graph of w against $\ln(d)$, and estimate the liquid limit w_{LL}. If the soil has a plastic limit of 22%, calculate the plasticity index and classify the soil using the chart given in Figure 1.15.

($w_{LL} \approx 65\%$; $PI \approx 43\%$; High plasticity clay (CH).)

Compaction

1.9   The following results were obtained from a standard (2.5 kg) Proctor compaction test:

Mass of tin empty (g)	14	14	14	14	14
Mass of tin + sample wet (g)	88	68	98	94	93
Mass of tin + sample dry (g)	81	62	87	82	80
Density (kg/m^3)	1730	1950	2020	1930	1860

Plot a graph to determine

  (i)   the maximum dry density
 (ii)   the optimum water content
(iii)   the actual density at the optimum water content.

If the particle relative density $G_s = 2.65$, calculate

 (iv)   the specific volume
  (v)   the saturation ratio at the maximum dry density.

((i) about 1770 kg/m^3; (ii) 14%; (iii) 2018 kg/m^3; (iv) 1.497; (v) 74.6%.)

Notes

1 Sedimentary rocks were originally particulate materials, which have been converted to rock by the application of vertical pressure as further material was deposited on top of them. Chalk and limestone are made up of the skeletal and shelly remains of tiny marine creatures, while sandstone is derived from sand, shale from clay and coal from peat. Crystalline rocks appear as a mass of interlocking crystalline units, fused together either by heat (e.g. in a volcano), in which case the rock is described as **igneous**; or by pressure and chemical processes occurring in the solid state, in which case the rock is described as **metamorphic**.

2 A **mineral** may be defined as a structurally homogeneous solid of definite chemical composition, formed by the inorganic processes of nature (Whitten and Brooks, 1972). This definition includes ice, but excludes coal. Mercury, though liquid at normal temperatures, is usually classed as a mineral. Minor variations in chemical composition between prescribed limits, which do not markedly alter the fundamental properties, are allowable, and indeed occur in many minerals.

3 The gauge pressure is the pressure above atmospheric, which is the pressure that would be measured using a pressure gauge. In soil mechanics, we are concerned almost exclusively with gauge pressures. The terms **pore water pressure** or even **pore pressure** are therefore used, and it is taken for granted that what is meant is actually the gauge, rather than the absolute, pressure.

4 That is, the maximum shear stress which the soil can withstand when it is sheared undrained at constant volume and constant water content. This **undrained shear strength** is not the same as the more fundamental effective frictional strength, because the undrained shear strength depends on the magnitude of the effective stress. In the liquid and plastic limit tests, the total stress is virtually zero, so that any effective stress must be attributable to the presence of a negative pore water pressure. It will be shown in Chapter 5 that, for clay soils deforming at constant volume, the effective (normal and shear) stresses at failure – and hence the undrained shear strength – depend entirely on the water content.

References

American Society for Testing Materials (1970) *Standard test method for classification of soils for engineering purposes*. ASTM document number D2487–1969.

Atterberg, A. (1911) Lerornas forhallande till vatten, deras plasticitetsgranser och plasticitetsgrader. *Kungl. Lantbruks akademiens Handlingar och Tidskrift*, **50**(2), 132–158.

Blyth, F.G.H. and de Freitas, M.H. (1984) *A Geology for Engineers*, 7th edn., Edward Arnold, London.

BS1377 (1990) *Methods of Test for Soils for Civil Engineering Purposes*, British Standards Institution, London.

BS5930 (1999) *Code of Practice for Site Investigations*, British Standards Institution, London.

Chandler, R.J., Crilly, M.S. and Montgomery-Smith, G. (1992) A low cost method of assessing clay desiccation for low-rise buildings. *Proceedings of the Institution of Civil Engineers (Civil Engineering)*, **92**(2), 82–89.

Clayton, C.R.I., Simons, N.E. and Matthews, M.C. (1995) *Site Investigation*, Blackwell Science, London.

Department of Transport (1993) *Manual of Contract Documents for Highway Works, Volume 1. Specification for Highway Works*, HMSO, London.

Driscoll, R.M.C. (1983) The influence of vegetation on the swelling and shrinking of clay soils in Britain, *Géotechnique*, **43**(2), 93–105.

Gillott, J.E. (1968) *Clay in Engineering Geology*, Elsevier, London.

Heyman, J. (1972) *Coulomb's Memoir on Statics: an Essay in the History of Civil Engineering*, Cambridge University Press, Cambridge.

Hight, D.W., Boese, R., Butcher, A.P., Clayton, C.R.I. and Smith, P.R. (1992) Disturbance of the Bothkennar Clay prior to laboratory testing, *Géotechnique*, **42**(2), 199–217.

Kenney, T.C., Chahal, R., Chiu, E., Ofoegbu, G.I., Omange, G.N. and Ume, C.A. (1985) Controlling constriction sizes of granular filters, *Canadian Geotechnical Journal*, **22**, 32–43.

Kolbuszewski, J.J. (1948) An experimental study of the maximum and minimum porosities of sand. *Proceedings of the Second International Conference on Soil Mechanics and Foundation Engineering*, Rotterdam, **1**, 158–165.

Marshall, T.J., Holmes, J.W. and Rose, C.W. (1996) *Soil Physics* (3rd edition), Cambridge University Press, Cambridge.

Mitchell, J.K. (1993) *Fundamentals of Soil Behavior*, 2nd edn., John Wiley & Sons, New York.

Preene, M., Roberts, T.O.L., Powrie, W. and Dyer, M. (2000) Groundwater control – design and practice. CIRIA Report C515. Construction Industry Research and Information Association, London.

Robinson, H. (1977) *Morphology and Landscape*, 3rd edn., University Tutorial Press, Slough.

Rowe, P.W. (1972) The relevance of soil fabric to site investigation practice. Twelfth Rankine Lecture, *Géotechnique*, **22**(2), 195–300.

Sherard, J.L., Dunnigan, L.P. and Talbot, J.R. (1984) Basic properties of sand and gravel filters, *American Society of Civil Engineers, Journal of the Geotechnical Engineering Division*, **102**, 69–85.

Skempton, A.W. (1953) The colloidal activity of clay. *Proceedings of the 3rd International Conference on Soil Mechanics and Foundation Engineering*, **1**, 57–61.

Skempton, A.W. (1960) Effective stress in soil, concrete and rocks. *Proceedings of the Conference on Pore Pressure and Suction in Soils*, 4–16, Butterworths, London.

Terzaghi, K. (1936) The shearing resistance of saturated soils. *Proceedings of the First International Conference on Soil Mechanics*, **1**, 54–56.

Whitten, D.G.A. and Brooks, J.R.V. (1972) *The Penguin Dictionary of Geology*, Penguin Books Ltd, Harmondsworth.

Whyte, I.L. (1982) Soil plasticity and strength: a new approach using extrusion, *Ground Engineering*, **15**(1), 16–24.

Winterkorn, H.F. and Fang, H.-Y. (1991) Soil technology and engineering properties of soil, in *Foundation Engineering Handbook*, 2nd edn (ed. H.-Y. Fang), Chapman & Hall/Van Nostrand Reinhold, New York.

Chapter 2

Soil strength

2.1 Introduction and objectives

One of the key questions concerning any geotechnical structure is 'is it safe?'. To begin to answer this question, the geotechnical engineer must address two points:

- How are the applied loads (due to external agencies and the soil's own weight) distributed within the soil mass as stresses?
- Is the soil strong enough to withstand these stresses?

This chapter is concerned primarily with the quantification and measurement of soil strength, using a laboratory testing apparatus known as a **shearbox**. It is assumed that the reader has an understanding of the concepts of stress and strain, and is familiar with the representation of the states of stress and strain within a plane by means of Mohr circles. A brief summary of some of the elements of stress analysis is given in section 2.2; further details will be found in, for example, Case *et al.* (1992).

Objectives

After having worked through this chapter, you should understand that

- soil strength is due primarily to interparticle friction (section 2.4)
- the ability of soil to resist shear on a particular plane depends on the normal effective stress acting on that plane (section 2.4)
- positive pore water pressures reduce the normal effective stress, and generally have a destabilizing effect on a geotechnical structure (section 2.4)
- when sheared, a soil will eventually reach a **critical state** in which unlimited shear strain could be applied without further changes in specific volume, normal effective stress or shear stress (sections 2.7 and 2.8)
- the combination of specific volume, normal effective stress and shear stress at the critical state lies on a unique line – the **critical state line** – on a three-dimensional plot with axes representing specific volume, normal effective stress and shear stress (section 2.8)
- a dense soil will develop a **peak strength** before reaching a critical state, because it must increase in volume or **dilate** to reach the critical state. The peak strength

cannot be sustained indefinitely, because dilation must cease when the critical state is reached (sections 2.7–2.9)

- for saturated clays sheared at constant volume, the shear stress at failure depends on the specific volume or the water content. This shear stress is known as the **undrained shear strength**, and can be used to define an alternative failure criterion in terms of total, rather than effective, stresses (section 2.10).

You should be able to

- process data from a shearbox test and present results in engineering terms and units (sections 2.5 and 2.6)
- determine the critical state strength ϕ'_{crit}, the peak strength ϕ'_{peak} and the undrained shear strength τ_u, from processed shearbox test data (sections 2.7–2.10)
- use these strength parameters to investigate the failure of simple geotechnical structures, such as pull-out anchors and friction piles (section 2.11).

Sections 2.12 and 2.13 contain material which might quite reasonably be omitted from an introductory-level soil mechanics course. If, however, these sections are included in your studies, they should help you to gain an appreciation of:

- the analysis of the shearbox test in terms of Mohr circles of stress and strain increment
- the uncertainties associated with the stresses and strains in a shearbox test sample, and the limitations of the conventional interpretation of the data obtained.

2.2   Stress analysis

2.2.1   General three-dimensional states of stress

The loads and forces applied to a solid body (such as a soil mass) are distributed within the body as **stresses**. Provided that there are no planes of weakness which might interrupt the transfer of stress, it is usually assumed that the stresses vary smoothly and continuously throughout the body, which is then described as a **continuum**.

If we were to imagine a cubical element within a three-dimensional body under a general state of stress, there could be up to three independent stresses acting on each pair of opposite faces, as shown in Figure 2.1. Two of the stresses on each face act parallel to the face, at right angles to each other: these are **shear stresses**, and are denoted by the symbol τ. The third stress acts perpendicular to the face of the cube: it is known as the **normal** (or **direct**) stress, and is given the symbol σ if it is a total stress, or σ' if it is an effective stress. To distinguish between the various shear and normal stresses, a system of double subscripts is used, for example τ_{xy}. The first subscript (in this case x) denotes the direction of the normal to the plane on which the stress acts, while the second subscript (in this case y) denotes the direction in which the stress acts.

In soil mechanics, it is usual to take compressive direct stresses and strains, and anticlockwise shear stresses and the associated shear strains, as positive. This is in contrast to structural mechanics, in which tensile direct stresses and strains, and

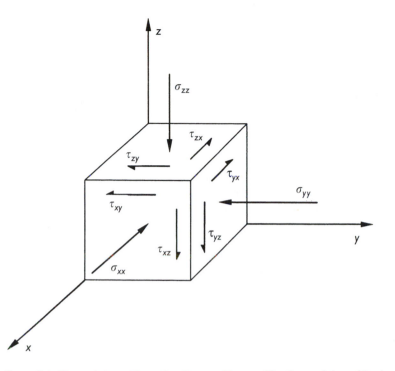

Figure 2.1 General three-dimensional state of stress. The faces of the cubic element shown are termed the **positive faces**, because the outward normals are in the positive directions of *x*, *y* and *z*. On the faces not shown (the **negative faces**), the directions of the stresses are reversed in order to satisfy to condition of equilibrium.

clockwise shear stresses, are conventionally taken as positive. The reason we adopt the 'compression is positive' sign convention in soil mechanics is that soils are essentially particulate materials, which cannot sustain tensile stresses unless the particles are cemented together. Stresses in soil mechanics are therefore almost always compressive. (A stress **increment**, however, can be tensile provided that the overall stress remains compressive. Tensile strains are also permissible, again provided that the overall stress remains compressive.)

2.2.2 Principal stresses

By rotating the cube shown in Figure 2.1, we should be able to find one particular orientation for which all of the shear stresses acting across the faces of the cube are zero. The three planes defined by the three pairs of parallel faces are the **principal planes,** and the normal stresses which act on the three planes are the **principal stresses**. By definition, the shear stress associated with a principal stress is zero, and a principal plane is a plane on which there is zero shear stress.

The largest principal stress is termed the **major principal stress,** the smallest principal stress is the **minor principal stress,** and the remaining principal stress is the **intermediate principal stress**. As an alternative to the system of double subscripts described in section 2.2.1, the major, intermediate and minor principal stresses are sometimes denoted by subscripts 1, 2 and 3 respectively.

2.2.3 Plane strain

In many geotechnical problems there is no need to carry out a full three-dimensional stress analysis. Many embankments and retaining walls, and some types of foundation, are long in comparison with their width and height. One cross-sectional plane must therefore be identical to any other in all respects, including the stresses acting within it. Also, if the embankment, foundation or retaining wall is effectively infinitely long, any cross-sectional plane must be a plane of symmetry.

Imagine a long embankment, cut in half along a cross-sectional plane. Suppose that, on the left-hand half of the embankment, there is a shear stress acting in the plane of the cross-section in some particular direction. The condition of equilibrium across the imaginary cut requires an equal and opposite shear stress to act on the right-hand side of the embankment, whereas the condition of symmetry requires the shear stress on the right-hand side to be acting in the same direction. These two requirements can only be met if the shear stress on the cross-sectional plane is zero.

This means that the plane of the cross-section is a principal plane. In fact, it is almost always the plane on which the intermediate principal stress acts. The planes on which the major and minor principal stresses act are therefore perpendicular to the plane of the cross-section, and the major and minor principal stresses are contained within the cross-sectional plane.

We shall see later in this chapter that the failure of soil is governed by the ratio of the major and minor principal effective stresses, or in certain conditions the difference between the major and minor principal total stresses. For geotechnical constructions that are long in comparison with their other dimensions, we can therefore concentrate our analysis on the plane of the cross-section, which contains the important major and minor principal stresses. The intermediate principal stress, which acts parallel to the length of the embankment, foundation or retaining wall, is generally of less importance and can often be ignored in analysis. This is particularly useful, because the intermediate principal stress is usually the most difficult to calculate.

A further consequence of the embankment, foundation or retaining wall being long in comparison with its height and width is that there is no strain in the longitudinal direction. If one section were to expand along its length, the adjacent section would have to contract so that the overall length remained the same. This would contravene the basic requirement that the stresses and strains must be identical at every crosssection. All deformation takes place within the cross-section of the structure, with the longitudinal (intermediate) principal stress taking up whatever value is necessary to ensure that the strain in the longitudinal direction is zero. This condition is known as **plane strain**.

In reality, geotechnical structures are of finite extent and there are bound to be differences in geometry and/or soil conditions along their length. Nonetheless, for retaining walls, embankments, foundations, excavations and other constructions which are long in comparison with their width and height, the assumption that deformation occurs in plane strain is a very useful and reasonable approximation, which simplifies the analysis enormously.

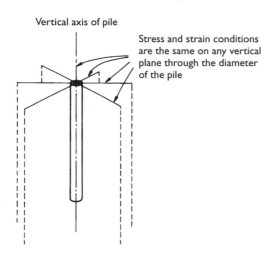

Vertical axis of pile

Stress and strain conditions
are the same on any vertical
plane through the diameter
of the pile

Figure 2.2 Axisymmetry.

2.2.4 Axisymmetry

In some cases, the stress analysis of a geotechnical structure can be simplified in a
slightly different way. For a single foundation formed from a circular cylinder of
reinforced concrete installed in the ground with its axis vertical (known as a **pile**), or
a pumped well, the conditions on any vertical plane which includes a diameter of the
pile or the well must be the same (Figure 2.2). Using the same arguments as above, the
diametral plane is the principal plane which contains both the major and the minor
principal stresses. The stress and strain conditions have rotational symmetry about the
vertical axis of the pile or well, and are termed **axisymmetric**. Stress analysis is focused
on a typical diametral plane, in the same way that a typical cross-sectional plane is
used in the analysis of a plane strain problem.

It is not always possible to idealize a real problem as either plane strain or axisymmetric. In these cases, a full three-dimensional analysis may be required. This must
usually be carried out numerically by means of a finite element or finite difference
analysis, as described in section 11.2.1.

2.2.5 Mohr circle of stress

The normal and shear stresses σ and τ acting on an imaginary cut within a typical
cross-sectional or diametral plane will depend on the orientation of the cut with respect
to the major and minor principal stress directions (Figure 2.3). If the cut is perpendicular to either the major or the minor principal stress, the shear stress acting in the
direction of the cut will be zero. In general, however, there will be a shear stress acting
along the cut, the magnitude of which increases as the cut is rotated away from the
direction of the planes of principal stress.

The stress state within a plane containing the major and minor principal stresses (or
indeed, within a plane containing any pair of principal stresses) is most conveniently
represented by means of a graphical construction known as the **Mohr circle of stress**

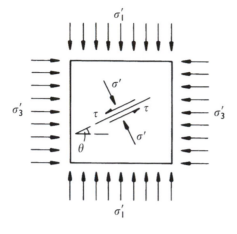

Figure 2.3 Normal and shear stresses acting on an imaginary cut within the cross-sectional plane of a long geotechnical construction.

(Figure 2.4). The Mohr circle is literally a circle, plotted on a graph of shear stress τ against normal stress, σ. The circle may be plotted for either total normal stresses, σ, or for effective normal stresses, σ'. The total and effective shear stresses are the same, because all shear must be carried by the soil skeleton as effective stress.

The plane containing the major and minor principal stresses σ_1 and σ_3 is often referred to as the 1–3 plane, the plane containing the major and intermediate principal stresses σ_1 and σ_2 as the 1–2 plane, and the plane containing the intermediate and minor principal stresses σ_2 and σ_3 as the 2–3 plane.

The Mohr circle passes through the points representing the major and the minor principal stresses, whose coordinates are $(\sigma_1', 0)$ and $(\sigma_3', 0)$ (for effective stresses) and $(\sigma_1, 0)$ and $(\sigma_3, 0)$ (for total stresses) respectively. The centre of the circle of effective stress is at $((\sigma_1' + \sigma_3')/2, 0)$, and the centre of the circle of total stress is at $((\sigma_1 + \sigma_3)/2, 0)$. Recalling that $\sigma = \sigma' + u$ (where u is the pore water pressure), the centres of the circles of effective and total stress are separated by a distance equal to u along the normal stress axis. $(\sigma_1' + \sigma_3')/2$ is the average of the major and minor principal effective stresses, and is conventionally given the symbol s'. Similarly, $(\sigma_1 + \sigma_3)/2$ is the average of the major and minor principal total stresses, and is given the symbol s.

The radius of the circle of effective stress is $(\sigma_1' - \sigma_3')/2$, while the radius of the circle of total stress is $(\sigma_1 - \sigma_3)/2$. These are identical, because the pore water pressure u is cancelled out in the subtraction of the two principal effective stresses. $(\sigma_1' - \sigma_3')/2$ (or $(\sigma_1 - \sigma_3)/2$) is equal to the maximum shear stress acting within the 1–3 plane, and is conventionally referred to by the symbol t.

The stresses acting on an imaginary cut at an angle θ anticlockwise from the plane on which the major principal stress acts are found by drawing a line from the centre of the Mohr circle to the circumference, which makes an angle 2θ (measured anticlockwise) with the normal stress (σ or σ') axis. The stresses on the 'cut' (in effective stress terms, (σ', τ)) are given by the point where this diameter meets the circumference of the circle (Figure 2.4(b)).

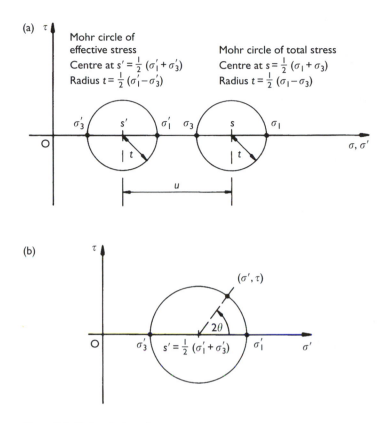

Figure 2.4 Mohr circles of stress.

The Mohr circle of stress shows that, unless the major and minor principal stresses are equal, there must be some shear stress acting somewhere within the plane under consideration. The maximum shear stress within the plane is equal to the radius of the Mohr circle, $(\sigma_1' - \sigma_3')/2 = (\sigma_1 - \sigma_3)/2$. It occurs at angles of $\pm 90°$ to the normal stress axis on the Mohr diagram, indicating that in reality, the shear stress is largest on planes that are at $\pm 45°$ to the planes on which the major and minor principal stresses act.

2.2.6 Mohr circle of strain

Strain, like stress, may be categorized as either **direct** or **shear** (Figure 2.5). Direct strain is given the symbol ε. Engineering shear strain, defined in Figure 2.5(b), is given the symbol γ. (The engineering shear strain γ is the overall decrease in the angle between the positive directions of two perpendicular lines – in the case of Figure 2.5, the x- and y-coordinate axes.) The same systems of subscripts are used as for stresses. If a material is elastic, the directions of the principal strain increments will coincide with the directions of the principal stress increments during an increment of loading, but otherwise this may not be the case.

(a)

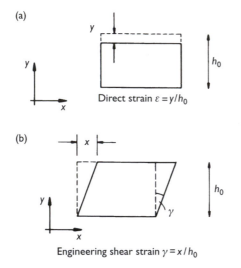

Direct strain $\varepsilon = y/h_0$

(b)

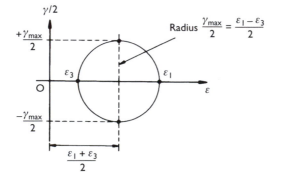

Engineering shear strain $\gamma = x/h_0$

Figure 2.5 (a) Direct strain and (b) engineering shear strain.

Figure 2.6 Mohr circle of strain.

The state of strain in a plane containing two principal strains may be represented by the **Mohr circle of strain,** as shown in Figure 2.6. This is exactly analogous to, and is used in the same way as, the Mohr circle of stress. However, to obtain a circle on the Mohr diagram of strain, the vertical ordinate is $\gamma/2$, not γ.

It is the behaviour of the material perpendicular (**normal**) to a plane that determines the strain related to that plane. For consistency with the soil mechanics sign convention on stresses, the direct strain associated with a plane is taken as positive if the material in the normal direction compresses. The shear strain is positive if this material rotates anticlockwise relative to the plane.

For example, for the x-plane in Figure 2.7, the rotation of the material normal to the plane is anticlockwise relative to the plane, and the associated shear strain γ_{xy} plots as positive on the Mohr circle of strain. For the y-plane in Figure 2.7, the rotation of the material normal to the plane is clockwise relative to the plane, and the

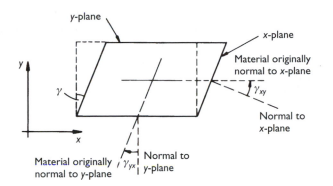

Figure 2.7 True shear strain.

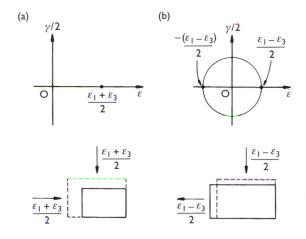

Figure 2.8 (a) Uniform compression; (b) pure shear.

associated shear strain γ_{yx} plots as negative on the Mohr circle of stress. This, too, is consistent with the soil mechanics sign convention for stresses. The deformation shown in Figure 2.7 would be caused by clockwise shear stresses τ_{yx} – which plot as negative on the Mohr circle of stress – acting on the top and bottom faces of the element. To fulfil the condition of equilibrium, the shear stresses on the vertical faces τ_{xy} must be anticlockwise, plotting as positive on the Mohr circle of stress.

Figure 2.6 shows the Mohr circle of strain for an element of soil, subjected to compressive strains of ε_1 in the vertical direction and ε_3 in the horizontal. We can consider the overall strain pattern as being made up of a uniform compression (in both directions) of $(\varepsilon_1 + \varepsilon_3)/2$ (Figure 2.8(a)), onto which is superimposed a shear deformation involving a further compression of $(\varepsilon_1 - \varepsilon_3)/2$ in the vertical direction and expansion of the same magnitude in the horizontal (Figure 2.8(b)).

The strain pattern shown in Figure 2.8(a) causes a change in the area of the cross-section, without shear in the 1–3 plane. (In plane strain, the prevention of out-of-plane strains will lead to the development of shear strains in the 1–2 and 2–3 planes).

Conversely, the strain pattern shown in Figure 2.8(b) involves pure shear in the 1–3 plane, without changing the area of the cross-section. Figure 2.6 shows that the 'volumetric' strain component $(\varepsilon_1 + \varepsilon_3)/2$ gives the location of the centre of the Mohr circle of strain, while the 'distortional' strain component $(\varepsilon_1 - \varepsilon_3)/2$ is equal to the radius. As with the Mohr circle of stress, the position of the centre is associated with direct effects, and the radius with shear.

Magnitudes of stresses are absolute, but strains must always be expressed with reference to some arbitrary and non-reproducible datum condition. In the analysis of a soil test it is sometimes necessary to distinguish between the strain that has occurred since the start of the test and the strain that occurs between two stages, both of which are some way into the test. The strain that has occurred since the start of the test is generally described as simply the **strain**, but sometimes, in order to remove any possibility of confusion, the term **cumulative strain** may be used. The strain between two stages, both some way into the test, is described as the **incremental strain** or a **strain increment**. Generally, an incremental quantity will be prefixed by a d, δ or Δ, as in $d\varepsilon$, $\delta\varepsilon$ or $\Delta\varepsilon$. Mohr circles of **strain increment** are used in the stress analysis of the shearbox sample in section 2.12.

Full details of the analysis of stresses and strains, including the mathematical arguments associated with their Mohr circle representations, will be found in Case *et al.* (1992) and other textbooks on stress analysis.

2.3   Soil strength

In soil mechanics and geotechnical engineering, strength may be defined as the ability to resist shear. It is the ability of a material to resist shear which enables the principal stresses to be different in different directions. This is indicated by the Mohr circle of stress for the plane containing the major and minor principal stresses, shown in Figure 2.9(a). Fluids such as water and treacle cannot sustain shear stresses when they are stationary. The stress within a stationary fluid must therefore be the same in all directions, and the Mohr circle of stress is effectively a single point (Figure 2.9(b)). Unless a material can withstand shear stresses, it will not be possible to form and maintain non-horizontal surfaces, such as embankment and cutting slopes.

Soil is able to withstand shear stresses, while water is not. This is why it is necessary to distinguish the component of stress carried by the soil skeleton from the component carried by the pore water, according to the principle of effective stress

$$\sigma' = \sigma - u \tag{1.13}$$

Soil is able to withstand shear stresses due to interparticle friction. Although in some natural soils the particles may be lightly cemented, the cementitious bonds are brittle: once broken, their strength is gone forever. As mentioned in Section 1.4, surface forces in clays may be significant at low effective stresses, but at depths of more than a few centimetres they are generally small in comparison with the self-weight of the soil. Thus interparticle friction is the main source of strength for nearly all soils, whether they are predominantly sand, silt or clay.

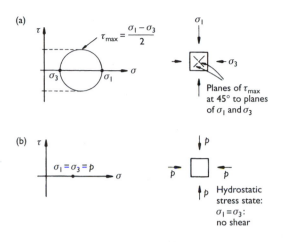

Figure 2.9 Mohr circles of stress for the plane containing the major and minor principal stresses. (a) Material which is able to resist shear stress; (b) material which is unable to resist shear stress.

2.4   Friction

Imagine a wooden block on a wooden table, as shown in Figure 2.10. An experiment is carried out in which the normal load N is kept constant, and the sideways force F is gradually increased until the block starts to slide. The value of F needed to make the block slide is recorded. The experiment is then repeated with different values of N. If the results are plotted as a graph of N against F at sliding, it will be found that there is a linear relationship between them, which may be expressed in the form

$$F = \mu N \tag{2.1a}$$

where μ is the **coefficient of friction** between the block and the table.

Alternatively, the coefficient of friction μ can usefully be expressed in terms of the **angle of friction** ϕ. ϕ is the angle of inclination of the resultant force R on the sliding interface, measured from the normal (Figure 2.11(a)). It is also the slope of the line joining possible combinations of F and N at sliding (Figure 2.11(b)). Substituting $\mu = \tan \phi$ into equation (2.1a),

$$F = N \tan \phi \tag{2.1b}$$

The wooden block could be taken to represent an **earth retaining wall** (i.e. a wall built to retain a mass of earth) of weight N, required to resist the lateral thrust F of the retained soil without moving. The onset of continued sliding would then be regarded as the failure of the structure. The line represented by equation (2.1b) defines combinations of N and F which must not be approached if failure is to be avoided. If $F < N \tan \phi$, the block will remain stationary. If $F = N \tan \phi$, the block will be on the verge of sliding. If $F > N \tan \phi$, the block will accelerate away sideways under the action of a net lateral force. Equation (2.1b) is a simple **failure criterion**, which defines the relationship between F and N at which failure will just occur.

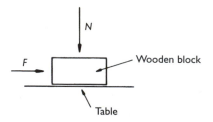

Figure 2.10 Wooden block on a wooden table.

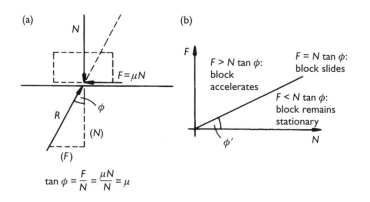

Figure 2.11 (a) Inclination of resultant force on interface; (b) relationship between *F* and *N* when the block starts to slide.

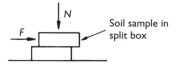

Figure 2.12 Sliding interface (shearbox) test to investigate frictional characteristics of a soil.

The principle of the sliding block could be used to investigate the frictional strength of a soil. It would be necessary to contain the sides of the block of soil to prevent it from collapsing into a heap. Steps would also have to be taken to prevent the 'block' from ploughing into the 'table'. The most practical way of testing a sample of soil in this way is to place the sample in a box divided horizontally into two halves, and to shear the top half of the sample across the bottom (Figure 2.12). The basic apparatus for doing this is known as a **shearbox**, and is illustrated in Figure 2.14.

The results of a shearbox test on a sample of soil are analysed in terms of stresses, rather than forces. Also, the analysis must be carried out in terms of effective, rather than total, stresses. The shear stress $\tau = F/A$ and the (compressive) normal total stress $\sigma = N/A$ are obtained by dividing F and N by the cross-sectional area of the soil sample A. The normal effective stress σ' is obtained by subtracting the pore water pressure at the interface u from the total normal stress σ, according to equation (1.13). If $u = 0$ (as is almost always the case in shearbox tests on sands, but not necessarily in shear box tests on clays), $\sigma = \sigma' = N/A$ and equation (2.1b) may be rewritten as

$$\tau = \sigma' \tan \phi' \tag{2.2}$$

where the prime (') is added to the angle of friction (ϕ') to indicate that it is a parameter which relates to effective stresses. A soil which relies solely on interparticle friction for its strength would be expected to obey the failure criterion given in equation (2.2).

Equation (2.2) defines combinations of shear stress τ and normal effective stress σ' which may not be exceeded. It is not possible for the soil to sustain a stress state whose Mohr circle of effective stress crosses the line $\tau = \sigma' \tan \phi'$ (Figure 2.13(a)). Permissible states of stress have Mohr circles either inside this line (Figure 2.13(b)), or – in the case of a soil which is just on the verge of failure – touching it (Figure 2.13(c)). A stress condition represented by a Mohr circle which just touches the line $\tau = \sigma' \tan \phi'$ is known as a **limiting stress state**. The line $\tau = \sigma' \tan \phi'$ is a tangent to the Mohr circles of stress representing all possible stress states at failure or limiting states of stress (Figure 2.13(d)): it is therefore termed a **failure envelope**. The condition $\tau = \sigma' \tan \phi'$ is sometimes referred to as a **Mohr–Coulomb failure criterion**.

According to the failure envelope shown in Figure 2.13(d), the soil has zero strength when the normal effective stress σ' is zero. If the soil grains are in reality cemented together, the soil will have some strength at $\sigma' = 0$, because some shear stress will be required to break the cementitious bonds. As mentioned before, however, the cementitious bonds are brittle: once broken, their strength is lost completely. For

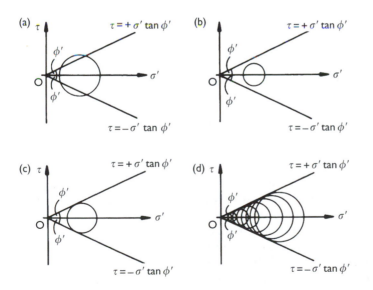

Figure 2.13 (a) Impossible stress state; (b) permissible stress state with soil not at failure; (c) permissible stress state with soil on verge of failure (limiting stress state); (d) the lines $\tau = \pm\sigma' \tan \phi'$ as an envelope to all possible limiting stress states.

this reason, it is generally unwise in geotechnical engineering design to rely on the strength of cementitious bonds between soil particles, unless the bonds are strong in comparison with the loads which are expected to be applied to the soil. If this is the case, the 'soil' is fairly heavily cemented, and is probably more appropriately classed as a weak rock.

We shall see in section 2.7 and later in this book that the behaviour of soil is somewhat more complicated than that of a wooden block sliding across a wooden table. One of the reasons for this is the tendency of the soil to change in volume as it is sheared. Along with a measure of the shear stress and a measure of the normal effective stress, the specific volume of the soil is an essential indicator of its state. It is important, therefore, to measure the changes in volume which occur during shear: the standard shearbox apparatus is designed to enable this.

As mentioned earlier, the frictional nature of soil is one of the reasons why it is necessary to distinguish between the effective stress and the pore water pressure. Water cannot resist shear, so that all of the applied shear stress must be carried by the soil skeleton. Furthermore, if the pore water pressure is positive, the normal effective stress σ' will be less than the applied normal total stress σ. This will reduce the ability of the soil to resist shear, according to equation (2.2) with $\sigma' = \sigma - u$. This is important in all geotechnical applications, but is illustrated particularly clearly in the assessment of the stability of a slope, as described in sections 8.10–8.11.

The angle of friction measured in a shearbox test on a soil sample is not the same as the true **interparticle** or **material** angle of friction. The true interparticle angle of friction could be measured by, for example, forcing two solid blocks, made from the same mineral as the soil particles, to slide across each other. In engineering applications, however, the angle of friction measured in the shearbox test, which is sometimes termed the **apparent** or **effective angle of friction** or the **angle of shearing resistance**, is much more significant. In fact, the interparticle friction angle may have surprisingly little influence on the effective angle of shearing resistance of an assemblage of particles that move relative to each other mainly by rolling other than sliding (Ni, 2003).

2.5   The shearbox or direct shear apparatus

The standard laboratory shearbox (or direct shear) apparatus, which is illustrated in Figure 2.14, is perhaps the simplest way of investigating the shear strength and the shear stress–strain behaviour of a soil.

In a shearbox test, a soil sample 20 mm thick is placed inside a split brass box, of internal dimensions 60 mm × 60 mm on plan. A vertical stress is applied to the sample by means of dead weights on a hanger, which bears on the lid of the shearbox. The shearbox lid covers the entire sample, and distributes the hanger load N uniformly across it. Usually, the shearbox lid simply rests on the upper surface of the soil sample, so that it is free to move up or down as the volume of the sample changes. Initially, the two halves of the box are held together by two screws. Before the test commences, these screws are removed. The top half of the box is raised slightly (using a second pair of screws), so that there is no metal-to-metal contact during shear.

If the soil to be tested is a sand, it is usually necessary to recompact the sample into the shearbox at an appropriate specific volume. The initial specific volume v_0 may be determined provided that the dry mass of soil m_s, the specific gravity of the particles

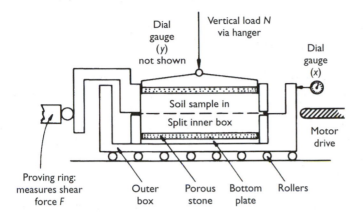

Figure 2.14 Standard shearbox apparatus.

G_s, and the initial total volume V_{t0} are known:

$$V_{t0} = V_s + V_{v0} = V_s(1 + [V_{v0}/V_s]) = V_s v_0$$

and (by definition)

$$V_s = m_s/\rho_s = m_s/(G_s\rho_w)$$

giving

$$v_0 = (V_{t0}\ G_s\rho_w)/m_s \qquad\qquad\qquad (2.3)$$

The test is carried out by shearing the two halves of the sample relative to each other, by means of a motor-driven ram acting on the bottom half. The lateral force F needed to hold the top half of the shearbox stationary is measured by means of a proving ring or a load cell. A proving ring is a stiff steel spring in the form of a ring, whose spring constant (i.e. force to deflection ratio) is determined by calibration prior to the shearbox test. In the shearbox test, the lateral force required to stop the lower half of the shearbox from moving is transmitted across the diameter of the proving ring. The shear load can then be deduced from the diametral compression of the proving ring, which is measured by means of a dial gauge or an electrical displacement transducer. The compression of the proving ring or load cell will result in a small displacement of the top half of the box. This is taken into account by calculating the shear strain from the relative displacement of the two halves of the shearbox (see section 2.6). The upward vertical displacement y of the lid of the shearbox, and the lateral displacement of the lower half of the shearbox, are measured using dial gauges or electrical displacement transducers known as **lvdts** (linearly variable differential transformers).

The displacement of the top half of the shearbox relative to the bottom half x is calculated as the absolute displacement of the bottom half of the shearbox, minus the displacement (in the same direction) of the top half, due to the compression of the proving ring or load cell. Alternatively, the lateral displacement transducer may be

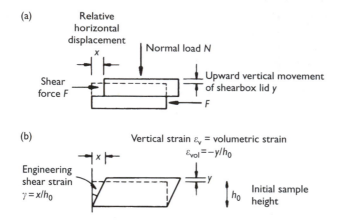

Figure 2.15 Schematic deformation of shearbox sample, showing quantities measured during shear test: (a) actual deformation; (b) idealized deformation.

positioned so as to measure the relative displacement of the top half of the shear box directly. The quantities measured during shear are shown schematically in Figure 2.15.

We shall see in Chapter 4 that, when a soil is subjected to an increase in load, additional pore water pressures are generated. These additional pore water pressures dissipate with time. In a sand, in which water can flow relatively easily between the particles, the additional pore water pressures usually dissipate almost instantaneously. In a clay, the additional or excess pore water pressures dissipate much more slowly. In a standard shearbox test, there is no facility to measure pore water pressures within the sample. To analyse the test in terms of effective stresses, the shearbox test must be carried out slowly enough to prevent the development of significant pore water pressures. This will ensure that the entire applied vertical load is carried by the soil skeleton as an effective stress.

A test carried out under these conditions is termed a **drained test**. Sands, being relatively free-draining, may be tested quite quickly without causing the build-up of excess pore water pressure. Some sands may be satisfactorily tested dry, which eliminates the problem of drainage entirely. Clay soils, on the other hand, cannot be tested dry, and pore water pressures take much longer to dissipate. A shearbox test on a clay might therefore need to be carried out so slowly that it takes several days. Bolton (1991) suggests suitable shear rates for drained tests of approximately 1 mm/min for a sand, 0.01 mm/min for a silt and 0.001 mm/min for a clay. If in doubt, it is better to err on the side of caution. The test results should not be affected if the sample is sheared more slowly, while the results of a test which has been carried out too quickly will be meaningless.

2.6 Presentation of shearbox test data in engineering units

Assuming that the pore water pressure within the sample is zero (either because the sample is dry, or because the test is being carried out sufficiently slowly), the vertical stress σ' and the shear stress τ acting on the central horizontal plane of the shearbox are obtained by dividing the forces N and F, respectively, by the cross-sectional area A of the sample:

$$\sigma' = N/A \tag{2.4}$$

$$\tau = F/A \tag{2.5}$$

According to the idealized mode of deformation shown in Figure 2.15(b), the engineering shear strain γ is given by

$$\gamma = x/h_0 \tag{2.6}$$

where x is the relative horizontal displacement, and h_0 is the initial sample height.

As the cross-sectional area of the sample remains constant, any increase in sample volume must result in an upward movement y of the lid of the shearbox. The increase in sample volume ΔV_t which corresponds to an upward movement y of the shearbox lid is

$$\Delta V_t = Ay$$

In soil mechanics, it is conventional to take compression as positive. The volumetric compression is $-\Delta V_t$, and the compressive volumetric strain ε_{vol} is given by

$$\varepsilon_{vol} = -\Delta V_t/V_{t0} = -(Ay)/(Ah_0) = -y/h_0 \tag{2.7}$$

The specific volume of the sample at any stage of the test may be calculated using the general form of equation (2.3),

$$v = (V_t G_s \rho_w)/m_s$$

where

$$V_t = V_{t0} + Ay \tag{2.8}$$

and y is measured positive upward.

Shearbox test data are conventionally plotted as graphs of shear stress τ or stress ratio τ/σ' against shear strain γ, and volumetric strain ε_{vol} or specific volume v against shear strain γ, as shown in Figure 2.16. The stress ratio (τ/σ') is used instead of the shear stress τ because, owing to the frictional nature of soil, the shear stresses generated in a shearbox test would be expected generally to increase in proportion to the applied normal stress. In dividing the shear stress by the normal effective stress at which the test is carried out, differences in shearbox test data due to the effect of differences in σ' alone are eliminated. The specific volume v is often preferred to the volumetric strain because it has an absolute value: the volumetric strain is measured relative to a datum which is arbitrarily set to zero at the start of the shearbox test.

An upward movement of the shearbox lid implies an increase in volume or dilation. This corresponds to a negative volumetric strain, according to the usual soil mechanics convention that compressive normal stresses and strains, and anticlockwise shear stresses and strains, are positive. Figure 2.16(b) is plotted with the positive volumetric strain axis pointing downward. This reflects the actuality of the shearbox test, and means that the volumetric shear–strain relationship shown in Figure 2.16(b) has the same shape as the path followed by the shearbox lid as the test progresses.

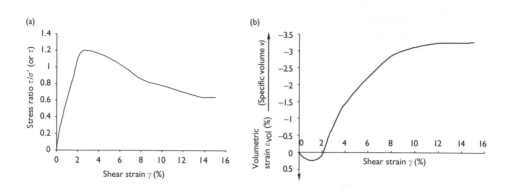

Figure 2.16 Shearbox test data plotted as (a) τ/σ' (or τ) vs. γ, and (b) ε_{vol} (or v) vs. γ.

2.7   Volume changes during shear

During a small time interval dt, the lid of the shearbox moves horizontally by a distance dx. If, during this time, it also moves vertically upward by distance dy, it is travelling at an angle $\psi = \tan^{-1}(dy/dx)$ to the horizontal (Figure 2.17). ψ is known as the **angle of dilation**, and is an indication of the rate at which the sample is changing in volume as it is sheared. If ψ is positive, the lid of the shearbox is moving upward and the sample is increasing in volume, or **dilating**. If ψ is negative, the lid of the shearbox is moving downward and the sample is reducing in volume, or **compressing**.

More formally, ψ is defined as the negative of the rate of increase of volumetric strain with shear strain,

$$\psi = \tan^{-1}(-d\varepsilon_{vol}/d\gamma) \tag{2.9}$$

The negative sign in equation (2.9) is required so that a negative (i.e. expansive) change in volume corresponds to a positive rate of dilation.

For the shearbox test, dε_{vol} (compression positive) $= -dy/h_0$ and d$\gamma = dx/h_0$ (writing equations (2.6) and (2.7) in incremental form), so that $\psi = \tan^{-1}(dy/dx)$, as in the text above.

So far, we have avoided the question of why the soil should change in volume (dilate or compress) as it is sheared. This behaviour arises because the soil is essentially a particulate material. The particles must take up a suitable arrangement of packing – corresponding to what is known as a **critical void ratio** – before continued shearing can take place. If the particles are initially more densely packed than the critical void ratio, some loosening will have to occur before steady shear can take place.

A loosening of the packing requires an increase in void ratio. This corresponds to an increase in the overall volume – that is, dilation – and an upward movement of the shearbox lid. If the particles are initially more loosely packed than the critical void ratio, densification of the sample will take place before the appropriate arrangement for steady shear is reached. Densification requires a reduction in void ratio, which corresponds to a compression of the sample and a downward movement of the shearbox lid.

The concept of a degree of packing at which steady shear can take place can be illustrated initially with reference to an imaginary assemblage of ball bearings. In

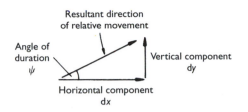

Resultant direction
of relative movement

Angle of
duration
ψ

Vertical component
dy

Horizontal component
dx

Figure 2.17 Dilation.

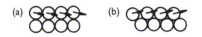

(a)   (b)

Figure 2.18 Conceptual model for (a) compression and (b) dilation during shear.

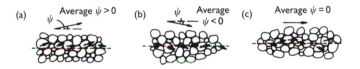

(a)   Average $\psi > 0$ (b) ψ Average   (c)   Average $\psi = 0$
      ψ $\psi < 0$

Figure 2.19 Visualization of rearrangement of soil particles during shear. (Redrawn with permission from Bolton, 1991.)

Figure 2.18(a), the ball bearings are initially loosely packed, with the ball bearings in the upper layer resting on the peaks of those in the lower layer. On shearing, the ball bearings in the upper layer will have to move downward into the troughs, resulting in a reduction in overall volume (compression). In Figure 2.18(b), the ball bearings are initially densely packed, with the upper layer nestling in the troughs between the ball bearings in the lower layer. On shearing, the ball bearings in the upper layer will have to climb out of the troughs, resulting in an increase in overall volume (dilation).

The diagrams shown in Figure 2.18 have only two layers of ball bearings in regular packing arrangements, and are therefore much too simple for anything except a basic illustration of the concept. A real assemblage of ball bearings might have only a single particle size, but will be randomly packed. A soil will be randomly packed with a variety of particle sizes. The result of this is that, after dilation or contraction to a critical void ratio, continued shearing of a soil (or an assemblage of ball bearings) can take place at constant volume, without the lumpiness (as the top layer of ball bearings continues to fall into and climb out of the troughs in the lower layer) implied by the simple model of Figure 2.18.

Figure 2.19 gives a more realistic visualization of the rearrangement of real soil particles during shear. Figure 2.19(a) shows an initially dense sample during the early stages of shear: the average particle velocity is inclined upwards, implying dilation. Figure 2.19(b) shows an initially loose sample: the average particle velocity is downward, implying compression. Figure 2.19(c) shows a sample deforming at the critical void ratio: some of the soil particles are moving upward and others downward, but the average movement is horizontal.

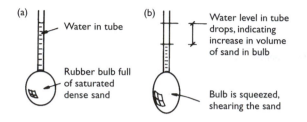

Figure 2.20 Demonstration of dilation.

The dilation of a dense sand when sheared can be demonstrated quite graphically by means of a rubber bulb, filled with sand, on the end of a tube (Figure 2.20). The sand is saturated, and some excess water is added so that the tube is initially about three-quarters full of clear water (Figure 2.20(a)). When the bulb is squeezed, the sand is sheared and dilates. This draws water into the bulb to fill the additional void space, resulting in a fall in the water level in the tube (Figure 2.20(b)). It follows from the behaviour shown in Figure 2.20 that, in a saturated soil, a tendency to dilate is accompanied by the development of negative pore water pressures. It is these negative pore water pressures which are responsible for the drawing in of water from the surroundings. Conversely, a tendency to contract on shearing will be associated with the generation of positive pore water pressures, which lead to the expulsion of water from the soil in order to attain the required reduction in specific volume.

2.8   Critical states

The achievement of a critical void ratio or specific volume, at which continued shear can take place without change in volume, is illustrated by the idealized results from shearbox tests on dense and loose samples of sand shown in Figure 2.21. The normal effective stress σ' is the same in each test.

In the test on the initially dense sample, the shear stress gradually increases with shear strain (Figure 2.21(a)). The shear stress increases to a peak at P, before falling to a steady value at C which is maintained as the shear strain is increased. The initially dense sample may undergo a small compression at the start of shear, but then begins to dilate (Figure 2.21(b)). The curve of ε_{vol} vs. γ becomes steeper, indicating that the rate of dilation $-d\varepsilon_{vol}/d\gamma$ is increasing. The slope of the curve reaches a maximum at p, but with continued shear strain the curve becomes less steep until at c it is horizontal. When the curve is horizontal $d\varepsilon_{vol}/d\gamma$ is zero, indicating that dilation has ceased. The peak shear stress at P in Figure 2.21(a) coincides with the maximum rate of dilation at p in Figure 2.21(b). The steady state shear stress at C in Figure 2.21(a) corresponds to the achievement of the critical specific volume at c in Figure 2.21(b).

The initially loose sample displays no peak strength, but eventually reaches the same critical shear stress as the initially dense sample (Figure 2.21(a)). Figure 2.21(b) shows that the initially loose sample does not dilate, but gradually compresses during shear until the critical specific volume is reached (i.e. the volumetric strain remains constant). Figure 2.21(c) shows that the critical specific volume is the same in each case.

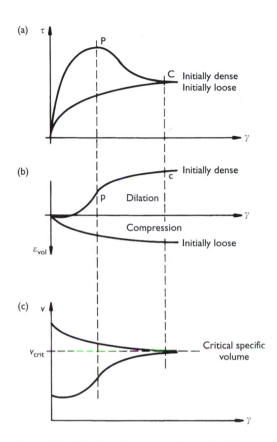

(a) τ

P

C   Initially dense
    Initially loose

γ

(b)

Initially dense

c

P   Dilation

Compression

Initially loose

ε_{vol}

γ

(c) v

v_{crit}

Critical specific
volume

γ

Figure 2.21 Idealized shearbox test results: (a) τ vs. γ; (b) ε_{vol} vs. γ; (c) v vs. γ.

The curves shown in Figure 2.21 indicate that, when sheared, a soil will eventually reach a critical void ratio, at which continued deformation can take place without further change in volume or stress. This condition, at which unlimited shear strain can be applied without further changes in specific volume v, shear stress τ and normal effective stress σ', is known as a **critical state**. The critical state concept was originally recognized by Casagrande (1936), and was developed by Roscoe *et al.* (1958).

Figure 2.21 also shows that the development of a peak strength is inextricably linked to the dilation of the soil: this is discussed in section 2.9.

It is worth emphasising here that, to define the state of a soil, three variables are required, quantifying the **specific volume**, the **shear stress** and the **normal effective stress**. We will see later in the book that the exact definition of the shear and normal stress parameters used may vary depending on circumstances. However, the underlying requirement to quantify the three state variables of specific volume, shear stress and normal effective stress remains the same.

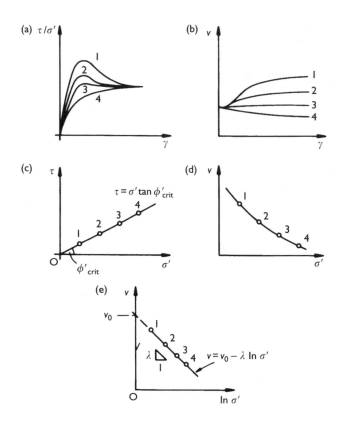

Figure 2.22 Idealized results from shearbox tests, carried out at different normal effective stresses, on four samples having the same initial void ratio. (a) Stress ratio τ/σ' vs. γ; (b) specific volume v vs. γ; (c) critical states (end points of tests); τ vs. σ'; (d) critical states; v vs. σ'; (e) critical states; v vs. $\ln \sigma'$.

The critical state eventually reached depends on the normal effective stress at which the soil is sheared. Figure 2.22 shows idealized results of shearbox tests carried out at different normal effective stresses, on soil samples having the same initial void ratio. Test 1 was carried out at the lowest normal effective stress and Test 4 at the highest. As the normal effective stress is different in each case, Figure 2.22(a) plots stress ratio τ/σ', rather than simply the shear stress τ against shear strain γ. Also, Figure 2.22(b) is in terms of the specific volume rather than the volumetric strain, because the volumetric strain tells us nothing about the soil state in absolute terms.

Figure 2.22(a) shows that, as the normal effective stress σ' is increased, the *maximum* or *peak* stress ratio achieved is reduced, while the *critical* stress ratio is unaffected.

Figure 2.22(b) shows that the specific volume at the critical state is reduced as the normal effective stress is increased.

Figure 2.22(c) shows that the critical states lie on a straight line of gradient $\tan \phi'_{crit}$ on a graph of τ against σ'. The equation of this line may be written as

$$\tau = \sigma' \tan \phi'_{crit} \tag{2.10}$$

Equation (2.10) is, as would be expected for an essentially frictional material, similar in form to equation (2.2). However, we have now stated explicitly that the apparent angle of friction ϕ' is measured at the critical state, rather than at the peak which might otherwise be described as the initial point of failure. In fact, we could not describe the peak strengths by means of an expression as simple as equation (2.10), because the peak strength varies with the normal effective stress σ'.

Figure 2.22(e) shows that the critical states also lie on a straight line on a graph of v against $\ln(\sigma')$. This line may be described by an equation of the form

$$v = v_0 - \lambda \ln \sigma' \tag{2.11}$$

where v_0 is the intersection of the line with the v-axis (i.e. the value of v on the critical state line at $\ln \sigma' = 0$ or $\sigma' = 1$ kPa), and $-\lambda$ is its slope.

The lines shown in Figures 2.22(c) and (d) are lines joining all possible critical states. They are in fact projections (onto the τ, σ' and v, σ' planes) of a single critical state line in three-dimensional (σ', τ, v) space (Figure 2.23).

The implication of Figure 2.23 is that, although the number of possible critical states is unlimited, there is only one critical state for any given value of σ' (or τ or v). In other words, if we have carried out sufficient tests to locate the critical state line (or its projections onto the τ, σ' and $v, \ln \sigma'$ planes), we can predict the values of any two of the parameters σ', τ and v at the critical state, provided that the third is known. Usually, in dealing with a sand, σ' will be known so that τ and v can be predicted.

Sometimes, a clay soil is sheared quickly, so that (assuming it is saturated) it reaches the critical state without change in specific volume. In these circumstances, v can be determined from the water content, using equation (1.10) with $S_r = 1$:

$$v = 1 + wG_s \tag{2.12}$$

and the critical state model represented by equations (2.10) and (2.11) can be used to predict the values of σ' and (more importantly) τ at the critical state. This point is discussed in section 2.10.

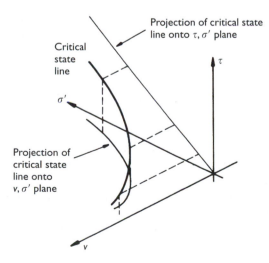

Figure 2.23 Critical state line in (σ', τ, v) space with projections onto (τ, σ') and (v, σ') planes.

In Chapter 5, the critical state model is developed in terms of different stress para-meters, to encompass more general stress states. The underlying principles, however, remain the same. These are:

---

1. The state of a soil must be described by three parameters, which give an indication of the specific volume, the normal effective stress and the shear stress.
2. When sheared, the soil will eventually reach a critical state in which unlimited shear strain could be applied without further changes in specific volume, normal effective stress or shear stress.
3. The combination of specific volume, normal effective stress and shear stress at the critical state lies on a unique line – the **critical state line** – when plotted on a three-dimensional graph with axes representing specific volume, normal effective stress, and shear stress.

2.9   Peak strengths and dilatancy

Figures 2.21 and 2.22 show that the ability of a soil to develop a peak strength before reaching a critical state depends on its ability to increase in volume or dilate. In Figure 2.21, the maximum rate of dilation of the dense sample corresponded to the development of the peak shear stress τ_{peak}. The shear stress then fell as the rate of dilation became smaller, until the critical state was reached. The loose sample did not dilate, and did not exhibit a peak strength.

In Figure 2.22(a), the peak stress ratio becomes smaller as the normal effective stress is increased. This is because the critical state void ratio reduces with increasing effective stress, resulting in a diminishing potential for dilation. The terms 'dense' and 'loose' should therefore be used in relation to the critical state void ratio at the normal effective stress under consideration. A soil of a certain initial specific volume may be dense at a normal effective stress of 10 kPa (because its initial void ratio is less than the critical state void ratio at 10 kPa), but loose at an effective stress of 100 kPa (because its initial void ratio is greater than the critical state void ratio at 100 kPa).

Although this might seem paradoxical, it is possible to turn a 'dense' sand into a 'loose' sand by increasing the normal effective stress at which it is tested. This is, in effect, what is happening in Figure 2.22(a). In test 1, the soil is effectively dense and displays a peak strength. In test 4, the soil is sheared from the same initial void ratio but at a higher normal effective stress. It displays no peak strength, and behaves as if it were loose.

When the soil in a shearbox test dilates, the lid of the shearbox moves upward at the angle of dilation ψ (Figure 2.17). It might be argued that, although the macroscopic plane of shearing is horizontal, the microscopic planes of shearing are in effect inclined at an angle of ψ to the horizontal. This can be visualized by imagining the two halves of the shearbox to be sliding along a series of sawteeth, as indicated in Figure 2.24(a). If the actual angle of shearing resistance mobilized on the surfaces of the saw teeth is ϕ'_{crit}, the angle of shearing resistance $\phi'_{current}$ measured at some stage in the shearbox

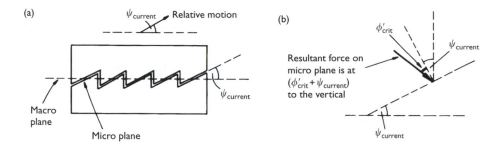

Figure 2.24 Sawtooth analogy for dilation. (Redrawn with permission from Bolton, 1991.)

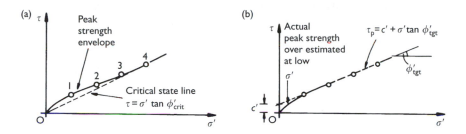

Figure 2.25 (a) Peak strength data, plotted as τ vs. σ', showing curved failure envelope; (b) how not to interpret peak strength data.

test (on the basis of assumed relative sliding on a horizontal plane) is given by

$$\phi'_{current} = \phi'_{crit} + \psi_{current} \tag{2.13}$$

(Figure 2.24(b)), where $\psi_{current}$ is the current angle of dilation.

Experimental evidence shows that equation (2.13) overestimates the effect of dilation on peak strength. On the basis of a considerable volume of data from shear tests on sands, Bolton (1986) shows that in plane strain, the contribution of dilation to peak strength is more closely represented by the expression

$$\phi'_{peak} = \phi'_{crit} + 0.8\psi_{max} \tag{2.14}$$

In some respects, the peak strength of a soil should be of academic interest only. A peak strength is a transient thing, sustainable only while the soil is dilating. A soil cannot carry on dilating for ever, and sooner or later its strength must fall to the critical state value, ϕ'_{crit}. Furthermore, the peak strength is not a soil property in the same way that the critical state strength is. The peak strength depends on the potential for dilation, and will therefore (for a soil of given initial specific volume) decrease as the normal effective stress increases, as shown in Figure 2.22(a).

This point may be reinforced by plotting the peak strength data from Figure 2.22 on a graph of τ against σ'. Figure 2.25(a) shows that the envelope formed by the peak strength data is curved. In contrast to the critical state strength shown in Figure 2.22(c), the peak strength data cannot accurately be described by means of a simple equation. Unfortunately, this does not stop some people from trying. In some books, you will find peak strength data described by means of a 'best-fit' straight line, having an equation of the form

$$\tau_p = c' + \sigma' \tan \phi'_{tgt} \qquad (2.15)$$

(Figure 2.25(b)), where c' is the intersection of the line with the τ-axis, and $\tan \phi'_{tgt}$ is its slope. Such an approach is conceptually flawed. For at least three reasons, it is also potentially dangerous:

1. It can lead to the overestimation of the actual peak strength at either low (Figure 2.25(b)) or high effective stresses, depending on where the 'best fit' straight line is drawn. (The position of the best fit straight line will depend on the available data points, which – if the testing programme has not been specified with care – may be predominantly at either high or low normal effective stresses.)
2. The designer of a geotechnical engineering structure cannot guarantee that the peak strength will be uniformly mobilized everywhere it is needed at the same time. It is much more likely that only some soil elements will reach their peak strength first. If any extra strain is imposed on these elements, they will fail in a brittle manner as their strength falls towards the critical value. In doing so, they will shed load to their neighbours, which will then also become overstressed and fail in a brittle manner. In this way, a progressive collapse can occur, which – like the propagation of a crack through glass – may be sudden and catastrophic. (Experience suggests that progressive failure is potentially particularly important with embankments and slopes, and perhaps less of a problem with retaining walls and foundations.)
3. Quite apart from the possibility of progressive collapse, many of the design procedures used in geotechnical engineering assume that the soil can be relied on to behave in a ductile manner. When a ductile material fails, it will undergo continued deformation at constant load. This is in contrast to a brittle material, which at failure breaks and loses its load carrying capacity entirely. At the critical state, the behaviour of the soil is ductile: the definition of the critical state is that unlimited shear strain can be applied *without further changes in stress* or specific volume. Between the peak strength and the critical state, however, the soil loses strength with continued strain.

In design, it is safer to use the critical state strength. It is also simpler mathematically. It must be appreciated, however, that this has not always been standard practice. This means that procedures which have led to acceptable designs on the basis of the peak strength might err on the conservative side when used unmodified with the critical state strength.

In cases where it is really necessary, the peak strength can be quantified with reference to the slope of the line joining the origin to the peak stress state (τ_{peak}, σ'):

$$\phi'_{peak} = \tan^{-1}(\tau/\sigma')_{peak} \qquad (2.16)$$

The value of ϕ'_{peak} defined in this way will gradually diminish with increasing normal effective stress σ', until it eventually falls to ϕ'_{crit} (e.g. at test 4 in Figure 2.25(a)). At a given normal effective stress σ', the value of ϕ'_{peak} (calculated according to equation (2.16)) may be useful as an empirical indicator of density, and hence of the relative stiffness of the soil. This point is discussed further, in the context of retaining wall and foundation design, in Chapters 7 and 8.

Example 2.1 Analysis of shearbox test data and presentation in engineering units

Table 2.1 gives data from a standard shearbox test on a sample of 125 g of dry sand. The initial dimensions of the sample were 60 mm × 60 mm on plan × 20 mm in height. The test was carried out at a constant normal effective stress of 50 kPa.

Plot graphs of (a) shear stress τ against shear strain γ; (b) volumetric strain ε_{vol} against shear strain γ; and (c) specific volume v against shear strain γ. Comment on these graphs, and estimate the peak and critical state effective angles of friction of the soil.

Take the specific gravity of the soil grains $G_s = 2.65$.

Table 2.1 Shearbox test data

Relative horizontal displacement x (mm)	Upward vertical movement of shearbox lid y (mm)	Shear stress τ (kPa)
0.00	0.000	0
0.02	0.002	19
0.04	0.008	34
0.06	0.016	43
0.08	0.026	47
0.20	0.064	56
0.32	0.128	51
0.48	0.192	46
0.64	0.256	41
0.80	0.288	37
0.96	0.320	34
1.12	0.321	33

Solution

The tabulated values of x and y are converted to shear strain γ and volumetric strain ε_{vol} by dividing them by the initial sample height h_0:

$$\gamma = x/h_0 \tag{2.6}$$

$$\varepsilon_{vol} = -y/h_0 \tag{2.7}$$

where compressive stresses and strains are taken as positive.

The initial specific volume of the sample v_0 may be calculated using equation (2.3):

$$v_0 = (V_{t0}G_s\rho_w)/m_s = \{(0.06 \times 0.06 \times 0.02)m^3 \times 2.65 \times 1000\,kg/m^3\}$$
$$\div 0.125\,kg$$

$$v_0 = 1.526$$

The specific volume v at a general stage of the test is given by

$$v = (V_t G_s \rho_w)/m_s$$

where

$$V_t = V_{t0} + Ay$$

or

$$v = \{(V_{t0} + Ay)G_s\rho_w\}/m_s = v_0 + (AyG_s\rho_w)/m_s$$

With $A = (0.06 \times 0.06)\text{m}^2$, $G_s = 2.65$, $\rho_w = 1000\,\text{kg/m}^3$ and $m_s = 0.125\,\text{kg}$,

$$v = v_0 + (0.07632y)$$

with y in mm.

The calculated values of γ, ε_{vol} and v are given in Table 2.2. Graphs of shear stress τ against shear strain γ; volumetric strain ε_{vol} against shear strain γ; and specific volume v against shear strain γ are shown in Figure 2.26.

Table 2.2 Processed shearbox test data

x (mm)	γ (%)	y (mm)	ε_{vol} (%)	v	τ (kPa)
0.00	0.0	0.000	0.00	1.526	0
0.02	0.1	0.002	0.01	1.526	19
0.04	0.2	0.008	0.04	1.527	34
0.06	0.3	0.016	0.08	1.528	43
0.08	0.4	0.026	0.13	1.528	47
0.20	1.0	0.064	0.32	1.531	56
0.32	1.6	0.128	0.64	1.536	51
0.48	2.4	0.192	0.96	1.541	46
0.64	3.2	0.256	1.28	1.546	41
0.80	4.0	0.288	1.44	1.548	37
0.96	4.8	0.320	1.60	1.551	34
1.12	5.6	0.321	1.60	1.551	33

The shear stress rises rapidly to a peak of 56 kPa at a shear strain of 1%, before falling to a steady value of about 33 kPa (Figure 2.26(a)). The peak shear stress corresponds approximately to the maximum rate of dilation as indicated by Figures 2.26(b) and (c). By the end of the test, dilation has ceased and deformation is taking place at constant shear stress, indicating that the critical state has probably been reached. From Figure 2.26(a),

$$\tau_{peak} = 56\,\text{kPa}$$

giving

$$\phi'_{peak} = \tan^{-1}(56/50) = 48°$$

$$\tau_{crit} = 33\,\text{kPa}$$

giving

$$\phi'_{crit} = \tan^{-1}(33/50) = 33°$$

Note that ϕ'_{peak} here has been calculated from the slope of the line joining the origin with the peak stress state, as suggested at the end of section 2.9 (equation (2.16)).

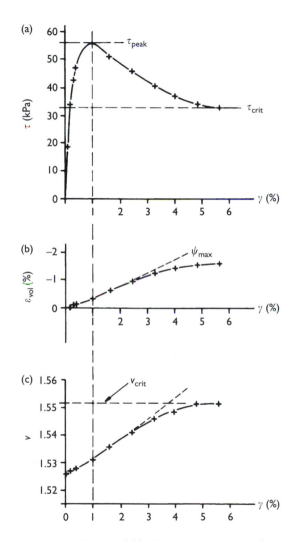

Figure 2.26 Graphs of (a) shear stress τ against shear strain γ; (b) volumetric strain ε_{vol} against shear strain γ; and (c) specific volume v against shear strain γ for Example 2.1.

From Figure 2.26(b), the maximum rate of dilation $-d\varepsilon_{vol}/d\gamma$ is approximately $(0.75/1.5)$, giving an angle of dilation $\psi = \tan^{-1}(0.5) = 26°$. For $\phi'_{crit} = 33°$ and $\psi = 26°$, equation (2.14) suggests a peak angle of friction of

$$\phi'_{peak} = \phi'_{crit} + 0.8\psi = 33° + 21° = 54°$$

which is somewhat in excess of the observed value, $\phi'_{peak} = 48°$.

2.10 Shearbox tests on clays

Shearbox tests on clays which are carried out sufficiently slowly to prevent the generation of non-equilibrium or excess pore water pressures may be analysed in terms of effective stresses, in exactly the same way as shearbox tests on sands.

At the other extreme, it might be possible to test a low permeability clay so quickly that it reaches the critical state without having time to expel or draw in water in response to the non-equilibrium pore water pressures generated during shear. In these circumstances, the simple critical state model given in Figure 2.22 predicts that the effective stress state at failure (i.e. at the critical state) will depend only on the specific volume, which is related (by equation 2.12) to the water content.

Although the normal total stress differs from the normal effective stress by the pore water pressure, the total shear stress is the same as the effective shear stress because all of the applied shear load must be carried by the soil skeleton. Thus, whatever the applied normal total stress, the applied shear stress cannot exceed a certain limit, which is dictated by the specific volume or the water content of the soil. This limit is known as the **undrained shear strength**, and defines an alternative failure criterion in terms of total stresses (Figure 2.27).

The undrained shear strength is conventionally given the symbol c_u: for the reason given in section 5.12, I have used the symbol τ_u. The undrained shear strength failure criterion is only applicable to a low permeability soil of a particular water content, which is brought rapidly to failure without changing the specific volume.

As the undrained shear strength depends on the water content of the soil, it may vary significantly (most usually, increasing with depth) within a single soil deposit. It is nonetheless used in the analysis of the short-term undrained stability of geotechnical

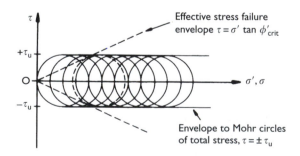

Figure 2.27 Envelope to all possible Mohr circles of total stress at failure for a clay sheared at constant volume.

structures, in terms of total – rather than effective – stresses. One of the reasons for this is that the initial, non-equilibrium pore water pressures generated by construction (loading) and excavation (unloading) processes in clay soils can be difficult to predict. If the pore water pressures are not known, the effective stresses cannot be calculated, and an effective stress analysis cannot be carried out.

In attempting to measure the undrained shear strength of a clay in the shearbox, the usual uncertainties of disturbance during sampling, and whether or not the test specimen is large enough to be representative of the soil and its fabric *in situ* will arise. An additional risk is that, if the test is not carried out quickly enough, partial drainage may occur. If the test is truly undrained, the lid of the shearbox should move neither up nor down: however, any tendency of the lid to rotate could mask whether this is in fact the case. Alternative methods of measuring the undrained shear strength of a clay include the triaxial test (Chapter 5), in which the sample is surrounded by a rubber membrane so that drainage may be physically prevented, and *in situ* using a field vane (section 11.3.4).

Undrained shear strengths are discussed more fully in section 5.12. Methods of total stress analysis on the basis of the undrained shear strength are described in Chapters 7–11. Also, it was mentioned in section 1.11 that the liquid and plastic limit tests in essence measure the undrained shear strength, and that the undrained shear strength at the plastic limit is approximately 70 times that at the liquid limit.

Example 2.2 Development of a simple critical state model from shearbox test data

(a) Explain briefly what is meant by a critical state in soil mechanics.

(b) Data obtained from three slow shearbox tests on samples of saturated clay are given in Table 2.3. Comment on these data and use them to construct a simple critical state model in terms of the shear and normal effective stresses τ and σ' on the horizontal plane of the apparatus, and the specific volume v.

Table 2.3 Shearbox test data

Parameter	Sample		
	A	B	C
Vertical effective stress σ' (kPa)	25	50	100
Peak shear stress τ_{peak} (kPa)	11.7	21.2	36.4
Critical state shear stress τ_{crit} (kPa)	9.1	18.2	36.4
Specific volume at start of test	2.209	2.209	2.209
Specific volume at end of test	2.417	2.313	2.209

(c) Estimate the values of undrained shear strength which would have been obtained if the tests had been conducted very quickly. Account for any difference between the shear stresses at the critical state in the quick and the slow tests.

[*University of London 1st year BEng (Civil Engineering) examination,*
King's College]

Solution

(a)   When sheared, a soil will eventually reach a critical state in which continued deformation takes place while the shear and normal effective stresses and the specific volume remain constant in the zone(s) of shear.

(b)   Samples A and B both have a larger specific volume at the end of the test than at the start. From this, it can be inferred that these samples dilate during shear, and that they are initially dense of their respective critical states. Both samples exhibit peak shear strengths while dilation is taking place. Sample C has the same specific volume at the start of the test as at the critical state. It does not dilate during shear, and therefore does not show a peak shear strength.

The critical state data are plotted as τ against σ' in Figure 2.28(a), and as v against $\ln \sigma'$ in Figure 2.28(b).

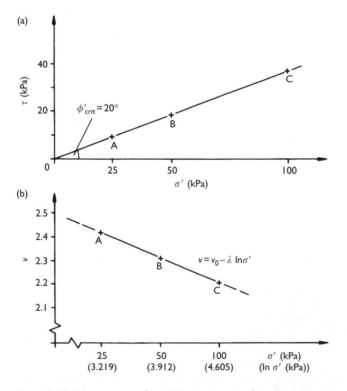

Figure 2.28 (a) τ against σ' and (b) v against $\ln \sigma'$ for Example 2.2.

From Figure 2.28(a), the graph of τ against σ' is a straight line of slope 20°, so that the equation of the critical state line in τ, σ' space is

$$\tau = \sigma' \tan 20°$$

that is, $\phi'_{crit} = 20°$ (cf. equation (2.10) and Figure 2.22(c)).

From Figure 2.28(b), the graph of v against $\ln \sigma'$ is a straight line of slope $-\lambda$ and intercept v_0 (at $\ln \sigma' = 0$, $\sigma' = 1\,\text{kPa}$), cf. equation (2.11) and Figure 2.22(e). The slope is given by

$$\lambda = (2.417 - 2.209) \div (\ln 100 - \ln 25), \quad \text{or} \quad \lambda = 0.15$$

v_0 may be calculated from the specific volume at $\sigma' = 100\,\text{kPa}$, v_{100}, as

$$v_0 = v_{100} + \lambda(\ln 100 - \ln 1), \text{ giving } v_0 = 2.9$$

(c) If the tests had been carried out very quickly, the samples would have been forced to deform without change in water content and would all have failed at a specific volume of 2.209. From Figure 2.28(b), this implies an effective stress $\sigma' = 100\,\text{kPa}$ at the critical state. From Figure 2.28(a), the corresponding shear stress – the undrained shear strength – is $\tau_u = 36\,\text{kPa}$.

The difference between the critical state shear stresses in the slow and the quick tests for samples A and B is due to the generation of negative pore water pressures during rapid shear. The negative pore water pressures increase σ' at the critical state and hence also the shear stress τ. In the slow tests, the tendency to develop negative pore water pressure results in the drawing in of water by the clay, which swells and softens to a critical state at a greater specific volume v, with correspondingly smaller values of σ' and τ.

2.11 Applications

The basic concepts of frictional strength and a frictional failure criterion may be used to calculate the collapse loads of simple geotechnical structures, as illustrated by Examples 2.3 and 2.4.

Example 2.3 Calculation of the pull-out resistance of a grouted ground anchor

Figure 2.29 shows a cross-section through a grouted ground anchor, installed in a sandy gravel behind a retaining wall. Estimate the tensile load which will cause the anchor to fail by pulling out of the soil (a) if the soil is dry, and (b) if the pore water varies along the anchor from 40 kPa at a depth of 5.5 m to 70 kPa at a depth of 9 m. Assume that failure will occur by slippage between the grout and the surrounding soil, rather than between the steel tendon and the grout; that the angle of friction between the soil and the grout is the same as the critical state angle of friction of the soil; and that the effective stress at any depth is the same in all directions.

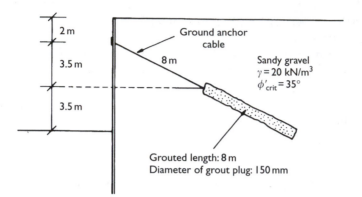

Figure 2.29 Cross-section through grouted ground anchor.

Solution

(a) Dry soil, pore water pressures zero.
The total tensile load at pull-out is given by

$$T = \tau_{av} \times A$$

where A is the total area of the soil/grout interface and τ_{av} is the average shear stress acting on it.

From equation (2.10), $\tau_{av} = \sigma'_{av} \tan \phi'_{crit}$. ϕ'_{crit} is uniform along the grouted anchor length, while σ' and hence τ increase with depth.

At the top of the grouted length (depth $z = 5.5$ m), $\sigma' = 5.5$ m $\times$ $20 \text{ kN/m}^3 = 110$ kPa.

At the bottom of the grouted length (depth $z = 9$ m), $\sigma' = 9$ m $\times$ $20 \text{ kN/m}^3 = 180$ kPa.

Assuming that σ' increases linearly with depth, and at any given depth is the same in all directions,

$$\sigma'_{av} = (110 + 180)/2 = 145 \text{ kPa}$$

Hence

$$\tau_{av} = 145 \text{ kPa} \times \tan \phi'_{crit} = 145 \text{ kPa} \times \tan 35° = 101.5 \text{ kPa}$$

The surface area of the grout cylinder is

$$\pi \times 0.15 \text{ m} \times 8 \text{ m} = 3.77 \text{ m}^2$$

Thus

$$T = 101.5 \text{ kPa} \times 3.77 \text{ m}^2 \approx 380 \text{ kN}$$

(b)   Saturated soil, pore water pressures increase from 40 kPa to 70 kPa along the grouted length.

At the top of the grouted length (depth $z = 5.5$ m), the normal total stress $\sigma = 5.5$ m $\times$ 20 kN/m^3 = 110 kPa. The normal effective stress $\sigma' = \sigma - u = $ 110 kPa $-$ 40 kPa = 70 kPa.

At the bottom of the grouted length (depth $z = 9$ m), the normal total stress $\sigma = 9$ m $\times$ 20 kN/m^3 = 180 kPa. The normal effective stress $\sigma' = \sigma - u = $ 180 kPa $-$ 70 kPa = 110 kPa.

Again, assuming that σ' increases linearly with depth, and at any given depth is the same in all directions,

$$\sigma'_{av} = (70 + 110)/2 = 90 \text{ kPa}$$

Hence

$$\tau_{av} = 90 \text{ kPa} \times \tan 35° = 63 \text{ kPa}$$

and

$$T = 63 \text{ kPa} \times 3.77 \text{ m}^2 \approx 237 \text{ kN}$$

The presence of the pore water pressure has led to a 38% reduction in the ultimate capacity of the anchor. In reality, the assumption that the interface between the grout and the soil is more critical than the interface between the grout and the tendon would need to be checked. Also, the horizontal effective stress in a sandy soil is usually somewhat less than the vertical effective stress at the same depth: in this respect, the pull-out force T may have been overestimated.

Example 2.4 Short- and long-term capacity of a friction pile in clay

A pile of circular cross-section, 15 m long and 0.5 m in diameter, is installed in a soft clay of unit weight 18 kN/m^3. The pile is required to resist an applied upward vertical load by friction between the pile and the surrounding soil. Estimate the ultimate capacity of the pile:

(a)   in the short term, if the undrained shear strength of the clay at the clay/pile interface increases linearly from zero at the soil surface to 30 kPa at the base of the pile;

(b)   in the long term, if the clay/pile interface has an effective angle of friction of 20°, pore water pressures are hydrostatic below a water table at the soil surface, and the ratio of the horizontal effective stress to the vertical effective stress is 0.5 at all depths.

Solution

(a) The total shear resistance of the clay/pile interface is given by

T = average shear stress × surface area of pile

The average shear stress is the average undrained shear strength on the interface, so that

$T = [(0 + 30\,\text{kPa}) \div 2] \times [(\pi \times 0.5\,\text{m}) \times 15\,\text{m}] = 353\,\text{kN}$

(b) In the long term, the ultimate or limiting shear stress on the interface is given by

$\tau_{\text{ult}} = \sigma_\text{h}' \tan \delta$

where $\sigma_\text{h}' = 0.5 \times \sigma_\text{v}'$ is the horizontal effective stress and δ is the effective angle of friction between the clay and the pile.

At a depth z,

$\sigma_\text{v}(\text{kPa}) = 18(\text{kN/m}^3) \times z(\text{m})$

$u(\text{kPa}) = 9.81(\text{kN/m}^3) \times z(\text{m})$

$\sigma_\text{v}' = \sigma_\text{v} - u$

As before, T = average shear stress × surface area of pile.
The shear stress on the soil/pile interface is now

$0.5 \times \sigma_\text{v}' \tan \delta$

which increases linearly from zero at the top of the pile to

$0.5 \times [(18\,\text{kN/m}^3 \times 15\,\text{m}) - (9.81\,\text{kN/m}^3 \times 15\,\text{m})] \times \tan 20°$
$= 22.36\,\text{kPa}$ at the base

Hence

$T = [(0 + 22.36\,\text{kPa}) \div 2] \times [(\pi \times 0.5\,\text{m}) \times 15\,\text{m}] = 263\,\text{kN}$

The ultimate capacity of the pile in the long term is approximately 25% less than in the short term. This suggests that the clay has a tendency to generate negative pore water pressures and/or soften when sheared, that it is initially dense of the appropriate critical state.

2.12 Stress states in the shearbox test

2.12.1 Conventional interpretation

In the analysis of a shearbox test, it is conventionally assumed that the stress state is uniform throughout the sample, and that the ratio of shear to normal effective stress (τ/σ') is greatest on the horizontal plane. (The plane of maximum stress ratio is sometimes called the **plane of maximum stress obliquity**, because it is the plane on which the resultant stress direction is furthest from the normal to the plane – that is,

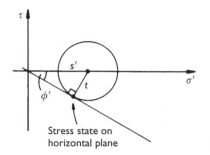

Figure 2.30 Mohr circle of stress for shearbox sample, assuming that the horizontal plane is the plane of maximum stress ratio. For the direction of shear shown in Figure 2.15, τ_{yx} is clockwise and therefore plots as negative on the Mohr circle of stress.

most oblique.) These two assumptions enable the Mohr circle of effective stress to be drawn as shown in Figure 2.30, and the effective angle of friction or angle of shearing resistance ϕ' of the soil to be calculated. In addition to the particular cases of ϕ'_{crit} and ϕ'_{peak} already examined, the parameter ϕ'_{mob}, the **mobilized angle of friction**, may be used to describe the stress ratio at any stage of the shearbox test, $\phi'_{mob} = \tan^{-1}(\tau/\sigma')$, where τ is not necessarily either τ_{peak} or τ_{crit}.

A useful result which follows from the geometry of the Mohr circle of stress (Figure 2.30) is that

$$\sin \phi'_{mob} = t/s'$$

or

$$\phi'_{mob} = \sin^{-1}(t/s') \qquad (2.17)$$

2.12.2 *Alternative interpretation*[1]

Some engineering materials – particularly those used in structures, such as mild steel – are **isotropic** (i.e. they have the same properties in all directions) and – under normal working conditions – elastic (i.e. the strain is proportional to the applied stress). For these materials, the principal axes of the strain increment which results from an increase in applied load are coincident with the principal axes of the stress increment. In other words, an elastic material stretches or compresses in the direction in which it is pulled or pushed.

Materials at failure do not behave elastically. For example, a ductile material at failure will undergo increasing strain at constant stress. This type of behaviour is described as plastic. In an ideal plastic material, the principal axes of **plastic** strain increment coincide with the principal axes of stress (Hill, 1950). This means that when an ideal plastic material is brought to failure, it strains not in the direction of the last push, but in a direction which is governed by the overall loading pattern when failure occurs.

The behaviour of soil prior to failure is not elastic, but is sometimes assumed to be approximately so (e.g. in the methods used to estimate settlements of foundations, described in Chapter 6). The behaviour of soil at failure, however, can in many circumstances be reasonably regarded as approximately plastic. This has led people to believe that a more realistic assessment of the stress state in the shearbox could be made by assuming that the directions of the principal stresses are coincident with the directions of the principal plastic strain *increments* (Rowe, 1969; Jewell and Wroth, 1987).

Let us assume that the shearbox test has reached a stage where the soil is deforming essentially plastically. During a short time interval δt, the relative horizontal displacement is δx, and the upward movement of the lid of the shearbox is δy. The vertical effective stress is σ'_{yy}, and the shear stress on the horizontal plane is τ_{yx}. In each case, the first subscript denotes the direction of the normal to the plane on which the stress acts, while the second subscript denotes the direction in which the stress acts. The plastic vertical strain increment $\delta \varepsilon_{yy}$ is $-\delta y/h_0$, and the plastic shear strain increment associated with the horizontal plane is $\delta \gamma_{yx} = -\delta x/h_0$. These stresses and strain increments are shown schematically in Figure 2.31.

Figure 2.32 shows the Mohr circle of strain increment. The plane associated with the major principal strain increment is at an angle of $(90° + \psi)/2$ anti-clockwise from the horizontal plane, where ψ is the angle of dilation, $\psi = \tan^{-1}(\delta y/\delta x)$.

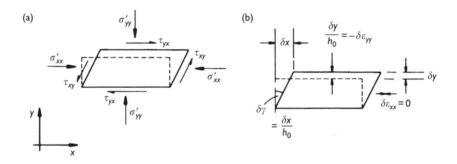

Figure 2.31 (a) Stresses and (b) plastic strain increments in a shearbox test.

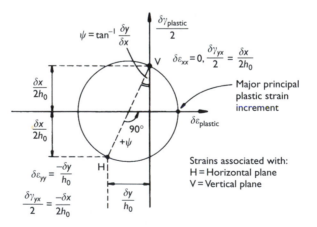

Figure 2.32 Mohr circle of plastic strain increment for shearbox test.

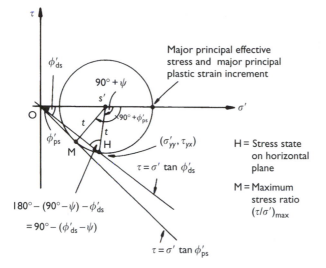

Figure 2.33 Mohr circle of effective stress for shearbox test, assuming that planes of major principal stress and major principal strain increment coincide.

Figure 2.33 shows the Mohr circle of effective stress corresponding to Figure 2.31(a), drawn on the assumption that the plane associated with the major principal strain increment is the same as that on which the major principal effective stress acts. The radius of the Mohr circle is t, and its centre is located at a distance s' from the origin. With reference to the triangle OMs',

$$t = s' \sin \phi'_{ps} \tag{2.18}$$

where ϕ'_{ps} (ps is for plane strain) is the effective angle of friction on the actual plane of maximum stress ratio, $(\tau/\sigma')_{max}$.

Applying the sine rule to the triangle OHs',

$$t / \sin \phi'_{ds} = s' / \sin(90° - [\phi'_{ds} - \psi]) \tag{2.19}$$

where ϕ'_{ds} (ds is for direct shear) is the apparent angle of friction on the horizontal plane.

Combining equations (2.18) and (2.19), and noting that $\sin(90° - \theta) = \cos\theta$,

$$t/s' = \sin \phi'_{ps} = \sin \phi'_{ds} / \cos(\phi'_{ds} - \psi)$$

But $\cos(\phi'_{ds} - \psi) = (\cos \phi'_{ds} \cos \psi) + (\sin \phi'_{ds} \sin \psi)$, so that $\sin \phi'_{ps} = \sin \phi'_{ds} / [(\cos \phi'_{ds} \cos \psi) + (\sin \phi'_{ds} \sin \psi)]$ or

$$\sin \phi'_{ps} = \tan \phi'_{ds} / [\cos \psi (1 + \tan \phi'_{ds} \tan \psi)] \tag{2.20}$$

At the critical state, the angle of dilation $\psi = 0$, giving

$$\sin \phi'_{ps} = \tan \phi'_{ds} \tag{2.21}$$

Table 2.4 Relationship between effective angle of friction on plane of maximum stress ratio ϕ'_{ps} and effective angle of friction on horizontal plane of shearbox ϕ'_{ds} according to equation (2.21) ($\psi = 0$)

ϕ'_{ps} (degrees)	ϕ'_{ds} (degrees)	$\tan \phi'_{ps}/\tan \phi'_{ds}$
20	18.9	1.06
25	22.9	1.10
30	26.6	1.15
35	29.8	1.22

According to this interpretation, the conventional assumption that the horizontal plane is the plane of maximum stress ratio will lead to the underestimation of the actual effective angle of friction in plane strain ϕ'_{ps}. Jewell and Wroth (1987) show that the discrepancy between ϕ'_{ds} and ϕ'_{ps} according to equation (2.20) is insensitive to the angle of dilation ψ. For $\psi = 0$, the values given in Table 2.4 apply (equation 2.21).

From the Mohr circle of effective stress shown in Figure 2.33, the actual plane of maximum stress ratio is at an angle of $(\phi'_{ps} - \psi)/2$ clockwise (for the sense of shear shown in Figure 2.31) from the horizontal plane.

The question which faces the geotechnical engineer is whether to use ϕ'_{ds} or ϕ'_{ps} in design. Equations (2.20) and (2.21) have been substantiated by comparing ϕ'_{ds} measured in shearbox tests with ϕ'_{ps} measured in different types of plane strain test (Rowe, 1969; Jewell and Wroth, 1987). However, many geotechnical analyses are based on the limiting equilibrium of a series of blocks of soil separated by presumed planes of failure. It might therefore be argued that in this context, ϕ'_{ds} – which is measured along a plane of failure in the shearbox test – is more appropriate than ϕ'_{ps}. As with many things in geotechnical engineering, the issue is not straightforward, and calls for the exercise of engineering judgement according to the circumstances of an individual case. If in doubt, it is safer to err on the side of caution and use the lower of the two values, ϕ'_{ds}.

2.12.3 Undrained tests on clays

The conventional interpretation of an undrained shearbox test on a clay soil is that the horizontal plane is the plane on which the shear stress is greatest, so that the measured shear stress on this plane is the undrained shear strength τ_u. The Mohr circle of total stress corresponding to this assumption is shown in Figure 2.34. The assumption that the horizontal plane is the plane of maximum shear stress is clearly incompatible with the assumption that it is the plane of maximum stress ratio. It is, however, consistent with the notion that the axes of principal stress and principal plastic strain increment coincide.

For a clay tested undrained, the vertical strain ε_{yy} is zero. The Mohr circle of plastic strain increment is therefore as shown in Figure 2.35. (This is essentially the same as that shown in Figure 2.32, with the angle of dilation $\psi = 0$.) Figure 2.35 shows that the plane associated with the major principal plastic strain increment is at an angle

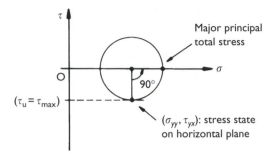

Figure 2.34 Mohr circle of total stress for an undrained shearbox test on a clay, assuming that the horizontal plane is the plane of maximum shear stress.

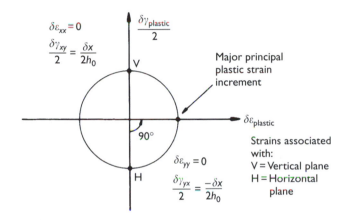

Figure 2.35 Mohr circle of plastic strain increment for an undrained shearbox test on a clay.

of $(90°)/2 = 45°$ anticlockwise from the horizontal plane. If this is also the plane on which the major principal total stress acts, the maximum shear stress acts on the horizontal plane, as conventionally assumed (Figure 2.34).

Unfortunately, things may not be this simple. Wroth (1987) quotes experimental data from simple shear tests (section 2.13) which suggest that the sequence of stress states followed by a sample of clay tested undrained is quite complex, and that the shear stress on certain planes at various stages of the test will be greater than that on the horizontal plane at the end of the test. Practically, however, these complications are probably of little significance. As Wroth (1987) himself notes, the important thing is the maximum shear stress experienced by the potential failure surface in the clay, in a test in which the principal axes are free to rotate.

2.13 The simple shear apparatus

The actual deformation imposed on a sample of soil in a conventional shearbox (Figure 2.15(a)) is rather different from the idealized deformation (Figure 2.15(b)), in that in reality, the relative displacement between the top and bottom of the shear-box is concentrated at the central horizontal plane. This means that the stress state within the sample will be highly non-uniform, with stress concentrations leading to the possible development of a rupture surface at the interface between the two sections (Ni *et al.*, 2000). Indeed, the conventional shearbox was originally developed to investigate the well-defined rupture surfaces which are frequently associated with landslips in clay (Collin, 1846).

The main difficulty lies in the use of the average state of the sample – as determined from measurements made at its boundaries – to describe the state of the soil, which may actually only be at failure in a thin rupture zone. This is a criticism which applies to other forms of soil test in which ruptures may develop – in particular, the triaxial test described in Chapter 5. Although the data from the initial stages of a test may be substantially unaffected, the development of a rupture at or near the peak stress would cast considerable doubt on the reliability of a critical state strength parameter determined from measurements made at the sample boundaries.

In a simple shear apparatus, the idealized deformation pattern shown in Figure 2.13(b) is actually imposed on the sample. This may be achieved by containing the sample within an arrangement of hinged and sliding platens (Roscoe, 1953), or within a stack of thin Teflon-coated aluminium plates which slide over each other as shown in Figure 2.36 (Kishida and Uesughi, 1987). Alternatively, a circular cylindrical sample may be enclosed within a rubber membrane, reinforced with a helical wire spring (Kjellman, 1951; Dyvik *et al.*, 1987). Stress cells can be incorporated into the boundaries of these devices in an attempt to measure the stresses directly.

Although a simple shear apparatus can offer a greater degree of controllability over sample drainage, and can incorporate more sophisticated measuring devices, there is conflicting evidence as to its success in imposing a uniform stress state on the soil sample (Airey and Wood, 1987). As in the case of the conventional shearbox, however, the non-uniformities do not materially affect the consistency and the engineering interpretation of the results. The term 'simple shear' relates to the mode of deformation imposed on the sample, rather than to the apparatus itself, which is mechanically quite complex. Perhaps, as a result of its mechanical complexity, the simple shear apparatus remains primarily a research tool, and will not be discussed further in this book.

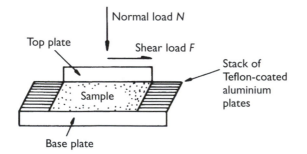

Figure 2.36 Simple shear apparatus. (Redrawn with permission from Kishida and Uesughi, 1987.)

Example 2.5 Analysis of stresses and strains in simple shear

Figure 2.37 shows a research shearbox, and defines the stresses and displacements measured during a test.

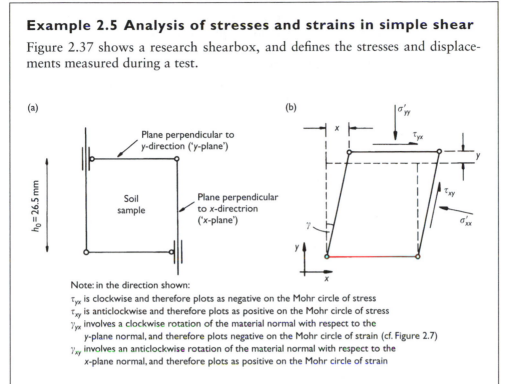

Note: in the direction shown:

τ_{yx} is clockwise and therefore plots as negative on the Mohr circle of stress

τ_{xy} is anticlockwise and therefore plots as positive on the Mohr circle of stress

γ_{yx} involves a clockwise rotation of the material normal with respect to the y-plane normal, and therefore plots negative on the Mohr circle of strain (cf. Figure 2.7)

γ_{xy} involves an anticlockwise rotation of the material normal with respect to the x-plane normal, and therefore plots as positive on the Mohr circle of strain

Figure 2.37 Research shearbox: (a) at start of test; (b) during test.

The stresses $\sigma'_{yy}, \tau_{yx}, \sigma'_{xx}$ and τ_{xy} are measured by means of load cells. The displacements x and y are measured using displacement transducers. The data in Table 2.5 were obtained during a drained test on a dense sand.

Table 2.5 Shear test data

Stage	x (mm)	σ'_{yy} (kPa)	τ_{yx} (kPa)	σ'_{xx} (kPa)	τ_{xy} (kPa)	y (mm)
a	0	70	0	30	0	0
b	0.30	70	−31.3	71	31.0	−0.50
c (peak)	2.50	70	−49.0	145.5	43.3	+0.60
d	3.00					+0.82
e	10.00	70	−32.0	90.6	24.5	+1.5

Sign conventions: positive direct stresses are compressive; positive shear stresses (τ_{xy}) are anticlockwise; negative shear stresses (τ_{yx}) are clockwise; y is measured positive upward; x is measured positive in the direction shown in Figure 2.37(b).

(i) Plot the Mohr circles of stress for stages a, b, c and e and determine the values of s' ($= [\sigma'_1 + \sigma'_3]/2$) and t ($= [\sigma'_1 - \sigma'_3]/2$) in each case.

(ii)   Determine also the 'volumetric' and 'shear' strain components $\varepsilon_{vol} = (\varepsilon_1 + \varepsilon_3)$ and $\gamma_{max} = (\varepsilon_1 - \varepsilon_3)$, and plot graphs of (τ/σ') against γ_{max} and ε_{vol} against γ_{max}.

(iii)   Plot the Mohr circle of strain increment between stages c and d, and estimate the peak angle of dilation ψ_{max}.

(iv)   From (ii), estimate ϕ'_{peak} and ϕ'_{crit}.

[*University of London 2nd year BEng (Civil Engineering) examination, King's College*]

Solution

Note: the 'x-plane' refers to the plane initially perpendicular to the x-direction, that is, the plane that is vertical at the start of the test, and the 'y-plane' refers to the plane initially perpendicular to the y-direction, that is, the plane that is horizontal at the start of the test.

(i)   For small displacements, the x and y planes remain approximately perpendicular. For larger displacements, this is not the case. The Mohr circles must therefore be constructed graphically. This is done by plotting the stress states associated with the x and y directions $(\sigma'_{xx}, \tau_{xy})$ and $(\sigma'_{yy}, \tau_{yx})$ on the (σ', τ) axes, and using a pair of compasses to locate (by trial and error) the centre of the Mohr circle. The centre of the circle must lie on the σ' axis, and it must be equidistant from the two stress points $(\sigma'_{xx}, \tau_{xy})$ and $(\sigma'_{yy}, \tau_{yx})$. The Mohr circles for each of the stages a, b, c and e are shown in Figure 2.38, and give the values of s' and t shown in Table 2.6.

Table 2.6 Values of s', t and t/s'

Stage	s' (kPa)	t (kPa)	t/s'	Change in angle between x and y planes (degrees)
a	50	20	0.40	0
b	70.5	31.2	0.44	~0
c (peak)	103	60	0.58	5.25
e	70	32	0.46	21

(ii)   The strains associated with the x and y planes are calculated using:

$\varepsilon_{yy} = -y/h_0$

$\varepsilon_{xx} = 0$

$\gamma_{xy} = -\gamma_{yx} = x/h_0$

(cf. Figure 2.31, ignoring the prefixes δ).

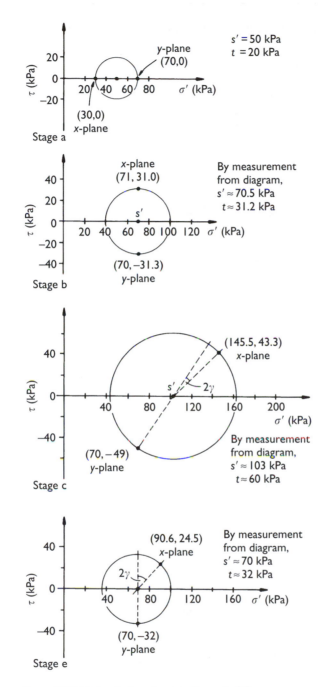

Figure 2.38 Mohr circles of stress, Example 2.5.

From Figure 2.39, the centre of the Mohr circle of strain is at $(\varepsilon_1 + \varepsilon_3)/2 = (y/2h_0)$, giving

$$\varepsilon_{\text{vol}} = (\varepsilon_1 + \varepsilon_3) = y/h_0$$

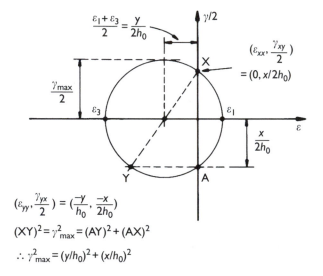

$$(\varepsilon_{yy}, \frac{\gamma_{yx}}{2}) = (\frac{-y}{h_0}, \frac{-x}{2h_0})$$

$$(XY)^2 = \gamma^2_{\text{max}} = (AY)^2 + (AX)^2$$

$$\therefore \gamma^2_{\text{max}} = (y/h_0)^2 + (x/h_0)^2$$

Figure 2.39 Geometry of Mohr circle of strain.

Also, the diameter of the Mohr circle is equal to $(\varepsilon_1 - \varepsilon_3) = \gamma_{\text{max}}$. Considering the triangle XAY,

$$(XY)^2 = (\gamma_{\text{max}})^2 = (AY)^2 + (AX)^2$$

But $AY = (y/h_0)$ and $AX = (x/h_0)$, giving

$$(\gamma_{\text{max}})^2 = (y/h_0)^2 + (x/h_0)^2$$

Using these expressions to calculate ε_{vol} and γ_{max} we obtain the data shown in Table 2.7. Sign convention: positive volumetric strain indicates compression.

Table 2.7 'Volumetric' and 'shear' strain components

Stage	$-y/h_0 (= \varepsilon_{yy})$	$x/h_0 (= -\gamma_{yx})$	ε_{vol}	γ_{max}
a	0	0	0	0
b	+1.89%	1.13%	+1.89%	2.20%
c	−2.26%	9.43%	−2.26%	9.70%
d	−3.09%	11.32%	−3.09%	11.74%
e	−5.66%	37.74%	−5.66%	38.16%

Graphs of (t/s') against γ_{max} and ε_{vol} against γ_{max} are plotted in Figure 2.40.

(iii)   From Figure 2.32, the angle of dilation $\psi = \tan^{-1}(dy/dx)$, where dx and dy are the *incremental* changes in x and y between stages c and d, rather than the *overall* changes since the start of the test. From the data for stages c and d, $dy = 0.22$ mm and $dx = 0.50$ mm, giving

$$\psi_{max} = \tan^{-1}(dy/dx) = \tan^{-1}(0.22/0.50)$$

or

$$\psi_{max} = 23.75°$$

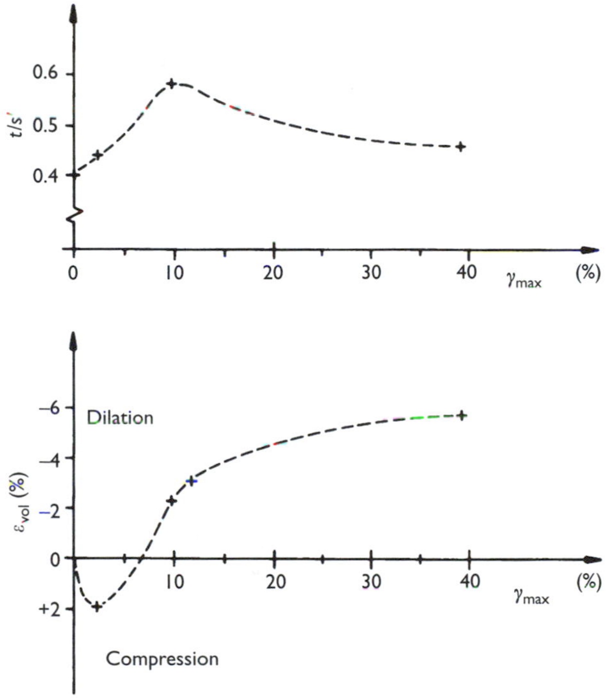

Figure 2.40 Plots of (a) (t/s') against γ_{max}; (b) ε_{vol} against γ_{max}.

(iv)   From the geometry of the Mohr circle of stress (Figure 2.30), the mobilized angle of shearing ϕ'_{mob} at any stage of the test is given by

$$\phi'_{mob} = \sin^{-1}(t/s')$$

Hence $\phi'_{peak} = \sin^{-1}(t/s')_{peak}$ and $\phi'_{crit} = \sin^{-1}(t/s')_{crit}$. From Figure 2.40, $(t/s')_{peak} \approx 0.583$, giving

$$\phi'_{peak} \approx 35.6°$$

and $(t/s')_{crit}$ (at the end of the test) ≈ 0.457, giving

$\phi'_{crit} \approx 27.2°$

As in Example 2.1, the value of ϕ'_{peak} calculated using equation (2.14) and the measured values of ϕ'_{crit} and ψ is somewhat high:

$\phi'_{peak} = \phi'_{crit} + 0.8\psi = 27.2° + (0.8 \times 23.75°) = 46.2°$

Key points

- The state of a soil must be described by three parameters, quantifying the **specific volume**, the **normal effective stress** and the **shear stress**.
- The strength of a soil is due primarily to interparticle friction. In terms of effective stresses, soils would be expected to obey a frictional failure criterion of the form

  $\tau = \sigma' \tan \phi'$

  where ϕ' is the effective **angle of friction** or the **angle of shearing resistance**.
- When sheared, a soil eventually reaches a **critical state** in which further deformation takes place at constant specific volume, normal effective stress and shear stress.
- Although the number of possible critical states is infinite, the combination of specific volume v, normal effective stress σ' and shear stress τ at each critical state is unique. If one of these parameters is known, the other two may be deduced using the equations that define the critical state line.
- The critical state line is the line which joins all possible critical states, on a three-dimensional plot with axes σ', τ and v. On a graph of τ against σ', the critical state line appears as a straight line with equation

  $\tau = \sigma' \tan \phi'_{crit}$

  and, on a graph of v against $\ln \sigma'$, as a straight line of equation

  $v = v_0 - \lambda \ln \sigma'$

- When a dense soil is sheared, it can develop a peak strength τ_{peak} before reaching a critical state. This is due to the effects of **dilation**, which cannot be sustained indefinitely. As the rate of dilation decreases, the shear stress falls until the critical state is reached. Loose soils do not dilate, and do not exhibit a peak strength.
- For a soil of a given initial specific volume, the peak stress ratio $(\tau/\sigma')_{peak}$ becomes smaller as the normal effective stress σ' is increased. This is because the critical state void ratio decreases with increasing effective stress, resulting in a diminishing potential for dilation. A soil should therefore be categorized as 'dense' or 'loose' with reference to the critical state void ratio at the normal effective stress under consideration.
- Clay soils sheared rapidly are forced to deform at constant specific volume. According to the critical state model, the specific volume will in this case define

the effective stress state at failure. The limiting shear stress, or the **undrained shear strength** τ_u, then defines an alternative failure criterion in terms of total stresses.

- The undrained shear strength failure criterion is only applicable to a low permeability soil, brought rapidly to failure at constant volume. The undrained shear strength is not a soil constant, since its value depends on the specific volume (or water content) of the soil as sheared.

Questions

The shearbox test

2.1   Describe with the aid of a diagram the essential features of the conventional shearbox apparatus. Stating clearly the assumptions you need to make, show how the quantities measured during the test are related to the stresses and strains in the soil sample.

[*University of London 2nd year BEng (Civil Engineering) examination, King's College (part question)*]

2.2   With the aid of sketches, describe, explain and contrast the results you would expect to obtain from conventional shearbox tests on samples of dry sand which were (a) initially loose and (b) initially dense. What factors would you take into account in selecting a soil strength parameter for use in design?

[*University of London 2nd year BEng (Civil Engineering) examination, King's College (part question)*]

Development of a critical state model

2.3   Mining operations frequently generate large quantities of fine, particulate waste known as tailings. Tailings are generally transported as slurries, and stored in reservoirs impounded by embankments or dams made up from the material itself. In order to investigate the geotechnical behaviour of a particular tailings material ($G_s = 2.70$), an engineer carried out three slow, drained shear tests – each over a period of one day – and three fast, undrained shear tests – each over a period of two minutes – in a conventional 60 mm × 60 mm shearbox apparatus.

The three samples in each group were initially allowed to come into drained equilibrium under the application of vertical hanger loads of 100 N, 200 N and 300 N. During each shear test, the hanger load was kept constant and the ultimate shear force F_{ult} recorded. Immediately after each test, a water content sample was taken from the centre of the rupture zone. All of the samples were initially saturated, and all of the tests were carried out with the sample underwater in the shearbox.

Use the results of the drained tests to construct a critical state model in terms of the normal effective stress σ' and shear stress τ on the shear plane, and the specific volume v. Give the values of ϕ'_{crit}, v_0 and λ. Deduce a relationship between the undrained shear strength τ_u and the normal effective stress at the start of the test, and compare its predictions with the experimental data from the undrained tests.

Test type	Vertical load V (N)	Shear load $F_{ult}/(N)$	Water content w (%)
Slow, drained	100	53	35.1
	200	105	31.3
	300	156	29.5
Fast, undrained	100	42	36.0
	200	80	32.6
	300	120	30.6

$(\phi'_{crit} = 28°$; $v_0 = 2.43$; $\lambda = 0.14$; critical state model predicts undrained shear strength $\tau_u = \exp\{(v_0 - v)/\lambda\} \tan \phi'_{crit}$, where $v = 1 + wG_s$, giving theoretical values of $\tau_u = 14\,\text{kPa}$ at $V = 100\,\text{N}$, $27\,\text{kPa}$ at $200\,\text{N}$ and $39.8\,\text{kPa}$ at $300\,\text{N}$. Measured values are smaller by about 16%, probably due to internal drainage and discontinuous sample behaviour.)

Determination of peak strengths

2.4   The following results were obtained from a shearbox test on a $60\,\text{mm} \times 60\,\text{mm}$ sample of dry sand of unit weight $18\,\text{kN/m}^3$.

Parameter	Reading on proving ring deflection dial gauge (divisions)
Zero force	91
Peak shear force for a hanger load of 3 kg	128
Peak shear force for a hanger load of 10 kg	162
Peak shear force for a hanger load of 20 kg	210

One division on the proving ring dial gauge corresponds to a force of $1.1\,\text{N}$ across the proving ring.

(a) Plot the data on a graph of shear stress against normal effective stress, and sketch the peak strength failure envelope.

(b) What is the peak resistance to shear on a horizontal plane at a depth of 3 m below the top of a dry embankment made from this soil?

(c) A model of the embankment is constructed from the same sand at a scale of 1:10. What is the peak resistance to shear on a horizontal plane at a depth of 300 mm below the top of the model?

(d) Would you expect the model to behave in the same way as the real embankment?

((a) The peak strength failure envelope i highly non-linear, with $\phi'_{peak} = 55°$ at $\sigma' \approx 8$ kPa, falling to $\phi'_{peak} = 34°$ at $\sigma' \approx 55$ kPa; (b) 36.4 kPa; (c) 7.7 kPa; (d) No, because the operational values of ϕ'_{peak} at corresponding depths in the model and the real embankment are quite different.)

Use of strength data to calculate friction pile load capacity

2.5 A friction pile, 300 mm in diameter, is driven to a depth of 25 m in dense sand of unit weight 19 kN/m^3. The ratio of horizontal to vertical effective stresses is 0.5. The angle of friction between the pile and the sand is 26° and the resistance offered at the base of the pile may be ignored. The natural water table, below which the pore water pressures are hydrostatic, is 5 m below ground level. During construction works, the water table is temporarily lowered to a depth of 16 m by pumping from wells. A load test on the pile is carried out even while pumping to lower the groundwater level is still in progress. Calculate the ultimate load capacity of the pile (a) observed in the test, and (b) after pumping from the wells has stopped and the water table has recovered to its natural level.

((a) 1273 kN; (b) 914 kN.)

2.6 The depth of the friction uplift pile described in Example 2.4 is increased to 20 m, where the undrained shear strength of the clay is 40 kPa. Calculate the short- and long-term uplift resistance of the 20 m pile.

(628 kN short-term; 468 kN long-term.)

Stress analysis and interpretation of shearbox test data

2.7 A drained shearbox test was carried out on a sample of saturated sand. The normal effective stress of 41.67 kPa was constant throughout the test, and the initial sample dimensions were 60 mm × 60 mm on plan × 30 mm deep. In the vicinity of the peak shear stress, the data recorded were:

Shear stress τ (kPa)	42.5	43.1	42.8
Relative horizontal displacement x (mm)	0.30	0.40	0.80
Upward movement of shearbox lid y (mm)	0.05	0.075	0.105

(a) Draw the Mohr circle of stress for the soil sample when the shear stress is a maximum, stating the assumption that you need to make. Determine ϕ'_{peak}, and the orientations of the planes of maximum stress ratio $(\tau/\sigma')_{max}$. Draw the Mohr circle of strain increment leading to the peak, and hence determine the maximum angle of dilation, ψ_{max}. Use an empirical relationship between ϕ'_{peak}, ψ_{max} and ϕ'_{crit} to estimate the critical state friction angle, ϕ'_{crit}.

(b) Three further drained tests on similar samples of the same soil were carried out, at different normal effective stresses. The peak and critical state shear stresses were:

Normal effective stress (kPa)	20	100	200
Peak shear stress (kPa)	23.8	83.9	132.0
Critical state shear stress (kPa)	12.6	63.2	126.4

For all four tests, plot the peak and critical state shear stresses τ_{peak} and τ_{crit} as functions of the normal effective stress σ'. Sketch failure envelopes for both peak and critical states, and comment briefly on their shapes. Which would you use for design, and why?

[University of London 2nd year BEng (Civil Engineering) examination, King's College (part question)]

$((a)\phi'_{peak} \approx 46°$; planes of maximum stress ratio $(\tau/\sigma')_{max}$ are at $\pm(90° - \phi'_{peak}) = \pm44°$ to the horizontal; $\psi_{max} \approx 14°$; estimate ϕ'_{crit} as $\phi'_{peak} - 0.8 \times \psi_{max} = 34.8°$. (b) $\phi'_{crit} = 32.5°$; peak strength failure envelope is curved, with lower peak strength at higher effective stresses, due to the suppression of dilation.)

2.8   In order to investigate the drained strength of a natural silt containing thin clay laminations at a spacing of approximately 6 mm, an engineer carried out a series of shearbox tests. The clay laminations were inclined at various angles θ to the horizontal. With the laminations horizontal ($\theta = 0$), the rupture formed entirely in the clay and the apparent angle of shearing resistance was 18°. With the laminations at an angle $\theta = 60°$, the rupture formed entirely in the silt and the apparent angle of shearing resistance was 30°. Stating clearly the assumptions you need to make, construct Mohr circles of stress at failure for various values of apparent angle of shearing resistance, marking on each the stress state corresponding to the clay laminations. (*Hint:* the mobilized strength on the clay laminations must never exceed 18°.) Plot a graph showing the relationship between the angle θ and the apparent angle of shearing resistance of the soil.

[*University of London 1st year BEng (Civil Engineering) examination,*
King's College (part question)]

($\phi'_{apparent} = 30°$ for $31.9° \leq \theta \leq 88.1°$ and $139.9° \leq \theta \leq 160.1°$, falling steeply between these plateaux to $\phi'_{apparent} = 18°$ at $\theta = 0°, 108°$ and $180°$.)

Note

1 As mentioned in the introduction to this chapter, the material covered in sections 2.12.2, 2.12.3 and 2.13 is unlikely to form part of the core syllabus of a general civil engineering course at first degree level. It may therefore be omitted by the first-time reader.

References

Airey, D.W. and Wood, D.M. (1987) An evaluation of direct simple shear tests on clay. *Géotechnique*, **37**(1), 25–33.

Bolton, M.D. (1986) The strength and dilatancy of sands. *Géotechnique*, **36**(1), 65–78.

Bolton, M.D. (1991) *A Guide to Soil Mechanics*, 232 Queen Edith's Way, Cambridge CB1 4NL.

Casagrande, A. (1936) Characteristics of cohesionless soils affecting the stability of slopes and earth fills. *Journal of the Boston Society of Civil Engineers*, **23**(1), 13–32.

Case, J., Chilver, A.H. and Ross, C.T.F. (1992) *Strength of Materials, and Structures*, 3rd edn. Edward Arnold, London.

Collin, A. (1846) *Recherches experimentales sur les glissements spontanes des terrains argileux*, Carilian-Goeurley et Dalmont, Paris. English translation by W. R. Schriever (1956) *Experimental Investigation on Sliding of Clay Slopes*, University of Toronto Press, Toronto.

Dyvik, R., Berre, T., Lacasse, S. and Raadim, B. (1987) Comparison of truly undrained and constant volume direct simple shear tests. *Géotechnique*, **37**(1), 3–10.

Hill, R. (1950) *The Mathematical Theory of Plasticity*. Oxford University Press, Oxford.

Jewell, R.A. and Wroth, C.P. (1987) Direct shear tests on reinforced sand. *Géotechnique*, **37**(1), 53–68.

Kishida, H. and Uesughi, M. (1987) Tests of the interface between sand and steel in the simple shear apparatus. *Géotechnique*, **37**(1) 45–52.

Kjellman, W. (1951) Testing the shear strength of clay in Sweden. *Géotechnique*, **2**(3), 225–35.

Ni, Q. (2003) The effects of particle properties and boundary conditions on soil shear behaviour: 3D numerical simulations. Ph D dissertation, University of Southampton.

Ni, Q., Powrie, W., Zhang, X. and Harkness, R.M. (2000) Effect of particle properties on soil behaviour: 3D numerical modelling of shearbox tests. American Society of Civil Engineers Geotechnical Special Publication, **96**, 58–70. ASCE, Reston, VA.

Roscoe, K.H. (1953) An apparatus for the application of simple shear to soil samples. *Proceedings of the 3rd International Conference on Soil Mechanics and Foundation Engineering*, Zurich, **1**, 186–191.

Roscoe, K.H., Schofield, A.N. and Wroth, C.P. (1958) On the yielding of soils. *Géotechnique*, **8**(1), 22–52.

Rowe, P.W. (1969) The relation between the shear strength of sands in triaxial compression, plane strain and direct shear. *Géotechnique*, **19**(1), 75–86.

Wroth, C.P. (1987) The behaviour of normally consolidated clay as observed in undrained direct shear tests. *Géotechnique*, **37**(1), 37–43.

Chapter 3

Groundwater, permeability and seepage

3.1   Introduction and objectives

We saw in Chapter 1 that a saturated soil is a two-phase material, comprising the soil particles and the pore water. The influence of the pore water on the overall behaviour of the soil and of geotechnical engineering structures cannot be overstated. In particular,

- The effective stress depends on both the total stress and the pore water pressure. The effective stress state of the soil governs both its stress–strain behaviour and its proximity to failure.
- Pore water pressures often represent a significant proportion of the total load which retaining walls and underground structures, such as tunnels and basements, must be able to withstand.

It might with some justification be argued that most problems in geotechnical engineering practice are associated with groundwater. Groundwater conditions may have been misunderstood, or not as originally envisaged on the basis of the site investigation data. Unfortunately, ignorance, and a failure to appreciate the potentially calamitous effects of uncontrolled groundwater in excavations, also play a large part.

To the engineer who has good site investigation data and a sound understanding of the principles of effective stress and groundwater flow and control, very few groundwater-related problems are truly unforeseeable. This chapter is concerned primarily with the study of steady state groundwater flow and the calculation of the associated pore water pressures and flowrates.

Objectives

After having worked through this chapter, you should know that:

- groundwater flow through the soil pores is driven by a **hydraulic gradient**, i, which is defined as the (negative of the) rate of change of total head with distance (section 3.3)
- the ease with which groundwater can flow through the soil pores is quantified by the soil **permeability**, k. Roughly, k is proportional to the square of the D_{10}

size of the soil. It therefore varies over an enormous range for the soils commonly encountered in geotechnical engineering practice (section 3.3)

- in most soil mechanics applications, the volumetric flowrate of water, q, through a soil element of cross-sectional area A is given by **Darcy's Law** (section 3.3),

$$q = Aki$$

You should understand:

- the principles of plane flownet sketching (sections 3.7–3.16)
- that the control of groundwater around an excavation is at least as much to do with pore water pressures and stability as with flowrates and the prevention of flooding (sections 3.10–3.12 and 3.16).

You should have an appreciation of:

- natural groundwater conditions, and the likely influence of construction activities such as excavation (section 3.2)
- the methods used to measure soil permeability, both in the laboratory and in the field, and the potential shortcomings of these methods (sections 3.4 and 3.5)
- the influence of soil structure and fabric on permeability (section 3.6)
- the mathematical equations governing groundwater flow (section 3.7).

You should be able to:

- sketch plane flownets for a variety of boundary conditions in both confined and unconfined situations (sections 3.8, 3.9, 3.12, 3.13 and 3.16), including cases where the soil is anisotropic (section 3.14)
- use the flownet to calculate flowrates (sections 3.8, 3.9 and 3.12) and pore water pressures (section 3.10), and to assess the stability of an excavation floor (section 3.11) or sideslope (sections 3.12 and 3.16).
- calculate the deflection of flowlines as they cross between two zones of soil of differing permeability (sections 3.15 and 3.16).

Sections 3.17–3.19 are probably outside the scope of many first-degree courses in civil engineering. They nonetheless contain potentially useful information of direct practical applicability, and if you read through them you should gain an appreciation of:

- the use of well pumping formulae to estimate the required capacity of a construction dewatering system (section 3.17)
- the use of numerical methods in the solution of groundwater flow problems (section 3.18)
- methods of groundwater control in practice (section 3.19).

3.2   Pore water pressures in the ground

3.2.1   Artesian conditions and underdrainage

We saw in section 1.7 that if the groundwater is stationary, pore water pressures would be hydrostatic below the level at which the gauge pore water pressure was zero – the **water table**.

Natural pore water pressures are not always hydrostatic. Figure 3.1 (a) shows an old river valley in which an **aquifer** (i.e. a porous, water-bearing stratum) is overlain by a clay soil. The aquifer extends beyond the edges of the clay, up into the surrounding hills. In the valley, the pore water pressures in the aquifer can be comparatively high, because the pore water can flow relatively easily through the aquifer from the high hills while the clay acts as a seal.

The pore water pressure at a particular point within the ground may be measured by means of a simple device known as a **standpipe piezometer**. A standpipe piezometer is essentially a small-bore pipe (usually about 20–25 mm in diameter) with a porous ceramic tip. After installation, the water level in the piezometer rises until the pressure at the bottom of the column of water in the standpipe is the same as the pore water pressure in the ground just outside the tip. (Owing to the volume of water that must flow into the standpipe before pressure equilibrium is achieved, the response time of the device increases as the permeability of the soil decreases. It is therefore unsuitable for measuring rapidly changing pore water pressures in a low-permeability soil, for which a different type of piezometer must be used.)

A standpipe piezometer driven through the clay may indicate a groundwater level in the aquifer which is above the ground surface in the valley. If the standpipe is not tall enough, it will overflow, bringing water from the aquifer to the surface. These conditions, in which the pore water pressure head in a buried or **confined** aquifer rises to a level above the ground surface, are described as **artesian**. A borehole drilled through the overlying clay layer to the aquifer for the purpose of obtaining a water supply is known as an **artesian well**.

In artesian conditions, the pore water pressure within the clay layer increases gradually with depth, from zero at (say) the ground surface. At the base of the clay layer, the pore water pressure is the same as that at the top of the aquifer (Figure 3.1(b)). Groundwater flows upward through the clay, but not more quickly than it can evaporate from the clay surface. (The pore water pressures in the aquifer may well be hydrostatic, below the apparent 'water table', or **piezometric level**, indicated by the standpipe piezometer.)

Figure 3.1(b) shows that the rate of increase of pore water pressure with depth in artesian conditions is (within the clay) greater than hydrostatic. If an excavation is made into the clay, a depth may be reached at which the weight of clay remaining is insufficient to hold back the water pressure in the aquifer, and the base of the excavation will fail.

Figure 3.1(c) also shows a clay layer underlain by a porous aquifer, but in this case there is a layer of gravel on top. The groundwater level in the gravel is maintained at the ground surface by recharge from the river. Pore water pressures in the underlying aquifer were originally artesian, but after many years of being pumped for water supply purposes, the groundwater level in the aquifer is now significantly below that in the

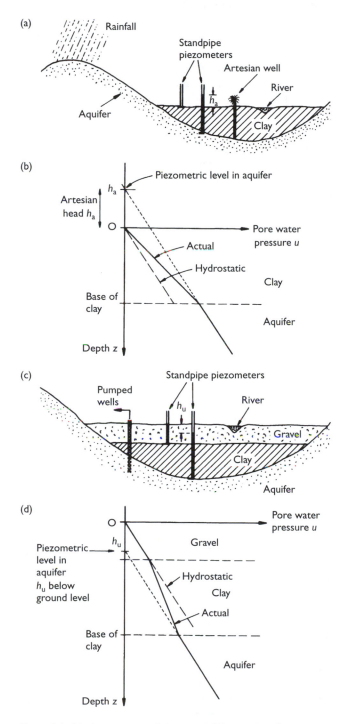

Figure 3.1 (a) Artesian conditions and (b) associated pore water pressures in clay; (c) under-drained conditions and (d) associated pore water pressures in clay.

overlying gravel. There is therefore a slow but steady flow of water downward through the clay from the gravel to the underlying aquifer. The increase in pore water pressure with depth within the clay is less than hydrostatic, as shown in Figure 3.1(d). In these conditions, the clay is described as **underdrained**.

Figure 3.1 is in effect an idealized representation of the hydrogeology of the London Basin in England. The underlying aquifer is the Chalk, the clay is the London Clay and the overlying gravels are the Thames Flood Plain or Terrace Gravels. 200 years ago, the pore water pressures in the chalk aquifer underlying the London Clay were artesian below the centre of the London Basin. With the industrialization and growth of London in the late nineteenth and early twentieth centuries, the water demand of the metropolis increased dramatically. The chalk aquifer was pumped to such an extent that the groundwater level within it fell to below that in the overlying Thames Gravel, and the London Clay became underdrained.

In the post-industrial age, pumping rates were reduced considerably, with the result that the groundwater level in the chalk aquifer is now gradually rising. The geotechnical implications of this may be very significant. Many foundations, and much of London's underground rail network, were constructed during the period when the pore water pressures in the London Clay were comparatively low, due to the effects of underdrainage.

The engineering implications of the rising groundwater beneath London are examined in detail by Simpson *et al.* (1989), who also cite evidence of rising groundwater levels beneath cities such as Paris, New York, Tokyo, Dohar (Qatar), Birmingham, Liverpool and Nottingham.

3.2.2 Effect of construction activities

Even if the natural groundwater conditions are hydrostatic, they are likely to be altered at least locally by the construction of geotechnical features such as excavations. Consider, for example, the retaining wall shown in Figure 3.2. Behind the wall, the water

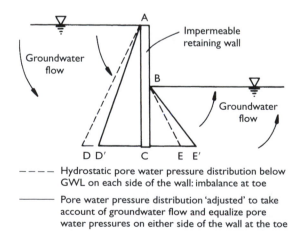

- - - - Hydrostatic pore water pressure distribution below GWL on each side of the wall: imbalance at toe

———— Pore water pressure distribution 'adjusted' to take account of groundwater flow and equalize pore water pressures on either side of the wall at the toe

Figure 3.2 Groundwater flow around an embedded retaining wall.

table is in its original position, level with the retained soil surface. In front of the wall, the water table must be maintained at the level of the excavated soil surface if the excavation is not eventually to flood. (This is usually achieved by means of drains, or by pumping from wells.) Assuming that the wall is impermeable, water will flow around the wall from the high groundwater level behind to the low groundwater level in front.

In the design of a retaining wall like this, it would be necessary to estimate the pore water pressures acting on the structure. We could assume that the pore water pressures are hydrostatic below the different groundwater levels on each side of the wall – lines AD and BE in Figure 3.2. This would create an imbalance in the pore water pressure at the toe of the wall C, and therefore seems unlikely. We could remove this imbalance by reducing the pore water pressures behind the wall and increasing them in front (lines AD' and BE'), so that the pore water pressure at the toe is the same on both sides of the wall. In doing this, the pore water pressures behind the wall have been reduced to below their hydrostatic values, while those in front have been increased.

At present, we have no theoretical justification for what we have done: however, the techniques introduced in this chapter can be used to show that the pore pressure distribution we would obtain is in many cases not unreasonable (see also section 7.8.1).

We should also note at this stage that it is not differences in pressure as such that drive groundwater flow. This can be demonstrated by the fact that the pore pressure at the soil surfaces behind and in front of the retaining wall (points A and B) is zero, yet water is flowing around the wall from A to B. In contrast, when the groundwater is stationary the pore water pressures increase hydrostatically with depth z below the water table, $u = \gamma_w z$ (section 1.7.1). From this we can infer that it is variations in pore water pressure away from the hydrostatic condition which cause groundwater to flow. This is discussed in more detail in section 3.3.

3.2.3 Pore water pressures above the water table

Pore water pressures can be negative (rather than positive), in which case, by equation (1.13), the effective stress will be greater than the total stress. There is, however, a limit to the negative pore water pressure a soil can sustain without drawing in air (at atmospheric or zero gauge pressure) through any surface which is exposed to the atmosphere. This limiting negative pore water pressure is known as the **air entry value**: it will increase as the soil pore size decreases. This can be shown by the analysis in Figure 3.3(a).

The gauge pore water suction U_e (i.e. the negative pore water pressure measured relative to ambient atmospheric pressure) at air entry may be estimated by considering the equilibrium of a hemispherical water meniscus in a circular pore of diameter d.

Force due to surface tension around = Force due to difference in pressure
    the rim of the meniscus   between the pore water and the air

$$\pi d T = (\pi d^2 / 4) U_e$$

or

$$U_e = 4T/d \qquad (3.1)$$

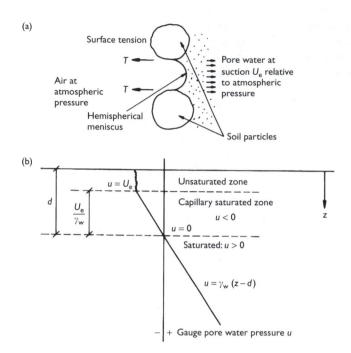

Figure 3.3 (a) Surface tension analysis for the gauge pore water suction at air entry; (b) pore water pressures in a fine soil above the water table (groundwater at rest).

where T is the surface tension of the water/air interface ($= 7 \times 10^{-5}$ kN/m at 10°C). The effective pore size of a real soil is generally assumed to be somewhere between D_{10} and $0.2 \times D_{10}$ (where D_{10} is the D_{10} particle size, defined in section 1.8). It may be shown (by substituting the appropriate values of D_{10} or $0.2 \times D_{10}$ into equation (3.1)) that the gauge pore water suction at air entry for a coarse sand with $D_{10} = 1$ mm lies in the range 0.28 to 1.4 kPa, while that for a clay with $D_{10} = 0.001$ mm is 1000 times greater. The consequence of this is that coarse soils above the water table or the **phreatic surface**, on which the gauge pore water pressure is zero, will tend to be unsaturated, with very little water retained in the pores by capillary action.

Fine soils (i.e. silts and clay) may remain saturated for several metres above the water table, with pore water pressures continuing to decrease until the air entry value is reached (Figure 3.3(b)). If the groundwater is at rest, the rate of decrease of pore water pressure with height above the water table will be approximately hydrostatic (i.e. 9.81 kPa/m). In reality, flow through the capillary saturated and unsaturated zone may take place by the infiltration of rainwater (downwards), and by evaporation (upwards). Water is also lost from the soil by transpiration from plants.

3.3   Darcy's Law and soil permeability

If the pore water is at rest, the distribution of pore water pressure must be hydrostatic. Conversely, any localized change in pore water pressure from the hydrostatic value will cause water to flow through the voids between the soil particles.

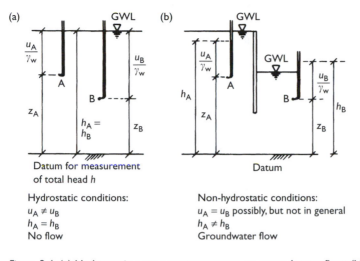

(a) Hydrostatic conditions:
$u_A \neq u_B$
$h_A = h_B$
No flow

(b) Non-hydrostatic conditions:
$u_A = u_B$ possibly, but not in general
$h_A \neq h_B$
Groundwater flow

Figure 3.4 (a) Hydrostatic pore water pressures, no groundwater flow. (b) Non-hydrostatic pore water pressures, groundwater flow.

Imagine two standpipe piezometers (see section 3.2.1) installed in the ground so that their lower ends are at the points A and B in Figure 3.4. If the groundwater is stationary, the absolute height to which water rises in each standpipe will be the same (Figure 3.4(a)). Water will only flow from A to B if there is a difference between the water levels in the standpipes (Figure 3.4(b)).

For any particular problem the standpipe piezometer rise may be measured from any convenient datum, but once the datum level has been chosen it must not be changed. The standpipe piezometer rise above an arbitrary datum is known as the **total head** or the **hydraulic potential**. In soil mechanics, it is often called the **excess head** because it is in effect a measure of the excess pore water pressure, over and above hydrostatic. (The term **excess head** may also be used to indicate a head difference over and above a set of steady state conditions which are not hydrostatic, particularly in the analysis of transient flow (Chapter 4). For this reason, the terms total head, head and potential will generally be used in this chapter.)

If the points A and B are at different levels and the groundwater is stationary, the pore water pressures at A and B will be different (Figure 3.4(a)). If there is a difference in total head (potential) between the two points, their pore water pressures will again in general be different, but could be the same (Figure 3.4(b)).

In 1836, Robert Stephenson used pumped wells to lower groundwater levels in order to enable the construction of the Kilsby tunnel on the London to Birmingham railway line in Northamptonshire. Stephenson observed that on pumping from one well, the water levels in adjacent wells dropped. He also recognized that the head difference between the wells was, for a given rate of pumping, an indication of the ease with which water could flow through the soil. However it was Henri Darcy (1856) who, on the basis of a series of experiments carried out at Dijon in France, proposed what is now known as **Darcy's Law**, which governs the flow of groundwater through

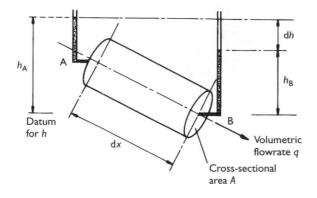

Figure 3.5 Darcy's Law.

soil (Figure 3.5):

$$q = Aki \tag{3.2}$$

where q (m³/s) is the volumetric flowrate of water, A (m²) is the cross-sectional area of the flow, i is the rate of decrease of total head (potential) with distance in the direction of the flow, $-dh/dx$, termed the **hydraulic gradient**, and k (m/s) is a soil parameter known as the **coefficient of permeability**.

Darcy's Law is one of the equations in soil mechanics that you should learn. The negative sign in the definition of the hydraulic gradient is mathematically necessary because the flow is always in the direction of decreasing head. If dh/dx is positive, the flowrate will be in the negative x direction. If dh/dx is negative, the flowrate will be in the positive x direction. In practice, the direction of flow is usually fairly obvious, and a pedantic insistence on the mathematical correctness of the equations can lead to confusion.

The Darcy coefficient of permeability depends on the properties of the permeant fluid as well as the soil matrix:

$$k = K\gamma_f/\eta_f \tag{3.3}$$

where K (m²) is the intrinsic permeability of the soil matrix, which does not depend on the properties of the permeant, γ_f (kN/m³) is the unit weight of the permeant, and η_f (kNs/m²) is the dynamic viscosity of the permeant. When the permeating fluid is water, the Darcy coefficient of permeability is more correctly termed the **hydraulic conductivity**, but in this book, the word 'permeability' is used.

Darcy's Law is sometimes written in the form

$$v_D = ki \tag{3.4}$$

where v_D is known as the **superficial** or **Darcy seepage velocity**. v_D is calculated by dividing the volumetric flowrate q by the total area (soil particles and voids) of the flow. The average *true* fluid velocity v_{true} is obtained by dividing the flowrate q by the cross-sectional area of the voids alone. Assuming that the void ratio for the cross-section is the same as the volumetric void ratio, the cross-sectional area of the voids is smaller than the total area by a factor $e/(1 + e)$. Thus

$$v_{\text{true}} = v_{\text{D}}(1 + e)/e \qquad\qquad (3.5)$$

The factor $e/(1 + e)$ is the porosity, n (section 1.5).

The coefficient of permeability used in Darcy's Law is a measure of the ease with which water can flow through the voids between the solid soil particles. For uniform soils, Darcy's coefficient of permeability depends on a number of factors, including the void size, the void ratio and the viscosity of the pore fluid (which for water varies by a factor of about 2 between temperatures of 20–60°C). Void size (which is related to particle size) is by far the most significant effect. This is shown by Hazen's (1892) empirical relationship for clean filter sand:

$$k\,(\text{m/s}) = 0.01D_{10}^2 \qquad\qquad (3.6)$$

where D_{10} is the sieve size in millimetres which just allows 10% by mass of the soil particles to pass through it. Characteristic permeabilities for various types of ground are shown in Figure 3.6: the range is enormous. This point is reinforced by comparing the difference in permeability between pebbles and clays (a factor of perhaps 10^{10}) with the difference in shear strength between high tensile steel and soft clay (about 10^5).

Darcy's Law can be applied to both the steady state seepage of water through an undeforming soil skeleton and the transient case of water being squeezed out from between soil particles which are moving closer together as a result of an increase in effective stress. The main proviso is that the flow must be laminar, rather than turbulent. In soils which have a particle size larger than a coarse gravel, groundwater velocities may be large enough for turbulent flow. In most other geotechnical applications, flow will be laminar. In this chapter, it will be assumed that the void ratio of the soil skeleton remains constant, and that conditions of steady state seepage have been reached. Transient flow is discussed in Chapter 4.

It has been assumed in the foregoing discussion of permeability that the soil is saturated. The permeability of an unsaturated or a partly saturated soil is altogether a different matter. Surface tension effects offer considerable resistance to flow, so that when a soil becomes unsaturated its permeability will fall by perhaps three orders of magnitude (McWhorter, 1985). For this reason, the surface of zero gauge pore water pressure (the phreatic surface) will generally represent a flow boundary in coarse soils

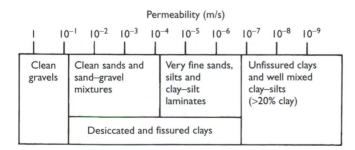

Figure 3.6 Range of soil permeability. (Redrawn with permission from BS8004, 1986.)

where the pore water suction at air entry is low – the soil above the phreatic surface being unsaturated and therefore effectively unpermeable.

Example 3.1 Effectiveness of a slurry trench cut-off wall

Darcy's Law may be applied directly in circumstances where the flow is one-dimensional. One example of one-dimensional flow is the constant head permeameter, which is used for the laboratory measurement of the permeability of recompacted samples of coarse-grained soils (section 3.4.1). A more directly practical application is the use of Darcy's Law to investigate the effectiveness of the slurry trench cut-off wall shown in Figure 3.7.

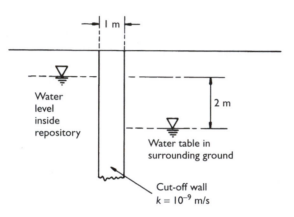

Figure 3.7 Use of Darcy's Law to estimate leakage through cut-off wall.

(A slurry trench cut-off wall is made by excavating a trench, usually 600 mm wide, to the required depth and filling it with a low-permeability cement–bentonite grout. They are often installed around landfills and other areas of contaminated land, in order to prevent the leakage of leachate or polluted groundwater into the surrounding soil.)

In Figure 3.7

$$\Delta h = \text{head drop across cut-off wall} = 2\,\text{m}$$
$$x = \text{thickness of cut-off wall} = 1\,\text{m}$$
$$\therefore \text{hydraulic gradient } i = \Delta h/x = 2$$
$$k = \text{permeability of cut-off wall material} = 10^{-9}\,\text{m/s}$$

Hence the rate of leakage $q = Aki = (1) \times 10^{-9} \times 2 = 2 \times 10^{-9}\,\text{m}^3/\text{s}$ per square metre, which is equal to the Darcy seepage velocity v_D (m/s).

If the void ratio of the cut-off wall material is uniform and constant and equal to 0.5, the actual average flow velocity is 3 ($= 1.5/0.5$) times greater and $v_{\text{true}} = 6 \times 10^{-9}\,\text{m/s}$. The time taken for the first particle of contaminated water to emerge on the clean side of the cut-off wall is thus $(6 \times 10^{-9})^{-1}$ seconds or about 5.3 years. While this may seem reasonably reassuring, the result of the calculation is critically dependent on the permeability of the cut-off wall. If this is increased by a factor of 10, the time taken is reduced by the same factor. If the wall is in reality cracked, the calculation may be worthless.

3.4   Laboratory measurement of permeability

3.4.1   The constant head permeameter

The constant head permeameter (Figure 3.8) is perhaps the simplest method of permeability measurement. The soil sample is contained within a perspex tube with inlet/outlets and filters at the top and bottom. Water flows one-dimensionally through the sample in the direction of its axis, and the hydraulic gradient required to maintain a flowrate q is determined from the head difference Δh indicated by manometers inserted at two points a distance L apart along the direction of flow.

Usually, the hydraulic gradient $i = (\Delta h/L)$ is found for a number of different flowrates q. The flowrate q is determined using a measuring cylinder and a stopwatch. For a sample of cross-sectional area A, Darcy's Law may be applied directly:

$$q = Aki = Ak\Delta h/L \tag{3.7}$$

so that a graph of q (m^3/s) against Δh (m) will have gradient Ak/L (m^2/s). The gradient is divided by A/L (m) to give the permeability k in m/s. The graph of q against Δh might not be a straight line if the volume (and hence the void ratio) of the soil sample changes during the test.

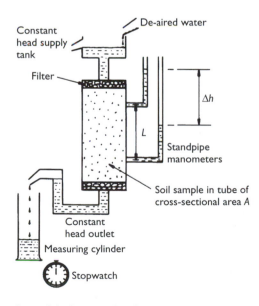

Figure 3.8 Constant head permeameter.

The soil can be tested in either upward or downward flow. In downward flow there may be a tendency for the soil to compact as the hydraulic gradient is increased. In upward flow, the soil sample will tend to decrease in density (become more loosely packed) until the hydraulic gradient reaches a critical value at which the soil grains are effectively buoyant and fluidization occurs (section 3.11).

Example 3.2 Interpretation of constant head permeameter test data

Table 3.1 gives data from a constant head permeameter test on a sample of initially dense sand in upward flow. Plot a graph of flowrate q against hydraulic gradient i, and estimate the initial permeability of the sample. Cross-sectional area of sample $= 8000\,\text{mm}^2$.

Table 3.1 Data from constant head permeameter test

Hydraulic gradient i	0	0.2	0.4	0.6	0.8
Flowrate q (cm³/s)	0	1.00	2.20	3.75	5.80

[*University of London 2nd year BEng (Civil Engineering) examination,*
Queen Mary and Westfield College (part question)]

Solution

The graph of q (y-axis) against i (x-axis) is shown in Figure 3.9. q is plotted in m³/s ($1\,\text{cm}^3/\text{s} = 1 \times 10^{-6}\,\text{m}^3/\text{s}$), and i is dimensionless.

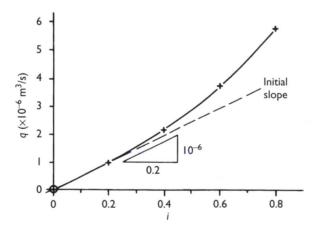

Figure 3.9 Flowrate q against hydraulic gradient i.

Applying Darcy's Law, $q = Aki$, to the permeameter test sample, the graph of q against i has gradient $S = dq/di = Ak$. (The slope of the graph increases with increasing flowrate, indicating an increasing permeability. This is due to the gradual loosening of the sample, as the upward hydraulic gradient is increased.)

At the beginning of the test, the slope of the graph S_i is approximately

$$S_i = dq/di \approx (1 \times 10^{-6}\,\text{m}^3/\text{s})/0.2 \approx 5 \times 10^{-6}\,\text{m}^3/\text{s} = Ak_i$$

The cross-sectional area $A = 8000\,\text{mm}^2 = 8 \times 10^{-3}\,\text{m}^2$, giving

initial permeability $k_i = S_i/A \approx 6.25 \times 10^{-4}\,\text{m/s}$

3.4.2 The falling head permeameter

The constant head permeameter is unsuitable for investigating the permeability of fine-grained (low permeability) soils where the flowrates are so small that evaporation from the measuring cylinder could lead to significant error. For fine soils, a falling head permeameter is used (Figure 3.10). Water flows from a small-bore tube of cross-sectional area A_2, through the soil sample which is contained within a larger tube of cross-sectional area A_1.

At the start of the test (time $t = 0$), the water level in the upper (small-bore) tube is at a height h_1 above the permeameter outlet. The water level in the upper tube then falls as water flows through the soil sample. At the end of the test (time $t = T$), the water level in the upper tube has fallen to a height h_2 above the permeameter outlet.

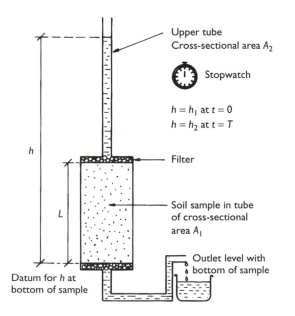

Figure 3.10 Falling head permeameter.

At a general time t $(0 < t < T)$, the water level is at a general height h $(h_1 > h > h_2)$. Applying Darcy's Law at a general time t to the soil sample in the large tube,

$$q = Aki = A_1 kh/L \qquad (3.8)$$

In the small-bore tube, the flowrate is given by the cross-sectional area multiplied by the velocity

$$q = A_2 v$$

but the velocity $v = -dh/dt$, so

$$q = -A_2 dh/dt \qquad (3.9)$$

(the negative sign is needed because v has been taken as positive downward, while h is measured as positive upward).
   Equating (3.8) and (3.9)

$$dh/dt = -(A_1/A_2)(k/L)h$$

Integrating between limits of $h = h_1$ at $t = 0$ and $h = h_2$ at $t = T$:

$$\int_{h1}^{h2} dh/h = -\int_0^T \left(\frac{A_1}{A_2} \frac{k}{L} \right) dt \qquad (3.10)$$

hence

$$\ln(h_2/h_1) = -(A_1/A_2)(k/L)T$$

or

$$k = (A_2 L/A_1 T) \ln(h_1/h_2) \qquad (3.11)$$

In general terms, a graph of $\ln(h_1/h)$ against t will have slope $S = (kA_1/A_2L)$ (Example 3.3).

Example 3.3 Interpretation of falling head permeameter test data

In an attempt to investigate the overall vertical permeability of a layered deposit, an engineer carries out a falling head permeability test on an artificial sample comprising 100 mm of silt overlying 100 mm of sand. The results from this test are given in Table 3.2.

(a)   Plot a graph of $\ln(h_1/h)$ against t, and explain its shape in terms of what happens to the overall vertical permeability of the sample during the test. What, physically, might be the explanation for this?
(b)   Estimate the overall vertical permeability at the start and at the end of the test.

Table 3.2 Data from falling head premeameter test

Time t since start of test (s)	0	40	100	190	330	600
Height of water in top tube h (m)	1.00	0.85	0.70	0.55	0.40	0.25

Cross-sectional area of sample $A_1 = 8000\,\text{mm}^2$
Cross-sectional area of top tube $A_2 = 10\,\text{mm}^2$
Overall sample length $= 200\,\text{mm}$

[*University of London 2nd year BEng (Civil Engineering) examination,
Queen Mary and Westfield College (part question)*]

Solution

(a) The processed data are given in Table 3.3 and plotted as a graph of $\ln(h_1/h)$ against t in Figure 3.11 ($h_1 = 1\,\text{m}$).

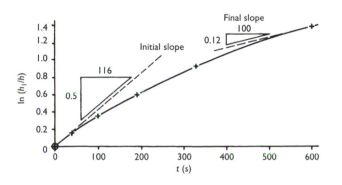

Figure 3.11 $\ln(h_1/h)$ against time.

Table 3.3 Processed data

t (s)	0	40	100	190	330	600
h_1/h	1	1.176	1.429	1.819	2.500	4.000
$\ln(h_1/h)$	0	0.163	0.357	0.598	0.916	1.386

The graph is curved, indicating a reduction in overall vertical permeability as the test progresses. Physically, this might be due to the migration of silt particles into the sand.

(b) Rewriting equation (3.11) with $T = t$, $h_1 = h_1$ and $h_2 = h$,

$$\ln(h_1/h) = (kA_1/A_2L) \times t$$

so that the graph of $\ln(h_1/h)$ (*y*-axis) against time t (*x*-axis) has slope

$$S = d[\ln(h_1/h)]/dt = (kA_1/A_2L)$$

or

$$k = (A_2L/A_1) \times S$$

From the graph, the initial slope $S_i = d \ln(h_1/h)/dt = 0.5/116 = 4.3 \times 10^{-3}\,\text{s}^{-1}$. Substituting $A_1 = 8000\,\text{mm}^2$, $A_2 = 10\,\text{mm}^2$ and $L = 200\,\text{mm}$,

$$(A_2 L/A_1) = (10 \times 200 \div 8000)\,\text{mm} = 0.25 \times 10^{-3}\,\text{m}$$

giving

$$k_i = (A_2 L/A_1)S_i = (0.25 \times 10^{-3}) \times (4.3 \times 10^{-3})\,\text{m/s}$$

$$\Rightarrow k_i = 1.075 \times 10^{-6}\,\text{m/s}$$

Similarly, the final slope $S_f = 0.12/100 = 1.2 \times 10^{-3}\,\text{s}^{-1}$, giving

$$k_f = 3.0 \times 10^{-7}\,\text{m/s}$$

3.5   Field measurement of permeability

The laboratory methods described above might give a reasonable estimate of the *in situ* permeability of a uniform isotropic soil, provided that the sample has been selected and taken with care and that some attempt has been made to replicate its field density in the permeameter. The loss of fine particles on sampling is often a problem, because it will lead to an overestimation, perhaps by more than an order of magnitude, of the permeability of the soil in the field. The soil structure and fabric (e.g. fissures, and anisotropy due to layering as described in section 3.6) may be destroyed during sampling and cannot be replicated in the laboratory. In many cases, these features will contribute significantly to the effective bulk permeability of even a homogeneous soil in the field (Rowe, 1972). Large-scale inhomogeneities, such as high permeability lenses, are an additional complicating factor.

In practice, the effective permeability of a soil stratum *in situ* may be investigated by means of a pumping test. Water is pumped from one well and the resulting fall in the groundwater level is monitored at a number of locations using standpipe piezometers or observation wells. Pumping tests can be expensive, but if properly planned and executed represent the most reliable method of determining a suitable value of permeability for design. A pumping test which has been carried out badly, however, may be worse than useless.

Observation wells differ from standpipe piezometers in that an observation well is open to the soil (via a slotted wellscreen and a granular filter) over a significant proportion of its length, whereas a standpipe piezometer is open to the soil over only a short length (typically 150–300 mm) at the bottom. In general, a standpipe piezometer measures the pore water pressure at a point, whereas an observation well might measure the maximum value of total head over its open length. For the pumping tests described in sections 3.5.1 and 3.5.2, the flow of groundwater is horizontal so that the head is the same along any vertical line. An observation well will, in these circumstances, record the same water level as a standpipe piezometer, but this will not in general be the case. The importance of using standpipe piezometers with a well-defined response zone, rather than observation wells, to measure pore pressures at discrete points in heterogeneous materials with complex flow fields is highlighted by Cox (2003).

In fine-grained soils, water levels may continue to fall for days or even weeks after the commencement of pumping. In coarse soils, steady state conditions may be reached quite quickly. Different methods of test and analysis are appropriate for different situations. The execution and analysis of well pumping tests is a subject in its own right, on which many papers and even whole books have been written (e.g. Cooper and Jacob, 1946; Clark, 1977; Kruseman and De Ridder, 1983; BS6316, 1992). As examples, we will here consider the steady state analysis of pumping tests in two highly idealized situations.

3.5.1 Well pumping test in an ideal confined aquifer

The aquifer shown in Figure 3.12 is described as **confined** because it is bounded at the top and the bottom by relatively impermeable strata. These prevent any water from entering or leaving the aquifer in vertical flow. The initial **piezometric surface**, which is defined by the levels to which water rises in an array of standpipe piezometers, is above the upper surface of the aquifer. In this respect, the piezometric surface is subtly different from the water table or **phreatic surface** mentioned in section 3.2.3, because the phreatic surface must always be contained within the body of the aquifer. The piezometric surface is the level to which the water in the confined aquifer would rise if it were free to do so.

For the following analysis to be valid, the piezometric surface must remain above the top of the aquifer during pumping. This requires that the water level in the well must not be drawn down below the top of the aquifer. It is also assumed that the well penetrates to the bottom of the aquifer, and that all flow is horizontal towards the well.

The flow area at a general radius r is $2\pi rD$, where D is the thickness of the aquifer. The hydraulic gradient $i = -dh/dr$. Applying Darcy's Law, the flowrate into the well is

$$q = Aki = 2\pi rDk\, dh/dr \qquad\qquad (3.12)$$

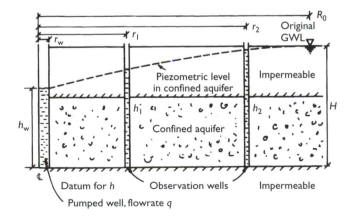

Figure 3.12 Well pumping test in a confined aquifer.

(The negative sign has been omitted from the hydraulic gradient in equation (3.12) because we are interested in the flow into the well, which is in the r negative direction.)

At the steady state, q is equal to the known pumped flowrate from the well. The **drawdown** (defined as the depth of the new piezometric surface below the initial piezometric surface) decreases with increasing distance from the well. Eventually, at a radius R_0, the drawdown is zero (i.e. the initial groundwater level beyond R_0 is unaffected by pumping). R_0 is termed the **radius** or **distance of influence** of the well. Rearranging equation (3.12) and integrating between limits of $(h = H, r = R_0)$ at the distance of influence, and $(h = h, r = r)$ at a general radius r

$$\int_r^{R_0} \frac{dr}{r} = (2\pi Dk/q) \int_h^H dh \tag{3.13}$$

Hence

$$\ln(R_0/r) = (2\pi Dk/q)(H - h) \Rightarrow (H - h) = [(q/2\pi Dk)\ln R_0] - [q/2\pi Dk]\ln r \tag{3.14}$$

A graph of the drawdown $(H - h)$ (on the y-axis) against the natural logarithm of the radial distance r (on the x-axis), plotted from the observation well data, will have a negative slope of magnitude $S = q/2\pi Dk$. The permeability k is given by

$$k = q/2\pi DS \tag{3.15}$$

The radius of influence R_0 may be read off from the graph at zero drawdown (Figure 3.13). It will generally be found that the drawdown in the soil given by the graph at $r = r_w$ (the radius of the well) is somewhat less than the measured drawdown inside the well. This is due to head losses at entry into, and vertical flow components in the vicinity of, the well, which are ignored in the idealized analysis shown in Figure 3.12.

If the graph of drawdown against radial distance is plotted directly onto $\log_{10}$–linear graph paper, equation (3.15) becomes

$$k = 2.3q/2\pi DS_{10} \tag{3.16}$$

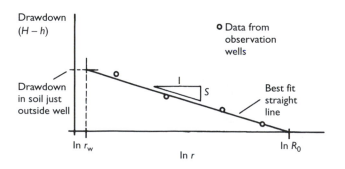

Figure 3.13 Analysis of data from a pumping test in a confined aquifer.

This is because $\log_{10} r = 2.3 \ln r$: the slope of the graph of $(H - h)$ against $\log_{10} r$ is therefore $S_{10} = 2.3q/2\pi Dk$.

3.5.2 Well pumping test in an ideal unconfined aquifer

If the aquifer is not overlain by an impermeable layer, the surface of zero gauge pore water pressure will be drawn down into the body of the aquifer to form a **phreatic surface** when pumping starts. In this case, the aquifer is described as an **unconfined** or a **water table aquifer**. It is still assumed that flow is essentially horizontal (which in practice means that the drawdown must be small in comparison with the thickness of the aquifer), and that the pumped well penetrates the entire depth of the aquifer. These conditions are shown in Figure 3.14.

The important difference between Figure 3.14 and Figure 3.12 is that in the case of the unconfined aquifer, the area available for flow diminishes more rapidly as the well is approached. This is because both the radius and the saturated aquifer thickness, rather than just the radius, decrease. (It is assumed that the soil above the surface of zero gauge pore pressure has become unsaturated, and there is therefore no flow within this zone. This assumption is reasonable for a coarse soil, but not necessarily appropriate for a fine-grained soil where flow may still take place in the capillary saturated zone.) At a general radius r, the area available for flow is $2\pi rh$, where h now varies with r. Applying Darcy's Law

$$q = Aki = 2\pi rhk \, dh/dr \tag{3.17}$$

(Again, flow into the well is taken as positive.) Rearranging equation (3.17) and integrating between limits of $h = H, r = R_0$ at the distance of influence, and $h = h, r = r$ at a general point

$$\int_r^{R_0} \frac{dr}{r} = (2\pi k/q) \int_h^H h \, dh \tag{3.18}$$

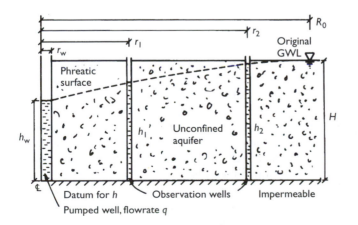

Figure 3.14 Well pumping test in an unconfined aquifer.

Hence

$$\ln(R_0/r) = (\pi k/q)(H^2 - h^2)$$

$$\Rightarrow (H^2 - h^2) = [(q/\pi k)\ln R_0] - [q/\pi k]\ln r \tag{3.19}$$

To estimate the permeability in this case, it is necessary to plot the data from the observation wells as a graph of $(H^2 - h^2)$ against the natural logarithm of the radial distance r. The slope is again negative, but of magnitude $S = q/\pi k$, so that the permeability k is given by

$$k = q/\pi S \tag{3.20}$$

If the data are plotted as $(H^2 - h^2)$ against $\log_{10} r$, $k = 2.3q/\pi S_{10}$ because $\log_{10} r = 2.3\ln r$, as before.

In practice, the well may not penetrate the full thickness of the aquifer. Correction factors which may be applied to take this into account are given in Leonards (1962). Also, vertical flow components will generally in reality be significant in the vicinity of the well. Equations (3.14) and (3.19) therefore tend to overestimate the drawdown at distances closer to the well than about 1.5 times the depth of the aquifer. Vertical flow does not seem to affect the flowrate significantly.

In three dimensions, the drawndown phreatic or piezometric surface around a well in an unconfined or confined aquifer takes up the shape of an inverted, concave-sided cone. This is known as the **cone of depression**. Consideration of the cone of depression is important in the design of an array of pumped wells to lower the groundwater level in the vicinity of an excavation. The wells must be installed close enough together to ensure that the required drawdown is achieved at the mid-points between every pair of adjacent wells. Generally, cones of depression are steeper in low-permeability soils than in high-permeability soils. Further details are given by Preene *et al.* (2000).

3.6 Permeability of laminated soils

We saw in section 1.3 that, due to the nature of the environment in which they were deposited, some soils exhibit a varved or layered structure, comprising alternating bands of fine and coarse material. This is particularly true of soils which were originally laid down over many years in glacial lakes. Each season's meltwater would bring with it soil particles of various sizes: the coarser particles would settle quite quickly to form one layer, and the finer particles would settle more slowly to form the next. A glacial lake deposit with this type of structure was encountered during the construction of the immersed tube tunnel crossing of the River Conwy in North Wales in 1989 (Figure 3.15).

The layered structure of a varved deposit can lead to an orders of magnitude difference between the bulk permeability in the vertical and horizontal directions. A material having different values of certain properties in different directions is described as **anisotropic**. Where the two horizontal directions are indistinguishable, as is often the case for soils as a result of their deposition in layers (even if a varved fabric is not readily identifiable), the term **cross-anisotropic** is sometimes used.

To construct the Conwy tunnel, three large excavations up to 15 m deep (below mean sea level) with a total perimeter of 2.6 km were made adjacent to the Conwy

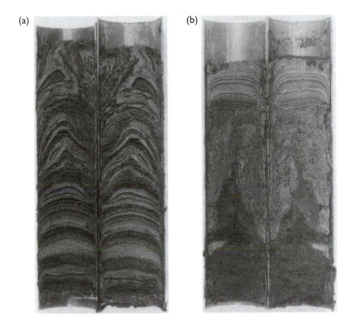

Figure 3.15 Laminated glacial lake deposits (100 mm diameter samples) from Conwy, North Wales. (a) 19.30–19.75 m below original ground level (silty clay with regular thin partings of silty fine sand). (b) 22.65–23.10 m below original ground level, showing a transition to a more uniform fine sand near the bottom of the stratum. Note disturbance due to sampling. (Reproduced with permission from Powrie and Roberts, 1990.)

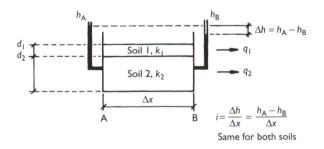

Figure 3.16 Horizontal flow through a layered soil.

estuary. The stability of these excavations depended on the maintenance of low pore water pressures in the slopes and the base throughout the construction period: this was achieved by means of pumped wells around each excavation. One of the factors influencing the design of the well systems was the anisotropic permeability of the Glacial Lake Deposits. The way in which the marked difference between the horizontal and vertical permeabilities of a layered deposit arises can be investigated using the idealized model shown in Figure 3.16.

In horizontal flow (i.e. flow parallel to the laminations), the hydraulic gradient between A and B is the same for both layers. The total flowrate q_T is the sum of the flowrates through the individual layers. We seek an expression of the form

$$q_T = A_T k_h i$$

where k_h is the overall (bulk) permeability in the horizontal direction and A_T is the total area available for flow. For a unit depth perpendicular to the plane of the paper,

$$A_T = d_1 + d_2$$

Applying Darcy's Law to each layer in turn,

$$q_1 = d_1 k_1 i \quad \text{and} \quad q_2 = d_2 k_2 i$$

Hence

$$q_T = q_1 + q_2 = (d_1 k_1 + d_2 k_2)i$$

and by comparison with the initial expression $k_h = q_T/(A_T i)$, the overall horizontal permeability is given by

$$k_h = (d_1 k_1 + d_2 k_2)/(d_1 + d_2) \tag{3.21}$$

In vertical flow (i.e. flow perpendicular to the laminations), the same flow passes through each layer and the overall head drop Δh_T is the sum of the head drops across the individual layers (Figure 3.17).

The hydraulic gradients across each layer are $i_1 = \Delta h_1/d_1$, and $i_2 = \Delta h_2/d_2$. The flow area A is the same for all layers, and we seek an expression of the form

$$q_T = A k_v i_T$$

where the overall hydraulic gradient $i_T = (\Delta h_1 + \Delta h_2)/(d_1 + d_2)$ and k_v is the overall vertical permeability. Since the flowrate through each layer is the same

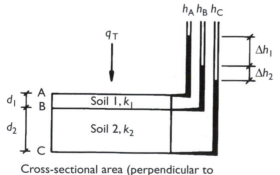

Cross-sectional area (perpendicular to direction of flow) is A.

Figure 3.17 Vertical flow through a layered soil.

(and equal to q_T),

$$q_T = Ak_1\Delta h_1/d_1 = A_1k_2\Delta h_2/d_2$$

and

$$\Delta h_1 + \Delta h_2 = (q_T/A)[(d_1/k_1) + (d_2/k_2)]$$

Hence

$$i_T = (\Delta h_1 + \Delta h_2)/(d_1 + d_2) = (q_T/A)[(d_1/k_1) + (d_2/k_2)]/(d_1 + d_2)$$

By comparison with the initial expression $k_v = q_T/Ai_T$, the overall vertical permeability is

$$k_v = (d_1 + d_2)/[(d_1/k_1 + d_2/k_2)] \tag{3.22}$$

For a system of n layers, each having a different thickness and permeability, the horizontal and vertical permeabilities are given by the generalized forms of equations (3.21) and (3.22):

$$k_h = \left(\sum_{r=1}^{r=n}(d_rk_r)\right)\left(\sum_{r=1}^{r=n}d_r\right)^{-1}$$

and

$$k_v = \left(\sum_{r=1}^{r=n}d_r\right)\left(\sum_{r=1}^{r=n}\frac{d_r}{k_r}\right)^{-1} \tag{3.23}$$

Example 3.4 Calculating the bulk horizontal and vertical permeabilities of a laminated deposit

A laminated soil consists of alternating bands of clay 15 mm thick and sandy silt 3 mm thick. These materials in isolation have permeability 10^{-8} m/s and 10^{-6} m/s respectively. Calculate the bulk permeability in the directions parallel and perpendicular to the laminations.

(i) k_h (parallel to the laminations) $= (d_1k_1 + d_2k_2)/(d_1 + d_2)$

$$= \frac{0.015 \times 10^{-8} + 0.003 \times 10^{-6}}{0.018}\ \text{m/s}$$

so

$$k_h = 1.75 \times 10^{-7}\ \text{m/s}$$

(ii) k_v (perpendicular to laminations) $= (d_1 + d_2)/[(d_1/k_1) + (d_2/k_2)]$

$$= 0.018/[(0.015/10^{-8})$$
$$+ (0.003/10^{-6})]\,\text{m/s}$$

so

$$k_v = 1.2 \times 10^{-8}\,\text{m/s}$$

3.7 Mathematics of groundwater flow

The differential equation governing the flow of groundwater is rarely used explicitly in geotechnical engineering. The assumptions which underlie it, however, must be appreciated, because it underpins the graphical and numerical methods that are used on a daily basis in the solution of groundwater flow problems. In this section, we will look at the derivation of the differential equation governing groundwater flow. Do not worry if you find yourself unable to remember this derivation: the important thing is to develop an understanding of the assumptions and limitations involved.

The differential equation governing the flow of groundwater is derived with reference to an element of soil having dimensions $\delta x, \delta y$ and δz in the directions of the x-, y- and z-axes, respectively (Figure 3.18). The space coordinates of the soil element are (x, y, z) and the total head or potential at this location is h.

The total flowrate into the soil element is

$$q_{in} = \sum (vA)_{in} = (v_x \delta y \delta z) + (v_y \delta x \delta z) + (v_z \delta x \delta y) \tag{3.24}$$

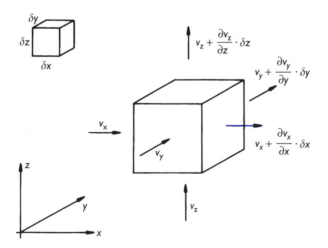

Figure 3.18 Flow through a soil element.

The total flowrate out of the soil element is

$$q_{out}= \sum (vA)_{out} = [v_x + (\partial v_x/\partial x)\delta x]\delta y\delta z + [v_y + (\partial v_y/\partial y)\delta y]\delta x\delta z$$

$$+ [v_z + (\partial v_z/\partial z)\delta z]\delta x\delta y$$

$$= q_{in} + (\delta x\delta y\delta z)(\partial v_x/\partial x + \partial v_y/\partial y + \partial v_z/\partial z) \qquad (3.25)$$

If steady state conditions have been reached the volume of the soil element is constant, so

$$q_{in} = q_{out}$$

or from equation (3.25)

$$\partial v_x/\partial x + \partial v_y/\partial y + \partial v_z/\partial z = 0 \qquad (3.26)$$

(The expressions $\partial v_x/\partial x$, $\partial v_y/\partial y$ and $\partial v_z/\partial z$ are **partial derivatives**. A partial derivative is used when a quantity – in this case, v_x, v_y or v_z – is a function of more than one variable. In this case, v_x, v_y, and v_z are each functions all three space coordinates, x, y and z. The notation $\partial v_x/\partial x$ is used to mean the rate of change of v_x with distance in the x-direction only, and similarly with $\partial v_y/\partial y$ and $\partial v_z/\partial z$.)

Darcy's Law applies in each of the three independent directions:

$$v_x = k_x i_x, \quad v_y = k_y i_y \quad \text{and} \quad v_z = k_z i_z \qquad (3.27)$$

where k_x is the soil permeability and i_x is the hydraulic gradient in the x-direction and so on. Also,

$$i_x = -\partial h/\partial x, \quad i_y = -\partial h/\partial y \quad \text{and} \quad i_z = -\partial h/\partial z \qquad (3.28)$$

Substitution of equations (3.27) and (3.28) into (3.26) yields

$$k_x\partial^2 h/\partial x^2 + k_y\partial^2 h/\partial y^2 + k_z\partial^2 h/\partial z^2 = 0 \qquad (3.29)$$

This is the differential equation governing the flow of groundwater through a rigid soil skeleton. If the permeability of the soil is the same in all directions, equation (3.29) can be simplified to **Laplace's Equation**,

$$\partial^2 h/\partial x^2 + \partial^2 h/\partial y^2 + \partial^2 h/\partial z^2 = 0 \qquad (3.30)$$

3.8 Plane flow

We have seen in Chapter 2 that in earthworks such as excavations, cuttings and embankments which are long in comparison with their other dimensions, the deformation in the longitudinal direction is (by symmetry) approximately zero. Exactly the same argument applies for seepage flow, with significant flow occurring only in the plane of the cross-section. The problem is therefore reduced to two dimensions, in which Laplace's equation may be solved graphically using a technique known as **flownet sketching**.

A **flownet** is a network of **flowlines**, which represent the trajectories of individual fluid particles, and **equipotentials**, along which the total head or potential is constant and therefore there is no flow. As well as providing a graphical representation of the flow pattern, the flownet may be used to calculate the seepage flowrate and pore water pressures at any point on the cross-section.

The flownet is constructed by trial and error so as to satisfy the following conditions:

1. Flowlines cross equipotentials at right-angles. This is because there is by definition no flow along an equipotential, which means that all of the flow must be at 90° to it.
2. Flowlines cannot cross other flowlines. Two molecules of water cannot occupy the same space at the same time, which is what flowlines crossing each other would imply.
3. Equipotentials cannot cross other equipotentials: one point cannot have two different values of total head.
4. Impermeable boundaries and lines of symmetry are flowlines: as there is no flow across them, all of the flow must be along them. (Flow across a line of symmetry would require flowlines from opposite sides to cross.)
5. Bodies of water such as reservoirs are equipotentials. (Try imagining standpipe piezometers inserted at a number of different locations within a lake: the water level in all of the standpipes will coincide with the water level in the lake.)
6. Although the number of flowlines and equipotentials which could be sketched is infinite, the flownet must be constructed so that each element is a **curvilinear square**. Although its sides may be curved, a curvilinear square is as broad as it is long, so that a circle may be inscribed within it. If this condition is fulfilled, the drop in head between any two consecutive equipotentials is the same, and the flowrate through all **flowtubes** is the same. (A flowtube is the channel defined by two adjacent flowlines.)

Consideration of the flow through a single flow element (Figure 3.19) shows that, for a flownet constructed according to these rules, the total flowrate q per metre run

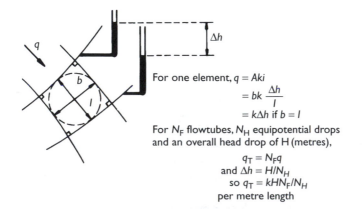

For one element, $q = Aki$

$$= bk\,\frac{\Delta h}{l}$$

$$= k\Delta h \text{ if } b = l$$

For N_F flowtubes, N_H equipotential drops and an overall head drop of H (metres),

$$q_T = N_F q$$
$$\text{and } \Delta h = H/N_H$$
$$\text{so } q_T = kHN_F/N_H$$
per metre length

Figure 3.19 Flow through a single flow element.

in the longitudinal direction is

$$q \, (\mathrm{m}^3/\mathrm{s} \text{ per metre}) = kHN_F/N_H \tag{3.31}$$

where k is the permeability of the soil (in m/s), H is the overall head drop (in metres) between the first and last equipotentials, N_F is the number of flowtubes (i.e. the number of spaces between flowlines) and N_H is the number of equipotential drops (i.e. the number of spaces between equipotentials). Equation (3.31) is quite straightforward, and is worth committing to memory.

3.9   Confined flownets

If all the boundaries to the flow regime are known at the outset, the flownet is described as **confined**. The construction of a confined flownet is illustrated below by means of a worked example based on a case study.

Example 3.5 Confined flownet for a long cofferdam

Figure 3.20 shows a cross-section through a long excavation in Norwich Crag, a fine sand of mean permeability $k = 1.5 \times 10^{-4}$ m/s. The sides of the excavation are supported by steel sheet piles, a structure known as a **cofferdam**. The purpose of the excavation is to enable the construction of a cooling water outfall pipe for a coastal power station. The excavation is to be made across a beach, so

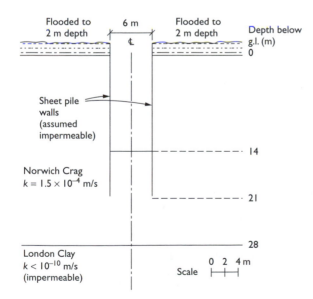

Figure 3.20 Geometry for example flownet: excavation for cooling water outfall pipe.

that the ground surface outside the sheet pile cofferdam must be assumed to be flooded with seawater to a depth of 2 m at high tide. Estimate the rate (per metre length of the excavation) at which water must be removed from the floor of the excavation in order to prevent flooding.

Newcomers to flownet sketching are often nervous about making a start, in case they make a mistake. It is at this stage that you must bear in mind that flownet sketching is an iterative process: it is only by making mistakes, which are then corrected, that a satisfactory solution is approached. It takes practice! The flownet should be constructed methodically, and the following procedure should enable you to make a reasonable start. Remember also that the person who never made a mistake never made anything: do not be afraid to try.

1. Is the problem geometry symmetrical about the centreline? If so – as in this case – it is only necessary to sketch one half of the flownet. Remember, however, that if only half the flownet is sketched, the number of flowtubes must be multiplied by two in order to calculate the flowrate per metre length.

2. Identify where the water is going to – the sink. Sketch in the bottom equipotential (i.e. the one with the smallest value of total head h) – in this case, the excavation floor, which we shall assume remains just covered in a shallow depth of water. Make this (or some other convenient point) the datum level from which the total head or potential is measured (Figure 3.21(a)).

3. Identify where the water is coming from – the source. Sketch in the top equipotential (i.e. the one with the greatest value of total head h) – in this case, the flooded beach on either side of the excavation (Figure 3.21(a): only the left-hand side is shown in the diagram).

   In Figure 3.21, the top equipotential has a value of 16 m relative to the floor of the excavation, even though the beach is only 14 m above the floor of the excavation. This is because the beach is flooded to a depth of 2 m. (Imagine a standpipe piezometer placed with its tip at the retained soil surface. Water will rise in the standpipe to the level of the free water surface, which is 2 m above the piezometer tip and 16 m above the floor of the excavation.)

4. Identify the bounding flowlines. In this case, one flowline runs from the the beach down the back of each of the sheet pile walls, round the bottom of the wall and into the excavation. The underlying London Clay has a permeability of less than 10^{-10} m/s and, in comparison with the Norwich Crag, is effectively impermeable. Bounding flowlines therefore follow the interface between the Norwich Crag and the London Clay, coming in from the left (and right) and turning through 90° to follow the centreline up to the floor of the excavation (Figure 3.21(a): only the left-hand side is shown).

5. Starting with zones where the flow pattern is reasonably well-defined (in this case between the sheet pile walls of the cofferdam), begin to sketch in equipotentials and flowlines within the boundaries you have now defined (Figure 3.21(b) and (c)). Keep it simple: in this case, one intermediate flowline is sufficient, at least for a start. You can always go back and subdivide large flow elements by sketching in further flowlines and equipotentials as a check, but if you start off by being too ambitious, you will get into a hopeless mess.

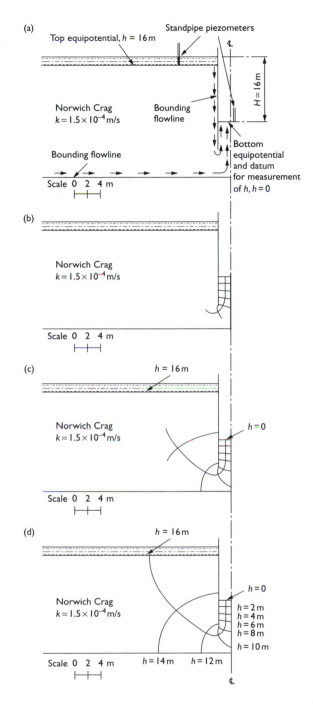

(a)

Standpipe piezometers

Top equipotential, $h = 16\,\mathrm{m}$

Norwich Crag
$k = 1.5 \times 10^{-4}\,\mathrm{m/s}$

Bounding flowline

$H = 16\,\mathrm{m}$

Bottom equipotential and datum for measurement of h, $h = 0$

Bounding flowline

Scale 0 2 4 m

(b)

Norwich Crag
$k = 1.5 \times 10^{-4}\,\mathrm{m/s}$

Scale 0 2 4 m

(c)

$h = 16\,\mathrm{m}$

Norwich Crag
$k = 1.5 \times 10^{-4}\,\mathrm{m/s}$

$h = 0$

Scale 0 2 4 m

(d)

$h = 16\,\mathrm{m}$

Norwich Crag
$k = 1.5 \times 10^{-4}\,\mathrm{m/s}$

$h = 0$
$h = 2\,\mathrm{m}$
$h = 4\,\mathrm{m}$
$h = 6\,\mathrm{m}$
$h = 8\,\mathrm{m}$
$h = 10\,\mathrm{m}$

Scale 0 2 4 m

$h = 14\,\mathrm{m}$ $h = 12\,\mathrm{m}$

Figure 3.21 Construction of flownet for cooling water outfall pipe excavation: (a) indentify the sink, the source and the bounding flowlines; (b), (c) start to sketch in intermediate flowlines and equipotentials, but keep it simple; (d) the finished flownet.

6. If some flowlines and equipotentials do not cross at right angles, or if some flow elements are not 'square', rub out the offending lines and redraw them so that the significant errors are gradually eliminated. The flownet does not have to be perfect. You will soon reach a stage where further improvements make very little practical difference.

7. When you are satisfied with the flownet (Figure 3.21(d)), count the number of flowtubes N_F (i.e. the spaces between the flowlines, not the flowlines themselves) and the number of head drops N_H (again, the spaces between the equipotentials, not the equipotentials themselves). The head on each equipotential is calculated using the fact that the head drop between adjacent equipotentials is H/N_H, where H is the overall head drop. Equation (3.31) is used to calculate the flowrate – remembering in this case to multiply the number of flowtubes by two because we have only sketched half the flownet.

The required flowrate is given by

$$q \text{ (m}^3/\text{s per metre)} = kHN_F/N_H$$

with

$$k = 1.4 \times 10^{-4} \text{ m/s}$$
$$H = 16 \text{ m}$$
$$N_F = 2 \times 2 \text{ for symmetry} = 4$$
$$N_H = 8$$
$$\Rightarrow q = (1.4 \times 10^{-4})(\text{m/s}) \times 16(\text{m}) \times (4/8)$$
$$= 1.2 \times 10^{-3} \text{ m}^3/\text{s per metre length}$$

Perhaps the two most common mistakes in the calculation of the flowrate from a completed flownet are

- Dimensional inconsistencies. Always work in SI units, i.e. metres and seconds, with permeability in m/s. If appropriate, the answer can be changed to more meaningful units (e.g. litres/second, litres/ hour, m³/hour or m³/day) at the end of the calculation.
- Using the number of flowlines (instead of the number of flowtubes) for N_F, and/or the number of equipotentials (instead of the number of head drops) for N_H. Think of fence posts and fence panels: the number of panels (flowtubes or head drops) is always one less than the number of posts (flowlines or equipotentials).

If the geometry of the cross-section is not symmetrical, the whole flownet will need to be sketched, as in Example 3.6.

Example 3.6 An asymmetric confined flownet

Figure 3.22(a) shows a cross-section through a long excavation, situated within the infilled channel of an old river. The infill material, in which the excavation is made, is a silty fine sand of permeability 1.5×10^{-5} m/s. The surrounding rock is of permeability 10^{-12} m/s. The natural groundwater level is at the original soil surface.

(a) Stating carefully the assumptions you make and the conditions you are attempting to fulfil, estimate by means of a carefully sketched flownet the capacity of the dewatering system required to keep dry a 100 m length of the excavation.

(b) If the pumps were to fail, how long would it take for the cofferdam to flood to a depth of 1 m?

[*University of London 2nd year BEng (Civil Engineering) examination,*
Queen Mary and Westfield College (modified)]

Solution

(a) The main assumptions are that:

- the rock is impermeable in comparison with the sand
- the sand is isotropic and homogeneous
- the water table outside the excavation is maintained (by rainfall and other sources of recharge) at its original level
- the groundwater level inside the excavation is maintained (by the pumped dewatering system) at the excavated soil surface
- the side walls are impermeable
- end effects may be neglected.

In addition to the boundary conditions that follow from the above assumptions, the flownet must be sketched in accordance with the usual rules:

- all flowlines must cross all equipotentials at right-angles
- flowlines must not merge with or cross other flowlines, and equipotentials must not merge with or cross other equipotentials
- each flow element should be a curvilinear square, that is, each element should be as broad as it is long.

The flownet is sketched following the procedure described for Example 3.5. In this case:

1. The problem geometry is *not* symmetrical about the centreline. The entire flownet must therefore be sketched.
2. Where is the water going to? The sink or lowest equipotential, and the datum for measurement of hydraulic head h, is the excavation floor, between the side walls (Figure 3.22(b)).

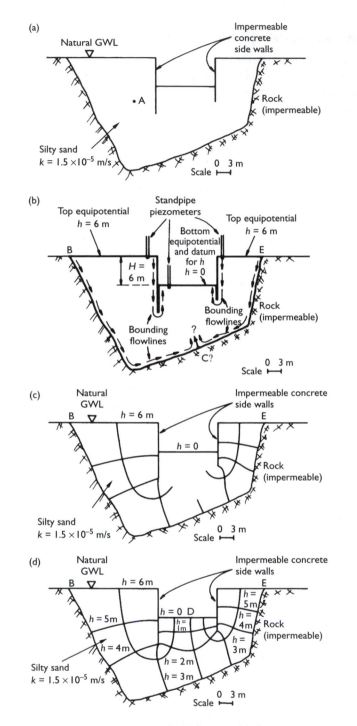

Figure 3.22 Asymmetric confined flownet. (a) Geometry; (b) identify the sink, the source and the bounding flowlines; (c) start to sketch in intermediate flowlines and equipotentials; (d) the finished flownet.

3. Where is the water coming from? As far as the soil is concerned, from the flooded soil surfaces on either side of the cofferdam. These are the top equipotentials, with the highest values of h. If we imagine a stand-pipe piezometer with its tip at any point along these equipotentials, the water level in the standpipe will not rise at all above the position of the tip. This is, however, 6 m above our datum for the measurement of h, and the value of h along the top equipotentials is therefore 6 m (Figure 3.22(b)).

4. Identify the bounding flowlines. As before, one flowline runs down the back of each of the side walls, round the bottom of the wall and into the excavation. Bounding flowlines also start from B and E, and follow the impermeable boundary between the sand and the rock. At some point C, they will turn through approximately 90° and run side-by-side up to the floor of the excavation at D (Figures 3.22(c) and (d)).

The flowpath CD separates the flow from the left and right sides of the cross-section, and must be located by trial and error. The criterion that must be fulfilled is that every flowline must cross the same number of equipo-tentials, whichever side of the flownet it starts from. This is because the change in head between any pair of equipotentials must be the same. In this case (Figure 3.22(d)), there are seven equipotentials (i.e. six equipotential drops) on each side of the cross-section. A flownet with, say, six equipo-tentials on the left side and seven on the right would be wrong, because it would be necessary to assign two conflicting values of head to the common equipotentials between the sheet piles.

The finished flownet, Figure 3.22(d), was arrived at gradually, start-ing with zones where the flow pattern is reasonably well defined (Figure 3.22(c)). Where, during initial attempts, flowlines and equipoten-tials did not cross at right-angles or flow elements were not 'square', the lines were rubbed out and redrawn until significant errors were eliminated.

The flowrate is given by

$$q \text{ (m}^3/\text{s per metre)} = kHN_F/N_H$$

with

$$k = 1.5 \times 10^{-5} \text{ m/s}$$
$$H = 6 \text{ m}$$
$$N_F = 4$$
$$N_H = 6$$
$$\Rightarrow q = (1.5 \times 10^{-5}) \text{ (m/s)} \times 6 \text{ (m)} \times (4/6)$$
$$= 6 \times 10^{-5} \text{ m}^3/\text{s per metre length}$$

For a 100 m length, $q_{100} = 6 \times 10^{-5}$ m^3/s per m $\times$ 100 m

$\Rightarrow q_{100} = 6 \times 10^{-3}$ m^3/s $= 21.6$ m^3/h

(b) If the pumps should fail, the water level within the excavation will start to rise. Provided that the excavated surface remains flooded, the geometry of the flownet does not change. What happens is that the value of h on the lowest equipotential starts to rise, so that the overall head drop H, and the corresponding flowrate q calculated according to equation (3.31), decrease.

Suppose that, at a time t after pumping ceased, the excavation is flooded to a height z above the excavation floor. The value of h on the lowest potential is then z (metres), and the overall head drop across the flownet $H = (6-z)$ m. The flowrate is still given by equation (3.31) with $k = 1.5 \times 10^{-5}$ m/s; $H = (6-z)$ m; $N_F = 4$ and $N_H = 6$:

$q = 10^{-5}(6 - z)$ m^3/s per metre

The flowrate q is related to the rate at which the water level within the excavation rises, dz/dt, because $dz/dt = q/A$, where A is the area of the excavation. For a 1 m length, A is equal to the width of the excavation, 12 m. Hence

$dz/dt = 10^{-5} \times (6 - z) \div 12$ (m/s)

This can be integrated between limits of $z = 0$ at $t = 0$, and a general depth z at a general time t, to give

$$\int_0^z \frac{dz}{(6 - z)} = \int_0^t \left(\frac{10^{-5}}{12} \right) dt$$

or

$$- [\ln(6 - z)]_0^z = \left[\frac{10^{-5}}{12} t \right]_0^t$$

$\Rightarrow t = (12 \times 10^5) \times \ln[6/(6 - z)]$ s

For $z = 1$ m, this gives

$t = (12 \times 10^5) \times \ln(1.2)$ s $= 60$ h

This might seem reassuring. However, the main danger with excavations in low permeability soils is not flooding, but instability due to uncontrolled pore water pressures. This is discussed in section 3.11.

In the flownets shown in Figures 3.21 and 3.22 we have taken the reduced water table inside the excavation to be level with the excavated soil surface. In reality, the water level inside the excavation will be lowered by means of a dewatering system, comprising an array of wells penetrating some way below formation level, and probably installed outside the excavation. The water level in each well will be lower than in the surrounding ground, due to vertical flow in the vicinity of the well; head losses at entry into the well; and – especially in less permeable soils – cone of depression effects (section 3.5), because there will be some recovery of groundwater levels in between the wells.

Experience suggests that (subject to a reasonable estimate of the effective soil permeability) the approach adopted in Figures 3.21 and 3.22 will in most circumstances lead to a realistic estimate of the pumping rate required to achieve an overall drawdown inside the excavation to formation level. The detailed flow pattern near the wells need not normally be considered. The potentially limiting factors of well capacity and performance should however be investigated in their own right. In some cases a test well or a pumping trial might be required.

3.10 Calculation of pore water pressures using flownets

Subject to the limitation outlined above concerning the detailed flow pattern in the vicinity of an individual well, the pore water pressure at any point may be calculated from a flownet by interpolation between equipotentials.

Example 3.7 Calculating pore water pressures from a flownet

Calculate the pore water pressure at the point marked A on the flownet shown in Figure 3.22.

Solution

The first step is to calculate the total head h at the point A, by linear interpolation between the equipotential lines on the flownet. By scaling from Figure 3.22, the point A is approximately 2 m from the 5 m equipotential ($h = 5$ m), in a region where the 5 m and 4 m equipotentials are separated by a distance of about 6 m (Figure 3.23(a)).

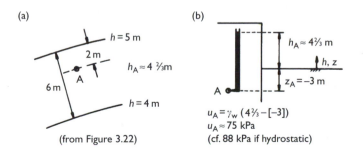

(a)

$h = 5$ m
2 m
$h_A \approx 4\,\tfrac{2}{3}$ m
6 m
A
$h = 4$ m

(from Figure 3.22)

(b)

$h_A \approx 4\tfrac{2}{3}$ m
h, z
$z_A = -3$ m
A

$u_A = \gamma_w\,(4\tfrac{2}{3} - [-3])$
$u_A \approx 75$ kPa
(cf. 88 kPa if hydrostatic)

Figure 3.23 Calculating pore water pressure from a flownet. (a) Determining the total head h_A; (b) relationship between total head and pore water pressure.

The potential at A is therefore given by

$$h_A = 5\,\text{m} - [(2/6) \times 1\,\text{m}] = 4.67\,\text{m}$$

This must now be converted to a pore pressure head u_A/γ_w, by subtracting the elevation of the point A above the datum for the measurement of h:

$$u_A = \gamma_w(h_A - z_A) \tag{3.32}$$

(Figure 3.23(b): u_A is the gauge pore water pressure at A, h_A is the total head at A and z_A is the elevation of A above the datum used for the calculation of total head.)

In this case, the point A is approximately 3 m below the datum for h, so z_A is negative: $z_A = -3\,\text{m}$. Hence

$$u_A = \gamma_w(h_A - z_A) = 9.81\,\text{kN/m}^3 \times [4.67\,\text{m} - (-3\,\text{m})]$$

$$\Rightarrow u_A \approx 75\,\text{kPa}$$

In hydrostatic conditions, the pore water pressure at a depth of 9 m below the water table would be approximately 88 kPa. This example illustrates the effect of downward seepage in reducing pore water pressures, compared with hydrostatic conditions. Conversely, upward seepage will increase pore water pressures, perhaps to such an extent that the soil **fluidizes** or 'boils'. This is discussed in section 3.11.

3.11   Quicksand

In regions of strong upward flow – for example in front of the retaining walls shown in Figures 3.21 and 3.22 – the soil may **fluidize** or **boil** if the uplift force due to seepage exceeds the weight of the soil. This condition is also known as **quicksand** if it occurs over a large area, and **piping** if it occurs in localized channels.

Fluidization takes place when the upward hydraulic gradient exceeds a **critical** value, i_{crit}. The **critical hydraulic gradient**, which is just sufficient to cause boiling, may be calculated by considering the forces acting on a block of soil which is on the verge of uplift (Figure 3.24).

Neglecting side friction, uplift will just occur when the upward force due to the pore water pressure acting on the base ($A\gamma_w[z + h_{crit}]$) begins to exceed the weight of the block of soil ($A\gamma z$):

$$A\gamma_w(z + h_{crit}) = A\gamma z$$

$$z(\gamma - \gamma_w) = \gamma_w h_{crit}$$

or

$$i_{crit} = h_{crit}/z = (\gamma - \gamma_w)/\gamma_w \tag{3.33}$$

For soils with $\gamma = 20\,\text{kN/m}^3 \approx 2\gamma_w, i_{crit} = 1$.

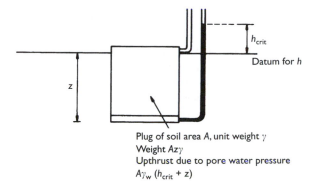

Figure 3.24 Forces on a block of soil on the verge of uplift.

Example 3.8 Assessing the potential for fluidization of the floor of an excavation

Use the flownet shown in Figure 3.21(d) to investigate the stability of the floor of the excavation with the proposed dewatering scheme. If there is a potential difficulty, suggest two ways in which it might be overcome.

Solution

The key question is, does the upward hydraulic gradient i exceed i_{crit} (assume here $i_{crit} = 1$) below the floor of the excavation? In the flownet shown in Figure 3.21(d), the embedment of the sheet pile below formation level is 7 m and the head drop between the toe of the sheet pile and the floor of the excavation is 8 m. This corresponds to an average hydraulic gradient $\Delta h/\Delta z$ in front of the wall 8 m/7 m = 1.14, so that in this case fluidization is quite likely.

The problem could be overcome by driving longer sheet piles into the London Clay to form a complete cut-off. Alternatively, it would be necessary to design the dewatering system so that it is capable of lowering the groundwater level inside the cofferdam to considerably below the floor of the excavation, to prevent the base from boiling or 'blowing'. This would require a higher pumping rate than that calculated using Figure 3.21. Typically, a dewatering system would be expected to limit the maximum upward hydraulic gradient to between 65–80% of the critical value calculated using equation (3.33) (i.e. a 'factor of safety' $F = i_{crit}/i_{actual}$ of between 1.25 and 1.5).[1]

Some upward movement of the excavated soil surface must still be anticipated, due to the reduction in vertical effective stess which results from the removal of overburden. In a fine soil or a laminated deposit, a system of vacuum wells could be installed in order to effect a reduction in pore water pressure which compensates for the removal of overburden, so that the effective stress remains substantially unchanged.

3.12   Unconfined flownets

In the examples given in section 3.9 (Figures 3.21 and 3.22), the boundaries to the flownet were known at the outset. In an unconfined or **water table** aquifer in which the soil surface does not remain flooded, this is not the case, because the surface of zero gauge pore water pressure – the phreatic surface – will be drawn down into the body of the aquifer. Neglecting capillary effects, there is no flow in the region above the phreatic surface, because the gauge pore water pressure would be less than zero in this zone, which as a result will become unsaturated and effectively impermeable. In cross-section, the phreatic surface therefore represents the upper flow boundary or the top flowline.

The position of the top flowline depends on the flow regime as represented by the flownet, while the flownet depends on the position of the top flowline. The only way forward is to sketch the flownet by trial and error, including the determination of the position of the top flowline or phreatic surface in the iterative process. Because of the additional requirement (sometimes called the **phreatic surface condition**) to locate the top flowline, unconfined flownets are generally more difficult to sketch than confined flownets. Sketching unconfined flownets will probably seem daunting at first, but – as with confined flownet sketching – becomes easier with practice and experience.

The position of the top flowline is located using the fact that the gauge pore water pressure at any point on it is zero. This means that the total head h at any point is equal to the elevation of that point above the datum for measurement of h. For example, the 3 m equipotential intersects the top flowline at an elevation of 3 m above datum level, and so on. Fulfilment of this additional condition (as with the rest of the flownet, by trial and error) will ensure that the top flowline or phreatic surface is correctly located.

The construction of an unconfined flownet is illustrated in Example 3.9.

Example 3.9 An unconfined flownet

The cross-section shown in Figure 3.25 is based on one of the excavations for the A55 Conwy Crossing in North Wales. Water flows through the earth bund from the river estuary to the line of wellpoints (here assumed to act as a continuous slot), so that the top and bottom equipotentials (i.e. the source and the sink) are clearly defined. The interface between the bund and the underlying impermeable stratum represents one bounding flowline, but the position of the top flowline is not initially known.

The unconfined flownet was constructed as follows:

- Can we use symmetry? In this case, no.
- In this case, the datum for the measurement of total head h is taken to coincide with Ordnance Datum OD. This is for consistency with the construction drawings and procedures for the job as a whole.
- The water is flowing from the river ($h = +3$ m OD) to the wellpoint line ($h = -7.5$ m OD). If we imagine a standpipe piezometer with its tip at any

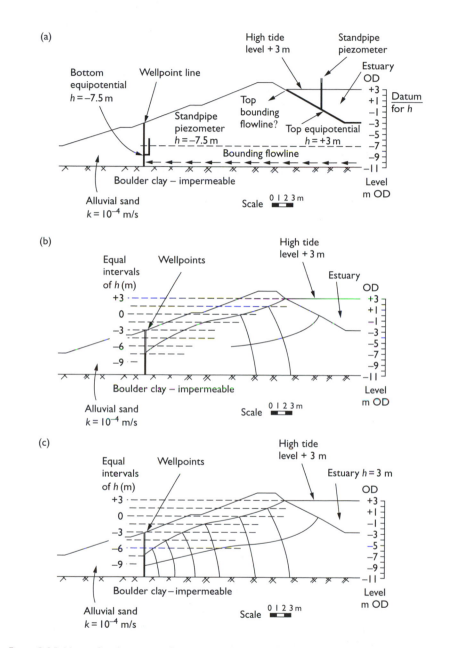

Figure 3.25 Unconfined seepage flownet. (a) Identify the source and the sink; identify the bottom boundary flowline; (b) draw horizontal lines at key elevations to assist in locating the phreatic surface; sketch in a trial phreatic surface and start to draw the flownet, modifying lines as required; (c) the finished flownet.

point along the submerged river bed, the water level in the standpipe will rise to the level of the water surface at +3 m OD. The river bed is the top equipotential (the source), and the wellpoint line is the bottom equipotential (the sink) (Figure 3.25(a)).

• Identify the bottom boundary flowline: this follows the impermeable boundary between the sand and the boulder clay, coming in from the right-hand side of the diagram towards the wellpoint line (Figure 3.25(a)).

• When starting to sketch an unconfined flownet, you will find it useful to draw horizontal lines at elevations corresponding to the total head along key equipotentials (Figure 3.25(b)). The intersection point between the top flowline and each horizontal line must also be the intersection point between the top flowline and the equipotential in question (Figures 3.25 (b) and (c)).

• The finished flownet, Figure 3.25(c), was again arrived at gradually. The initial guess for the shape of the phreatic surface was developed as the first trial flownet was sketched in, starting from the top equipotential. The phreatic surface in an embankment dam such as this is usually approximately parabolic in shape, with the convex side upward as shown in Figure 3.22(c). With experience, you can use this knowledge to make your initial guess at the top flowline quite close to the correct position. It does not really matter too much, however, what shape you start with: you will soon discover that you have to alter the shape of the top flowline, to fulfill the requirements of curvilinear square flownet elements, equipotentials crossing flowlines at right-angles, and the condition of zero gauge pore water pressure on the phreatic surface. The important point is to start off with something, which then forms a basis for iteration towards the correct solution.

As with the previous examples (3.5 and 3.6), errors were gradually eliminated by relocating flowlines and equipotentials that did not cross at right-angles or resulted in non-'square' flow elements or non-zero pore water pressures on the phreatic surface.

The purpose of the dewatering system shown in Figure 3.25 is threefold. It prevents instability of the slope by reducing pore water pressures and increasing effective stresses; it prevents erosion of the bund due to the emergence of the top flowline on the downstream face; and it also prevents the excavation from flooding.

The phreatic surface is a flow boundary in coarse-grained soils, which become unsaturated at gauge pore water pressures only slightly less than zero. In a fine-grained soil, in which capillary effects are not negligible, the surface of zero gauge pore water pressure will not in general represent a boundary to the flownet. In a fine-grained soil, an appropriate procedure might be to draw the flownet assuming that the upper surface of the soil represents the top flowline, and then to check that the resulting pore water suctions do not exceed the air entry value.

3.13 Distance of influence

In practice, the boundaries to most flownets are unlikely to be as well-defined as those in Figures 3.21 and 3.22. Often (but not in fact in the case of Figure 3.21), both the flooded soil surface which defines the source equipotential or recharge boundary and the impermeable stratum which defines the extent of the flownet are figments of a degree examiner's (or a text-book author's) imagination. A more realistic situation for a reasonably uniform soil is that the effective recharge boundaries will be located at some uncertain distance and depth from the excavation, beyond which the *in situ* pore water pressures are not affected – the **distance of influence.**

The distance (and depth) of influence will depend on the permeability of the soil and the drawdown at the excavation, and must be estimated before the flownet can be drawn. To estimate the lateral distance of influence, Sichardt's empirical formula

$$L_0 = C\Delta h \sqrt{k} \qquad (3.34)$$

may be used, where L_0 is the distance of influence in metres (measured from the edge of the excavation), k is the soil permeability in m/s, Δh is the drawdown in metres and C is a factor of between 1500 and 2000 $(m/s)^{-1/2}$ for plane flow: the units of L_0, k and h must be as stated. Generally, this seems to work reasonably satisfactorily in practice. Unfortunately, there is little or no guidance in the literature as to the depth of influence in cases where no natural boundary is present. This topic is discussed in more detail by White (1981).

3.14 Soils with anisotropic permeability

If the soil is anisotropic with k_x not equal to k_z, the differential equation governing plane flow is (from equation (3.29), ignoring the y-direction)

$$k_x \partial^2 h / \partial x^2 + k_z \partial^2 h / \partial z^2 = 0 \qquad (3.35)$$

This equation in the (x, z) plane can be reduced to the simpler Laplace equation in a destorted (x', z) plane by means of the transformation $x' = \alpha x$ where $\alpha = \sqrt{(k_z/k_x)}$.

$$x = x'/\alpha$$

so

$$\partial x = \partial x'/\alpha$$

Therefore

$$\partial h / \partial x = \alpha \partial h / \partial x' \quad \text{and} \quad \partial^2 h / \partial x^2 = \alpha^2 \partial^2 h / \partial x'^2 \qquad (3.36)$$

If $\alpha^2 = k_z/k_x$, substitution of equation (3.36) into equation (3.35) gives

$$k_x(k_z/k_x)(\partial^2 h/\partial x'^2) + k_z\partial^2 h/\partial z^2 = 0$$

or

$$\partial^2 h/\partial x'^2 + \partial^2 h/\partial z^2 = 0 \tag{3.37}$$

This enables the flownet sketching technique to be used for soils having k_x not equal to k_z, provided that the true cross-section is first redrawn in the destorted coordinate system (x', z), where $x' = x\sqrt{(k_z/k_x)}$. Usually, the permeability in the horizontal (x) direction is greater than that in the vertical (z) direction. Multiplication of horizontal distances by $\sqrt{(k_z/k_x)}$ then leads to the compression of the horizontal scale. The use of this transformation is illustrated in Example 3.10.

Example 3.10 Flownet in an anisotropic soil

A sheet-piled excavation, which is square on plan and whose actual dimensions are shown in Figure 3.26(a), is to be made in a laminated soil with horizontal permeability $k_x = 5 \times 10^{-5}$ m/s and vertical permeability $k_z = 5 \times 10^{-6}$ m/s. Estimate the capacity (per metre perimeter) of the dewatering system required to prevent the excavation from flooding. Assume that the distance of influence of the dewatered excavation is 100 m horizontally and 50 m vertically, measured from the side and the floor of the excavation respectively.

The transformation factor $\alpha = \sqrt{(k_z/k_x)} = \sqrt{(5 \times 10^{-6}/5 \times 10^{-5})} = \sqrt{(1/10)} = 0.316$. The transformed cross-section, with the horizontal dimensions reduced following multiplication by the transformation factor 0.316, is shown in Figure 3.26(b). (On the transformed cross-section, the scales for both true horizontal distances x and reduced horizontal distances x' are shown. A true horizontal distance of 12.65 m reduces to a transformed horizontal distance of 0.316×12.65 m $= 4$ m, as indicated by the horizontal scale).

The flownet was sketched on the transformed cross-section in the usual way, with flowlines crossing equipotentials at right angles and 'square' flow elements, and fulfilling the phreatic surface condition of zero gauge pore water pressure on the top flowline between A and B.

It would in reality be necessary to check that the aquifer really was deep enough to act as a recharge boundary at a depth of 50 m below the floor of the excavation. If the aquifer is deep but of decreasing permeability with depth, the assumption of an impermeable lower boundary to the flownet might be more appropriate. On the other hand, an underlying more permeable stratum would almost certainly act as a source of recharge. The sensitivity of the calculated flowrate to alternative assumptions concerning the depth and nature of the bottom boundary to a flownet is discussed with reference to an excavation in chalk by Powrie and Roberts (1995).

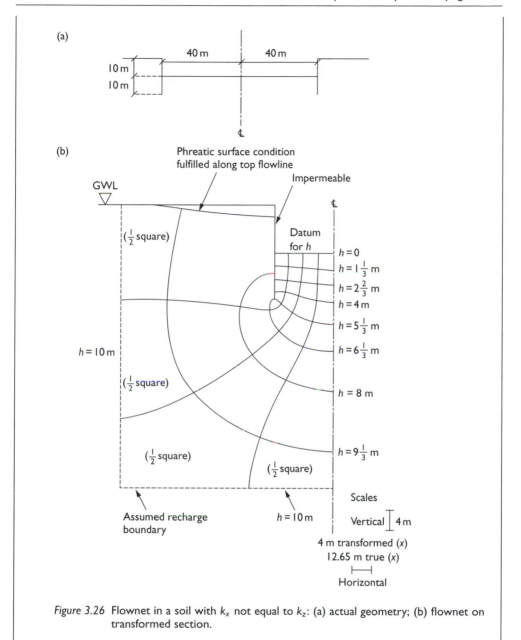

(a)

40 m   40 m

10 m
10 m

℄

(b)

Phreatic surface condition
fulfilled along top flowline

Impermeable

GWL
▽

($\frac{1}{2}$ square)

Datum
for h

℄

h = 0

h = 1$\frac{1}{3}$ m

h = 2$\frac{2}{3}$ m

h = 4 m

h = 5$\frac{1}{3}$ m

h = 6$\frac{1}{3}$ m

h = 10 m

($\frac{1}{2}$ square)

h = 8 m

($\frac{1}{2}$ square)

h = 9$\frac{1}{3}$ m

($\frac{1}{2}$ square)

Scales

Assumed recharge
boundary

h = 10 m

Vertical 4 m

4 m transformed (x)
12.65 m true (x)

Horizontal

Figure 3.26 Flownet in a soil with k_x not equal to k_z: (a) actual geometry; (b) flownet on
transformed section.

The flowrate is calculated from the transformed section in the usual way, using the
effective permeability of the transformed section k_t. A general expression for k_t in
terms of k_x and k_z may be deduced by equating the flowrate through a soil element
in transformed space to the flowrate through the corresponding element in real space
(Figure 3.27).

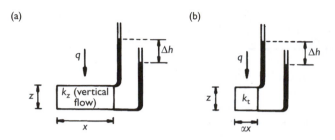

Figure 3.27 Calculation of equivalent permeability of a transformed cross-section: (a) true geometry; (b) transformed geometry.

Applying Darcy's Law to each element

$$q = \alpha x k_t h / z = x k_z h / z$$

so

$$k_t = k_z / \alpha$$

Recalling that $\alpha = \sqrt{(k_z / k_x)}$

$$k_t = \sqrt{k_x \cdot k_z} \qquad (3.38)$$

Example 3.11 Calculating the flowrate from the transformed-section flownet shown in Figure 3.26

For the excavation shown in Figure 3.26, the total flowrate is given by

$$q = k_t H (N_F / N_H) \times \text{perimeter}$$

with

$$k_t = \sqrt{(5 \times 10^{-5} \times 5 \times 10^{-6})}\,\text{m/s} = 1.58 \times 10^{-5}\,\text{m/s}$$

$$H = 10\,\text{m}$$

$$N_F = 4$$

$$N_H = 7.5$$

perimeter $= 4 \times 80\,\text{m} = 320\,\text{m}$, giving

$$q = (1.58 \times 10^{-5})\,(\text{m/s}) \times 10\,(\text{m}) \times (4/7.5) \times 320\,\text{m} = 0.027\,\text{m}^3/\text{s}$$

or

$$q \approx 97\,\text{m}^3/\text{hour}$$

For square, symmetrical excavations it is usual to calculate the flowrate in m^3/s per metre for one-half of the cross-section (as in Figure 3.26), which is then multiplied by the length of the perimeter to determine the overall flowrate, as in Example 3.11. In doing this, the additional flow to the corners has been neglected. Finite element analyses by Powrie and Preene (1992) suggest that this may be reasonable if the distance of influence L_0 (measured from the edge of the excavation) is small in comparison with the width of the excavation $a(L_0/a < 0.3)$. In the present case, $L_0/a = 100/80 = 1.25$, so that the neglect of corner effects will lead to an underestimation of the required flowrate.

3.15   Zones of different permeability

When flowlines pass from a zone of permeability k_1 into a zone of different permeability k_2, they are deflected through an angle $(\beta_1 - \beta_2)$ as shown in Figure 3.28. Applying Darcy's Law in zone 1

$$q_1 = A_1k_1i_1 = (\cos \beta_1)k_1(h_1 - h_2)/\sin \beta_1 \qquad (3.39)$$

and in zone 2

$$q_2 = A_2k_2i_2 = (\cos \beta_2)k_2(h_1 - h_2)/\sin \beta_2 \qquad (3.40)$$

Since flowlines cannot cross other flowlines, the flowrate through any flowtube remains constant and $q_1 = q_2$. Hence

$$k_1/\tan \beta_1 = k_2/\tan \beta_2$$

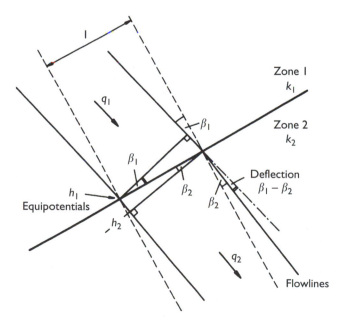

Figure 3.28 Deflection of flowlines on passing between zones of different permeability.

or

$$\tan \beta_1 / \tan \beta_2 = k_1 / k_2 \tag{3.41}$$

The direction of the deflection of the flowlines must be such that the equipotentials are closer together in the zone of lower permeability, because a higher hydraulic gradient is required in order to drive the same flowrate.

A comprehensive guide to flownet sketching for complex cross-sections is given by Cedergren (1989).

3.16   Boundary conditions for flow into drains

In the flownet shown in Figure 3.25, one of the purposes of the dewatering system was to ensure that the top flowline remained below the surface of the slope at all points. In practice a dewatering system of this type would comprise a line of small wells known as **well-points**, installed at a spacing of 1–3 m. This spacing is sufficiently close for the recovery of the groundwater level between the wellpoints to be small, so that the flownet analysis shown in Figure 3.25, in which the line of discrete wellpoints is treated as if it were a continuous pumped slot, is quite reasonable.

The earth bund shown in Figure 3.25 is essentially a dam, preventing the water in the Conwy estuary from flooding the excavation. In this case, the dam was needed for a period of only eighteen months or so, and the requirement to operate the dewatering system continuously during this time was not particularly onerous. Similar embankments are used worldwide as permanent dams: in these circumstances a dewatering system which must be continuously actively pumped is undesirable.

There are about 2000 dams in the UK alone, many of which are embankment or earth dams. Seepage patterns in earth dams are usually controlled by means of a horizontal toe drain, as shown in Figure 3.29. A passive toe drain like this could not have been installed at Conwy, partly because the slope was formed by the excavation of material rather than being built up, and partly because the excavation was below sea level, so that pumping would still have been required to remove the water from a passive toe drain in any case.

Figure 3.29 has been drawn on the assumption that the horizontal toe drain remains flooded, so that it acts as a reservoir or an equipotential. This is because the gravel drain is several orders of magnitude more permeable than the earth dam, enabling water to flow through it without any significant loss of head.

Vertical drains will not in general be flooded (unless they are blocked), and so must be treated somewhat differently. Figure 3.30 shows the flow pattern from a reservoir through an earth embankment, supported by a retaining wall with a vertical drain behind it. The gravel can be viewed as a medium of effectively infinite permeability in comparison with the material from which the embankment is constructed. On entering the drain, the flowlines will therefore experience a deflection consistent with equation (3.41).

However, the actual angle of deflection is indeterminate, because $\beta_2 \rightarrow 90°$ (so that $\tan \beta_2 \rightarrow \infty$). Furthermore, the flowlines in the drain will effectively merge: if the permeability is infinite, the width of the flowtubes must be infinitesimal, so that the flowrate through each flowtube is the same as it was in the soil. Finally, the phreatic surface condition must be fulfilled at the interface between the soil and the drain,

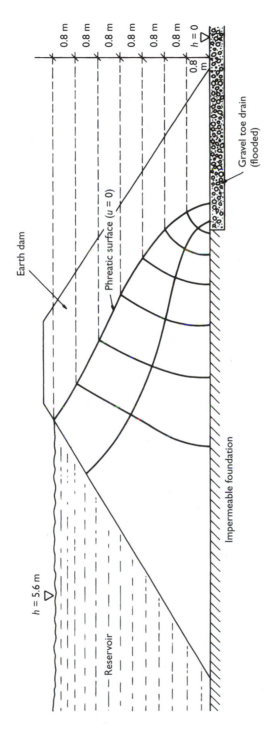

h = 5.6 m

Reservoir

Earth dam

Phreatic surface (*u* = 0)

0.8 m

0.8 m

0.8 m

0.8 m

0.8 m

0.8 m

h = 0

0.8 m

Gravel toe drain (flooded)

Impermeable foundation

Figure 3.29 Seepage control in an earth dam by means of a horizontal toe drain.

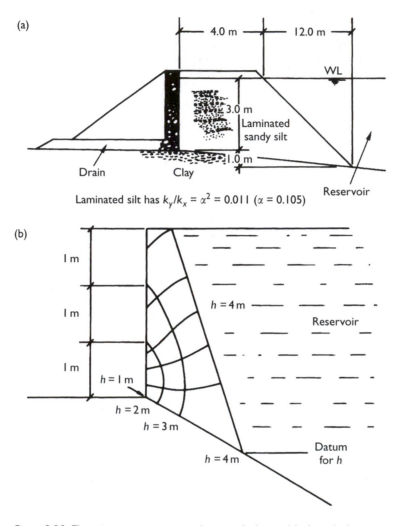

(a)

4.0 m   12.0 m

WL

3.0 m
Laminated
sandy silt

1.0 m

Drain   Clay

Reservoir

Laminated silt has $k_y/k_x = \alpha^2 = 0.011$ ($\alpha = 0.105$)

(b)

1 m

1 m

$h = 4\,m$

Reservoir

1 m

$h = 1\,m$

$h = 2\,m$

$h = 3\,m$

$h = 4\,m$

Datum
for h

Figure 3.30 Flow into an unsaturated vertical drain. (a) Actual dimensions (not to scale);
(b) flownet on transformed section. (This example is reproduced from a BSc (Eng)
in Civil Engineering second year examination, King's College London.)

where the gauge pore water pressure $u = 0$. The flownet is sketched by trial and error, in the usual way.

Figure 3.30 has been drawn assuming that water will emerge from the soil pores at atmospheric pressure. This is reasonable for coarse soils. However, there is some evidence (e.g. Hall, 1955) that in fine soils where the pore size is small, a pressure difference between the pore water and atmospheric air will be required for water to drain out of the soil. This is known as the **water exit effect**. It is analogous to the air entry effect illustrated in Figure 3.3, but its magnitude is rather smaller: the water exit pressure is perhaps 15–50% of the air entry pressure for the same soil (Preene, 1992).

3.17 Application of well pumping formulae to construction dewatering

Construction dewatering is the term applied to the process of lowering the groundwater level by means of a system of pumped wells, so that an excavation below the natural water table will remain dry and stable. The first stage in the design of a construction dewatering system is to estimate the flowrate which must be pumped to achieve the required drawdown. We have already seen how flownet sketching can be used to do this for plane flow problems in Figures 3.21, 3.22, 3.25 and 3.26.

In some circumstances, flow may be essentially radial rather than plane. If information on pore water pressures is not required, a rectangular dewatering system of plan dimensions $a \times b$ may then be idealized as a large single equivalent well of radius $r_e = (a + b)/\pi$. (The radius of the equivalent circular well r_e is chosen so that the perimeter of the equivalent well is $2(a + b)$, which is the same as that of the actual dewatering system). The standard well pumping formulae derived in section 3.5 can then be used to estimate the pumping rate required to achieve the specified drawdown. At the edge of the excavation, the required drawdown in the soil corresponds to a head $h = h_w$ at the equivalent radius, $r = r_e$. For a confined aquifer, substitution of these values into equation (3.14) yields

$$(H - h_w) = (q/2\pi Dk)\ln(R_0/r_e)$$

or

$$q = 2\pi Dk(H - h_w)/\ln(R_0/r_e) \tag{3.42}$$

For an unconfined aquifer, substitution of $h = h_w$ at $r = r_e$ into equation (3.19) gives

$$(H^2 - h_w^2) = (q/\pi k)\ln(R_0/r_e)$$

or

$$q = \pi k(H^2 - h_w^2)/\ln(R_0/r_e) \tag{3.43}$$

The approximate mode of flow (i.e. radial or plane) to a construction dewatering system of dimensions $a \times b$ will depend on both the aspect ratio a/b and the relative distance of influence L_0/a. Three possible idealizations are shown in Figure 3.31. L_0 is measured from the edge of the excavation, rather than from the centre of the well as it was in section 3.5.

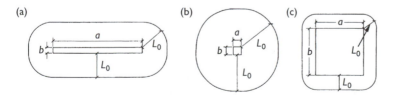

Figure 3.31 Idealized flow patterns towards excavations treated as single equivalent wells. (a) Flow to a long well; (b) radial flow to a rectangular well with a distant recharge boundary; (c) plane and radial flow to a rectangular well with a close recharge boundary. (Redrawn with permission from Powrie and Preene, 1992.)

For horizontal flow in a confined aquifer, Powrie and Preene (1992) suggested the following guidelines on the basis of finite element analyses:

1. For long excavations ($a/b > 10$) with close recharge boundaries ($L_0/a < 0.1$), plane flow to the long sides dominates and end effects may be neglected.
2. For rectangular excavations with $1 < a/b < 5$ and close recharge boundaries ($L_0/a < 0.3$), plane flow to the sides dominates and corner effects may be neglected.
3. For rectangular excavations with $1 < a/b < 5$ and more distant recharge boundaries ($L_0/a > 3$), flow is essentially radial and flowrates may be estimated using equation (3.42) with the radius of the equivalent well $r_e = (a + b)/\pi$ and $R_0 = L_0 + (a/2) \approx L_0$.

For other geometries, Powrie and Preene (1992) present a chart which enables the flowrate to be estimated at a glance. This chart is reproduced here as Figure 3.32. Figure 3.32 plots the dimensionless flowrate $q/kD(H - h_w)$ as a function of L_0/a for various aspect ratios a/b. Figure 3.32 is based on horizontal flow to a fully penetrating equivalent well in a uniform confined aquifer.

In geotechnical engineering, it is usually necessary to make gross simplifications to arrive at a conceptual model of a real situation which is amenable to analysis. The secret of success is not to ignore any factor that could destroy the applicability of the conceptual model adopted.

In groundwater flow problems, permeable strata and lenses which could act as close sources of recharge are likely to be very significant. Plane flownet sketching can be difficult if more than one stratum is present, unless the flow pattern of interest is contained within a single aquifer because the other soil layers present are effectively either impermeable or of infinite permeability. The calculated flowrate is enormously sensitive to the soil permeability k, which can be difficult to estimate even to within a factor of 3 either way. In many cases, it is also necessary to estimate the distance of influence of the dewatering system L_0 before the flownet can be sketched. If Sichardt's

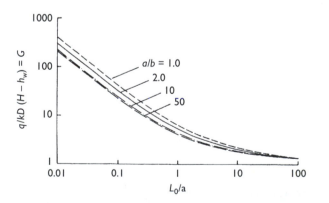

Figure 3.32 Flowrates by finite element analysis for equivalent wells of various geometries. (Redrawn with permission from Powrie and Preene, 1992.)

formula (equation 3.34) is used, the dependency of the calculation on the estimated permeability may be increased.

Flowrates calculated using the equivalent well approach are also heavily dependent on the soil permeability and the distance of influence. In addition, it is assumed that the aquifer is uniform and homogeneous, that all flow is horizontal, and that the equivalent well fully penetrates the aquifer. In short, the equivalent well approach probably involves a higher degree of idealization than flownet sketching.

An alternative possibility is to use the **principle of superposition** to predict the draw-down pattern produced by pumping at a certain flowrate from a number of wells, on the basis of the drawdown pattern produced by pumping at the same flowrate from a single test well. The attraction of this method is that there is no need to estimate the effective soil permeability, or indeed to worry too much about the ground conditions. However it is only theoretically correct if the relationship between flowrate and draw-down remains linear. In practical terms, this means that the wells must penetrate fully a confined aquifer of uniform thickness. The distance of influence must be independent of drawdown (which according to equation (3.34) it is not), and the drawdown must never be so large that the aquifer becomes unconfined.

The method of superposition seems to work quite well in practice, provided that these conditions are satisfied approximately. The tendency of the distance of influence to increase with drawdown means that the method should err on the conservative side, in that flowrates will be overestimated at higher drawdowns.

3.18 Numerical methods

The differential equation governing groundwater flow (equation 3.29) may be solved numerically for given boundary conditions using **finite element analysis** or **a finite difference** technique. In some cases, the simplifying assumptions that enable problems to be solved by techniques such as flownet sketching and equivalent well analysis cannot be justified, and a numerical approach is the only analytical alternative.

A finite difference solution involves the division of the flow field into a network of nodes. The variation in total head is assumed to be linear between adjacent nodes. The flowrate into a node is then calculated in terms of the head difference between it and the surrounding nodes, using equation (3.29) in finite difference form. In principle, values of total head at each node which satisfy equation (3.29) could be calculated iteratively by hand: it is more usual to assemble the nodal equations into matrix form and solve them using a computer.

In finite element analysis, the region of flow is divided into discrete (finite) elements with common nodes. An approximate solution to the differential equation govern-ing the flow of groundwater is obtained numerically for the specified geometry and boundary conditions. The solution in this case is defined over the entire flow field, in contrast to the finite difference method in which only nodal values are found. The variation in head over a finite element need not be linear.

A further advantage of finite element analysis over finite difference methods for groundwater flow is their ability to calculate the strains and deformations associated with the changes in effective stress which result from lowering the groundwater level. This is particularly useful in modelling the effects of dewatering systems in fine-grained soils, where compression or swelling due to changes in effective stress is likely to be important.

Example 3.12 Numerical analysis of the influence of a high permeability zone on dewatering system efficiency

Numerical methods can be useful in the analysis of dewatering systems in all soils, particularly where the ground conditions are complex or where the presence of a singularity such as a lens of higher permeability is suspected. Both the importance of such features and the application of finite element analysis may be illustrated with reference to the dewatering system shown in Figure 3.33.

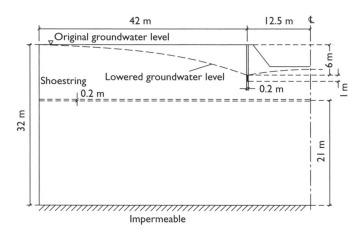

Figure 3.33 Excavation in uniform soil with lens of higher permeability at 11 m depth. (Redrawn with permission from Powrie *et al.*, 1989.)

There are two lines of wellpoints 25 m apart, and the depth of each wellpoint is 7 m. It is required to lower the groundwater level between the lines of wellpoints so that a trench may be excavated safely. The original water table is at ground level and the distance of influence of the wellpoint lines has been taken as 42 m. This is consistent with a soil permeability k_1 of 1.2–2.2×10^{-5} m/s according to equation (3.34), with $C = 1500$–$2000 (\text{m/s})^{-1/2}$ and a drawdown at the wellpoints of 6 m. A thin lens of higher permeability k_2 is located at a depth of 11 m below ground level. The soil below a depth of 32 m was assumed to be impermeable.

The steady state drawdown produced by the wellpoint system was investigated using the finite element method to take account of the effect of the lens of higher permeability. The finite element mesh is shown in Figure 3.34. This is an unconfined flow problem, and an iterative technique was used to determine the position of the phreatic surface or top flowline. The vertical height of each node was adjusted until the condition of zero gauge pore water pressure was fulfilled. In the analysis, the trench excavation was ignored. This reflects reality in that in practice the groundwater level would be lowered before the excavation commenced. If excavation proceeded below the phreatic surface the analysis would become invalid, since the boundary condition would be changed. This is an entirely theoretical point: in reality the excavation would collapse and flood.

The plane seepage flownet for the homogeneous case ($k_1 = k_2$) is shown in Figure 3.35. The phreatic surface is below the floor of the proposed excavation, and the performance of the dewatering system is satisfactory.

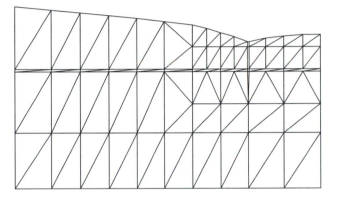

Figure 3.34 Finite element mesh. (Redrawn with permission from Powrie et al., 1989.)

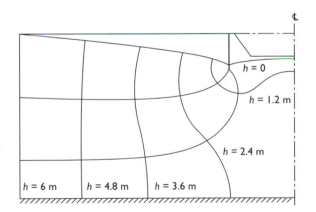

Figure 3.35 Seepage flownet for homogeneous case (no lens). (Redrawn with permission from Powrie et al., 1989.)

The effect of an increase in the permeability of the thin lens (k_2) is to raise the level of the phreatic surface at the centreline of the proposed excavation.

Figure 3.36 shows the equipotentials (calculated by finite element analysis) and the position of the phreatic surface for $k_2 = 400\,k_1$. It may be seen that the dewatering system is in this case inadequate, since the drawdown at the centre-line of the proposed excavation is insufficient. It would have been difficult to investigate this problem analytically, without using a numerical method.

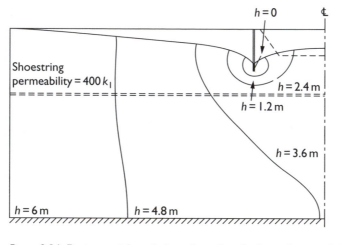

Figure 3.36 Equipotentials and phreatic surface for lens of permeability 400 times that of the soil. (Redrawn with permission from Powrie et al., 1989.)

3.19 Wells and pumping methods

It has already been mentioned that groundwater lowering is usually carried out by pumping from an array of wells known as a **construction dewatering system**. The number and type of wells, and the pumping devices used, will depend primarily on the drawdown required and the soil type and permeability.

In very high-permeability soils ($k > 5 \times 10^{-3}$ m/s or so), construction dewatering may not be feasible because the pumped flowrate required to achieve anything more than a modest drawdown will probably be excessive. In these circumstances, a combination of a physical cut-off (to prevent water ingress through the sides of the excavation) and a well system (to reduce pore water pressures below the base of the excavation and prevent piping) is often used. In soils of low permeability (k less than about 10^{-7} m/s), groundwater flow may be induced by the application of an electric potential rather than a hydraulic potential, using a technique known as **electro-osmosis** (section 11.4.1).

Tentative limits (in terms of soil permeability and drawdown) for the successful application of groundwater control techniques involving pumping are given in Figure 3.37.

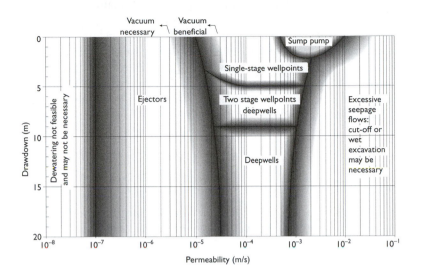

Figure 3.37 Range of application of dewatering techniques. (Adapted from Roberts and Preene, 1994 after Preene *et al.*, 2000.)

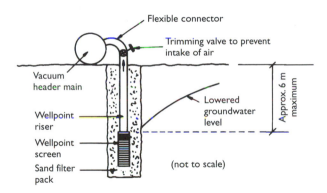

Figure 3.38 Cross-section through a typical wellpoint.

3.19.1 *Wellpoints*

Perhaps the most common method of ground-water control using pumped wells is the **wellpoint system**. Wellpoints are essentially small wells, with **wellscreens**[2] of approximately 50 mm diameter and up to 1 m in length connected by a **riser pipe** to a common **header main** from which water is removed by means of a vacuum pump (Figure 3.38).

In wellpoints, the vacuum is used to raise the water from the bottom of the well. It is not actually applied to the soil: water drains from the soil into the well mainly by gravity. Owing to the limitations of suction lift, the maximum drawdown achievable by a wellpoint system is approximately 6 m. If a drawdown in excess of 6 m is required, it may be possible to use a **multi-stage wellpoint system**. The first stage of wellpoints enables excavation to proceed to a certain depth, from which the second stage of wellpoints is installed. The feasibility of this procedure depends mainly on the available space within the excavation and on programming constraints.

Wellpoints are comparatively inexpensive, and are usually installed at a spacing of between 1 m and 3 m. Because drawdowns are comparatively small, and due to the fact that wellpoints are not normally used in soil of permeability less than about 10^{-5} m/s, well losses (i.e. the difference between the water levels in the well and in the ground immediately around it) are not usually significant. The close spacing of the wellpoints means that cone of depression effects (i.e. the recovery of groundwater levels in between extraction points; section 3.9) along the line of the wellpoints do not normally need to be considered.

3.19.2 Deep wells

For drawdowns in excess of 6 m in soils where the water will drain into the wells under gravity alone, and where a multi-stage wellpoint system is not feasible, a system of **deep wells** is often used (Figure 3.39).

Deep wells are of much larger diameter than wellpoints – 200 mm diameter liners and wellscreens inside 350 mm diameter bores are not untypical – and are therefore much more expensive to install. Each well is pumped using an individual submersible electric pump, lowered to within perhaps 2 m of the bottom of the well. Deep wells are spaced more widely than wellpoints: spacings of 15 m to 20 m are not uncommon. Both well losses and cone of depression effects are likely to be very significant, so that the groundwater level between the wells will probably be very much higher than the water levels inside the well casings. For this reason, deep wells must generally be installed to a depth considerably in excess of the required drawdown: a well depth of 20 m for a drawdown of 10 m would not be unusual.

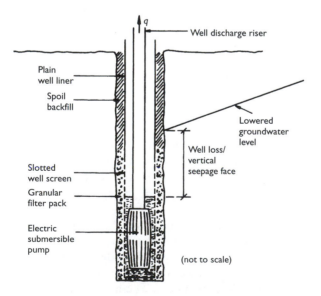

Figure 3.39 Cross-section through a typical deep well.

As the pore size of the soil is reduced, capillary (water exit) effects become more significant, and drainage of pore water under gravity alone becomes very slow or even non-existent. In soils where the pore size is small – which in practice means soils of permeability less than about 10^{-5} m/s – it may be necessary to apply a vacuum at the interface between the soil and the well in order to promote drainage. This can be achieved with a deep well system by sealing the top of the well and pumping all the air from the well using a second pump in order to create a vacuum. This is known as a **deep well system with vacuum**. Usually, all of the wells in the system would be connected to one or two vacuum pumps. Each well must still be provided with an individual submersible pump in order to remove the water.

3.19.3 Ejectors

If the soil permeability is low enough to warrant the application of vacuum, the flowrates into each well are likely to be very small. Since many types of submersible electric pump rely on a reasonable flow of water to cool the bearings and to support the impeller shaft, the practical operation of deep well systems with vacuum can be difficult. For groundwater and pore pressure control in low-permeability soils an ejector system may be preferable. An **ejector** (also known as an **eductor**) is a nozzle and venturi device which acts as a small water-driven jet pump (Figure 3.40).

Water is pumped at high pressure through a supply pipe to the nozzle. The supply stream emerges from the nozzle with a high velocity and an absolute static pressure close to zero (i.e. a gauge pressure of nearly -100 kPa). Groundwater is drawn from the well and entrained into the supply stream, and carried up out of the well through a return riser. If the water level in the well is drawn down to the intake of the ejector, the ejector pumps air as well as water to create a vacuum inside the well, provided that the well is sealed.

Ejector wells are of smaller diameter than deep wells pumped by electric submersible pumps, and are therefore cheaper to install. Cone of depression effects and well losses generally increase with decreasing soil permeability, so that ejector wells must usually be deep in comparison with the drawdown required. Spacings of 5–15 m are typical. Each well is connected to a supply main and a return main, with one pumping station for perhaps up to 60 or so individual ejector wells.

3.19.4 Well design

Typical construction details for wellpoints, deep wells and wells pumped using ejectors are shown in Figures 3.38, 3.39 and 3.40 respectively. The granular material used in the filter between the wellscreen and the natural soil must be selected so that it neither impedes the flow of water nor allows fine particles to be drawn into the well from the surrounding ground. Similarly, the slots or holes in the wellscreen must be large enough not to impede flow, but small enough to retain the material of the filter pack. Design rules relating to the particle size distribution curves for the natural soil and the filter were given in section 1.9: these must be used to ensure that both the filter and the wellscreen perform satisfactorily.

Further details of construction dewatering systems in theory and in practice are given by Powers (1992), Roberts and Preene (1994), Preene *et al.* (2000), and Cashman and Preene (2001).

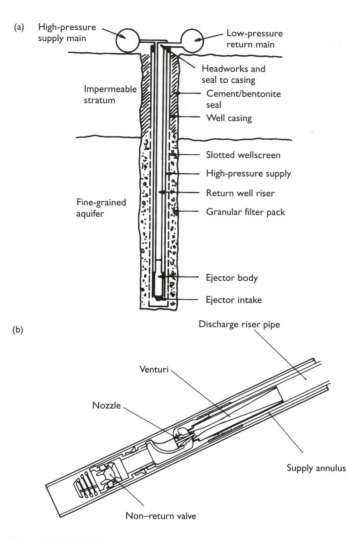

Figure 3.40 (a) Schematic cross-section through an ejector; (b) cross-section through typical single-pipe ejector body.

Notes
 Well casing and slotted wellscreen may be omitted; surface of zero gauge pore water pressure may be below the bottom of the well: flow can still take place through soil which has negative pore water pressures but remains saturated.

Key points

- Water in the ground may be stationary, in which case the pore water pressures will be hydrostatic. Otherwise, it will flow through the soil pores in response to a **hydraulic gradient**. The hydraulic gradient is defined as the (negative of the) rate of change of total head with distance, in the direction of flow.

- The ease with which water can flow through a soil is quantified by the soil **permeability**, k. The permeability can be measured in the laboratory or in the field. Laboratory tests involve disturbance to the soil structure and void ratio, and cannot take account of large-scale inhomogeneities such as high permeability lenses and soil fabric effects such as fissures.

- The flow of water through soil is governed by **Darcy's Law**,

$$q = Aki \tag{3.2}$$

where $q\,(\mathrm{m^3/s})$ is the volumetric flowrate, $A\,(\mathrm{m^2})$ is the cross-sectional area of the flow, k (m/s) is the soil permeability and i is the hydraulic gradient.

- The plane flow patterns associated with long earthworks and excavations can be investigated using the technique of **flownet sketching**. A flownet may be used to calculate pore water pressures, and also flowrates using the equation

$$q\,(\mathrm{m^3/s\ per\ metre}) = kHN_F/N_H \tag{3.31}$$

where H (m) is the overall head drop and N_F and N_H are the numbers of flowtubes and equipotential drops determined from the flownet.

- Flownets may be **confined** (if all of the flow boundaries are known at the outset) or **unconfined** (if the top flowline is defined by the **phreatic surface**, on which the gauge pore water pressure is zero). Flownet sketching can be applied in anisotropic soils, by means of an appropriate transformation of the cross-section.

- The groundwater in the vicinity of an excavation must be properly controlled by means of a suitable **construction dewatering system**. The purpose of a dewatering system is in general threefold. It prevents instability of a slope or the base of an excavation by reducing pore water pressures and increasing effective stresses; it prevents the erosion of soil due to uncontrolled seepage; and it prevents the excavation from flooding by groundwater.

Questions

Laboratory measurement of permeability; fluidization; layered soils

3.1 Describe by means of an annotated diagram the principal features of a constant head permeameter. Give three reasons why this laboratory test might not lead to an accurate determination of the effective permeability of a large volume of soil in the ground. Suggest how each of these problems might be overcome.

> [*University of London 2nd year BEng (Civil Engineering) examination, Queen Mary and Westfield College (part question)*]

3.2 Describe by means of an annotated diagram the principal features of a falling head permeameter.

Show that the water level in the top tube h would be expected to change with time t according to the following equation

$$\ln(h/h_1) = -(kA_1/A_2L)t$$

where h_1 is the initial water level in the top tube, A_1 is the cross-sectional area of the sample and L is its length, k is the soil permeability and A_2 is the cross-sectional area of the top tube.

Give two reasons why this laboratory test might not lead to an accurate determination of the effective permeability of a large volume of soil in the ground.

[University of London 2nd year BEng (Civil Engineering) examination, Queen Mary and Westfield College (part question)]

3.3   In the constant head permeameter test described in Example 3.2, the sample was found to fluidize in upward flow at a hydraulic gradient of 0.84. Estimate the unit weight of the soil in its loosest state.

[University of London 2nd year BEng (Civil Engineering) examination, Queen Mary and Westfield College (part question)]

$(18.05 \, \text{kN/m}^3.)$

3.4   An engineer wishes to investigate the bulk permeability of a layered soil comprising alternating bands of fine sand (5 mm thick) and silt (3 mm thick). The engineer makes a special constant head permeameter of square cross-section (internal dimensions 112 mm × 112 mm) and carries out two tests on undisturbed samples. In one test, the flow is parallel to the laminations; in the other test, the flow is perpendicular to the laminations. The data recorded in downward flow are as follows:

Hydraulic gradient i	0	1	2	5	10
Flowrate, test 1 (mm^3/s)	0	79	158	395	–
Flowrate, test 2 (mm^3/s)	0	–	–	16	33

Unfortunately, the engineer is not very careful in keeping a laboratory notebook, and omits to record the orientation of the sample in each test.

Estimate the permeability of the fine sand and the silt. Estimate also the flowrates at which fluidization would just occur in upward flow, both parallel and perpendicular to the laminations. Derive from first principles any formulae you use.

[University of London 1st year BEng (Civil Engineering) examination, King's College (part question)]

$(k_{\text{sand}} = 10^{-5}\,\text{m/s};\; k_{\text{silt}} = 10^{-7}\,\text{m/s};$ flowrates at fluidization in upward flow are $3.0\,\text{mm}^3/\text{s}$ perpendicular to the laminations, and $72.6\,\text{mm}^3/\text{s}$ parallel to the laminations, assuming $\gamma_{\text{soil}} = 2 \times \gamma_{\text{w}}$. You need to derive equations (3.21), (3.22) and (3.33).)

3.5 The following data were obtained from a constant head permeameter test in downward flow on a sample of medium sand.

Measured flowrate $q(\text{cm}^3/\text{s})$	2.0	3.0	4.0	5.0	6.0
Head difference between manometer tappings Δh (mm)	18.8	31.0	45.1	60.0	75.0
Sample height z (mm)	180	175	170	165	160

Relative density of soil particles (grain specific gravity) $G_s = 2.65$.
Cross-sectional area of permeameter $A = 8000\,\text{mm}^2$.
Distance between pressure tappings $L = 120\,\text{mm}$.

Prior to the test, the sample had been brought to its loosest possible state – corresponding to a sample height of 180 mm – by fluidization in upward flow. At fluidization, the upward flowrate was $11.725\,\text{cm}^3/\text{s}$ and the head difference between the manometer tappings was 109.9 mm.

Plot a graph of flowrate q against hydraulic gradient i for downward flow, and explain its shape. Estimate the maximum and minimum permeability k and specific volume v of the sample during this part of the test.

[*University of London 1st year BEng (Civil Engineering) examination,*
King's College (part question)]

$(k_{\text{max}} = 1.6 \times 10^{-3}\,\text{m/s}, v_{\text{max}} = 1.8$ at start of downward flow; $k_{\text{min}} = 1.2 \times 10^{-3}\,\text{m/s}, v_{\text{min}} = 1.6$ at the end of the test.)

Well pumping test (field measurement of permeability)

3.6 A well pumping test was carried out to determine the bulk permeability of a confined aquifer. The aquifer was overlain by a clay layer 4 m thick, the depth of the aquifer was 20 m, and the initial piezometric level in the aquifer was 2 m below ground level. After a period of pumping when steady state conditions had been reached, the following observations were made:

pumped flowrate $q = 1.637\,\text{l/s}$
well radius $= 0.1\,\text{m}$
drawdown just outside well $= 2\,\text{m}$
drawdown in piezometer at 100 m distance from well $= 0.2\,\text{m}$.

Deriving from first principles any equations you need to use, determine the bulk permeability of the aquifer. Would your analysis still apply for a drawdown in the well of 4 m?

> [*University of London 2nd year BEng (Civil Engineering) examination,*
> *Queen Mary and Westfield College (part question)*]

(You need to derive equation (3.12) and integrate between limits of ($h = 21.8$ m at $r = 100$ m) and ($h = 22.0$ m at $r = 0.1$ m); $k = 5 \times 10^{-5}$ m/s; no, because the aquifer would become unconfined.)

Confined flownets; quicksand

3.7   Figure 3.41 shows a cross-section through a square excavation at a site where the ground conditions are as indicated. Assuming that the water levels in the overlying gravels, the underlying fractured bedrock and the medium sand outside the excavation do not change, estimate by means of a carefully sketched flownet the capacity of the required dewatering system.

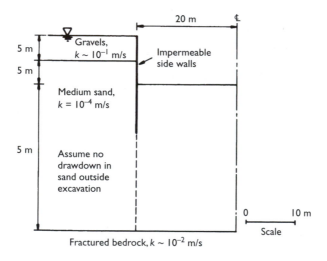

Figure 3.41   Cross-section through excavation.

What proportion of the extracted groundwater must be recirculated through the medium sand and the gravels in order to maintain the initial groundwater level outside the excavation if there is no other close source of recharge?

Do you foresee any problem concerning the stability of the base of the excavation?

> [*University of London 2nd year BEng (Civil Engineering) examination,*
> *Queen Mary and Westfield College*]

($q \approx 160\,1/s$ extraction; recharge $\approx 66\%$, based on the proportion of flowtubes starting from the sand; maximum upward hydraulic gradient is approximately 0.7, which should not be a problem.)

3.8   Figure 3.42 shows a **plan** view of an excavation underlain by a confined aquifer of uniform thickness 20 m. The aquifer is bounded on two sides by a river having a water level $h = 12$ m above datum level. On the third side, the effective recharge boundary to the aquifer is as indicated. A sheet pile cut-off wall is installed along the edge of the river adjacent to the excavation, extending for a certain distance on either side. The datum level for the measurement of hydraulic head is at the upper surface of the aquifer.

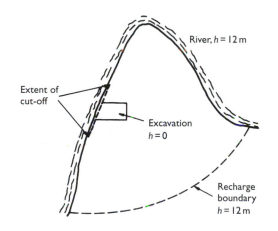

Figure 3.42 Plan view of excavation and surrounding aquifer.

Estimate by means of a carefully sketched flownet the rate at which water must be pumped from a dewatering system, in order to reduce the groundwater level at the excavation to datum level. (The permeability of the aquifer is 3.6×10^{-4} m/s).

Explain why your analysis would be invalid for drawdowns at the excavation to below datum level.

[University of London 2nd year BEng (Civil Engineering) examination, Queen Mary and Westfield College]

($q \approx 238\,1/s$ from a plan flownet. The aquifer would become unconfined if the groundwater level were drawn down to below datum level.)

Unconfined flownet

3.9   Figure 3.43 shows a cross-section through a long canal embankment. Explaining carefully the conditions you are attempting to fulfil, estimate by means of a flownet the rate at which water must be pumped from the drainage ditch back into the canal, in litres per hour per metre length.

Describe qualitatively what might happen if the drain beneath the toe of the embankment became blocked.

[*University of London 2nd year BEng (Civil Engineering) examination,
Queen Mary and Westfield College*]

($q \approx 29$ l/h per metre length; if toe drain becomes blocked, the top flowline would emerge on the surface of the embankment leading to erosion and failure.)

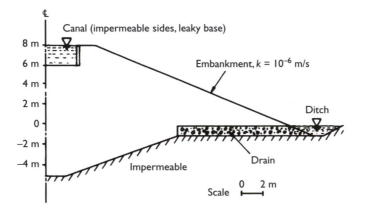

Figure 3.43 Cross-section through canal embankment.

Flownets in anisotropic soils

3.10   Figure 3.44 shows a true cross-section through a long cofferdam. It is proposed to dewater the cofferdam by lowering the water level inside it to the floor of the excavation. Investigate the suitability of this proposal by means of a carefully sketched flownet on an appropriately transformed cross-section (horizontal scale factor $\alpha = \sqrt{(k_v/k_h)}$). How might the stability of the base be ensured?

[*University of London 2nd year BEng (Civil Engineering) examination,
Queen Mary and Westfield College*]

(Transformation factor $\alpha = 0.5$; transformed permeability $k_t = 5 \times 10^{-5}$ m/s; flowrate $q \approx 0.3$ l/s per metre length; hydraulic gradient below formation level ≈ 1, so there is a danger of base instability. Need to drive sheet piles into impermeable clay, or pump at a higher flowrate to reduce groundwater level within excavation to below formation level.)

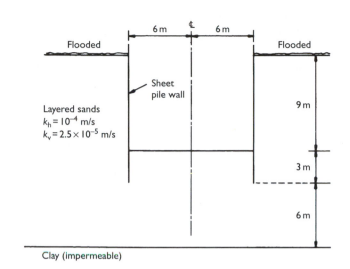

Figure 3.44 Cross-section through cofferdam.

3.11 Figure 3.45 shows a true cross-section through a sheet-piled excavation in a laminated soil of permeability $k_v = 10^{-6}$ m/s (vertically) and $k_h = 1.6 \times 10^{-5}$ m/s (horizontally). The laminated soil is overlain by 4 m of highly permeable gravels, and the natural groundwater level is 2 m below the soil surface. By means of a flownet sketched on a suitably modified cross-section estimate:

(a) the minimum capacity required of the dewatering system
(b) the pore water pressure at the point A.
      Comment briefly on the stability of the base of the excavation.
      (Transformation factor $\alpha = \sqrt{(k_v/k_h)}$.)

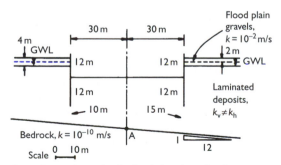

Original groundwater level is 2 m below the soil surface

Figure 3.45 Cross-section through excavation.

[*University of London 2nd year BEng (Civil Engineering) examination, Queen Mary and Westfield College*]

(Transformation factor $\alpha = 0.25$; transformed permeability $k_t = 4 \times 10^{-6}$ m/s; flowrate $q \approx 0.025$ l/s per metre length; pore water pressure at A is approximately 300 kPa; upward hydraulic gradient between the sheet piles below formation level is ≈ 0.42, so there should be no danger of base instability.)

Notes

1 Note, however, that the hydraulic gradient determined from a flownet is not an appropriate criterion for deciding whether a pumped well dewatering system is needed. The flownets in this chapter have all been drawn on the assumption that a dewatering system capable of lowering the groundwater level within the excavation to below formation level is already in place.
2 Well bores are usually lined or cased to prevent the sides of the borehole from falling in. Where it is intended that water should be able to flow into the casing, the casing is slotted and possibly wrapped round with a fine-meshed plastic material known as a **geotextile**. The slotted portion of the casing is termed the **wellscreen.**

References

British Standards Institution (1992) *Code of Practice for Test Pumping of Water Wells* (BS6316), BSI, London.

British Standards Institution (1986) *Code of Practice for Foundations* (BS8004), BSI, London.

Cashman, P.M. and Preene, M. (2001) *Groundwater Lowering in Construction – a Practical Guide*, Spon Press, London.

Cedergren, H. (1989) *Seepage, Drainage and Flownets* (3rd edn). John Wiley, New York.

Clark, L. (1977) The analysis and planning of step drawdown tests. *Quarterly Journal of Engineering Geology*, **10**, 125–43.

Cooper, H.H. and Jacob, C.E. (1946) A generalized graphical method for evaluating formation constants and summarizing well field history. *Transactions of the American Geophysical Union*, **27**, 526–34.

Cox, S.E. (2003) *The use of horizontal wells for leachate and gas control in landfills*. PhD dissertation, University of Southampton.

Darcy, H. (1856) *Les Fontaines Publiques de la ville de Dijon*, Dalmont, Paris.

Hall, H.P. (1955) An investigation of steady flow toward a gravity well. *Houille Blanche*, **10** (8), 8–35.

Hazen, A. (1892) *Physical properties of sands and gravels with reference to their use in filtration*. Report of the Massachusetts State Board of Health.

Kruseman, G.P. and De Ridder, N.A. (1983) *The Analysis and Evaluation of Pumping Test data*, International Institute for Land Reclamation and Improvement, Wageningen.

Leonards, G.A. (ed.) (1962) *Foundation Engineering*, McGraw-Hill, New York.

McWhorter, D.B. (1985) Seepage in the unsaturated zone: a review, in *Seepage and leakage from dams and empoundments*, ASCE Geotechnical Engineering Division symposium, Denver, Colorado, 200–19.

Powers, J.P. (1992) *Construction Dewatering: New Methods and Applications* (2nd edn), John Wiley, New York.

Powrie, W. and Preene, M. (1992) Equivalent well analysis of contruction dewatering systems. *Géotechnique*, **42** (4), 635–9.

Powrie, W. and Roberts, T.O.L. (1995) Case history of a dewatering and recharge system in chalk. *Géotechnique*, **45** (4), 599–609.

Powrie, W., Roberts, T.O.L. and Moghazy, H.E.-D. (1989) Effects of high permeability lenses on the efficiency of wellpoint dewatering systems. *Géotechnique*, **39** (3), 543–7.

Preene, M. (1992) *The design of pore pressure control systems in fine soils*. PhD dissertation, University of London (Queen Mary and Westfield College).

Preene, M., Roberts, T.O.L., Powrie, W. and Dyer, M. (2000) *Groundwater control – design and practice*. CIRIA Report 0515 Construction Industry Research and Information Association, London.

Roberts, T.O.L. and Preene, M. (1994) Range of application of construction dewatering systems, in *Groundwater Problems in Urban Areas* (ed. W.B. Wilkinson), pp. 415–23, Thomas Telford, London.

Rowe, P.W. (1972) The relevance of soil fabric to site investigation practice. Twelfth Rankine Lecture, *Géotechnique*, **22**(2), 195–300.

Simpson, B., Blower, T., Craig, R.N. and Wilkinson, W.B. (1989) *The engineering implications of rising groundwater levels in the deep aquifer below London*, Construction Industry Research and Information Association, Special Publication 69, CIRIA, London.

White, J.K. (1981) On the analysis of dewatering systems. *Proceedings of the 10th International Conference on Soil Mechanics and Foundation Engineering*, Stockholm, **1**, 511–16.

Chapter 4

One-dimensional compression and consolidation

4.1 Introduction and objectives

One-dimensional compression, in which deformation takes place in the direction of loading only, has a particular significance in soil mechanics and foundation engineering. The natural loading and unloading of a soil stratum – for example during the deposition and erosion of overlying material – generally take place under conditions of one-dimensional compression, because lateral strains at any point are prevented by the surrounding soil. This mode of deformation is often assumed to be approximately appropriate for soil subjected to vertical loads from pad, strip and especially raft foundations (Figure 4.1).

If the soil is of low permeability, the application of a surface load results initially in an increase in the pore water pressure. This gives rise to a hydraulic gradient, in response to which pore water flows out of the soil and the soil deforms. As the water flows out of the soil, the pore water pressures gradually return to their equilibrium values, after which no further deformation takes place. The time-related process of soil deformation due to the dissipation of non-equilibrium pore water pressures is described as **consolidation**. The term **compression** is used more generally to describe changes in volume due to changes in effective stress, without reference to the timescale over which they occur.

To estimate the settlement of a foundation beneath which the soil can be assumed to deform in one-dimensional compression, a relationship between vertical effective

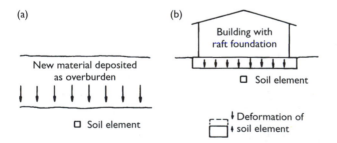

Figure 4.1 Field situations in which the mode of soil deformation approximates to one-dimensional compression. (a) Geological deposition; (b) surface loads from wide foundations.

stress σ_v' and vertical strain ε_v is required. To predict how long these settlements will take to occur, information concerning the consolidation characteristics of the soil is needed. These parameters are traditionally obtained by testing soil elements in the laboratory in conditions of one-dimensional compression, in an apparatus called the **oedometer** (from the Greek word *oedema*, meaning swelling). The device appears to have been developed by J. Frontard in France, in or around 1910; and Karl Terzaghi's discovery of the principle of effective stress was based on the results of a series of experiments using oedometers, carried out in the early 1920s at Robert College in Istanbul, Turkey (Skempton, 1960; Clayton *et al.*, 1995).

In this chapter, the oedometer test is used as a basis for the examination of the behaviour of soil in one-dimensional compression and swelling and for the analysis of consolidation. Several examples are given of the application of oedometer test data and consolidation theory to field problems where deformation occurs primarily in the vertical direction. Particular emphasis is placed on the time-dependent process of consolidation, with the calculation of ultimate settlements being covered in more detail in Chapter 6.

Objectives

After having worked through this chapter, you should understand that

- on first loading, the main component of soil deformation is **irrecoverable** or **plastic**, while on unloading or reloading, deformation is primarily **recoverable** or **elastic** (section 4.2)
- the apparent stiffness of a soil depends on its stress history and stress state, and the change in stress to which it is subjected (section 4.2)
- following a change in total stress or boundary pore water pressure, a low-permeability soil responds over a period of time by **consolidation** (section 4.3)
- the distribution of **non-equilibrium** or **excess** pore water pressure within a clay layer during consolidation is represented by an **isochrone** (sections 4.3 and 4.4)
- the timescale for consolidation increases with reducing soil permeability k and reducing soil stiffness E_0', and with increasing drainage path length (sections 4.3–4.5).

You should be able to

- determine the properties relevant to the one-dimensional compression and consolidation of a soil from oedometer test data (sections 4.2 and 4.6)
- use one-dimensional compression and consolidation theory, together with oedometer test data, to estimate ultimate settlements and rates of settlement in field situations corresponding approximately to one-dimensional conditions (sections 4.7.1 and 4.7.2).

You should have an appreciation of

- the use of vertical drains to reduce consolidation times in the field by encouraging horizontal (radial) flow and reducing the drainage path length (section 4.9)

- the limitations of the conceptual models introduced in this chapter (section 4.10).

The detail of sections 4.5, 4.7.2, 4.7.3, and 4.7.4 may be, and section 4.8 (which is included primarily for reference purposes) almost certainly is, outside the scope of a first course in soil mechanics. However, if you do work through these sections, you should be able to

- analyse the one-dimensional consolidation process in an oedometer sample, assuming that the isochrones are parabolic in shape (section 4.5)
- apply the parabolic isochrone approximation to other field conditions (sections 4.7.2–4.7.4).

You should also gain an understanding of

- the general analysis of one-dimensional consolidation problems, starting with the derivation of the governing differential equation (section 4.8).

4.2   One-dimensional compression: the oedometer test

The oedometer test is used to investigate the stress–strain behaviour of a low-permeability soil (i.e. a clay or a silt), in one-dimensional vertical compression and swelling. A soil sample, usually approximately 75 mm in diameter and 20 mm in height, is retained in a steel confining ring and immersed in a water bath. It is subjected to a compressive stress by the application of a vertical load, which is assumed to act uniformly over the area of the sample. Two-way drainage is permitted through porous discs at the top and bottom (Figure 4.2).

The soil does not respond instantaneously to an increase in vertical total stress, but continues to compress or settle for some time after the load is applied (Figure 4.3). This is because any increase in effective stress must be accompanied by the compression of

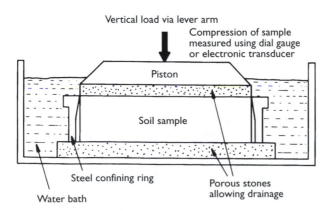

Figure 4.2 Schematic diagram of oedometer test sample.

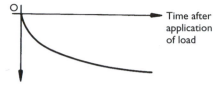

Figure 4.3 Response of a soil element to an increase in vertical total stress.

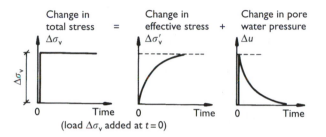

Figure 4.4 Corresponding rates of increase in vertical effective stress and decrease in pore water pressure during one-dimensional consolidation at constant vertical total stress.

the soil skeleton, that is, by a decrease in the void ratio or specific volume. If the soil is saturated, water must escape from the pores for a decrease in void ratio to take place. The rate of drainage – and hence the rate at which the soil deforms – is controlled by the permeability of the soil k and the length of the maximum drainage path d. In an oedometer test with two-way drainage, d is equal to half the sample thickness.

As the void ratio cannot change instantaneously, neither can the effective stress. The increase in the total vertical stress that is applied to the oedometer sample at the start of a loading stage is therefore taken initially entirely by an increase in the pore water pressure. As water then gradually bleeds out from between the pores and the soil skeleton consolidates, the additional or **excess** pore water pressure dissipates. The vertical effective stress increases in accordance with equation (1.13), which may be rewritten in terms of *changes* in vertical effective stress $\Delta\sigma_v'$, total stress $\Delta\sigma_v$ and pore water pressure Δu to give equation (4.1) (Figure 4.4):

$$\Delta\sigma_v' + \Delta u = \Delta\sigma_v \qquad (4.1)$$

The transient process of pore water pressure dissipation and change in void ratio of the sample at constant total stress is known as **consolidation**. Only when consolidation in response to an increment of external load has ceased, and the pore water pressures have returned to their equilibrium values, is the vertical effective stress known.

Transient flow occurs in all saturated soils subjected to a change in external load or a change in the boundary pore water pressures. In sands and gravels of permeability greater than 10^{-4} m/s, the process is usually very rapid – unless perhaps the drainage path length is extremely long. Also, the settlements which occur due to changes in volume in sands and gravels are generally very much smaller than those which occur

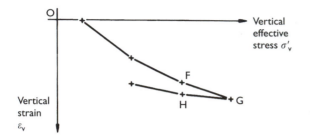

Figure 4.5 Oedometer test data plotted as σ'_v vs. ε_v.

in clays and silts, because sands and gravels are usually (but not always) comparatively stiff. The term **consolidation** is therefore used almost exclusively with clays and possibly silts in mind, and oedometer tests are usually carried out on these comparatively compressible, slow-draining soils.

An oedometer test normally consists of a number of stages in which the vertical load on the sample is increased (or decreased), and the resulting settlement (or swelling) is monitored. Two types of data are obtained from an oedometer test:

- Following each load change, a graph may be plotted of settlement against time. This will tell us about the consolidation characteristics of the soil, and the data may be used to estimate how long it will take for settlements to develop in reality when the soil in the field is subjected to an increase in external load. The analysis of consolidation processes is discussed in sections 4.3–4.9.
- The end points from several increments (or decrements) in load may be plotted on one graph of vertical strain against vertical effective stress, to give a vertical effective stress–strain relationship for the soil in one-dimensional compression. These data may be used to estimate the eventual magnitude of settlements in the field, but give no information as to how long the settlements will take to develop.

The end points from a number of loading and unloading stages of an oedometer test are shown plotted as a conventional stress–strain curve in Figure 4.5. The **increment** of vertical strain $\Delta\varepsilon_v$ which occurs during each loading stage is given by

$$\Delta\varepsilon_v = \Delta h / h_0 \tag{4.2}$$

where Δh is the **incremental** settlement, that is, the change in sample height, due to an increase in vertical effective stress of $\Delta\sigma'_v$; and h_0 is the height of the sample at the start of the loading or unloading stage. (It does not usually make much difference if h_0 is taken as the sample height at the start of the entire test.)

In Figure 4.5, the *cumulative* strain (i.e. the total strain since the start of the test) is plotted as a function of the current vertical effective stress. The strain datum may be set at some convenient point – in this case, the sample height at the first known effective stress state is taken to correspond to zero strain. However, the curve does not start from zero effective stress, but from the initial load under which the sample is allowed to come into equilibrium at the start of the oedometer test.

A sample in which the effective stress was zero would be a slurry with no strength, and could not be tested in a standard oedometer. Usually, the sample used in an oedometer test has been prepared from a core taken from a site investigation borehole. The total stresses acting on the sample as prepared are zero, and the pore water pressures within it are negative but unknown. The effective stress is equal in magnitude to the unknown pore water suction. The sample must be allowed to swell or compress to a known effective stress state, with zero pore water pressures and a vertical effective stress provided by a known external load, as the first stage of the oedometer test.

The apparent stiffness of the soil in one-dimensional compression E'_0 over each load increment (or decrement) may be calculated from the incremental vertical strain which results from the increase (or decrease) in the vertical effective stress. By definition,

$$E'_0 = \Delta\sigma'_v/\Delta\varepsilon_v, \text{ where } \Delta\varepsilon_v = \Delta h/h_0 \tag{4.3}$$

E'_0 is sometimes known as the **constrained modulus** and given the symbol M'_0.

Figure 4.5 demonstrates quite clearly that the concept of an elastic stiffness modulus for a soil is no more than a convenient fiction. Soil is not an elastic material (at least, not in the conventional sense) because

- its stiffness increases with effective stress;
- its stiffness on unloading/reloading will be different from its stiffness during first loading, even over the same stress range.

In addition, the stiffness of the soil may change if the stress or strain path is changed: in the oedometer test, the mode of deformation (i.e. one-dimensional compression) does not vary. Despite all this, it might in some circumstances be acceptable to model the soil as an elastic material:

- The stiffness modulus must have been determined in a test in which the soil has been subjected to the same changes in stress as are anticipated in the field.
- Ideally, the recent stress history of the soil in the field will also have been reproduced in the laboratory test.
- The soil should be on an unload/reload line (or there should be no reversal in the direction of the stress path, which could give rise to a sudden change in stiffness: compare the stiffness during loading from F to G in Figure 4.5 with the stiffness during unloading over the same stress range from G to H).
- The changes in stress and strain must be small.

It was mentioned earlier that the strain datum (at which the strain is zero) could be set at any convenient point. In Figure 4.5, the height of the sample in its first known effective stress state at the start of the oedometer test has been taken to correspond to zero strain. We might equally have decided to set the strain datum to correspond to the state of the soil when it was first laid down as a loose deposit in prehistoric times. In this case, the strain at the start of the oedometer test – which will usually be chosen to represent the current state of the soil in the ground – could easily be 50% or more. For the purpose of engineering calculations such as soil settlements, this would not be a particularly sensible approach. It is more appropriate to describe the volumetric state of the soil in terms of the void ratio e or the specific volume v. This is because these parameters – unlike strain – do not depend on the selection of some arbitrary reference point.

The specific volume v may be related to the height of the oedometer test sample as follows. The total sample volume V_t at any stage of the test is equal to the sample area A multiplied by the current sample height h:

$$V_t = Ah$$

Also, the total volume is equal to the volume of voids V_v plus the volume of soil grains or solids V_s:

$$V_t = V_s + V_v = V_s(1 + V_v/V_s) = V_s v$$

Hence

$$V_t = V_s v = Ah$$

or

$$v/h = A/V_s = \text{constant} \tag{4.4}$$

Assuming that the sample is fully saturated at the end of the test, the final sample height h_f can be related to the final specific volume v_f by measurement of the final moisture content w_f:

$$e_f = w_f G_s$$

Hence

$$v_f = 1 + w_f G_s$$

and

$$v = h(v_f/h_f) = h[(1 + w_f G_s)/h_f] \tag{4.5}$$

It is conventional to plot specific volume as a function of the natural logarithm (ln) of the vertical effective stress (Figure 4.6), because the soil tends to become stiffer as the vertical effective stress is increased.

For a soil which is being compressed for the first time, there should be a unique relationship between the specific volume $v(= 1 + e)$ and $\ln \sigma'_v$,

$$v = v_{\lambda_0} - \lambda_0 \ln \sigma'_v \tag{4.6}$$

This is a straight line on the graph of v against $\ln \sigma'_v$ with slope $-\lambda_0$. It is known as the **normal compression line for one-dimensional compression** (or **1-d ncl** for short).

On unloading (and also on reloading) the soil will be found to be much stiffer, following a hysteresis loop on a graph of v against $\ln \sigma'_v$ which is usually idealized to a straight line of slope $-\kappa_0$. Unlike the normal compression line, there is no unique unload/reload line: an unload/reload line can begin from any point on the normal compression line at which the sample starts to be unloaded.

When the soil moves from a reload line onto the normal compresssion line there will be a marked change in slope on the graph of v against $\ln \sigma'_v$. This point should indicate the maximum previous vertical effective stress to which the sample has been subjected, which is termed the **preconsolidation pressure**. In the field, the soil may at this point exhibit a marked change in stiffness: settlements calculated on the basis of the stiffness

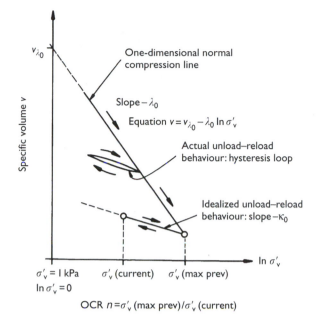

Figure 4.6 Oedometer test data plotted as v against $\ln \sigma'_v$.

in reloading might be a factor of perhaps 5 too small. This is yet another reason why it is important that soil parameters used in engineering calculations should be relevant to the stress history, stress state and anticipated changes in stress and strain of the soil in the field.

You might see oedometer test data plotted as a graph of void ratio e against $\log_{10}(\sigma'_v)$, rather than specific volume v against $\ln \sigma'_v$. A unique one-dimensional normal compression line and an infinite number of possible unload/reload lines are still obtained, but the slopes are conventionally given the symbols C_c and C_s respectively. C_c is known as the **compression index**, and is numerically equal to $2.3 \times \lambda_0$ (because $\lambda_0 = -dv/d(\ln \sigma'_v)$ and $C_c = -de/d(\log_{10} \sigma'_v)$; $dv = de$ and $d(\ln \sigma'_v) = 2.3 \times d(\log_{10} \sigma'_v)$). C_s is known as the **swelling index**, and is (by a similar argument) numerically equal to $2.3 \times \kappa_0$.

A soil which is on the normal compression line has never before been subjected to a vertical effective stress higher than the current value. In this state, a soil is described as **normally consolidated**. A soil which has previously been consolidated to a higher vertical effective stress than that which currently acts is **overconsolidated**, with an **overconsolidation ratio** (OCR or n) given by

$$n = \sigma'_{v(\text{max prev})}/\sigma'_{v(\text{current})} \qquad (4.7)$$

The overconsolidation ratio is a simple but important indicator of the stress state of the soil in relation to its previous stress history.

During normal compression, the greater part of the deformation is due to slippage of the soil particles as the soil skeleton rearranges itself to accommodate higher loads.

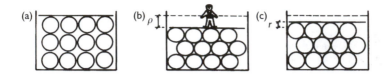

Figure 4.7 Rubber balls analogy. (a) Rubber balls are loosely packed, no load applied; (b) compression ρ in first loading due to (i) particle rearrangement and (ii) particle distortion; (c) rebound r on unloading due to recovery of particle shape. Component of compression due to particle rearrangement is irrecoverable.

This component of deformation is **irrecoverable** or **plastic**. On an unload/reload line, changes in stress can be accommodated without the need for a rearrangement of the soil skeleton. Deformation is primarily due to distortion of the soil particles. It is recovered on unloading, and may in this sense be described as **elastic**.

The behaviour of soil in one-dimensional compression and unloading can be illustrated with reference to a tub full of rubber balls. Initially, the rubber balls are quite loosely packed (Figure 4.7(a)). If someone stands on a platform resting on the upper surface of the rubber balls, the platform will move downward as the rubber balls (i) rearrange themselves to a more dense packing which enables the applied load to be carried by inter-ball contact forces, and (ii) distort in shape, without change in volume (Figure 4.7(b)).

If the person steps off the platform, the platform will move back up, but not by so much as to return to its first position. This is because although the rubber balls will rebound to their original shapes, they will not rearrange themselves back into their initial packing (Figure 4.7(c)). The component of deformation due to particle distortion without slip is elastic in the sense that it is recoverable on unloading, while the component of deformation due to rearrangement of the particles (re-packing) is plastic, or not recoverable on unloading.

If the person then steps back onto the platform, it will move to its previous position under the same load. The settlement due to distortion is re-established, but there is no need for further rearrangement of the rubber balls because their packing is already such that the load due to one person can be carried by interparticle forces.

If a second person steps onto the platform in addition to the first, further rearrangement of the rubber balls will be necessary in order to carry the increased applied load. However, the amount of additional particle rearrangement will not be as great as that which was required to carry the weight of the first person, because the more densely packed the particles are, the more difficult it is to push them closer together. A graph of settlement against load for a tub of rubber balls is shown in Figure 4.8. In essence, this behaviour is the same as that observed in clays during one-dimensional compression and unloading.

In summary, soils are stiffer in unloading/reloading because the deformation is primarily elastic or recoverable. An initially overconsolidated soil, which is loaded past its maximum previous effective stress so that it becomes normally consolidated, may exhibit a marked reduction in stiffness as it moves from a reload line back onto the normal compression line.

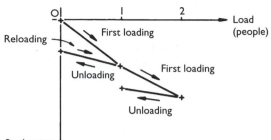

Figure 4.8 Settlement against load for a tub of rubber balls.

Example 4.1 Analysis and interpretation of one-dimensional compression test data

Table 4.1 gives data from an oedometer test on a sample of Weald clay.

Table 4.1 Oedometer test data (equilibrium states)

Vertical effective stress σ_v' (kPa)	50	100	150	200	250	200	150
Equilibrium sample height (mm)	20.23	19.89	19.70	19.35	19.07	19.18	19.32

Initial sample height: 20 mm

$G_s = 2.75$

Moisture content data at the end of the test:
mass of tin empty   4.97 g
mass of tin + wet sample   23.85 g
mass of tin + dry sample   20.52 g

Calculate the water content and the void ratio at the end of the test, assuming that the sample is then fully saturated. Show that the specific volume v is related to the sample height h by the expression

$v/h = \text{constant}$

Hence plot a graph of specific volume against the natural logarithm (ln) of the vertical effective stress. Explain the shape of this graph. Calculate the preconsolidation pressure and the slopes of the one-dimensional normal compression line and the unloading/reloading lines. For each loading/unloading step, calculate the apparent one-dimensional modulus E_0'.

Comment briefly on the significance of these results for the selection of parameters for use in design. Explaining your choice of E_0', estimate the compression of a 2 m thick layer of clay located at a depth of 4–6 m below ground level, which results from the application of a uniform increase in vertical effective stress of 50 kPa.

[*University of London 2nd year BEng (Civil Engineering) examination,*
Queen Mary and Westfield College (part question)]

Solution

$$\text{Water content at end of test} = \frac{m_w}{m_s} = \frac{(m_t + m_w + m_s) - (m_t + m_s)}{(m_t + m_s) - m_t}$$

$$= \frac{(23.85\,\text{g} - 20.52\,\text{g})}{(20.52\,\text{g} - 4.97\,\text{g})} = 21.415\%$$

(m_t = mass of tin, m_w = mass of pore water, m_s = mass of soil particles).
If saturated, $e = wG_s = 0.21415 \times 2.75 = 0.589$.

Total volume $V_t = V_s + V_v = V_s(1 + V_v/V_s) = V_s v$

Hence

$$V_t = V_s v = Ah$$

or

$$v/h = A/V_s = \text{constant (equation (4.4))}$$

$$A/V_s = [v/h]_{\text{at end of test}} = [1.589/19.32]\text{mm}^{-1} \text{ so } v = \frac{1.589}{19.32}h$$

The processed data are given in Table 4.2 and plotted, as a graph of v against $\ln(\sigma_v')$, in Figure 4.9.

Table 4.2 Processed oedometer test data

σ_v'(kPa)	50	100	150	200	250	200	150
v	1.664	1.636	1.620	1.591	1.568	1.577	1.589
$\ln(\sigma_v')$	3.912	4.605	5.011	5.298	5.521	5.298	5.011

AB: Reloading to preconsolidation stress at B, approximately 150 kPa. 'Elastic' compression of soil matrix, without particle slip.

BC: One-dimensional normal compression line, with plastic deformation taking place due to particle slip and rearrangement of soil skeleton to carry higher loads, in addition to 'elastic' deformation of particles.

CD: Unloading: recovery of 'elastic' deformations which occurred during phase BC.

From the graph, the slope of the unload/reload lines AB and CD is

$$\Delta v/\Delta \ln \sigma_v' = (0.620 - 0.664)/(5.011 - 3.912) = -0.04 (= -\kappa_0)$$

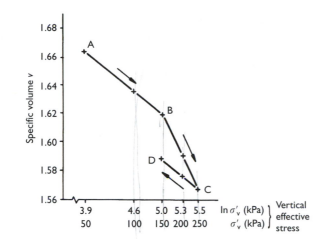

Figure 4.9 v against $\ln(\sigma_v')$.

(check that AB and CD have the same slope). Similarly, the slope of the one-dimensional normal compression line BC is

$$\Delta v / \Delta \ln \sigma_v' = (0.568 - 0.620)/(5.521 - 5.011) = -0.102(= -\lambda_0)$$

The change in slope occurs at B, indicating that the preconsolidation pressure

$$\sigma_{v,\text{max prev}}' \approx 150 \, \text{kPa}$$

The one-dimensional modulus E_0' is given by the ratio $\Delta\sigma_v'/\Delta\varepsilon_v$ for each load increment or decrement, where $\Delta\varepsilon_v = \Delta h/h_0$. Values are given in Table 4.3.

Table 4.3 Values of one-dimensional modulus for each loading and unloading stage

Load (kPa)	50–100	100–150	150–200	200–250	250–200	200–150
$\Delta\sigma_v'$ (kPa)	50	50	50	50	−50	−50
Δh (mm)	0.34	0.19	0.35	0.28	−0.11	−0.14
h_0 (mm)	20.23	19.89	19.70	19.35	19.07	19.18
$\Delta\varepsilon_v$ (%)	1.681	0.955	1.777	1.447	−0.577	−0.730
E_0' (kPa)	2974	5236	2814	3455	8666	6849

Notes

Δh = change in sample height during load increment/decrement; h_0 = sample height at start of load increment/decrement; negative signs denote reductions in stress, and heave (i.e. upward movement) rather than settlement.

The data show that the soil is stiffer in unloading or reloading than during first (normal) compression, and that provided the soil remains on either the normal compression line (slope $-\lambda_0$) or on an unload/reload line (slope $-\kappa_0$), it becomes stiffer as σ_v' increases. In design calculations, parameters (in this case, the value of E_0') appropriate to the stress state, stress history, and anticipated stress and strain paths should be used.

The centre of the clay layer is at 5 m below ground level. Assuming that the groundwater level is at the soil surface, that pore water pressures are hydrostatic, and that the unit weight of the clay and the overlying soil is $20\,\text{kN/m}^3$, the initial vertical effective stress at the centre of the clay layer is

$$(5\,\text{m} \times 20\,\text{kN/m}^3) - (5\,\text{m} \times 10\,\text{kN/m}^3) = 50\,\text{kPa}$$

(taking the unit weight of water as $10\,\text{kN/m}^3$).

The eventual increase in vertical effective stress is $50\,\text{kPa}$, so that the appropriate stress range for the measurement of E_0' is 50–100 kPa, $E_0' = 2974 \approx 3000\,\text{kPa}$.

The eventual settlement of the clay layer ρ_{ult} is obtained from the definition of E_0',

$$E_0' = \Delta\sigma_v'(\rho_{\text{ult}}/h_0)$$
$$\Rightarrow \rho_{\text{ult}} = \Delta\sigma_v'h_0/E_0'$$

Hence

$$\rho_{\text{ult}} = 50(\text{kPa}) \times 2(\text{m})/2800(\text{kPa}) = 0.0357\,\text{m} \approx 36\,\text{mm}$$

4.3 One-dimensional consolidation

It is often important to be able to estimate not only the settlements that will eventually result from an increase in the load applied to a clay or a silt, but also how long these settlements will take to occur. An example of this is found in the construction of an embankment on soft clay (Figure 4.10(a)). It may not be possible to construct such an embankment all in one operation, because the loads which would be applied to the surface of the soft clay would cause a failure of the foundation.

This problem can be circumvented by constructing the embankment in a number of stages (Figure 4.10(b)). The additional load imposed at each stage is insufficient to cause failure, but the soft clay must be allowed to consolidate between stages so that

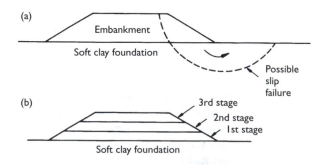

Figure 4.10 Stage-constructed embankment on a soft clay foundation. (a) Embankment constructed quickly; (b) embankment constructed in three stages, with clay foundation allowed to consolidate in between each stage.

its void ratio is reduced and its undrained shear strength (which governs rapid failure) is increased. The period between stages (which could be several months) will be vital to the design of the embankment and the programming of its construction sequence.

Although there are many problems in attempting to predict the behaviour of a large mass of soil in the field from the results of laboratory tests on small soil elements, laboratory test data usually represent at least a starting point for design. Parameters relevant to the processes of transient flow and consolidation are obtained from graphs of settlement against time for the appropriate stages of an oedometer test.

Before the application of an increment of external load, the oedometer test sample is in equilibrium with the pore water pressures hydrostatic. As the sample is only 20 mm thick, the variation in pore water pressure between the top and the bottom is only $1000(\text{kg/m}^3) \times 9.81 \times 10^{-3}(\text{kN/kg}) \times 0.02(\text{m}) = 0.2\,\text{kPa}$. This is negligible in comparison with the applied loads (which are usually tens or even hundreds of kPa), and it is conventional to assume that in an oedometer test sample the pore water pressures at equilibrium are effectively zero.

Before the start of a load increment, the vertical effective stress σ'_v throughout the sample is equal to the vertical total stress applied at the surface. Effective stresses cannot increase unless there is a compression of the soil skeleton. This requires water to flow from the pores, which cannot occur instantaneously because the permeability of the soil is finite.

As the effective stresses cannot change instantaneously, the increase in external load at the start of a loading increment must result initially in an increase in pore water pressure throughout the sample. Because there now exists a pressure gradient over and above hydrostatic – the pressure in the water bath surrounding the sample has not changed – water begins to flow from the soil pores. As water flows out of the sample, the pore water pressures start to fall and the effective stresses start to increase, and the process of consolidation begins.

Figure 4.11 shows a conceptual model of the consolidation process. The piston runs smoothly in the cylinder, but a seal around its circumference prevents water from leaking between the piston and the cylinder wall. Water can only escape through the

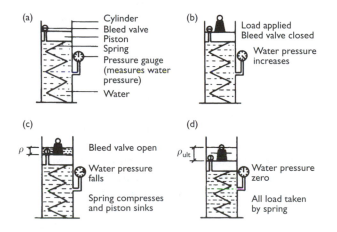

Figure 4.11 Conceptual model for consolidation.

piston via the bleed valve. The space below the piston is filled with water, and there is also a compression spring between the underside of the piston and the base of the cylinder.

Initially, the spring is unstressed, the gauge pressure in the water is zero and the bleed valve is closed (Figure 4.11(a)). A load is applied to the top of the piston. As no water can escape, the piston cannot move down and the spring cannot compress. All of the applied load must therefore be carried by an increase in the water pressure (Figure 4.11(b)). Now the bleed valve is opened so that water can slowly escape (Figure 4.11(c)). As the water escapes, the water pressure below the piston is reduced and the piston sinks. The spring compresses as it takes up the load. Eventually, the water pressure returns to zero and the applied load is carried entirely by the compression of the spring (Figure 4.11(d)). To enable the spring to take the load, there has been a reduction in the volume below the piston.

The time taken to reach the steady conditions shown in Figure 4.11(d) will depend on the size of the bleed valve hole, which governs the rate at which water can escape and is analogous to the soil permeability, and the stiffness of the spring, which governs how much compression must take place and is analogous to the stiffness of the soil in one-dimensional compression, E'_0. The time taken to reach the steady state does not, however, depend on the magnitude of the applied load. Although more compression is needed for the spring to carry a larger applied load, the water can escape more quickly because the pressure gradient across the bleed valve is increased in proportion to the applied load.

The consolidation of soil is actually rather more complicated than suggested by Figure 4.11. Even if the soil is reasonably uniform and the directions of drainage and compression are well defined (as in the oedometer test), the drop in water pressure or head which in Figure 4.11 occurs across the bleed valve will in a soil be distributed through the sample. Consolidation is inextricably linked to the changes in effective stress which result from changes in pore water pressure as water flows out of the soil. To analyse quantitatively the consolidation process, it is therefore necessary to consider these changes in pore water pressure in some detail.

At any stage of the consolidation process, the pore water pressures will vary within the thickness of the sample. We are really interested in the *excess* pore water pressures u_e[1] (i.e. the pressures over and above hydrostatic), because it is these that cause seepage flow, according to Darcy's Law. The **excess head**, defined as $h_e = u_e/\gamma_w$, is in effect the same as the hydraulic total head h used in the analysis of groundwater flow in Chapter 3.

In the oedometer test, analysis is relatively straightforward, because the hydrostatic equilibrium pore water pressures are small enough in comparison with the applied loads to be neglected. In field problems, however, excess pore water pressures must be calculated by subtracting the hydrostatic component from the actual pore water pressure at each point.

The distribution of excess pore water pressure within an oedometer sample at any given time after an increase in external load is represented by a line known as an **isochrone**. The excess pore water pressure varies with depth and time only: it does not vary over the cross-section of the sample. Each isochrone is in effect a graph of excess pore water pressure (which, in the case of the oedometer test, is for practical purposes the same as the actual pore water pressure) against depth at a fixed time t.

4.4   Properties of isochrones

For the upper half of an oedometer test sample of total thickness $2d$ with two-way drainage (or for a sample of thickness d with one-way drainage), Figure 4.12 depicts schematically the succession of isochrones which indicate the progress of consolidation and excess pore water pressure dissipation in response to an increase in the applied vertical stress of $\Delta\sigma_v$.

Isochrones have two important properties:

1. The slope of the isochrone at any point is $\partial u_e/\partial z$. This is equal to γ_w times the magnitude of the hydraulic gradient $\partial h_e/\partial z$, where γ_w is the unit weight of water (h_e is the excess head, $h_e = u_e/\gamma_w$).[2]
2. The compression $\delta\rho$ of a layer of thickness δz at a depth z below the surface of the sample and at a time t after the start of consolidation is obtained from the expression

$$\varepsilon_v = \delta\rho/\delta z = \Delta\sigma_v'/E_0' \tag{4.8}$$

where $\varepsilon_v = \delta\rho/\delta z$ is the vertical strain in the layer, $\Delta\sigma_v'$ is the increase in vertical effective stress at time t, and E_0' is the stiffness of the soil in one-dimensional compression.

The increase in effective stress is equal to the applied increase in vertical total stress minus the excess pore water pressure remaining at time t (Figure 4.13):

$$\Delta\sigma_v' = \Delta\sigma_v - u_e \tag{4.9}$$

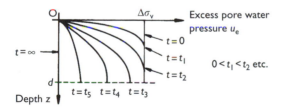

Figure 4.12 Isochrones of excess pore water pressure in an oedometer test sample at various times during consolidation.

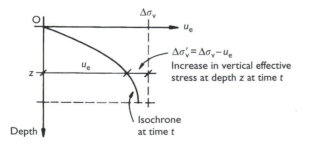

Figure 4.13 Increase in effective stress at a depth z.

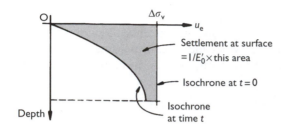

Figure 4.14 Calculation of settlement from isochrone.

Substituting equation (4.9) into equation (4.8) and rearranging,

$$\delta\rho = (1/E_0')\,(\Delta\sigma_v' - u_e)\delta z \tag{4.10a}$$

The overall compression, measured at the surface of the sample, at an elapsed time t is given by integration of equation (4.10a), over the half-depth of the sample $(0 \le z \le d)$,

$$\rho = \frac{1}{E_0'}\int_0^z (\Delta\sigma_v' - u_e)\partial z \tag{4.10b}$$

The integral on the right-hand side of equation (4.10b) is equal to $1/E_0'$ times the area swept out by the isochrone at time t (i.e. the area bounded by the isochrones at time zero and time t). This area is shown shaded in Figure 4.14.

The first of these properties arises by definition, and is always true. The second applies in this form only in cases of vertical consolidation with vertical drainage, where settlements are uniform across the entire soil surface. A more general statement of the second property is that the volume change per unit area of outflow is equal to $1/E_0'$ times the area bounded by the isochrone at $t = 0$ and the current isochrone. An example of the use of the second property in its more general form is given in section 4.7.3.

4.5   One-dimensional consolidation: solution using parabolic isochrones

The exact analysis of the process of one-dimensional consolidation is presented in section 4.8. It is rather mathematical, and probably lies outside the scope of many first degree courses in soil mechanics. As an alternative, the consolidation of an oedometer test sample (and also a number of other consolidation problems) may be analysed approximately by assuming that the isochrones are parabolic in shape. The justification for this is that the results match the exact solution very closely, with errors generally less than 5% of the ultimate settlement for many plane flow problems. In comparison with the other assumptions made in both analyses – in particular that the soil is uniform and that the soil parameters k and E_0' are constant – an error of 5% is insignificant.

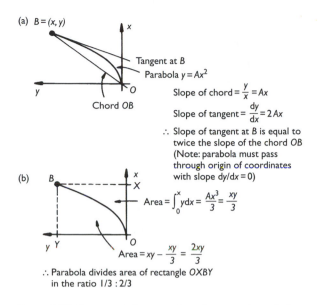

(a) $B = (x, y)$

Tangent at B
Parabola $y = Ax^2$

Slope of chord $= \dfrac{y}{x} = Ax$

Slope of tangent $= \dfrac{dy}{dx} = 2Ax$

∴ Slope of tangent at B is equal to twice the slope of the chord OB
(Note: parabola must pass through origin of coordinates with slope $dy/dx = 0$)

(b)

Area $= \displaystyle\int_0^x y\,dx = \dfrac{Ax^3}{3} = \dfrac{xy}{3}$

Area $= xy - \dfrac{xy}{3} = \dfrac{2xy}{3}$

∴ Parabola divides area of rectangle $OXBY$ in the ratio $1/3 : 2/3$

Figure 4.15 Geometrical properties of a general parabola.

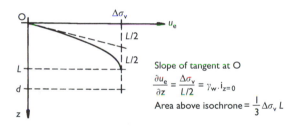

Slope of tangent at O

$\dfrac{\partial u_e}{\partial z} = \dfrac{\Delta\sigma_v}{L/2} = \gamma_w \cdot i_{z=0}$

Area above isochrone $= \dfrac{1}{3}\Delta\sigma_v L$

Figure 4.16 Characterization of first set of isochrones, upper half of an oedometer test sample with two-way drainage.

Once it is assumed that the isochrones are parabolic, the properties described in section 4.4 can be quantified very easily by considering the geometry of a general parabola, as shown in Figure 4.15.

For the consolidation of an oedometer sample, the initial isochrones are characterized by the distance L from the drainage boundary beyond which the excess pore water pressures have not yet begun to dissipate (Figure 4.16).

Assuming that the soil particles and the pore water are incompressible, changes in volume must be due to changes in void ratio as water flows out of (or into) the soil sample. The rate of reduction of volume due to water flowing from the top half of the sample is given by Darcy's Law:

$$q = Aki \tag{4.11}$$

where i is the hydraulic gradient[3] at the drainage boundary (which is equal to $1/\gamma_w \times \Delta\sigma_v/(L/2)$ – Figure 4.16) and A is the cross-sectional area of the sample.

As the surface settlement is uniform, the rate of reduction in volume is equal to $A \times$ the rate of settlement $\partial \rho / \partial t$. Hence

$$\partial \rho / \partial t = k i_{z=0} = (k \Delta \sigma_v)/(\gamma_w L/2) = 2k \Delta \sigma_v / \gamma_w L \tag{4.12}$$

The settlement at time t is given by $1/E_0' \times$ the area swept out by the isochrone (Figure 4.16),

$$\rho = (1/E_0')(\Delta \sigma_v L/3) = \Delta \sigma_v L / 3E_0' \tag{4.13}$$

Differentiating equation (4.13) and setting the result equal to equation (4.12),

$$\partial \rho / \partial t = (\Delta \sigma_v / 3E_0')\partial L / \partial t = 2k \Delta \sigma_v / \gamma_w L \tag{4.14}$$

Hence

$$L\, \partial L = (6kE_0'/\gamma_w)\partial t$$

Integrating between limits of ($L = 0$ at $t = 0$) and a general point ($L = L$ at $t = t$),

$$(L^2)/2 = (6kE_0'/\gamma_w)t$$

or

$$L = \sqrt{(12c_v t)} \tag{4.15}$$

where $c_v = kE_0'/\gamma_w$ is known as the **coefficient of consolidation**. Substituting this expression for L into equation (4.13) gives

$$\rho = (\Delta \sigma_v / E_0')\sqrt{(4c_v t/3)} \tag{4.16}$$

which applies until $L = d$ or $t = d^2/12c_v$.

For a sample of depth $2d$ with two-way drainage, this is the settlement due to the compression of the upper half of the sample only. To obtain the total settlement of the whole sample, we must multiply the expression given in equation (4.16) by 2.

After this time ($t = d^2/12c_v$), the isochrones are characterized by the excess pore water pressure u_{ed} remaining at the central horizontal plane – or, in the case of a sample with only one-way drainage, at the base of the sample (Figure 4.17).

The hydraulic gradient at the central horizontal plane is zero (i.e. the isochrones are vertical), because by symmetry there can be no flow across it. Similarly the hydraulic gradient at the impermeable base of a sample with one-way drainage is zero, because again there is no flow.

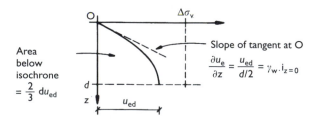

Figure 4.17 Characterization of second set of isochrones, upper half of oedometer test sample with two-way drainage.

For one of the second set of isochrones, the hydraulic gradient at the drainage boundary is $u_{ed}/(\gamma_w d/2)$ (Figure 4.17), so that at time t the rate of water outflow is

$$q = 2Aku_{ed}/(\gamma_w d)$$

and the rate of settlement is

$$\partial \rho/\partial t = (2ku_{ed})/\gamma_w d \tag{4.17}$$

The area swept out is $(d\Delta\sigma_v - 2du_{ed}/3)$ (Figure 4.17), so the settlement at time $t(t > d^2/12c_v)$ is

$$\rho = (1/E_0')(d\Delta\sigma_v - 2du_{ed}/3) \tag{4.18}$$

Differentiating equation (4.18) and setting the result equal to equation (4.17),

$$\partial \rho/\partial t = -(2d/3E_0')\partial u_{ed}/\partial t = 2ku_{ed}/\gamma_w d \tag{4.19}$$

Hence

$$(1/u_{ed})\partial u_{ed} = -(3kE_0'/\gamma_w d^2)\partial t$$

Integrating between limits of ($u_{ed} = \Delta\sigma_v$ at $t = d^2/12c_v$), and a general point ($u_{ed} = u_{ed}$ at $t = t$),

$$\ln(u_{ed}/\Delta\sigma_v) = 1/4 - (3c_v t/d^2)$$

or

$$u_{ed} = \Delta\sigma_v \exp(1/4 - 3c_v t/d^2) \tag{4.20}$$

Substituting equation (4.20) into equation (4.18)

$$\rho = (d\Delta\sigma_v/E_0')[(1 - 2/3)\exp(1/4 - 3c_v t/d^2)] \tag{4.21}$$

for $t > d^2/12c_v$.

Again, this result is for the upper half of the sample only, and for a sample of total thickness $2d$ with two-way drainage we must multiply it by 2.

Equations (4.16) and (4.21) may be rewritten in terms of dimensionless parameters R and T. R is the **proportional settlement** ρ/ρ_{ult}, where ρ_{ult} is the settlement at $t = \infty$. (For the upper half only of a sample of thickness $2d$ with two-way drainage, $\rho_{ult} = d\Delta\sigma_v/E_0'$.) R is also known as the **degree of consolidation**. T is the **time factor**, given by $T = c_v t/d^2$.

Substitution of the expressions $R = \rho/\rho_{ult}$, $\rho_{ult} = d\Delta\sigma_v/E_0'$ and $T = c_v t/d^2$ into equations (4.16) and (4.21) yields

$$R = \sqrt{(4T/3)} \text{ for } 0 < T < 1/12 \tag{4.22}$$

and

$$R = 1 - 2/3\exp(1/4 - 3T) \text{ for } T > 1/12 \tag{4.23}$$

R is plotted as a function of T in Figure 4.18.

We have derived Figure 4.18 for the case of one-dimensional vertical consolidation of a layer of soil in response to an initial, non-equilibrium excess pore water pressure distribution which is uniform with depth. Drainage may be either one-way, in which case the drainage path length is equal to the whole sample thickness, or two-way, in which case the drainage path length is equal to half the sample thickness. It will be shown in section 4.8 that Figure 4.18 also applies to the consolidation of a soil

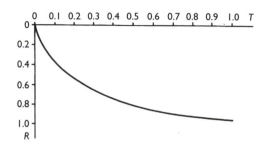

Figure 4.18 Non-dimensional settlement R as a function of time factor T during one-dimensional vertical consolidation: uniform change in excess pore pressure with one- or two-way drainage, and any linear-with-depth change in excess pore pressure with two-way drainage.

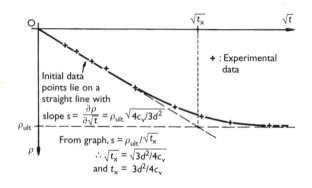

Figure 4.19 Determination of consolidation coefficient c_v from oedometer test data.

sample with two-way drainage, in which the initial non-equilibrium excess pore water pressure distribution is not uniform, but varies linearly with depth.

The applicability of Figure 4.18 depends on the validity of our assumption that the isochrones are parabolic in shape. Comparison with the 'exact' solutions developed in section 4.8 shows that the error in R is always less than approximately 0.05, which in soil mechanics terms – given the likely uncertainties concerning the values of permeability k and stiffness E_0' – is quite acceptable.

4.6   Determining the consolidation coefficient c_v from oedometer test data

The behaviour of the soil element during one-dimensional consolidation in response to a single load increment in an oedometer test is investigated by plotting a graph of settlement against time or – more usefully – settlement against the square root of time. According to the analysis of the consolidation process presented in section 4.5, a graph of settlement ρ against $\sqrt{t}$ should have initial slope $s = \rho_{ult}\sqrt{(4c_v/3d^2)}$ (substitute $\rho_{ult} = d\Delta\sigma_v/E_0'$ into equation 4.16). This can be used to determine the consolidation coefficient experimentally (Figure 4.19 and Example 4.2). Remember that $d = h/2$ for a sample of height h with two-way drainage.

Example 4.2 Analysis of settlement/time data from an oedometer test

(a) Table 4.4 gives data from an oedometer test on a sample of peat. Plot a graph of settlement against $\sqrt{time}$, and determine the values of c_v, E_0' and k for the soil over this load increment.

Table 4.4 Oedometer test data: consolidation phase

Time (min)	0	0.32	0.64	1.28	2.40	4.80	9.60	16.00
Settlement (mm)	0	0.16	0.23	0.33	0.45	0.65	0.86	0.96

Vertical stress increment 10 kPa.
Initial sample thickness 20 mm, with two-way drainage through porous stones.

$$c_v = 3d^2/4t_x$$

(b) Explain briefly, without attempting the calculation, how you could use the data given in Table 4.4 to estimate the rate of consolidation of a 4 m thick stratum of the same peat, following a uniform increase in vertical stress of 10 kPa, with two-way vertical drainage from the peat layer. What factors would you take into account in assessing the applicability of the laboratory parameters to the field situation?

[*University of London 2nd year BEng (Civil Engineering) examination, Queen Mary and Westfield College (part question)*].

Solution

(a) The data given in Table 4.4 are re-presented in Table 4.5, and plotted in Figure 4.20, as settlement against $\sqrt{time}$.

Table 4.5 Processed settlement data

Time (min)	0	0.32	0.64	1.28	2.40	4.80	9.60	16.00
$\sqrt{Time}$ (min$^{1/2}$)	0	0.57	0.80	1.13	1.55	2.19	3.10	4.00
Settlement (mm)	0	0.16	0.23	0.33	0.45	0.65	0.86	0.96

From Figure 4.20, $\sqrt{t_x} = 3.42 \text{ min}^{1/2}; t_x = 11.7 \text{ min}$.

$t_x = 3d^2/4c_v; \quad d = 20/2 = 10 \text{ mm}$

$c_v = 3d^2/4t_x = 3 \times 10^2/4 \times 11.7 = 6.41 \text{ mm}^2/\text{min} (= 1.07 \times 10^{-7} \text{ m}^2/\text{s})$

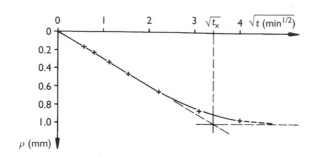

Figure 4.20 Settlement against √*time* for oedometer test.

At the end of the load increment, the vertical strain $\varepsilon_v = 1\,\text{mm}/20\,\text{mm} = 0.05$ (5%). For this load increment, the one-dimensional modulus E'_0 is given by

$$E'_0 = \Delta\sigma'_v/\varepsilon_v = 10\,\text{kPa}/0.05 = 200\,\text{kPa}$$

Calculate the permeability k by inference from $c_v = kE'_0/\gamma_w; k = \gamma_w c_v/E'_0$:

$$k = \frac{(9.81\,\text{kN/m}^3 \times 1.07 \times 10^{-7}\text{m}^2/\text{s})}{200\,\text{kN/m}^2} = 5.2 \times 10^{-9}\text{m/s}$$

(b)   The graph of $R(= \rho/\rho_{\text{ult}})$ against $T(= c_v t/d^2)$ will in theory be the same for the oedometer test sample as for the 4 m thick peat layer in the field. In the field, $d = 2\,\text{m}$ (half the layer thickness, because there is two-way drainage) and ρ_{ult} is given by

$$\varepsilon_v = \rho_{\text{ult}}/2d = \Delta\sigma_v/E'_0 \text{ or}$$
$$\rho_{\text{ult}} = 2d\Delta\sigma_v/E'_0 = 4\,\text{m} \times 10\,\text{kPa} \div 200\,\text{kPa} = 0.2\,\text{m}$$

The development of settlement with time in the field may be investigated by re-plotting the experimental data as R against T, then calculating corresponding values of ρ and t for the field situation using $\rho_{\text{ult}} = 0.2\,\text{m}$ and $d = 2\,\text{m}$. In effect, the data from the oedometer test are multiplied by the ratio of the sample thicknesses (4000 mm ÷ 20 mm = 200) for the settlement ρ, and by the ratio of the drainage path lengths squared ($200^2 = 40\,000$) for the time t, to arrive at the corresponding field values (see also section 4.7.1).

We have assumed that the values of c_v and E'_0 are the same in the field as those measured in the laboratory test. This would require that the stress history and stress state of the sample in the laboratory test are the same as those of the soil in the field, and that the laboratory test sample is subjected to the same changes in stress and strain as are expected to occur in the field.

4.7 Application of consolidation testing and theory to field problems

In many field situations, drainage and consolidation are not unidirectional, so that the application of one-dimensional consolidation theory is at best an approximation. However, the main difficulty with field consolidation problems is often the identification of boundaries, drainage path lengths and soil parameters; so that the error introduced in assuming one-dimensional consolidation may not be that significant.

In some cases, one-dimensional consolidation theory will be entirely appropriate and straightforward to apply, as in the case studies and examples that follow. In the first example, the soil in the field is subjected to the same drainage conditions and changes in effective stress as an oedometer sample, and the solution developed in section 4.5 may be applied directly. In the remaining examples, either the drainage boundary conditions or the changes in effective stress (or both) are different, but the principles used to derive appropriate relationships between settlement and time are the same.

4.7.1 Consolidation due to an increase in effective stress following groundwater lowering

In the oedometer, increases in effective stress result from increases in total stress while the steady state pore water pressure regime remains unaltered. In a field problem, changes in effective stress are just as likely to result from changes in the steady state pore water pressure regime, while the total stress remains approximately constant. This is illustrated in Example 4.3.

Example 4.3 Using one-dimensional consolidation theory and oedometer test data to estimate field rates of settlement due to construction dewatering

(a) Figure 4.21 shows a cross-section through a site at which it is proposed to lower the groundwater level by 2 m as indicated. Sketch the initial and final distributions of pore water pressure (whose variation with depth may be assumed to be hydrostatic) in the soft clay layer. Sketch also the initial and final distributions of excess pore pressure, together with 3 or 4 isochrones in between. Take the unit weight of water $\gamma_w = 10\,\text{kN/m}^3$, and the datum for the measurement of excess head as shown.

(b) Table 4.6 gives data from an oedometer test on a sample of the soft clay. Plot a graph of settlement against $\sqrt{\text{time}}$, and determine the values of c_v, E_0' and k for the soft clay over this load increment.

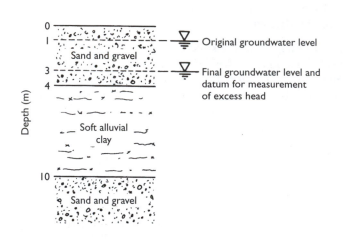

Figure 4.21 Soil profile at site for groundwater lowering.

Table 4.6 Oedometer test data

Time(s)	0	6	24	48	120	240	600
Settlement (mm)	0	0.096	0.184	0.264	0.424	0.592	0.768

Vertical stress increment 20 kPa

Initial sample thickness 20 mm, with two-way drainage through porous stones

$$c_v = 3d^2/4t_x$$

Estimate the final compression of the soft clay layer shown in Figure 4.21 that will result from the proposed reduction in groundwater level. It is estimated that a compression of the clay layer in excess of 40 mm might cause damage to existing buildings. For how long could the dewatering system be operated before this occurs?

[University of London 2nd year BEng (Civil Engineering) examination, Queen Mary and Westfield College]

Solution

Initial and final pore pressure distributions, together with isochrones of excess pore pressure referred to the datum indicated in Figure 4.21, are shown in Figure 4.22. The pattern of isochrones of excess pore water pressure is exactly the same as for an oedometer test sample.

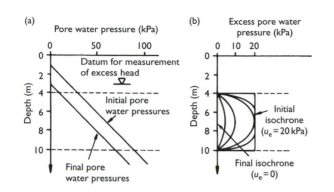

Figure 4.22 (a) Initial and final pore water pressure distributions and (b) isochrones of excess pore water pressure.

The settlement vs. $\sqrt{\text{time}}$ data are given in Table 4.7 and plotted in Figure 4.23.

Table 4.7 Processed settlement data

Time (s)	0	6	24	48	120	240	600
$\sqrt{\text{time}}$ (s$^{1/2}$)	0	2.45	4.89	6.93	10.95	15.5	24.5
ρ (mm)	0	0.096	0.184	0.264	0.424	0.592	0.768

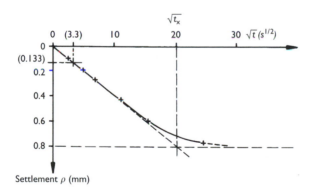

Figure 4.23 Settlement vs. $\sqrt{\text{time}}$.

Take $\rho_{\text{ult}} = 0.8$ mm:

$$E_0' = \Delta\sigma_v'/\varepsilon_v = 20\,\text{kPa}/(0.8/20) = 500\,\text{kPa}$$

From Figure 4.23, $\sqrt{t_x} \approx 20\,\text{s}^{1/2}$ and $c_v = 3d^2/4t_x$:

$$d = 10\,\text{mm (maximum drainage path length)}$$

$$c_v = 0.188\,\text{mm}^2/\text{s} \approx 2 \times 10^{-7}\,\text{m}^2/\text{s}$$

$$c_v = kE_0'/\gamma_w$$

so

$$k = 4 \times 10^{-9}\,\text{m/s}$$

Final compression of clay layer in the field is

$$\Delta\sigma_v' 2d/E_0' = (20\,\text{kPa} \times 6000\,\text{mm})/500\,\text{kPa} = 240\,\text{mm}$$

(Alternatively, the settlements scale according to the thickness of the clay, giving $(6000/20) \times 0.8\,\text{mm} = 240\,\text{mm}$. $(6000/20)$ is the ratio of the thickness of the clay layer in the field to the thickness of the oedometer test sample, and $0.8\,\text{mm}$ is the settlement of the sample in the oedometer under the same increase in vertical effective stress. Either way, we have assumed that the one-dimensional modulus of the soil is the same in the field as in the laboratory test.)

When the compression in the field is 40 mm, the proportional settlement $R = \rho/\rho_{\text{ult}}$ is $40/240 = 0.167$. The corresponding settlement in the oedometer test is $0.167 \times 0.8\,\text{mm} = 0.133\,\text{mm}$. This occurs at an elapsed time $t = 3.3^2\,\text{s} = 10.9\,\text{s}$ (scaling from Figure 4.23). The time factor $T = c_v t/d^2$ will be the same in the field as in the laboratory. Assuming that c_v is the same in the laboratory as in the field, the ratio t/d^2 must also be the same. Thus

$$t_{\text{field}}/d_{\text{field}}^2 = t_{\text{lab}}/d_{\text{lab}}^2 \quad \text{or} \quad t_{\text{field}} = (d_{\text{field}}/d_{\text{lab}})^2 \times t_{\text{lab}}$$

$$d_{\text{field}} = 3000\,\text{mm},\, d_{\text{lab}} = 10\,\text{mm} \text{ and } t_{\text{lab}} = 10.9\,\text{s}$$

Therefore

$$t_{\text{field}} = (3000/10)^2 \times 10.9\,\text{s} = 11.3\,\text{days}$$

This is unlikely to be a very accurate prediction, not least because of the inevitable errors at the start of the consolidation increment due to the impossibility of increasing the load on the oedometer sample instantaneously, and the magnification of any discrepancy by a scaling factor of 300^2.

4.7.2 Underdrainage of a compressible layer

In Example 4.3, the alluvial clay layer remained submerged throughout the consolidation process. The final pore water pressures were therefore still hydrostatic, below the reduced groundwater level. For this reason, the change in excess pore water pressure in the alluvial clay during consolidation was uniform with depth, exactly as in the oedometer test.

If the groundwater level were lowered to the base of the alluvial clay layer, the situation would be somewhat different. Steady state conditions in the clay would no longer be hydrostatic, but would correspond to downward seepage into the underlying sand and gravel. The clay would eventually become underdrained, as discussed in section 3.2.1 (Figures 3.1(c) and (d)).

The underdrainage of a compressible layer is actually quite an interesting application of one-dimensional consolidation theory, which is discussed in Example 4.4 with reference to a case study.

Example 4.4 Underdrainage of a compressible layer

A firm of civil engineering contractors install a wellpoint dewatering system at one of their sites in order to lower the groundwater level during the construction in open excavation of a sub-surface pumping chamber. After the dewatering system has been in continuous operation for a period of 4 weeks or so, the owners of buildings up to a distance of 500 m from the site file claims against the civil engineering contractor for stuctural damage arising from ground movements which they allege are the result of the dewatering operation. Advise the contractor. A borehole log is given in Figure 4.24.

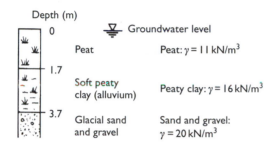

Figure 4.24 Borehole log: one-dimensional consolidation due to underdrainage.

We will assume that the pore water pressure at the upper surface of the compressible layer is maintained (by rainfall and natural or artificial recharge) at zero. The initial pore water pressures may be assumed to be hydrostatic below the water table indicated in Figure 4.24. At the start of pumping, the pore water pressure in the sand and gravel aquifer at the base of the compressible peat and clay will be reduced very quickly to zero. Eventually, the pore water pressures within the peat and clay will also fall to zero, corresponding to steady downward seepage with a hydraulic gradient of unity. Thus the steady state pore pressure regime is not hydrostatic, and does not correspond to conditions of zero flow. As the total vertical stress remains constant, the changes in pore water pressure must be accompanied by changes in vertical effective stress. These can only occur as the clay and peat consolidate.

The analysis of the consolidation process for a two-layer system is quite complicated, and is beyond the scope of this book. However, an approximate analysis, in which the discontinuity in hydraulic gradient between the two layers (which will occur because they have different permeabilities and compressibilities) is ignored, will still give some insight into the problem. Furthermore, the available soils data are somewhat limited, and do not justify the use of a sophisticated analysis. Unfortunately, this is often the case, especially where the geotechnical engineer is called upon to investigate something which went wrong. In back-analysis – but not in design – it is quite reasonable to estimate parameters for which measured values are not available, provided that this limitation is borne in mind in assessing the significance of the calculations.

Assuming that the peat and the clay can be considered to behave as a single layer, isochrones of pore water pressure u and excess pore water pressure u_e as functions of depth z are shown in Figure 4.25. The excess pore water pressure u_e is related to the pore water pressure u by the expression

$$u_e = u - \gamma_w z \tag{4.24}$$

where the datum for and the direction of z are as shown in Figure 4.25.

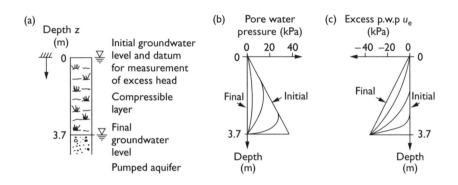

Figure 4.25 (a) Idealized geometry and isochrones of (b) pore water pressure and (c) excess pore water pressure.

We could solve the problem from first principles, by assuming that the curved parts of the isochrones of excess pore water pressure are parabolic in shape. Alternatively, we might save ourselves some work by considering very carefully the drainage boundary conditions. Although pore water drains from the two compressible strata in downward flow only, the important point is that the boundary conditions are such that if water wanted to escape from the top of the layer, it could. This means that we are dealing in effect with the one-dimensional consolidation of a compressible layer with two-way drainage, in response to a change in excess pore water pressure which varies linearly with depth. We may therefore use Figure 4.18, with the maximum drainage path length $d =$ half the thickness of the compressible layer. (If you do not think that this is obvious, it is demonstrated mathematically in section 4.8.3, case (c).) All we need to do is to calculate the ultimate settlement of the peat–clay system, and use Figure 4.18 to investigate how it develops with time.

Take the unit weight of the clay as $16\,\text{kN/m}^3$, the unit weight of the peat as $11\,\text{kN/m}^3$, and the unit weight of water as $10\,\text{kN/m}^3$. Assume also that the soil remains saturated even though the pore water pressure is zero, and that the unit weights remain approximately constant. Calculate the ultimate changes in vertical effective stress at key depths (i.e. at the top and bottom of each stratum), from the total vertical stress and the initial and final pore water pressures (Table 4.8).

The average long-term increases in vertical effective stress are $(0 + 17)/2 = 8.5\,\text{kPa}$ in the peat, and $(17 + 37)/2 = 27\,\text{kPa}$ in the clay.

Table 4.8 Changes in vertical effective stress with depth

Depth (m)	Stratum	σ_v (kPa)	Initial u (kPa)	Final u (kPa)	Initial σ'_v (kPa)	Final σ'_v (kPa)	$\Delta\sigma'_v$ (kPa)
0	Peat	0	0	0	0	0	0
1.7	Peat/clay	18.7	17.0	0	1.7	18.7	17.0
3.7	Clay	50.7	37.0	0	13.7	50.7	37.0

Oedometer tests on each soil gave the E'_0 values in Table 4.9.

Table 4.9 E'_0 from oedometer tests

Stratum	Range of σ'_v (kPa)	E'_0 (kPa)
Peat	16–21	194
Clay	16–21	416
Clay	21–27	435
Clay	27–38	488
Clay	38–59	666

Taking $E'_0 = 200\,\text{kPa}$ for the peat, and $E'_0 = 500\,\text{kPa}$ for the clay, the ultimate settlement ρ may be calculated. For each layer,

$$\varepsilon_v = \Delta\sigma'_{v,\text{av}}/E'_0 \text{ and } \varepsilon_v = \rho/h$$

where h is the thickness of the layer, so

$$\rho = \Delta\sigma'_{v,\text{av}}h/E'_0$$

For the peat, $E'_0 = 200\,\text{kPa}$, $\Delta\sigma'_{v,\text{av}} = 8.5\,\text{kPa}$ and $h = 1.7\,\text{m}$, giving $\rho = 0.072\,\text{m}$ or 72 mm. For the clay, $E'_0 = 500\,\text{kPa}$, $\Delta\sigma'_{v,\text{av}} = 27\,\text{kPa}$ and $h = 2\,\text{m}$, giving $\rho = 0.108\,\text{m}$ or 108 mm.

Thus the total ultimate settlement is 180 mm. The next question is how quickly this will occur – and in particular, how much settlement will have occurred after four weeks, which is when the alleged damage to nearby buildings began to occur. Unfortunately, values of consolidation coefficient were not included in the site investigation report. However, if it is accepted that the rate of consolidation is controlled by the clay (as the less permeable of the two strata, and the stratum closest to the drainage boundary), the consolidation coefficient may be estimated by assuming a value for the permeability. The permeability of the clay is likely to be in the range 10^{-7} to 10^{-10} m/s. Rather than assuming just one value, it is more useful to carry out the calculation for a number of values in the likely range, in order to investigate the sensitivity of the result to this uncertainty.

The values of consolidation coefficient $c_v (= kE'_0/\gamma_w)$, and time factor $T(= c_v t/d^2)$ after $t = 4$ weeks $(= 2.42 \times 10^6\,\text{s})$ corresponding to permeabilities of $10^{-7}, 10^{-8}, 10^{-9}$ and 10^{-10} m/s, are indicated in Table 4.10. Remember that in this case drainage is (or could be) two-way, so that the maximum drainage path length $d = 3.7\,\text{m}/2 = 1.85\,\text{m}$. The proportional settlements $R(= \rho/\rho_{\text{ult}})$

corresponding to the values of T according to Figure 4.18 are also shown, as are the actual settlements calculated from R using $\rho_{ult} = 180\,\text{mm}$.

Table 4.10 Results of consolidation analysis

Permeability k (m/s)	Consolidation coefficient $c_v\,(\text{m}^2/s)$	Time factor T after 4 weeks	Proportional settlement R	Settlement ρ (mm) after 4 weeks
10^{-7}	5.1×10^{-6}	3.6	1	180
10^{-8}	5.1×10^{-7}	0.36	0.71	128
10^{-9}	5.1×10^{-8}	0.036	0.22	39
10^{-10}	5.1×10^{-9}	0.0036	0.069	12

The likely settlement after 4 weeks is at least 12 mm, and probably nearer to 40 mm. Buildings are more susceptible to damage from differential than from uniform settlements. The settlements we have calculated should in theory extend uniformly across a wide area. This is because of the boundary conditions to the compressible layer in this case, which make it in effect like a huge oedometer test. (If the peat and the clay had been pumped directly, however, instead of via the underlying aquifer, this would not be so. Pumping from the peat and clay directly would lead to vertical compression due to horizontal flow, which would result in non-uniform settlements. This is discussed in section 4.7.3.)

In practice, there are several reasons why differential settlements might occur in the present case. It is extremely unlikely that the thickness of compressible soil is uniform across the entire site. If the thickness of the compressible layer is reduced, then so is the surface settlement. Different parts of the same building might have different types of foundation, resting either on the clay (which will be affected by the dewatering-induced settlements), or on the underlying sand and gravel (which is comparatively stiff, and will not settle significantly). Thus on balance, it seems likely that the alleged damage to nearby buildings is probably attributable to the dewatering system.

Should the dewatering system be switched off to prevent further damage? If the permeability of the clay is closer to 10^{-7} m/s, there is probably no point, because consolidation has ceased. If on the other hand the permeability of the clay is closer to 10^{-9} m/s or 10^{-10} m/s, there is every point because perhaps 80% or 90% of the settlement has yet to occur. Unfortunately, in the absence of the appropriate soils data (or better still, the monitoring of ground movements or buildings during construction activities), there is no way of knowing. This demonstrates a salutary point concerning the need to obtain soils data before work on site commences, rather than waiting until after a problem has arisen.

A further complication concerns our initial assumption that the long-term pore pressures would be maintained at zero due to recharge and downward percolation: this may not be correct. The amount of recharge required per unit area of the surface is equal to the permeability of the soil (from Darcy's Law, $q/A = ki$ with the hydraulic gradient $i = 1$). A permeability of 10^{-7} m/s corresponds to a rate of recharge of 9 mm/day or 3.15 m per annum.

Apart from in mountainous areas, the average annual rainfall in the UK does not generally exceed 1500 mm which corresponds to a permeability of approximately 5×10^{-8} m/s. In many parts of southeast England, the annual rainfall is less than 635 mm or 2×10^{-8} m/s. Thus, where the vertical permeability of the soil is greater than about 10^{-8} m/s, rainfall alone is unlikely to maintain long-term conditions of downward percolation with zero pore water pressures. What may happen instead is that the pore water pressures near the surface become negative, increasing the effective stresses (and hence the settlements) still further.

If it is assumed that the steady state pore water pressures vary linearly with a hydrostatic gradient from zero at the bottom of the compressible layer to some negative value (which does not exceed in magnitude the air entry value of pore water suction derived in section 3.2.3) at the top, then the average increase in vertical effective stress within the compressible layer is doubled (Figure 4.26). So too is the ultimate settlement. The isochrones are now exactly the same as those in Figure 4.12, so to use Figure 4.18 to predict the rate of settlement we have to assume that there is only one-way drainage, with the maximum drainage path length equal to the full thickness of the layer, 3.7 m in this case.

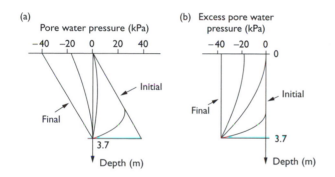

Figure 4.26 Isochrones of (a) pore water pressure and (b) excess pore water pressure with steady state suctions and no recharge.

Interestingly, the rate at which settlement begins to occur is the same in each case. This is because initially $R = \sqrt{(4T/3)}$. If the effective drainage path length d is doubled, $T (= c_v t/d^2)$ is reduced by a factor of 4 at a given elapsed time t. R, being initially proportional to the square root of T, is reduced by a factor of 2. However, ρ_{ult} has also been doubled. Thus the actual settlement ρ after an elapsed time t remains the same, provided that T (based on the increased drainage path length) is less than 1/12.

Unfortunately, it is not possible to be sure what will happen after $T = 1/12$ until after the event, unless positive steps (such as the installation of a recharge system) are taken to ensure that the first set of boundary conditions considered is actually imposed.

One final point concerns the very wide area over which the alleged settlement damage occurred in this case. The distance of influence of the dewatering system would have been controlled by the sand and gravel aquifer, rather than by the peat and clay. Using Sichardt's formula (equation (3.34)) with $C = 1500\text{--}2000$ $(\text{m/s})^{-1/2}$, $\Delta h = 3.7\,\text{m}$ and $k = 4 \times 10^{-3}\,\text{m/s}$, the distance of influence would have been expected to have been in the range 350–470 m. In fact, piezometers installed to investigate the problem (after the event) indicated no significant variation in drawdown up to a distance of 250 m from the site, suggesting an effective distance of influence substantially in excess of 500 m.

The reason for this is probably that the aquifer was confined laterally by impermeable boundaries, rather than being of effectively infinite extent as assumed by the use of equation (3.34). The slow rate of recovery of the groundwater level in the aquifer after the dewatering system was switched off would tend to confirm this. The unexpectedly wide area over which a significant drawdown was produced in the sand and gravel aquifer, together with the fact that many of the buildings had mixed or otherwise unsuitable foundations, undoubtedly contributed to the extent of the alleged damage to property in this case.

Is the contractor liable for the damage that might have been caused? Surprisingly, the answer under current English law is probably not. The owner of property has a right to support from the underlying soil, but not to support from water which flows in undefined channels through the soil. Thus a contractor who excavated a trench too close to a building, which removed or reduced the support of the soil, would be liable for any damage so caused. The operator of an incorrectly installed dewatering system, which damaged buildings by removing soil particles as well as water from the ground (e.g. due to inadequate filters or wellscreens, section 3.19.4), would probably be similarly liable. However, if the dewatering system merely removes water flowing in undefined channels (i.e. groundwater), the contractor is unlikely to be held legally responsible for any damage to property which results.

This position has developed over the last 150 years or so on the basis of cases (originally concerned with pumping rights for water supply) brought to court. The most recent ruling was in the Court of Appeal, and there is a possibility that the general legal position could one day be reversed by a House of Lords decision in some future case. Contractors – or at least their insurers – are probably aware of this, because in practice *bona fide* claims for subsidence damage due to groundwater extraction seem often to elicit a sympathetic response.

4.7.3 *Vertical compression due to plane horizontal flow*

The Conwy Crossing project was mentioned in Chapter 3, in connection with the control of pore water pressures and groundwater by means of pumped wells. Figure 3.25, which shows an idealized cross-section through one of the large temporary excavations adjacent to the Conwy estuary and is concerned with the control of groundwater in the alluvial sands and gravels, tells only part of the story.

Below the 'impermeable' boulder clay lies the stratum of laminated glacial lake deposits, illustrated in Figure 3.15. A full soil profile is given in Figure 4.27(a). The glacial lake deposits are strongly anisotropic, with estimated permeabilities of

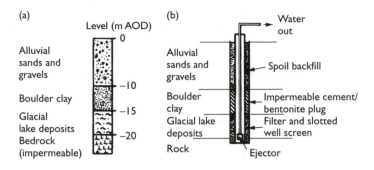

Figure 4.27 (a) Ground conditions and (b) ejector well installation at Conwy, North Wales.

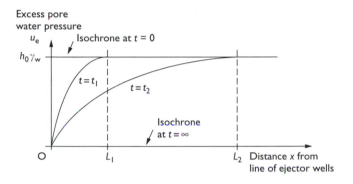

Figure 4.28 Isochrones for horizontal plane flow to a pumped slot.

$k_h \approx 2 \times 10^{-7}$ m/s horizontally and $k_v \approx 10^{-10}$ or 10^{-11} m/s vertically. This horizontal permeability was high enough to necessitate the installation of a pore pressure control system in order to ensure that the base of the excavation remained stable throughout the approximately two-year construction period of the Conwy tunnel. Calculated flowrates were very low, and an ejector well system was therefore used (section 3.19.3; Figure 4.27(b)).

The stratum of lake deposits was expected to remain saturated and to drain by consolidation rather than by air entry. One of the key questions which had to be addressed during the design of the dewatering system and the programming of the project in general was the length of time it would take for the ejector system to reduce pore water pressures sufficiently for excavation to proceed.

If it is assumed that flow within the lake deposits is entirely horizontal, it follows that the equipotentials are vertical and that the excess pore water pressure within the stratum at any time after the start of pumping varies only with horizontal distance, and not with depth. Assuming also that ejector wells are close enough together to behave as a pumped slot, the distribution of excess pore water pressure u_e within the stratum at any time may be represented by an isochrone showing u_e as a function of distance from the line of ejector wells (Figure 4.28).

The datum for the measurement of excess head $h_e = u_e/\gamma_w$ is taken as the reduced (final) piezometric level at the line of ejector wells. If this is below the top of the stratum of glacial lake deposits, then the assumption of purely horizontal flow will not be valid, at least in the vicinity of the wells.

Although the lake deposits drain by horizontal flow, the accompanying change in volume results in vertical settlements. The surface settlements will be non-uniform, due to the variation of excess pore water pressure (and hence of vertical effective stress) with distance from the ejector line.

Assume that the isochrones are parabolic in shape, and extend a distance L (at a given time t) into the stratum of lake deposits. At $x = L$, the slope of the isochrone $\partial u_e/\partial x$ is zero. Consider a thin horizontal layer at a depth z, of thickness δz. Then (per unit length of the slot perpendicular to the plane of the paper), the rate of outflow of water is given by Darcy's Law:

$$q = A k_h i_0$$

where i_0 is the hydraulic gradient at $x = 0$, k_h is the horizontal permeability of the soil and A is the area through which flow takes place: $A = 1 \times \delta z$ in this case. From the geometric properties of the parabolic isochrones (Figure 4.15),

$$i_0 = h_0/(L/2) = 2h_0/L$$

where $h_0 = u_0/\delta_w$ is the drawdown imposed at the ejector line.

Hence the rate of change in volume is given by

$$q = A k_h i_0 = \delta z k_h 2 h_0/L = -\partial V/\partial t \tag{4.25}$$

where the negative sign indicates that the volume is reducing as the soil compresses.

The reduction in volume of an element of the lake deposits of thickness δz, width δx and unit length in the horizontal direction parallel to the pumped slot, located at a distance $x(0 < x < L)$ from the slot, is given by

$$\delta(\Delta V) = \delta\rho\delta x$$

where $\delta\rho$ is the vertical compression, $\delta\rho = \Delta\sigma_v'\delta z/E_0'$.

The increase in vertical effective stress $\Delta\sigma_v'$ at time t and distance x is equal to $(h_0\gamma_w - u_e)$ (Figure 4.29).

In the limit as $\delta V \rightarrow \partial V$ and $\delta x \rightarrow \partial x$, the total change in volume in the layer of thickness δz which has occurred by a time t after the start of pumping is given by

$$\Delta V = \left(\frac{\delta z}{E_0'}\right) \times \int_0^L \Delta\sigma_v'\partial x = \left(\frac{\partial z}{E_0'}\right) \times \int_0^L (h_0\gamma_w - u_e)\partial x$$

which is equal to $(\delta z/E_0') \times$ the area above the isochrone at time t (Figure 4.29).

From the general properties of a parabola (Figure 4.15), the area above the isochrone at time t is $(\gamma_w h_0 L/3)$. Thus

$$\Delta V = \gamma_w h_0 L \delta z/3E_0'$$

and the total volume of the layer at time t is

$$V = V_0 - \Delta V = V_0 - \gamma_w h_0 L \delta z/3E_0' \tag{4.26}$$

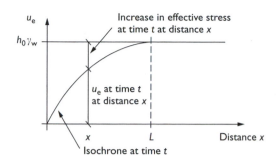

Figure 4.29 Calculation of volume change at a time t after the start of pumping, horizontal plane flow to a pumped slot.

where V_0 is the volume at $t = 0$.

Differentiating equation (4.26) and setting the result equal to equation (4.25), we have

$$\partial V/\partial t = -(\gamma_w h_0 \delta z/3E_0')\partial L/\partial t = -\delta z k_h 2h_0/L$$

or

$$L\partial L = (6k_h E_0'/\gamma_w)\partial t$$

Integrating between limits of ($L = 0$ at $t = 0$) and a general point ($L = L$ at $t = t$), we have

$$(L^2)/2 = (6k_h E_0'/\gamma_w)t$$

or

$$L = \sqrt{(12c_{hv}t)} \qquad (4.27)$$

where $c_{hv} = (k_h E_0'/\gamma_w)$ is the coefficient of consolidation in one-dimensional vertical compression due to horizontal drainage.

The key question in a dewatering application is how long it will take for the pore water pressure to fall by a given amount at a certain distance from the line of wells. In the case of vertical compression due to horizontal flow, the general form of the equation of the parabolic isochrone at time t is

$$u_e = Ax^2 + Bx + C$$

The boundary conditions are

$$u_e = 0 \text{ at } x = 0$$
$$u_e = r_w h_0 \text{ at } x = L$$

and

$$\partial u_e/\partial x = 0 \text{ at } x = L$$

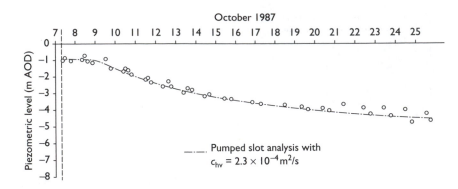

Figure 4.30 Piezometric level against time at a distance of 18 m from a line of ejector wells. (Data from the A55 Conwy Crossing, with permission from Powrie and Roberts, 1990.)

which give $A = -\gamma_w h_0/L^2, B = 2\gamma_w h_0/L$ and $C = 0$. Thus the equation of the isochrone of excess pore water pressure u_e at time t is

$$u_e = \gamma_w h_0(2x/L - x^2/L^2) \text{ where } L = \sqrt{(12c_{hv}t)} \tag{4.28}$$

The pore water pressure response at a distance x from the line of wells may be calculated as a function of time using equation (4.28). Equation (4.28) is not valid for $x > L$, because the pore water pressure at a distance x from the line of wells will not begin to fall until $x = \sqrt{(12c_{hv}t)}$ or $t = x^2/12c_{hv}$.

Figure 4.30 compares the measured response of a piezometer at a distance of 18 m from the ejector line at Conwy with that calculated using equation (4.28) with $h_0 = 6.73$ m and $c_{hv} = 2.3 \times 10^{-4}$ m²/s. In Figure 4.30, excess head $h_e = u_e/\gamma_w$ is plotted rather than excess pore water pressure: this is common in dewatering applications, because the drawdown achieved provides the most obvious indicator of the performance of the dewatering system. Also, the excess head is plotted relative to or above ordnance datum (AOD). The initial water level was – 1 m AOD, with a drawdown to -7.73 m AOD along the line of ejector wells.

4.7.4 Self-weight consolidation: hydraulic fill

In some situations, a soil or a soil-like material may be deposited as a slurry, in which the soil particles are not initially in contact and the effective stress is zero. The total weight of the slurry must therefore be carried by the pore water pressure, which as a result will be greater than hydrostatic. Initially, the particles will settle through the water at constant velocity, according to equation (1.15), until they begin to come into contact with each other. After this time, the excess pore water pressures will dissipate as the soil consolidates to carry part of its own weight via interparticle contact forces as effective stress. Soil placed in this way is sometimes known as **hydraulic fill**.

Materials such as mine wastes and pulverized fuel ash from coal-fired power stations are often pumped as slurries into storage lagoons, where they are left to consolidate under their own weight. In any of these cases, it may be important to be able to estimate how long the consolidation process will take. If the base of the lagoon is permeable, pore water may escape through it so that consolidation takes place by two-way drainage, and Figure 4.18 may be used. If, however, the base of the lagoon is impermeable, drainage takes place only through the upper surface of the material: as the change in pore water pressure which occurs during consolidation varies linearly with depth, and there is only one-way drainage (Figure 4.31), Figure 4.18 does not apply.

Until the soil particles come into contact, the vertical effective stress must be zero. At the start of consolidation, the initial pore water pressure u_i must at any depth z be equal to the vertical total stress,

$$u_i = \sigma_v = \gamma z$$

where γ is the unit weight of the slurry. Eventually, at the end of consolidation, the pore water pressures will be hydrostatic below the surface of the consolidated material,

$$u_f = \gamma_w z$$

(assuming that any surplus water is drained off). The final **excess** pore water pressures (which are related to the pore water pressures using equation (4.24)) are zero, so the initial excess pore water pressure distribution is $u_{ei} = (\gamma - \gamma_w)z$ (Figure 4.31). (We have tacitly assumed that the overall settlement of the soil surface is small, so that changes in geometry – and soil properties – during consolidation can be neglected. The solution therefore applies only to situations in which the strains are small.)

Because the base of the lagoon is impermeable, the slope of the isochrone $\partial u_e/\partial z$ must be zero at this depth, at all stages during the consolidation process. The general geometrical properties given in Figure 4.15 show that a parabola OP which has a slope $\partial u_e/\partial z = (\gamma - \gamma_w)$ at the surface ($z = 0$), and zero slope at the base, will intersect the bottom of the lagoon at $u_e = 0.5(\gamma - \gamma_w)d$, where d is the depth of the consolidating layer (Figure 4.32). This means that only the lower part of the initial isochrones can be parabolic, and that the slope of the isochrone at the surface remains unchanged until the isochrone OP applies. Until this time, the isochrones may be characterized by the

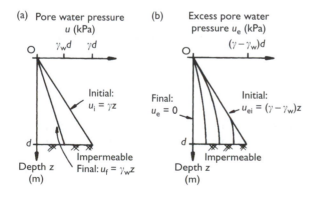

Figure 4.31   (a) Initial and final distributions of pore water pressure and (b) isochrones of excess pore water pressure: self-weight consolidation with drainage through the upper surface only.

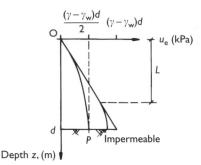

Figure 4.32 Characterization of initial isochrones during self-weight consolidation.

depth L below the surface at which the hydraulic gradient begins to change (i.e. the transition between the straight and parabolic sections of the isochrone: Figure 4.32).

The flowrate of water out of the layer is given by Darcy's Law, $q = Aki$:

$$q = -\partial V/\partial t = Aki_0 \tag{4.29}$$

where i_0 is the hydraulic gradient at the surface, $i_0 = (1/\gamma_w) \cdot (\gamma - \gamma_w)$ initially, A is the surface area, and $-\partial V/\partial t$ is the rate of change of volume (as in section 4.7.3, the negative sign denotes that the volume of the soil is decreasing).

As the surface settlement is uniform, the rate of reduction of sample volume is equal to $A\times$ the rate of settlement $\partial\rho/\partial t$. Thus

$$\partial\rho/\partial t = (k/\gamma_w)(\gamma - \gamma_w) \tag{4.30}$$

Rearranging equation (4.30) and integrating between limits of ($\rho = 0$ at $t = 0$) and ($\rho = \rho$ at $t = t$),

$$\rho - k\{(\gamma - \gamma_w)/\gamma_w\}t \tag{4.31}$$

Equation (4.31) applies until the isochrone OP (Figure 4.32) is reached, at which time the settlement is $1/E_0'\times$ the area between the isochrone at $t=0$ and the isochrone OP

$$\rho = 1/E_0' \times \{[d^2(\gamma - \gamma_w)/2] - [(2/3)d^2(\gamma - \gamma_w)/2]\}$$
$$= d^2(\gamma - \gamma_w)/6E_0' \tag{4.32}$$

As the rate of settlement until this point is uniform, the time at which the isochrone OP is reached may be calculated by dividing equation (4.32) by Equation (4.31):

$$t = d^2\gamma_w/6kE_0' = d^2/6c_v \tag{4.33}$$

After this time, the slope of the isochrone at the surface of the soil layer begins to decrease. The second set of isochrones may be characterized by the excess pore water pressure u_{ed} remaining at the base of the lagoon (Figure 4.33).

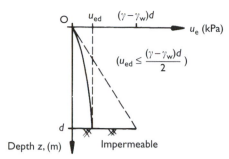

Figure 4.33 Characterization of second set of isochrones during self-weight consolidation.

The slope of the isochrone at $z=0$ is $2u_{ed}/d$, so the rate of settlement is

$$\partial\rho/\partial t = ki_0 = (k/\gamma_w)(2u_{ed}/d) \tag{4.34}$$

The settlement which has occurred up until this time is given by $1/E_0' \times$ the area between the isochrone at $t = 0$ and the current isochrone,

$$\rho = (1/E_0') \times \{d^2(\gamma - \gamma_w)/2\} - \{(2/3)du_{ed}\} \tag{4.35}$$

Differentiating equation (4.35) and setting the result equal to equation (4.34),

$$\partial\rho/\partial t = (k/\gamma_w)(2u_{ed}/d) = -(2/3E_0')d\partial u_{ed}/\partial t$$

Rearranging and integrating between limits of ($u_{ed} = (\gamma - \gamma_w)d/2$ at $t = d^2\gamma_w/6kE_0'$) and ($u_{ed} = u_{ed}$ at $t = t$), we have

$$\int_{(\gamma-\gamma_w)d/2}^{u_{ed}} \frac{\partial u_{ed}}{u_{ed}} = -\int_{d^2\gamma_w/6kE_0'}^{t} \frac{3kE_0'}{\gamma_w d^2}\partial t$$

or

$$u_{ed} = \{d(\gamma - \gamma_w)/2\}\ \exp\{(1/2) - 3c_v t/d^2\} \tag{4.36}$$

where $c_v = kE_0'/\gamma_w$ as before.

Substituting equation (4.36) into equation (4.35)

$$\rho = \{d^2(\gamma - \gamma_w)/E_0'\}\{(1/2) - (1/3)\exp[(1/2) - 3c_v t/d^2]\} \tag{4.37}$$

which applies for $t \geq d^2/6c_v$.

Noting that the ultimate settlement ρ_{ult} is $(1/E_0') \times \{d^2(\gamma - \gamma_w)/2\}$, equations (4.31) and (4.37) may be written in terms of the non-dimensional quantities $R = \rho/\rho_{ult}$ and $T = c_v t/d^2$,

$$R = 2T \qquad\qquad \text{for } 0 \leq T \leq 1/6 \tag{4.38}$$

$$R = 1 - \tfrac{2}{3}\exp\left(\tfrac{1}{2} - 3T\right) \quad \text{for } T \geq 1/6 \tag{4.39}$$

Equations (4.38) and (4.39) are plotted in Figure 4.34. Although in this example, consolidation was due to the self weight of the soil deposited as an hydraulic fill, this is not a necessary assumption for the solution presented. Figure 4.34 applies,

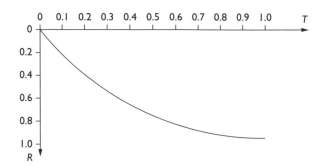

Figure 4.34 Non-dimensional settlement as a function of time factor for consolidation during a change in excess pore water pressure which increases with depth, with one-way drainage through the upper surface only.

therefore, to all cases of one-dimensional consolidation due to a change in excess pore water pressure which increases linearly with depth, with one-way drainage through the upper surface only.

4.8   One-dimensional consolidation: exact solutions

As mentioned in the introduction to this chapter, the 'exact' solution of the one-dimensional consolidation problem is somewhat mathematical, and might very reasonably be omitted from a first course in soil mechanics. In a wider context, the classical consolidation theory presented in this section is important for three reasons:

- It forms the basis of the dimensionless plots of proportional settlement R against time factor T commonly used to predict rates of consolidation in practice. (These are given in Figure 4.37.)
- It provides the justification for the approximate solutions based on parabolic isochrones, presented in sections 4.5 and 4.7.
- In its generalized three-dimensional form (Biot, 1941), it is used extensively in the numerical analysis of geotechnical engineering problems.

This section is included primarily for reference purposes. Unless you are particularly adept at mathematical analysis, it recommended that you do not attempt to work through it in detail, especially if you are new to the subject of soil mechanics.

4.8.1   Derivation of differential equation governing one-dimensional consolidation

In this section, we shall assume that consolidation is taking place with vertical settlements and vertical flow.

Consider a thin layer of saturated soil of thickness δz at a depth z below the soil surface, during a time increment δt at a time t after the start of consolidation (Figure 4.35).

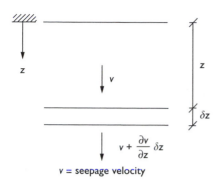

Figure 4.35 Geometry of one-dimensional consolidation for derivation of governing differential equation.

The reduction in thickness of the layer $\delta\rho$ caused by an increase in effective stress $\delta\sigma'_v$ is

$$\delta\rho = (\delta\sigma'_v\delta z)/E'_0 \tag{4.40}$$

where $\delta\sigma'_v = (\partial\sigma'_v/\partial t)\delta t$, and $\partial\sigma'_v/\partial t$ is the rate of change of effective stress with time. For a unit area perpendicular to the direction of flow ($A = 1$), $\delta\rho$ is also the reduction in volume, and the rate of settlement $\partial\rho/\partial t$ is equal to the rate of reduction of volume or the net outward flowrate of pore water, $q_{out,net}$.

The flowrate into the layer per unit area is equal to the seepage velocity v ($= q/A$). The change in seepage velocity v across the layer is $(\partial v/\partial z)\delta z$, so that the net rate of outflow from the layer per unit area is given by

$$q_{out,net} = q_{out} - q_{in} = [v + (\partial v/\partial z)\delta z] - v$$

$$= (\partial v/\partial z)\delta z \tag{4.41}$$

Dividing both sides of equation (4.40) by δt and setting the result equal to equation (4.41), and noting that in the limit $\delta t \to \partial t$, $\delta\rho \to \partial\rho$ and $\delta\sigma'_v \to \partial\sigma'_v$,

$$q_{out,net} = (\partial\rho/\partial t) = (\partial\sigma'_v/\partial t)\delta z/E'_0 = (\partial v/\partial z)\delta z$$

Hence

$$\partial\sigma'_v/\partial t = E'_0\partial v/\partial z \tag{4.42}$$

Now at constant total stress the increase in effective stress $\delta\sigma'_v$ is equal to the reduction in excess pore water pressure δu_e, so that

$$\partial\sigma'_v/\partial t = -\partial u_e/\partial t$$

and using Darcy's Law

$$v = ki$$

where

$$i = -(1/\gamma_w)\partial u_e/\partial z$$

(the negative sign is required, $i = -\partial h_e/\partial z$, because v is defined as positive in the z positive direction in Figure 4.35) so that

$$v = (k/\gamma_w)\partial\sigma'_v/\partial z \tag{4.43}$$

Substituting equation (4.43) into equation (4.42) to eliminate v,

$$\partial\sigma'_v/\partial t = E'_0\partial\{(k/\gamma_w)\partial\sigma'_v/\partial z\}/\partial z = (kE'_0/\gamma_w)\partial^2\sigma'_v/\partial z^2$$

or

$$\partial u_e/\partial t = c_v\partial^2 u_e/\partial z^2 \tag{4.44}$$

where c_v is the consolidation coefficient, kE'_0/γ_w. This is the differential equation governing one-dimensional consolidation and the dissipation of excess pore water pressures.

4.8.2 General solution to the consolidation equation

Equation (4.44) can be solved if it is assumed that the excess pore water pressure u_e can be expressed as a separable function of the two variables z and t, that is

$$u_e = f(z)g(t) \tag{4.45}$$

where $f(z)$ is a function of z and not t, and $g(t)$ is a function of t and not z. Substitution of equation (4.45) into equation (4.44) yields

$$f(\partial g/\partial t) = c_v g\partial^2 f/\partial z^2$$

or

$$(1/f)\partial^2 f/\partial z^2 = (1/[c_v g])\,(\partial g/\partial t) \tag{4.46}$$

The left-hand side is now a function of f alone, and the right-hand side is a function of g alone. If equation (4.46) is valid for all values of z and t, then each side must be a constant. Let the constant be equal to $-\lambda^2$. Then

$$(1/f)\partial^2 f/\partial z^2 = -\lambda^2 = [1/(c_v g)](\partial g/\partial t)$$

or

$$\partial^2 f/\partial z^2 + \lambda^2 f = 0 \tag{4.47a}$$

and

$$\partial g/\partial t + \lambda^2 c_v g = 0 \tag{4.47b}$$

The solutions to equations (4.47) are

$$f = A\sin(\lambda z) + B\cos(\lambda z)$$

and

$$g = \exp\{-c_v\lambda^2 t\}$$

so that

$$u_e = [A\sin(\lambda z) + B\cos(\lambda z)]\exp\{-c_v\lambda^2 t\}$$

Equation (4.44) is also satisfied by the solution

$$u_e = Cz + D$$

where C and D are constants. The general solution to the one-dimensional consolidation equation is therefore

$$u_e = [A \sin(\lambda z) + B \cos(\lambda z)] \exp\{-c_v \lambda^2 t\} + Cz + D \tag{4.48}$$

4.8.3 Solutions for particular boundary conditions using Fourier series

The values of the constants A, B, C, D and λ in equation (4.48) depend on the drainage boundary conditions and the initial and final distributions of excess pore water pressure. We will now consider the three cases of one-dimensional vertical consolidation examined using parabolic isochrones in sections 4.5 (and 4.7.1), 4.7.2 and 4.7.4.

(a) The oedometer test sample: consolidation during an increment in effective stress that is uniform with depth and of magnitude $\Delta\sigma_v$, with one-way or two-way drainage (cf. section 4.5, the oedometer test, and section 4.7.1).

Considering again the top half of a soil layer of thickness $2d$ with two-way drainage from the top and the bottom, or a soil layer of thickness d with one-way drainage through the upper surface only, the succession of isochrones is as indicated in Figure 4.12. The boundary conditions which apply to the isochrones at all times are

$$u_e = 0 \text{ at } z = 0 \quad \text{and} \quad \partial u_e/\partial z = 0 \text{ at } z = d$$

If the first of these conditions is to apply for all values of t, then B and D in equation (4.48) must be zero. Differentiating equation (4.48) and setting $B = D = 0$,

$$\partial u_e/\partial z = A\lambda \cos \lambda z \exp\{-c_v \lambda^2 t\} + Cz$$

If the second boundary condition is to apply for all values of t, C must be zero and (assuming A is non-zero) $\cos \lambda d = 0$. This implies that $\lambda d = \pi/2, 3\pi/2, 5\pi/2$ etc., or in general $\lambda = n\pi/2d$ where n is odd. Thus, taking into account all possible solutions,

$$u_e = \sum_{\substack{n=1 \\ n \text{ odd}}}^{\infty} \left\{ A_n \sin\left(\frac{n\pi z}{2d}\right) \exp\left[\frac{-c_v n^2 \pi^2 t}{4d^2}\right] \right\}$$

$$= \sum_{\substack{n=1 \\ n \text{ odd}}}^{\infty} \left\{ A_n \sin\left(\frac{n\pi z}{2d}\right) \exp\left[\frac{-n^2 \pi^2 T}{4}\right] \right\} \tag{4.49}$$

where $T = c_v t/d^2$. The values of the constants A_n may be found by considering the distribution of excess pore water pressure at $t = 0$,

$$u_{e,t=0} = \Delta\sigma_v = \sum_{\substack{n=1 \\ n \text{ odd}}}^{\infty} \left\{ A_n \sin\left(\frac{n\pi z}{2d}\right) \right\} \tag{4.50}$$

Any function $f(z)$ may be represented over its range $0 \le z \le l$ by a summation of sine waves known as a **Fourier series**,

$$f(z) = \sum_{n=1}^{\infty} \left\{ A_n \sin \left[\frac{n\pi z}{l} \right] \right\},$$

where

$$A_n = \frac{2}{l} \int_0^l f(z) \sin \left[\frac{n\pi z}{l} \right] dz \qquad (4.51)$$

Equation (4.51) is the same as equation (4.50) with $f(z) = \Delta\sigma_v$ and $l = 2d$, so that

$$
\begin{aligned}
A_n &= \left(\frac{2}{2d} \right) \int_0^{2d} \left(\Delta\sigma_v \sin \left[\frac{n\pi z}{2d} \right] \right) dz \\
&= \left(\frac{\Delta\sigma_v}{d} \right) \left(\frac{2d}{n\pi} \right) \left[-\cos \left(\frac{n\pi z}{2d} \right) \right]_0^{2d} \\
&= \left(\frac{2\Delta\sigma_v}{n\pi} \right) (1 - \cos n\pi)
\end{aligned}
$$

Substituting this into equation (4.49), and noting that because $(1 - \cos n\pi) = 2$ when n is odd and zero when n is even there is now no need to specify that the summation is carried out for odd values of n only,

$$u_e = \sum_{n=1}^{\infty} \left\{ \left(\frac{2\Delta\sigma_v}{n\pi} \right) (1 - \cos n\pi) \sin \left(\frac{n\pi z}{2d} \right) \exp \left(\frac{-n^2 \pi^2 T}{4} \right) \right\} \qquad (4.52)$$

For a given time factor $T(= c_v t / d^2)$, equation (4.52) may be used to calculate the excess pore water pressure remaining at any depth $z(0 \le z \le d)$. In this way, isochrones showing the variation of excess pore water pressure with depth may also be produced. Unfortunately, the summation must be carried out to calculate each individual value of pore water pressure. Although the terms in the summation decay quite rapidly with n (because of the term in $\exp(-n^2)$), this is a time-consuming process to undertake by hand.

The calculation of the surface settlement ρ is less tedious. From equation (4.10b) (for the top half of a sample of thickness $2d$ with two-way drainage, or for the whole of a sample of thickness d with one-way drainage),

$$\rho = \frac{1}{E_0'} \int_0^d (\Delta\sigma_v' - u_e) \partial z \qquad (4.10b)$$

Substituting equation (4.52) into equation (4.10b) and carrying out the appropriate integration gives

$$\rho = \left(\frac{d\Delta\sigma_{\mathrm{v}}}{E_0'}\right)\left[1 - \sum_{n=1}^{\infty}\left\{\left(\frac{4}{n^2\pi^2}\right)(1 - \cos n\pi)\left(1 - \cos\frac{n\pi}{2}\right)\exp\left(\frac{-n^2\pi^2 T}{4}\right)\right\}\right]$$

(4.53)

Recalling that the ultimate settlement $\rho_{\mathrm{ult}} = d\Delta\sigma_{\mathrm{v}}/E_0'$ and evaluating the significant terms of the summation,

$$R = \rho/\rho_{\mathrm{ult}} = 1 - (8/\pi^2)[\exp\{-\pi^2 T/4\} + (1/9)\exp\{-9\pi^2 T/4\}$$

$$+ (1/25)\exp\{-25\pi^2 T/4\}]$$

(4.54)

(b) Consolidation during an increment in effective stress that increases linearly with depth (triangular distribution, with one-way drainage towards the thin end of the triangle: cf. section 4.7.4, self-weight consolidation/hydraulic fill).

The boundary conditions in this case are exactly the same as in (a) above, so that the solution presented above applies as far as equation (4.49).

The difference between the two situations is that the initial excess pore water pressure distribution used to evaluate the constants A_n is now triangular. This means that the isochrones during the first part of the consolidation process will be different in shape (Figure 4.31). Also, when it comes to evaluate the coefficients A_n, it is necessary to define the initial pore water pressure distribution – $f(z)$ in equation (4.51) – over a depth $2d$, even though the depth of the clay layer in this case is only d. As the slope of any isochrone $\partial u_{\mathrm{e}}/\partial z$ is zero at $z = d$, $f(z)$ must be symmetrical about $z = d$. In the case of self-weight consolidation, the initial excess pore water pressure distribution is for, $0 \leq z \leq d$

$$f(z) = u_{\mathrm{e},t=0} = (\gamma - \gamma_{\mathrm{w}})z$$

so that to meet the requirement of symmetry,

$$f(z) = u_{\mathrm{e},t=0} = (\gamma - \gamma_{\mathrm{w}})z \qquad \text{for } 0 \leq z \leq d$$
$$f(z) = u_{\mathrm{e},t=0} = (\gamma - \gamma_{\mathrm{w}})(2d - z) \quad \text{for } d \leq z \leq 2d$$

$$A_n = \left(\frac{2}{2d}\right)\int_0^{2d} f(z)\sin\left(\frac{n\pi z}{2d}\right)dz$$

$$= \left(\frac{1}{d}\right)\int_d^{2d}(\gamma - \gamma_{\mathrm{w}})(2d - z)\sin\left(\frac{n\pi z}{2d}\right)dz$$

$$+ \left(\frac{1}{d}\right)\int_0^{d}(\gamma - \gamma_{\mathrm{w}})z\sin\left(\frac{n\pi z}{2d}\right)dz$$

which must be integrated by parts to give

$$A_n = [8(\gamma - \gamma_{\mathrm{w}})d/(n^2\pi^2)]\sin(n\pi/2)$$

(4.55)

so that

$$u_e = \sum_{n=1}^{\infty} \left\{ \left(\frac{8(\gamma - \gamma_w)d}{n^2 \pi^2} \right) \sin \left(\frac{n\pi}{2} \right) \sin \left(\frac{n\pi z}{2d} \right) \exp \left(\frac{-n^2 \pi^2 T}{4} \right) \right\} \qquad (4.56)$$

The surface settlement at time t is

$$\rho = \frac{1}{E_0'} \int_0^d [(\gamma - \gamma_w)z - u_e]\partial z \qquad (4.57)$$

Substituting equation (4.56) into equation (4.57), integrating, and dividing both sides by the ultimate settlement $\rho_{ult} = (\gamma - \gamma_w)d^2/2E_0'$ we obtain

$$R = 1 - \left(\frac{32}{\pi^3} \right) \sum_{n=1}^{\infty} \left\{ \left(\frac{1}{n^3} \right) \left(\sin \frac{n\pi}{2} \right) \exp \left(\frac{-n^2 \pi^2 T}{4} \right) \right\} \qquad (4.58)$$

Evaluating the significant terms,

$$R = 1 - (32/\pi^3)\{\exp(-\pi^2 T/4) - (1/27) \exp(-9\pi^2 T/4)$$
$$+ (1/125) \exp(-25\pi^2 T/4)\} \qquad (4.59)$$

Equation (4.59) applies to all cases of one-dimensional vertical consolidation by one-way drainage to the surface, during a change in effective stress that increases linearly with depth.

(c) Consolidation due to underdrainage of a stratum with initially hydrostatic pore water pressures, in response to an increment of effective stress that increases with depth (triangular distribution, with apparently one-way drainage through the base of the stratum towards the thick end of the triangle: cf. section 4.7.2, underdrainage of a compressible stratum).

The boundary conditions in this case are different from those in (a) and (b). A further slight complication is that the excess pore water pressures decay from zero to negative values in the steady state, as indicated in Figure 4.25. This is due to the definition of excess pore water pressure used in this book, which is that excess pore water pressure drives seepage flow in general, rather than consolidation in particular. In the situations described in sections 4.8.3(a) (and 4.5 and 4.7.1), 4.8.3(b) (and 4.7.4) and 4.8.3(d), the definition of excess pore water pressure used makes no difference, because the steady state pore water pressures are hydrostatic, so that zero seepage flow corresponds to the end of consolidation. In this case, the definition of excess pore water pressure, once chosen, must be followed with care if mistakes in the analysis are to be avoided.

Taking the thickness of the soil layer as $2d$, the boundary conditions which apply at all times are

$$u_e = 0 \text{ at } z = 0, \quad \text{and} \quad u_e = -2\gamma_w d \text{ at } z = 2d$$

If the first of these conditions is to apply for all values of t, then B and D in equation (4.46) must be zero. The second boundary condition then requires that

$$u_{e,z=2d} = -2\gamma_w d = A \sin 2\lambda d \exp\{-c_v\lambda^2 t\} + 2Cd$$

If the second boundary condition is to apply for all values of t, C must be equal to $-\gamma_w$ and (assuming A and λ are non-zero) $\sin 2\lambda d = 0$. This implies that $2\lambda d = \pi, 2\pi, 3\pi$ etc., or in general $\lambda = n\pi/2d$. Thus

$$u_e = \sum_{n=1}^{\infty} \left\{ A_n \sin\left(\frac{n\pi z}{2d}\right) \exp\left(\frac{-c_v n^2 \pi^2 t}{4d^2}\right) \right\} - \gamma_w z$$

$$= \sum_{n=1}^{\infty} \left\{ A_n \sin\left(\frac{n\pi z}{2d}\right) \exp\left(\frac{-n^2 \pi^2 T}{4}\right) \right\} - \gamma_w z \tag{4.60}$$

where $T = c_v t/d^2$. The values of the constants A_n may again be found by considering the distribution of excess pore water pressures at $t = 0$,

$$u_{e,t=0} = 0 = \sum_{n=1}^{\infty} \left\{ A_n \sin\left(\frac{n\pi z}{2d}\right) \exp\left(\frac{-n^2 \pi^2 T}{4}\right) \right\} - \gamma_w z \tag{4.61}$$

so that

$$f(z) = (u_{e,t=0} + \gamma_w z)$$

$$= \sum_{n=1}^{\infty} \left\{ A_n \sin\left(\frac{n\pi z}{2d}\right) \exp\left(\frac{-n^2 \pi^2 T}{4}\right) \right\}$$

From equation (4.24),

$$u_e = u - \gamma_w z$$

so that $f(z)$ is the actual pore water pressure, $u = u_e + \gamma_w z$, at $t = 0$. Evaluation of the Fourier coefficients A_n is carried out as before,

$$A_n = \left(\frac{2}{2d}\right) \int_0^{2d} \left(\gamma_w z \sin\left[\frac{n\pi z}{2d}\right]\right) dz$$

Integration by parts gives

$$A_n = (-2\gamma_w z/n\pi)[\cos(n\pi z/2d)]$$
$$+ (-4d\gamma_w/n^2\pi^2)[\sin(n\pi z/2d)]$$
$$= (-4d\gamma_w/n\pi)\cos n\pi \tag{4.62}$$

Substituting equation (4.62) into equation (4.61),

$$u_e = \sum_{n=1}^{\infty} \left\{ \left(\frac{-4d\gamma_w}{n\pi}\right)(\cos n\pi)\sin\left(\frac{n\pi z}{2d}\right) \times \exp\left(\frac{-n^2\pi^2 T}{4}\right) \right\} - \gamma_w z \tag{4.63}$$

The surface settlement ρ is given by

$$\rho = \frac{1}{E'_0} \int_0^d \Delta\sigma'_v \partial z \tag{4.64}$$

In this case, the increase in effective stress $\Delta\sigma'_v$ is equal to the decrease in pore water pressure, which is equal to $-u_e$ (Figure 4.25). Thus

$$\rho = \frac{1}{E'_0} \int_0^d -u_e \partial z \tag{4.65}$$

Substituting equation (4.63) into equation (4.65) and integrating,

$$\rho = \left(\frac{2\gamma_w d^2}{E'_0}\right)\left[1 + \sum_{n=1}^{\infty}\left\{\left(\frac{4}{n^2\pi^2}\right)(1 - \cos n\pi)(\cos n\pi) \times \exp\left(\frac{-n^2\pi^2 T}{4}\right)\right\}\right] \tag{4.66}$$

Dividing by the ultimate settlement $\rho_{ult} = 2\gamma_w d^2/E'_0$ (remember that the layer has thickness $2d$) and evaluating the significant terms of the summation,

$$R = \rho/\rho_{ult} = 1 - (8/\pi^2)[\exp\{-\pi^2 T/4\}$$
$$+ (1/9)\exp\{-9\pi^2 T/4\} + (1/25)\exp\{-25\pi^2 T/4\}] \tag{4.67}$$

which is exactly the same as equation (4.54).

Because we have calculated T using a maximum drainage path length of half the sample thickness, this demonstrates that the underdrained layer does indeed behave as if the drainage were two-way. (In section 4.7.2 it was stated that the important point was that two-way drainage could occur if it needed to: it was just that all of the pore water drained to the bottom of the stratum under the particular conditions of consolidation due to underpumping.) It also shows that the relationship between non-dimensional surface settlement and time factor for a layer with two-way drainage is the same whether the change in excess pore water pressure driving consolidation is uniform, or varies linearly with depth. It therefore follows that Figure 4.18 and equation (4.54) may be applied to the consolidation of a layer with two-way drainage following any trapezoidal change in excess pore water pressure. This is because a trapezoidal excess pore water pressure distribution may be derived by the superposition of a distribution which varies with depth onto a distribution which is uniform.

(d) Consolidation of a sample with one-way upward drainage during an increment in effective stress that decreases with depth (triangular distribution, with upward drainage towards the thick end of the triangle).

There is one further case of one-dimensional consolidation which it is worth mentioning briefly. It will be seen in Chapter 6 that the vertical stresses imposed on the soil by a surface load tend to decrease with depth. A load applied near the surface of a layer of clay soil, for example by an earth embankment or the foundation of a building, would therefore be expected to give rise to an initial distribution of excess pore water pressure which decreases with depth. If the compressible layer is underlain by an impermeable stratum, the excess pore water pressures must dissipate by upward

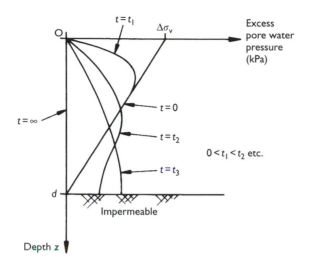

Figure 4.36 Isochrones of excess pore water pressure: initially triangular excess pore pressure distribution that decreases linearly with depth, with drainage towards the upper surface only.

drainage alone, at least near the centre of the loaded area. (Away from the centre, lateral drainage will probably be more significant, because the horizontal permeability is likely to be greater than the vertical permeability. Also, the horizontal drainage path lengths may be shorter.)

Assuming vertical flow and one-dimensional vertical consolidation, initial and final distributions of excess pore water pressure for this case are shown in Figure 4.36, together with some intermediate isochrones. Drainage is towards the thick end of the triangle which represents the initial excess pore water pressure distribution at the start of consolidation. This is different from the case of hydraulic fill (sections 4.7.4, 4.8.3(b) and Figure 4.31), where drainage was to the thin end of the triangle. It is also different from sections 4.7.2, 4.8.3(c) and Figure 4.25, because these cases had effectively two-way drainage. The boundary condition at the thin end of the triangle in sections 4.7.2 and 4.8.3(c) was $u_e = 0$, whereas in the present case it is $\partial u_e / \partial z = 0$.

It may be seen from Figure 4.36 that during the initial stages of consolidation there is some downward flow, which results in a transient increase in excess pore water near the impermeable base of the stratum. This gradually decays, as eventually the pore water must drain through the top of the layer. The shapes of the intermediate isochrones in this case are such that the problem is not really amenable to the parabolic isochrones approximation. In the solution of equation (4.44) using the Fourier series approach, the same boundary conditions apply as in section 4.8.3(b). The initial pore water pressure distribution must again be defined symmetrically over the range $0 \leq z \leq 2d$:

$$f(z) = u_{e,t=0} = \Delta\sigma_v(1 - [z/d]) \quad \text{for } 0 \leq z \leq d$$
$$f(z) = u_{e,t=0} = \Delta\sigma_v([z/d] - 1) \quad \text{for } d \leq z \leq 2d$$

Setting $l = 2d$ in equation (4.51), and integrating by parts to evaluate the constants A_n; using equation (4.64) with $\Delta\sigma_v' = \Delta\sigma_v(1 - [z/d]) - u_e$ and integrating between limits of

$0 \leq z \leq d$ to evaluate the surface settlement; and noting that $\rho_{ult} = d\Delta\sigma_v/2E'_0$ where $\Delta\sigma_v$ is the eventual increase in vertical effective stress at the surface of the clay layer,

$$R = \frac{\rho}{\rho_{ult}} = 1 - \sum_{n=1}^{\infty} \left\{ \left(\frac{8}{n^3\pi^3}\right) \left[n\pi(1-\cos n\pi) - 4\sin\left(\frac{n\pi}{2}\right)\right] \exp\left(\frac{-n^2\pi^2 T}{4}\right) \right\}$$

(4.68)

Evaluating the significant terms,

$$\begin{aligned} R = \rho/\rho_{ult} = 1 & - (16/\pi^3)[(\pi - 2)\exp(-\pi^2 T/4) \\ & + (1/27)(3\pi - 2)\exp(-9\pi^2 T/4) \\ & + (1/125)(5\pi - 2)\exp(-25\pi^2 T/4)] \end{aligned}$$

(4.69)

If the consolidating layer were underlain by a permeable stratum, equation (4.54) and Figure 4.18 would apply, with the maximum drainage path length equal to half the thickness of the layer.

Equations (4.54), (4.59) and (4.69), which represent the exact mathematical solutions to the one-dimensional consolidation equation (equation (4.44)) for the various situations considered, are shown graphically in Figure 4.37. Comparison with Figures 4.18 and 4.34 shows that, for an increase in effective stress which is uniform with depth and the case of hydraulic fill, the results obtained using the parabolic isochrones approximation are practically identical.

The solutions (using both the parabolic isochrones approximation and the Fourier series approach) to most of the problems of one-dimensional consolidation presented in sections 4.5–4.8 (and many others) are given by Terzaghi and Fröhlich (1936). This, however, is a text for the enthusiast: it is difficult to obtain and is written in German.

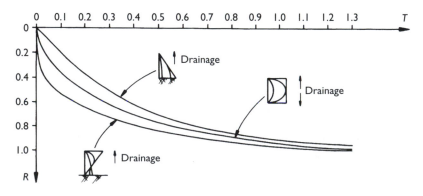

Figure 4.37 Non-dimensional settlement R as a function of time factor T for consolidation in response to different patterns of effective stress increment and different boundary conditions (exact solutions).

4.9   Radial drainage

In section 4.7.3, the problem of vertical compression due to plane horizontal flow to a pumped slot was examined. In some cases, vertical compression will occur due to horizontal flow which is not plane but radial, for example to an individual pumped well. The mathematics for radial flow is much more complicated than for plane flow, because the fact that the flow area decreases with radius must be taken into account. A solution for the flowrate to a single pumped well withdrawing water from a soil stratum which consolidates was presented by Jacob and Lohman (1952), but their solution is not in a suitable form for estimating how long it will take for a given drawdown to be achieved at a certain distance from the well.

In the case of a construction dewatering system analysed as a single equivalent well (as in section 3.17), the time taken to achieve the required drawdown over the area of the site is likely to be a more important design consideration than the flowrate. The consolidation equation was solved numerically by Rao (1973), whose isochrones of drawdown (i.e. reduction in head) h as a function of radial distance r from the edge of the excavation or equivalent well (normalized with respect to the drawdown at the well h_0 and the radius of the well r_w respectively) may be used for this purpose (Figure 4.38). The time factor used in radial flow is $T_r = c_{hv}t/r_w^2$, where $c_{hv} = k_h E_0'/\gamma_w$ is the coefficient of consolidation with vertical compression and horizontal flow.

Where a stratum of soil of low permeability is to be loaded, for example by the construction of an embankment, it is quite common to install a large number of vertical drains in a grid pattern, so that vertical consolidation takes place primarily by the horizontal flow of pore water to the drains. This results in a significant reduction in the consolidation time, both because the maximum drainage path length is reduced (to approximately half the distance between the vertical drains), and because the horizontal permeability is usually appreciably higher than that in the vertical direction. Originally, the vertical drains were boreholes backfilled with sand, but prefabricated porous plastic band drains, which are easier and cheaper to install, are now more common.

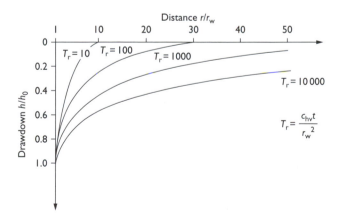

Figure 4.38 Dimensionless isochrones for radial flow to a single well. (Redrawn with permission from Rao, 1973.)

Isochrones of the form shown in Figure 4.38 will apply initially, but only until the patterns of excess pore pressure reduction due to adjacent vertical drains begin to interfere. In this application, it is not usually necessary to know the pore water pressure distribution between individual drains. What is important is the time taken for significant consolidation (say 90%) to occur. This is expressed by the **proportional average settlement** $R = \rho_{av}/\rho_{av,ult}$, which is really the **proportional volume change** $\Delta V/\Delta V_{ult}$ or the **degree of consolidation**. (The average settlement has to be used because the settlement at the surface during consolidation is not uniform. The excess pore water pressure remaining increases, and hence the increase in effective stress decreases with radial distance from each drain.)

Solutions for the degree of consolidation as a function of time factor T_r for radial horizontal flow from a cylinder of clay of radius r_0 to a central vertical drain of radius r_w are given in Figure 4.39, for various values of radius ratio $n = r_0/r_w$. (This solution is due to Barron (1948), who used a time factor $T_D = c_{hv}t/D^2$, where $D = 2r_0$ is the diameter of the clay cylinder. For this reason, the values on the horizontal axis of Barron's original graph are a factor of 4 smaller than in Figure 4.39.)

In the analysis which led to Figure 4.39, it was assumed that the initial excess pore water pressure is uniform with depth; that $u_e = 0$ at $r = r_w$ for $t > 0$; that no flow occurs across the external boundary at $r = r_0$, (i.e. $\partial u_e/\partial r = 0$ at $r = r_0$); and that the soil surface is free to settle non-uniformly with time (cf. section 4.7.3). Barron states (but does not prove) that where excess pore water pressures are able to dissipate by both vertical and radial flow, the excess pore water pressure u_e remaining at any point within the soil mass at any time is given by

$$u_e = (u_{e,r}\, u_{e,z})/u_{e0} \qquad\qquad (4.70)$$

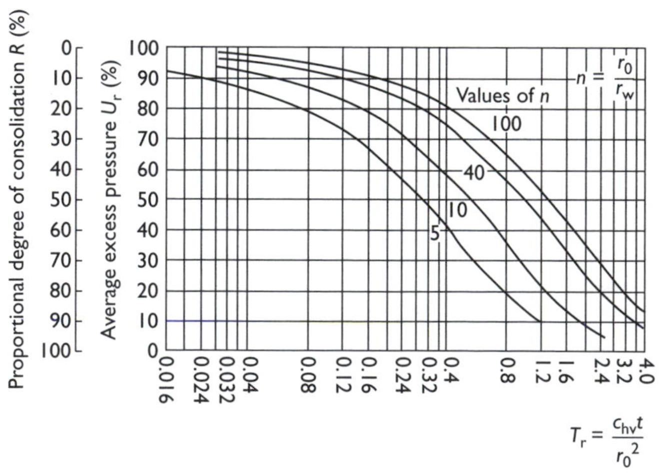

Figure 4.39 Degree of consolidation as a function of time factor T_r for radial flow to vertical drains draining a circular cylinder of soil of radius r_0. (Redrawn with permission from R.A. Barron, Consolidation of fine-grained soil by drain wells, Transactions of the ASCE, 1948.)

where $u_{e,r}$ is the excess pore water pressure which would remain in the case of radial drainage alone, $u_{e,z}$ is the excess pore water pressure which would remain in the case of vertical drainage alone, and u_{e0} is the initial excess pore water pressure. The same applies to the average excess pore water pressure U,

$$U = (U_r \cdot U_z)/U_0 \qquad (4.71)$$

As the initial excess pore water pressure distribution is uniform with depth, the average settlement ρ_{av} is given by

$$\rho_{av} = (1/E'_0)d(\Delta\sigma_v - U)$$

and the ultimate settlement ρ_{ult} by

$$\rho_{av,ult} = (1/E'_0)d\Delta\sigma_v$$

where $\Delta\sigma_v$ is the applied surface load and d is the depth of the clay layer. Hence, the degree of consolidation $R = \rho_{av}/\rho_{av,ult}$ is equal to $1 - U/\Delta\sigma_v$. As $\Delta\sigma_v = U_0$,

$$R = 1 - U/U_0, \quad \text{or} \quad U/U_0 = 1 - R \qquad (4.72)$$

Similarly, $U_z/U_0 = 1-R_z$, and $U_r/U_0 = 1-R_r$, where R_z is the degree of consolidation due to vertical flow alone and R_r is the degree of consolidation due to radial flow alone. Dividing both sides of equation (4.71) by U_0 and rewriting it in terms of R, R_r and R_z,

$$(1 - R) = (1 - R_r)(1 - R_z) \qquad (4.73)$$

where R_r may be obtained using Figure 4.39 and R_z using Figure 4.18 or Figure 4.37 (equation 4.54).

Vertical drains are usually installed in a square grid, in which case the cylinder of clay they drain is square on plan, or in a triangular grid, in which case the cylinder is hexagonal on plan. In either case, the radius of the equivalent circular cylinder r_0 is calculated on the assumption that its plan area is the same as that of the real cylinder. Thus for a square grid, $r_0 = S\sqrt{(1/\pi)} = 0.564S$, and for a triangular grid $r_0 = S\sqrt{(6/(4\pi \tan 60°))} = 0.525S$, where S is the distance between adjacent drains.

Barron (1948) also gives solutions for the case where the load is rigid so that uniform strains are imposed. This analysis is complicated by the fact that it can no longer be assumed that all volume change is due to vertical settlement, and radial strains must be permitted to develop within the soil mass (see also Al-Tabbaa and Muir-Wood, 1991). Barron presents further analyses, taking account of the effects of smear during well drain installation. Smear might result in a zone of reduced permeability in the immediate vicinity of the drain, and a consequent slowing of consolidation.

4.10 Limitations of the simple models for the behaviour of soils in one-dimensional compression and consolidation

One of the aims of the various examples presented in this book, particularly in Chapters 3 and 4, is to show that simple ideas and models can be used successfully in the analysis of real geotechnical problems. However, the model must encapsulate the dominant aspects of field behaviour, and it is equally important that nothing significant is overlooked.

To be certain that a model is genuinely suitable for a potential real application, you must be aware of its shortcomings and inherent assumptions. In this section, some of the main limitations of the models for the behaviour of clay soils in one-dimensional compression and consolidation are reiterated.

4.10.1 Model for one-dimensional compression and swelling: specific volume against ln σ'_v

It was mentioned in section 4.2 that the unload/reload path in v, ln σ'_v space is in reality a hysteresis loop (Figure 4.6), and the slope κ_0 is not constant. The one-dimensional normal compression line also displays some variation in slope at low values of σ'_v. This can be substantially eliminated by plotting ln v against ln σ'_v (Butterfield, 1979). However, the curvature of the one-dimensional normal compression line is very much less pronounced than that of the unload/reload lines, and for most practical purposes the model described herein is quite adequate.

The soil particles and the pore water are assumed to be incompressible in themselves, so that any change in volume results from a rearrangement of the soil skeleton and an accompanying change in void ratio. For most soils, this assumption is reasonable, unless stresses are high enough to cause significant particle crushing.

In practice, sample disturbance may result in a curved transition from a reload line to the one-dimensional normal compression line: this makes the identification of the preconsolidation pressure somewhat subjective.

4.10.2 One-dimensional consolidation solutions

The underlying assumptions in the analyses of one-dimensional consolidation presented in this chapter are that the physical properties of the soil (k, E'_0, c_v and the unit weight γ) remain constant while consolidation takes place; and that the compression which occurs does not significantly affect the geometry (primarily the thickness of the sample and the drainage path length). It is well known that the physical properties of the soil depend on the effective stress: thus they will change during consolidation. In many cases, however, the effect of this is not too significant, provided that the effective stress is not increased or decreased by a factor of more than approximately two. In the self-weight consolidation of a slurry or an hydraulic fill, the initial effective stresses are close to zero, and the effective stress is almost certain to increase during consolidation by a factor of considerably more than two. Furthermore, the settlements during self-weight consolidation will bring about significant changes in geometry. In these circumstances, the theories presented in this chapter must be viewed as even more than usually approximate. A rigorous mathematical treatment of the consolidation process, in which changes in geometry and soil properties are taken into account, is given by Gibson et al. (1967).

The use in calculations of a consolidation coefficient based on oedometer tests in vertical flow must be expected to underestimate consolidation times considerably in field conditions where horizontal flow is important. This is demonstrated by the probable three orders of magnitude difference between the vertical and the horizontal permeabilities of the glacial lake deposits at Conwy (section 4.7.3). The value of $c_{hv} = k_h E'_0/\gamma_w$ used to calculate the piezometer response shown in Figure 4.30 was reasonably consistent with values of k_h measured in a field pumping test and E'_0 measured in an oedometer. The value of $c_v = k_v E'_0/\gamma_w$ obtained from the oedometer test would have been a factor of 1000 or so too small, because of the difference between k_h and k_v.

Consolidation in the field may also be accelerated by macro-fabric effects (such as fissures and sand partings), which represent preferential drainage paths that cannot be reproduced in small-scale laboratory tests.

4.10.3   Horizontal stresses in one-dimensional compression and swelling

One significant limitation of a conventional oedometer is that it has no facility for the measurement of horizontal stress. This is mainly because of the additional complexity that this would entail, and the considerable difficulties associated with obtaining reliable measurements of stresses in soils. Nonetheless, the horizontal stress in a natural soil deposit due to a geological stress history of one-dimensional compression and swelling can be an important consideration in the analysis of underground structures such as *in situ* retaining walls (Chapter 10).

One-dimensional compression tests carried out in modified oedometers equipped with lateral stress transducers (Brooker and Ireland, 1965; Mayne and Kulhawy, 1982) have generally shown that, for normally consolidated soils, the horizontal effective stress σ'_h increases in proportion to the vertical effective stress σ'_v, with

$$\sigma'_h = (1 - \sin \phi') \times \sigma'_v \tag{4.74}$$

as originally suggested by Jaky (1944). (ϕ' is the effective angle of friction of the soil.)

The ratio σ'_h/σ'_v is conventionally given the symbol K, and is known as the **earth pressure coefficient**. The subscript 0 (K_0) is used to indicate the natural, undisturbed earth pressure coefficient in a soil deposit *in situ*. The earth pressure coefficient of a normally consolidated soil is sometimes denoted K_{nc}: from equation (4.74), $K_{nc} = (1 - \sin \phi')$.

On unloading, the horizontal effective stress tends to remain 'locked in' as the vertical effective stress is reduced. Thus, the earth pressure coefficient gradually increases with overconsolidation ratio (OCR) as the soil is unloaded. On the basis of experimental data, Mayne and Kulhawy (1982) suggest the relationship

$$K_{oc} = (1 - \sin \phi') \times (OCR)^{\sin \phi'} \tag{4.75}$$

in one-dimensional swelling. Equation 4.75 applies up to a limit of $K_p = (1 + \sin \phi')/(1 - \sin \phi')$, at which the full strength of the soil is mobilized (section 7.3). The subscript oc denotes that the soil is overconsolidated. In natural clay deposits, K_{oc} can be – and indeed often is – greater than one.

Key points

- During natural deposition and unloading due to the removal of overburden, beneath many foundations and in certain other circumstances, soil compresses or swells primarily in the vertical direction. This mode of deformation, in which direct strain occurs only in the direction of loading or unloading, is described as **one-dimensional**.

- The behaviour of a soil in one-dimensional loading and unloading is investigated in an **oedometer test**. Two types of data are obtained from an oedometer test, relating to (a) equilibrium states, after consolidation has ceased; and (b) the consolidation or swelling phase of each loading or unloading stage.

- For a soil which is being compressed for the first time, equilibrium states lie on a unique straight line – the **one-dimensional normal compression line** – on a graph of specific volume v against the natural logarithm of the vertical effective stress, $\ln \sigma_v'$:

$$v = v_{\lambda_0} - \lambda_0 \ln \sigma_v' \tag{4.6}$$

- On unloading and reloading, the soil is much stiffer, following a hysteresis loop in v, $\ln \sigma_v'$ space, which is usually idealized to a straight, **unload/reload line** of slope $-\kappa_0$.

- On a graph of void ratio e against $\log_{10} \sigma_v'$, the one dimensional compression line has slope $-C_c$ (where C_c is the compression index, $C_s = 2.3\lambda_0$) and an unload/reload line has slope $-C_s$ (the swelling index, $C_s = 2.3\kappa_0$).

- Deformation in first loading is mainly **plastic** or **irrecoverable**, while deformation in unloading and reloading is essentially **elastic** or **recoverable**.

- A soil on the normal compression line is described as **normally consolidated**. On an unload/reload line, the soil is described as **overconsolidated**, with an **overconsolidation ratio** n defined as

$$n = \sigma_{v(\text{maximum previous})}' / \sigma_{v(\text{current})}' \tag{4.7}$$

- The soil stiffness may be characterized by means of the **one-dimensional modulus**, E_0'. Overconsolidated soils are stiffer than normally consolidated soils over the same stress range, and E_0' increases with effective stress, unless the soil moves from a reloading line onto the normal compression line.

- The idealization of a soil as an almost elastic material requires that the changes in stress and strain are small; the stiffness modulus has been determined over an appropriate stress range and from a realistic initial stress state; and that the soil is overconsolidated (or there is no reversal in the direction of the stress path).

- A low-permeability soil does not respond immediately to an increase in total stress, because the effective stress cannot increase unless the soil skeleton compresses. For the soil skeleton to compress, water must flow from the pores. The rate at which this can occur depends on the permeability of the soil.

- As the effective stresses cannot change instantaneously, an increase in boundary stress is carried initially by an increase in pore water pressure. This sets up a non-equilibrium pore water pressure gradient, in response to which water begins to flow from the soil pores. As water flows from the pores, the pore water pressure starts to fall, the effective stresses start to increase, and the soil **consolidates**.

- The rate of consolidation of a soil increases as the permeability k and one-dimensional modulus E'_0 increase. In addition to the soil properties, consolidation times in the field depend on the problem geometry, increasing in proportion to the square of the drainage path length d.

Questions

Analysis and interpretation of one-dimensional compression test data

4.1 (a) What factors govern the relevance to a given design situation of the parameters obtained from an oedometer test?

(b) Data from an oedometer test are given below. Show that the specific volume v is related to the sample height h by the expression $(v/h) = $ constant. Plot a graph of specific volume v against the natural logarithm of the vertical effective stress, $\ln \sigma'_v$, and explain its shape. Calculate the values of κ_0 and λ_0.

σ'_v (kPa)	25	50	100	200	100	50
Equilibrium sample height h (mm) (after consolidation has ceased)	19.86	19.56	19.27	18.48	18.79	19.08

Water content of sample at the end of the test ($\sigma'_v = 50\,\text{kPa}$, $h = 19.08\,\text{mm}$): 20.88%.

Particle relative density (grain specific gravity) $G_s = 2.75$.

(c) Figure 4.40 shows the ground conditions at the site of a proposed new office building. The office building will have a raft foundation, the effect of which will be to increase the vertical effective stress in the clay layer by 50 kPa throughout its depth. The oedometer test sample was taken from the mid-depth of the clay layer, that is, 5 m below ground level. Explaining your choice of one-dimensional modulus E'_0, estimate the eventual settlement of the clay layer. What, qualitatively, would be the effect if the foundation load were to be increased by a further 50 kPa?

[University of London 2nd year BEng (Civil Engineering) examination, Queen Mary and Westfield College]

((b) $\kappa_0 = 0.035$; $\lambda_0 = 0.094$; (c) 74 mm.)

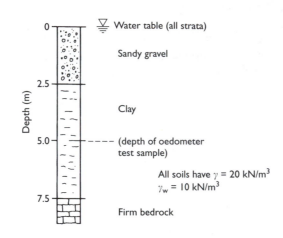

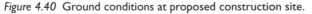

Figure 4.40 Ground conditions at proposed construction site.

4.2 (a) Describe with the aid of a diagram the important features of a conventional oedometer, and define the parameters that this apparatus is
used to measure.

(b) Data from an oedometer test on a sample of clay are given below.

σ'_v (kPa)	50	100	200	400	800	600	400
Equilibrium sample height h (mm)	17.123	16.912	16.701	15.496	14.300	14.390	14.521

Cross-sectional area of sample $= 80\,000\ \text{mm}^2$
$G_s = 2.61$
Two-way sample drainage.
Calculate the specific volume at the end of the test, assuming $S_r = 1$
and $w = 32.84\%$ at this stage. What was the saturation ratio at the
start of the test, if the initial water content was 45.14%?
    Show that the specific volume is related to the sample height by the
expression $v/h = A/V_s = $ constant, where A is the cross-sectional
area of the sample and V_s is the volume occupied by the soil grains.
    Plot a graph of the specific volume against the natural logarithm of
the vertical effective stress. Explain the shape of this graph. Calculate
the preconsolidation pressure and the slopes of the one-dimensional
normal compression line and unloading/reloading lines.

[*University of London 2nd year BEng (Civil Engineering) examination,*
King's College (part question)]

((b) $v = 1.857$; $S_r = 99\%$ initially; preconsolidation pressure $\sigma'_{v,\text{max prev}} = 200\,\text{kPa}$; $\lambda_0 = 0.221$; $\kappa_0 = 0.039$.)

4.3   For the oedometer test described in question 4.2, plot a graph of vertical effective stress σ'_v against vertical strain ε_v. For each of the loading and unloading steps, calculate the one-dimensional modulus $E'_0 = \Delta\sigma'_v/\Delta\varepsilon_v$ ($\Delta\sigma'_v$ and $\Delta\varepsilon_v$ are the changes in vertical effective stress and strain that occur during the loading or unloading step). Comment briefly on the significance of these results in the context of the selection of parameters for design.

Load increment/decrement (kPa)	50–100	100–200	200–400
E'_0 (MPa)	4.06	8.02	2.77
Load increment/decrement (kPa)	400–800	800–600	600–400
E'_0 (MPa)	5.18	31.78	21.97

Analysis of data from the consolidation phase

4.4   (a)   Data from one stage of an oedometer test are given below.

Time (min)	0.25	1	4	9	16	25
Settlement (mm)	0.063	0.075	0.103	0.133	0.160	0.185
Time (min)	36	49	64	81	100	196
Settlement (mm)	0.210	0.228	0.240	0.250	0.258	0.265

Load increment 25–50 kPa
Sample diameter 76 mm
Initial sample thickness 20 mm
Two-way drainage.
  For this load increment, estimate the one-dimensional modulus E'_0, the consolidation coefficient c_v, and the vertical permeability of the soil k_v. (It may be assumed that the initial slope of a graph of proportional settlement $R = \rho/\rho_{\text{ult}}$ against the square root of the time factor $T = c_v t/d^2$ is equal to $\sqrt{(4/3)}$, that is, $R = \sqrt{(4T/3)}$.)
  What factors would you take into account in the laboratory determination of E'_0, c_v and k_v for use in design? What difficulties might you encounter in attempting to use oedometer test results to predict rates of settlement in the field?

[*University of London 2nd year BEng (Civil Engineering) examination, Queen Mary and Westfield College (part question)*]

($E'_0 = 2.22$ MPa; $c_v = 1.265$ mm^2/min; $k_v = 3.35 \times 10^{-7}$ m/s.)

4.5   An engineer carries out an oedometer consolidation test on a sample of stiff clay, in connection with the design of a proposed grain silo. The results from this test are as follows:

Time (min)	0	0.5	1	2	4	8	16	32	64
Settlement (mm)	0.020	0.044	0.052	0.066	0.086	0.110	0.150	0.192	0.2

Load increment 100–200 kPa
Initial sample thickness 20 mm
Two-way drainage.

Suggest a reason for the initial settlement of 0.02 mm. Estimate the one-dimensional modulus E'_0 and the consolidation coefficient c_v for the clay over the stress range under consideration. (It may be assumed that a graph of the consolidation settlement ρ against the square root of the elapsed time t has an initial slope of $\rho_{ult}\sqrt{(4c_v/3d^2)}$, that is, that $\rho = \rho_{ult}\sqrt{(4c_v t/3d^2)}$.)

[*University of London 2nd year BEng (Civil Engineering) examination, King's College (part question)*]

Initial settlement is probably due to trapped air (careless sample preparation), therefore use a datum of 0.02 mm for calculating the true consolidation settlements. $E'_0 \approx 10$ MPa; $c_v \approx 2.1$ mm²/min.

Application of one-dimensional compression and consolidation theory to field problems

4.6 Figure 4.41 shows a cross-section through a long sheet-piled excavation. The width of the excavation is b, its depth is h and the sheet piles penetrate a further depth d to a permeable aquifer. A standpipe piezometer is driven into the aquifer, and the water in the standpipe rises to a height H above the bottom of the sheet piles.

(a) Show that the base of the excavation will become unstable if $H > H_{crit}$, where $H_{crit} = d\gamma/\gamma_w$.
(*Note:* this is a quicksand problem: see section 3.11.)

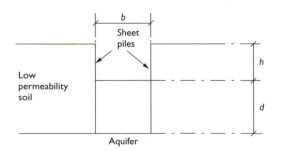

Figure 4.41 Sheet-piled cofferdam.

(b)   Some time after the excavation has been made, and steady state seepage from the aquifer to the excavation floor has been established, it is found that H is indeed very close to H_{crit}. In order to reduce the risk of base instability it is decided to reduce the head in the aquifer to $H/2$ by pumping. This is done very rapidly. Explain why the pore water pressures in the soil between the sheet piles cannot respond instantaneously.

(c)   Taking $d = 10\,m$, $\gamma = 20\,kN/m^3$ and $\gamma_w = 10\,kN/m^3$, draw diagrams to show the initial and final distributions of pore water pressure with depth in the soil between the sheet piles. Draw also the initial and final distributions of excess pore water pressure with depth, and sketch in three or four isochrones at various stages in between.

(d)   The soil between the sheet piles has a one-dimensional modulus E'_0 that increases linearly with depth from 5 MPa at the excavated surface to 45 MPa at the interface with the aquifer. Estimate the settlement which ultimately results from the reduction in pore water pressure due to pumping.

[University of London 2nd year BEng (Civil Engineering) examination, King's College (part question)]

((c)   Initial pore water pressure = 0 at excavation floor, increasing linearly to 200 kPa at the upper surface of the aquifer. Final pore water pressure = 0 at excavation floor, increasing linearly to 100 kPa at the upper surface of the aquifer. Excess pore water pressures are calculated by subtracting hydrostatic component: initial excess pore water pressure increases linearly from 0 at the excavation floor to 100 kPa at the upper surface of the aquifer. Final pore water pressures are hydrostatic below excavation floor, so final excess pore water pressures are zero. (d) Settlement = 19 mm (by integration).)

4.7   Figure 4.42 shows the ground conditions at the site of the Jubilee Line Extension station at Canary Wharf, in East London. During construction of the station, it was necessary to lower the groundwater level at the top of the Thanet Sands to 84 m above site datum.

(a)   Assuming that the groundwater level in the Thames Gravels is unaffected, and that the groundwater level at the top of the chalk is reduced by only 35 kPa, construct a table for the soil behind the retaining wall, showing the initial and final vertical effective stresses at the ground surface and at the interface levels between each of the strata.

(b) Using the geotechnical data given in Figure 4.42, estimate

    (i)   the immediate settlement of the soil surface,
    (ii) the long-term settlement of the soil surface, and
    (iii) using the relation between R and T given in Figure 4.18, the settlement after a period of 18 months.

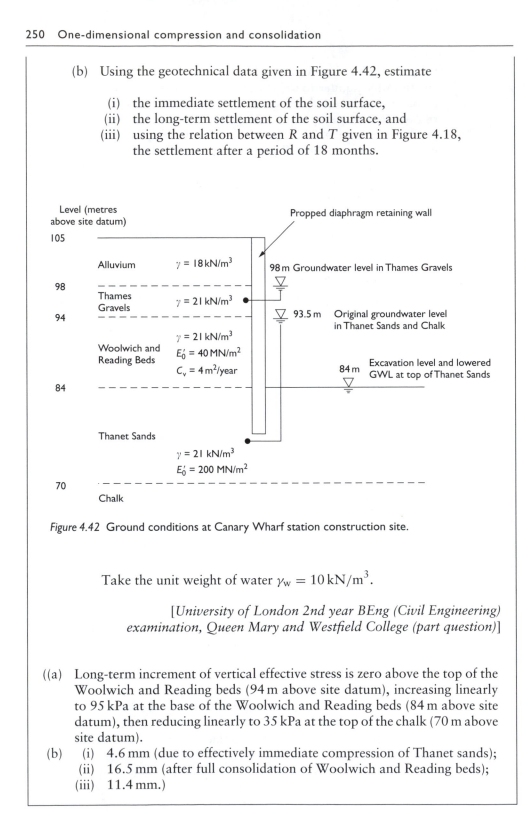

Figure 4.42 Ground conditions at Canary Wharf station construction site.

Take the unit weight of water $\gamma_w = 10\,kN/m^3$.

[University of London 2nd year BEng (Civil Engineering) examination, Queen Mary and Westfield College (part question)]

((a) Long-term increment of vertical effective stress is zero above the top of the Woolwich and Reading beds (94 m above site datum), increasing linearly to 95 kPa at the base of the Woolwich and Reading beds (84 m above site datum), then reducing linearly to 35 kPa at the top of the chalk (70 m above site datum).

(b) (i)   4.6 mm (due to effectively immediate compression of Thanet sands);
    (ii) 16.5 mm (after full consolidation of Woolwich and Reading beds);
    (iii) 11.4 mm.)

Notes

1 As mentioned in Chapter 3, there are unfortunately two different definitions of the term excess pore water pressure. In this chapter, the term is used to mean pore water pressures over and above hydrostatic. This means that zero excess pore water pressures correspond to zero flow. If the excess pore water pressures are non-zero, flow may be either transient (in which case consolidation is taking place) or steady (in which case it is not). The equivalence between excess head h_e and hydraulic total head h is dependent on the adoption of this definition of excess pore water pressure.

   In some texts, the term excess pore water pressure is used to mean pore water pressures over and above those at the steady state, which may or may not be hydrostatic. Using the second definition, zero excess pore water pressures will indicate that consolidation has ceased. Steady state flow, however, may still be taking place.

   The fact that there are two alternative definitions of the term excess pore pressure is confusing, but is unlikely to lead to errors except in the analysis of problems in which the steady state conditions are not hydrostatic. An example of such a problem is given in section 4.7.2. A full discussion of the implications of each definition is given by Gibson *et al.* (1989).

2 In the analysis of consolidation problems, the excess pore water pressure u_e and the quantities related to it are functions of both depth z and time t. It is therefore appropriate to express differentials using partial notation, for example, $\partial u_e/\partial z$ rather than du_e/dz. The term $\partial u_e/\partial z$ signifies the rate of change of u_e with z at constant t.

3 Note that there is no need to specify the hydraulic gradient as $-\partial h_e/\partial z$, because the flowrate has been taken as positive upwards, which is in the z-negative direction.

References

Al-Tabbaa, A. and Muir-Wood, D. (1991) Horizontal drainage during consolidation: insights gained from analyses of a simple problem. *Géotechnique*, **41**(4), 571–85.

Barron, R.A. (1948) Consolidation of fine-grained soil by drain wells. *Transactions of the ASCE*, **113**, 718–42.

Biot, M.A. (1941) General theory of three-dimensional consolidation. *Journal of Applied Physics*, **12**, 155–64.

Brooker, E.W. and Ireland, H.O. (1965) Earth pressures at rest related to stress history. *Canadian Geotechnical Journal*, **2**(1), 1–15.

Butterfield, R. (1979) A natural compression law for soils (an advance on e-log p′). *Géotechnique*, **29**(4), 469–80.

Clayton, C.R.I., Muller-Steinhagen, H. and Powrie, W. (1995) Terzaghi's theory of consolidation, and the discovery of effective stress. *Proceedings of the Institution of Civil Engineers, Geotechnical Engineering*, **113**(4), 191–205.

Gibson, R.E., England, G.L. and Hussey, M.J.L. (1967) The theory of one-dimensional consolidation of saturated clays. I: Finite non-linear consolidation of thin homogeneous layers. *Géotechnique*, **17**(3), 261–73.

Gibson, R.E., Schiffman, R.L. and Whitman, R.V. (1989) On two definitions of excess pore water pressure. *Géotechnique*, **39**(1), 169–71.

Jacob, C.E. and Lohman, S.W. (1952) Non-steady flow to a well of constant drawdown in an extensive aquifer. *Transactions of the American Geophysical Union*, **33**(4), 559–69.

Jaky, J. (1944). The coefficient of earth pressure at rest. *Journal of the Union of Hungarian Engineers and Architects*, 355–8 (in Hungarian).

Mayne, P.W. and Kulhawy, F.M. (1982) K_0–OCR relationships for soils. *Journal of Geotechnical Engineering Division, American Society of Civil Engineers*, **108**(6), 851–872.

Powrie, W. and Roberts, T.O.L. (1990) Field trial of an ejector well dewatering system at Conwy, North Wales. *Quarterly Journal of Engineering Geology*, **23**, 169–85.

Rao, D.B. (1973) Construction dewatering by vacuum wells. *Indian Geotechnical Journal*, 217–24.

Skempton, A.W. (1960) Terzaghi's discovery of effective stress, in *From Theory to Practice in Soil Mechanics: Selections From the Writings of Karl Terzaghi*, 42–53, John Wiley, New York.

Terzaghi, K. von and Fröhlich, O.K. (1936) *Theorie der Setzung von Tonschichten: eine Einfuhrung in die Analytische Tonmechanik*. Franz Deuticke, Leipzig.

Chapter 5

The triaxial test and soil behaviour

5.1 Objectives

The **triaxial test** is widely used, both in industry and research, to investigate the stress–strain behaviour of soils. In this chapter, the triaxial test apparatus and standard testing procedures, and the analysis of triaxial test data, are described. The triaxial test then is used to introduce the more general behaviour of soils, especially saturated clays. At the end of the chapter, issues such as anisotropy, creep and partial saturation, which are not taken into account in simple models of soil behaviour, are mentioned.

Some of the later sections – and possibly the numerical parts of section 5.12 on state paths during shear – may be outside the scope of some courses in soil mechanics at first degree level.

Objectives

After having worked through this chapter, you should understand

- the standard procedures adopted in the triaxial testing of soils, in particular the distinctions between the **isotropic compression** and **shear** stages of a test, and between **drained** and **undrained** shear tests (sections 5.2–5.9)
- that in the triaxial test, the appropriate indicators of soil state are the **specific volume** v, **average principal effective stress** p' and **deviator stress** q (section 5.3)
- the methods of analysis used, and their underlying assumptions and limitations (sections 5.4–5.9)
- the distinction between **failure** and **yield**, and the concept of a **yield locus** in three-dimensional v, p', q space (sections 5.10 and 5.11)
- that when sheared, a **loose** or **lightly overconsolidated** soil will tend to contract or generate positive pore water pressures following yield. It will eventually reach a **critical state** in which further shear strain occurs at constant v, p' and q (sections 5.10–5.12)
- that a **dense** or **heavily overconsolidated** soil will tend to dilate or generate negative pore water pressures following yield. It will develop a **peak strength**, and will then probably fail by rupture before reaching a uniform critical state. The possible combinations of v, p' and q at rupture represent a three-dimensional surface – the **Hvorslev surface** – in (v, p', q) space (section 5.13).

You should be able to

- process data from a triaxial test, and present results as graphs of q, q/p', mobilized strength ϕ'_{mob}, and pore water pressure change Δu or volumetric strain ε_{vol} against axial strain ε_a or shear strain γ (section 5.3 and 5.4)
- plot and interpret state paths on graphs of v against p'; v against $\ln p'$; q against p'; and q against p (sections 5.3 and 5.4)
- draw Mohr circles of stress and strain for triaxial test samples, sketch peak and critical state strength failure envelopes, and determine the mobilized strength at any stage ϕ'_{mob}, the critical state strength ϕ'_{crit}, the peak strength ϕ'_{peak}, and the undrained shear strength τ_u (sections 5.4–5.6, 5.10 and 5.13)
- predict the state paths followed by triaxial test samples, using the Cam clay model and/or other triaxial test data (sections 5.12–5.13).

You should have an appreciation of

- the limitations of the Cam clay and critical state models, with regard to the development of **ruptures** (sections 5.13–5.14), soil structure effects and **sensitivity** (section 5.15), **creep** (section 5.17) and **anisotropy** (section 5.18)
- the correlation between critical state parameters and index tests (section 5.16)
- the difficulties associated with partly saturated soils (section 5.19)
- the application of critical state models to sands (section 5.20)
- the availability of more sophisticated and realistic models, developed from Cam clay, for use in numerical analysis (section 5.21).

5.2   The triaxial test

5.2.1   Apparatus

The triaxial test apparatus is illustrated in Figure 5.1. The test is based on a cylindrical sample with a height : diameter ratio of 2. Commonly, samples are either 38 mm or 100 mm in diameter, but other sizes are sometimes used. In general, larger samples provide a better representation of the *in situ* behaviour of the soil, as they contain more of it. Samples of 38 mm diameter are only really suitable for homogeneous soils, without significant fabric or fissuring. A 38 mm diameter sample may be small enough to fit between the fissures, giving an unduly optimistic indication of the bulk soil strength (Marsland, 1972).

The sample is enclosed by a thin rubber membrane and placed inside the triaxial cell, where it can be subjected independently to an all-round isotropic pressure from the cell fluid (known as the **cell pressure**), and a vertical force acting through a piston. If the force through the piston (the **ram load**) is compressive, as is usually the case, its effect is to increase the axial (vertical) stress to a value greater than the radial (horizontal) stress applied by the cell pressure. The axial (vertical) stress will then be the major principal stress, and the radial (horizontal) stress the minor. A ram load which is tensile will reduce the axial (vertical) stress to a value which is less than the cell pressure, but is still compressive: in this case, the radial stress is the major principal stress, and the axial stress the minor.

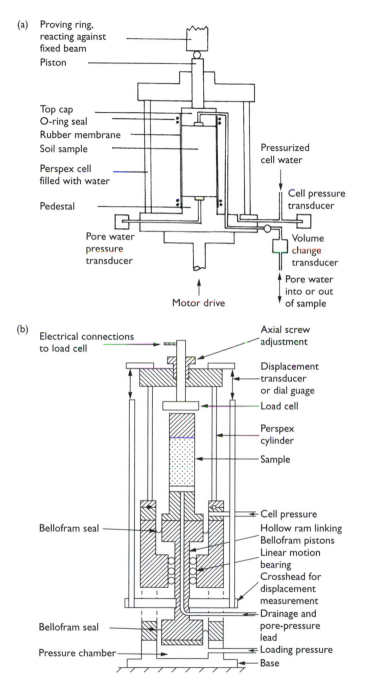

(a) Proving ring, reacting against fixed beam
Piston

Top cap
O-ring seal
Rubber membrane
Soil sample
Perspex cell filled with water
Pedestal

Pressurized cell water

Cell pressure transducer

Pore water pressure transducer

Volume change transducer

Pore water into or out of sample

Motor drive

(b) Electrical connections to load cell

Axial screw adjustment

Displacement transducer or dial guage

Load cell

Perspex cylinder

Sample

Bellofram seal

Cell pressure
Hollow ram linking
Bellofram pistons
Linear motion bearing
Crosshead for displacement measurement
Drainage and pore-pressure lead

Bellofram seal

Pressure chamber

Loading pressure
Base

Figure 5.1 Triaxial test apparatus: (a) schematic; (b) hydraulically operated cell. (Redrawn with permission from Bishop and Wesley, 1975.)

Modern versions of the triaxial apparatus are computer-controlled. The cell pressure, pore water pressure, axial load, axial strain, volumetric strain and combinations thereof can be varied so as to subject the sample to almost any desired pattern of loading or unloading. However, the imposed stress state is always axisymmetric (i.e. has rotational symmetry about the vertical axis of the sample: section 2.2.4), whereas conditions of plane strain would be more suitable for many civil engineering applications. Plane strain conditions are found (at least approximately) in geotechnical engineering constructions such as embankments, cuttings, trenches and retaining walls, which may reasonably be considered to be long in comparison with their width and height (section 2.2.3).

5.2.2 Procedures

There are normally two stages in a conventional triaxial test. In the **compression** or **consolidation** stage, the cell pressure alone is increased to some desired value which represents the starting point for the **shear test**. In the shear test, an additional axial stress (which may be either compressive or tensile) is applied vertically via the ram. Before the development of computer control, the triaxial apparatus was usually used to carry out shear tests in which the cell pressure was kept constant while the axial stress was increased until the sample failed. For this largely historical reason, such a procedure is known as a **conventional triaxial compression test**. Less commonly, a **triaxial extension** test might be carried out, in which the cell pressure is kept constant while the axial stress is reduced (by means of a tensile ram load) until the sample fails.

During a triaxial shear test, the drainage taps to the sample can be kept either open or closed. With the drainage taps closed, water cannot move out of or into the sample: such a test is described as **undrained**. An undrained test in which the sample has been allowed to consolidate or swell to its equilibrium moisture content during the application of the cell pressure prior to shear, is known as a **consolidated–undrained** test. A test in which no drainage into or out of the sample prior to shear has been allowed (i.e. the drainage taps have remained closed during the application of the cell pressure, preventing consolidation from taking place) is termed an **unconsolidated–undrained** test. Unconsolidated–undrained triaxial tests are usually carried out – if at all – on 'undisturbed' samples of soil from the field, in an arguably misguided attempt to estimate the undrained shear strength of the soil *in situ*.

In an undrained test, deformation takes place at constant volume (making the usual soil mechanics assumption that the soil particles and the pore water are incompressible), and excess or non-equlibrium pore water pressures will be generated during shear. These pore water pressures should be measured using an appropriate transducer, so that the results of the test may be analysed in terms of effective stresses. Undrained tests should be carried out sufficiently slowly for the distribution of pore water pressure within the sample to remain uniform. Typically, a high-quality undrained shear test on a 38 mm diameter clay specimen would be carried out over a working day.

With the drainage taps open, water is free to flow out of or into the test specimen: this is known as a **drained** test. Drained tests are always carried out on samples which have been allowed to consolidate or swell to a moisture content in equilibrium with the test cell pressure. Drained tests must be carried out slowly enough not to allow significant excess pore water pressures to develop within the sample, at the point furthest from the drainage boundary. The rate of testing will depend on the permeability and size of the sample and on the drainage arrangements: typically, a high-quality conventional drained shear test on a 38 mm diameter clay sample would take a week or more. Research-quality tests, in which the sample is prepared in the triaxial cell and subjected to a variety of complex stress or strain paths before being brought finally to failure, may take months.

Analysis of the behaviour of soils which are only partly saturated is possible, but difficult (section 5.19). Below the water table in temperate zones, an unsaturated specimen is probably the result of poor sampling and preparation rather than representative of reality. To avoid difficulties with unsaturated conditions by ensuring that small amounts of air which may be present remain dissolved in the pore water, samples are often prepared and tested at an increased pore water pressure known as a **back pressure**. The cell pressure must be increased by the same amount as the pore water pressure if the effective confining stress is to remain unaltered.

Following consolidation of the sample in the triaxial cell, it is good practice to check the degree of saturation by measuring the increase in pore water pressure Δu in response to an increase in cell pressure $\Delta \sigma_c$ with the drainage taps closed. If the sample is fully saturated, the ratio $\Delta u / \Delta \sigma_c$ (known as the **B-value** should be equal to one. A B-value of less than one indicates that air is present in the soil pores, and the back pressure should be increased to dissolve it. The relation between the B-value and the degree of saturation S_r depends on the stiffness of the soil. A B-value of 0.95 might indicate a saturation ratio S_r of 99.9% in a stiff clay, but a saturation ratio of only 96% in a soft clay (Black and Lee, 1973).

5.3   Stress parameters

5.3.1   Stress invariants

We have already seen (in section 2.8) how a complete description of the state of a soil requires information concerning the **shear stress**, the **normal effective stress** and the **void ratio** or the **specific volume**. In the case of the shearbox test, the shear stress and the normal effective stress acting on the central horizontal shear plane were deduced from the external normal and shear loads applied to the sample. The specific volume or void ratio was determined from the total volume of the sample at any stage (monitored by measuring the sample height) and the known volume of soil grains.

A knowledge of the stress state on just one plane is not sufficient in itself to define the stress state of a sample completely, even if we confine our attention to the major and minor principal effective stresses σ'_1 and σ'_3, and neglect the intermediate principal effective stress σ'_2. In the analysis of the shearbox test in Chapter 2, it was necessary to assume, for example, that the shear plane is the plane of maximum stress ratio – that is, the plane on which the ratio of shear to normal effective stress (τ/σ') is greatest. This enabled the Mohr circle of stress to be drawn as shown in Figure 5.2.

In general to draw a Mohr circle of stress, the major and the minor principal effective stresses σ'_1 and σ'_3 – or alternatively the circle radius and the location of its centre on the σ' axis – must be known. From the geometry of the Mohr circle shown in Figure 5.2, the **radius** is equal to $(\sigma'_1 - \sigma'_3)/2$ and the centre is at $(\sigma'_1 + \sigma'_3)/2$. These parameters are known as the **two-dimensional stress invariants**, and are conventionally given the

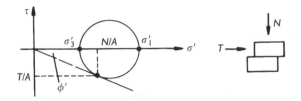

Figure 5.2 Mohr circle of stress for a shearbox sample.

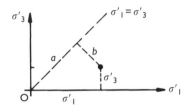

Figure 5.3 Representation of a plane stress state on a graph of σ_3' against σ_1'.

symbols t and s' respectively:

$$t = (\sigma_1' - \sigma_3')/2 \tag{5.1}$$

$$s' = (\sigma_1' + \sigma_3')/2 \tag{5.2}$$

s' is the **average effective stress** and t is the **maximum shear stress** within the sample. Since the parameters t and s' provide indications of shear and normal effective stresses, respectively, they may be used (together with the void ratio e or the specific volume v) to define the state of a soil.

State paths, indicating how the state of a soil element changes as it is loaded or unloaded, may be plotted as graphs of t against s' and s' against v. These graphs are in fact two-dimensional projections of a three-dimensional state path being followed in (t, s', v) space. The stress parameters t and s' are particularly appropriate to conditions of plane strain, in which the intermediate principal effective stress is often considered to be of only secondary importance.

The stress state in a plane may be defined completely either by the principal stresses σ_1' and σ_3', or by the parameters t and s'. Another way of looking at the parameters t and s' is to consider a stress state represented as a point on a graph of σ_3' against σ_1' (Figure 5.3). This point may be reached either by travelling a distance σ_1' in the x-direction, followed by a distance σ_3' in the y-direction, or by travelling a distance a along the line $\sigma_3' = \sigma_1'$, followed by a distance b along a line at 90°. It may be shown that the distance a along the line $\sigma_3' = \sigma_1'$ is equal to $(\sqrt{2})s'$, while the distance b is equal to $(\sqrt{2})t$.

Conditions in the triaxial test do not correspond to plane strain, and the intermediate principal effective stress is known. In a conventional triaxial compression test, the intermediate principal total stress σ_2 is equal to the minor principal total stress σ_3, which is provided by the cell pressure. Effective stresses are calculated by subtracting the pore water pressure: since this is the same in all directions, $\sigma'_2 = \sigma'_3$.

In an extension test, the ram load is tensile so that the vertical stress on the sample is less than the cell pressure σ_c (but still positive, i.e. compressive, overall). The intermediate principal stress is again equal to the cell pressure. However, in this case, the cell pressure is the major principal stress, so that $\sigma'_2 = \sigma'_1 = \sigma_c - u$.

In general, if the intermediate principal effective stress is known, and the mode of deformation is not plane strain, it is appropriate to use the **three-dimensional stress invariants** q and p':

$$q = (1/\sqrt{2})\sqrt{\{(\sigma'_1 - \sigma'_2)^2 + (\sigma'_1 - \sigma'_3)^2 + (\sigma'_2 - \sigma'_3)^2\}} \tag{5.3}$$

and

$$p' = (1/3)(\sigma'_1 + \sigma'_2 + \sigma'_3) \tag{5.4}$$

The parameters q and p' provide an indication of the **shear** and **normal** stresses respectively, and so may be used (together with e or v) to define stress states and to plot state paths. p' is the **mean** or **average principal effective stress**, and is directly analogous to s'. The interpretation of q, which is termed the **deviator stress**, is more difficult. A three-dimensional stress state can be defined by the three principal stresses, so that stress states may be represented by points in a three-dimensional coordinate system whose axes are σ'_1, σ'_2 and σ'_3 (Figure 5.4(a)).

Referring to Figure 5.4, the stress point may be reached by travelling along the **hydrostatic axis** $\sigma'_1 = \sigma'_2 = \sigma'_3$ for a distance a, and then along a line perpendicular to the hydrostatic axis for a distance b. It may be shown that the distance a along the hydrostatic axis is equal to $(\sqrt{3})p'$, while the distance b along the perpendicular is equal to $(\sqrt{2/3})q$. A third parameter is needed to specify the stress state completely: this is the Lode angle θ, which indicates the orientation of the line perpendicular to the hydrostatic axis (Figure 5.4(b)). For the axisymmetric conditions imposed on the soil sample in the triaxial test, σ'_2 is equal to either σ'_1 or σ'_3, and the general expression

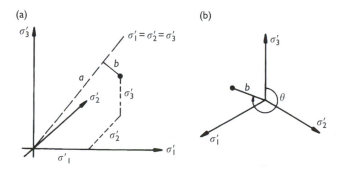

Figure 5.4 (a) Representation of a three-dimensional stress state in $\sigma'_1, \sigma'_2, \sigma'_3$ space (orthogonal view); (b) view along the hydrostatic $\sigma'_1 = \sigma'_2 = \sigma'_3$.

for q (equation (5.3)) becomes

$$q = \sigma'_1 - \sigma'_3 \tag{5.5}$$

It is conventional to use the state parameters q, p' and v to describe the behaviour of the soil sample in the triaxial test. State paths are plotted as graphs of q against p', and v against p' or the natural logarithm of p'. Again, these are two-dimensional projections of three-dimensional state paths being followed by the soil in (q, p', v) space. In addition, the imposed total stress path is plotted in terms of q and p. Section 5.4 is concerned with converting the raw data measured in the triaxial test to the state parameters q, p' and v. A clear and concise summary of the various stress parameters, and their application to common civil engineering situations, is given by Wood (1984).

5.3.2 Notation

It is perhaps worth mentioning at this point the systems of subscripts used to describe the stresses. In the ground, we will usually be interested in the vertical and horizontal effective stresses, which are denoted by the subscripts v and h respectively. In the triaxial test, the relevant stresses are axial and radial, which are given the subscripts a and r respectively. Usually, if a sample is taken from the ground and placed in the triaxial cell, the vertical stress in the ground will correspond to the axial stress in the triaxial cell. The radial stress in the triaxial cell represents the horizontal stress in the ground.

In some cases, it is convenient to refer to the effective stresses by the numerical subscripts 1, 2 and 3. This system is used when the stresses are principal stresses (i.e. they act on planes on which there is no shear stress). The numbers do not in general correspond to unique physical directions. In the triaxial test, however, the axial and radial stresses will always be principal stresses, because there is no facility to apply direct shear to the sample. In a conventional triaxial compression test, the axial effective stress is the major principal effective stress, $\sigma'_1 = \sigma'_a$, while the cell pressure provides both the intermediate and the minor principal effective stresses, $\sigma'_2 = \sigma'_3 = \sigma'_r = \sigma_c - u$. In a triaxial extension test, the axial effective stress is the minor principal effective stress, $\sigma'_3 = \sigma'_a$, while the cell pressure provides both the intermediate and the major principal stresses, $\sigma'_2 = \sigma'_1 = \sigma'_r = \sigma_c - u$.

A further source of potential confusion arises in the exact definition of the deviator stress q. Equation (5.5) defines q in terms of major and minor principal effective stresses. This means that q is always positive, because (by definition) $\sigma'_1 \geq \sigma'_3$. In interpreting triaxial tests and soil behaviour in general, it is often useful to distinguish between conditions of triaxial compression (with $\sigma'_a \geq \sigma'_r$) and triaxial extension (with $\sigma'_a \leq \sigma'_r$). (These correspond to conditions in the field where $\sigma'_v \geq \sigma'_h$ and $\sigma'_v \leq \sigma'_h$ respectively.) It is therefore quite common to define the deviator stress q as

$$q = \sigma'_a - \sigma'_r \tag{5.6}$$

so that state paths on a graph of q against p' plot above the p' axis (q positive) for triaxial compression, and below the p' axis (q negative) for triaxial extension.

5.4   Stress analysis of the triaxial test

5.4.1   Assumptions

The principal assumptions made in the stress analysis of the triaxial test are that the sample deforms as an upright circular cylinder with stresses and strains uniform and continuous, and that the principal stresses are axial (vertical) and radial (horizontal). In practice, the sample may **barrel** or **rupture**, invalidating these assumptions.

Barrelling might occur due to friction on the end platens. Inaccuracies resulting from this effect can be eliminated by the measurement of strains over a short gauge length at the centre of the sample, using instruments mounted on the sample inside the cell. The onset of rupture will unavoidably invalidate the continuum analysis, because stresses and strains are no longer uniform throughout the sample. This is particularly important in the interpretation of measurements made at the sample ends, which relies on the validity of a continuum analysis.

5.4.2   Measured quantities

The quantities measured in a triaxial test are

- the cell pressure σ_c (kPa)
- the ram load Q (kN)
- the pore water pressure u (kPa)
- the change in sample height Δh, or the movement over a small gauge length at the centre of the sample (mm)
- the change in sample volume ΔV_t (mm^3).

The **cell pressure** σ_c provides the minor principal total stress σ_3 in a triaxial compression test, or the major principal total stress σ_1 in an extension test, as already discussed.[1]

The **ram load** Q is used to determine the difference between the axial stress σ_a and the cell pressure σ_c, which is the deviator stress defined according to equation (5.6), $q = (\sigma_a - \sigma_c) = (\sigma'_a - \sigma'_c) = Q/A$, where A is the current cross-sectional area of the sample. σ'_c is the effective cell pressure, defined as $\sigma'_c = \sigma_c - u$ (i.e. the cell pressure minus the pore water pressure).[2]

The **pore water pressure** u is needed to calculate the effective stresses acting on the soil skeleton.

The **axial strain** ε_a is calculated as $\Delta h/h_0$, and the **volumetric strain** ε_{vol} as $\Delta V_t/V_{t0}$, where h_0 and V_{t0} are the height and total volume respectively of the sample at the start of the shear test.

The **radial strain** ε_r may be calculated from the axial and volumetric strains, assuming that the sample remains cylindrical in shape (equation (5.14)). Alternatively, it might be attempted to measure ε_r directly, using a strain-gauged clip designed for this purpose.

5.4.3   Converting measurements to stress, strain and state parameters

The sample height and volume at the start of the shear test (h_0 and V_{t0}) will not in general be the same as the height and volume of the sample as prepared (h_i and

V_{ti}), because of the change in volume which occurs during the application of the cell pressure. (Unconsolidated–undrained triaxial tests are an exception, because the drainage taps are closed while the cell pressure is applied, preventing any change in volume from taking place.) V_{t0} may be calculated provided that V_{ti} and the volume of water expelled during consolidation ΔV_{tc} are known:

$$V_{t0} = V_{ti} - \Delta V_{tc} \tag{5.7}$$

The volumetric strain during consolidation is $\Delta V_{tc}/V_{ti}$. Assuming that the sample deforms isotropically during this stage, the linear strain in all three directions must be the same and equal to $(\Delta V_{tc}/V_{ti})/3$. This is because for small strains, $\varepsilon_{vol} = \varepsilon_1 + \varepsilon_2 + \varepsilon_3$ where $\varepsilon_1, \varepsilon_2$ and ε_3 are the principal strains. Thus

$$h_0 = h_i(1 - [\Delta V_{tc}/3V_{ti}]) \tag{5.8}$$

To convert the readings of axial force or ram load Q into deviator stress $q(= \sigma_a - \sigma_c)$, it is necessary to divide by the current true area of the sample A. Taking compression positive,

$$V_t = (V_{t0} - \Delta V_{tq}) = A(h_0 - \Delta h)$$

and

$$q = Q/A = Q(h_0 - \Delta h)/(V_{t0} - \Delta V_{tq})$$

where ΔV_{tq} is the volume of water expelled during shear. Since

$$h_0 - \Delta h = h_0(1 - \Delta h/h_0) = h_0(1 - \varepsilon_a),$$

and

$$(V_{t0} - \Delta V_{tq}) = V_{t0}(1 - \Delta V_{tq}/V_{t0}) = V_{t0}(1 - \varepsilon_{vol})$$

the expression for the deviator stress q becomes

$$q = Qh_0(1 - \varepsilon_a)/V_{t0}(1 - \varepsilon_{vol})$$

and since $V_{t0} = A_0 h_0$,

$$q = (Q/A_0)(1 - \varepsilon_a)/(1 - \varepsilon_{vol}) \tag{5.9}$$

For a drained test, the volumetric strain ε_{vol} may be found from the volume of water expelled, which is measured in a burette or a volume gauge. For an undrained test, $\varepsilon_{vol} = 0$ and $q = (Q/A_0)(1 - \varepsilon_a)$.

It is also necessary to calculate from the measured quantities the average effective stress p', which is in general terms equal to $(\sigma'_1 + \sigma'_2 + \sigma'_3)/3$. In the triaxial test, since two of the principal stresses are given by the cell pressure, $p' = [(\sigma_a + 2\sigma_c)/3] - u$. As $\sigma_a = \sigma_c + q$,

$$p' = \sigma_c - u + q/3 \tag{5.10}$$

The average total stress $p = (\sigma_1 + \sigma_2 + \sigma_3)/3$ is given by $p' + u$, hence

$$p = \sigma_c + q/3 \tag{5.11}$$

The specific volume at the end of the test v_f may be determined from the dry mass of the sample m_s and its final total volume V_{tf} (which should be equal to the volume of the sample as prepared V_{ti} minus the volume of water expelled during consolidation

ΔV_{tc} and shear ΔV_{tq}, $V_{tf} = V_{ti} - \Delta V_{tc} - \Delta V_{tq}$), provided that the relative density of the soil particles (the grain specific gravity) G_s is known:

$$V_{tf} = V_s + V_{vf} = V_s(1 + V_{vf}/V_s) = V_s(1 + e_f) = V_s v_f$$

where V_s is the volume of soil solids and V_{vf} is the final volume of voids. Also,

$$V_s = m_s/\rho_s = m_s/(G_s\rho_w)$$

hence

$$v_f = V_{tf}G_s\rho_w/m_s \tag{5.12}$$

Measurement of the final water content w_f provides a check, assuming that the sample is saturated:

$$e_f = w_f G_s$$

so

$$v_f = 1 + w_f G_s \tag{5.13}$$

In practice, it can be difficult to remove the sample cleanly from the membrane to measure the dry sample mass m_s, so that equation (5.13) will usually give a more reliable result than equation (5.12).

In addition to plotting the state paths followed in terms of q, p' and v – usually, as graphs of q against p' and specific volume v against $\ln(p')$ – test data may also be presented as graphs of q (or stress ratio $\eta = q/p'$) against axial strain ε_a, and volumetric strain ε_{vol} (for a drained test) or change in pore water pressure Δu (for an undrained test) against ε_a. The usual sign convention is that compressive volumetric strains and increases in pore water pressure are positive, but plotted below the horizontal axis (Figure 5.5).

5.4.4 Mohr circles of stress

It has already been stated that the stress state imposed on the soil sample in a triaxial test is radially symmetric rather than plane strain. However, two-dimensional Mohr circles of stress and strain can still be plotted, representing the conditions on any vertical plane containing a diameter (as in Figure 2.2). The Mohr circle of effective stress is shown in Figure 5.6.

In a conventional triaxial compression test, σ_3 is equal to the cell pressure and $\sigma_1 = \sigma_3 + q$. The effective stresses are $\sigma_1' = \sigma_1 - u$ and $\sigma_3' = \sigma_3 - u$: in an undrained test, or in a drained test carried out against a back pressure, u is not equal to zero. The maximum stress ratio (τ/σ') mobilized in the sample corresponds to a **mobilized strength** or a **mobilized angle of shearing** $\phi'_{mob} = \sin^{-1}(\sigma_1' - \sigma_3')/(\sigma_1' + \sigma_3')$. ϕ'_{mob} is defined by the tangent to the Mohr circle which passes through the origin, and is a measure of the strength used or **mobilized** to enable the soil to carry the applied stresses. If the soil had only this strength, it would be on the verge of failure. The concept of mobilized strength is discussed further in sections 7.6, 7.7, 8.5–8.7, 8.10, 8.11 and 10.2–10.5.

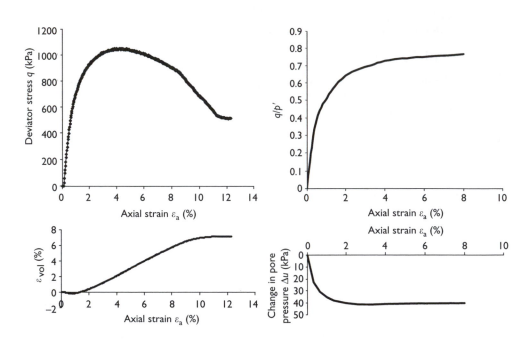

Figure 5.5 Typical triaxial test data: (a) deviator stress q and volumetric strain ε_{vol} against axial strain for a drained test on dense sand; (b) stress ratio q/p' and change in pore water pressure Δu against axial strain for an undrained test on a lightly overconsolidated clay.

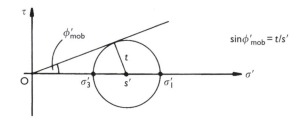

Figure 5.6 Mohr circle of effective stress for a triaxial compression test.

5.4.5 Mohr circles of strain

In an undrained test, $\varepsilon_{vol} = \varepsilon_1 + \varepsilon_2 + \varepsilon_3 = \varepsilon_a + 2\varepsilon_r = 0$, hence

$$\varepsilon_r = -(1/2)\varepsilon_a \qquad (5.14a)$$

and the maximum shear strain is $\gamma_{max} = 1.5\varepsilon_a$, as shown by the Mohr circle of strain in Figure 5.7.

In a drained test, ε_{vol} is non-zero, so that

$$\varepsilon_r = (1/2)(\varepsilon_{vol} - \varepsilon_a) \qquad (5.14b)$$

and the maximum shear strain γ_{max} is equal to $(1/2)(3\varepsilon_a - \varepsilon_{vol})$ (Figure 5.8).

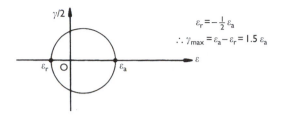

$$\varepsilon_r = -\tfrac{1}{2}\varepsilon_a$$
$$\therefore \gamma_{max} = \varepsilon_a - \varepsilon_r = 1.5\,\varepsilon_a$$

Figure 5.7 Mohr circle of strain for an undrained triaxial compression test specimen.

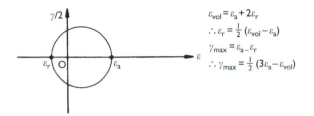

$$\varepsilon_{vol} = \varepsilon_a + 2\varepsilon_r$$
$$\therefore \varepsilon_r = \tfrac{1}{2}\left(\varepsilon_{vol} - \varepsilon_a\right)$$
$$\gamma_{max} = \varepsilon_a - \varepsilon_r$$
$$\therefore \gamma_{max} = \tfrac{1}{2}\left(3\varepsilon_a - \varepsilon_{vol}\right)$$

Figure 5.8 Mohr circle of strain for a drained triaxial compression test specimen.

5.4.6 Other ways of presenting shear test data

Triaxial shear test data can also usefully be presented as graphs of mobilized angle of shearing ϕ'_{mob}, and volumetric strain ε_{vol} (in a drained test) or change in pore water pressure Δu (in an undrained test), as functions of shear strain γ, as shown in Figure 5.9. It is shown in section 5.10.1 that, in triaxial compression, the mobilized angle of shearing ϕ'_{mob} is related to the stress ratio q/p' by

$$\sin\phi'_{mob} = \frac{3q/p'}{6 + q/p'} \tag{5.15}$$

An alternative shear strain parameter which is sometimes used is the **triaxial shear strain** ε_q, which is defined as

$$\varepsilon_q = \tfrac{2}{3}(\varepsilon_a - \varepsilon_r) \tag{5.16}$$

The triaxial shear strain ε_q is defined in this way so that the increment of work done ΔW when a triaxial test specimen is subjected to small increments of shear strain $\delta\varepsilon_q$ and volumetric strain $\delta\varepsilon_{vol}$ is given by

$$\Delta W = q \cdot \delta\varepsilon_q + p' \cdot \delta\varepsilon_{vol} \tag{5.17}$$

(see e.g. Muir Wood, 1990).

It is shown in section 6.2.3 that, if the soil is assumed to be isotropic, the Young's modulus can be obtained from a graph of q against ε_a. For both isotropic and anisotropic soils, the shear modulus may be obtained from a graph of q against γ or ε_q measured in an undrained test, and the bulk modulus from a graph of ε_{vol} against p' in compression or swelling without shear (section 6.10).

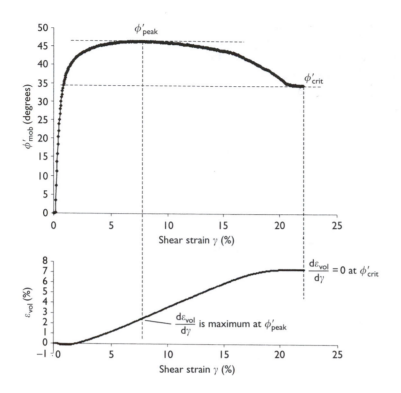

Figure 5.9 Mobilized angle of shearing ϕ'_{mob} and volumetric strain ε_{vol} against shear strain γ for a drained triaxial test on a dense sand.

Example 5.1 Mohr circle analysis of triaxial test data

(a) Stating the assumptions you need to make, show that the deviator stress q is related to the quantities measured in a drained triaxial compression test by the expression

$$q = Q(h_0 - \Delta h)/(V_{t0} - \Delta V_{tq})$$

where Q is the ram load, h_0 and V_{t0} are initial sample height and total volume at the start of the shear test, and Δh and ΔV_{tq} are the changes in sample height and total volume measured during the shear test.

(b) An engineer wanted to investigate the drained strength of a natural sandy silt having thin laminations of clay at intervals of approximately 20 mm. The engineer decided to take cores through the deposit at various angles, so that the laminations were inclined at different angles β to the horizontal when each sample was installed in the triaxial cell. A number of very slow drained tests at zero pore water pressure were carried out at a cell pressure of 100 kPa. The maximum measured values of Q for each test are given in Table 5.1, along with the corresponding values of β. Samples 2 and 3 ruptured along a clay lamination at peak strength: otherwise the samples remained cylindrical.

Table 5.1 Triaxial test data (conditions at failure)

Test no.	β (degrees)	$Q(N)$	$\Delta h(mm)$	$\Delta V_{tq}(cm^3)$
1	0	2230	18.5	67.5
2	45	1410	8.2	32.2
3	60	1170	3.0	11.8
4	90	2200	18.0	62.0

Sample dimensions at start of shear: height $h_0 = 200$ mm, diameter $= 100$ mm.

On one diagram, plot Mohr circles of effective stress at failure for each of the tests, and mark the points that indicate the stress states on the clay laminations.

What are the individual values of ϕ' at failure for the sandy silt and the clay laminations?

[*University of London 2nd year BEng (Civil Engineering) examination, Queen Mary and Westfield College (part question)*]

Solution

(a) Assume that the sample deforms as a right circular cylinder with stresses and strains uniform and continuous, up till the moment of rupture, and that the principal strains and stresses are radial and axial. Ignore any tendency to distortion which would probably in practice result from anisotropy of the soil.

The deviator stress $q = Q/A$, where A is the current cross-sectional area of the sample. The current volume $= V_{t0} - \Delta V_{tq} = A(h_0 - \Delta h)$, hence

$$q = Q/A = Q(h_0 - \Delta h)/(V_{t0} - \Delta V_{tq})$$

(b) For the test samples, $h_0 = 200$ mm and $V_{t0} = [\pi 100^2/4 \times 200 = 1\,570\,796$ mm^3.

The calculated deviator stresses at failure in each test are given in Table 5.2.

Table 5.2 Processed data (conditions at failure)

	Test number			
	1	*2*	*3*	*4*
q (kPa)	269.2	175.8	147.8	265.4
β (degrees)	0	45	60	90

These are drained tests so $u = 0$. The Mohr circles of stress are drawn with the minor principal effective stress σ_3' equal to the cell pressure (100 kPa), and the major principal effective stress σ_1' equal to the cell pressure plus the deviator stress, $\sigma_c + q$. The centre of the Mohr circle is at $(100 + q/2)$ kPa and the radius is $(q/2)$ kPa.

The major principal effective stress acts on the horizontal plane. The stress state on the clay laminations (which are at an inclination β to the horizontal) is found by turning through an angle 2β at the centre of the Mohr circle. (In Figure 5.10, β is taken as anticlockwise.)

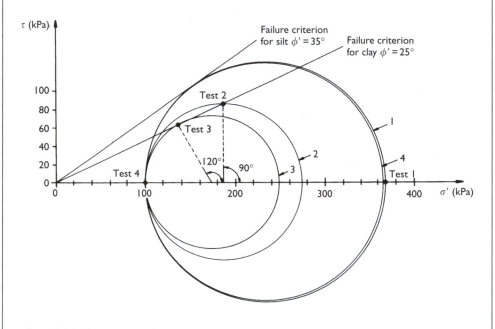

Figure 5.10 Mohr circles of stress for Example 5.1.

Samples 1 and 4 give $\phi' = 35°$ for the sandy silt; samples 2 and 3 give $\phi' = 25°$ for the clay laminations.

5.5   Determining the effective angle of shearing resistance ϕ' from triaxial shear tests

Figure 5.9 shows that the mobilised angle of shearing ϕ'_{mob} may increase with shear strain γ to a peak value ϕ'_{peak} before falling to a lower steady state value ϕ'_{crit} which is then maintained with continued shearing. Examination of the associated volumetric strain data shows that the peak strength coincides with the maximum rate of dilation (defined as the rate of change of volumetric strain with shear strain, $d\varepsilon_{vol}/d\gamma$), and that there is no further change in sample volume once the steady state has been reached.

The steady state corresponds to the critical state already seen in shearbox tests in section 2.8, and discussed further with respect to the triaxial test in section 5.10. For the time being, the important point is that when we come to determine the strength of a soil from triaxial test data, we must be very careful to specify whether we are talking about the strength at the peak or at the critical state. (In fact, for clay soils that have been extensively sheared along a slip plane on which the particles have become aligned with the direction of shear, a third, lower strength – termed the residual strength – may operate. This is discussed in section 5.14.)

As in the shearbox test, the peak strength can only be maintained while the sample is dilating. Also as in the shearbox test, the potential for a sample to dilate depends on the initial specific volume relative to that at the critical state. The specific volume at the critical state depends in turn on the confining stress (the normal effective stress in the shearbox or the effective cell pressure in the triaxial test) at which the test is carried out. For a saturated soil, the specific volume v is linked to the water content w by equation (2.12): $v = 1 + w \cdot G_s$. Thus, the peak angle of shearing resistance of a saturated clay must be expected to depend on both the water content of the sample as tested and the cell pressure at which the test is carried out: it is not a soil property or constant. In contrast, the steady or critical state angle of shearing resistance should be purely frictional in nature and hence independent of both the water content of the clay as tested and the cell pressure at which the test is carried out.

Apart from their inherent variability, peak strengths can be unreliable for use in design because the soil strength falls with continuing strain after the peak has been passed. This leads to the possibility of **progressive failure**, in which the first zone of soil to fail sheds load to its neighbour, causing the next zone of soil to fail and shed load and so on throughout the soil mass. Having said this, the numerical values of factors of safety recommended in many of the traditional codes of practice used in design are specified on the assumption that they will be applied to the peak strength: in this case, the application to the critical state strength of high factors of safety intended for use with peak strengths could lead to an uneconomical design. Strength parameters for use in design are discussed in Chapters 7–11.

Soil strengths expressed as angles of shearing resistance have traditionally been determined by carrying out three conventional triaxial compression tests on similar samples at different cell pressures; plotting the Mohr circles of effective stress at either the peak or the critical state angle of shearing (or stress ratio q/p'); drawing the best fit tangent; and using this to define a **failure envelope** (Figure 5.11).

At the critical state, the failure envelope represented by the tangent to the Mohr circles of effective stress should be a straight line through the origin, with equation

$$\tau = \sigma' \cdot \tan \phi'_{crit} \tag{2.10}$$

(Figure 5.11(a)). This corresponds exactly with the shearbox test data discussed in section 2.8.

Determination of the corresponding failure envelope in terms of peak strengths is more problematic, because the peak angle of shearing resistance will depend on the ability of the sample to dilate, which in turn depends on the specific volume of the sample in relation to that at the eventual critical state. If three samples of the same soil, each having the same initial specific volume, are tested at different cell pressures, the peak angle of shearing resistance will be found to decrease with increasing cell

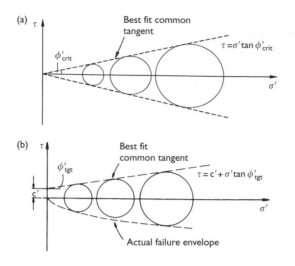

Figure 5.11 Failure envelopes determined from Mohr circles of stress at (a) critical state and (b) peak stress ratios in triaxial tests.

pressure because the ability of the soil to dilate is reduced. Thus the failure envelope in terms of peak strengths will be curved as indicated below the σ' axis in Figure 5.11(b).

Despite this, it is quite common practice to attempt to describe the peak strength failure envelope by means of a 'best fit' straight line having an equation of the form

$$\tau_p = c' + \sigma' \cdot \tan \phi'_{tgt} \tag{2.15}$$

(Figure 5.11(b)).
Equation (2.15) has a number of serious shortcomings:

- The physical interpretation of equation (2.15), with c' as a cohesion (i.e. an ability to withstand shear stresses at zero effective stress) and ϕ'_{tgt} as a friction angle, is incorrect. In the absence of particle bonding, the soil cannot withstand shear stresses at zero effective stress. For unbonded soils, therefore, $c' = 0$.
- Equation 2.15 takes no account of differences in stress history and specific volume/water content, which would be expected to alter the potential for dilation and hence the peak strength achieved.
- As a consequence of this, the scatter in the values of c' and ϕ'_{tgt} obtained from similar sets of samples can be very wide.
- Although it represents a peak strength failure envelope, the parameter ϕ'_{tgt} is actually smaller than the critical state strength ϕ'_{crit}. Thus, there will be a point to the right of the graph of τ against σ' at which the peak strength failure envelope intersects, and then lies below, the critical state failure envelope $\tau = \sigma' \cdot \tan \phi'_{crit}$.
- It is too easy to apply equation (2.15) outside the range of stresses for which it has been determined. At the left-hand end, the peak strength failure envelope is limited at best by the Mohr circle corresponding to a minor principle effective stress $\sigma'_3 = 0$. This is because an unbonded soil is unable to carry tension so that a

Mohr circle crossing the τ axis (i.e. having $\sigma'_3 < 0$) represents an impossible stress state. At the right-hand end, the peak strength failure envelope is limited by its point of intersection with the critical state line $\tau = \sigma' \cdot \tan \phi'_{crit}$.

On a Mohr diagram, the peak strength is best defined by the straight line tangent, passing through the origin, to the Mohr circle representing the stress state at the peak effective angle of shearing resistance. This is equivalent to joining the point of maximum stress obliquity $(\tau/\sigma')_{max}$ to the origin, as in section 2.9:

$$\phi'_{peak} = \tan^{-1}(\tau/\sigma')_{peak} \tag{2.16}$$

A more satisfactory interpretation of peak strength data from a number of tests, which uses a normalization procedure to take account of differences in the specific volume of samples relative to their critical states, is given in section 5.13.3.

Example 5.2 Interpretation of strength data from triaxial tests using Mohr circles of stress

Table 5.3 gives data from consolidated–undrained triaxial compression tests on three samples of a particular type of clay. Explaining your reasoning, identify the peak and the critical state strengths for each test. Why is it not possible to identify a critical state strength in test 3? On one diagram, plot the Mohr circles of stress at the peak strength for all three tests, and sketch in the peak strength failure envelope. On a separate diagram, plot the Mohr circles of stress at the critical state for tests 1 and 2, and estimate the critical state strength ϕ'_{crit}.

Table 5.3 Triaxial test data

	Axial strain ε_a (%)	Deviator stress q (kPa)	Pore water pressure u (kPa)
Test 1	3.89	207.9	235.1
Cell pressure 410 kPa,	4.56	219.2	230.5
Initial back pressure 200 kPa	5.68	232.5	222.7
	6.74	240.5	215.8
	7.91	241.0	216.2
Test 2	6.87	132.4	349.6
Cell pressure 450 kPa,	7.36	138.8	346.7
Initial back pressure 310 kPa	8.44	145.3	339.8
	9.40	151.3	333.0
	10.38	152.6	329.0
	11.63	152.8	329.2
Test 3	2.05	72.7	371.4
Cell pressure 420 kPa,	3.34	87.4	367.0
Initial back pressure 340 kPa	4.18	95.9	363.3
	5.80	91.8	351.1
	6.60	83.5	345.7

Solution

To draw the Mohr circles of stress, we must first calculate the values of major and minor principal effective stress σ_1' and σ_3' at each stage. As this is an undrained compression test, the minor principal effective stress σ_3' is given by

$$\sigma_3' = \sigma_c - u$$

where σ_c is the cell pressure, and the major principal effective stress σ_1' is given by

$$\sigma_1' = \sigma_3' + q$$

(section 5.3).

The mobilized strength ϕ_{mob}' is given by

$$\phi_{mob}' = \sin^{-1}(\sigma_1' - \sigma_3')/(\sigma_1' + \sigma_3') \text{ (Figure 5.6).}$$

Tha calculated values of σ_3', σ_1' and ϕ_{mob}' for each stage of all three tests are given in Table 5.4.

Table 5.4 Processed triaxial test data, Sample A

	Axial strain ε_{ax} (%)	Deviator stress q (kPa)	Pore water pressure u (kPa)	Minor principal effective stress σ_3' (kPa) ($= \sigma_c - u$)	Major principal effective stress σ_1' (kPa) ($= \sigma_3' + q$)	ϕ_{mob}' (degrees)
Test 1	3.89	207.9	235.1	174.9	382.8	21.9
$\sigma_c = 410$ kPa	4.56	219.2	230.5	179.5	398.7	22.3
	5.68	232.5	222.7	187.3	419.8	22.5 (peak)
	6.74	240.5	215.8	194.2	434.7	22.5
	7.91	241.0	216.2	193.8	434.8	22.5 (crit)
Test 2	6.87	132.4	349.6	100.4	232.8	23.4
$\sigma_c = 450$ kPa	7.36	138.8	346.7	103.0	242.1	23.7 (peak)
	8.44	145.3	339.8	110.2	255.5	23.4
	9.40	151.3	333.0	117.0	268.3	23.1
	10.38	152.6	329.0	121.0	273.6	22.8
	11.63	152.8	329.2	120.8	273.6	22.8 (crit)
Test 3	2.05	72.7	371.4	48.6	121.3	25.3
$\sigma_c = 420$ kPa	3.34	87.4	367.0	53.0	140.4	26.9
	4.18	95.9	363.3	56.7	152.6	27.3 (peak)
	5.80	91.8	351.1	68.9	160.7	23.6
	6.60	83.5	345.7	74.3	157.8	21.1

The peak strengths are given by the maximum values of ϕ_{mob}' in each test. These are indicated (peak) in Table 5.4. The peak strength, expressed as a ratio of shear to normal effective stress, does not necessarily coincide with the peak deviator stress q_{max}.

The critical state should be indicated by a period of shearing at a constant stress state, that is, the values of σ_1' and σ_3' should not change. From Table 5.4, it would appear that the samples in tests 1 and 2 do reach critical states, because the stress states for the last two data points in each case are more or less the same.

In test 3, however, a critical state is not achieved, probably because the sample ruptures at or near the peak strength. With continued shearing, deformation occurs by relative sliding on the rupture surface, rather than by straining as a continuum. The soil in the vicinity of the rupture softens as it draws in water from the rest of the sample. Eventually, the rupture surface will become polished, and its strength will fall to a residual value as described in section 5.14.

All three samples are overconsolidated and dry of critical, as evidenced by the reducing pore water pressures during shear as failure is approached.

Mohr circles of stress for the three tests at the point of the development of the peak strength are plotted in Figure 5.12(a). The peak strength failure envelope is curved, indicating a reducing peak strength (defined as $\phi'_{peak} = \tan^{-1}(\tau/\sigma')_{peak}$; Equation (2.16)) with increasing effective stress σ', due to the reduced potential for dilation.

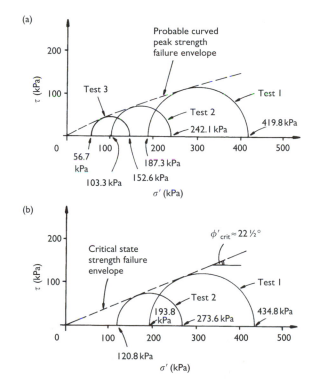

Figure 5.12 Mohr circles of stress: (a) peak strengths; (b) at critical states.

Mohr circles of stress at critical states are plotted for tests 1 and 2 (but not test 3, for which a critical state cannot be identified) in Figure 5.12(b). This is a straight line through the origin, with a critical state angle of shearing resistance $\phi'_{crit} \approx 22.5°$.

5.6   Undrained shear strengths of clay soils

So far, we have discussed failure envelopes for soils in terms of effective stresses. If these failure criteria are to be applied to a practical situation, it is necessary to determine the effective stresses, which means that the pore water pressures must be known or calculated.

Steady state pore water pressures can be estimated using techniques such as flownet sketching (Chapter 3). However, we have seen in Chapter 4 that clay soils will take some time to respond to a change in loading or in hydraulic boundary conditions. Changes in external loads are taken initially by changes in pore water pressure. The pore water pressures move slowly towards their long-term steady state values as the clay consolidates or swells, allowing changes in effective stress to occur. Except for the simple geometry of one-dimensional consolidation, the non-equilibrium pore water pressures – and hence the effective stresses – in a clay soil are difficult to predict using simple methods.

Fortunately, the critical state framework provides an alternative way of investigating the strength of a clay soil in the short term, assuming that the clay is sheared quickly in comparison with the time it takes for drainage of pore water to occur. Such conditions are conventionally termed **undrained**, as there is no drainage of pore water into or out of the soil skeleton, and hence (provided that the soil is saturated) no change in specific volume during shear.

In section 2.8 we saw that, when sheared, a soil will eventually reach a critical state in which continued shear may take place without any further changes in shear stress, normal effective stress or specific volume. Possible combinations of shear stress, normal effective stress and specific volume at critical states form a unique line – the critical state line – when plotted on a three-dimensional graph with axes representing shear stress, normal effective stress and specific volume. If the value of any one of these parameters at a critical state is known, the other two may be determined from the equations that define the critical state line.

For a clay soil sheared undrained, there is no change in specific volume, and the initial specific volume of the sample (which, assuming that the soil is saturated, may be determined from the water content – equation (2.12)) fixes the point at which failure on the critical state line is reached, and hence the shear stress at failure τ_c. The effective stress state at failure on the critical state line will be the same, irrespective of the external changes in stress applied, and the pore water pressure u will take up the difference between the total stress σ and the effective stress σ'.

For a sample of clay sheared undrained to failure, there is only one possible Mohr circle of effective stress. The diameter of this Mohr circle – and hence its position on the σ' axis, given that it must touch the failure envelope $\tau = \sigma' \tan \phi'_{crit}$ – is equal to twice the shear stress at failure τ_c, and is a function of the specific volume of the clay as sheared. The position of the Mohr circle of total stress depends on the applied loading and the pore water pressures (which may be negative or positive) generated. However, the diameter of the Mohr circle of total stress must be the same as the diameter of the Mohr circle of effective stress. The envelope to all possible Mohr circles of total stress is given by

$$\tau_{max} = \pm \tau_u \tag{5.18}$$

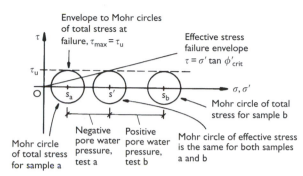

Figure 5.13 Mohr circle representation of undrained shear strength failure criterion, for clay sheared undrained at constant specific volume.

where τ_{max} is the maximum shear stress within the sample, and τ_u is the undrained shear strength of the clay[3] (Figure 5.13). From the geometry of the Mohr circle of total stress, the maximum shear stress within the sample τ_{max} is related to the major and minor principal total stresses σ_1 and σ_3 by

$$\tau_{max} = (\sigma_1 - \sigma_3)/2 = q/2$$

where q is the deviator stress. Hence

$$\tau_u = q_c/2 \tag{5.19}$$

where q_c is the deviator stress at (undrained) failure.

The τ_u model for failure in terms of total stresses represents a very special case, and is applicable only to clay soils sheared at constant volume. Shearing at constant volume may take place in an undrained triaxial test, or in a real construction where failure occurs rapidly. In either case, the development of a rupture surface would result in local drainage, as the clay in the immediate vicinity of the rupture surface softens preferentially by taking in water from the surrounding soil.

With clays, it is often necessary to investigate the possibility of failure in both the short term and the long term. The long-term calculation must be carried out in terms of effective stresses and pore water pressures. Where the applied load results in an increase in average effective stress p', for example, in the case of a foundation, the short-term condition is likely to be more critical. It will be shown in Example 5.6 that for normally consolidated or lightly overconsolidated clays, the deviator stress at failure q_c is much smaller in undrained loading than in drained loading, where the clay is given time to consolidate. Where the applied change in load results in a decrease in the average effective stress p', as in the case of an excavation, the long-term condition may be more critical, because the clay will swell and q_c is reduced. The failure of free-draining soils must always be analyzed in terms of effective stresses and pore water pressures.

The undrained shear strength τ_u of a clay of a given specific volume is traditionally investigated by carrying out unconsolidated–undrained triaxial tests at three different cell pressures. When the Mohr circles of total stress are plotted, they should all have the same radius τ_u. Their common tangent should be horizontal, intercepting the τ axis at $\tau = \tau_u$ (Figure 5.14(a)).

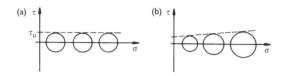

Figure 5.14 How and how not to determine the undrained shear strength τ_u of a clay of given specific volume. (a) Circles all the same size, radius τ_u; (b) circle diameter increases with confining pressure; samples not saturated or of different specific volumes.

If the common tangent is not horizontal (Figure 5.14(b)), the inference is that the samples as tested did not have the same specific volume at failure. This would occur if the samples had been allowed to consolidate to equilibrium at the test cell pressure before the start of shear (i.e. the tests carried out were consolidated–undrained, rather than unconsolidated–undrained). In this case, the test results provide an indication of the increase in τ_u with decreasing water content, due to increasing initial stress, perhaps corresponding to increasing depth in the field.

Alternatively, the samples may not have been fully saturated. In this case, the air is compressed and dissolved in the pore water as the cell pressure is increased, so that changes in sample volume and void ratio occur without the passage of pore fluid (air or water) into or out of the sample.

The notion that test results such as those shown in Figure 5.14(b) may be described in terms of an undrained friction angle ϕ_u and an undrained cohesion intercept c_u, though regrettably still found in some textbooks, is fundamentally flawed.

Example 5.3 Determining the undrained shear strength from triaxial test data

For each of the three tests described in Example 5.2, determine the undrained shear strength. Explain why this is different for each case. Suggest a relationship between undrained shear strength and depth, in a uniform bed of this clay having unit weight $18\,kN/m^3$ and a water table at the surface of the deposit.

Solution

The maximum shear stress in the sample at any stage of the test is given by the radius of the Mohr circle of stress, $\tau_{max} = (\sigma_1' - \sigma_3')/2$. In a triaxial test, the deviator stress $q = (\sigma_1' - \sigma_3')$, so that the maximum shear stress is equal to $q/2$. In tests 1 and 2, the maximum deviator stress occurs at the critical state, $q_{max} = 241\,kPa$ (test 1), and $q_{max} = 152.8\,kPa$ (test 2). In test 3, there is no obvious critical state, but the peak deviator stress is $95.9\,kPa$. (In tests 1 and 2, the peak deviator stress coincides with the critical state, rather than the peak stress ratio $(q/p')_{peak}$ or ϕ_{peak}'.) Dividing these values of q_{max} by 2 to obtain the undrained shear strength τ_u in each case

$\tau_u = 120.5\,kPa$ (test 1, critical state)

$\tau_u = 76.4\,\text{kPa (test 2, critical state)}$

$\tau_u = 48.0\,\text{kPa (test 3, peak)}$

The undrained shear strengths are different because each sample has been tested at a different water content. The trend of increasing undrained shear strength with increasing initial effective stress p' or decreasing water content is exactly as would be expected.

The depth to which each sample corresponds depends on its stress state at the start of the test. (The final effective stress state, both in the triaxial tests and in the field, depends – according to the critical state model – on the water content.) Assuming that the stress state within the deposit is such that $\sigma'_h = \sigma'_v$ (due to the effects of overconsolidation: section 4.10.3), the *in situ* variation in average effective stress p' with depth is given by

$$\sigma'_v = \sigma'_h = p' = (\gamma \cdot z - \gamma_w \cdot z) = (18z - 9.81z)\,\text{kPa} = 8.19z\,\text{kPa}$$

The initial value of p' in each test is given by subtracting the initial back pressure from the cell pressure. The corresponding field depth z (in m) is obtained by dividing the initial value of p' (in kPa) by 8.19 kPa/m. Values of τ_u, initial p' and corresponding depth z for each test are given in Table 5.5. τ_u is plotted against depth in Figure 5.15, which gives the relationship

$$\tau_u\,(\text{kPa}) = 4.73\,(\text{kPa/m}) \times z\,(\text{m})$$

Table 5.5 Undrained shear strengths, initial p'
and corresponding sample depths

Test	τ_u (kPa)	p' initially (kPa)	z (m)
I	120.5	210	25.6
2	76.4	135	16.5
3	48.0	80	9.8

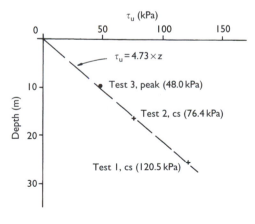

Figure 5.15 Undrained shear strength against equivalent depth.

5.7   Isotropic compression and swelling

Traditionally, triaxial test samples were brought to the starting point for the shear test by increasing (and in some cases then reducing) the cell pressure, without the application of deviator stress q through the ram. Provided that the drainage taps are left open, the sample is able to expel or take in water until its specific volume is in equilibrium with the applied cell pressure. As the applied stress is the same in all three directions, the sample undergoes **isotropic** compression or swelling as the cell pressure is increased or reduced. State paths during isotropic compression and swelling may be plotted on a graph of v against $\ln p'$ (Figure 5.16). For clay samples, these are analogous to the graphs of v against $\ln \sigma'_v$ plotted for equilibrium states in the oedometer (Figure 4.6).

For a clay which is being compressed for the first time, there is a unique relationship between the specific volume v and the mean normal effective stress p'. This is represented by a straight line on the graph of v against $\ln p'$ known as the **isotropic normal compression line (INCL)**,

$$v = N_p - \lambda \ln p' \tag{5.20}$$

The behaviour of the soil on unloading and reloading is generally idealized as straight line of slope $-\kappa$ (Figure 5.16(b)).

During first loading, strains are substantially **plastic** and **irrecoverable**: during unloading and reloading, the strains are **elastic** (but not linearly so) and **recoverable**. This behaviour is essentially the same as already seen in the oedometer, except that the modes of deformation are different.

In isotropic compression and swelling, the **overconsolidation ratio** or OCR n_p is defined in terms of p' rather than σ'_v:

$$n_p = p'_0/p' \tag{5.21}$$

where p' is the current value of the average effective stress, and p'_0 is the maximum value of average effective stress to which the sample has previously been subjected. As in one-dimensional compression and swelling, a **normally consolidated** soil (which

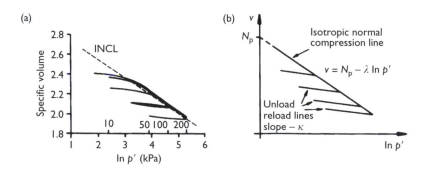

Figure 5.16 Normal compression line and elastic unload/reload lines plotted as v against $\ln p'$: isotropic compression and swelling. (a) Actual data (Redrawn with permission from Allman and Atkinson, 1992); (b) idealization.

has $n_p = 1$) has never before been subjected to a higher value of p' than that which currently acts. A soil that has in the past been subjected to a higher value of p' than the current value has $n_p > 1$, and is described as **overconsolidated**.

5.8 Sample preparation by one-dimensional compression and swelling: K_0 consolidation

The preparation of triaxial test samples by isotropic compression and swelling in the triaxial cell, so that the shear test starts from a deviator stress of zero, is a convention which arose primarily because it was a convenient procedure for a triaxial apparatus without computer control.

In high-quality tests, samples are now often prepared by increasing (and then possibly decreasing) the cell pressure and the deviator stress in a ratio which results in zero horizontal strain (i.e. one-dimensional compression), so as to mimic the stress history and stress state of an undisturbed soil element in the ground – or in an oedometer.[4] This technique is known as K_0 **consolidation**, K_0 being the symbol used to denote the *in situ* earth pressure coefficient (i.e. the ratio of the horizontal to vertical effective stress, σ_h'/σ_v') in undisturbed soil (section 4.10.3). For a sample prepared in this way, the deviator stress is non-zero at the start of the shear test.

The stress path followed in terms of q and p' during K_0 consolidation may be estimated, by means of Jaky's (1944) empirical relationship between horizontal and vertical effective stress in one-dimensional normal compression (section 4.10.3):

$$\sigma_h' = (1 - \sin \phi')\sigma_v' \tag{4.74}$$

Taking the radial effective stress σ_r' to represent the horizontal effective stress σ_h' in the ground, and the axial effective stress σ_a' to represent the vertical effective stress σ_v',

$$\sigma_r' = (1 - \sin \phi')\sigma_a' \tag{5.22}$$

Hence

$$q = \sigma_a' - \sigma_r' = \sigma_a'[1 - (1 \sin \phi')] = \sigma_a' \sin \phi' \tag{5.23a}$$

and

$$p' = (\sigma_a' + 2\sigma_r')/3 = \sigma_a'(3 - 2 \sin \phi')/3 \tag{5.23b}$$

giving

$$q/p' = (3 \sin \phi')/(3 - 2 \sin \phi') \tag{5.23c}$$

(Figure 5.17(a)).

Since during one-dimensional compression, p' is proportional to $\sigma_a'(\sigma_v')$ (from equation (5.23b)), the slope of the one-dimensional normal compression line (section 4.2: Figure 4.6) should be the same as that of the isotropic normal compression line (in either a $v, \ln p'$ or a $v, \ln \sigma_v'$ plot). However, the intercept of the normal compression line with the v-axis (at $\ln p' = 0$ or $p' = 1$ kPa), will be different. In general, the normal compression line at any constant stress ratio $\eta (= q/p')$ will on a graph of v against $\ln p'$ have slope $-\lambda$, but the intercept with the v-axis will depend on the stress ratio η. The one-dimensional and isotropic compression lines in $v, \ln p'$ space are therefore separated by a constant vertical distance (Figure 5.17(b)).

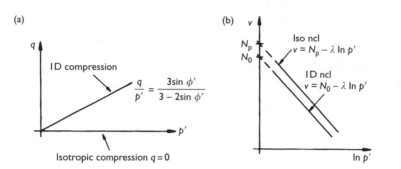

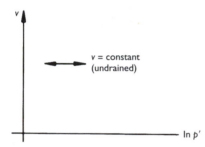

Figure 5.17 Comparison of state paths on graphs of q against p' and v against $\ln p'$ during isotropic and one-dimensional compression (K_0 consolidation).

Figure 5.18 State path on a graph of v against $\ln p'$ for undrained triaxial compression.

In one-dimensional swelling, the ratio σ_h'/σ_v' is not constant, but increases as unloading continues. Hence the ratio σ_v'/p' changes during unloading, and the one-dimensional and isotropic unload/reload lines are not parallel when plotted on a graph of v against $\ln p'$.

5.9   Conditions imposed in shear tests

In a conventional undrained triaxial compression test, the sample deforms with changes in deviator stress q and average effective stress p' at constant specific volume v (assuming that it is saturated), so that the state path on a graph of v against $\ln p'$ is horizontal (Figure 5.19). On a graph of q against p, the total stress path ascends with gradient $dq/dp = 3$ from the starting point ($p = \sigma_c, q = 0$) according to equation (5.11). The average effective stress p' must be calculated by subtracting the pore water pressure from p. In an extension test, equation (5.11) is still valid, but the changes in q and p are negative. Some common total stress paths are illustrated in Figure 5.19.

In a drained triaxial test, the change in pore water pressure is zero. The effective stress path is therefore parallel to the total stress path, with slope $dq/dp' = dq/dp = 3$ from the starting points ($p = \sigma_c, p' = \sigma_c - u_0, q = 0$), where u_0 is the constant back pressure. If $u_0 = 0$, the effective and total stress paths are coincident. To plot the state path in terms of v against $\ln p'$, the specific volume v must be determined at each stage of the test from the total volume of the sample at the start of the shear

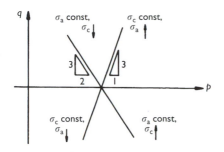

Figure 5.19 Total stress paths in q, p plot for triaxial compression and extension.

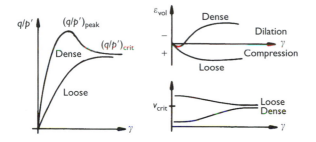

Figure 5.20 q/p', ε_{vol} and v against γ for drained triaxial compression tests on clays at different initial water contents.

test V_{t0} (equation (5.7)) and the volume of water expelled since the start of the shear test, ΔV_{tq}. Equation (5.12) may be written in general terms to relate the total volume $V_t = V_{t0} - \Delta V_{tq}$ and the specific volume v at any stage of the test:

$$v = (V_{t0} - \Delta V_{tq})G_s\rho_w/m_s \tag{5.24}$$

The term $(G_s\rho_w/m_s)$ is calculated from the final water content w_f, using equations (5.12) and (5.13):

$$(G_s\rho_w/m_s) = v_f/V_{tf} = (1 + w_f G_s)/V_{tf} \tag{5.25}$$

Substituting equation (5.25) into equation (5.24),

$$v = (V_{t0} - \Delta V_{tq})(1 + w_f G_s)/(V_{tf}) \tag{5.26}$$

5.10 Critical states

Figure 5.20 shows idealized graphs of stress ratio q/p', ε_{vol} and v as functions of shear strain γ for two drained triaxial compression tests on samples of clay having different initial water contents, corresponding to dense and loose states of particle packing. Although the denser sample (i.e. the one with the lower initial moisture content) displays a higher peak strength, this is only sustainable while the clay is **dilating** (i.e. the volume is increasing and ε_{vol} is becoming more negative). The stress ratio q/p' at the end of each test is the same, and the difference between the peak and

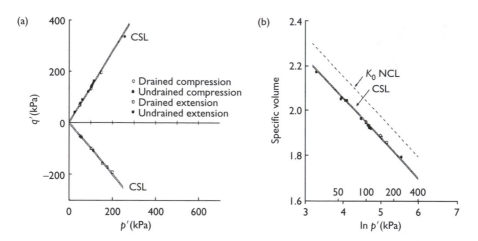

Figure 5.21 End points of triaxial compression and extension tests in q, p' and $v, \ln p'$ plane. (Redrawn with permission from Allman and Atkinson, 1992.)

the final strength at any stage of the test is related to the rate of dilation, $d\varepsilon_{vol}/d\gamma$. The end points of both tests lie on straight lines on graphs of q against p' and v against $\ln p'$. The end points of other tests on the same type of soil – whether drained or undrained – would also lie on these lines (Figure 5.21).

The end points of the shear tests represent **critical states,** where the soil continues to deform at constant stress ratio q/p' and constant specific volume. The line joining critical states (often referred to as the **critical state line** or **CSL**) has equations

$$q = Mp' \tag{5.27}$$

and

$$v = \Gamma - \lambda \ln p' \tag{5.28}$$

where M, Γ and λ are parameters whose values depend on the soil type, and are determined from triaxial tests.

Figure 5.21 shows two-dimensional projections (onto the q, p' and v, p' planes) of what is in reality a single line in three-dimensional q, p', v plot. The critical state line is parallel to the isotropic normal compression line on the $v, \ln p'$ plot, and is located a distance $(\lambda - \kappa)$ vertically below it. Recalling that the isotropic normal compression line has equation $v = N_p - \lambda \ln p'$ (equation (5.20)), and considering the separation of the iso-NCL and the CSL at $p' = 1$ kPa ($\ln p' = 0$), we have

$$N_p - \Gamma = \lambda - \kappa$$

or

$$N_p = \Gamma + \lambda - \kappa \tag{5.29}$$

The critical state line shown in Figure 5.21 is exactly analogous to that developed in Chapter 2 from the results of shearbox tests. The deviator stress q is a measure of the shear stress acting on the soil, which was characterized in the shearbox test by the shear stress τ on the central horizontal plane. The average effective stress p' is analogous to the effective stress σ' on the central horizontal plane of the shearbox. The critical state parameter M is a measure of the ratio of shear to normal effective stresses at failure, and is therefore related to the effective soil friction angle ϕ'_{crit}.

5.10.1   Relation between M and ϕ'_{crit}

From Figure 5.6,

$$\sin \phi'_{mob} = t/s' = (\sigma'_1 - \sigma'_3)/(\sigma'_1 + \sigma'_3) \tag{5.30}$$

In triaxial compression,

$$\sigma'_a = \sigma'_1, \quad \sigma'_r = \sigma'_3$$
$$q = (\sigma'_a - \sigma'_r) = (\sigma'_1 - \sigma'_3)$$

and

$$p' = (\sigma'_a + 2\sigma'_r)/3 = (\sigma'_1 + 2\sigma'_3)/3$$

Hence

$$(\sigma'_1 - \sigma'_3) = q \tag{5.31a}$$

and

$$(\sigma'_1 + \sigma'_3) = (6p' + q)/3 \tag{5.31b}$$

Substituting equations (5.31) into equation (5.30), with $\phi'_{mob} = \phi'_{crit}$ and $q/p' = M$ at the critical state,

$$\sin \phi'_{crit} = 3M/(6 + M) \tag{5.32a}$$

or

$$M = 6 \sin \phi'_{crit}/(3 - \sin \phi'_{crit}) \tag{5.32b}$$

At a general stage of the test prior to failure when the stress ratio $q/p' = \eta$, the mobilized effective angle of shearing ϕ'_{mob} is given by

$$\sin \phi'_{mob} = \frac{3\eta}{6 + \eta} \tag{5.32c}$$

In triaxial extension,

$$\sigma'_a = \sigma'_3, \quad \sigma'_r = \sigma'_1$$
$$q = (\sigma'_a - \sigma'_r) = (\sigma'_3 - \sigma'_1)$$

and

$$p' = (\sigma'_a + 2\sigma'_r)/3 = (\sigma'_3 + 2\sigma'_1)/3$$

Hence

$$(\sigma'_1 - \sigma'_3) = -q \tag{5.33a}$$

and

$$(\sigma'_1 + \sigma'_3) = (6p' + q)/3 \tag{5.33b}$$

Substituting equations (5.33) into equation (5.30), with $\phi'_{mob} = \phi'_{crit}$ and $q/p' = -M$ at the critical state,

$$\sin \phi'_{crit} = 3M/(6 - M) \tag{5.34a}$$

or

$$M = 6 \sin \phi'_{crit}/(3 + \sin \phi'_{crit}) \tag{5.34b}$$

At a general stage of the test prior to failure, when the stress ratio $q/p' = \eta$, the mobilized effective angle of shearing ϕ'_{mob} is given by

$$\sin \phi'_{mob} = \frac{3\eta}{6 - \eta} \tag{5.34c}$$

Comparison of equations (5.34) with equations (5.32) shows that the relation between M and ϕ'_{crit} depends on the mode of deformation, and is different for triaxial compression and extension. More specifically, the relation between M and ϕ'_{crit} depends on the value of the intermediate principal effective stress σ'_2. Starting with the full definitions of q and p' given in equations (5.3) and (5.4), and writing the intermediate principal effective stress in terms of the parameter

$$b = (\sigma'_2 - \sigma'_3)/(\sigma'_1 - \sigma'_3) \tag{5.35}$$

Bishop (1971) showed that

$$\sin \phi'_{crit} = 3M/\left\{\left[6\sqrt{(1 - b + b^2)}\right] - [(2b - 1)M]\right\} \tag{5.36}$$

In triaxial compression, $\sigma'_2 = \sigma'_3$ and $b = 0$, so that equation (5.36) reduces to equation (5.32a). In triaxial extension, $\sigma'_2 = \sigma'_1$ and $b = 1$, so that equation (5.36) reduces to equation (5.34a). The relation between ϕ'_{crit} and M in plane strain is often estimated using equation (5.36) with $b = 0.5$.

5.11 Yield

A material **yields** when its stress–strain behaviour changes from being purely elastic to partly plastic – in other words, when the deformation stops being recoverable in unloading. **Yield** and **failure** are not synonymous, and in general are not coincident.

Figure 5.22 illustrates the stress–strain behaviour of a copper wire during a cyclic tensile test. At very small strains, the behaviour of the material is perfectly elastic. If it is loaded to a point such as A and then unloaded, all of the deformation is recovered. This is what is meant by the term **elastic**.

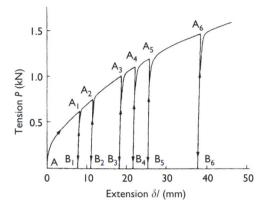

Figure 5.22 Tension test on copper wire. (Redrawn with permission from Muir Wood, 1990; data from Taylor and Quinney, 1931, reworked by D. Muir Wood from original data published in the Proceedings of the Royal Society.)

If the wire is loaded to a point such as A_1, not all of the deformation is recovered on unloading (to B_1). The material has deformed **plastically**, so that it retains a **permanent set**. However, on second loading, the behaviour of the copper wire is purely elastic until the initial loading point A_1 is reached. Only as the load is increased beyond this point (to A_2, say) is further plastic strain induced, increasing the permanent set on unloading (e.g. from A_2 to B_2). On third loading, yield does not occur until A_2. As the load is increased further, the stress–strain curve becomes gradually flatter. Eventually a point may be reached at which the material is **fully plastic,** and continues to deform at constant stress. This point represents the failure of the sample.

Between yield and failure, the capacity of the material to sustain an increasing stress with increasing plastic strain is known as **work hardening**. (Mild steel, one of the most common engineering materials, is perhaps unusual in that it does not work harden to any significant extent, with the result that yield and failure occur effectively simultaneously. Metals such as mild steel which undergo ductile failure continue to deform under constant load, in contrast with brittle materials such as cast iron or glass which fail by fracture, with the sudden loss of all their load-carrying ability.)

In soils, the onset of a **critical state** at which deformation continues at constant stress ratio and volume represents the **failure** of the material. The failure of the soil is usually undesirable in geotechnical engineering. It is therefore important to be able to predict combinations of q and p' (or shear and normal effective stress) that will cause failure. The critical state concepts described in Chapter 2 and section 5.10 provide the means to do this. It is also necessary to be able to predict combinations of q and p' that will cause yield, at which rates of deformation may be expected to increase significantly.

We have seen in sections 4.2 and 5.7 that a soil which is loaded, unloaded and reloaded either one-dimensionally or isotropically will deform plastically during loading if it is **normally consolidated** (i.e. has an OCR of unity, and has not previously been subjected to a higher value of p' or σ'_v than that which currently acts). During unloading and reloading, when the soil is **overconsolidated** (i.e. it has an OCR of more than 1, and has previously been subjected to a higher value of p' or σ'_v than that which now acts), it behaves elastically, in the sense that deformation is substantially recoverable.

Figure 5.22 looks rather like Figure 4.6 or Figure 5.11 turned on its side, except that the slope of the normal compression lines does not flatten off because failure cannot be achieved in either one-dimensional or isotropic compression. In isotropic (or one-dimensional) loading, an overconsolidated soil will yield when the preconsolidation pressure is exceeded (Figure 5.23).

Most real stress paths will involve the application of a non-zero deviator stress q. It is therefore necessary to be able to predict the combination of stresses (p' and q) that will cause a soil, which has previously been isotropically (or one-dimensionally) compressed to a preconsolidation pressure p'_0 (or σ'_{v0}), to yield. This may be investigated experimentally by subjecting similar samples of soil with the same stress history to different exploratory stress paths in the q, p' plane (Figure 5.24).

The need to predict combinations of types of stress or load which will cause a material to yield is not confined to soil mechanics. In 1931, Taylor and Quinney published the results of tests in which thin-walled copper tubes were subjected to combined tension P and torsion Q, in order to determine the line representing combinations of P and Q which would cause yield (Figure 5.25). This line is known as the **yield locus**.

For soils, Roscoe and Schofield (1963) proposed a model incorporating yield, known as **Cam clay**.[5] In Cam clay, the yield locus defining combinations of deviator stress q

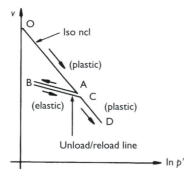

Figure 5.23 Yield of a soil under isotropic loading.

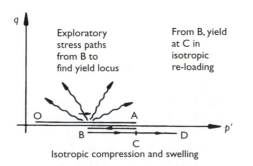

Figure 5.24 Exploratory stress paths to determine combinations of q and p' that will cause an overconsolidated soil to yield.

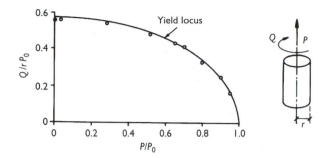

Figure 5.25 Yield of thin walled copper tubes under combined tension and torsion. P_0 is the axial load which, acting alone, causes yield. (Redrawn with permission from Muir Wood, 1990; data from Taylor and Quinney, 1931, reworked by D. Muir Wood from original data published in the Proceedings of the Royal Society.)

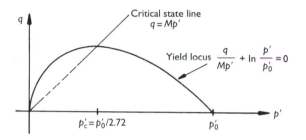

Figure 5.26 Cam clay yield locus and critical state line on a graph of q vs p'.

and average effective stress p' which would cause yield of a soil preconsolidated isotropically to an average effective stress of p_0' is given by the expression (Figure 5.26).

$$q/Mp' + \ln(p'/p_0') = 0 \qquad (5.37)$$

Soils are more complex than most metals, not least because their specific volume must be taken into consideration. The yield locus defined by equation (5.37) is associated not with a constant specific volume, but with the particular unload/reload line that intersects the isotropic normal compression line at $p' = p_0'$. The yield locus may be imagined as suspended above the unload/reload line in three-dimensional q, p', v space: the succession of yield loci associated with the infinite number of possible unload reload lines represents a three-dimensional yield surface (Figure 5.27).

The Cam clay model comprises the yield locus (equation (5.37)), the isotropic normal compression line (equations (5.20) and (5.29)) and the critical state line (equations (5.27) and (5.28)), and is fully defined by the soil properties Γ, λ, κ and M.

Equation (5.37) may be derived, following Roscoe and Schofield (1963), by considering the energy dissipated when a soil sample is loaded slowly in a drained triaxial test. First note that the work done ΔW in straining a unit cube of material by small amounts of $\delta\varepsilon_1$, $\delta\varepsilon_2$ and $\delta\varepsilon_3$ in the directions of the major, intermediate and minor

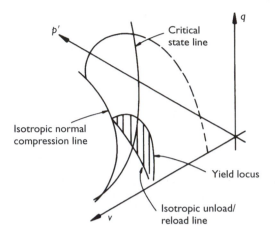

Figure 5.27 Three-dimensional yield surface and critical state line in q, p', v space.

principal effective stresses σ'_1, σ'_2 and σ'_3 respectively is given by

$$\Delta W = \sigma'_1 \cdot \delta\varepsilon_1 + \sigma'_2 \cdot \delta\varepsilon_2 + \sigma'_3 \cdot \delta\varepsilon_3 \tag{5.38}$$

(This is because work = force × distance, and for a unit cube with a principal stress acting on each face, the force is equal to the stress and the distance moved in the direction of the force is equal to the increase in strain, $\delta\varepsilon$, associated with that stress. The changes in strain must be small enough not to affect significantly the areas of the faces of the cube. To cause the changes in strain, the principal effective stresses σ'_1, σ'_2 and σ'_3 will themselves have had to be increased: provided the increases in principal effective stress $\delta\sigma'_1$, $\delta\sigma'_2$ and $\delta\sigma'_3$ are small, equation (5.38) still holds because the work associated with them is the product of two small quantities ($\delta\sigma'_1 \cdot \delta\varepsilon_1$ etc.) and may be neglected.) In a triaxial compression test, the intermediate and minor principal effective stresses are equal, $\sigma'_2 = \sigma'_3$. Hence, equation (5.38) can be re-written in terms of the triaxial stress parameters $p' (= [\sigma'_1 + \sigma'_2 + \sigma'_3]/3)$ and $q \ (= \sigma'_1 - \sigma'_3)$, the volumetric strain increment $\delta\varepsilon_{\text{vol}} \ (= \delta\varepsilon_1 + 2\delta\varepsilon_3)$ and the triaxial shear strain increment $\delta\varepsilon_q \ (= 2/3[\delta\varepsilon_1 - \delta\varepsilon_3]$ – see section 5.4.6 and equation (5.16)) as

$$\Delta W = q \cdot \delta\varepsilon_q + p' \cdot \delta\varepsilon_{\text{vol}} \tag{5.17}$$

This can be demonstrated by substituting the above expressions for p', $\delta\varepsilon_{\text{vol}}$, q and $\delta\varepsilon_q$ in terms of σ'_1, σ'_3, ε_1 and ε_3 into equation (5.17) and multiplying out the terms, which should give equation (5.38) with $\sigma'_2 = \sigma'_3$. As with equation (5.38), it is assumed that the changes in strain $\delta\varepsilon_{\text{vol}}$ and $\delta\varepsilon_q$ and the stress increments $\delta p'$ and δq applied to cause them are very small.

Now, let us assume that all of the work done in shear, the deviatoric component $q \cdot \delta\varepsilon_q$, is dissipated. This is equivalent to assuming that shear strains are irreversible. Volumetric strains, on the other hand, are partly elastic (recoverable) and partly plastic (irrecoverable). Writing the plastic or irrecoverable component of the volumetric strain as $\delta\varepsilon_{\text{vol,plastic}}$ we obtain an expression for the component of the work done on the soil element that is dissipated,

$$\Delta W_{\text{dissipated}} = q \cdot \delta \varepsilon_q + p' \cdot \delta \varepsilon_{\text{vol,plastic}} \tag{5.39}$$

Let us also assume that the work dissipated is lost in overcoming interparticle friction to cause shear. The work dissipated $\Delta W_{\text{dissipated}}$ will then be proportional to the amount of shear that has taken place $\delta \varepsilon_q$: $\Delta W_{\text{dissipated}} \propto \delta \varepsilon_q$. As interparticle frictional forces increase the harder the particles are pushed together, the work dissipated in friction will also increase in proportion to the average effective stress p' : $\Delta W_{\text{dissipated}} \propto p'$. If the overall constant of proportionality is M, the work dissipated in overcoming friction during shear is

$$\Delta W_{\text{dissipated}} = M \cdot p' \cdot \delta \varepsilon_q \tag{5.40}$$

Combining equations (5.39) and (5.40) and dividing through by ($p' \cdot \delta \varepsilon_{\text{vol,plastic}}$), we have

$$\frac{q}{p'} \cdot \frac{\delta \varepsilon_q}{\delta \varepsilon_{\text{vol,plastic}}} + 1 = M \cdot \frac{\delta \varepsilon_q}{\delta \varepsilon_{\text{vol,plastic}}}$$

or, writing q/p' as the stress ratio η,

$$(M - \eta) \cdot \frac{\delta \varepsilon_q}{\delta \varepsilon_{\text{vol,plastic}}} = 1 \tag{5.41}$$

The final assumption required to derive the equation of the Cam clay yield locus (equation (5.37)) is that the soil exhibits a property known as **normality**. This means that the direction of the plastic strain increment vector is normal (i.e. at right angles) to the yield locus when the stress and strain axes are superimposed. In the present case, the relevant stress axes are p' and q, and the corresponding plastic or irrecoverable strain axes are ε_q and $\varepsilon_{\text{vol,plastic}}$ (Figure 5.28).

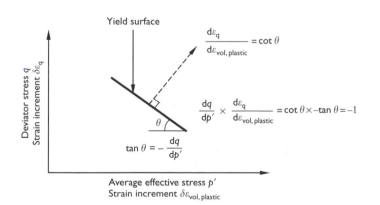

Figure 5.28 Normality of the strain increment vector to the yield locus (redrawn, with permission, from Bolton, 1991).

The normality condition leads to the relationship

$$\frac{dq}{dp'} \cdot \frac{\delta\varepsilon_q}{\delta\varepsilon_{vol,plastic}} = -1 \tag{5.42}$$

sometimes termed 'associated flow'. Substitution of equation (5.42) into equation 5.41 allows the strain increments to be eliminated, giving

$$(M - \eta) = \frac{\delta\varepsilon_{vol,plastic}}{\delta\varepsilon_q} = -\frac{dq}{dp'}$$

or

$$\eta = M + \frac{dq}{dp'} \tag{5.43}$$

Writing $q = \eta p'$ and differentiating with respect to p' we obtain

$$\frac{dq}{dp'} = \eta + p' \cdot \frac{d\eta}{dp'} \tag{5.44}$$

Substituting equation (5.44) into equation (5.43),

$$\eta = M + \eta + p' \cdot \frac{d\eta}{dp'}$$

or

$$\frac{dp'}{p'} = -\frac{d\eta}{M} \tag{5.45}$$

Integrating equation (5.45) between limits of $p' = p'_0$ and $\eta = 0$ in isotropic normal compression (i.e. the top of the yield locus) and a general point (p', η),

$$\int_{p'}^{p'_0} \frac{dp'}{p'} = -\int_{\eta}^{0} \frac{d\eta}{M}$$

or

$$\ln\left(\frac{p'_0}{p'}\right) = \frac{\eta}{M}$$

Substituting $\eta = q/p'$ and rearranging,

$$\frac{q}{M \cdot p'} + \ln\left(\frac{p'}{p'_0}\right) = 0 \tag{5.37}$$

More detailed derivations of the equation for the Cam clay yield locus may be found in Schofield and Wroth (1968) and Bolton (1991).

We will in sections 5.12 and 5.13 use the concept of the yield locus represented by equation (5.37) and Figure 5.26 to predict the state paths followed by triaxial test samples during shear. The methods used to predict these state paths are equally applicable to soil elements in the ground, which undergo changes in stress as a result of construction or excavation activities. Their application in this way is more difficult to visualize, because of the large number of soil elements being subjected to a variety of different stress paths. In practice, predictions of the behaviour of geotechnical engineering structures using Cam clay-type models are usually made by means of a finite element analysis (e.g. Britto and Gunn, 1987).

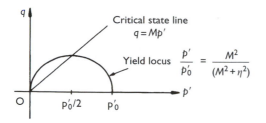

Figure 5.29 Modified Cam clay yield locus.

The Cam clay model is undoubtedly elegant, and represents an important contribution to modern soil mechanics. It does have some shortcomings. It is unable to distinguish between one-dimensional and isotropic compression, and the shape of the yield locus bears little resemblance to those obtained experimentally for many natural soils (Muir Wood, 1990). The first of these was addressed by Roscoe and Burland (1968), who proposed a variation to the original model, known as Modified Cam clay. Modified Cam clay has an elliptical yield locus, given by

$$p'/p_0' = M^2/(M^2 + \eta^2) \tag{5.46}$$

where p_0' is the equivalent isotropic preconsolidation pressure and η is the stress ratio q/p' (Figure 5.29).

The imperfections of the Cam clay model are, however, of only minor importance in comparison with the concepts it introduces and embodies.

Example 5.4 Determination of Cam clay parameters from experimental data

(a) Show that the total stress path followed in q, p space during a conventional triaxial compression test at constant cell pressure is given by

$$dq/dp = 3$$

from any starting point $(p, 0)$.

(b) A sample of saturated kaolin ($G_s = 2.61$) undergoes isotropic normal compression in a triaxial cell to a cell pressure of 200 kPa. At this stage, its total volume is 86×10^3 mm^3 and its water content is 61.28%. The cell pressure is increased to 400 kPa, and 5956 mm^3 of water is expelled. The cell pressure is then reduced to 300 kPa, and the volume of the sample increases by 476 mm^3. Use these data to determine the Cam clay parameters Γ, λ and κ.

[*University of London 2nd year BEng (Civil Engineering) examination,*
King's College (modified, part question)]

Solution

(a) The average total stress $p = (\sigma_1 + \sigma_2 + \sigma_3)/3 = (2\sigma_r + \sigma_a)/3$ and the deviator stress $q = (\sigma_a - \sigma_r)$.

In a conventional triaxial compression test, the cell pressure σ_r is constant and changes in p and q are related to changes in σ_a by

$$dp = d\sigma_a/3; \quad dq = d\sigma_a$$

so that

$$dq/dp = 3$$

(b) It is necessary to relate changes in overall volume ΔV_t to changes in the specific volume Δv.

$$V_t = V_s + V_v = V_s(1 + V_v/V_s) = V_s(1 + e) = V_s v$$

Therefore $\Delta V_t = V_s \Delta v$.

At a cell pressure of 200 kPa, $w = 0.6128$ and $e = wG_s$, so that $e = 0.6128 \times 2.61 = 1.599$. Thus, $v = 1 + e = 2.599$.

Hence, the volume of solids $V_s = V_t/v = 86\,000/2.599 = 33\,090 \text{ mm}^3$.

Compressing to 400 kPa, the change in total volume is 5956 mm^3. Hence, the change in specific volume Δv is

$$\Delta v = \Delta V_t/V_s = 5956/33\,090 = 0.180$$

From the Cam clay model, on the isotropic normal compression line[6]

$$\Delta v = \lambda \Delta \ln p' = \lambda \ln(400/200)$$
$$\lambda = 0.18/(\ln 2) = 0.26$$

Swelling back from 400 to 300 kPa on an unloading line,

$$\Delta v = \Delta V_t/V_s = 476/33\,090 = 0.01439$$
$$\Delta v = \kappa \Delta \ln p' = \kappa \ln(400/300)$$
$$\kappa = 0.01439/(\ln 4/3) = 0.05$$

On the isotropic normal compression line, $v = N_p - \lambda \ln p'$ and $N_p = (\Gamma + \lambda - \kappa)$, so

$$2.599 = N_p - 0.26 \ln 200$$

and

$$N_p = (\Gamma + 0.26 - 0.05) = 3.977, \text{ so}$$
$$\Gamma = 3.767$$

5.12 State paths during shear: normally consolidated and lightly overconsolidated clays

5.12.1 Drained tests

During a conventional drained triaxial compression test, the effective stress path in terms of q and p' has no option but to follow the imposed changes in total stress. Assuming that there is no back pressure, $u = 0$ and (from equations (5.10) and (5.11)),

$$p' = p = \sigma_c + q/3$$

This stress path is shown in Figure 5.30.

If the sample is initially normally consolidated, the stress path starts from the tip of the initial yield locus at $(p' = p'_0, q = 0)$, as shown in Figure 5.30(a). The stress state of the sample can never lie outside the current yield locus: as the deviator stress is increased, the yield locus expands so that the current stress state is always on the current yield locus. The soil **work hardens** (in that it is able to withstand higher and higher stress ratios q/p') as it deforms plastically, until it eventually **fails** when the stress path intersects the critical state line. By **failure**, we mean that the sample has reached a **critical state**, in which deformation continues at **constant specific volume** and **constant stress**.

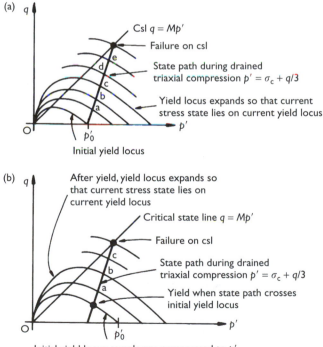

Figure 5.30 State paths in the q, p' plane during conventional drained triaxial compression tests on (a) normally consolidated and (b) lightly overconsolidated clay samples.

For a sample which is initially lightly overconsolidated, the early stages of the stress path lie within the initial yield locus, as shown in Figure 5.30(b). While the soil remains within the yield locus, its behaviour is described as **elastic** in the sense that deformations are recoverable. If we chose to describe the behaviour of the soil using conventional elastic parameters such as Young's modulus, the values of these would *not* be constant.

Increases in q after reaching the initial yield locus cause the yield locus to expand, so that again the current stress state lies on the current yield locus. Eventually, failure occurs when the stress path reaches the critical state line.

Using the Cam clay model, it is possible to predict the changes in specific volume that occur during a drained test. In the v, $\ln p'$ plane, the state path for a normally consolidated sample crosses a succession of κ-lines (i.e. unload/reload lines), each of which corresponds to a yield locus in the plane q,p' (Figure 5.31(a)). At any given deviator stress q, the current value of p' may be calculated from equation (5.10). The current value of p'_0 (at the tip of the yield locus) may be found using the equation of the yield locus in the q,p' plane (equation (5.37) for Cam clay; equation (5.46) for Modified Cam clay). The current specific volume can then be calculated from the Cam

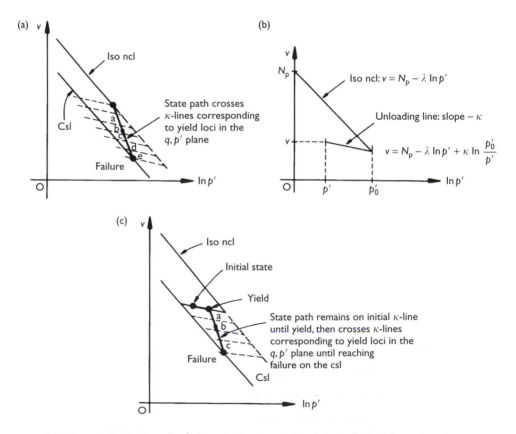

Figure 5.31 State paths in the v, $\ln p'$ plane during conventional drained triaxial compression tests on (a) normally consolidated and (c) lightly overconsolidated clay samples. Part (b) shows the calculation of the specific volume.

clay model in the v, $\ln p'$ plane (Figure 5.31(b)):

$$v = N_p - \lambda \ln p_0' + \kappa \ln(p_0'/p') \qquad (5.47)$$

The volume of water expelled at any stage since the start of shear may be calculated from the change in specific volume, using equation (5.12) written in difference form:

$$\Delta V_{tq} = (m_s \Delta v)/(G_s \rho_w)$$

For a sample that is initially lightly overconsolidated, the state path in the v, $\ln p'$ plane remains on the original κ-line until the initial yield locus is reached as in Figure 5.30(b). After yield, the yield locus expands so that the soil state is always on the current locus, as for a normally consolidated sample (Figure 5.31(c)).

5.12.2 Undrained tests

In an undrained test, the state path in the v, $\ln p'$ plane must be horizontal, because there can be no change in specific volume (Figure 5.18). However, the sample is able to generate either positive or negative pore water pressures, which means that in the q, p' plane the effective stress path followed differs from the imposed total stress path, $p = \sigma_c + q/3$.

The pore water pressure is given by the distance between the total (q, p) and effective (q, p') stress paths when these are drawn on the same diagram (Figure 5.33(c)).

As the test progresses, the state path in the v, $\ln p'$ plane crosses a succession of κ-lines as it moves towards the critical state line (Figure 5.32(a)). Each κ-line corresponds to a yield locus in the q, p' plane which has an associated value of p_0', given by the intersection of the κ-line with the isotropic normal compression line. The initial value of p_0' (p_0' in Figure 5.32) is known from the stress history of the sample. The value of p_0' which corresponds to the critical state (p_{0f}' in Figure 5.32) may be calculated from the equation of the critical state line (equation 5.28) and the known, constant specific volume of the sample being sheared. For intermediate values of p_0', the corresponding values of p' and q may be calculated using equation (5.47) (with the as-tested specific volume) and equation (5.37) respectively. The effective stress path followed on a graph of q against p' by an initially normally consolidated sample is shown in Figure 5.32(b).

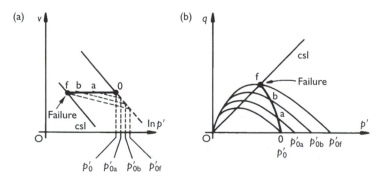

Figure 5.32 State paths in the $v, \ln p'$ and q, p' planes during a conventional undrained triaxial compression tests on a normally consolidated clay sample.

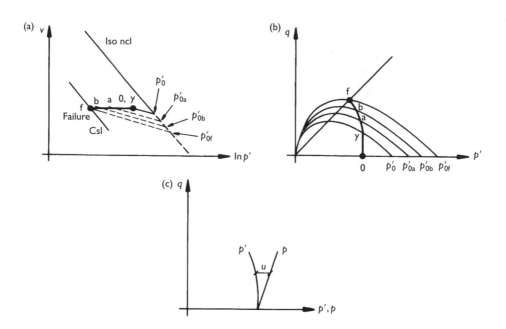

Figure 5.33 State paths in the $v, \ln p'$ and q, p' planes during a conventional undrained triaxial compression test on a lightly overconsolidated clay sample.

For a sample which is initially lightly overconsolidated, the early part of the state path lies inside the initial yield locus in the q, p' plane. This means that in the $v, \ln p'$ plane, the state path remains on the initial κ-line. As the test is undrained, the state path in the $v, \ln p'$ plane must also be horizontal. These two requirements can only be fulfilled if the state path stays in the same place in the $v, \ln p'$ plane that is, has $p' = $ constant until the initial yield locus is reached (Figure 5.33). If p' is not to change, it follows from equation (5.10) ($p' = \sigma_c - u + q/3$) that, if the test is carried out at constant cell pressure, the increase in pore water pressure Δu must be equal to one third of the increase in deviator stress, $\Delta u = \Delta q/3$.

After yield, the sample follows the same state path to the critical state as an initially normally consolidated sample of the same specific volume. Between yield and failure, the increase in pore water pressure with deviator stress du/dq accelerates, that is, $du > dq/3$ (Figure 5.33(b)). In construction works (such as embankments) which involve the application of a large surcharge to a normally consolidated or lightly overconsolidated clay, the pore water pressures within the clay are usually monitored. A rise in the rate of increase in pore water pressure with surcharge load would indicate that the clay had begun to yield, providing a warning to halt or at least slow down the rate of construction in order to avoid the undrained failure of the clay.

The state path followed between yield and failure is a projection of the three-dimensional yield locus onto a $v = $ constant plane, and is known as the **undrained state boundary** (Figure 5.34). The position of the undrained state boundary in the q, p' plane depends on the specific volume of the sample, but it always has the same shape. (This follows because the Cam clay yield loci associated with individual κ-lines are

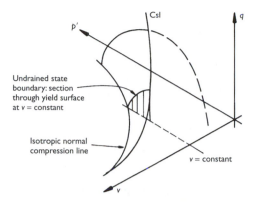

Figure 5.34 Three dimensional view of yield locus in q, p', v space showing undrained state boundary surface as projection onto a $v =$ constant plane.

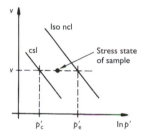

Figure 5.35 Definition of p'_c and equivalent consolidation pressure p'_e.

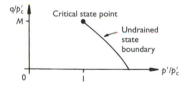

Figure 5.36 Undrained state boundary surface and critical state point in the $q/p'_c, p'/p'_c$ plane.

geometrically similar.) Thus, if q and p' are normalized with respect to some suitable reference stress, the undrained state boundary may be described as a single line.

Two alternative reference stresses are commonly used. These are the value of p' on the critical state line at the current specific volume p'_c, and the value of p' at the current specific volume on the isotropic normal compression line, p'_e (Figure 5.35). p'_e is known as the **equivalent consolidation pressure**. Atkinson (1993) argues that it is preferable to normalize with respect to p'_c, because the critical state line is unique, whereas the position of the normal compression line depends on the type of loading (e.g. isotropic or one-dimensional). The normalized undrained state boundary and the critical state are shown in Figure 5.36. On this normalized plot, the critical state reduces to a point at $p'/p'_c = 1, q/p'_c = M$.

During conventional triaxial compression, a normally consolidated or lightly over-consolidated sample tends to generate positive pore water pressures at an increasing rate between yield and failure. In a drained test, where these pore water pressures are not allowed to build up, the sample will expel water as it reduces in volume to the critical state. The initial states of normally consolidated and lightly overconsolidated samples lie between the critical state line and the isotropic normal compression line in the v, $\ln p'$ plane and are sometimes described as being on the **wet side** of the critical state.

Example 5.5 Determination of Cam clay parameters; prediction of state paths during shear using Cam clay concepts; comparison between normally consolidated and lightly overconsolidated samples

(a) Define in terms of principal stresses the parameters q, p and p'. Indicate how they relate to the quantities measured during a conventional undrained triaxial compression test.

(b) Table 5.6 gives data from both consolidation and shear stages of an undrained triaxial compression test on a sample (Sample A) of reconstituted London clay.

Table 5.6 Triaxial test data, Sample A

CP (kPa)	50	75	100	100	100	100	100	100
q (kPa)	0	0	0	13.4	25.5	35.9	44.1	48.9
u (kPa)	0	0	0	14.5	28.5	42.0	54.7	62.4
v		2.129	2.064	2.018	2.018	2.018	2.018	2.018	2.018

CP: cell pressure; q: deviator stress; u: pore water pressure; v: specific volume.

Plot the state paths followed by the sample on graphs of q against p' and v against $\ln p'$, and explain their shapes. Stating clearly the assumptions you need to make, estimate the soil parameters M, Γ and λ.

It is intended to prepare an identical sample (Sample B) in exactly the same way as Sample A, but the cell pressure is accidentally increased during isotropic compression to a maximum of 115 kPa. To bring Sample B to the same specific volume as Sample A ($v = 2.018$), it is found necessary to reduce the cell pressure to 80 kPa. Estimate the parameter κ.

Sketch and explain the shape of the effective stress path followed by Sample B on the graph of q against p', when it is subjected to an undrained triaxial compression test from a cell pressure of 80 kPa.

[University of London 2nd year BEng (Civil Engineering) examination, Queen Mary and Westfield College (modified, part question)]

Solution

(a) $q =$ deviator stress $= \sigma'_1 - \sigma'_3 = \sigma_1 - \sigma_3$

$p =$ mean total stress $= (\sigma_1 + \sigma_2 + \sigma_3)/3$

$p' =$ mean effective stress $= p - u$

   In a conventional triaxial compression test, $\sigma_2 = \sigma_3 =$ the cell pressure, the pore water pressure u is measured directly, and $q = Q/A$, where Q is the ram load and A is the current cross-sectional area of the sample.

   The area A is inferred from the axial strain and the assumed deformation of the sample as a right circular cylinder of constant total volume V_{t0}:

$A = V_{t0}/h = A_0 h_0/h$ and $h/h_0 = (h_0 - \Delta h)/h_0 = (1 - \varepsilon_a)$

so $A = A_0/(1 - \varepsilon_a)$, where ε_a is the axial strain.

(b) From the data in Table 5.6, calculate p' using

$p' = q/3 + CP - u$

Also, $v =$ constant during undrained shear. The calculated values of p', v and $\ln p'$ at each stage of the test are given in Table 5.7.

Table 5.7 Processed triaxial test data, Sample A

q (kPa)	0	0	0	13.4	25.5	35.9	44.1	48.9
p' (kPa)	50	75	100	90	80	70	60	53.9
v		2.129	2.064	2.018	2.018	2.018	2.018	2.018	2.018
$\ln p'$		3.912	4.317	4.605	4.500	4.382	4.248	4.094	3.987

The state paths followed in the q, p' and $v, \ln p'$ planes are shown in Figure (5.37).

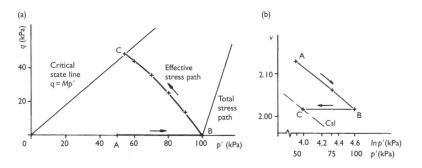

Figure 5.37 State paths in q, p' and $v, \ln p'$ planes (Sample A).

A to B: Isotropic normal compression, with no shear stress applied. The sample is being loaded for the first time, so that changes in specific volume are mainly plastic.

B to C: Undrained compression to failure, with gradually increasing shear stress. Undrained test, therefore v = constant. As the sample is initially normally consolidated, it yields immediately shear is applied, and remains on the current yield locus until the critical state is reached. This path is a section through the three-dimensional state boundary surface at v = constant, and will be followed between yield and failure by any sample on the wet side of the critical state having a specific volume of 2.018.

Assuming that the shear test ends on the critical state line, $q = Mp'$ at the end of the test,

$$M = 48.9/53.9 = 0.91$$

λ is the slope of the isotropic normal compression line:

$$\lambda = |\Delta v/\Delta \ln p'| = (2.129 - 2.018)/(4.605 - 3.912)$$
$$\lambda = 0.16$$

Γ may be calculated from the equation of the critical state line in the $v, \ln p'$ plane:

$$v = \Gamma - \lambda \ln p'$$

so that

$$2.018 = \Gamma - 0.16 \times \ln(53.9)$$
$$\Gamma = 2.656$$

The state path followed by Sample B in the $v, \ln p'$ plane is shown in Figure 5.38(a). The increase in specific volume on swelling back to $p' = 80 \, \text{kPa}$ from the overstress point B' (115 kPa) is equal to the reduction in specific volume following the isotropic normal compression line from 100 to 115 kPa. Thus

$$\Delta v = \lambda \ln(115/100) = \kappa \ln(115/80)$$

so that

$$\kappa = \lambda \ln(115/100)/\ln(115/80) = 0.385 \times \lambda$$
$$\kappa = 0.062$$

The state paths followed by Sample B are shown in Figure 5.38.

A' to B': Isotropic normal compression, as before.

B' to C': Isotropic unloading: only the 'elastic' component of the deformation during isotropic normal compression (from 80 to 115 kPa) is recovered.

C' to Y: Application of deviator stress q at constant specific volume, within existing yield locus; p' = constant.

Y: Soil yields on reaching initial yield locus.

Y to F: Soil follows undrained state boundary (i.e. the same path as Sample A) from yield to failure on the critical state line.

F: The critical state is reached, and any further deformation would take place at constant stresses q and p'.

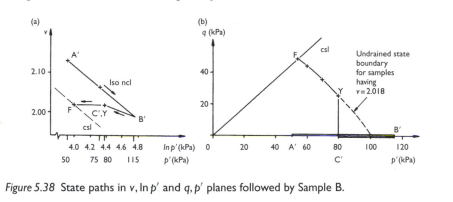

Figure 5.38 State paths in v, $\ln p'$ and q, p' planes followed by Sample B.

Example 5.6 Numerical prediction of state paths using the Cam clay model; contrast between undrained and drained state paths

(a) With the cell pressure at 300 kPa, the drainage taps are closed, and the sample in Example 5.4 is subjected to a conventional undrained compression test. Use the Cam clay model with $M = 1.02$ to sketch the state paths (in $q, p'; q, p$ and $v, \ln p'$ planes) followed by the sample during the test. Give the values of q, p' and u at yield and at failure, and obtain some numerical values along the state path between these points.

(b) A second, identical sample is treated similarly, except that it is subjected to a drained (rather than an undrained) shear test from an effective cell pressure of 300 kPa. Show that yield should occur at $q = 70$ kPa. Predict the values of q and p' at failure.

(c) What is the engineering significance of the difference between the two results?

(d) Explain briefly why you would expect the Cam clay predictions to be less realistic if the samples had been allowed to swell back to 50 kPa before the start of the shear tests.

[*University of London 2nd year BEng (Civil Engineering) examination, King's College (modified, part question)*]

Solution

(a) The state path followed in the q, p plane is given by the expression $dq/dp = 3$. The state paths followed in q, p' and $v, \ln p'$ planes are as shown in Figure 5.33. While the state of the sample remains within the initial yield locus, the requirement that deformation takes place on the same κ-line (in the $v, \ln p'$ plane) and at constant specific volume means that $p' = $ constant ($= 300 \, kPa$) until yield. The equation of the Cam clay yield locus is

$$q/Mp' + \ln(p'/p'_0) = 0$$

with $p'_0 = 400 \, kPa$ initially. Thus the value of q at yield q_y may be calculated:

$$q_y/(1.02 \times 300) + \ln(300/400) = 0$$

Hence

$$q_y = 88 \, kPa (p'_y = 300 \, kPa; p_y = 300 + q_y/3 = 329.3 \, kPa; \text{ and}$$

$$u_y = p_y - p'_y = 29.3 \, kPa)$$

Eventually, the sample reaches the critical state line and fails. On the critical state line, $q = Mp'$ and $v = \Gamma - \lambda \ln p'$. We can calculate p' at the critical state from the equation of the critical state line in the $v, \ln p'$ plane, because v, Γ and λ are known.

The specific volume during undrained shear is (from Example 5.4)

$$v = 2.599 - 0.18 + 0.01439 = 2.433$$

Using the subscript c to denote the critical state, the equation of the critical state line in the $v, \ln p'$ plane is $v_c = \Gamma - \lambda \ln p'_c$, so

$$\lambda \ln p'_c = 3.767 - 2.433$$

Hence

$$p'_c = 169.1 \, kPa$$

From the equation of the CSL in the q, p' plane,

$$q_c = 1.02 p'_c = 172.5 \, kPa$$

The value of p_c is given by $p_c = 300 + q_c/3 = 357.5 \, kPa$, and $u_c = p_c - p'_c = 188.4 \, kPa$.

At any stage between yield and failure on the critical state line, the current state of the sample lies on a κ-line (in the $v, \ln p'$ plane) and on a yield locus associated with a particular value of p'_0. The value of p'_0 at the critical state, p'_{0c}, is given by

$$q_c/Mp'_c + \ln(p'_c/p'_{0c}) = 0$$

or

$$\ln(p'_{0c}/p'_c) = 1$$

that is, $p'_{0c} = 2.72 p'_c = 460 \, kPa$ in this case.

The easiest way of calculating a state point between yield and failure is to choose a value of p'_0 between the initial value of 400 kPa and the value at the critical state, 460 kPa. The corresponding value of p' may then be calculated from the relation between v, p'_0 and p' in the $v, \ln p'$ plane

$$v = (\Gamma + \lambda - \kappa) - \lambda \ln p'_0 + \kappa \ln(p'_0/p')$$

(Figure 5.30(b)), with $v = 2.433$.

Knowing p' and p'_0, the value of q may be calculated from the expression for the current yield locus in the q, p' plane

$$q/Mp' + \ln(p'/p'_0) = 0$$

The average total stress is $p = 300 + q/3$, and the pore water pressure is $u = p - p'$. Values of p'_0, p', q, p and u between yield and failure are given in Table 5.8.

Table 5.8 Values of p'_0, p', q, p and u between yield and failure

	Yield						Failure
p'_0 (kPa)	400	410	420	430	440	450	460
p' (kPa)	300	273.7	247.4	224.1	203.5	185.1	169.1
q (kPa)	88.1	112.8	133.6	149.0	160.1	167.7	172.5
p (kPa)	329.3	337.6	344.5	349.7	353.4	355.9	357.5
u (kPa)	29.3	63.9	97.1	125.6	149.9	170.8	188.4

The state paths followed in the q, p', q, p and $v, \ln p'$ planes are plotted in Figure 5.39.

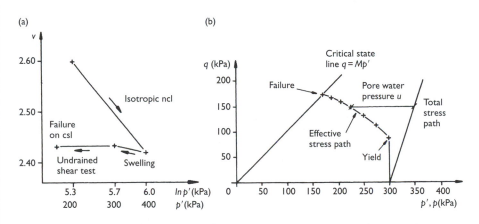

Figure 5.39 State paths according to Cam clay model.

(b)   For a sample subjected to a drained shear test, the pore water pressures are zero and the effective stress path is the same as the total stress path, $p' = 300 + q/3$. The sample will yield when the initial yield locus is reached, so that

$$p'_y = 300 + q_y/3$$

and

$$q_y/Mp'_y + \ln(p'_y/p'_0) = 0$$

or

$$q_y/[1.02 \times (300 + q_y/3)] + \ln\{(300 + q_y/3)/400\} = 0$$

By trial and error, this equation is satisfied when $q_y = 70\,\text{kPa}$ ($p'_y = 300 + q_y/3 = 323.3\,\text{kPa}$).

At failure, $q_c = Mp'_c$ and $p'_c = 300 + q_c/3$:

$$q_c = 1.02 \times 300 + 1.02q_c/3$$

$$= 463.6\,\text{kPa}; p'_c = 454.5\,\text{kPa at drained failure.}$$

(c) q_c for drained failure is much greater than in the undrained case. The engineering significance of this is that constructions involving rapid loading of a clay soil might cause undrained failure, whereas if time were allowed for pore water pressures to dissipate, failure could be avoided. This is particularly applicable to embankments on soft clay, which are often stage-constructed (cf. Figure 4.10) for this reason.

(d) A sample which had been allowed to swell back to a cell pressure of 50 kPa would have an overconsolidation ratio n_p of $400/50 = 8$. This would bring the soil onto the dry side of the critical state. Failure would be expected to occur by rupture (section 5.13), rather than by continuum yield as assumed in the Cam clay model.

5.13   Peak strengths

5.13.1   Predictions using Cam clay

The state paths followed by a more heavily overconsolidated sample during conventional triaxial compression may in principle be predicted using the Cam clay model, by means of exactly the same procedures as for a normally consolidated or a lightly overconsolidated sample (section 5.12). The state paths followed in the q, p' and $v, \ln p'$ planes during drained and undrained triaxial tests are shown in Figures 5.40 and 5.41.

For a heavily overconsolidated sample, the Cam clay model predicts that yield will occur at a stress ratio q/p' in excess of the critical state value M. The strength of the sample will then reduce as it 'softens' to the critical state (Figures 5.40 and 5.41). Between yield and failure, the state path crosses a series of κ-lines which intersect the isotropic normal compression line at decreasing values of p'_0, so that the associated current yield loci viewed in the q, p' plane appear to contract.

During undrained triaxial compression, a heavily overconsolidated sample generates positive pore water pressures given by $\Delta u = \Delta q/3$ until yield, because of the

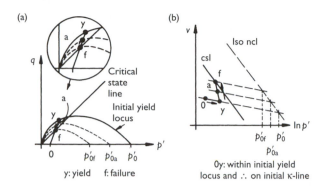

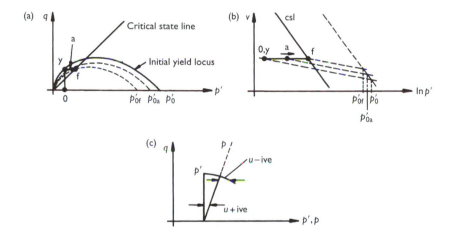

Figure 5.40 State paths in q, p' and $v, \ln p'$ planes during a conventional drained triaxial compression test on a heavily overconsolidated clay sample.

Figure 5.41 State paths in the q, p' and $v, \ln p'$ planes during a conventional undrained triaxial compression test on a heavily overconsolidated clay sample.

requirement $p' =$ constant. After yield, the pore water pressures decrease dramatically, and may well become negative (Figure 5.41(c)). In a drained test, the sample would suck in water to enable it to dilate to the critical state. The initial states of such samples lie to the left of the critical state line in the $v, \ln p'$ plane, and are described as **dry** of critical.

The apparent overconsolidation ratio at failure that divides wet or lightly overconsolidated samples from dry or heavily overconsolidated samples is given by $p'_{0c}/p'_c = 2.72$ (from Figure 5.26). For undrained tests, which follow the undrained state boundary to the critical state, the ratio between the average effective stress at failure p'_c and the **initial** preconsolidation pressure p'_{0i} is approximately 1.86: Figure 5.36.

The Cam clay models are based on the assumption that the soil deforms as a continuum, with stresses and strains which are uniform throughout the sample. In practice, the progress of a heavily overconsolidated sample towards the initial yield locus is likely to be interrupted by the formation of a rupture zone. As the test continues, deformations become concentrated in the rupture zone, and the state of the soil within the rupture zone may not be the same as that in the rest of the sample. This makes the analysis of the postrupture behaviour of a triaxial sample somewhat difficult, although it might be argued that the continuum analysis is applicable until the rupture occurs. Postpeak stresses and strains in the triaxial test cannot reliably be calculated on the basis of loads and displacements measured at the boundaries of a ruptured sample: such data should therefore be treated with caution.

5.13.2 Hvorslev rupture and tensile fracture

Muir Wood (1990) shows that data of peak strengths from carefully conducted tests on overconsolidated clays fall on a line on a normalized q, p' plot of the form

$$(q/p_c') = H[(p'/p_c') + g] \tag{5.48}$$

(Figure 5.43). This may be used to estimate peak strengths of overconsolidated samples that are dry of critical. In q, p', v space, equation (5.48) represents a three-dimensional surface defining possible failure states. It is known as the **Hvorslev surface**, after the geotechnical engineer M.J. Hvorslev who carried out pioneering work on the peak strengths of overconsolidated clays in the 1930s.

Three important points concerning the Hvorslev surface line should be noted:

- It applies only to the left of the critical state point on the $(q/p_c', p'/p_c')$ diagram. If the stress state of the sample lies to the right of the critical state point, the sample is either normally consolidated or lightly overconsolidated. It will not then exhibit a peak strength, but will yield and move along the undrained state boundary towards failure on the critical state line.
- The extent of the Hvorslev line on the left-hand side of the $(q/p_c', p'/p_c')$ diagram is also limited. A drained triaxial compression test starting from an effective cell pressure of zero would follow the line $q = 3p'$, from equation (5.11) with $\sigma_c - u = 0$ (Figure 5.42). A stress state to the right of this line implies an effective cell pressure of greater than zero, while a stress state to the left of this line implies an effective cell pressure which is negative. As soil cannot withstand tensile stresses, this is impossible. Thus the line $q = 3p'$ represents a third possible failure condition (in addition to the critical state and the Hvorslev line), corresponding to tensile

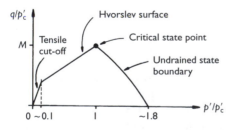

Figure 5.42 Tensile cut-off, Hvorslev line, critical state point and Cam clay undrained state boundary in a normalized $(q/p_c', p'/p_c')$ plot.

fracture. (In conditions other than triaxial compression, the equation of the tensile cut-off would be slightly different, but the same principle applies.)

- In Figure 5.42, the stress state (q, p') has been normalized with respect to p'_c (Figure 5.35). p'_c depends on the specific volume of the sample. Thus peak strengths measured in practice must also be expected to depend on the specific volume (or the water content) of the sample.

Figure 5.42 depicts the limiting states of stress to which a soil element can be subjected, defined by the mode of failure appropriate to the stress history and stress state of the soil in-normalized $(q/p'_c, p'/p'_c)$ terms. Normally consolidated and lightly overconsolidated samples will display continuum yield as predicted by a Cam clay-type model, eventually reaching failure at a critical state. More heavily overconsolidated samples will probably rupture on planes of maximum stress ratio, according to a Hvorslev-type failure criterion (equation 5.48). At very low effective stresses, failure would be expected to occur by tensile fracture. This behavioural regime was expounded by Schofield (1980).

5.13.3 Interpreting peak strength data

A traditional interpretation of peak strengths, based on drawing a best-fit common tangent to a series of Mohr circles of stress, was described in section 5.5. Muir Wood (1990) shows that the interpretation of peak strength data obtained by plotting q/p'_e against p'/p'_e at peak stress ratio q/p' is much more satisfactory. (p'_e is the equivalent consolidation pressure defined in Figure 5.35, but normalization with respect to p'_c would be equally acceptable.) In this way, the dependence of peak strength on stress history and specific volume/water content is to some extent taken into account, and the scatter in the results is considerably reduced (Figure 5.43). Figure 5.43 also demonstrates quite clearly the departure of the peak strength envelope from equation (5.48) at low effective stresses, as it curves round to the no-tension cut-off and passes through the origin.

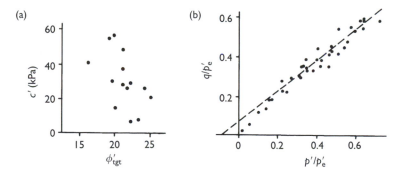

Figure 5.43 How and how not to interpret peak strength data from tests on clay soils. (a) Range of values of c' and ϕ'_{tgt} from drained triaxial tests on London Clay, interpreted according to Figure 5.11(b); (b) failure data as (a) plotted in normalized $(q/p'_e, p'/p'_e)$ terms. (Redrawn with permission from Muir Wood, 1990.)

5.14 Residual strength

The Cam clay model and critical state soil mechanics are concerned with soil that deforms as a continuum. Peak strengths (by which we really mean peak stress ratios τ/σ' or q/p') will be observed due to the effects of dilation, which occurs in dense (heavily overconsolidated) soils because the particles need more room to move relative to one another. Dilation cannot continue indefinitely, so that the soil eventually reaches a condition in which deformation continues at constant stress ratio and constant specific volume – the critical state.

In dense soils, deformation may become concentrated in a very thin band as a rupture surface develops. Stress states at rupture are indicated by the Hvorslev line shown in Figure 5.42. Once a rupture surface has developed, the sample no longer behaves as a continuum. Movement occurs by relative sliding across the rupture, and the shearing of the rupture zone generates negative pore water pressures, enabling the rupture zone to soften by drawing in water from the surrounding soil. With continued shearing, the plate-like clay particles become aligned along the direction of the rupture surface. The rupture surface becomes polished, and the shear strength gradually falls to the **residual value**, ϕ_r'.

The relative movement required to polish a rupture to the extent that the shear strength falls to ϕ_r' is rather more than can be accommodated in a shearbox or a triaxial apparatus. Residual strengths are investigated using a device known as a **ring shear apparatus** (Figure 5.44). This is rather like a shearbox in which the sample is in the shape of a ring, and is sheared by applying a torque which twists the top half of the sample relative to the bottom. Unlike the shearbox, the ring shear apparatus does not run out of travel, and the relative displacement of the two halves can be continued indefinitely. Typical data from a ring shear test, including both the peak and residual states, are shown in Figure 5.45.

The movements associated with the reduction of the strength of a clay from the critical to the residual state would be unacceptably large for a geotechnical structure in practice. Unfortunately, it cannot be guaranteed that the clay at a particular site does not have pre-existing shear surfaces, on which the residual strength has already been reached due to ground movements (such as landslides) in the past. Skempton (1964) showed that the stability of many natural slopes in London Clay is governed by the residual strength ($\phi_r' \approx 16°$) on historic slip surfaces, rather than the critical state strength ($\phi_{crit}' \approx 20°$) of the intact material. This significant reduction in strength must be taken into account in the design of cuttings, embankments or foundations for buildings on ground which is already of marginal stability. (Although it might seem

Figure 5.44 Schematic diagram of ring shear apparatus.

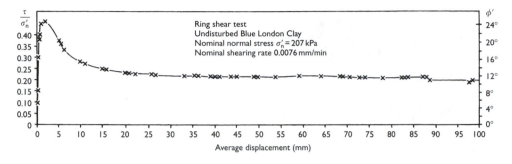

Figure 5.45 Typical data from the ring shear apparatus. (Redrawn with permission from Bishop *et al.*, 1971.)

odd to build a house on the site of an old landslide, many areas of marginal stability, such as the Kent and Essex coasts in England, are attractive places to live, with pleasant sea views.)

Undulations in the surface of the ground are sometimes indicative of historic slip surfaces. Careful examination of soil cores retrieved intact from trial boreholes may reveal to the experienced eye the presence of ruptures, as greasy-wet or 'slickensided' discontinuities.

5.15   Sensitive soils

Real clays are usually sampled from the ground and tested in the triaxial apparatus with as little disturbance as possible. Although a degree of disturbance is inevitable, at least some of the natural structure of the soil will remain. The soil parameters determined from tests on natural or undisturbed samples may be different from those determined from tests on the material after it has been remoulded or reconstituted (i.e. dried to a powder, mixed with water to form a slurry, and recompressed to give the desired stress history), because the natural structure is destroyed in remoulding or reconstitution.

Structured natural soils may be more compressible than remoulded soils when tested in an oedometer, because of the additional deformation associated with the breakdown of their structure (Burland, 1990). The slow deposition of soil particles through still water tends to give rise to an open structure, and a relation between specific volume and $\ln \sigma_v'$ – the **sedimentation compression** line – which lies above and approximately parallel to the one-dimensional normal compression line for reconstituted soils. If a natural soil having an open structure is sampled and then tested in an oedometer, its state will move from the sedimentation compression line to the one-dimensional normal compression line as the vertical effective stress is increased, leading to an increased apparent compressibility.

Strength parameters determined from natural samples may be unreliable, because the components of shear resistance associated with soil structure and interparticle bonds are probably only transitory. Neglecting the effects of sample disturbance, the strength parameters determined from tests on remoulded soils will usually err

Figure 5.46 Effect of remoulding on the undrained shear strength of a sensitive soil (Leda Clay). (From Crawford, 1963, with permission.)

on the safe side, because the contribution of any natural structure or cementing of the particles is neglected.

The **sensitivity** S of a clay is a measure of the ratio of the undrained shear strength of the structured material to that in the remoulded state, at the same water content,

$$S = \tau_{u,peak} / \tau_{u,remoulded} \tag{5.49}$$

If $S \approx 1$, the clay is insensitive, and the peak and remoulded strengths are similar. However, the sensitivity S can be as high as 1000. Figure 5.46 (from Crawford, 1963) illustrates quite graphically the effect of remoulding on a sensitive clay. An undisturbed sample capable of supporting a reasonable load is reduced to a slurry by remoulding at the same water content, due to the loss of its natural structure.

Sensitive clays are often geologically recent (postglacial), and normally consolidated (i.e. they have never been subjected to the removal of overburden). A sensitive structure might arise from a cementing of the soil particles in the natural condition, or the leaching out of some component of the material as deposited – for example the replacement of salt water by fresh water. The differences between natural and remoulded soils are discussed in more detail by Mitchell (1993) and Burland (1990).

5.16   Correlation of critical state parameters with index tests

The liquid and plastic limit tests described in section 1.11 are in effect indicators of the undrained shear strength of the soil. It is generally accepted that the undrained shear strength at the liquid limit $\tau_{u,LL}$ is of order 1.6 kPa. At the plastic limit, $\tau_{u,PL}$ is approximately 110 kPa, or 70 times greater (Whyte, 1982; Skempton and Northey,

1953). Critical states of remoulded soils are given by equations (5.27) and (5.28),

$$q = Mp' \tag{5.27}$$

and

$$v = \Gamma - \lambda \ln p' \tag{5.28}$$

Recalling that the water content is related to the specific volume by

$$v = (1 + e) = (1 + wG_s)$$

and assuming that the relation between q_c and τ_u in the triaxial test, $\tau_u = q_c/2$, is appropriate to the liquid and plastic limit tests, then

$$(1 + wG_s) = \Gamma - \lambda \ln(2\tau_u/M)$$

so that

$$wG_s = \Gamma - 1 + \lambda \ln(M/2) - \lambda \ln(\tau_u),$$

or

$$w = C - (\lambda/G_s) \ln(\tau_u) \tag{5.50}$$

where C is a constant. Noting that the plasticity index I_P is equal to $w_{LL} - w_{PL}$, we have (using equation 5.50):

$$w_{LL} = C - (\lambda/G_s) \ln(\tau_{u,LL})$$
$$w_{PL} = C - (\lambda/G_s) \ln(\tau_{u,PL})$$

so that

$$w_{LL} - w_{PL} = (\lambda/G_s) \ln(\tau_{u,PL}/\tau_{u,LL})$$

and

$$I_P = (\lambda/G_s) \ln 70, \lambda = I_P G_s / \ln 70, \text{ or}$$
$$\lambda \approx 0.63 \times I_P, \text{taking } G_s = 2.7 \tag{5.51}$$

Equation (5.51) may be used to estimate the slope of the normal compression line in the $v, \ln p'$ (or $v, \ln \sigma_v'$) plane from the results of the index tests described in section 1.11.

Muir Wood (1990) shows that, if the liquid limit tests are carried out using the fall cone apparatus, there is a further correlation between the water content of the sample and the cone penetration d. For a given cone mass and geometry and for soils of the same type, dimensional analysis may be used to show that $\tau_u \propto 1/d^2$

(Wood and Wroth, 1978). Thus

$$\ln(\tau_u) = D - 2\ln d \tag{5.52}$$

where D is a constant. Substituting equation (5.52) into equation (5.50) to eliminate τ_u,

$$w = C - (\lambda/G_s)(D - 2\ln d)$$
$$\Rightarrow w = E + (2\lambda/G_s)\ln d \tag{5.53}$$

where E is a constant ($E = C - [D\lambda/G_s]$). Thus a graph of water content w against the logarithm of the cone penetration d may be used to determine the parameter λ from the slope, which should be equal to $2\lambda/G_s$, or (taking $G_s = 2.7$) approximately 0.74λ.

Further correlations that can be made from the results of fall cone tests in which cones of different mass are used are detailed by Muir Wood (1990).

5.17   Creep

Throughout this book, and in most conventional soil mechanics theory, it is generally assumed that deformation of soil occurs only as a result of changes in effective stress. In reality, some soils continue to deform when the effective stresses are not changing: this behaviour is known as **creep**.

Creep may be investigated experimentally by maintaining a triaxial test sample at a constant stress state (q and p'), and observing the development of continuing deformation (e.g. axial strain) with time. Data from creep tests presented by Bishop and Lovenbury (1969) are shown in Figure 5.47. All except the uppermost of these curves take the form of a straight line when the strain is plotted as a function of the

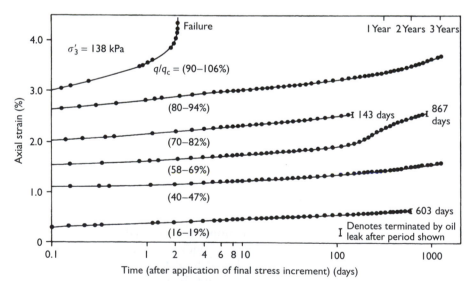

Figure 5.47 Creep test data. The ranges given for q/q_c by each line reflect the range in q_c from a number of conventional shear tests. (Redrawn from Bishop and Lovenbury, 1969.)

logarithm of the elapsed time, indicating that the same creep strain occurs in each log cycle of time.

In general, the development of creep strains with time (at constant effective stress) may be described mathematically by an equation of the form

$$\varepsilon_{creep} = [At_1/(1 - m)] \exp(\alpha q/q_c)[t/t_1]^{(1-m)} (m \neq 1) \tag{5.54a}$$

or

$$\varepsilon_{creep} = At_1 \exp(\alpha q/q_c) \ln[t/t_1] \quad (m = 1) \tag{5.54b}$$

(Mesri *et al.*, 1981; Mitchell, 1993), where t_1 is a reference time (e.g. 1 minute or 1 hour), q_c is the deviator stress at failure, and A, m and α are soil parameters.

Most of the data shown in Figure 5.47 illustrate a reasonably stable condition, in which the rate of increase of shear strain (in real time) is decreasing. In some situations, particularly with high shear stress ratios q/q_c and/or sensitive, structured soils, the stress state is unsustainable and creep strains accelerate, leading to creep rupture. This is indicated by the uppermost curve in Figure 5.47, and would be associated with a value of the parameter $m < 1$ in equation (5.54). Equation (5.54) suggests that creep is likely to be more significant when the soil is closer to shear failure, that is, at higher values of q/q_c: this is confirmed by the data of Figure 5.47.

Creep may also affect the results of oedometer tests, and triaxial tests carried out at a constant rate of strain. The slower the test, the more opportunity there is for creep strains to develop, resulting in a softer response (i.e. more deformation at a given load). If the strain rate is changed during a test, the state of the soil will jump to the stress–strain relation appropriate to the new strain rate (Figure 5.48).

The effects of creep can lead to the indication of a false pre-consolidation pressure in oedometer tests where the rate of testing is faster than the rate of deposition in the field, as indicated in Figure 5.49.

In practice, creep is not a significant problem with many soils. If it were, the surface of the earth would be locally completely flat. The influence of creep on stress–strain relationships of soils can usually be overcome by not carrying out tests too quickly. For example, the load on an oedometer test sample might be increased at daily, rather than at hourly intervals. Creep is only usually troublesome with soils which are unstable due to their structure (e.g. sensitive soils), or are subjected to unstable (close to peak) stress states.

5.18 Anisotropy

We saw in section 3.6 that some soils exibit a layered structure, due to the way in which they were originally deposited. Soils such as the glacial lake deposits shown in Figure 3.15 will be **anisotropic** (i.e. they will not have the same properties in every direction) in terms of their stress–strain response, in addition to their permeability. The properties associated with the two horizontal directions are different from the properties associated with the vertical direction, but indistinguishable from each other. The term **cross-anisotropic** is used to describe this special form of anisotropy.

In fact, most natural soils will be cross-anisotropic to some extent, even if their structure is not as obviously layered as that shown in Figure 3.15. This is because the particles will tend to become aligned with the plane of deposition. The effect is more

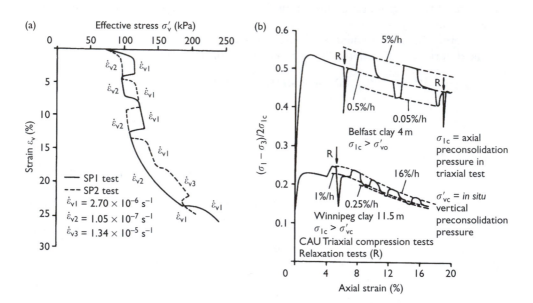

Figure 5.48 Effect of strain rate on constant rate of strain oedometer and triaxial test results. (a) Oedometer test (Redrawn with permission from Leroueil et al., 1985.); (b) anisotropically consolidated undrained (CAU) triaxial test. (Redrawn with permission from Graham et al., 1983.)

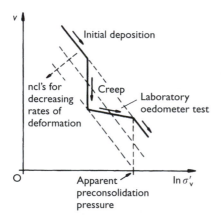

Figure 5.49 Apparent overconsolidation of a normally consolidated soil exhibiting creep.

pronounced with flat particles (i.e. clays) than with more rotund particles (i.e. sands): clay particles deposited through water settle onto their flat faces, not their edges.

Few of the simple conceptual models for the behaviour of soils take anisotropy into account directly. One reason for this is that it is difficult to obtain the parameters which would be required to describe it. When a sample which has been cored vertically is tested in the triaxial cell, the horizontal stresses imposed during the test are aligned with the directions which were horizontal during deposition in the field. The axes of structural anisotropy (due to the direction of deposition in the field) and stress anisotropy in the triaxial cell are coincident, and conditions of axisymmetry are preserved. If a sample is cored with its axis horizontal, then one of the horizontal directions in the triaxial cell is aligned with what was originally the vertical direction in the field. Complete symmetry is then lost, and the sample will almost certainly deform in a non-cylindrical manner.

Anisotropy is probably usually neglected because it is difficult to quantify and model, rather than because it is unimportant. In soil deposits where the structure is not obviously anisotropic, however, it seems that structural anisotropy due to the alignment of the particles with the plane of deposition affects the stiffness more than the soil strength. Thus calculations based on soil strength (which are discussed in Chapters 7–10) are unlikely to be affected significantly by depositional anisotropy. In calculations based on soil stiffness (which are discussed in Chapter 6), the direction of loading and deformation may be reasonably well-defined in advance (e.g. vertical compression), so that the parameters can be determined and used accordingly.

In soils which have an obvious layered structure, layers of softer materials (such as normally consolidated clay) represent planes of potential weakness, which must be taken into account in the assessment of strength parameters for design. Old slip surfaces, and fissures or joints in rocks, represent perhaps more serious planes of weakness, because their orientation is unlikely to be primarily horizontal. (The example problem given in section 5.4 indicates that planes of weakness will have a more significant effect on the strength of the soil mass when they are inclined at angles nearer 45° or so to the direction of the principle effective stress.)

5.19 Partly saturated soils

It is usually assumed in conventional soil mechanics theory that the soil is either saturated or dry. In practice, there are situations where the soil is only partly saturated: some of the voids are filled with water and some with air.

A saturated soil is a two-phase material, comprising solid soil particles and water. Its behaviour is controlled by the effective stress, defined as $\sigma - u$ where σ is the total stress and u is the pore water pressure. A partly saturated soil is a three-phase material, comprising soil particles, water and air. Its behaviour is controlled by two stress parameters, $(\sigma - u_a)$ and $(u_a - u_w)$, where σ is the total stress, u_w is the pore water pressure and u_a is the pore air pressure.

Bishop and Donald (1961) were able to describe the shear behaviour of a partly saturated soil in terms of a stress σ_i', where

$$\sigma_i' = (\sigma - u_a) + \chi(u_a - u_w) \tag{5.55}$$

and χ is a parameter whose value depends on the saturation ratio, S_r. Unfortunately, the parameter χ also depends on the way in which it is measured. Jennings and Burland (1962) showed that below a critical degree of saturation, there was no unique relationship between specific volume and σ_i'. σ_i' is not therefore an effective stress, in the sense that an effective stress controls completely both the shear and volumetric behaviour of a soil.

The ideal of a single effective stress parameter which controls the behaviour of a partly saturated soil remains elusive. Toll (1990) presents a critical state framework for the behaviour of unsaturated soils in terms of five state variables. These are the deviator stress q, the specific volume v, the saturation ratio S_r, and the isotropic stress parameters $(p - u_a)$ and $(u_a - u_w)$. Critical states are defined by the equations,

$$q = M_a(p - u_a) + M_w(u_a - u_w) \tag{5.48}$$

and

$$v = \Gamma_{aw} - \lambda_a \ln(p - u_a) - \lambda_w \ln(u_a - u_w) \tag{5.49}$$

where the five soil parameters $M_a, M_w, \Gamma_{aw}, \lambda_a$ and λ_w depend on the saturation ratio S_r. The dependence of $M_a, M_w, \Gamma_{aw}, \lambda_a$ and λ_w on S_r must be determined experimentally.

Wheeler (1991) points out that Toll's (1990) critical state framework for unsaturated soils cannot be used as a predictive tool, because S_r (and hence the values of the parameters $M_a, M_w, \Gamma_{aw}, \lambda_a$ and λ_w) will change as the soil is sheared. Wheeler (1991) proposes an alternative critical state framework that overcomes this shortcoming, but with some loss of simplicity.

One of the difficulties of formulating mathematical descriptions of the behaviour of unsaturated soils concerns the effects of soil fabric and structure. Jennings and Burland (1962) describe how, at low saturation ratios, granular soils may develop interparticle 'bonds' at the contact points, due to the presence of high-curvature menisci. As the saturation ratio is increased on wetting, these 'bonds' will disappear, leading to the collapse of the rather open structure their presence has sustained. Compacted clay soils tend to aggregate, forming packets or lumps with air voids in between.

A review of the concepts appropriate to partly saturated soils is given by Fredlund (1979). Soils are rarely unsaturated except above the groundwater level. In temperate regions (such as the UK), the soil above the water table is usually avoided for engineering purposes, and the concepts of conventional (saturated) soil mechanics are generally sufficient. In warmer and more arid climates, the water table may be very deep, and the need to consider the effects of non-saturation is rather more important. Non-saturation may also be important in attempting to describe the mechanical behaviour of biologically active sediments (Sills *et al.*, 1991) and wastes (Hudson *et al.*, 2004) in which on going degradation leads to the generation of methane gas within the deposit.

5.20   The critical state model applied to sands

Although there is no doubt that a critical state for sands exists in much the same way as the critical state for clays (witness the shearbox tests of Chapter 2), the full Cam clay-type model is difficult to apply. This is primarily because it is not possible to identify a normal compression line (i.e. a unique relation between specific volume and p' or σ'_v in first compression), in the same way that it is for clays. There has also been some discussion in the literature as to whether the critical state line, which is usually determined on the basis of drained tests on dense samples, is actually the same as the so-called **steady state line**, which is usually determined on the basis of undrained tests on loose samples. For practical purposes, it seems that it probably is (Been *et al.*, 1991).

The critical state line determined for Erksak 330/0.7 sand by Been *et al.* (1991) is shown in Figure 5.50. The change in slope of the critical state line, which occurs at $p' \approx 1000\,\text{kPa}$, is due to the onset of particle crushing at high confining pressures.

The behaviour of a sand sample during shear before reaching the critical state depends on its specific volume relative to the specific volume on the critical state line at the current average effective stress p'. Samples that are initially loose will compress, while samples that are initially dense will dilate. In this context, the terms 'loose' and 'dense' must be defined with respect to the specific volume at the critical state, at the confining pressure at which the shear test is carried out. You can turn a dense sand into a loose sand (relative to its critical state) by testing it at a high enough confining stress.

Been and Jefferies (1985) propose the use of a parameter $\Psi = v - v_c$ to describe the state of a sand, where v_c is the specific volume at the critical state (Figure 5.51). This is more satisfactory that the density index $I_D = (e - e_{min})/(e_{max} - e_{min})$, because Ψ takes into account the effects of the confining pressure p'. Muir Wood (1990) suggests that the influence of mineralogy, angularity and particle size could be eliminated by normalizing Ψ by dividing it by $(v_{max} - v_{min})$.

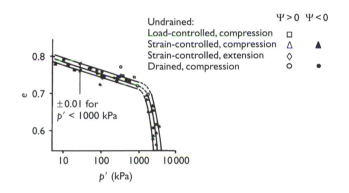

Figure 5.50 Critical state line for Erksak 330/0.7 sand. (Redrawn with permission from Been *et al.*, 1991.)

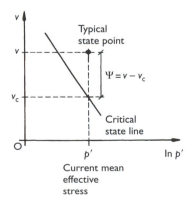

Figure 5.51 Definition of state parameter for sand.

5.21   Non-linear soil models

Soils only behave truly elastically at very small strains (Simpson *et al.*, 1979). Furthermore, the stiffness of soil decreases rapidly with increasing strain following a change in the direction of the stress path (e.g. loading followed by unloading): Jardine *et al.* (1984); Atkinson *et al.* (1990). If first-class predictions of the deformations associated with complex geotechnical constructions are required, it is necessary to take these factors into account in the model used to describe the stress–strain behaviour of the soil.

One possibility is to relate the shear stiffness of the soil to the current stress and the strain following the last reversal in the direction of the stress path, by means of an empirical curve fitted to triaxial stress–strain data (Duncan and Chang, 1970; Jardine *et al.*, 1986). This is an oversimplification, in that the stress paths followed by soil elements in the vicinity of a geotechnical engineering construction will not all be the same. Also, the stiffness at a given strain may depend on the magnitude of the rotation of the stress path direction (Figure 6.2).

A more general and fundamental approach involves a development of the Cam clay model with three special surfaces (Stallebrass, 1990; Atkinson and Stallebrass, 1991), as shown in Figure 5.52.

While the stress state of the soil lies within the innermost surface, its behaviour is truly elastic. The innermost surface may therefore be termed the yield surface. The middle surface helps to define (in conjunction with the stress path) the region within which the stress state is influenced by the recent stress history of the soil (e.g. a change in the stress path direction). The outermost surface is established by the geological or overall stress history of the soil, in the same way as the yield surface in the Cam clay model.

The three surfaces are similar in shape (elliptical, as in Modified Cam clay), and the two inner surfaces expand or contract with the outer surface, so that their sizes are always in the same proportion. In addition, the inner surfaces can move around within the outer surface, depending on the stress state and stress history of the soil: for this reason they are termed **kinematic**.

Figure 5.53 shows the translation of the two inner surfaces as an overconsolidated soil is loaded along the stress path OH (Figure 5.53(d)). The stress state O has been reached via the stress path AO (Figure 5.53(b)). The stress state of the soil at O lies on the boundaries of both of the inner surfaces, which have a common tangent at

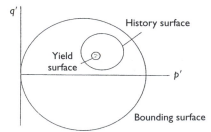

Figure 5.52 Three-surface soil model. (Redrawn with permission from Atkinson and Stallebrass, 1990.)

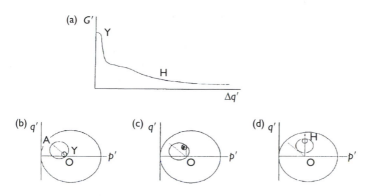

Figure 5.53 Translation of inner kinematic surfaces following a change in the direction of the stress path. (Redrawn from Stallebrass, 1992.)

this point. (Whether or not this is the case in general depends on the length of the stress path AO.) As the stress state moves towards H, the innermost yield surface does not start to move until the stress path has traversed it completely (the point Y: Figure 5.53(b)).

While the stress state is still inside the innermost yield surface, the behaviour of the soil is truly elastic. As the stress path continues from Y towards H, the innermost (yield) surface moves with it until the middle surface is reached (Figure 5.53(c)). As loading continues, both surfaces move and the influence of the previous stress path AO (i.e. the recent stress history effect) is apparent until the two inner surfaces are aligned with their common tangent at some angle to the current stress path OH (Figure 5.53(d)). Figure 5.53(a) shows the variation in shear stiffness modulus G with deviator stress q predicted by the model.

The three-surface model uses the same parameters as the Cam clay model, together with three new parameters. The three additional parameters are the size ratios of the kinematic surfaces, and a parameter which relates the change in shear stiffness G to the movement of the kinematic surfaces. An example of the use of the three surface model in an axisymmetric finite element analysis is given by Stallebrass and Taylor (1997), who compare the calculated soil movements with those measured in a centrifuge model test on a circular footing and an analysis based on modified Cam clay.

At present, multi-surface kinematic models are developed and used primarily in research. It is likely, however, that models like this – together with PC-based finite element programs – will be used increasingly in leading-edge commercial geotechnical analyses.

Key points

- The **triaxial test** is used extensively in industry and in research to investigate the stress–strain behaviour of soils.
- The sample is usually first consolidated by increasing the cell pressure to a known **isotropic** stress state, from which it is sheared by the application of a **deviator stress** q.

- Triaxial shear tests may be carried out either **drained**, in which case the sample expels or takes in water and there is no build-up of pore water presure during shear, or **undrained**, in which case the sample is prevented from changing in volume so that either positive or negative pore water pressures are generated.
- Triaxial test data may be presented as graphs of **deviator stress** q and either **pore water pressure change** Δu (in an undrained test) or **volumetric strain** $\varepsilon_{\mathrm{vol}}$ (in a drained test) against the **axial strain** ε_{a} or the **shear strain** γ. As alternatives to q, the **stress ratio** q/p' or the **mobilized strength** ϕ'_{mob} may be used. The point of maximum deviator stress q will not in general correspond to the peak mobilized strength ϕ'_{peak} or the maximum stress ratio $(q/p')_{\mathrm{max}}$.
- In addition, state paths may be plotted on graphs of q against p' (and p), and specific volume v against the natural logarithm of the average effective stress, $\ln p'$. The deviator stress q quantifies the shear stress, the average effective stress p' quantifies the normal effective stress, and the specific volume v quantifies the volumetric state of the soil.
- Peak and critical state strength envelopes can be determined by plotting Mohr circles of stress on a diagram of shear stress τ against normal effective stress σ'. For an uncemented soil, the peak strength envelope is curved, and passes through the origin. Peak strengths may be represented as $\phi'_{\mathrm{peak}} = tan^{-1}(\tau/\sigma')_{\mathrm{peak}}$. At a given specific volume, the peak strength increases at low effective stresses due to the increased potential for dilation.
- For clays which are sheared undrained, the deviator stress at failure depends on the specific volume or the water content. The limiting shear stress, or the **undrained shear strength** τ_{u}, is equal to half the deviator stress at failure, and defines an alternative failure criterion in terms of total stresses.
- The **undrained shear strength** failure criterion is applicable only to a low permeability soil, brought rapidly to failure at constant volume. The undrained shear strength is not a soil constant, since its value depends on the specific volume (or water content) of the soil as sheared.
- When sheared, a lightly overconsolidated sample will behave elastically (in the sense that deformation is recoverable, but not in the sense that the elastic parameters are constant) until it **yields**. This requires $p' = $ constant, and in a conventional undrained compression test the change in pore water pressure $\Delta u = \Delta q/3$. The combinations of q, p' and v that cause yield define a three-dimensional **yield surface**, whose magnitude depends on the stress history of the soil. A normally consolidated soil is already on the yield surface, and will continue to yield immediately on further loading.
- Provided that the sample does not rupture, it should on being sheared eventually fail by reaching a **critical state**, at which deformation continues at constant q, p' and v.
- Between yield and failure, the pore water pressure in an undrained shear test on a lightly overconsolidated sample increases more rapidly than during the initial stage of the shear test, prior to yield. In a drained test, the sample will compress or reduce in volume to reach the critical state. Such a sample is described as being on the **wet side** of the critical state.
- In an undrained shear test on a heavily overconsolidated sample, the pore water pressures will start to decrease at yield. In a drained test, the sample will dilate or

increase in volume to reach the critical state. Such a sample is described as being on the **dry side** of the critical state.

- In practice, a heavily overconsolidated soil will probably rupture before yield, or between yield and the critical state. Combinations of q, p' and v at rupture define a three-dimensional surface in v, p', q plane, known as the **Hvorslev surface**. At very low effective stresses, failure will occur by **tensile fracture**.
- The comparatively simple soil behavioural models described in this chapter do not take into account the effects of soil structure and sensitivity, creep, anisotropy and partial saturation, each of which may be important in some soils in some situations.

Questions

Interpretation of triaxial test results

5.1   Data from a conventional, consolidated–undrained triaxial compression test, carried out at a constant cell pressure of 400 kPa, are given below.

Axial strain ε_a (%)	0	0.05	0.09	0.18	0.39	0.69
Deviator stress q (kPa)	0	10.9	22.3	33.5	45.0	53.5
Pore water pressure u (kPa)	274.6	280.3	284.6	290.8	300.0	307.6
Axial strain ε_a (%)	1.51	3.22	4.74	6.13	7.89	9.39	11.03
Deviator stress q (kPa)	65.4	79.0	85.7	89.6	91.4	93.9	94.0
Pore water pressure u (kPa)	314.4	317.0	315.1	312.6	312.1	312.7	312.8

Plot graphs of mobilized strength ϕ'_{mob} and pore water pressure change Δu against shear strain γ. Plot also the total and effective stress paths in the q, p and q, p' planes. Comment on these curves, and estimate the critical state strength ϕ'_{crit}. Is the sample lightly or heavily overconsolidated?

($\phi'_{crit} \approx 20.5°$. No peak strength, and sample has positive pore water pressures at failure: sample is therefore wet of critical, that is, lightly overconsolidated.)

5.2   Two further consolidated–undrained triaxial compression tests are carried out on samples of the same clay as in Question 5.1. These gave the following results.

	s' at ϕ'_{peak}	t at ϕ'_{peak}	s' at ϕ'_{crit}	t at ϕ'_{crit}
Test 2	88 kPa	35.8 kPa	90 kPa	31.5 kPa
Test 3	43 kPa	19.5 kPa	45 kPa	15.8 kPa

Using data from all three tests, plot peak and critical state strength failure envelopes on a graph of τ against σ', and comment on the data.

($\phi'_{crit} = 20.5°$; ϕ'_{peak} decreases from $27°$ in test 3 to $20.5°$ in test 1.)

5.3   Using the data from Questions 5.1 and 5.2, determine the equations of the critical state line in the q, p' and v, $\ln p'$ planes. (The as-tested water contents were 41.7% for sample 1; 45.5% for sample 2; and 52.0% for sample 3. Take $G_s = 2.65$.) Hence predict the undrained shear strength of a fourth sample of the same clay, which is subjected to a conventional undrained triaxial compression test at a water content of 35%.

($\Gamma = 3.3$; $\lambda = 0.25$; $M = 0.793$; τ_u of fourth sample $= 96.1$ kPa.)

Determination of critical state and Cam clay parameters

5.4   Define in terms of principal stresses and the quantities measured during a conventional undrained compression test the parameters q, p and p'.

Data from both consolidation and shear stages of an undrained triaxial compression test on a sample of reconstituted London Clay are given below. Plot the state paths followed by the sample on graphs of q vs p', q vs p and v vs $\ln p'$, and explain their shapes.

CP (kPa)	50	100	200	150	150	150	150	150
q (kPa)	0	0	0	0	21	39	61	86
u (kPa)	0	0	0	0	7	13	43	82
v	2.228	2.116	2.005	2.023	2.023	2.023	2.023	2.023

Notes
CP: cell pressure; q: deviator stress; u: pore water pressure; v: specific volume.

Stating clearly the assumptions you need to make, estimate the soil parameters M, λ, κ and ϕ'_{crit}.

[*University of London 2nd year BEng (Civil Engineering) examination, Queen Mary and Westfield College*]

($M = 0.89$; $\lambda = 0.161$; $\kappa = 0.063$; $\phi'_{crit} = 22.8°$.)

5.5   Define the parameters q, p and p', in terms of principal stresses. Show also how q, p and p' are related to the quantities measured during a conventional undrained compression test.

Data from the shear stage of an undrained triaxial compression test on a sample of kaolin clay are given below. Plot the state paths followed by the sample in the q, p' and q, p planes, and explain their shapes. Stating clearly the assumptions you need to make, estimate the slope of the critical state line M and the corresponding value of ϕ'_{crit}.

q (kPa)	0	13.8	27.5	41.3	53.0	59.5	63.0
u (kPa)	0	4.6	9.2	13.8	33.6	48.0	59.3

Notes
Cell pressure = 100 kPa; q: deviator stress; u: pore water pressure.

A second, identical, sample is subjected to a drained compression test starting from a cell pressure of 100 kPa. Estimate the value of q at failure, and show the effective stress path followed (in the q, p' plane) on the diagram you have already drawn for the first sample.

[*University of London 2nd year BEng (Civil Engineering) examination, Queen Mary and Westfield College*]

($M = 1.02; \phi'_{crit} = 25.8°; q_c$ for drained test = 154.5 kPa.)

Prediction of state paths from triaxial test data using Cam clay concepts

5.6   A sample of saturated kaolin ($G_s = 2.61$) was compressed isotropically in a triaxial cell to an effective cell pressure of 300 kPa. In this state, the cylindrical sample had a height of 80 mm and a diameter of 38 mm. The drainage taps were closed and the sample was subjected to a conventional undrained compression test to failure at a constant cell pressure of 300 kPa. The following values of deviator stress q and pore water pressure u were recorded.

Deviator stress $q(= \sigma'_1 - \sigma'_3)$ (kPa)	0	24.5	45.4	63.2	78.3	101.6	117.7	136.5
Pore water pressure u (kPa)	0	30.2	53.6	76.6	97.3	132.8	161.8	211.8

At the end of the test, the water content of the sample was found to be 57.2%.

(a)   Plot and explain the significance of the stress paths followed in the q, p' and q, p planes.
It was intended to prepare a second sample of kaolin in an identical manner, but the sample was accidentally over-stressed to an effective cell pressure of 320 kPa during isotropic compression. To make the water content of the second sample the same as that of the first, it was necessary to reduce the effective cell pressure to 229 kPa. During

swelling from $p' = 320\,$kPa to $229\,$kPa, the sample took in $618\,$mm^3 of water.

(b)   Use all of these data to calculate the parameters $\Gamma, \lambda, \kappa, M$ and ϕ'_{crit}.

(c)   The second sample was subjected to a conventional undrained compression test from an effective cell pressure of $229\,$kPa. Sketch the stress paths followed in terms of q vs p' and q vs p, giving the values of q at yield and at failure.

(d)   If the first sample had been subjected to a drained (rather than an undrained) test, what would have been the value of q at failure? Comment briefly on the engineering significance of this result.

[*University of London 2nd year BEng (Civil Engineering) examination, King's College*]

((b) $\Gamma = 3.766; \lambda = 0.26; \kappa = 0.05; M = 1.02; \phi'_{crit} = 25.8°$. (c) For test 2, $q_{yield} = 78.3\,$kPa and $q_c = 136.5\,$kPa. (d) For drained test $q_c = 464\,$kPa.)

5.7   (a)   Define the triaxial invariant stress parameters p' and q in terms of the principal stresses and the quantities measured in a conventional triaxial test.

Two saturated triaxial test samples, each containing $116.3\,$g of dry clay powder ($G_s = 2.70$), were prepared for a shear test by isotropic compression in the triaxial cell. For sample A, the cell pressure was gradually raised from $25\,$kPa to $174\,$kPa, with full drainage occurring throughout the process. At $174\,$kPa, the sample had a diameter of $40\,$mm and a height of $120\,$mm. The drainage taps were then closed, the cell pressure was increased to $274\,$kPa and the sample was subjected to an undrained compression test to failure. The data recorded during consolidation were:

Cell pressure (kPa)	25	50	75	100	150	174
Pore water pressure (kPa)	0	0	0	0	0	0
Volume of water expelled (cm^3) (cumulative)	0	22.4	34.47	43.08	56.01	60.31

The data recorded during shear were:

Cell pressure (kPa)	274	274	274	274	274	274
Pore water pressure (kPa)	100	104	114	132	162	189
Deviator stress q (kPa)	0	10	20	30	40	45

(b)   Plot the state paths followed by sample A in the q, p' and $v, \ln p'$ planes, and comment on their significance.

Sample B was consolidated in the same manner as sample A, but at the last increment of cell pressure was inadvertently overstressed to

200 kPa. To achieve the same void ratio at the start of the shear test, the cell pressure was reduced to 140 kPa and the sample was allowed to swell slightly as indicated below. The drainage taps were then closed, the cell pressure was increased to 240 kPa, and the undrained shear test was commenced.

Cell pressure (kPa)	150	200	140	240
Pore water pressure (kPa)	0	0	0	100
Volume of water expelled (cm^3) (cumulative)	56.01	64.62	60.31	–

(c)   Predict the state paths followed by sample B in terms of q vs p' and v vs $\ln p'$ during the shear test, giving values of q, p' and pore water pressure u at yield and at failure.

(d)   If sample B had been subjected to a drained shear test at a constant cell pressure of 140 kPa, estimate the values of q and p' at which failure would have occurred, and the volume of water that would have been expelled during the shear test.

[*University of London 2nd year BEng (Civil Engineering) examination, King's College*]

((c) For sample B, $q_{yield} = 35$ kPa; $p'_{yield} = 140$ kPa; $u_{yield} = 117.7$ kPa and $q_c = 45$ kPa; $p'_c = 100$ kPa; $u_c = 155$ kPa. (d) For drained test $q_c = 74.1$ kPa; $p'_c = 164.7$ kPa, $\Delta V_t = 15.5$ cm^3.)

Prediction of triaxial state paths using the Cam clay model

5.8   (a)   Describe by means of an annotated diagram the main features of the conventional triaxial compression test apparatus.

(b)   A sample of London clay is prepared by isotropic normal compression in a triaxial cell to an average effective stress $p' = 400$ kPa, at which point its total volume is 86×10^3 mm^3. The drainage taps are then closed and the sample is subjected to a special compression test in which the cell pressure is reduced as the deviator stress is increased so that the average total stress p remains constant. Sketch the state paths followed, in the q, p', q, p and $v, \ln p'$ planes. Give values of cell pressure, q, p, u, p' and specific volume v at the start of the test and at failure. (You must also calculate some intermediate values in order to sketch the state paths satisfactorily.)

How do the values of undrained shear strength τ_u and pore water pressure at failure compare with those that would have been measured in a conventional compression test?

> Use the Cam clay model with numerical values $\Gamma = 2.759, \lambda = 0.161, \kappa = 0.062, M = 0.89$ and $G_s = 2.75$.
>
> > [*University of London 2nd year BEng (Civil Engineering) examination, King's College*]
>
> (At the start of the test, $CP = 400\,\text{kPa}$; $q = 0$; $p = 400\,\text{kPa}$; $u = 0$; $p' = 400\,\text{kPa}$; $v = 1.893$. At failure, $CP = 335.7\,\text{kPa}$; $q = 192.9$; $p = 400\,\text{kPa}$; $u = 183.2\,\text{kPa}$; $p' = 216.8\,\text{kPa}$; $v = 1.893$. Undrained shear strength would be the same (because it is assumed that the critical state reached depends only on water content). Pore water pressure would be $q/3$ greater, so that effective stresses p' at failure were the same in each case, giving $u = 247.5\,\text{kPa}$.)

Notes

1 As the sample is compressed, the rubber membrane stretches. The resulting hoop tension $\sigma_t = \varepsilon_r E_m$ applies an additional radial pressure of $2t_m\sigma_t/d$, where E_m is the Young's modulus of the membrane, t_m is the thickness of the membrane, ε_r is the radial strain and d is the current diameter of the sample. In accurate work or at large strains, the cell pressure must be corrected to allow for this effect. Full details are given by Bishop and Henkel (1962).

2 Q is conventionally taken as the ram load needed to produce a difference between the axial stress σ_a and the cell pressure σ_c. If $Q = 0, \sigma_a = \sigma_c$. To calculate the axial stress, the cell pressure must be added to the deviator stress $q, \sigma_a = \sigma_c + q$.

3 The undrained shear strength was traditionally denoted by the symbol c_u. More recent authors, disliking the use of the symbol c_u because of its connotation of cohesion, have used the symbol s_u where the s stands for strength and the u for undrained. Unfortunately, the letter s is already associated with the average principal total stress, $(\sigma_1 + \sigma_3)/2$: to label the radius of a Mohr circle of total stress s_u and the distance to its centre s is potentially quite confusing. I have adopted the symbol τ_u, because τ is associated with shear stress, which is what τ_u is.

4 In the oedometer, the horizontal stresses needed to prevent horizontal strains are provided by the steel confining ring. The oedometer test sample has a **strain-controlled** horizontal boundary, while the triaxial test sample has a **stress-controlled** horizontal boundary. Provided that the horizontal stresses applied to the triaxial sample are the same as those which would be experienced in an oedometer, conditions of zero lateral strain can still be obtained.

5 You cannot go out into the field and dig up a sample of Cam clay: it is *not* a real soil. Cam clay is the name given to the theoretical model for soil behaviour, developed by Roscoe and Schofield and their colleagues at Cambridge University. The model is named after the River Cam, which flows past the engineering laboratories in Cambridge where Roscoe's group carried out their work.

6 Between $p' = p_1'$ and $p' = p_2'$ $(p_2' > p_1')$, $\Delta(\ln p') = \ln(p_2') - \ln(p_1') = \ln(p_2'/p_1')$.

References

Allman, M.A. and Atkinson, J.H. (1992) Mechanical properties of reconstituted Bothkennar soil. *Géotechnique*, **42**(2), 531–40.

Atkinson, J.H. (1993). *An Introduction to the Mechanics of Soils and Foundations*, McGraw-Hill, London.

Atkinson, J.H. and Stallebrass, S.E. (1991) A model for recent stress history and non-linearity in the stress–strain behaviour of overconsolidated soil. *Proceedings of the 7th IACMAG*, Cairns, 555–60.

Atkinson, J.H., Richardson, D. and Stallebrass, S.E. (1990) Effect of recent stress history on the stiffness of overconsolidated soil. *Géotechnique*, **40**(3), 531–40.

Been, K. and Jefferies, M.G. (1985) A state parameter for sands. *Géotechnique*, **35**(2), 99–112.

Been, K., Jefferies, M.G. and Hachey, J. (1991) The critical state of sands. *Géotechnique*, **41**(3), 365–81.

Bishop, A.W. (1971) Shear strength parameters for undisturbed and remoulded soil specimens, in *Stress–strain Behaviour of Soils: Proceedings of the Roscoe Memorial Symposium*, Cambridge (ed. R.H.G. Parry), pp. 3–8, Foulis, Yeovil.

Bishop, A.W. and Donald, I.B. (1961) The experimental study of partly saturated soil in the triaxial apparatus. *Proceedings of the V International Conference on Soil Mechanics*, Paris, **1**, 13–21.

Bishop, A.W. and Henkel, D.J. (1962) *The Measurement of Soil Properties in the Triaxial Test* 2nd edn, Edward Arnold, London.

Bishop, A.W. and Lovenbury, H.T. (1969) Creep characteristics of two undisturbed clays. *Proceedings of the VII International Conference on Soil Mechanics & Foundation Engineering*, Mexico City, **1**, 29–37.

Bishop, A.W. and Wesley, L.D. (1975) A hydraulic triaxial apparatus for controlled stress path testing. *Géotechnique*, **25**(4), 657–70.

Bishop, A.W., Green, G.E., Garga, V.K., Andresen, A. and Brown, J.D. (1971) A new ring shear apparatus and its application to the measurement of residual strength. *Géotechnique*, **21**(4), 273–328.

Black, D.K. and Lee, K.L. (1973) Saturating laboratory samples by back pressure. *Proceedings of the American Society of Civil Engineers: Journal of the Soil Mechanics and Foundations Division*, SM1, January, 75–93.

Bolton, M.D. (1991) *A Guide to Soil Mechanics*, M.D. & K. Bolton, Cambridge.

Britto, A. and Gunn, M.J. (1987) *Critical State Soil Mechanics via Finite Elements*. Ellis Horwood, Chichester.

Burland, J.B. (1990) On the strength and compressibility of natural soils. 30th Rankine Lecture. *Géotechnique*, **40**(3), 329–78.

Crawford, C.B. (1963) Cohesion in an undisturbed sensitive clay. *Géotechnique*, **13**(2), 132–46.

Duncan, J.M. and Chang, C.-Y. (1970) Non-linear analysis of stress and strain in soils. *Journal of ASCE Soil Mechanics & Foundation Engineering Division*, **96**, 1629–53.

Fredlund, D.G. (1979) Appropriate concepts and technology for unsaturated soils. *Canadian Geotechnical Journal*, **16**, 121–39.

Graham, J., Crookes, J.H.A. and Bell, A.L. (1983) Time effects on the stress–strain behaviour of natural soft clays. *Géotechnique*, **33**(3), 327–40.

Hudson, A.P., White, J.K., Beaven, R.P. and Powrie, W. (2004) Modelling the compression behaviour of landfilled domestic waste. *Waste Management*, **24**, 259–69.

Jaky, J. (1944) The coefficient of earth pressure at rest. *Journal of the Union of Hungarian Engineers and Architects*, 355–8 (in Hungarian).

Jardine, R.J., Symes, M.J. and Burland, J.B. (1984) Measurement of soil stiffness in the triaxial apparatus. *Géotechnique*, **34**(3), 323–40.

Jardine, R.J., Potts, D.M., Fourie, A.B. and Burland, J.B. (1986) Studies of the influence of non-linear stress–strain characteristics in soil–structure interaction. *Géotechnique*, **36**(3), 377–96.

Jennings, J.E.B. and Burland, J.B. (1962) Limitations to the use of effective stresses in partly saturated soils. *Géotechnique*, **12**(2), 125–44.

Leroueil, S., Kabbaj, M., Tavenas, F. and Bouchard, R. (1985) Stress–strain–strain rate relation for the compressibility of sensitive natural clays. *Géotechnique*, **35**(2), 159–80.

Marsland, A. (1972) The shear strength of stiff fissured clays, in *Stress–strain behaviour of soils: Proceedings of the Roscoe Memorial Symposium*, Cambridge (ed. R.H.G. Parry), 59–68, Foulis, Yeovil.

Mesri, G., Febres-Cordero, E., Shields, D.R. and Castro, A. (1981) Shear stress–strain–time behaviour of clays. *Géotechnique*, 31(4), 537–52.

Mitchell, J.K. (1993) *Fundamentals of Soil Behavior*, 2nd edn., John Wiley, New York.

Muir Wood, D. (1990) *Soil Behaviour and Critical State Soil Mechanics*, Cambridge University Press, Cambridge.

Roscoe, K.H. and Burland, J.B. (1968) On the generalized stress–strain behaviour of 'wet' clay in *Engineering Plasticity* (eds. J. Heyman and F.A. Leckie), 535–609, Cambridge University Press, Cambridge.

Roscoe, K.H. and Schofield, A.N. (1963) Mechanical behaviour of an idealized 'wet' clay. *Proceedings of European Conference on Soil Mechanics and Foundation Engineering*, Wiesbaden, 1, 47–54.

Schofield, A.N. (1980) Cambridge geotechnical centrifuge operations (20th Rankine lecture). *Géotechnique*, 30(3), 227–68.

Schofield, A.N. and Wroth, C.P. (1968) *Critical State Soil Mechanics*, McGraw-Hill, London.

Sills, G.C., Wheeler, S.J., Thomas, S.D. and Gardner, T.N. (1991) Behaviour of offshore soils combining gas bubbles. *Géotechnique*, 41(2), 227–241.

Simpson, B., O'Riordan, N.J. and Croft, D.D. (1979) A computer model for the analysis of ground movements in London Clay. *Géotechnique*, 29(2), 149–75.

Skempton, A.W. (1964) Long-term stability of clay slopes. 4th Rankine Lecture. *Géotechnique*, 14(2), 75–102.

Skempton, A.W. and Northey, R.D. (1953) The sensitivity of clays. *Géotechnique*, 3(1), 30–53.

Stallebrass, S.E. (1990) Modelling small strains for analysis in geotechnical engineering. *Ground Engineering*, 22(9), 26–29.

Stallebrass, S.E. (1992) Modelling the deformation of overconsolidated soils using finite element analysis. *Proceedings of a Workshop on Experimental Characterization and Modelling of Soils and Soft Rocks*, 111–30. CUEN, Naples.

Stallebrass, S.E. and Taylor, R.N. (1997) The development and evaluation of a constitutive model for the prediction of ground movements in overconsolidated clay. *Géotechnique*, 47(2), 235–253.

Taylor, G.I. and Quinney, H. (1931) The plastic distortion of metals. *Phil. Trans. Roy. Soc.*, A230, 323–62.

Toll, D.G. (1990) A framework for unsaturated soil behaviour. *Géotechnique*, 40(1), 31–44.

Wheeler, S.J. (1991) An alternative framework for unsaturated soil behaviour. *Géotechnique*, 41(2), 257–61.

Whyte, I.L. (1982) Soil plasticity and strength: a new approach using extrusion. *Ground Engineering* 15(1), 16–24.

Wood, D.M. (1984) On stress parameters. *Géotechnique*, 34(2), 282–7.

Wood, D.M. and Wroth, C.P. (1978) The use of the cone penetrometer to determine the plastic limit of soils. *Ground Engineering*, 11(3), 37.

Chapter 6

Calculation of soil settlements using elasticity methods

6.1 Introduction

Geotechnical engineers must design foundations, retaining walls and other soil constructions which are not only safe (in the sense that they will not collapse) but also serviceable, in the sense that they do not deform excessively under working conditions. The definition of 'excessively' will depend on the nature of the construction under consideration. In general, a need to limit soil movements will arise where there is a likelihood of damage to structures made of comparatively brittle materials, which may crack at tensile strains as small as 0.075%.

Damage to buildings is more likely to arise from differential than uniform settlements. Guidelines developed by Burland and Wroth (1975) relate the maximum permissible deflection ratio (defined in Figure 6.1) to the length: height ratio of the building L/H and the nature of its construction (infilled frame, load-bearing wall in sagging, load-bearing wall in hogging). This approach is developed further by Burland (2001) and summarized by Gaba et al. (2003). Uniform settlements are usually much less damaging, and in many cases several tens of millimetres might be acceptable.

One way in which designers can attempt to ensure that the in-service deformations of a building are small is to apply a **factor of safety** to either the failure load or the soil strength. This is advantageous in that the calculation does not require a detailed knowledge of the stress–strain behaviour of the soil, but disadvantageous in that the required value of the factor of safety depends largely on previous experience. There is always the danger that, if they are applied in circumstances outside the knowledge base, the accepted procedures may prove inappropriate. Calculations based on conditions at collapse, and their application to design, are discussed in Chapters 7–10 for a number of geotechnical constructions, including foundations.

As an alternative to the factor of safety approach, the stress–strain or **constitutive** relationship for the soil is used in a calculation based on the stresses under working conditions, in an attempt to estimate the associated strains and deformations directly. Unfortunately, stress–strain relationships for soils cannot easily be described in simple yet accurate terms. We have already seen in Chapters 4 and 5 that soil does not behave as an elastic material.

It is nonetheless tempting to assign to the soil a Young's modulus and a Poisson's ratio. This enables the vast number of solutions for the stresses and displacements due to the application of various patterns of load to the surface of an elastic material of

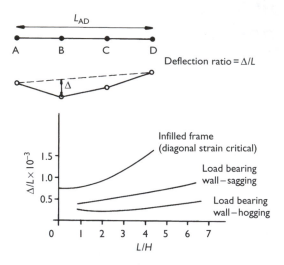

Figure 6.1 Limits to the deflection ratio Δ/L of a building foundation as a function of the length–height ratio L/H in order to prevent structural damage. (Redrawn with permission from Burland and Wroth, 1975.)

infinite depth and lateral extent (known as an **elastic half-space**) to be accessed by the geotechnical engineer (see e.g. Poulos and Davis, 1974).

This chapter is concerned with the applicability of elasticity methods in soil mechanics, and their use to calculate the vertical stress increases and soil settlements associated with shallow foundations and other surface loads.

Objectives

After having worked through this chapter, you should understand that

- although soil is not an elastic material, it is often treated as such for the purpose of calculating stress increases and settlements due to surface loads such as shallow foundations. One of the attractions of doing this is the large number of standard solutions and methods of analysis for elastic materials that can then be used
- the settlements predicted using elasticity methods depend on the elastic parameters used to characterize the soil – in particular, the stiffness modulus. The stiffness modulus depends on the stress history and the stress state of the soil, and on the applied stress path. All of these should be taken into account in selecting appropriate soil stiffness values for use in settlement calculations (section 6.2)
- there is a distinction between the **immediate** settlement due to shear at constant volume, and the **ultimate** settlement which includes a component due to consolidation (sections 6.2 and 6.8)
- the vertical stress increment caused by the application of a surface surcharge decreases with depth, due to the spreading out of the load into the surrounding soil (section 6.3).

You should be able to

- estimate increases in vertical stress at any depth, due to any pattern of applied surface load, using **Newmark's chart** (section 6.4), or – for rectangular surcharges – **Fadum's chart** (section 6.6)
- use these increases in vertical stress to estimate the associated long-term settlements, assuming that compression is predominantly one-dimensional (section 6.5)
- use standard formulae and charts to estimate stress increases and settlements due to surface loads of simple geometry (sections 6.3, 6.7 and 6.9).

You should have an appreciation of

- the potentially much more damaging effects on buildings of **differential**, as opposed to **uniform**, settlements (section 6.1)
- the main shortcomings of an elasticity-based approach, especially for the calculation of settlements (sections 6.2–6.10)
- the effect on calculated settlements of a soil stiffness that increases with depth (section 6.9)
- the interdepencence or **cross-coupling** of shear and volumetric effects in an anisotropic soil (section 6.10).

6.2   Selection of elastic parameters

6.2.1   Approximations and shortcomings of a simple elastic model

The main difficulty in ascribing a unique, 'elastic' stiffness modulus to a soil is illustrated by Figure 6.2, which shows the steady decrease in shear stiffness modulus G (which for an elastic material $= E/2(1 + \nu)$, where E is Young's modulus and ν is Poisson's ratio) with increasing triaxial shear strain ε_q (defined in section 5.4.6), measured in triaxial tests. Soil is clearly not an elastic material: if it were, G would remain constant with increasing shear strain. Nonetheless, elastic calculations, combined with judiciously selected elastic parameters, can often lead to reasonable estimates of the soil settlements associated with foundations and other near-surface loads. In particular:

> The elastic modulus should be appropriate to the stress and strain paths to which the soil is subjected, as well as the initial stress state and the expected changes in stress and strain. The soil should be on an unload/reload line (sections 4.2 and 5.7) – or there should be no reversal in the direction of the stress path – and the changes in stress and strain must be small.

The elastic modulus depends on the change in strain, and also on the direction of the new stress path relative to the direction of the immediately preceding stress path. The second of these is known as the **recent stress history** effect (Atkinson *et al.*, 1990).

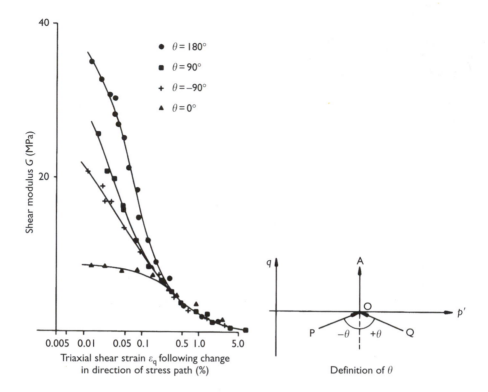

Figure 6.2 Stiffness of reconstituted London Clay following a change in the direction of the stress path. (Redrawn with permission from Atkinson *et al.*, 1990.)

The extreme example of this is when the stress path is reversed, giving a stress path rotation of 180°. (Stress path reversals are not infrequent in practice. For example, during the construction of a basement, the soil below the building is first unloaded vertically during excavation, and then re-loaded as the basement is built.) Figure 6.2 shows that the stiffness of the soil immediately following a change in the direction of the stress path ($\theta = 180°$) can be comparatively very high.

 If the stress–strain curve is non-linear, there are two alternative definitions of the effective stiffness modulus that may be used. These are illustrated in Figure 6.3. The **secant modulus** is given by the line joining the origin to the current stress–strain point (σ, ε), and is defined as the ratio of the change in stress to the change in strain measured from the datum point, $E_{secant} = \sigma/\varepsilon$. (Similarly, the **secant shear modulus** G_{secant} is defined as τ/γ, where τ is the shear stress and γ is the corresponding shear strain.) The **tangent modulus**, which is equal to the slope of the stress–strain curve at the point under consideration, is defined as the rate of change of stress with strain, $E_{tangent} = d\sigma/d\varepsilon$. (The **tangent shear modulus** $G_{tangent}$ is defined as the rate of change of shear stress with shear strain, $G_{tangent} = d\tau/d\gamma$.) For an elastic material, $G/E = $ constant, as shown in section 6.2.2.

 In an elastic calculation, the strain depends on both the stress and the modulus. If the soil modulus depends on the strain, an iterative procedure must be adopted, in

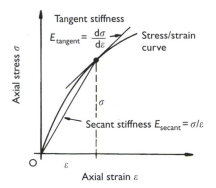

Figure 6.3 Definition of secant and tangent moduli.

which the calculation is repeated until the strain is consistent with the value of the modulus used to obtain it. For this purpose, it is easier to use the secant, rather than the tangent, modulus.

It is also necessary to take account of the fact that soil stiffness is likely to increase with increasing effective stress. This effect is often modelled by assuming that the soil stiffness is proportional to the depth. In reality, there may be a sequence of different strata of different stiffnesses, all of which may contribute to the observed settlement at the ground surface. Most of the standard solutions for stresses and displacements are for an elastic material which is uniform and isotropic, and so do not take into account the effect of a modulus which varies with depth. Some solutions for an elastic half-space whose stiffness increases linearly with depth are presented by Gibson (1974), and are discussed briefly in section 6.9.

A further approximation inherent in the use of nearly all of the standard formulae and methods is that the foundation is assumed either to be rigid in comparison with the soil (so that the settlement is uniform), or to be perfectly flexible (so that the stress distribution is uniform).

The concept of consolidation discussed in Chapter 4 is also relevant here. Indeed, one-dimensional consolidation is a special case of the use of elasticity to estimate soil settlements. It is special in that there is no opportunity for the load to spread out into the surrounding ground, because the soil is subjected to a uniform increase in stress over its entire surface area.

The application of a surface load causes an immediate increase in total stress, and an eventual increase in effective stress when the pore water pressures have returned to their equilibrium values. If the soil beneath a foundation is a clay, settlements will develop over a period of time as the excess pore water pressures generated by the application of the load dissipate and the clay consolidates. Elasticity calculations may be used to estimate either the initial (undrained) or the long-term (fully drained) soil movements. To estimate the immediate (undrained) soil movements, the elastic parameters E (Young's modulus) and ν (Poisson's ratio) must be obtained from undrained tests and defined in terms of total stresses. They are given the subscript 'u' to indicate that they are **undrained**, or **total stress**, parameters. To estimate the long-term (drained) soil movements, the elastic parameters E and ν must be obtained from drained tests and defined in terms of effective stresses. They are given a prime (') to indicate that they are **effective stress** parameters.

6.2.2 Relationships between elastic constants

Writing Hooke's Law for an isotropic elastic material in terms of undrained parameters E_u and ν_u, and changes in principal total stress $\Delta\sigma$ and changes in principal strain $\Delta\varepsilon$:

$$\Delta\varepsilon_1 = (1/E_u)(\Delta\sigma_1 - \nu_u\Delta\sigma_2 - \nu_u\Delta\sigma_3)$$

$$\Delta\varepsilon_2 = (1/E_u)(\Delta\sigma_2 - \nu_u\Delta\sigma_1 - \nu_u\Delta\sigma_3) \qquad (6.1)$$

$$\Delta\varepsilon_3 = (1/E_u)(\Delta\sigma_3 - \nu_u\Delta\sigma_1 - \nu_u\Delta\sigma_2)$$

Assuming that the principal strain increments are small, they may be added to obtain the volumetric strain increment:

$$(\Delta\varepsilon_1 + \Delta\varepsilon_2 + \Delta\varepsilon_3) = \Delta\varepsilon_{vol} = [(1 - 2\nu_u)/E_u](\Delta\sigma_1 + \Delta\sigma_2 + \Delta\sigma_3) \qquad (6.2)$$

But

$$(\Delta\sigma_1 + \Delta\sigma_2 + \Delta\sigma_3) = 3\Delta p$$

(where Δp is the change in the average principal total stress), so that

$$\Delta p/\Delta\varepsilon_{vol} = E_u/[3(1 - 2\nu_u)]$$

which is equal to the undrained bulk modulus K_u.

It is generally assumed that the soil can only compress by means of a reduction in the void ratio. In a saturated soil, this cannot happen unless water is expelled from the pores. In the short term as the soil deforms without drainage of pore water, there is no change in volume ($\Delta\varepsilon_{vol} = 0$), and the undrained bulk modulus K_u is infinite. This requires that the undrained Poisson's ratio $\nu_u = 0.5$.

The effective and total stress Young's modulus and Poisson's ratio are linked by the shear modulus G. As the pore water is unable to take shear, all of the shear stress is carried by the soil skeleton. Thus the effective shear stress is the same as the total shear stress, and the effective stress shear modulus G' is equal to the total stress shear modulus G_u.

In a conventional triaxial compression test, there is no change in the cell pressure so that $\Delta\sigma_2 = \Delta\sigma_3 = 0$. Substituting these values into equations (6.1),

$$\Delta\varepsilon_1 = (1/E_u)\,(\Delta\sigma_1) \qquad (6.3a)$$

$$\Delta\varepsilon_2 = \Delta\varepsilon_3 = (1/E_u)(-\nu_u\Delta\sigma_1) \qquad (6.3b)$$

From the Mohr circles of stress and strain increment shown in Figure 6.4 (with $\Delta\sigma_3 = 0$),

$$\Delta\tau = \Delta\sigma_1/2$$

$$\Delta\gamma = (\Delta\varepsilon_1 - \Delta\varepsilon_3) = [(1 + \nu_u)/E_u]\Delta\sigma_1$$

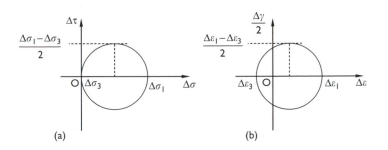

Figure 6.4 Mohr circles of (a) stress and (b) strain increment for a conventional undrained triaxial compression test.

so that

$$G_u = \Delta\tau/\Delta\gamma = E_u/2(1 + \nu_u) \qquad (6.4)$$

Similarly, in terms of effective stresses,

$$G' = \Delta\tau/\Delta\gamma = E'/2(1 + \nu') \qquad (6.5)$$

so that

$$G = G_u = G' = E_u/2(1 + \nu_u) = E'/2(1 + \nu') \qquad (6.6)$$

6.2.3 *Determining elastic moduli from triaxial compression test data*

In a conventional undrained triaxial compression test, the change in deviator stress Δq is equal to the change in major principal total stress $\Delta\sigma_1$. In a drained test, the change in deviator stress Δq is equal to the change in major principal effective stress $\Delta\sigma'_1$. Thus from equation (6.3a), a graph of q (y-axis) against axial strain ε_1 (x-axis) for an undrained test has slope E_u. If the data are from a drained test, the slope of the graph is E'.

As $\Delta\sigma_1 = \Delta q$, the change in shear stress $\Delta\tau = \Delta q/2$. From equation (6.4), a graph of deviator stress q (y-axis) against shear strain γ (x-axis) has slope $2G$, whether the test is drained or undrained. (A graph of q against triaxial shear strain $\varepsilon_q = 2\gamma/3$ has slope $3G$, i.e. $G = (dq/d\varepsilon_q)/3$.)

The effective stress bulk modulus K' may be obtained from the slope of a graph of average effective stress p' (y-axis) against volumetric strain ε_{vol} (x-axis), since by definition $K' = dp'/d\varepsilon_{vol}$.

In all cases, the tangent modulus is obtained from the local slope of the graph, while the secant modulus is given by slope of the line joining the point in question to the origin (Figure 6.3).

The form of Hooke's Law given in equations (6.1) applies only to isotropic elastic materials. Most soils are anisotropic, which results in an interdependence or **coupling** of volumetric and shear effects (section 6.10). This has important implications for the determination of elastic parameters for a soil: in particular, the rather simplistic approach described above is invalidated as explained in section 6.10.

6.3 Boussinesq's solution

Boussinesq (1885) used the conditions of equilibrium and compatibility, together with the stress–strain relationship for the material, to determine the stresses and strains at any point within an **isotropic** (i.e. having the same properties in all directions), **homogeneous** (i.e. having the same properties at any point) elastic **half-space** (i.e. a body of infinite depth and lateral extent, so that it occupies half of all possible space) resulting from the application of a point load at the surface (Figure 6.5).

Boussinesq's solution may be integrated for a collection of point loads acting over any given area of the surface of the half-space to determine the stresses and strains that result from the application of any pattern of applied load. In some cases, it is possible to derive a mathematical expression for the stresses and strains at any point, but it is often necessary to calculate them numerically.

In the case of a uniform surcharge of magnitude q^1 acting over a circular area of radius R at the surface of the half-space, integration of Boussinesq's solution gives the expression

$$\Delta\sigma_z = q\{1 - [1 + (R/z)^2]^{-3/2}\} \tag{6.7}$$

for the increase in vertical stress $\Delta\sigma_z$ at a depth z below the centreline. This expression is plotted non-dimensionally in Figure 6.6. It may be seen that the increase in vertical stress falls to below 10% of the surface value at a depth of approximately twice the diameter of the footing. This is due to the spreading of the loaded area with depth: in the case of a strip footing, where spreading can only occur in the direction perpendicular to the line of the foundation, the increase in vertical stress falls to 10% of the surface value at a depth of approximately six times the width of the footing (Figure 6.7).

Although it would be necessary to use the soil stiffness to calculate strains and displacements, the stiffness of the material does not feature in the expressions for increases in stress. It is perhaps not unreasonable, therefore, to suppose that the vertical

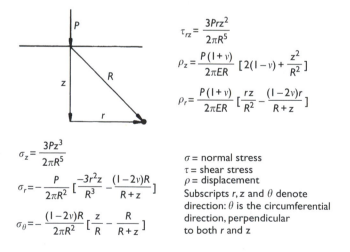

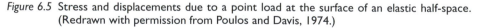

Figure 6.5 Stress and displacements due to a point load at the surface of an elastic half-space. (Redrawn with permission from Poulos and Davis, 1974.)

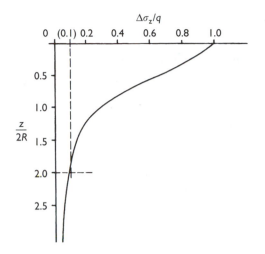

Figure 6.6 Increase in vertical stress below the centreline of a uniform circular surcharge.

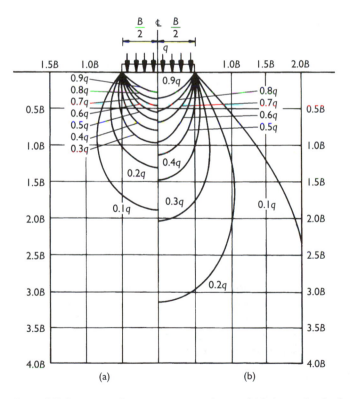

Figure 6.7 Contours of increase in vertical stress (a) below a circular footing of diameter B, and (b) below a strip footing of width B, subjected to a uniform vertical surcharge q. (Redrawn with permission from Whitlow, 1995.)

effective stresses within the soil mass which result from the application of some pattern of loading at the surface will remain substantially unaffected, even if the soil is not truly elastic.

Burland *et al.* (1977) quote ample evidence to show that the distribution of vertical stress within a soil mass due to the application of a surface load does not differ enormously from the Boussinesq solution in cases where the stress–strain relationship of the soil is nonlinear, or where the stiffness of the soil increases with depth. Thus the Boussinesq-type stress distributions can still be useful if there is a general increase in stiffness with depth, or if several distinct horizontal layers of different types of soil are present.

Significant deviation from the Boussinesq vertical stress distributions would be expected to occur if the applied loads were high enough to cause extensive yielding; or in the case of an inclusion or lens of soil of higher or lower stiffness, which was not continuous across an entire horizon. This is because the stiffer inclusion would attract more than its fair share of the load. The vertical stress distribution will also differ from the Boussinesq solution in cases where a stiff surface layer overlies a more compressible soil stratum, because the stiff layer will redistribute the load to some extent. For practical purposes, however, the Boussinesq distribution of vertical stress is reasonably accurate in most ground conditions.

Formulae and charts based on the idealization of the soil as a uniform isotropic elastic half-space can also be used to estimate settlements beneath surface loads and footings. However, the settlements depend on the elastic parameters used to characterize the soil (i.e. its Young's modulus and Poisson's ratio), and also on the calculated horizontal stresses, which are much more sensitive to factors such as changes in soil stiffness with depth. The settlements calculated in this way are not, therefore, as reliable as the vertical stress distributions. A more versatile and general approach to the calculation of settlements, based on the *ad hoc* conversion of stresses to vertical strains by dividing the soil into discrete layers and using a different stiffness modulus at each depth, is outlined in sections 6.4–6.6.

6.4   Estimation of increases in vertical stress at any depth due to any pattern of surface load, using Newmark's chart

The Newmark chart (originally devised by Newmark (1942); Figure 6.8) is essentially a plan view of the surface of an infinite half-space, divided into n elements. Each element will, when loaded with a uniform stress q, result in an increase in vertical stress of q/n at a depth Z below the centre of the chart. The depth Z is linked to the scale of the chart, and must be shown on it.

The chart shown in Figure 6.8 is divided by 9 concentric circles into 10 annuli, each of which is subdivided into 20 segments, giving 200 elements in total. If any element is loaded by a uniform stress of q, this will result in an increase in vertical stress of $q/200$ or $0.005q$ at a depth Z below the centre of the chart.

The radii of the concentric circles were calculated using equation (6.7). If there is a total of 10 annuli (the outermost of which actually extends to infinity), the loading of each annulus by a uniform stress q must result in an increase in vertical stress $0.1q$ at

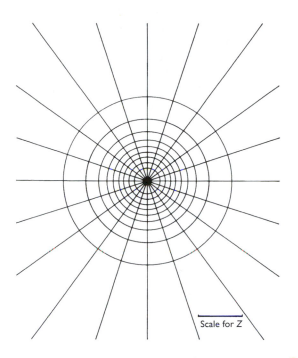

Scale for Z

Figure 6.8 Influence chart for vertical stress increment. (Redrawn from Newmark, 1942.)

the depth Z. The innermost circle must have radius R_1/Z such that (from equation 6.7)

$$\Delta\sigma_Z = 0.1q = q\{1 - [1 + (R_1/Z)^2]^{-3/2}\}$$

or

$$[1 + (R_1/Z)^2]^{-3/2} = 0.9$$

or

$$R_1/Z = \sqrt{(0.9^{-2/3} - 1)} = 0.27$$

Similarly, the second circle must have radius R_2/Z such that

$$\Delta\sigma_Z = 0.2q = q\{1 - [1 + (R_2/Z)^2]^{-3/2}\}$$

or

$$[1 + (R_2/Z)^2]^{-3/2} = 0.8$$

or

$$R_2/Z = \sqrt{(0.8^{-2/3} - 1)} = 0.40$$

and so on. To construct the chart, a scale must be chosen for Z (e.g. $Z \approx 12$ mm in Figure 6.8; $Z = 40$ mm is convenient for an A4 sheet), which must be shown on the finished diagram.

The Newmark chart may be used to calculate the increase in vertical stress at any location within the soil mass due to the application of any surface load, applied over an area of any size and shape. This is done by drawing a plan view of the loaded area on the chart to the appropriate scale (which is found by setting $Z =$ the depth under investigation), positioned with the point of interest (which may be either below or to the side of the loaded area) above the centre of the chart. The increase in vertical stress is estimated from the number of elements covered by the loaded area and the stress acting on each, as in Example 6.1.

Example 6.1 Calculating increases in vertical stress at depth below a surface surcharge using Newmark's chart

Figure 6.9 shows a plan and a cross-section of a proposed reinforced earth road embankment. Use the Newmark chart to estimate the increase in vertical stress along the existing pipeline A–A', which runs perpendicular to the new road at a depth of 8 m below original ground level. The embankment may be treated as a uniform surcharge of 100 kPa acting over a width of 8 m. State one major potential shortcoming of your analysis.

[University of London 2nd year BEng (Civil Engineering) examination,
Queen Mary & Westfield College]

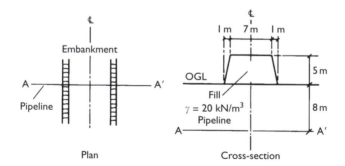

Figure 6.9 Problem geometry.

Solution

It is necessary to determine the increases in vertical stress at a number of points along the line A–A', both beneath and on either side of the embankment. Making use of symmetry, draw scale views of the embankment (on plan) on the lower part of the Newmark chart, as shown in Figure 6.10. In each case, the point of interest (A, B, C, D, E or F in Figure 6.10) is at the centre of the chart, and the 'scale for Z' has been set to the depth in question, which is 8 m.

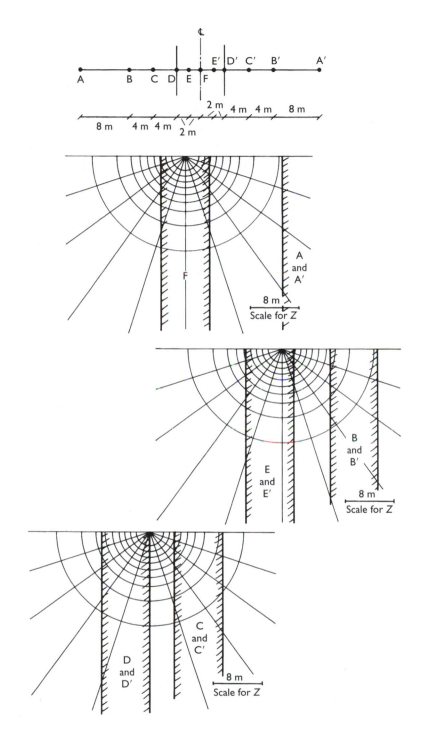

Figure 6.10 Use of Newmark charts.

The increase in vertical stress at each point is given by

$$\Delta\sigma_Z = (n/200) \times q$$

where q is the surface surcharge (100 kPa) and n is the number of squares covered in each case.

In plan view, the embankment (idealized as a uniform surcharge over a width of 8 m) runs from the top to the bottom of the Newmark chart. Only half of the Newmark chart is shown in Figure 6.10: the total number of squares covered by the embankment is therefore twice that counted from any of the half-charts shown in Figure 6.10.

Values of n and $\Delta\sigma_z$ are given for each of the points A–F (see Figure 6.10) in Table 6.1, and the distribution of increase in vertical stress $\Delta\sigma_z$ along the line A–A' is shown in Figure 6.11. (The calculation is approximate, and some degree of variability – say ±5% – is inevitable, due to differing estimates of the contributions of partly covered 'squares'. Do not worry, therefore, if the answers you obtain to the worked examples and the problems at the end of the chapter are slightly different from those given.)

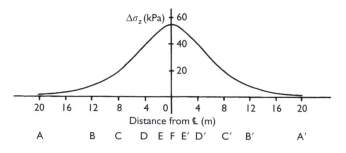

Figure 6.11 Increase in vertical stress along pipeline A–A'.

Table 6.1 Values n and increase in vertical stress

	Point					
	A	B	C	D	E	F
n	3	16	40	81	100	110
$\Delta\sigma_z$(kPa)	1.5	8	20	40.5	50	55

Apart from the reservations about the use of elastic stress distributions already mentioned, the calculation assumes that the presence of the pipeline does not alter the stress distribution within the soil. Since the pipeline is likely to be stiffer than the surrounding soil, it may well attract higher loads than indicated in Figure 6.11.

6.5   Estimation of settlements due to surface loads and foundations

Charts and tables are available for the calculation of surface settlements due to the application of various loads to the surface of an elastic half-space. It has already been mentioned, however, that the effect of heterogeneity (i.e. the likely increase in soil stiffness with depth) on settlements is much more significant than it is on stresses, so that such charts and tables may be of only limited value.

Terzaghi (1943) suggested that settlements could be calculated by assuming that soil movements are predominantly vertical, so that the relationship between vertical stress and strain increments is governed by the one-dimensional modulus, E_0'. Variations in stiffness with depth can then be taken into account by dividing the soil into a number of layers, according to the following procedure.

1. Identify a number of different layers within the soil, each of which may be characterized approximately by a uniform one-dimensional modulus E_0', and a uniform increase in vertical stress $\Delta\sigma_v$. Even if there is only one soil type present, this procedure can be used to take account, in a step-wise fashion, of an increase in modulus (or a decrease in stress increment) with depth. Since stresses change quite rapidly at shallow depths, the layers used in the analysis should in general be thinner near the soil surface.
2. Using the Newmark chart as described in sections 6.4 and 6.5, calculate a representative increase in vertical stress $\Delta\sigma_v$ for each layer at each point (on plan) where the settlement is to be estimated. This might be the increase in vertical stress at the centre of the layer, or the average of the increase in vertical stress at the top and at the bottom.
3. Assuming that strains are primarily one-dimensional, estimate the settlement of each layer $\delta\rho$ from

$$E_0' = \Delta\sigma_v'/\varepsilon_v; \quad \varepsilon_v = \delta\rho/h$$

Hence

$$\delta\rho = \Delta\sigma_v'h/E_0' \tag{6.8}$$

where h is the thickness of the layer.
4. For each point, add together the settlement of each layer to give an estimate of the settlement at the soil surface.

The one-dimensional modulus E_0' is defined as the ratio of the change in vertical effective stress $\Delta\sigma_v'$ to the change in vertical strain $\Delta\varepsilon_v$ during one-dimensional loading or unloading. The value of E_0' for a soil may be obtained directly from an oedometer test stage over the correct increment or decrement of vertical stress (section 4.2). Strictly, however, a value of Young's modulus E' obtained from a conventional triaxial compression test as described in section 6.2.3 must be corrected for the Poisson's ratio effect as follows.

From equations (6.1) in terms of effective stresses with $\sigma_1' = \sigma_v'$ and $\sigma_2' = \sigma_3' = \sigma_h'$,

$$\Delta\varepsilon_v = (1/E')(\Delta\sigma_v' - 2v'\Delta\sigma_h')$$
$$\Delta\varepsilon_h = (1/E')[(1-v')\Delta\sigma_h' - v'\Delta\sigma_v'] \tag{6.9}$$

In one-dimensional compression the horizontal strain increment $\Delta\varepsilon_h = 0$, so that

$$(1-v')'\Delta\sigma_h' = v'\Delta\sigma_v'; \quad \Delta\sigma_h' = [v'/(1-v')]\Delta\sigma_v'$$
$$\Delta\varepsilon_v = (\Delta\sigma_v'/E')[(1-2v'^2)/(1-v')]$$
$$E_0' = \Delta\sigma_v'/\Delta\varepsilon_v = E'(1-v')/[(1+v')(1-2v')] \tag{6.10}$$

For undrained conditions, substitution of the undrained total stress parameters E_u and v_u into equation (6.10) in place of the drained, effective stress parameters E' and v' indicates that the undrained one-dimensional modulus is infinite, because $v_u = 0.5$ (section 6.2.2). This is consistent with the fact that no volume change can occur without drainage of pore water. In one-dimensional compression, in which the horizontal strain is by definition zero, a change in volume must be accompanied by the development of vertical strain. Thus only the long-term or ultimate settlements, calculated from the long-term changes in effective stress, can be found using the one-dimensional modulus. In reality, some immediate settlement due to shear deformation will probably occur, because shear deformation takes place without any change in volume. This confirms the approximate nature of the approach. Methods of estimating immediate settlements are discussed in section 6.8.

In many cases, the loading pattern exerted by the building on its foundation is different from the loading pattern transmitted by the foundation to the soil, because the foundation itself has an appreciable stiffness. The foundation will tend to redistribute the stresses so that the settlements beneath it are more uniform. The settlements calculated using the pattern of loads imposed by the superstructure on the foundation (rather than by the foundation on the soil) correspond to a perfectly flexible foundation, which is in practice unrealistic. The settlement of a rigid foundation could be estimated by determining the settlement of the flexible foundation at a number of points, and taking the average. More rigorously, the stiffness of a raft of plan dimensions $L \times B$ and thickness t_r made of a material having Young's modulus E_r and Poisson's ratio v_r on soil of Young's modulus E_s and Poisson's ratio v_s may be characterized by the raft–soil stiffness ratio,

$$K_{rs} = \frac{4E_r B t_r^3 (1-v_s^2)}{3\pi E_s L^4 (1-v_r^2)}$$

which in practice may range in value from 0.001 to 10 (Clancy and Randolph, 1996).

Methods of correcting the settlements calculated using the one-dimensional approach for the effects of lateral strain have been proposed, and in some cases are reasonably widely used (e.g. Skempton and Bjerrum, 1957). However, more significant errors are likely to arise from the careless determination and/or selection of soil parameters than from the approximations inherent in the one-dimensional approach. In their review of the behaviour of foundations and structures, Burland et al. (1977)

concluded that for soils which are approximately elastic in their response to monotonically increasing stresses, total settlements obtained from the one-dimensional method of analysis compare very favourably with values obtained using more sophisticated methods.

Example 6.2 Estimation of soil settlements below a shallow foundation

Figure 6.12 shows a plan view of the loads transmitted to the soil by the raft foundation of a new industrial building, together with the soil profile at the proposed construction site. Use the Newmark chart to estimate the increase in vertical stress at depths of 6 m and 16 m below the points marked X and Y on Figure 6.12.

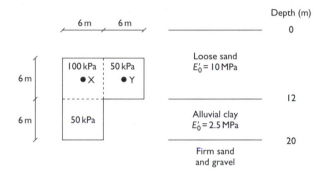

Figure 6.12 Plan view of proposed industrial building and soil profile at the construction site.

Hence estimate the expected eventual settlement of the foundation at the points X and Y. Suggest two possible shortcomings of your analysis.

[*University of London 2nd year BEng (Civil Engineering) examination, Queen Mary and Westfield College*]

Solution

Draw plan views of the foundation with the scale for Z set to (a) 6 m and (b) 16 m, with the points (i) X and (ii) Y located over the centre of the chart (see Figure 6.13).

The increase in vertical stress at each point is given by

$$\Delta\sigma_v = \{(n_{50}/200) \times 50\,\text{kPa}\} + \{(n_{100}/200) \times 100\,\text{kPa}\}$$

where n_{50} is the number of elements covered by the area over which the surface load is 50 kPa, and n_{100} is the number of elements covered by the area over which the surface load is 100 kPa.

The values of n_{50}, n_{100} and the increases in vertical stress at each depth and location are given in Table 6.2.

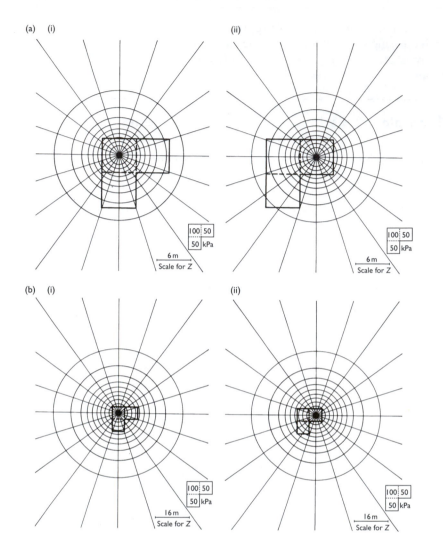

Figure 6.13 Newmark charts: (a) 6 m depth, (i) the point X, (ii) the point Y; (b) 16 m depth, (i) the point X, (ii) the point Y.

Table 6.2 Values of *n* and increase in vertical stress

	No. of elements @ 50 kPa (n_{50})	No. of elements @ 100 kPa (n_{100})	$\Delta\sigma_v = \{(n_{50}/200) \times 50\} + \{(n_{100}/200) \times 100\}$ kPa
X, depth 6 m	$18 \times 2 = 36$	$17.25 \times 4 = 69$	43.5
Y, depth 6 m	$69 + 8 = 77$	18	28.25
X, depth 16 m	$9 \times 2 = 18$	$4 \times 4 = 16$	12.5
Y, depth 16 m	$16 + 8 = 24$	9	10.5

The values of n_{50}, n_{100} and the increases in vertical stress at each depth and location are given in Table 6.2.

The settlement of each layer is given by equation (6.8):

$$\delta\rho = \Delta\sigma'_v h / E'_0$$

Taking the layers as 0–12 m below ground level (loose sand, $E'_0 = 10$ MPa) and 12–20 m below ground level (alluvial clay, $E'_0 = 2.5$ MPa), the overall settlement at X is

$$\rho = (12\,\text{m} \times 43.5\,\text{kPa} \div 10\,000\,\text{kPa}) + (8\,\text{m} \times 12.5\,\text{kPa} \div 2500\,\text{kPa})$$

$$= 92\,\text{mm}$$

The overall settlement at Y is

$$\rho = (12\,\text{m} \times 28.25\,\text{kPa} \div 10\,000\,\text{kPa}) + (8\,\text{m} \times 10.5\,\text{kPa} \div 2500\,\text{kPa})$$

$$= 68\,\text{mm}$$

The potential shortcomings of the analysis are as follows:

1. It has been assumed that the predominant mode of deformation is vertical compression, so that short term settlements (due to shear at constant volume) cannot be calculated.
2. The division of the soil into only two layers is somewhat crude, and the use of more layers (each with a different value of E'_0 and $\Delta\sigma_v$) would be preferable.

6.6   Influence factors for stresses

Although the Newmark chart is versatile, its use can be time consuming, particularly in analyses where the soil is divided into several layers. In many practical cases, foundations are rectangular on plan, at least to a reasonable approximation. Increases in vertical stress at a depth Z below the corner of a uniformly loaded rectangular surcharge of length L and breadth B may be calculated using a chart originally presented by Fadum (1948).

Fadum's chart (Figure 6.14) gives values of an influence factor, I_σ, in terms of the dimensionless parameters $m = L/Z$ and $n = B/Z$. The increase in vertical stress $\Delta\sigma_v$ at a depth Z below the corner of a uniform surcharge of length L and breadth B is

$$\Delta\sigma_v = qI_\sigma \tag{6.11}$$

One of the attractive features of elasticity calculations is that the principle of superposition applies. This means that Fadum's chart can be used to calculate the increase in vertical stress below any point within the loaded area by dividing the loaded area into four and adding the increases in vertical stress due to each of the four small rectangles individually (Figure 6.15(a)). Similarly, the judicious superposition of positive and negative loads enables Fadum's chart to be used to calculate the increase in vertical stress at points outside the loaded area (Figure 6.15(b)).

The use of Fadum's chart is illustrated in Example 6.3, which is an idealization of a case record presented by Somerville and Shelton (1972).

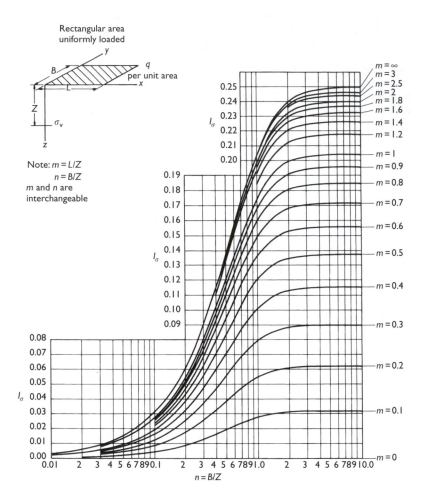

Figure 6.14 Influence factors for the increase in vertical stress below the corner of a uniform rectangular surcharge. (Redrawn from Fadum, 1948.)

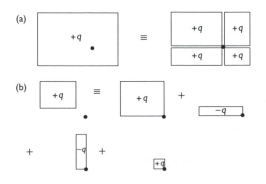

Figure 6.15 Use of superposition to calculate the increase in stress at any point using Fadum's chart.

Example 6.3 Settlement of high-rise buildings in Glasgow

Five 15-storey blocks of flats were constructed during 1968 in the Parkhead and Bridgeton districts of Glasgow. The foundations are underlain by deep alluvial deposits consisting of laminated clays and silts, and coal measures rocks. In view of the compressibility of the alluvial deposits, piled foundations would normally have been adopted for high-rise buildings of this nature. However, because the coal measures had been extensively mined during the 1800s, it was considered inadvisable to concentrate foundation loads using piles terminating on or just above the rockhead.

It was therefore decided to use semi-buoyant raft foundations to reduce the bearing pressure on the clays and silts to a value which provided an adequate margin of safety against failure. Nonetheless, there was considerable concern over the magnitude of possible settlements, and detailed settlement analyses were carried out. Settlements were monitored during and for some time after construction, and compared with the predicted values.

The foundation of each block of flats may be idealized as a rectangle 33.55 m × 22.22 m on plan. The gross bearing pressure is approximately 135 kPa, but by utilizing the buoyancy of cellular rafts placed 4.3 m below ground level, the net effective bearing pressure was reduced to 53.5 kPa. (The net effective bearing pressure is calculated from the weight of the building and the raft divided by the area of the foundation, minus the upthrust on the base of the foundation due to the pore water pressure at that depth, minus the effective surcharge due to the soil above founding level.) A soil profile, giving the variation in coefficient of compressibility m_v as a function of depth as used by the designers, is given in Figure 6.16. (The coefficient of compressibility m_v is equal to the reciprocal of the one-dimensional modulus, $m_v = 1/E_0'$.) Estimate the eventual settlement at the centre of the raft and at the middle of the longer side.

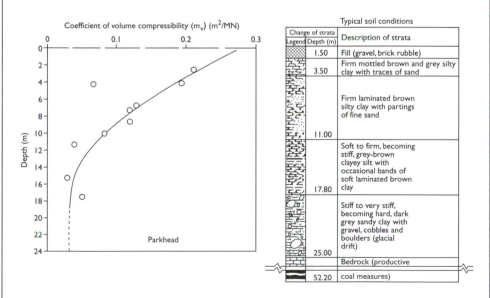

Figure 6.16 Soil profile and $m_v(= 1/E_0')$ as a function of depth. (Redrawn with permission from Somerville and Shelton, 1972.)

Solution

Divide the compressible strata into 2 m thick layers. Use Fadum's chart (Figure 6.14) to estimate the increase in vertical stress at the centre of each layer, below the centre of the foundation and the middle of the longer side. (Using superposition, the increase in stress below the centre of the raft is four times the increase in stress below the corner of a foundation of dimensions 16.78 m × 11.11 m, and the increase in stress below the middle of the longer side is twice the increase in stress below the corner of a foundation of dimensions 16.78 m × 22.22 m – Figure 6.17.) Read off the value of m_v at the centre of each 2 m layer from Figure 6.16, and calculate the settlement of each layer using equation (6.8):

$$\delta\rho = \Delta\sigma'_v h / E'_0 = \Delta\sigma'_v h m_v$$

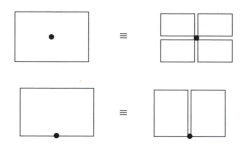

Figure 6.17 Use of symmetry.

The overall settlement is the sum of the settlements of each individual layer. The ground has been assumed to be effectively incompressible below a depth of 18.3 m, where the clay becomes very stiff with gravel, cobbles and boulders. Bedrock is reached at 25 m below ground level. The calculation is detailed in Tables 6.3 and 6.4.

Table 6.3 Calculation of settlement below centre of raft

Depth below ground level (m)	Depth Z below founding level (m)	$m_v (= 1/E'_0)$ (m^2/MN)	$B/Z = n$	$L/Z = m$	I_σ (from Figure 6.14)	$\delta\rho = q \times 4I_\sigma \times m_v \times h$ (m)
5.3	1.0	0.166	11.11	16.78	0.250	0.0178
7.3	3.0	0.129	3.70	5.59	0.248	0.0137
9.3	5.0	0.094	2.22	3.36	0.242	0.0097
11.3	7.0	0.066	1.59	2.40	0.229	0.0065
13.3	9.0	0.048	1.23	1.86	0.211	0.0043
15.3	11.0	0.039	1.01	1.53	0.194	0.0032
17.3	13.0	0.033	0.85	1.29	0.175	0.0025
					Total	58 mm

Table 6.4 Calculation of settlement below middle of longer side

Depth below ground level (m)	Depth below founding level Z (m)	$m_v(=1/E_0')$ (m^2/MN)	$B/Z = n$	$L/Z = m$	I_σ (from Figure 6.14)	$\delta\rho = q \times 2I_\sigma \times m_v \times h$ (m)
5.3	1.0	0.166	16.78	22.22	0.250	0.0089
7.3	3.0	0.129	5.59	7.41	0.250	0.0069
9.3	5.0	0.094	3.36	4.44	0.247	0.0050
11.3	7.0	0.066	2.40	3.17	0.242	0.0034
13.3	9.0	0.048	1.86	2.47	0.235	0.0024
15.3	11.0	0.039	1.53	2.02	0.225	0.0019
17.3	13.0	0.033	1.29	1.71	0.212	0.0015
					Total	30 mm

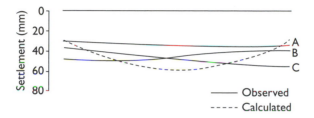

Figure 6.18 Comparison of measured and calculated settlements: blocks of flats at Parkhead and Bridgeton.

The calculated and measured settlements are compared in Figure 6.18. Although the calculated maximum settlement is somewhat greater than those observed, the agreement generally is reasonably close. However, the effect of the stiffness of the foundation (which was neglected in the calculations) in reducing differential settlements is clear. Further details given by Somerville and Shelton (1972) include the settlement contours, and the observed settlement–time curves.

Ground conditions cannot in reality be expected to be perfectly uniform, and as a result of variations in ground conditions across each site the settlement contours are not in general focused on the centre of the building. The rates of settlement were such that 90% of the ultimate movement had occurred after 12–15 months, whereas one-dimensional consolidation theory using an oedometer test value of c_v would have suggested a period of 25 years. The reason for this discrepancy was the laminated nature of the deposit. As discussed in sections 3.6, 3.14 and 4.7.3, a laminated soil must be expected to be significantly more permeable in the horizontal direction than in the vertical, so that excess pore water pressures dissipate primarily by horizontal flow.

6.7   Standard solutions for surface settlements on an isotropic, homogeneous, elastic half-space

Formulae and charts are available for the settlements which result from various patterns of surface load, calculated assuming that the soil behaves as a uniform, isotropic elastic material (e.g. Steinbrenner, 1934; Giroud, 1968). It has already been mentioned that settlements calculated in this way are considerably less robust than distributions of vertical effective stress, being much more sensitive to the elastic parameters assigned to the soil and factors such as anisotropy and the likely increase in soil stiffness with depth. It is, however, sometimes convenient to make use of these solutions because of the limitations of the one-dimensional approach – for instance in the calculation of initial undrained settlements due to shear rather than the eventual settlements following consolidation. As examples, Table 6.5 gives the settlements below the centre and edge of a uniform circular surcharge of diameter B and the settlement of a rigid circular plate.

Table 6.5 Formulae for soil surface settlements: circular foundation of diameter B on a uniform, isotropic elastic half-space

Uniform vertical stress q, settlement below the centre	$\rho_c = qB(1 - v^2)/E$	equation (6.12)
Uniform vertical stress q, settlement below the edge (at radius $B/2$)	$\rho_e = (2/\pi)qB(1 - v^2)/E$	equation (6.13)
Rigid circular plate, total load $Q(=\pi B^2 q/4)$	$\rho = Q(1 - v^2)/EB = \pi(1 - v^2)qB/4E$	equation (6.14)

The surface settlement below the corner of a uniform rectangular surcharge of magnitude q and plan dimensions $L \times B$ (where $L \geq B$), on a uniform isotropic elastic half-space, is given by equation (6.15), with the appropriate value of influence factor I_ρ from Table 6.6.

$$\rho_c = [qB(1 - v^2)/E]I_\rho \qquad (6.15)$$

Table 6.6 Influence factors for calculating the settlement below the corner of a uniformly loaded rectangular area at the surface of a homogeneous, isotropic elastic half-space (for use with equation (6.15))

L/B	I_ρ	L/B	I_ρ	L/B	I_ρ	L/B	I_ρ
1.0	0.561	1.6	0.698	2.4	0.822	5.0	1.052
1.1	0.588	1.7	0.716	2.5	0.835	6.0	1.110
1.2	0.613	1.8	0.734	3.0	0.892	7.0	1.159
1.3	0.636	1.9	0.750	3.5	0.940	8.0	1.201
1.4	0.658	2.0	0.766	4.0	0.982	9.0	1.239
1.5	0.679	2.2	0.795	4.5	1.019	10.0	1.272

Reproduced with permission from J.P. Giroud, Settlement of a linearly loaded rectangular area. *Proc. ASCE (SM4)*, 1968.

Many more formulae, tables and charts, for foundations of different shapes, taking account of non-uniform loading, a finite soil depth, anisotropy, inhomogeneity, and in some cases the stiffness of the foundation, are given – together with original references – by Poulos and Davis (1974).

6.8   Estimation of immediate settlements

It is often useful to be able to separate the component of the overall settlement which occurs immediately (due to shear, without change in volume) from that which occurs in the long term, as the soil consolidates.

$$\rho_t = \rho_i + \rho_c \tag{6.16}$$

where ρ_t is the overall (total) settlement, perhaps calculated using the one-dimensional method, ρ_i is the immediate undrained settlement and ρ_c is the settlement due to consolidation which develops in the long term. One reason for making this distinction is that it is only the deformation that develops *after* a brittle component has been fixed to a building which is likely to lead to damage.

Immediate settlements may be estimated by using values representing the undrained Young's modulus and Poisson's ratio ($\nu_u = 0.5$) in formulae such as those given in Tables 6.5 and 6.6. Alternatively, the proportion of the overall settlement (calculated using the one-dimensional method) which occurs effectively instantaneously (i.e. over the time period of construction) might be estimated empirically, on the basis of previous experience.

For stiff, overconsolidated clays, case data presented by Simons and Som (1970) indicate values of the ratio of settlement at the end of construction to total settlement in the range 0.32–0.74, with an average of 0.58. Morton and Au (1975) give a range of 0.4–0.82, with an average of 0.63, for buildings on London Clay.

For soft, normally consolidated clays, Simons and Som (1970) present further data with ratios of settlement at the end of construction to total settlement in the range 0.08–0.21, with an average of 0.16. Since in all cases the settlement at the end of construction will probably include some settlement due to consolidation, Burland *et al.* (1977) suggest that the ratio ρ_i/ρ_t for soft clays will normally be less than 0.1.

6.9   Effect of heterogeneity

Gibson (1974) discusses the effect of an increase in soil stiffness with depth on the stresses and settlements calculated using elasticity theory. Such a soil is described as **heterogeneous** (as opposed to homogeneous), and is sometimes known as a **Gibson soil**.

The Young's modulus of the soil at a depth z below the ground surface is given by

$$E_z = E_0 + \lambda z \tag{6.17}$$

The degree of heterogeneity is characterized by the parameter $E_0/\lambda R$, where R is a dimension representative of the loaded area. Settlement patterns (normalized with respect to the settlement at the centre of the loaded area, ρ_c) due to a uniformly loaded circular area of radius R at the surface of an elastic half-space are shown in Figure 6.19, for various values of $E_0/\lambda R$.

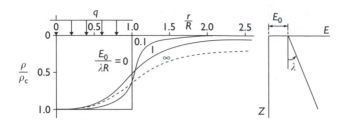

Figure 6.19 Normalized surface settlement profiles due to a uniformly loaded circular area of radius R on a heterogeneous elastic half space. (Redrawn with permission from Gibson, 1974.)

For a homogeneous soil, whose stiffness does not increase with depth (i.e. $\lambda = 0, E_0/\lambda R = \infty$), the zone of significant surface settlement extends well beyond the loaded area. As the degree of heterogeneity is increased (i.e. $E_0/\lambda R$ decreases), the extent of the settlement zone is reduced. When $E_0 = 0$, the settlements beneath the loaded area are uniform and one-dimensional. Figure 6.19 is presented by Gibson (1974) for an incompressible soil with Poisson's ratio $\nu = 0.5$, which corresponds to undrained conditions. He states, however, that the results obtained for other values of ν are practically indistinguishable. Perhaps the most important point demonstrated by Figure 6.19 is that the pattern of soil surface settlements around a shallow foundation is likely to be very sensitive to the stiffening of the soil with depth.

Gibson (1974) also shows that when $E_0 = 0$ and $\nu = 0.5$, the settlement beneath a uniformly loaded area of any shape is given by

$$\rho = 3q/2\lambda \tag{6.18}$$

(where q is the applied surface load), and is zero elsewhere. The settlement at the centre of a uniformly loaded circular area of radius R when E_0 is non-zero and Poisson's ratio $\nu = 0.5$ is given approximately by

$$\rho = 3qR/2(E_0 + \lambda R) \tag{6.19}$$

where q is the applied surface load (Poulos and Davis, 1974). On substituting $E_0 = 0$ into equation (6.19), equation (6.18) is recovered.

6.10   Cross-coupling of shear and volumetric effects due to anisotropy

Hooke's Law (equation (6.1)) may be written in terms of the bulk modulus K' and the shear modulus G' (rather than Young's modulus E' and Poisson's ratio ν'), in order to separate shear and volumetric effects. For an isotropic elastic material under the stress conditions imposed in the triaxial test Muir Wood (1990) shows that the stress–strain relationship may be written in matrix form

$$\begin{bmatrix} \Delta\varepsilon_p \\ \Delta\varepsilon_q \end{bmatrix} = \begin{bmatrix} 1/K' & 0 \\ 0 & 1/3G' \end{bmatrix} \begin{bmatrix} \Delta p' \\ \Delta q \end{bmatrix} \tag{6.20}$$

where $\Delta\varepsilon_p = \Delta\varepsilon_a + 2\Delta\varepsilon_r$ is the incremental volumetric strain, $\Delta\varepsilon_q = (2/3)(\Delta\varepsilon_a - \Delta\varepsilon_r)$ is the incremental triaxial shear strain, K' is the effective stress bulk modulus, G' is the shear modulus, $\Delta p' = \Delta\sigma_a + 2\Delta\sigma_r$ is the increment of average effective stress, and $\Delta q = \Delta\sigma_a - \Delta\sigma_r$ is the increment of deviatoric stress. The off-diagonal zeros in the 2×2 **compliance matrix** of equation (6.20) show that shear and volumetric deformations may be considered independently.

Two independent parameters are required to describe the stress–strain response of an isotropic elastic material. A material which is **cross-anisotropic** (i.e. has the same properties in both horizontal directions but different properties in the vertical direction, as do many soils) requires five independent parameters to describe its behaviour, while a material which is fully anisotropic requires 21. This is discussed by Muir Wood (1990) and Graham and Houlsby (1983), who also show that the stiffness of a cross-anisotropic elastic material may be written in the form

$$\begin{bmatrix} \Delta p' \\ \Delta q \end{bmatrix} = \begin{bmatrix} K^* & J \\ J & 3G^* \end{bmatrix} \begin{bmatrix} \Delta\varepsilon_p \\ \Delta\varepsilon_q \end{bmatrix} \tag{6.21}$$

where K^* and G^* are modified bulk and shear moduli. The non-zero off-diagonal terms J indicate that shear and volumetric effects are now **coupled** (i.e. that each has some influence on the other), and cannot be considered in isolation.

In practice, most soils are cross-anisotropic, and must be expected to display some degree of coupling between shear and volumetric strains. One practical result of this is that the modified shear modulus G^* should be obtained from an undrained triaxial shear test in which the change in volumetric strain $\Delta\varepsilon_p$ is zero, so that $\Delta q = 3G^*\Delta\varepsilon_q$. The modified bulk modulus K^* should be obtained from a drained test in which the applied changes in p' and q cause no resultant shear strain, $\Delta\varepsilon_q = 0$, so that $\Delta p' = K^*\Delta\varepsilon_p$.

Key points

- To calculate the settlements of shallow foundations, it is common to assign isotropic elastic properties such as stiffness to the soil. However, soil is usually anisotropic, and its stiffness depends on its stress history, stress state and the stress path to which it is subjected. Stiffness increases with increasing effective stress and decreases with strain following the last change in the direction of the strain path. Over a given stress range, overconsolidated soils are stiffer than normally consolidated soils. In the selection of parameters for an elastic analysis, all of these points should be borne in mind.

- The vertical stresses within an elastic body of infinite depth and lateral extent – **an infinite elastic half-space** – resulting from a surface load are comparatively insensitive to the elastic material properties. Experience has shown that the distribution of vertical stress within a soil mass due to the application of a surface load does not differ enormously from the elastic solution in cases where the stress–strain relationship of the soil is nonlinear, or where the stiffness of the soil increases with depth.

- Vertical stress distributions calculated using elasticity-based methods are reasonably applicable to layered soils and soils whose stiffness increases with depth. Significant deviation from the elastic solution may occur if the applied loads are high enough to cause extensive yielding or if there is a discontinuous inclusion of higher or lower stiffness.

- Increases in vertical stress at any depth, due to any pattern of applied surface load, can be calculated relatively easily using **Newmark's chart**, or – for rectangular buildings – **Fadum's chart**.

- Standard solutions are available for **settlements** calculated on the basis that the soil behaves as an elastic half-space. However, these are much more sensitive to the elastic properties assigned to the soil. Except possibly for solutions which allow for an increase in soil stiffness with depth (i.e. a **Gibson soil**), they are probably of limited value.

- Settlements may be calculated by assuming that soil movements are predominantly vertical, so that the relationship between vertical stress and strain increments is given by the one-dimensional modulus, E_0'. Variations in stiffness with depth can be taken into account by dividing the soil into a number of layers. The contribution of each layer to the overall settlement is then calculated on the basis of the average stiffness and stress increment within that layer.

- For soils which are approximately elastic in their response to monotonically increasing stresses, total settlements calculated using the one-dimensional approach are generally similar to those obtained using more sophisticated methods.

Questions

Note: Questions 6.2–6.4 may be answered using either the Newmark chart (Figure 6.8), or Fadum's chart (Figure 6.14), or both. Question 6.5 is based on Example 6.3, and should therefore be answered with the aid of a Newmark chart.

Determining elastic parameters from laboratory test data

6.1 (a) Write down Hooke's Law in incremental form in three dimensions and show that for undrained deformations Poisson's ratio $\nu_u = 0.5$. Assuming that the behaviour of soil can be described in terms of conventional elastic parameters, show that in undrained plane compression (i.e. $\Delta\varepsilon_2 = 0$), the undrained Young's modulus E_u is given by $0.75 \times$ the slope of a graph of deviator stress q (defined as $\sigma_1 - \sigma_3$) against axial strain ε_1. Show also that the maximum shear strain is equal to twice the axial strain, and that the shear modulus $G = 0.25 \times (q/\varepsilon_1)$.

(b) Figure 6.20 shows graphs of deviator stress q and pore water pressure u against axial strain ε_1 for an undrained plane compression test carried out at a constant cell pressure of 122 kPa. Comment on these curves and explain the relationship between

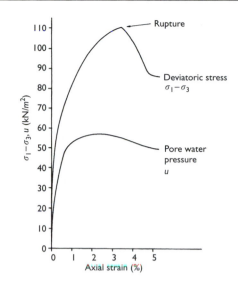

Figure 6.20 Undrained plane compression test data.

them. Calculate and contrast the shear and Young's moduli at 1% shear strain and at 10% shear strain. Which would be the more suitable for use in design, and why?

> *[University of London 2nd year BEng (Civil Engineering)*
> *examination, King's College (part question)]*

((b) Secant shear moduli are 3.4 MPa at $\gamma = 1\%$ and 0.4 MPa at $\gamma = 10\%$. (Corresponding Young's moduli are three times these values; tangent moduli at 10% shear strain are negative.))

Calculation of increases in vertical effective stress below a surface surcharge

6.2   The foundation of a new building may be represented by a raft of plan dimensions 10 m × 6 m, which exerts a uniform vertical stress of 50 kPa at founding level. A pipeline AA′ runs along the edge of the building at a depth of 2 m below founding level, as indicated in plan view in Figure 6.21.

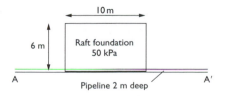

Figure 6.21 Plan view of foundation and pipeline.

Estimate the increase in vertical stress at a number of points along the pipeline AA′, due to the construction of the new building. Present your results as a graph of increase in vertical stress against distance along the pipeline AA′, indicating the extent of the foundation on the graph.

What is the main potential shortcoming of your analysis?

[University of London 2nd year BEng (Civil Engineering)
examination, Queen Mary and Westfield College]

(Increase in vertical stress calculated using a Newmark chart is 24.5 kPa beneath the mid-side, reducing to 12.5 kPa below the corner, and to less than 1 kPa at a distance of 4 m from the corner.)

Calculation of increases in vertical effective stress and resulting soil settlements

6.3 (a) In what circumstances might an elastic analysis be used to calculate the changes in stress within the body of the soil due to the application of a surface surcharge?

(b) Figure 6.22 shows a cross-section through a long causeway. Using the Newmark chart or otherwise, sketch the long-term settlement profile along a line perpendicular to the causeway. Given time, how might your analysis be refined?

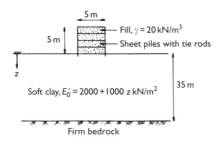

Figure 6.22 Cross-section through causeway.

[University of London 2nd year BEng (Civil Engineering)
examination, King's College]

(Settlement calculated using a Newmark chart is approximately 3 mm at a distance of 22.5 m from the centreline, increasing to 21 mm at 7.5 m from the centreline; 85 mm at the edge of the causeway, and 130 mm at the centre.)

6.4 (a) When might an elastic analysis reasonably be used to calculate the settlement of a foundation? Briefly outline the main difficulties encountered in converting stresses into strains and settlements.

(b) A square raft foundation of plan dimensions $5\,m \times 5\,m$ is to carry a uniformly distributed load of $50\,kPa$. A site investigation indicates that the soil has a one-dimensional modulus given by $E'_0 = (10 + 6z)$ MPa, where z is the depth below the ground surface in metres. Use a suitable approximate method to estimate the ultimate settlement of the raft.

[*University of London 2nd year BEng (Civil Engineering) examination, King's College*]

(Settlements calculated using a Newmark chart with the soil divided horizontally into three layers are approximately 9 mm below centre, 3 mm below corner, and 5 mm below mid-side, giving an average (for a rigid foundation) of about 5 mm.)

6.5   The foundations of a new building may be represented by a raft of plan dimensions $24\,m \times 32\,m$, which exerts a uniform vertical stress of $53.5\,kPa$ at founding level. The soil at the site comprises laminated silty clay underlain by firm rock. The estimated stiffness in one-dimensional compression E'_0 increases with depth as indicated below.

Depth below founding level (m)	E'_0 (MPa)
0–4	5
4–10	10
10–20	25
Below 20	Very stiff

Use the Newmark chart (Figure 6.8) to estimate the increase in vertical stress at depths of 4 m, 10 m and 20 m below the centre of the raft. Hence estimate the expected eventual settlement of the centre of the foundation. Suggest two possible shortcomings of your analysis.

[*University of London 2nd year BEng (Civil Engineering) examination, Queen Mary and Westfield College*]

(Increases in stress are $53.5\,kPa$ at $4\,m$ depth; $43.6\,kPa$ at $10\,m$, and $25.7\,kPa$ at $20\,m$. Total settlement $\approx 86\,mm$.)

Use of standard formulae in conjunction with one-dimensional consolidation theory (Chapter 4)

6.6   (a) To estimate the ultimate settlement of the grain silo described in question 4.5, the engineer decides to assume that the soil behaviour is elastic, with the same properties in loading and unloading. In what circumstances might this be justified?

(b) The proposed silo will be founded on a rigid circular foundation of diameter 10 m. Under normal conditions, the net or additional load imposed on the soil by the foundation, the silo and its contents will be 5000 kN. What is the ultimate settlement due to this load? (It may be assumed that the settlement ρ of a rigid circular footing of diameter B carrying a vertical load Q at the surface of an elastic half space of one-dimensional modulus E_0' and Poisson's ratio ν' is given by $\rho = (Q/E_0'B)[(1 - \nu')^2/(1 - 2\nu')]$. Take $\nu' = 0.2$.)

(c) To reduce the time taken for the settlement to reach its ultimate value, it is proposed to overload the foundation initially by 5000 kN, the additional load being removed when the settlement has reached 90% of the predicted ultimate value. In practice, this occurs after six months has elapsed, and the additional load is then removed. Giving two or three actual values, sketch a graph showing the settlement of the silo as a function of time. (Assume that the principle of superposition can be applied, and use the curve of R against T given in Figure 4.18.) State briefly the main shortcomings of your analysis.

[University of London 2nd year BEng (Civil Engineering) examination, King's College (part question)]

($\rho_{ult} = 53$ mm; occurrence of 90% settlement after 6 months suggests drainage path length of 1.9 m, assuming one-dimensional vertical consolidation. Super-impose one-dimensional consolidation solutions for loads of $+10\,000$ kN at $t = 0$ and -5000 kN at $t = 6$ months to obtain settlement–time plot; resultant settlement is 45 mm after 1 year, 50 mm after 2 years and 51 mm after 3 years.)

Note

1 Surcharge loads are traditionally given the same symbol q as the deviator stress in the triaxial test. This is not unreasonable, because the effect in each case is to increase the stress difference $\sigma_v' - \sigma_h'$.

References

Atkinson, J.H., Richardson, D. and Stallebrass, S.E. (1990) Effect of recent stress history on the stiffness of overconsolidated soil. *Géotechnique*, **40**(4), 531–40.

Boussinesq, J. (1885) *Applications des Potentiels à L'étude de L'équilibre et du Mouvement des Solides Élastiques*, Gauthier-Villars, Paris.

Burland, J.B. (2001) Assessment methods used in design. In *Building Response to Tunnelling: Case Studies from Construction of the Jubilee Line Extension*, London, Volume 1: Projects and methods (eds J.B. Burland, J.R. Standing and F.M. Jardine), pp 23–43. Special Publication 200. Construction Industry Research and Information Association and Thomas Telford Ltd., London.

Burland, J.B. and Wroth, C.P. (1975) Settlement of buildings and associated damage. *Proceedings, British Geotechnical Society Conference on Settlement of Structures*, Cambridge, 611–54, Pentech Press, London.

Burland, J.B., Broms, B. and de Mello, V.F.B. (1977) Behaviour of foundations and structures. State-of-the-art review. *Proc. IX Int. Conf. on Soil Mech. & Fndn. Eng.*, Tokyo, **2**, 495–546.

Clancy, P. and Randolph, M.F. (1996) Simple design tools for piled raft foundations. *Géotechnique*, **46**(2), 313–28.

Fadum, R.E. (1948) Influence values for estimating stresses in elastic foundations. *Proc. 2nd Int. Conf. on Soil Mech. & Fndn. Eng.*, Rotterdam, **3**, 77–84.

Gaba, A.R., Simpson, B., Powrie, W. and Beadman, D.R. (2003) Embedded retaining walls-guidance for economic design. Report C580, Construction Industry Research and Information Association, London.

Gibson, R.E. (1974) The analytical method in soil mechanics: 14th Rankine Lecture. *Géotechnique*, **24**(2), 115–40.

Giroud, J.P. (1968) Settlement of a linearly loaded rectangular area. *Proc. ASCE (SM4)*, **94**, July, 813–31.

Graham, J. and Houlsby, G.T. (1983) Elastic anisotropy of a natural clay. *Géotechnique*, **33**(2), 165–80.

Morton, K. and Au, E. (1975) Settlement observations on eight structures in London. *Proceedings British Biotechnical Society Conference Settlement of Structures*, Cambridge, 183–203, Pentech Press London.

Muir Wood, D. (1990) *Soil Behaviour and Critica State Soil Mechanics*, Cambridge University Press, Cambridge.

Newmark, N.M. (1942) Influence charts for computation of stresses in elastic foundations. *Bulletin No. 338*, University of Illinois Engineering Experimental Station.

Poulos, H.G. and Davis, E.H. (1974) *Elastic Solution for Soil and Rock Mechanics*, Wiley, New York.

Simons, N.E. and Som, N.N. (1970) *Settlement of structures on clay with particular emphasis on London Clay*. Report 22, Construction Industry Research & Information Association, London.

Skempton, A.W. and Bjerrum, L. (1957) A contribution to the settlement analysis of foundations or clay. *Géotechnique*, **7**(4), 168–78.

Somerville, S.H. and Shelton, J.C. (1972) Observed settlements of multi-storey buildings on laminated clays and silts in Glasgow. *Géotechnique*, **22**(3), 513–20.

Steinbrenner, W. (1934) Tafeln zur Setzungsberechnung. *Strasse* **1**, 121–4.

Terzaghi, K. (1943) *Theoretical Soil Mechanics*, John Wiley, New York.

Whitlow, R. (1995) *Basic Soil Mechanics*, 3rd edn Longman Scientific & Technical, Harlow.

Chapter 7

The application of plasticity and limit equilibrium methods to retaining walls

7.1 Engineering plasticity

Many of the traditional methods of geotechnical engineering design are based on the concepts of **plasticity** rather than **elasticity**. Typically, conditions at collapse are investigated, and a factor is applied to one of the parameters involved in the calculation to arrive at the final design. This factor, usually given the symbol F, is conventionally known as the **factor of safety**. This is something of a misnomer, because its purpose is not only to ensure that collapse will not occur under working conditions, but also that deformations will be small.

In structural engineering, we would call this procedure **plastic design**. The justifications for its use are that (particularly for complex frames) it is easier than an elastic calculation, and that an elastic calculation is in any case unwarranted because of the effects of, for example, foundation settlements and partially fixed joints, which are difficult or impossible to quantify. The factor F is usually applied to the strength of the frame or to the loads; so that either the permissible maximum bending moment under working conditions is M_p/F, or the permissible loads are W_c/F, where M_p is the fully plastic moment for the frame and W_c is the set of loads which causes collapse.

We have already seen in Chapters 4 to 6 that the engineering behaviour of soil is rather more complex than that of structural materials such as mild steel. The constitutive (stress–strain) relationship for a soil element depends on both its geological and recent stress history and the stress and strain paths to which it is subjected. It is probable that the soil elements at different locations in the vicinity of a geotechnical structure such as a retaining wall or a foundation will all have different stress–strain responses.

In these circumstances, **plasticity solutions** have the additional attraction that only the soil strength need be known, rather than the full stress–strain relation. Critical state soil strengths are comparatively easy to determine, since they are relatively insensitive to state path and stress history. The principal disadvantage of classical plasticity methods is that they give no information concerning deformations. It is therefore necessary to rely on experience and engineering judgement in choosing an appropriate numerical value for the factor of safety.

In geotechnical engineering, the factor of safety F is now usually applied to the soil strength. Unfortunately, it is also sometimes applied to other parameters – for example, the depth of embedment of an embedded retaining wall. This can give highly misleading results, and should be avoided.

In this chapter the concepts of engineering plasticity are summarized. They are then used to calculate horizontal soil stresses and in the analysis of simple retaining walls. The main emphasis is on conditions at collapse, but the concept of a factor of safety on soil strength is introduced where appropriate.

Objectives

After having worked through this chapter, you should understand that

- the concepts of **engineering plasticity** can be used to analyse a structure such as a retaining wall on the verge of collapse (sections 7.1–7.7)
- a **lower bound** solution, based on a system of stresses which is in **equilibrium** and does not violate the **failure criterion** for the soil, tends to err on the safe side (section 7.2)
- an **upper bound** solution, based on an assumed **mechanism** of collapse, tends to err on the unsafe side (section 7.2)
- soils fail according to a **frictional failure criterion** expressed in terms of effective stresses: $(\tau/\sigma')_{max} = \tan\phi'$ (section 7.3.1)
- as a special case, the undrained failure of clay soils which are sheared at constant volume may be analysed in terms of a **total stress failure criterion**: $\tau_{max} = \tau_u$ (section 7.3.3)

You should be able to:

- calculate, for a given vertical effective or total stress, the minimum possible (active) and maximum possible (passive) horizontal effective or total stress that can be applied to an element of soil (section 7.3)
- use the concept of active and passive zones to calculate the lateral stress distributions on retaining walls, and the resulting prop loads and bending moments, when the soil is at active and/or passive failure (section 7.5)
- calculate the depth of embedment and prop load required to maintain an embedded cantilever retaining wall in equilibrium for a specified mobilized strength in the surrounding soil (sections 7.6, 7.8.2 and 7.8.3)
- investigate the stability of mass retaining walls, in both the short term and the long term, by analysing the statical equilibrium of assumed sliding wedges both behind and in front of the wall, and then of the wall itself (section 7.7)
- estimate the long-term seepage pore water pressures acting on an embedded retaining wall, using the **linear seepage approximation** (section 7.8.1)
- use, where appropriate in stress field calculations, tabulated earth pressure coefficients which take account of the effects of soil/wall friction or adhesion (section 7.10)

You should have an appreciation of:

- the various common types of retaining wall, and the way in which each resists the lateral thrust of the retained soil (section 7.4)

- the shortcomings of calculations based on conditions at collapse (sections 7.5–7.7)
- the use of a strength mobilization factor M in design calculations, together with other devices to distance the working state of a real wall from collapse (sections 7.5–7.7)
- the shortcomings of the linear seepage approximation for steady state pore water pressures (section 7.8.1)
- the uncertainty concerning the degree of soil/wall friction or adhesion actually mobilized (section 7.9)

7.2 Upper and lower bounds (safe and unsafe solutions)

The rigorous analysis of soil constructions such as foundations, slopes and retaining walls at collapse is based on the **upper** and **lower bound theorems** of **plasticity** (see e.g. Baker and Heyman (1969); and for a full treatment of their use in soil mechanics, Atkinson (1981)).

The **lower bound theorem** states that if a system of stresses within the soil mass can be found which is in *equilibrium* with the external loads and body forces (i.e. self-weight), and nowhere violates the *failure criterion* for the soil, then the external loads and body forces represent a lower bound to those that will actually cause collapse. Any error will be on the safe side, since it may be that a different system of stresses will carry even higher loads without violating the failure criterion. In short, if it can be shown that the soil can carry the loads, then it will.

The **upper bound theorem** states that if a *mechanism* can be found such that the work done by the external loads and body forces is equal to the energy dissipated within the soil mass as it deforms, then the external loads and body forces represent an upper bound to those that will cause collapse. Any error will be on the unsafe side, since it may be that a different mechanism will already have been responsible for collapse at a lower load. In short, if it can be shown that the soil can fail, then it will.

The terms upper and lower bound relate to the collapse load, and are potentially ambiguous in soil mechanics applications such as an embedded retaining wall where a collapse load not the output of the calculation. For this reason, upper and lower bounds are sometimes referred to as unsafe and safe solutions, respectively.

The bound theorems apply strictly to **perfectly plastic** materials whose stress–strain relationship exhibits **plastic plateau**, that is there is no brittleness or tendency to shed load as deformation continues after failure (Figure 7.1). In soil mechanics, this would suggest the use of critical state, rather than peak, strengths.

A further requirement for the validity of the bound theorems is that the material exhibits a property known as **normality**. This was mentioned in section 5.11 and is discussed in section 7.7.

In the solution of problems in engineering mechanics using elasticity-based methods, we can use the conditions of *equilibrium* (of forces and stresses), *compatibility* (of strains and displacements) and the *stress–strain* relationship for the material. (If the problem can be solved using the condition of equilibrium alone, it is said to be **statically determinate**). In plasticity, a lower bound ('safe') solution will satisfy the condition of *equilibrium*, and a material property-based condition (that the failure criterion is not

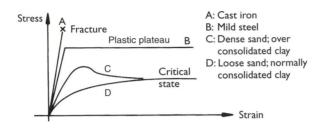

Figure 7.1 Brittle and plastic materials.

violated), which is analogous to the *stress–strain* relationship in an elastic analysis. An upper bound ('unsafe') solution involving a kinematically admissible mechanism satisfies a displacement-based criterion which is analogous to *compatibility*.

If we are able to find upper and lower bounds that give the same answer, we have satisfied all three criteria (equilibrium, kinematic admissibility/compatibility, and conformance with the failure criterion/stress–strain law), and the solution is theoretically correct. Unsafe and safe solutions which do not coincide will give an indication of the range within which the correct answer lies. So long as the range is reasonably narrow, this is likely to be sufficient in many cases.

7.3   Failure criteria for soils

7.3.1 $(\tau/\sigma')_{\max} = \tan \phi'$ *failure criterion*

All soils derive their strength from interparticle friction. In natural soils, cementing between the particles may also be present, but such bonds will be brittle and enhance the peak strength, not that at the critical state. In terms of effective stresses, most soils obey the purely frictional failure criterion developed in Chapter 2,

$$\tau = \sigma' \tan \phi' \tag{7.1}$$

on the plane of maximum stress ratio (Figure 7.2). In equation (7.1), the frictional failure criterion is expressed in terms of the stresses in a plane. This is more useful for the analysis of essentially plane problems such as long retaining walls or foundations than the analogous expression $q = Mp'$ in terms of the triaxial stress parameters, which was developed in Chapter 5.

From the geometry of the Mohr circle of stress for an element of soil obeying the $(\tau/\sigma')_{\max} = \tan \phi'$ failure criterion on the verge of failure, the maximum possible ratio of principal effective stresses is:

$$\sigma_1'/\sigma_3' = [1 + \sin \phi']/[1 - \sin \phi'] \tag{7.2}$$

For a given vertical effective stress σ_v', the smallest horizontal effective stress σ_h' that must be applied to the soil to prevent the Mohr circle from crossing the failure envelope occurs when σ_v' is the major principal effective stress σ_1', and σ_h' is the minor principal

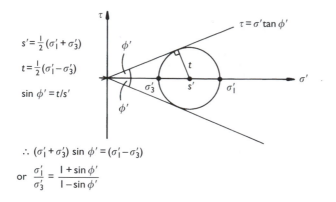

Figure 7.2 Frictional failure criterion in terms of effective stresses, showing the Mohr circle of effective stress for an element of soil at failure.

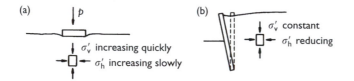

Figure 7.3 Zones of soil which tend towards active failure ($\sigma_v' > \sigma_h'$). (a) Below a foundation; (b) behind a retaining wall.

effective stress σ_3'. Substituting $\sigma_1' = \sigma_v'$ and $\sigma_3' = \sigma_h'$ into equation (7.2),

$$\sigma_h'/\sigma_v' = ([1 - \sin\phi']/[1 + \sin\phi']) = K_a \qquad (7.3)$$

This state is described as the **active condition,** and K_a is termed the **active earth pressure coefficient.** Active failure will eventually occur if the vertical effective stress is increased while the horizontal effective stress remains approximately constant or increases less quickly (e.g. in the soil beneath a foundation); or where the horizontal stress is reduced while the vertical stress remains constant (e.g. behind a retaining wall): Figure 7.3.

For a given vertical effective stress σ_v', the maximum horizontal effective stress σ_h' that can be applied before the Mohr circle touches the failure envelope occurs when σ_v' is the minor principal effective stress σ_3', and σ_h' is the major principal effective stress σ_1'. Substituting $\sigma_1' = \sigma_h'$ and $\sigma_3' = \sigma_v'$ into equation (7.2),

$$\sigma_h'/\sigma_v' = ([1 + \sin\phi']/[1 - \sin\phi']) = K_p \qquad (7.4)$$

This state is described as the **passive condition,** and K_p is the **passive earth pressure coefficient.** Passive failure will eventually occur if the horizontal effective stress is increased while the vertical effective stress remains constant or is reduced (e.g. in the soil to one side of a foundation, or in front of a retaining wall), or where the vertical stress is reduced while the horizontal stress remains approximately constant: Figure 7.4.

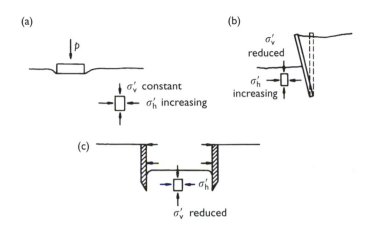

Figure 7.4 Zones of soil which tend towards passive failure ($\sigma'_h > \sigma'_v$). (a) On either side of a foundation; (b) in front of a retaining wall; (c) below the floor of an excavation.

7.3.2 ϕ'_{crit} or ϕ'_{peak}?

In using the failure criterion $(\tau/\sigma')_{max} = \tan\phi'$, the value of ϕ' could be taken to represent either the peak or the critical state strength. The main reasons for using the critical state strength are:

- for a given soil, the value of ϕ'_{crit} is a constant. ϕ'_{peak}, on the other hand, depends on the potential for dilation, which in turn depends on the soil density and the average effective stress at failure.
- ϕ'_{peak} can only be maintained while the soil continues to dilate. With continued deformation, ϕ'_{peak} falls and the soil **strain softens**. The upper and lower bound theorems of plasticity, which require that the material exhibits a plastic plateau (Figure 7.1), do not apply to a strain-softening material. A strain-softening material is prone to progressive failure, as described in section 2.9.
- For embedded retaining walls, the use of ϕ'_{crit} in calculations seems to give a reasonable indication of the onset of large deformations (Bolton and Powrie, 1987; Powrie, 1996).

The one possible advantage of using the peak strength is that it gives an indication of the stiffness of the soil. At a given effective stress, denser soils have both a higher stiffness and a higher peak strength. This is particularly relevant when retaining walls are designed by the application of a factor of safety to the soil strength. If critical state strengths are used in the collapse calculation, a higher factor of safety would be needed for a retaining wall in a loose soil than for an identical retaining wall in a dense soil if the deformations under working conditions were to be the same.

The dilemma is reflected in the current British code of practice for the design of earth retaining structures BS 8002 (BSI, 2001). BS 8002 recommends that the **design** strength, ϕ'_{design} which is the actual soil strength ϕ' divided by a factor of safety or a **mobilization factor** M, is taken as the lesser of

(a) $\tan^{-1}\{(\tan\phi'_{\text{peak}})/M\}$ with $M = 1.2$, and

(b) ϕ'_{crit}.

This means that, if the soil is very dense, with $\phi'_{\text{peak}} >$ about $1.2 \times \phi'_{\text{crit}}$, the design strength is ϕ'_{crit}. If the soil is not quite so dense, with $\phi'_{\text{crit}} < \phi'_{\text{peak}} < 1.2 \times \phi'_{\text{crit}}$, the design strength is less than ϕ'_{crit}, and the factor of safety on $\tan\phi'_{\text{crit}}$ increases as ϕ'_{peak} is reduced. If the soil is loose, with $\phi'_{\text{peak}} = \phi'_{\text{crit}}$, then the design soil strength is taken as $\tan^{-1}\{(\tan\phi'_{\text{crit}})/1.2\}$, and the factor of safety on $\tan\phi'_{\text{crit}}$ is 1.2. In this way, an increasing factor of safety on ϕ'_{crit} with reducing peak strength (and implied soil stiffness) is achieved. (There are other provisions made in BS 8002 which ensure that even in a dense soil, where the factor of safety on ϕ'_{crit} is only 1, the wall is sufficiently remote from collapse not to deform excessively under working conditions: these are discussed in sections 7.6.1 and 10.2.)

If a peak strength based on laboratory test data is to be used in design, it must be measured at the maximum effective stress experienced by the soil in the field. The sample must have a representative void ratio and/or stress history. With deep retaining walls, it may be that the effective stresses below the base are high enough to suppress dilation completely, so that the peak strength is the critical state strength.

7.3.3 $\tau_{\text{max}} = \tau_{\text{u}}$ failure criterion

Clay soils can be brought to failure quickly, so that there is no change in specific volume (sections 2.10 and 5.6). The critical state model expressed in terms of the plane stress parameters s' and t (Figure 7.5(a)) shows that the effective stresses at failure are then determined by the initial specific volume. The radius t_{f} and the centre s'_{f} of the Mohr circle of effective stress at failure are independent of the total stress path, with the difference between the applied average total stress at failure s_{f} and the average effective stress at failure s'_{f} being made up by the pore water pressure u_{f} (which may be either positive or negative), $s'_{\text{f}} = s_{\text{f}} - u_{\text{f}}$ (Figure 7.5(b)).

The radii of the Mohr circles of total and effective stress are the same, because water cannot take shear. Since the excess pore water pressures generated during undrained loading can be difficult to quantify, it is convenient to use an alternative failure criterion for the short-term (undrained) failure of clay soils, based on total stresses. On the plane of maximum shear stress,

$$\tau_{\text{max}} = \tau_{\text{u}} \qquad (7.5)$$

where τ_{u} (often denoted c_{u} and sometimes s_{u}) is the undrained shear strength (Figure 7.5(c)). As discussed in section 5.6, τ_{u} is not a soil property in the way that ϕ'_{crit} is. τ_{u} is a function of the specific volume, and hence the stress state and stress history of the soil.

For a soil obeying the 'maximum shear stress' failure criterion $\tau_{\text{max}} = \tau_{\text{u}}$, the maximum possible difference between the vertical and horizontal total stresses may be obtained by considering Mohr circles of total stress at failure (Figure 7.6). For a given

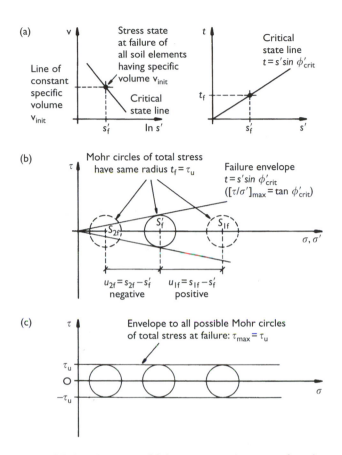

Figure 7.5 Development of failure criterion in terms of total stresses for the undrained loading or unloading of clay soils. (a) Lines joining critical states in terms of $(v, \ln s')$ and (t, s'); (b) Mohr circles of effective and total stress at undrained failure for two different loading paths; (c) 'Maximum shear stress' failure criterion.

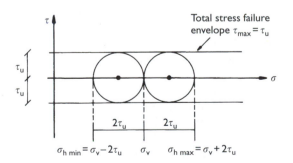

Figure 7.6 Mohr circles of total stress showing active and passive states for a soil obeying the 'maximum shear stress' failure criterion $\tau_{max} = \tau_u$.

vertical total stress σ_v, the horizontal total stress σ_h cannot be less than $(\sigma_v - 2\tau_u)$, nor can it exceed $(\sigma_v + 2\tau_u)$. These values are the **active** and **passive** limits to the total lateral stress:

$$\sigma_{h,min} = \sigma_v - 2\tau_u \quad \text{(active)} \tag{7.6a}$$

$$\sigma_{h,max} = \sigma_v + 2\tau_u \quad \text{(passive)} \tag{7.6b}$$

The undrained shear strength failure criterion and equations (7.6) apply only in the case of clay soils brought rapidly to undrained failure at constant volume.

7.4   Retaining walls

Figures 7.7–7.11 show some of the more common types of earth retaining structure, grouped according to the way in which they function.

- **Gravity** or **mass retaining walls** (Figure 7.7) rely primarily on their weight to prevent failure by sliding or overturning (toppling).
- **L-** and **T-cantilever walls** (Figure 7.8) resist the lateral pressure of the retained soil by bending. Sliding and overturning are prevented by the weight of the retained soil acting on a platform behind the wall which projects beneath the backfill (Figures 7.8(a), (b)), or by means of a piled foundation to the base (Figure 7.8(c)).
- **Embedded retaining walls**, which may be made of steel sheet piles, reinforced concrete piles or reinforced concrete diaphragm wall panels (Figure 7.9), are supported by the lateral pressure of the soil in front of the wall, and possibly by props or tie-back anchors at one or more levels.
- The facing panels of **reinforced soil retaining walls** (Figure 7.10) are held in place by the weight of the backfill acting on ties at various depths: these structures are described in detail in section 10.10.

Figure 7.11 indicates two ways in which bending moments in L-, T- and embedded cantilever walls (all of which, as structures, carry the loads imposed by the retained soil in bending) can be reduced. The downward load on a relieving platform behind an L- or T-cantilever (Figure 7.11 (a)), and the upward load on a platform in front of an embedded wall (Figure 7.11 (b)), will both give bending moments in the opposite sense to those caused by the lateral stress of the retained soil. The relieving platform behind the L- or T-cantilever has the additional advantage of reducing the lateral stresses acting on the wall stem below it.

A further distinction may be made between walls where the difference in level is created by removing soil from in front (sometimes called **excavated walls**), and walls where the difference in level is created by building up the retained height of the soil behind (**backfilled walls**). Gravity walls, L- and T-cantilever walls, and reinforced soil walls are usually backfilled, while embedded walls are installed from the original ground surface and the difference in level is formed by excavating soil from in front. Variants or combinations of retaining wall type may be used for particular applications, depending on the individual circumstances on site.

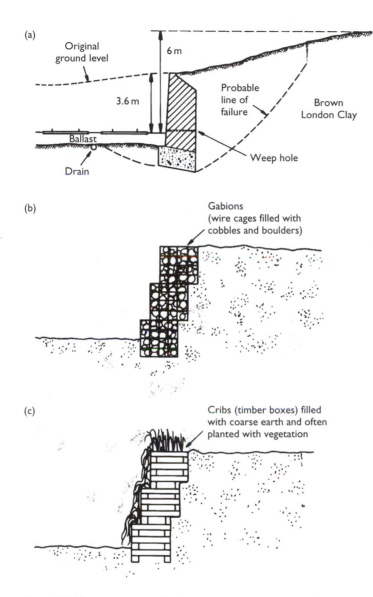

Figure 7.7 Mass retaining walls (masonry or concrete, or built up from gabions or timber cribs). (a) Kensal Green, London; built in 1913, failed 1942 (Reproduced from Institution of Structural Engineers, 1951); (b) gabions; (c) timber cribs.

The type of retaining structure selected for a given design application will depend on a number of criteria. For example, L-cantilevers with piled foundations are often used for bridge abutments so that relative vertical movement between the bridge deck and the retained soil (which would create an unacceptable step) is effectively eliminated. Embedded walls made from steel sheet piles are often used to form **cofferdams** to retain the sides of excavations in granular soils.

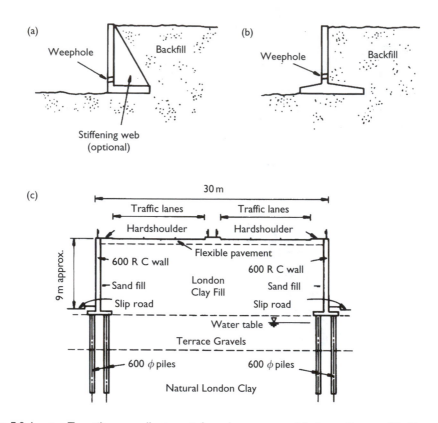

Figure 7.8 L- or T-cantilever walls in reinforced concrete. (a) L-cantilever; (b) T-cantilever; (c) T-cantilever with piled foundation: section through the M3 motorway near Junction 1. (Redrawn with permission from Mawditt, 1989.)

For deep excavations in clay soils in built-up areas, where a stiff wall must be constructed with the minimum of disturbance to the surrounding ground, *in situ* concrete walls made from bored piles or diaphragm wall panels are often used, although heavy-section steel sheet piles are available for this purpose. Embedded walls are frequently unpropped over retained heights of less than 4–5 m, but for deep excavations props or anchors are often installed at one or more levels. Temporary props may be used to support the wall at a high level, until the excavation has reached a sufficient depth for lower level permanent props to be installed.

For backfilled walls where relative movement between the wall and the retained soil is unimportant, reinforced soil walls may be the most economical form of construction.

Whatever form of construction is adopted, it should be checked against the possibility of at least three classes of collapse:

- Failure of the wall as a single rigid block (a monolith), for example sliding or overturning of a gravity wall or a cantilever L- or T-wall; and rotational or vertical instability of an embedded cantilever wall.

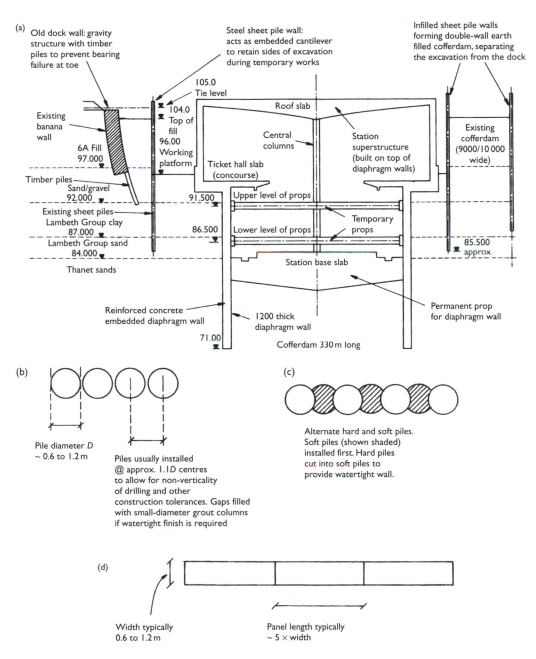

Figure 7.9 Embedded wall. May act as a free cantilever, or be propped or anchored at one or more levels. Constructed *in situ* using reinforced concrete (contiguous or secant piles or diaphragm wall panels/barrettes), or from steel sheet piles. (a) Typical application of *in situ* and sheet pile embedded retaining walls: construction of Canary Wharf station, Jubilee Line extension, east London (London Underground Limited.); (b) contiguous piles (plan view); (c) secant piles (plan view); (d) diaphragm wall, excavated and cast in panels or barrettes.

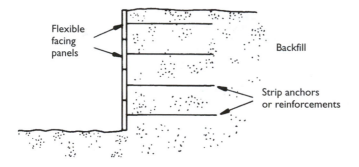

Figure 7.10 Reinforced soil wall.

(a)

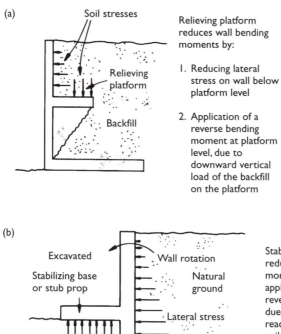

Relieving platform reduces wall bending moments by:

1. Reducing lateral stress on wall below platform level

2. Application of a reverse bending moment at platform level, due to downward vertical load of the backfill on the platform

(b)

Stabilizing base reduces wall bending moment by the applicaton of a reverse bending moment due to the upward reaction of the soil below it

Figure 7.11 Walls with moment-reduction platforms. (a) Backfilled wall with relieving platform (Tsagareli, 1967); (b) *in situ* embedded wall with stabilizing base or stub prop.

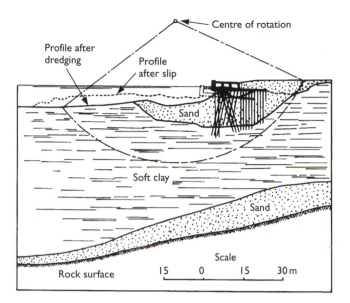

Figure 7.12 Failure of the quay wall at Gothenburg harbour. (Redrawn with permission from Skempton, 1946; after Petterson, 1916 and Fellenius, 1916.)

- Local structural failure, for example yield of props or anchors, or the failure in bending of reinforced concrete and sheet pile walls. For reinforced soil walls, the internal stability of the wall (i.e. the resistance of the reinforcement strips to breakage or pull-out) must be assessed: this is discussed in section 10.10.
- Inclusion in a landslide (Figure 7.12), which could be triggered by the changes in stress (loading and/or unloading) resulting from the construction of the retaining wall. This is particularly important for walls in sloping ground.

Further information on earth retaining systems, including many practical examples, may be found in Puller (1996).

7.5   Calculation of limiting lateral earth pressures

By identifying zones of soil in either the active or the passive condition, a **stress field** around a retaining wall, which satisfies the requirements of *equilibrium* and does not violate the *failure criterion* for the soil (and therefore represents a *lower bound* or inherently safe solution), can be developed. This approach was originally proposed by Rankine (1857), and is illustrated in Example 7.1.

Example 7.1 is relatively straightforward, because we need only consider the soil behind the retaining wall, which at failure will be in the active state (Figure 7.3(b)). There are, however, two different soil types – a sand and a clay – to which different short-term failure criteria are applied.

Example 7.1 Lateral stresses on a well-propped retaining wall

(a) Figure 7.13 shows a cross-section through an excavation supported by a braced retaining wall. What is the minimum thrust (in kN/m) due to the retained ground that the retaining wall must, in the short term, be able to resist? (Assume that the retaining wall is frictionless, and that the pore water pressures in the sandy soil are hydrostatic. Take the unit weight of water as $10 \, \text{kN/m}^3$). Is your answer reliable for use in design?

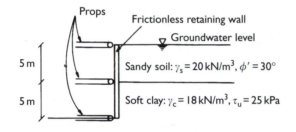

Figure 7.13 Cross-section through braced excavation.

(b) If the middle prop were accidentally removed, what would be the loads (in kN/m) in the remaining props?
(c) Why would it in reality be necessary to embed the wall to some depth below formation level?

[*University of London 2nd year BEng (civil engineering) examination,*
Queen Mary and Westfield College]

Solution

(a) Assume active conditions behind the retaining wall. This will give the minimum lateral thrust that must be resisted if collapse is to be avoided. It is a short-term calculation, so that the undrained shear strength failure criterion ($\tau_{max} = \tau_u$) may be used in the soft clay. In the overlying sand, the effective stresses and pore water pressures must be considered separately. Remember, however, that it is the total stress ($\sigma'_h + u$) that acts on the wall.

In the overlying sand, $\sigma'_h = K_a \sigma'_v$ where $K_a = (1 - \sin\phi'_{crit})/(1 + \sin\phi'_{crit})$ with $\phi'_{crit} = 30°$, that is $K_a = 1/3$. $\sigma'_v = \sigma_v - u$, where at depth z, $\sigma_v = \gamma_s z$ and $u = \gamma_w z$, assuming that the pore water pressures are hydrostatic. σ_h is calculated as $\sigma_h = \sigma'_h + u$.

In the clay, σ_h is calculated directly as $\sigma_h = \sigma_v - 2\tau_u$, where $\sigma_v = 5 \times \gamma_s + (z - 5) \times \gamma_c$ at a depth $z(z > 5\,\text{m})$ below original ground level (γ_s is the unit weight of the sandy soil, and γ_c is the unit weight of the clay).

Table 7.1 gives the values of $\sigma_v, u, \sigma_v', \sigma_h'$ and σ_h at key depths, where there is a change in soil type or groundwater conditions. Between these key depths, the lateral stress varies linearly.

Table 7.1 Calculated active stresses

	Depth z (m)	σ_v (due to overburden) (kPa)	u (kPa)	$\sigma_v'(= \sigma_v - u)$ (kPa)	σ_h' (kPa)	σ_h (kPa)
Sandy soil	0	0	0	0	0	0
Sandy soil	5	100	50	50	16.67	66.67
Clay	5	100	(50)	–	–	50
Clay	10	190	(100)	–	–	140

The active stress in the clay is greater than or equal to the hydrostatic water pressure at all depths. If this were not the case, a flooded tension crack could develop, and the thrust on the wall would have to be calculated on that basis.

The distribution of total lateral stress is shown in Figure 7.14.

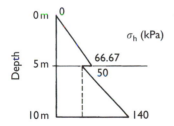

Figure 7.14 Calculated distribution of lateral total stress.

The total lateral thrust is given by:

$$[66.67\,\text{kPa}/2] \times 5\,\text{m} + [(50\,\text{kPa} + 140\,\text{kPa})/2] \times 5\,\text{m} = 641.7\,\text{kN/m}$$

Although the answer has been obtained from a stress field in which the effect of soil/wall friction has been ignored, it may be too low for use in design because:

1. It is the minimum thrust that must be applied to the soil to prevent failure: a larger thrust might be necessary to prevent excessive movements (i.e. we have not applied a factor of safety to the calculation).

2. The period of time for which the undrained shear strength of the clay can be relied upon is uncertain. The clay has ready access to water from the overlying more permeable sandy soil, and may well begin to swell and soften quite quickly. This would increase the lateral stresses to above those calculated assuming that the clay retains its original undrained shear strength.

(b) With three props the system is statically indeterminate, because there are three unknowns (the prop loads) and only two equations of equilibrium that can be used. Thus the prop loads cannot be calculated.

If the middle prop were removed, there would be only two unknowns. The system would be statically determinate, and the prop loads can be calculated.

Taking moments about the position of the top prop in order to calculate the load P_L in the lower prop:

$$\{P_L \times 10\,\text{m}\} = \{[(66.67\,\text{kPa}/2) \times 5\,\text{m}] \times [(2/3) \times 5\,\text{m}]\}$$

$$+ \{[50\,\text{kPa} \times 5\,\text{m}] \times [7.5\,\text{m}]\}$$

$$+ \{[(90\,\text{kPa}/2) \times 5\,\text{m}] \times [5\,\text{m} + (2/3 \times 5\,\text{m})]\}$$

$$\Rightarrow P_L = 430.6\,\text{kN/m}$$

From the condition of horizontal equilibrium, the sum of the prop loads must be equal to the total lateral thrust per metre run,

$$P_U + P_L = 641.7\,\text{kN/m}$$

$$\Rightarrow P_U = 641.7 - 430.6\,\text{kN/m} = 211.1\,\text{kN/m}$$

As a check, P_U can be calculated by taking moments about the position of the lower prop:

$$\{P_U \times 10\,\text{m}\} = \{[(66.67\,\text{kPa}/2) \times 5\,\text{m}] \times [5\,\text{m} + (5\,\text{m}/3)]\}$$

$$+ \{[50\,\text{kPa} \times 5\,\text{m}] \times [2.5\,\text{m}]\}$$

$$+ \{[(90\,\text{kPa}/2) \times 5\,\text{m}] \times [5\,\text{m}/3]\}$$

$$\Rightarrow P_U = 211.1\,\text{kN/m}$$

(c) It would in practice be necessary to embed the wall to some depth below formation level in order to prevent the soil below the toe of the wall from undergoing a **bearing failure**, due to the weight of the soil outside the excavation, as described in section 7.6.

7.6   Development of simple stress field solutions for a propped embedded cantilever retaining wall

In Example 7.1, the retaining wall was supported entirely by the props, and did not penetrate below the floor or **formation level** of the excavation. A more common form of structure is the embedded wall, as shown in Figure 7.9. Embedded walls are often designed to act as propped cantilevers. Although supported at or near the top by stiff props, the passive resistance of the soil in front of the wall below excavation level is required to maintain the wall in moment equilibrium. The passive resistance must be sufficient to prevent the wall from failing by rotation about the prop, and one of the key questions facing the designer is the depth of wall embedment required to achieve this.

In Examples 7.2 and 7.3, the concept of active and passive zones is used to calculate the depth of embedment required just to prevent a propped cantilever retaining wall from failing by rotation about the prop. These are *not* design calculations, because the calculated depth of embedment is such that the wall is on the verge of failure, or at the **collapse limit state**. In other words, we are not yet applying the factor of safety F.

7.6.1 $(\tau/\sigma')_{max} = \tan\phi'$ failure criterion

Figure 7.15 shows a stiff, frictionless embedded wall propped at the crest, retaining a height h of dry sandy soil. A conservative estimate (i.e. one which errs on the safe side) of the depth of embedment d required just to prevent collapse can be obtained by means of a stress field solution, based on zones of soil at active and passive failure.

The first step is to imagine that the soil is divided into four zones by vertical frictionless planes on either side of the wall, and a horizontal frictionless plane at the level of the toe of the wall, as indicated in Figure 7.15. These frictionless planes are called **stress discontinuities** because they divide regions in which the stress states (as represented by Mohr circles of stress) are different: in other words, the stresses are discontinuous across them.

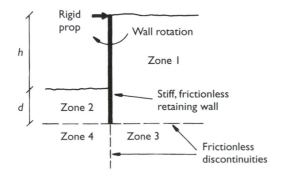

Figure 7.15 Stiff, frictionless embedded cantilever retaining wall propped at the crest.

A stiff wall will tend to fail by rotation about the prop. In the soil behind the wall (zone 1), the horizontal stress is reduced as the wall moves away, while in front of the wall (zone 2) the horizontal stress is increased as the wall moves into the soil. At failure, the soil in zone 1 will be in the active state and the soil in zone 2 in the passive state. The resulting stress field can be used to calculate the depth of embedment and prop load just to maintain equilibrium of the wall, as shown in Example 7.2.

Example 7.2 Calculating the depth of embedment at collapse of a propped cantilever retaining wall

By dividing the soil into active and passive zones, calculate the depth of embedment d required just to maintain the stability of the propped embedded cantilever retaining wall shown in Figure 7.15. The soil has $\phi'_{crit} = 30°$ and $\gamma = 20\,\text{kN/m}^3$, and the retained height $h = 5\,\text{m}$. Calculate the corresponding prop load F.

Solution

The vertical total stress at any point on either side of the wall is given by the unit weight of the soil γ multiplied by the depth below the appropriate soil surface. The vertical effective stress is obtained by subtracting the pore water pressure, which in this case is zero because the soil is dry. The horizontal effective stresses at a depth z below either soil surface are $K_a\sigma'_v$ in the active zone (behind the wall) and $K_p\sigma'_v$ in the passive zone (in front of the wall). K_a and K_p are the active and passive earth pressure coefficients, which for a frictionless wall are as defined in equations (7.3) and (7.4).

It follows that, in this case, both the vertical and the horizontal effective stresses increase linearly with depth on each side of the wall, as shown in Figure 7.16.

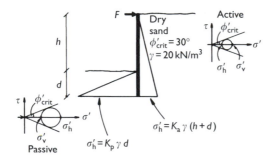

Figure 7.16 Lateral stress distributions for the calculation of limiting depth of embedment for a stiff, propped embedded wall.

Behind the wall, the vertical effective stress at the bottom of the wall (usually called the **toe**) is $\gamma \times (h+d) = 20\,\text{kN/m}^3 \times (5+d)\,\text{m}$, and the horizontal effective stress σ'_h is K_a (the active earth pressure coefficient) times this, $\sigma'_h = K_a\gamma(h+d)$.

In front of the wall, the vertical effective stress at the toe is $\gamma \times d = 20 \, \text{kN}/m^3 \times d$ m (in kPa), and the horizontal effective stress σ'_h is K_p (the passive earth pressure coefficient) times this, $\sigma'_h = K_p \gamma d$.

The horizontal stresses acting on each side of the wall are shown, together with schematic Mohr circles of stress, in Figure 7.16.

The depth of embedment d required just to prevent collapse is calculated by taking moments about the prop. The prop load then follows from the condition of horizontal force equilibrium.

In soil mechanics, stress distributions are often either triangular (increasing linearly with depth), or rectangular (uniform with depth). For the purpose of taking moments, the resultant force acts through the centre of pressure, which is at the geometric centroid of the shape defining the stress distribution. In the case of a triangular stress distribution of overall depth D, the centre of pressure is at a depth of $2D/3$ from the tip of the triangle (Figure 7.17(a)). In the case of a rectangular stress distribution of overall depth D, the centre of pressure is at mid-depth, $D/2$ (Figure 7.17(b)).

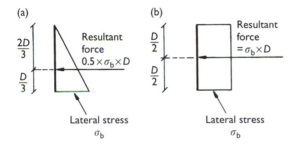

Figure 7.17 Taking moments: (a) triangular pressure distribution; (b) rectangular pressure distribution.

For the stress distribution shown in Figure 7.16, and considering a 1 m length of the wall perpendicular to the plane of the paper, taking moments about the prop gives:

$$[0.5 \, K_a \gamma (h + d)^2] \times [2(h + d)/3] = [0.5 \, K_p \gamma d^2] \times [h + (2d/3)]$$

[Force due to active stresses] × [Lever arm]

$\quad$ = [Force due to passive stresses] × [Lever arm]

For $\phi'_{\text{crit}} = 30°$ and soil/wall friction $\delta = 0, K_a = 0.333$ and $K_p = 3$. With $h = 5$ m and $\gamma = 20 \, \text{kN}/m^3$,

$$[0.5 \times 0.333 \times 20 \, \text{kN}/m^3 \times (5 + d)^2 \, m^2] \times [2 \times (5 + d)/3]m$$

$$= [0.5 \times 3 \times 20 \, \text{kN}/m^3 \times d^2 m^2] \times [5 + (2d/3)]m$$

$$\Rightarrow (5 + d)^3 \times (2/9) = 150d^2 + 20d$$

(with d in metres). Solution by trial and error gives

$$d \approx 2\,\mathrm{m}$$

For horizontal force equilibrium,

$$[0.5\,K_{\mathrm{a}}\gamma(h+d)^2] = [0.5K_{\mathrm{p}}\gamma d^2] + F$$

[Force due to active lateral stresses]

$$= [\text{Force due to passive lateral stresses}] + \text{Prop load}$$

For the values of $K_{\mathrm{a}}, K_{\mathrm{p}}, \gamma$ and h given, and $d = 2\,\mathrm{m}$ as calculated,

$$[0.5 \times 0.333 \times 20\,\mathrm{kN/m}^3 \times 7^2\,\mathrm{m}^2] = [0.5 \times 3 \times 20\,\mathrm{kN/m}^3 \times 2^2\mathrm{m}^2]$$

$$+F\,\mathrm{kN/m}$$

$$\Rightarrow 163.33\,\mathrm{kN/m} = 120\,\mathrm{kN/m} + F\,\mathrm{kN/m}$$

$$\Rightarrow F = 43.33\,\mathrm{kN/m}$$

The prop load is expressed in kN per metre length of wall, kN/m. Similarly, bending moments calculated from a plane strain analysis would be in kNm per metre length of wall, kNm/m.

A full lower bound plasticity solution requires us to show that the failure condition is not violated in the soil in zones 3 and 4 (below the toe of the wall). For equilibrium across the vertical frictionless stress discontinuity separating these two zones, the horizontal stress at any level must be the same in each zone, even though the vertical stress at any level is different in each zone.

At a general depth z below the retained soil surface, the vertical effective stress in zone 3 (behind the wall) is γz. The horizontal stress is σ'_{h}, which must be greater than or equal to the active limit, $K_{\mathrm{a}}\gamma z$.

At the same level in zone 4, the vertical effective stress is $\gamma(z-h)$. The horizontal stress σ'_{h} must be less than or equal to the passive limit, $K_{\mathrm{p}}\gamma(z-h)$. Thus equilibrium can be achieved without violating the failure condition provided that

$$\sigma'_{\mathrm{h}} \geq K_{\mathrm{a}}\gamma z$$

and

$$\sigma'_{\mathrm{h}} \leq K_{\mathrm{p}}\gamma(z-h)$$

This means that

$$K_{\mathrm{p}}\gamma(z-h) \geq K_{\mathrm{a}}\gamma z$$

or

$$z \geq [K_{\mathrm{p}}/(K_{\mathrm{p}} - K_{\mathrm{a}})]h$$

With the current values of $K_{\mathrm{a}} = 1/3$ and $K_{\mathrm{p}} = 3$ (corresponding to $\phi' = 30°$), this requires $z(= h + d) \geq 1.125\,h$, or $d \geq 0.125\,h$. If $h = 5\,\mathrm{m}$, $d \geq 0.625\,\mathrm{m}$.

If this condition is not fulfilled, the soil in zones 3 and 4 will be unable to support the weight of the soil behind the wall in zone 1, and will undergo what is termed a **bearing** (or **bearing capacity**) **failure**. This will result in large settlements behind the wall, while the soil in front of the wall will be pushed upward into the excavation. In Example 7.2, the depth of embedment required to prevent a bearing failure ($d = 0.625$ m) is less than that required to prevent rotational failure, $d = 2$ m as already calculated. With multi-propped walls, or walls with low level props, this is not always the case, and it is in general important to investigate the possibility of a bearing as well as a rotational failure.

It is worth setting out the shortcomings of the calculation we have just done.

- Pore water pressures have been assumed to be zero. In practice it will usually be necessary to estimate pore water pressures, perhaps corresponding to steady state seepage from a high water table behind the wall to a lower level in front. This can be done by means of a flownet (Chapter 3). A common alternative approach, which is useful in analysis, is to assume that the overall drop in excess head is distributed linearly around the wall (section 7.8.1). However, this method can lead to significant errors, particularly in terms of the prediction of base instability due to fluidization (section 3.11), when the embedded wall forms one side of a narrow cofferdam (Williams and Waite, 1993).

It is important to remember that the failure of the soil depends on the effective stress state. The earth pressure coefficients K_a and K_p are therefore applied to the **vertical effective stresses**. The horizontal effective stress at a given depth z is equal to K_a or K_p times the vertical effective stress. In the absence of wall friction, the vertical effective stress is equal to $\gamma z - u$, where u is the local pore water pressure.

The wall, however, experiences the effect of both the pore water pressures and the effective stresses, and is unable to distinguish between them. The equilibrium of the wall is therefore maintained by the **total** horizontal stress distributions – that is, the effective stresses and the pore water pressures acting together.

- Wall friction has been neglected. This is inherently conservative, but will in reality lead to uneconomical designs. Stress field solutions can be obtained which allow for the rotation in the direction of the principal stresses between the soil surface and the wall, which must occur when wall friction is present (see Chapter 9).

  Alternatively, tables are available (e.g. Caquot and Kerisel (1948), Kerisel and Absi (1990) and section 7.10.1) which give modified numerical values of the coefficients K_a and K_p for given angles of soil friction ϕ' and soil/wall friction δ, taking account of a non-horizontal backfill and a retaining wall with a sloping or **battered** back if desired. Such tables can be very useful, but care should be taken to interpret them correctly. In Caquot and Kerisel (1948), for example, the tabulated earth pressure coefficients are $\sigma'_h/(\gamma z - u)$, not σ'_h/σ'_v (σ_v is *not* equal to γz if the wall is frictional): and values of the *resultant* thrust (inclined at the angle of wall friction δ) and its horizontal component are tabulated separately.

Some of the published earth pressure coefficients result from true lower bound calculations, but others may be based on some degree of empiricism. A further problem is that considering soil/wall friction makes it more difficult to investigate rigorously the stress state in the soil below the wall (zones 3 and 4 in Figure 7.15). As a result, the answers obtained can no longer be classed as true lower bounds: they are known as **limit equilibrium solutions**.

- The possibility of a structural failure of the prop or of the wall itself has not been considered.
- Bending of the wall has been neglected, the assumed mode of wall deformation being rigid body rotation about the prop.
- The calculation tells us nothing about the deformations needed to reach the collapse condition. It would certainly not be acceptable to design a wall to be on the verge of collapse, and it is necessary to introduce a factor of safety into the calculation. For walls designed in accordance with BS 8002 (BSI, 2001), the current British code of practice for earth retaining structures, this is achieved by the combined effects of three modifications to the collapse calculation:

(a) a reduced soil strength is used to calculate the active and passive pressures (section 7.3.2),

(b) the retained height is increased to represent 'a depth of unplanned excavation in front of the wall. The depth of excavation should be not less than 10% of the retained height for [unpropped] cantilever walls or [10%] of the height retained below the lowest support level for propped or anchored walls, but the depth of the excavation may be limited to 0.5 m.' Eurocode 7 (BSI, 1995) is rather clearer that the additional depth of (unplanned) excavation is limited to a maximum of 0.5 m.

(c) an additional uniform surcharge of 10 kPa is assumed to act on the retained soil surface, although this value may, under certain circumstances, be reduced for walls with a retained height of less than 3 m.

Design calculations for embedded retaining walls are discussed in sections 10.2 and 10.3.

- The stress field solution could easily be modified to take account of the effect of a uniform surcharge on one or both of the free soil surfaces. The vertical stress in the underlying soil is simply increased by the amount of the surcharge. The effect of a line load or a strip load is more easily taken into account by analysing potential mechanisms of failure, although one possible approach using the stress field type of calculation is given in section 10.8.

Although the calculation would provide a lower bound to a set of collapse loads, it actually gives an upper limit to the depth of embedment required just to prevent collapse. For this reason, it is probably less ambiguous to refer to a rigorous stress-field calculation as a 'safe' solution, rather than a 'lower bound'.

7.6.2 $\tau_{max} = \tau_u$ failure criterion

A similar stress field may be constructed from the limiting active and passive pressures for a soil obeying the 'maximum shear stress' failure criterion $\tau_{max} = \tau_u$. This would be applicable to intact clays, of low bulk permeability, in the short term. It was shown in Figure 7.6 that, for a given vertical *total* stress σ_v, the horizontal *total* stress σ_h must lie between $(\sigma_v - 2\tau_u)$ and $(\sigma_v + 2\tau_u)$ (equations (7.6)). These values give the active and passive limits to the total lateral stress.

In the absence of a surface surcharge, the active stress $\sigma_v - 2\tau_u$ is equal to $\gamma z - 2\tau_u$, which is in theory negative for depths z less than $2\tau_u/\gamma$. For design purposes, the interface between the soil and the wall would not normally be relied upon to transmit tensile stresses. This 'no-tension' criterion requires that $\sigma_h \geq 0$. A dry tension crack, in which there is no contact between the wall and the soil so that $\sigma_h = 0$, could develop to a depth of $2\tau_u/\gamma$ below the soil surface on the active side of the wall.

If there is any possibility that the tension crack might flood, the lateral stress in the tension crack should be taken as $\gamma_w z$, where γ_w is the unit weight of water. Sources of water to flood the crack might include a burst pipe or a more permeable stratum of soil, such as a layer of gravel. A flooded tension crack can remain open to the depth at which $\sigma_h = \gamma_w z = \gamma z - 2\tau_u$ or $z = 2\tau_u/(\gamma - \gamma_w)$. If τ_u is constant and $\gamma = 2\gamma_w$, this is twice the depth to which a dry tension crack will remain open. If τ_u increases with depth (which is more common in practice than $\tau_u = $ constant), the difference between the dry and the flooded tension crack depths will be even greater. Criteria for deciding whether and how to allow for the possibility of flooded tension cracks in design are discussed by Gaba *et al.* (2003).

Example 7.3 Calculating the depth of embedment at undrained collapse of a propped cantilever retaining wall in a clay

Assuming that tension cracks remain dry, calculate the depth of embedment required just to prevent rotational failure of a frictionless, rigid, embedded cantilever wall, propped at the top, in a uniform clay soil of undrained shear strength $\tau_u = 40\,kPa$ and unit weight $\gamma = 20\,kN/m^3$. The retained height is 5 m, and there is a uniform surface surcharge of $q = 40\,kPa$ on the retained side.

Solution

Behind the wall, the vertical total stress σ_v is given by the unit weight of the soil γ multiplied by the depth z below the soil surface, plus the surface surcharge $q, \sigma_v = q + \gamma z$. The active horizontal total stress σ_h is $(\sigma_v - 2\tau_u) = (q + \gamma z - 2\tau_u)$ (equation (7.6a)), subject to the requirement that $\sigma_h \geq 0$. A dry tension crack can remain open to the depth at which

$$\sigma_h = q + \gamma z - 2\tau_u = 0$$

The dry tension crack depth is therefore $(2\tau_u - q)/\gamma$.

At the toe of the wall, $\sigma_h = [q + \gamma(h+d) - 2\tau_u] = [40 + 20 \times (5+d) - 2 \times 40]$ kPa. The total horizontal stress behind the wall increases linearly from zero at a depth of $(2\tau_u - q)/\gamma$ to $(q + \gamma z - 2\tau_u)$ at the toe.

At a depth z below the soil surface in front of the wall, the vertical total stress σ_v is given by γz (kPa), and $\sigma_h = \sigma_v + 2\tau_u$ (equation (7.6b)). At the toe, $\sigma_v = \gamma \times d = 20\,\text{kN/m}^3 \times d\,\text{m}$ (kPa), and $\sigma_h = (20d + 2\tau_u)$ kPa.

The horizontal total stresses acting on each side of the wall are shown, together with schematic Mohr circles of stress, in Figure 7.18.

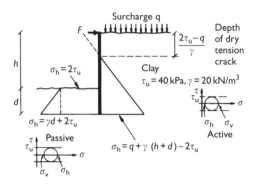

Figure 7.18 Total stress distributions for calculating the limiting depth of embedment of a stiff, propped embedded wall in a clay soil: short-term (undrained) conditions.

Taking moments about the prop to find the depth of embedment d:

$$[0.5(q + \gamma(h+d) - 2\tau_u)(h + d - \{(2\tau_u - q)/\gamma\})]$$
$$\times\, [(2\tau_u - q)/\gamma + 2(h + d - \{(2\tau_u - q)/\gamma\})/3]$$
$$= [0.5\gamma d^2] \times [h + (2d/3)] + [2\tau_u d] \times [h + (d/2)]$$

[Force due to active stresses] × [Lever arm]

    = [Force due to passive stresses] × [Lever arm]

(To take moments, the passive stress block must be split into two components, one rectangular (of magnitude $2\tau_u d$ and lever arm about the prop $h + d/2$), and one triangular (of magnitude $0.5\gamma d^2$ and lever arm $h + 2d/3$).)

For $\tau_u = 40$ kPa, $q = 40$ kPa, $h = 5$ m and $\gamma = 20\,\text{kN/m}^3$, the maximum depth of a dry tension crack $(2\tau_u - q)/\gamma = 2$ m and

$$[0.5 \times (40 + 100 + 20d - 80)\,\text{kPa} \times (5 + d - 2)\,\text{m}]$$
$$\times\, [2 + 2(5 + d - 2)/3]\,\text{m}$$
$$= [0.5 \times 20d^2]\,\text{kPa m} \times [5 + 2d/3]\text{m} + [80d]\,\text{kPa m} \times [5 + d/2]\,\text{m}$$
$$\Rightarrow [10(3 + d)^2] \times [2(6 + d)/3] = [10d^2] \times [5 + 2d/3] + [80d] \times [5 + d/2]$$

(with d in metres)

$\Rightarrow (20/3) \times (3+d)^2 \times (6+d) = 20d^3/3 + 90d^2 + 400d$

$\Rightarrow (3+d)^2 \times (6+d) = d^3 + 13.5d^2 + 60d$

Solution by trial and error gives

$d = 2.8\,\mathrm{m}$

For horizontal equilibrium,

$[0.5(q + \gamma(h+d) - 2\tau_u)(h + d - \{(2\tau_u - q)/\gamma\})] = [0.5\gamma d^2] + [2\tau_u d] + F$

[Force due to active lateral stresses]

$= $ [Force due to passive lateral stresses] + Prop load

For $\tau_u = 40\,\mathrm{kPa}$, $q = 40\,\mathrm{kPa}$, $h = 5\,\mathrm{m}$, $\gamma = 20\,\mathrm{kN/m}^3$, $(2\tau_u - q)/\gamma = 2\,\mathrm{m}$ and $d = 2.8\,\mathrm{m}$ as just calculated,

$[0.5 \times (40 + \{20 \times 7.8\} - 80)\,\mathrm{kPa} \times (5 + 2.8 - 2)\,\mathrm{m}]$

$= [0.5 \times 20 \times 2.8^2]\,\mathrm{kN/m} + [2 \times 40 \times 2.8]\,\mathrm{kN/m} + F\,\mathrm{kN/m}$

$\Rightarrow 336.4\,\mathrm{kN/m} = 302.4\,\mathrm{kN/m} + F\,\mathrm{kN/m}$

$\Rightarrow F = 34\,\mathrm{kN/m}$

For equilibrium across the frictionless stress discontinuity separating zones 3 and 4 in Figure 7.15, the horizontal total stress at any level must be the same in each zone. Again, the vertical total stress at any level is different in each zone.

At a general depth z below the retained soil surface, the vertical total stress in zone 3 (behind the wall) is $q + \gamma z$. The horizontal total stress is σ_h, which must be greater than or equal to the active limit, $\sigma_h \geq q + \gamma z - 2\tau_u$. At the same level in zone 4, the vertical total stress is $\gamma(z - h)$. The horizontal total stress σ_h must be less than or equal to the passive limit in zone 4, $\sigma_h \leq \gamma(z - h) + 2\tau_u$. Thus equilibrium can be achieved without violating the failure criterion provided that

$\sigma_h \geq q + \gamma z - 2\tau_u$

and

$\sigma_h \leq \gamma(z - h) + 2\tau_u$

This means that

$\gamma(z - h) + 2\tau_u \geq q + \gamma z - 2\tau_u$

or

$h \leq (4\tau_u - q)/\gamma$ \hfill (7.7)

With the current values of $\tau_u = 40\,\mathrm{kPa}$, $q = 40\,\mathrm{kPa}$ and $\gamma = 20\,\mathrm{kN/m}^3$, $h \leq 6\,\mathrm{m}$, irrespective of the depth of embedment d. Owing to the neglect of wall friction and the true strength of the soil along the vertical discontinuity separating zones 3 and 4, this solution will err on the conservative side. Nonetheless, it illustrates an important point: in a soft clay, the stability of the base of an excavation may well be more critical than the rotational stability of the sidewalls.

For the general case of an embedded wall propped at the crest, of retained height h and depth of embedment d in a uniform clay of undrained shear strength τ_u with unit weight γ in the range $1.6\gamma_w$–$2.0\gamma_w$ (i.e. $\gamma/\gamma_w = 1.6$–2.0, where γ_w is the unit weight of water) but no surface surcharge, Figure 7.19 shows how the embedment ratio d/h at collapse varies with the normalized undrained shear strength $\tau_u/\gamma h$, and how the normalized prop load $F/0.5\gamma h^2$ varies with the embedment ratio d/h at collapse.

Figure 7.20 shows the corresponding relationship between the normalized undrained shear strength $\tau_u/\gamma h$ and the embedment ratio d/h at collapse for unpropped embedded cantilever walls. Figures 7.19 and 7.20 should err on the conservative side, as they result from true lower bound calculations, in which soil/wall friction has been ignored. The increase in the depth of embedment required when tension cracks are flooded

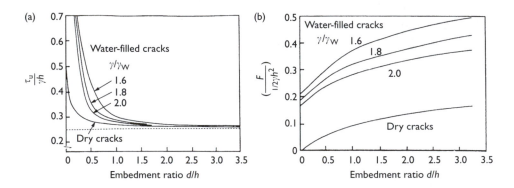

Figure 7.19 (a) Normalized undrained shear strength ($\tau_u/\gamma h$) and (b) prop load ($F/0.5\gamma h^2$) against embedment ratio d/h for short-term limiting equilibrium of embedded walls propped at the crest in uniform clay soil. (Redrawn with permission from Bolton *et al.* 1990.)

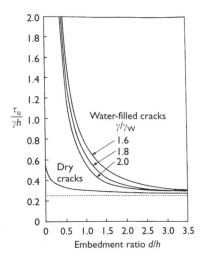

Figure 7.20 Normalized undrained shear strength ($\tau_u/\gamma h$) against embedment ratio d/h for short-term limiting equilibrium of unpropped embedded walls in uniform clay soil. (Redrawn with permission from Bolton *et al.*, 1990.)

rather than dry is clear, particularly for unpropped walls. In both cases, the limiting retained height in the absence of a surface surcharge (i.e. $q = 0$ in equation (7.7)), is $h_{max} = 4\tau_u/\gamma$ ($\tau_u/\gamma h \geq 0.25$ on the graph).

In Example 7.3, adhesion between the wall and the soil τ_w (which is analogous to soil/wall friction δ in the analysis based on the frictional failure criterion $(\tau/\sigma')_{max} = \tan\phi'_{crit}$) has been neglected. As in section 7.6.1, the possibility of a structural failure of the prop or the wall itself has not been considered. A uniform surcharge has been easily taken into account, because it simply increases the vertical total stresses below it. The effect of a line load, however, is again more easily dealt with in an analysis based on potential mechanisms of collapse (section 7.7). The extension of the stress field method to include the effects of wall adhesion, a non-horizontal backfill and a sloping or battered wall back is addressed in Chapter 9.

A further important point concerning an analysis based on undrained shear strengths is that it is only valid for the period during which the specific volume and hence the water content of the clay does not change. In problems involving the excavation of clay, for example an *in situ* retaining wall, the long-term effective stresses are generally smaller than those in the undisturbed ground. This means that the clay will tend to take in water, swell and soften, leading to a reduction in undrained shear strength with time. The rate of softening in practice will depend on

- the bulk permeability of the soil (which may be increased by the effect of fissures which open up on unloading)
- the soil stiffness (which affects the consolidation coefficient c_v, and might be quite high following a change in the direction of the stress path)
- the availability of water (potential sources include rainfall, surface run-off, natural water courses, leaking pipes, gravel aquifers and sand lenses, waste-water from hosing down construction plant etc.).

In assessing the applicability of the undrained shear strength model, all of these factors must be considered in relation to the time-scale over which the excavation is expected to remain open.

The depth L below the excavation to which the soil is affected by softening after an elapsed time t, is the same as the depth to which the isochrone of excess pore water pressure dissipation has penetrated. The parabolic isochrones approximation described in sections 4.5 and 4.7.3 may be used (equation (4.15)) to estimate the depth L:

$$L = \sqrt{12 c_v t} \tag{7.8}$$

where c_v is the consolidation coefficient. This assumes – perhaps pessimistically – that the excavated soil surface acts as a recharge boundary. In London Clay, the effects of softening are sometimes allowed for by applying a factor of 0.7–0.8 to the value of τ_u, and assuming that τ_u falls linearly to zero over a disturbed zone immediately below the excavated soil surface. Gaba *et al.* (2003) suggest that the depth of this disturbed zone could be taken as $\sqrt{12 c_v t}$ (equation (7.8)) where groundwater recharge may occur at excavation level but not within the soil, or 0.5 m in the absence of any recharge. They also note that the 20–30% reduction in the value of τ_u in London Clay often applied globally in the restraining soil might be too severe in some cases, and stress the importance of drawing on relevant previous experience wherever possible.

It is worth emphasizing again that so far we have investigated depths of embedment which are just sufficient for the wall to be on the verge of outright collapse. In design, it is necessary to factor the calculation so that deformations under working conditions are not excessive. In the design of retaining walls according to BS 8002 (BSI, 2001), the undrained shear strength is reduced by a mobilization factor of at least 1.5, in addition to the requirements for an increased excavation depth and an additional 10 kPa surcharge acting on the retained soil surface.

7.7 Mechanism-based kinematic and equilibrium solutions for gravity retaining walls

In addition to the non-brittle type of stress–strain behaviour shown in Figure 7.1, a further requirement for the validity of the upper and lower bound theorems is that the material exhibits a property known as **normality**. This means that the direction of the plastic strain increment vector must be normal to the surface which defines the failure criterion when the deformation and stress axes are superimposed (Figure 7.21).

If the normality condition applies, a material obeying the maximum shear stress failure criterion $\tau_{max} = \tau$ will move parallel to the surface along which the failure condition is reached (Figure 7.21(a)). This implies that the soil is deforming at constant volume, which is consistent with the notion of a critical state, and suggests that clay soils brought rapidly to failure at constant volume/water content can in this respect reasonably be regarded as perfectly plastic materials.

For a soil obeying the maximum stress ratio failure criterion $(\tau/\sigma')_{max} = \tan \phi'$, however, the normality condition requires relative movement at an angle of ϕ' to the failure surface (Figure 7.21(b)): that is, the soil has an angle of dilation ψ equal to ϕ'. This means that **kinematically admissible** rupture surfaces (i.e. rupture surfaces that allow the necessary soil movement physically to take place) may be straight lines and circles for $\tau_{max} = \tau_u$ materials, while for $(\tau/\sigma')_{max} = \tan \phi'$ materials they are straight lines and logarithmic spirals (Figure 7.22).

The assumption that the angle of dilation ψ is equal to the soil strength ϕ' at failure is unrealistic. The whole point about the critical state is that the soil is deforming at constant volume, that is, $\psi = 0$. However, the assumption $\psi = \phi' = \phi'_{crit}$ can be

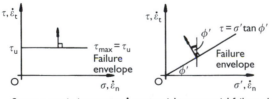

Figure 7.21 Normality. (a) $\tau_{max} = \tau_u$ failure criterion; (b) $(\tau/\sigma')_{max} = \tan \phi'$ failure criterion.

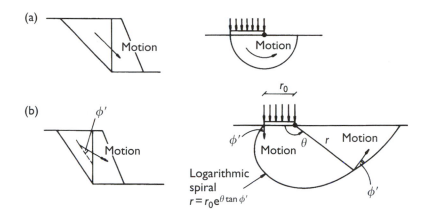

Figure 7.22 Kinematically admissible mechanisms for materials exhibiting normality. (a) $\tau_{max} = \tau_u$, failure criterion (motion parallel to slip surface); (b) $(\tau/\sigma')_{max} = \tan\phi'$ failure criterion (motion at ϕ' to slip surface).

shown to lead to true upper bounds, in the sense that loads causing collapse calculated on this basis will be greater than or equal to those for a real soil for which $\psi = 0$ at the critical state (e.g. Atkinson, 1981). Although there is no proof, it is usually just assumed that lower bounds (which apply strictly to soils with $\psi = \phi'$) are also lower bounds for soils with $\psi = 0$ at the critical state.

Strictly, there is no requirement that a plasticity solution based on an assumed mechanism of collapse should necessarily correspond to an equilibrium state. However, a non-equilibrium mechanism will have two **degrees of freedom** (i.e. it will be necessary to specify two independent parameters in order to define its position at any stage) and will therefore be unusable for the purpose of plasticity analysis (e.g. Baker and Heyman, 1969).

The upper bound theorem as stated in section 7.2 implies that the mechanism of collapse should be analysed in terms of a work balance. This requires the construction of a **velocity diagram** or **hodograph** – a vector diagram showing the relative velocities of all the components of the mechanism in terms of some reference velocity v_0, and providing a check on **kinematic admissibility** (i.e. that the deformations associated with the assumed mechanism can actually physically occur). Drescher and Detournay (1993) show that the same answer may be obtained using the condition of statical equilibrium, provided that the assumed mechanism really is a mechanism (i.e. that it is kinematically admissible). This is the approach that will be adopted in this section. Its main drawback is that there is no automatic check on kinematic admissibility (as there is with the work balance approach because of the requirement to construct a hodograph). Thus it is possible to apply the method to a system of sliding bocks that either does not constitute a mechanism at all or requires one or more of the blocks to distort (rather than remain rigid) as movement takes place. An example of a mechanism that includes a distorting or shearing zone is given in section 9.9.

Solutions that are not true upper or lower bounds (e.g. stress fields that are not extended to infinity, or systems of blocks that do not represent kinematically admissible mechanisms) are known as **limit equilibrium** analyses. They are actually much more common in geotechnical engineering than true plasticity solutions.

The analysis of retaining walls based on an assumed mechanism of failure was originated by the military engineer C. A. Coulomb (Coulomb, 1776; Heyman, 1972). As an extended worked example, we will consider the mass concrete retaining wall shown in Figure 7.23(a). This is an idealization of a real wall on the St Pancras–Sheffield railway line at Cricklewood, North London (Figure 7.23(b)), which was described by Skempton (1946). It is also mentioned, together with a number of other interesting case histories, in Appendix E of the former British Code of Practice on retaining wall design, CP2 (Institution of Structural Engineers, 1951).

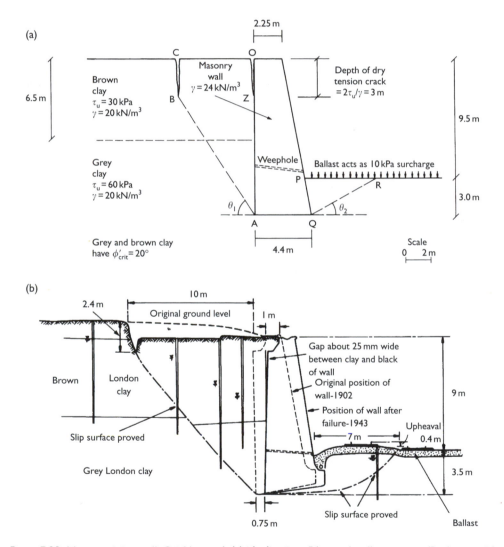

Figure 7.23 Mass retaining wall, Cricklewood: (a) idealization; (b) actual wall geometry. (Redrawn with permission from Skempton, 1946.)

In the worked examples that follow, we will determine the minimum (i.e. active) lateral thrust that the wall shown in Figure 7.23(a) must be able to withstand, and the maximum (i.e. passive) resistance available in front of the wall, both in the 'as-built' condition. We will then investigate the stability of the wall in terms of (i) sliding along the base and (ii) toppling or overturning about the front corner. If the wall is satisfactory in both these respects, we will go on to calculate the pressures exerted by the wall on the soil below it – the first step in checking that the bearing capacity of the foundation is adequate. Both the short-term condition (using the $\tau_{max} = \tau_u$ failure criterion in terms of total stresses) and the long-term condition (using the $(\tau/\sigma'_{max}) = \tan \phi'_{crit}$ failure criterion in terms of effective stresses) will be considered. Because we are analysing and possibly trying to explain the failure of a real wall, we will use the actual soil strengths in our calculations. If we were designing a new wall, we would reduce the soil strengths by a factor of safety to ensure that the wall we design remains stable and serviceable, in both the short and the long term. In the examples that follow, the differences between a design calculation and our analysis are highlighted as appropriate.

Example 7.4 Determining the forces exerted by the soil on the Cricklewood retaining wall in the short term, using the undrained shear strength ($\tau_{max} = \tau_u$) failure criterion

In general, retaining wall problems such as these are best tackled by scale drawing on a sheet of graph paper. The retained ('active') and excavated ('passive') sides of the wall will initially be investigated separately.

(a) Behind the wall (active side)

The first step is to assume a failure mechanism. This will probably involve a block of soil such as OABC sliding down a slip plane such as AB, which is at a general angle θ (θ_1 in Figure 7.23(a)) to the horizontal (Figures 7.23(a), 7.24(a)).

The second step is to investigate the forces acting on the block of soil OABCO. These are (Figure 7.24(a)):

1. The weight W of the block acting vertically downward. For a one metre run perpendicular to the plane of the paper, the weight in kN/m is simply the area of the block (m^2) multiplied by the unit weight of the soil (kN/m^3).
2. Any known external forces, for example surcharges or water in a flooded tension crack. In this case, there is no surcharge on the active side, and we will assume that the tension cracks (which may form to a depth $2\tau_u/\gamma$) remain dry.
3. The shear force T_R along the assumed rupture plane, which is known in both direction (parallel to the slip surface, acting so as to oppose motion) and magnitude (τ_u multiplied by the length of the slip line BA below the tension crack CB if τ_u is uniform, or the summed effect if – as in this case – τ_u is not uniform over the entire depth).

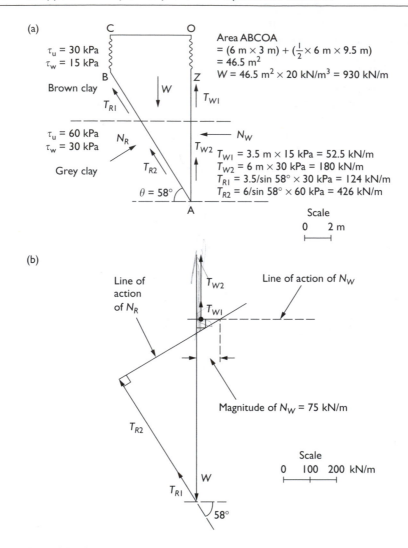

(a)

$\tau_u = 30$ kPa
$\tau_w = 15$ kPa

Brown clay

$\tau_u = 60$ kPa
$\tau_w = 30$ kPa

Grey clay

Area ABCOA
$= (6\ \text{m} \times 3\ \text{m}) + (\frac{1}{2} \times 6\ \text{m} \times 9.5\ \text{m})$
$= 46.5\ \text{m}^2$
$W = 46.5\ \text{m}^2 \times 20\ \text{kN/m}^3 = 930\ \text{kN/m}$

$\theta = 58°$

$T_{WI} = 3.5\ \text{m} \times 15\ \text{kPa} = 52.5\ \text{kN/m}$
$T_{W2} = 6\ \text{m} \times 30\ \text{kPa} = 180\ \text{kN/m}$
$T_{RI} = 3.5/\sin 58° \times 30\ \text{kPa} = 124\ \text{kN/m}$
$T_{R2} = 6/\sin 58° \times 60\ \text{kPa} = 426\ \text{kN/m}$

Scale
0   2 m

(b)

Line of action of N_R

Line of action of N_W

Magnitude of $N_W = 75$ kN/m

Scale
0   100 200 kN/m

58°

Figure 7.24 (a) Forces acting on the sliding block of soil OABCO on the retained side of the wall (assumed failure surface is the plane AB); (b) polygon of forces (force vector diagram) to determine N_R and N_W.

4. The shear force T_W between the wall and the sliding block, which is also known in both direction (parallel to the interface, acting so as to oppose motion) and magnitude (τ_w multiplied by the length of the interface ZA below the tension crack OZ if the wall adhesion τ_w is uniform, or the summed effect of wall adhesion along the length of the slip line if – as in this case – τ_w is not uniform).

5. The normal component N_R of the reaction at the assumed slip surface, which is known in direction (it acts at right-angles to the slip surface), but not in magnitude.

6. The normal component N_W of the reaction between the wall and the soil, which is again known in direction (right-angles to the wall) but not in magnitude.

Since this is a total stress analysis, pore water pressures are not considered separately.

The forces acting on the active-side sliding block of soil OABCO are shown in Figure 7.24(a).

In design, it is usual to use values of wall adhesion τ_w which are less than the undrained shear strength of the soil τ_u. This is to allow for the possible effects of softening due to wall construction effects, and for the fact that full wall friction may not be mobilized – for example, if there is insufficient relative movement between the soil and the wall. In this example (which is an after-the-event back-analysis rather than a design), we will assume that $\tau_w = 0.5\,\tau_u$. This is consistent with BS 8002 (BSI, 2001) and Gaba *et al.* (2003) for stiff clay.

The main difference between the analysis which we are about to do and a design analysis is that a design analysis would be carried out with the undrained shear strength of the soil reduced by a factor of safety F. In BS 8002, the factor of safety on soil strength is applied as a **mobilization factor** M, where

$$\tau_{u,design} = \tau_{u,actual} \div M \tag{7.9}$$

(The term **mobilization factor** is used because its main purpose is to reduce the soil strength mobilized under working conditions, in order to control wall movement.) Both Gaba *et al.* (2003) and BS 8002 (2001) recommend the use of $M = 1.5$ in an undrained analysis. In BS 8002, the design value of τ_w is given as $0.75 \times$ the design value of τ_u, giving

$$\tau_{w,design} = 0.75 \times \tau_{u,design} = 0.75 \times \tau_u/1.5 = 0.5\tau_u$$

The third step is to construct the polygon of forces for the block of soil taking part in the assumed failure mechanism.

The force polygon is shown in Figure 7.24(b): in practice it is best drawn to scale on graph paper. The lateral thrust on the retaining wall which corresponds to the assumed failure mechanism is given by the horizontal component of the total stress reaction between the sliding wedge and the wall, which is in this case normal to the back of the wall and – scaling from the force polygon – equal to 75 kN/m.

The answer we have just found will only be correct if the failure mechanism we have assumed is also correct. Furthermore, the error in our answer will be on the unsafe side, i.e. we have probably *underestimated* the minimum lateral thrust that the wall must be able to resist to maintain the stability of the retained soil. This is because it may be that a larger lateral thrust must be provided to prevent failure from occurring by sliding on a *different* slip plane. We must therefore repeat the calculation for different failure planes, such as AD and AF (Figure 7.25), until the failure plane which gives the greatest lateral thrust is identified. In practice, this is not too tedious if it is done methodically.

The triangles ABD, ADF etc., have the same base distance (BD = DF etc. = 2 m in this case) and height (AZ = 9.5 m), so that each extends the width of the sliding block of soil by the same amount. Each new volume ABCEDA, ADEGFA etc. therefore adds the same weight (465 kN/m in this case) to the original block. The shear force T_W at the soil/wall interface will not change, and the shear force on the rupture is given by $[30\,\text{kPa} \times (3.5\,\text{m}/\sin\theta)] + [60\,\text{kPa} \times (6\,\text{m}/\sin\theta)] = [465/\sin\theta]\,\text{kN/m}$. Force polygons for each new mechanism can be constructed quickly by scaling angles from the space diagram (Figure 7.25). Force polygons may be superimposed as shown in Figure 7.26, so that the failure surface which leads to the maximum lateral thrust is easily identifiable. The values of W, θ and T_R used in the construction of the force polygons in Figure 7.26, and the resulting values of active thrust N_W, are given in Table 7.2.

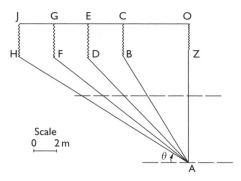

Figure 7.25 Succession of trial blocks for retaining wall analysis (active side).

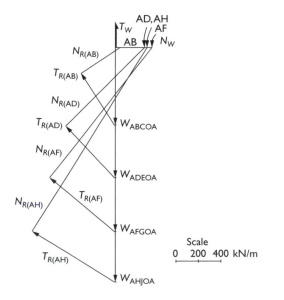

Figure 7.26 Graphical construction for 'least optimistic' sliding wedge failure mechanism: superposition of force polygons (active side).

Table 7.2 Numerical values used in and obtained from Figure 7.26 (active side of wall, short-term analysis)

Block	Weight W (kN/m)	$\tan\theta$	θ (degrees)	$T_R = (465/\sin\theta)$ (kN/m)	N_W (kN/m) from Figure 7.26
ABCOA	930	9.5 ÷ 6	57.7	550	75
ADEOA	1395	9.5 ÷ 9	46.5	641	295
AFGOA	1860	9.5 ÷ 12	38.4	749	335
AHJOA	2325	9.5 ÷ 15	32.3	869	295

From Figure 7.26 and Table 7.2, it may be seen that the maximum active lateral thrust is approximately 335 kN/m, occurring when $\theta \approx 38.4°$.

For a vertical wall back and a horizontal retained soil surface with no surcharge, it can be shown relatively easily using trigonometry and calculus that the maximum lateral thrust occurs when the assumed slip plane is at an angle α to the horizontal, where $\cot\alpha = \sqrt{[1 + (\tau_w/\tau_u)]}$ (e.g. Whitlow, 1995). For more complex geometries, the mathematical approach becomes rather more difficult. In the present case with $(\tau_w/\tau_u) = 0.5$, this equation gives $\alpha = 39.2°$, which is close enough to the approximate value of 38.4° just calculated.

(b)   In front of the wall (passive side)

The procedure must now be repeated for the soil that is pushed up in front of the retaining wall as the wall moves forwards. The forces acting on the assumed wedge (Figure 7.27(a)) are in principle the same as on the retained side, but there is now no need to consider tension cracks: as the wall is being pushed into the soil, there is no possibility that a tension crack will form. The shear forces on the rupture plane and at the soil/wall interface again act so as to resist the supposed soil movement, but while the block of soil behind the wall slides downward, the wedge of soil in front is pushed upward.

The shear force on the wall is the same for all wedges, and is given by $T_W = 30\,\text{kPa} \times (3/\cos 9.8°)\text{m} = 91\,\text{kN/m}$. The shear force on the rupture surface is $T_R = 60\,\text{kPa} \times (3/\sin\theta)\text{m}$. The ballast (a coarse granular aggregate, with a typical particle size of 40 mm or more, which is traditionally used as a bedding material for railway tracks) that supports the railway tracks has been modelled by a uniform surcharge of 10 kPa (i.e. a 0.5 m depth × a unit weight of 20 kN/m³). The forces acting on a typical wedge PQRP are shown in Figure 7.27(a), and the succession of trial wedges PQRP, PQSP etc. is shown in Figure 7.27(b).

As the soil in front of the wall helps to prevent failure, it is necessary to find the **minimum** lateral thrust that the passive wedge can be relied upon to provide.

The force polygons associated with each of the trial wedges are superimposed in Figure 7.28, from which the critical rupture plane can be identified. The values of W, θ, T_R and N_W used in and calculated from Figure 7.28 are given in Table 7.3.

From Figure 7.28 and Table 7.3, it may be seen that the minimum passive lateral thrust is approximately 480 kN/m, occurring when $\theta \approx 45°$.

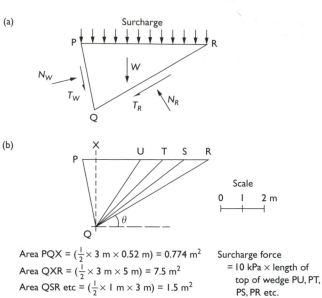

(a) Surcharge

(b)

Area PQX = $(\frac{1}{2} \times 3 \text{ m} \times 0.52 \text{ m}) = 0.774 \text{ m}^2$

Area QXR = $(\frac{1}{2} \times 3 \text{ m} \times 5 \text{ m}) = 7.5 \text{ m}^2$

Area QSR etc = $(\frac{1}{2} \times 1 \text{ m} \times 3 \text{ m}) = 1.5 \text{ m}^2$

Surcharge force
= 10 kPa × length of
top of wedge PU, PT,
PS, PR etc.

Figure 7.27 (a) Forces acting on a typical wedge and (b) succession of trial wedges for retaining wall analysis (passive side).

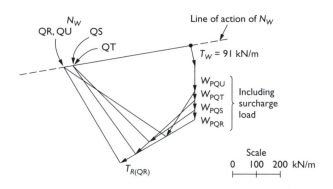

Figure 7.28 Graphical construction for 'least optimistic' sliding wedge failure mechanism: superposition of force polygons (passive side).

Table 7.3 Numerical values used in and obtained from Figure 7.28 (passive side of wall, short-term analysis)

Block	Weight W (kN/m)	Force due to surcharge (kN/m)	$\tan\theta$	θ (degrees)	$T_R = (180/\sin\theta)$ (kN/m)	N_W (kN/m) from Figure 7.28
PQRP	165	55	3 ÷ 5	31.0	350	525
PQSP	135	45	3 ÷ 4	36.9	300	495
PQTP	105	35	3 ÷ 3	45.0	255	480
PQUP	75	25	3 ÷ 2	56.3	216	500

Having determined the minimum lateral thrust that the wall must be able to withstand for stability (which is the largest of the values calculated, 335 kN/m), and the maximum lateral resistance that can be contributed by the passive wedge in front of the wall (which is the smallest of the values calculated, 480 kN/m), we need to investigate the stability of the wall in terms of failure by toppling (i.e. rotation about the point Q), and by sliding along the base AQ. It is also necessary to consider the possibility of a **bearing capacity failure**, due to the application of an excessive vertical stress to the soil beneath the base of the wall AQ.

Example 7.5 Investigating the short-term stability of the Cricklewood wall

The first step in our investigation of the stability of the wall is to draw a **free body diagram**, showing all of the forces acting on the wall, in order to check that the wall is or can be in equilibrium (Figure 7.29). The normal stress distribution on the base of the wall is usually assumed to be of the form shown. The values of the stresses σ_A (at A) and σ_Q (at Q) are needed to assess the possibility of a bearing failure, but not to investigate sliding and toppling. The calculation of σ_A and σ_Q is described in (c) below.

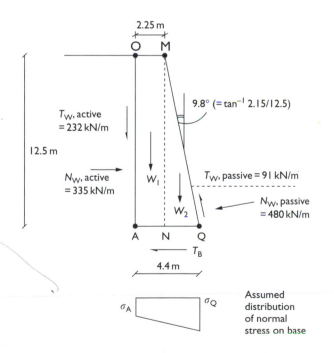

Figure 7.29 Free body diagram for the wall, showing calculated limiting soil thrusts and the assumed total normal stress distribution along the base.

(a) Sliding

In our back-analysis, we can investigate the safety of the wall against sliding by comparing the magnitude of the lateral active thrust with the total available resistance, which comes from the passive pressure of the soil in front of the wall and the shear force T_B along the base.

The lateral active thrust is 335 kN/m. The maximum base shear force is given by

$$T_{B\,max} = (4.4\,m \times \tau_w) = (4.4\,m \times 30\,kPa) = 132\,kN/m$$

The lateral component of the passive pressure forces $T_{W,passive}$ and $N_{W,passive}$ is calculated by resolving in the horizontal direction:

$$(N_{W,passive} \times \cos 9.8°) + (T_{W,passive} \times \sin 9.8°) = 489\,kN/m$$

This gives a maximum possible resisting force of 489 kN/m + 132 kN/m = 621 kN/m, which is considerably in excess of the destabilizing active thrust of 335 kN/m.

Although the wall should not fail by sliding along the base in the short term, it must be remembered that the calculated active and passive thrusts represent minimum and maximum values respectively, which might only be mobilized following an unacceptable amount of wall movement. In a design analysis following BS 8002 (BSI, 2001) one of the ways by which displacements under working conditions are limited is by calculating the active and passive thrusts using the reduced mobilized shear strength $\tau_{mobilized} = \tau_u/M$. This leads to an increase in the destabilizing active thrust and a decrease in the available passive resistance. The maximum available base resistance is calculated using a base adhesion $\tau_w = 0.75 \times \tau_{mobilized}$, that is, $0.5 \times \tau_u$ as in the present case.

By assembling the equations representing the forces shown in Figures 7.26 and 7.28 (in terms of the slip plane angles θ on each side of the wall) into a spreadsheet, the value of strength mobilization factor M needed to maintain the wall in horizontal equilibrium may be investigated. The mobilized shear strength τ_{mob} is then τ_u/M in each of the soil strata on both sides of the wall, and the mobilized soil/wall adhesion τ_w is $0.75 \times \tau_{mob}$ on both sides of the wall and on the base. Solution by trial and error gives a value of M in the case of the Cricklewood wall of 1.36, which is rather less than the value of 1.5 required by BS8002. Even in the short term, the Cricklewood wall would not therefore have an adequate factor of safety according to a modern design code. However, the values of undrained shear strength used in the calculation are probably on the conservative side for London Clay: Watson (1956) cites values of τ_u in the range 80–140 kPa in connection with another retaining wall failure at Uxbridge.

BS 8002 also requires that the water pressure regime used in the design calculation is the most onerous that could reasonably possibly occur in reality. In general, this might involve a flooded tension crack between the wall and the soil, in the absence (or following a blockage) of the weepholes. (**Weepholes** are drainage pipes set into the wall as shown in Figure 7.23(a) to prevent the build-up of pore water pressures in the retained soil.)

In this case, if the weepholes were to become blocked a flooded tension crack could remain open to a depth of approximately $2\tau_u/(\gamma - \gamma_w) = 12$ m, exerting a lateral thrust of $0.5 \times 10\,\text{kN/m}^3 \times 12\,\text{m} \times 12\,\text{m} = 720\,\text{kN/m}$, which is greater than the maximum available resistance of 621 kN/m.

(b) Toppling

If the wall were on the verge of toppling by rotation about Q, the vertical stresses on the base AQ would be zero. For the purpose of our back-analysis, the safety of the wall against toppling may be investigated by comparing the disturbing and resisting moments Q, ignoring the base bearing pressure.

To take moments about Q, it is necessary to know the point of action of the active and passive normal soil thrusts N_W.

It is generally assumed that the component of thrust which results from the weight of the soil is associated with a stress distribution that is triangular over the depth of soil in contact with the wall. On the passive side, the effect of the surcharge will be to raise the centre of pressure slightly.

In this case, the active thrust is due entirely to the weight of the soil, and may therefore be considered to act at a height AZ/3 above A.

In front of the wall, we will assume that the component of pressure resulting from the surcharge is uniformly distributed with depth. We will also assume that the components of the normal force N_W due to both the weight of soil and the surcharge load are proportional to the vertical forces W and V which cause them. The centre of pressure z^* of the overall normal stress distribution can then be found by taking moments about the toe of the wall Q:

$$z^* \times (W + V) = (d/2)V + (d/3)W$$

$$\Rightarrow (z^*/d) = [(V/2 + (W/3)]/(V + W)$$

where d is the depth of soil in front of the wall. In the present case, $V = 35$ kPa and $W = 105$ kPa, giving $z^* = 0.375d$.

For taking moments, the weight of the wall must be considered in two parts:

$$W_1 = 2.25\,\text{m} \times 12.5\,\text{m} \times 24\,\text{kN/m}^3$$

$$= 675\,\text{kN/m}, \quad \text{acting at a distance of } [2.15 + (2.25/2)]\,\text{m}$$

$$= 3.275\,\text{m} \quad \text{horizontally from Q}$$

$$W_2 = 0.5 \times 2.15\,\text{m} \times 12.5\,\text{m} \times 24\,\text{kN/m}^3$$

$$= 322.5\,\text{kN/m}, \quad \text{acting at a distance of } [2.15 \times (2/3)]\,\text{m}$$

$$= 1.43\,\text{m} \quad \text{horizontally from Q}$$

The disturbing moment about Q is due to $N_{W,\text{active}}$, and is equal to

$$N_{W,\text{active}} \times (9.5\,\text{m}/3) = 335\,\text{kN/m} \times 3.167\,\text{m} = 1061\,\text{kNm/m}$$

The resisting moment is due to the weight of the wall, $N_{W,passive}$ and $T_{W,active}$, and is equal to

$$(W_1 \times 3.175\,\text{m}) + (W_2 \times 1.43\,\text{m}) + (N_{W,passive} \times 3\,\text{m}/\cos 9.8° \times 0.375)$$

$$+ (T_{W,active} \times 4.4\,\text{m})$$

$$= (675\,\text{kN/m} \times 3.275\,\text{m}) + (322.5\,\text{kN/m} \times 1.43\,\text{m})$$

$$+ (480\,\text{kN/m} \times 1.142\,\text{m}) + (232\,\text{kN/m} \times 4.4\,\text{m})$$

$$= 4241\,\text{kNm/m}$$

The resisting moment of 4241 kNm/m is comfortably in excess of the disturbing moment of 1061 kNm/m, but the proviso concerning the wall movement required to mobilize the full soil strength of the soil still applies.

If a 12 m deep tension crack between the wall and the retained soil is assumed to fill with water, $T_{W,active}$ is substantially eliminated and the value of $N_{W,active}$ is increased to $(0.5 \times 12\,\text{m} \times 10\,\text{kN/m}^3 \times 12\,\text{m}) = 720\,\text{kN/m}$, acting at a height of $0.5\,\text{m} + 12\,\text{m}/3 = 4.5\,\text{m}$ above A. The effect of this is to increase the overturning moment to $720\,\text{kN/m} \times 4.5\,\text{m} = 3240\,\text{kNm/m}$. (The stresses between the wall and the 0.5 m of soil which remains in contact with the wall below the bottom of the tension crack are negligible in comparison.)

The ratio of the potential resisting moment to the disturbing moment could be viewed as a measure of the remoteness of the wall from toppling failure. In this case it is equal to $4241/1061 = 4$ if the tension crack remains dry, but is reduced to $4241/3240 = 1.31$ if the tension crack floods.

(c)   Calculating the base bearing pressures

The third possible mode of failure involves the **bearing failure** of the soil below the base of the retaining wall if the imposed vertical stresses are too high. To investigate this, it is necessary to calculate the stresses σ_A and σ_Q (Figure 7.29).

The wall must be in equilibrium under the action of the forces acting on it, which are as indicated in Figure 7.29. If it is assumed that the active and passive thrusts calculated in Example 7.4 act simultaneously, then the value of the shearing force T_B on the base of the wall required for horizontal equilibrium is given by

$$T_B = N_{W,active} - (N_{W,passive} \times \cos 9.8°) + (T_{W,passive} \times \sin 9.8°)$$

$$= 335 - 489 = -154\,\text{kN/m}$$

which is (a) in the opposite direction from that shown in Figure 7.29, and (b) slightly in excess of the limiting value of $0.5 \times \tau_u \times 4.4\,\text{m} = 132\,\text{kN/m}$ with $\tau_u = 60\,\text{kPa}$. Thus it is not possible for the calculated active and passive thrusts to be acting on the wall at the same time with the wall in a plausible equilibrium state.

There are two relatively straightforward adjustments we could make to ensure that the forces on the wall represent a potentially reasonable equilibrium state enabling the base bearing pressures to be calculated. We could either

(i) multiply the minimum possible (fully active) disturbing force $N_{W,active}$, and divide the maximum available resisting forces $N_{W,passive}$, $T_{W,passive}$ and T_B by the same factor F to distance them all roughly equally from failure. The factor we would need to use in this case would be the square root of the ratio of the maximum possible resisting force to the minimum possible disturbing force, that is $F = \sqrt{1.854} = 1.362$. This is (by coincidence) practically the same as the mobilization factor $M = 1.36$ that, when applied to the undrained shear strength in the soil on both sides of the wall, will maintain the wall in horizontal equilibrium with a soil/wall adhesion of $0.75 \times \tau_{mob}$, or

(ii) we could assume that the lateral force behind the wall has fallen to the active limit, and divide the passive forces in front of the wall and the maximum available base sliding resistance by a factor equal to the ratio of the maximum possible (fully passive) resisting force to the minimum possible (fully active) disturbing force, that is $(489 + 132) \div 335 = 1.854$. This might be considered reasonable on the basis that the wall movement required to reduce the lateral stresses to the active limit is very much smaller than that needed to increase them to the passive limit in many soils (but not necessarily in overconsolidated clays – this point is discussed in section 10.4). This would leave the value of $N_{W,active}$ unaltered at 335 kN/m, while reducing $N_{W,passive}$ and $T_{W,passive}$ by the factor 1.854 to values of 259 kN/m and 49 kN/m respectively, so that $(N_{W,passive} \times \cos 9.8°) + (T_{W,passive} \times \sin 9.8°) = 264$ kN/m. The base friction force T_B is 1/1.854 times the maximum possible value of 132 kN/m, giving $T_B = 71$ kN/m acting from right to left as shown in Figure 7.29.

Taking the first option as being nearer in spirit to the modern approach of applying a factor of safety uniformly to the soil strength, the values of σ_A and σ_Q which define the total stress distribution on the base of the wall are calculated as follows. For vertical equilibrium:

$$(T_{W,active} \div F) + W_1 + W_2 + (N_{W,passive} \times \sin 9.8° \div F)$$

$$-(T_{W,passive} \times \cos 9.8° \div F) = (\sigma_A \times 4.4\,\text{m}) + (0.5\{\sigma_Q - \sigma_A\} \times 4.4\,\text{m})$$

(Note that, viewing the factor F as it is not a destabilizing force. If we were to view the factor F as a strength mobilization factor or factor of safety on soil strength, the value of $T_{W,active}$ has been divided by F to account for the smaller adhesion mobilized on the soil/wall interface, $\tau_w = 0.5 \times \tau_{u,mobilized} = 0.5 \times \tau_u/F$). Substituting in the numerical values for F, $T_{W,active}$, W_1, W_2, $N_{W,passive}$ and $T_{W,passive}$:

$$(232\,\text{kN/m} \div 1.362) + 675\,\text{kN/m} + 322.5\,\text{kN/m}$$

$$+ (480\,\text{kN/m} \times 0.17 \div 1.362) - (91\,\text{kN/m} \times 0.985 \div 1.362)$$

$$= 1161.9\,\text{kN/m} = 4.4\,\text{m} \times \sigma_A + 2.2\,\text{m} \times (\sigma_Q - \sigma_A)$$

$$\Rightarrow \sigma_A + 0.5(\sigma_Q - \sigma_A) = 264.1\,\text{kPa} \tag{7.10}$$

Taking moments about Q,

$$(T_{W,active} \div F \times 4.4 \,\text{m}) + (W_1 \times 3.275 \,\text{m}) + (W_2 \times 1.43 \,\text{m})$$

$$+ (N_{W,passive} \div F \times [3 \,\text{m}/\cos 9.8°] \times 0.375) - [N_{W,active} \times F \times 9.5 \,\text{m} \div 3]$$

$$= [(\sigma_A \times 4.4 \,\text{m}) \times 2.2 \,\text{m}] + [0.5 \times \{\sigma_Q - \sigma_A\} \times 4.4 \,\text{m} \times 4.4 \,\text{m} \div 3]$$

Substituting in the numerical values,

$$(232 \,\text{kN/m} \div 1.362 \times 4.4 \,\text{m}) + (675 \,\text{kN/m} \times 3.275 \,\text{m})$$

$$+ (322.5 \,\text{kN/m} \times 1.43 \,\text{m}) + (480 \,\text{kN/m} \div 1.362 \times 1.142 \,\text{m})$$

$$- (335 \,\text{kN/m} \times 1.362 \times 3.167 \,\text{m})$$

$$= 2378.7 \,\text{kN} = (\sigma_A \times 9.68 \,\text{m}^2) + [\{\sigma_Q - \sigma_A\} \times 3.225 \,\text{m}^2]$$

$$\Rightarrow \sigma_A + 0.333(\sigma_Q - \sigma_A) = 245.7 \,\text{kPa} \tag{7.11}$$

Subtracting equation (7.11) from equation (7.10),

$$0.167(\sigma_Q - \sigma_A) = 18.37 \,\text{kPa}, \quad \text{or} \quad (\sigma_Q - \sigma_A) = 110.2 \,\text{kPa}$$

From equation (7.10),

$$\sigma_A = 209 \,\text{kPa}$$

Hence

$$\sigma_Q = 319 \,\text{kPa}$$

If the same calculation is carried out without reducing the active side soil/wall interface force $T_{W,active}$ by the factor $F = 1.362$, the calculated base bearing pressures are $\sigma_A = 265 \,\text{kPa}$ and $\sigma_Q = 291 \,\text{kPa}$.

If we had taken the first option of assuming that the forces behind the wall had reached the active limit (with fully mobilized soil strengths τ_u and wall adhesion $\tau_w = 0.5 \times \tau_u$), and reduced the forces in front of the wall $N_{W,passive}$ and $T_{W,passive}$ and the available base friction T_B by the factor 1.854, the calculated base bearing pressures would have been $\sigma_A = 350.4 \,\text{kPa}$ and $\sigma_Q = 206.4 \,\text{kPa}$.

The ability of the soil beneath the retaining wall to sustain the calculated base stresses without suffering a bearing capacity failure must be investigated using the methods described in sections 8.3–8.5 and 9.5. Note that the wall exerts a moment (evidenced by the non-uniform distribution of vertical stress) and a horizontal force on the underlying soil, as well as a vertical load. These must be taken into account in the calculation, as they may seriously reduce the bearing capacity of the wall foundation compared with the case of a purely vertical load. This point is addressed in sections 8.5 and 9.5.

It is possible that σ_A will be found to be negative (i.e. tensile) as initially calculated. If this occurs, the calculation must be repeated using the base pressure distribution shown in Figure 7.30, in which the unknowns are σ_Q and the extent of wall/soil separation x. For example, the base pressure distribution shown in Figure 7.30 will be found to apply if $T_{W,active}$ is set to zero and $N_{W,active}$ to 720 kN/m (corresponding to a flooded tension crack).

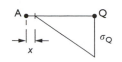

Figure 7.30 Amended base bearing pressure distribution when there is a tendency to tension near A.

The calculations just carried out have indicated that the wall is most vulnerable to failure or excessive movement by sliding. The Cricklewood retaining wall shown in Figure 7.23(b) did not actually fail until 1943, 41 years after its contruction in 1902. We will now investigate the long-term stability of the Cricklewood wall in terms of effective stresses, with pore water pressures corresponding to a steady-state seepage regime.

Example 7.6 Determining the forces exerted by the soil on the Cricklewood retaining wall in the long term, using the effective stress $(\tau/\sigma')_{max} = \tan\phi'_{crit}$ failure criterion

(a)   Behind the wall (active side)

The procedure is essentially similar to that for the short-term analysis, but in this case the pore water pressures and effective stresses must be considered separately. As in Example 7.4, the first step is to assume a failure mechanism, such as sliding along the plane AB, which is at a general angle θ (θ_1 in Figure 7.31) to the horizontal. The second step is to investigate the forces acting on the wedge of soil OAB. These now are:

1. The weight W of the wedge of soil itself (including the weight of water in the pores where present), which acts vertically downward.
2. The pore water reaction from the wall, U_W. This must be calculated from the pore water pressure distribution. U_W acts at right-angles to the wall, as water cannot take shear.

    At Cricklewood, weepholes were incorporated into the wall at a depth of approximately 9 m below ground level, in order to draw down the groundwater level in the vicinity of the wall, reducing the pore water pressures against it. This is in general sound engineering practice, provided that the weepholes are sufficiently close together to be effective, and can be relied on to continue to function over the design life (perhaps 120 years or more) of the wall. The reduction in pore water pressure in the retained soil will aid wall stability, but at the expense of increased effective stresses and possibly consolidation settlements. At a greenfield site, consolidation settlements due to groundwater lowering are unlikely to be of primary importance, but in an urban area the effect on nearby buildings must be considered.

In the present case, we will use the pore water pressures measured at Cricklewood to define the piezometric surface (along which the gauge pore water pressure is zero), as shown in Figure 7.31. This indicates that the weepholes were effective in reducing the water table at the wall to approximately 7 m below ground level. From Figure 7.31, assuming hydrostatic conditions below the water table, the pore water reaction from the wall is in this case equal to $0.5 \times (5.5\,\text{m} \times 10\,\text{kN/m}^3) \times 5.5\,\text{m} = 151.25\,\text{kN/m}$, taking the unit weight of water as $10\,\text{kN/m}^3$.

In general, long-term pore water pressures should be calculated by drawing a steady state seepage flownet. Also, pore water pressures are unlikely to be hydrostatic below the water table, due to the effects of seepage. In the case of the Cricklewood wall, however, the difference in hydraulic head between the two sides of the wall is not great, so the error introduced by assuming hydrostatic conditions behind the wall should be small.

3. The pore water reaction from the assumed rupture surface, U_R. In general, this must also be calculated from the steady state flownet, but in the present case we will use the idealized water table shown in Figure 7.31.

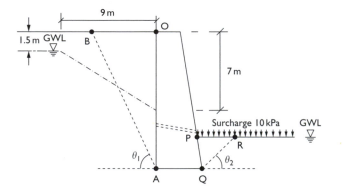

Figure 7.31 Idealized water table for Cricklewood retaining wall, showing trial wedges for long-term analysis.

Again assuming that pore water pressures are approximately hydrostatic below the water table, the pore water pressure at any point on the assumed slip surface is given by $\gamma_w \times$ the depth of the slip surface below the piezometric surface at that point.

For wedges such as OABO, where the rupture plane intersects the reduced water table before it has recovered to its original level, the pore water reaction from the assumed rupture surface is equal to $0.5 \times 55\,\text{kPa} \times$ the length of the rupture surface below the water table, which can be scaled off the space diagram (Figure 7.31 or Figure 7.33(a)). For wedges where the rupture plane intersects the water table at a distance of more than 9 m from the wall, account must be taken of the bilinear pore water pressure distribution on the rupture surface (see Figure 7.33(a) and Table 7.4 for details). The pore water reaction from the rupture surface acts at right-angles to the rupture surface.

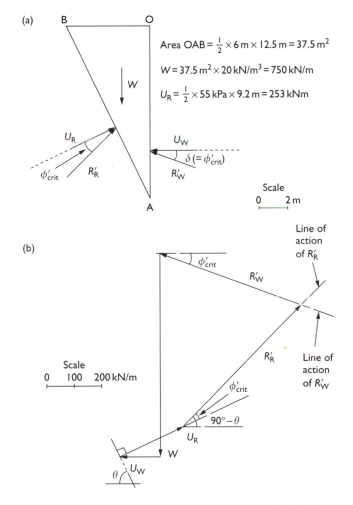

(a)

Area OAB $= \frac{1}{2} \times 6\,m \times 12.5\,m = 37.5\,m^2$

$W = 37.5\,m^2 \times 20\,kN/m^3 = 750\,kN/m$

$U_R = \frac{1}{2} \times 55\,kPa \times 9.2\,m = 253\,kNm$

Scale
0   2 m

(b)

Scale
0   100   200 kN/m

Figure 7.32 Forces acting on the sliding block of soil OABO on the retained side of the wall, long-term analysis (assumed failure surface is the plane AB). $\phi'_{crit} = 20°$.

4. The effective stress reaction from the wall, R'_W, which is unknown in magnitude but acts at an angle δ to the normal to the wall, where δ is the angle of soil/wall friction. We will here assume that the angle of soil/wall friction is equal to the critical state strength of the soil, ϕ'_{crit} (=20° in this case).

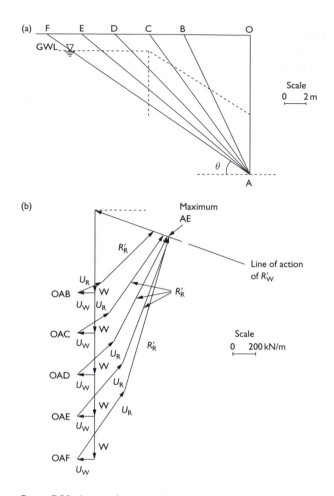

Figure 7.33 Array of potential sliding wedges and corresponding force polygons, retained side of wall, long-term analysis.

Table 7.4 Numerical values used in and obtained from Figure 7.33(b) (active side of wall, long-term analysis)

Block	Weight W (kN/m)	$\tan \theta$	θ (degrees)	$90° - \theta$ (degrees)	U_R (kN/m)	R'_W (kN/m) from Figure 7.33(b)
ABOA	750	12.5 ÷ 6	64.4	25.6	253	550
ACOA	1125	12.5 ÷ 9	54.2	35.8	330	640
ADOA	1500	12.5 ÷ 12	46.2	43.8	479[a]	670
AEOA	1875	12.5 ÷ 15	39.8	50.2	625[b]	700
AFOA	2250	12.5 ÷ 18	34.8	55.2	758[c]	690

Notes
a calculated as $[13\,\text{m} \times 0.5 \times (55\,\text{kPa} + 16\,\text{kPa})] + [2.2\,\text{m} \times 0.5 \times 16\,\text{kPa}]$
b calculated as $[11.8\,\text{m} \times 0.5 \times (55\,\text{kPa} + 35\,\text{kPa})] + [5.4\,\text{m} \times 0.5 \times 35\,\text{kPa}]$
c calculated as $[11\,\text{m} \times 0.5 \times (55\,\text{kPa} + 47\,\text{kPa})] + [8.4\,\text{m} \times 0.5 \times 47\,\text{kPa}]$ to allow for bilinear pore water pressure distribution on rupture surface.

This is quite reasonable for a wall which is rough in comparison with the representative soil particle size (D_{50}), provided that there is sufficient relative movement at the interface, in the appropriate direction. However, codes of practice generally recommend the use of soil/wall friction angles less than ϕ' in design. Soil/wall friction is discussed more fully in sections 7.9 and 7.10.

5. The effective stress reaction from the rupture surface, R'_R, which is again unknown in magnitude but acts at an angle ϕ' (=20° in this case) to the normal to the rupture surface.

For the long-term analysis, the forces acting on the active-side wedge of soil OABO are shown in Figure 7.32(a). The polygon of forces is shown in Figure 7.32(b).

As before, the most critical failure surface is that associated with the largest lateral thrust, and it is necessary to repeat the calculation with different trial wedges until this has been found.

Figure 7.33(a) shows the array of sliding wedges investigated, and Figure 7.33(b) shows the corresponding force polygons. The values of W, θ, U_R and R'_W used in and obtained from each force polygon are detailed in Table 7.4.

From Figure 7.33 and Table 7.4, the maximum active lateral effective thrust is approximately 700 kN/m, occurring when $\theta \approx 39.8°$. (For a vertical, frictionless wall retaining dry soil with a horizontal retained surface and no surcharge, the critical slip plane in active conditions is at an angle of $45° + \phi'_{crit}/2$ (=55° in this case) to the horizontal: e.g. Bolton (1991).)

(b)   In front of the wall (passive side)

As with the undrained analysis, the procedure must be repeated for the soil in front of the retaining wall.

The shear forces on the rupture plane and at the soil/wall interface again act so as to resist the supposed soil movement, giving effective stress reactions R'_W and R'_R acting in the directions shown in Figure 7.34(a). The ballast that supports the railway tracks has again been modelled by a uniform surcharge of 10 kPa.

The pore water pressure reaction from the wall U_W is $0.5 \times \gamma_w \times 3 \text{ m} \times (3 \text{ m}/\cos 9.8°) = 45.7 \text{ kN/m}$, taking $\gamma_w = 10 \text{ kN/m}^3$. The pore water reaction from the rupture U_R is $30 \text{ kPa} \times 0.5 \times$ the length of the rupture, which is $3 \text{ m}/\sin\theta$ where θ is the angle of inclination of the rupture to the horizontal (Figure 7.34(a)). Figure 7.34(b) shows that the horizontal components of U_R and U_W cancel each other out exactly: this must be the case if the pore water pressures are hydrostatic below a level water table. The forces acting on a typical wedge PQRP are shown in Figure 7.34(a), and the corresponding polygon of forces is shown in Figure 7.34(b).

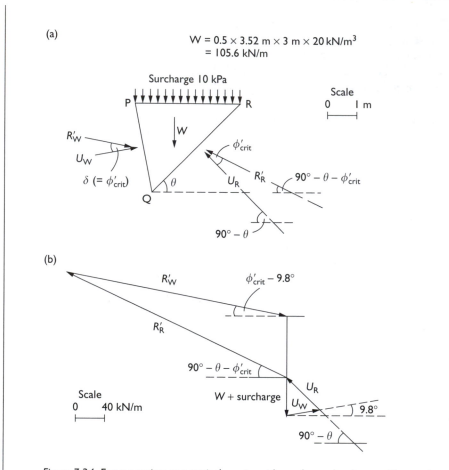

(a)

$$W = 0.5 \times 3.52\,\text{m} \times 3\,\text{m} \times 20\,\text{kN/m}^3$$
$$= 105.6\,\text{kN/m}$$

Surcharge 10 kPa

Figure 7.34 Forces acting on a typical passive-side wedge and polygon of forces, long-term analysis ($\phi'_{crit} = 20°$).

It is again necessary to find the minimum lateral thrust that the passive wedge can be relied upon to provide.

The sequence of trial wedges and the force polygons associated with each wedge are shown in Figure 7.35, from which the critical rupture plane can be identified. The values of W, θ, U_R and R'_W used in and calculated from Figure 7.35 are given in Table 7.5.

From Figure 7.35(b) and Table 7.5, the minimum passive thrust is approximately $210\,\text{kN/m}$, occurring when $\theta \approx 23°$. (For a vertical, frictionless wall in dry soil with a level surface and no surcharge, the critical passive-side slip plane would be at an angle of $45 - \phi'_{crit}/2$ ($=35°$ in this case) to the horizontal: e.g. Bolton (1991).)

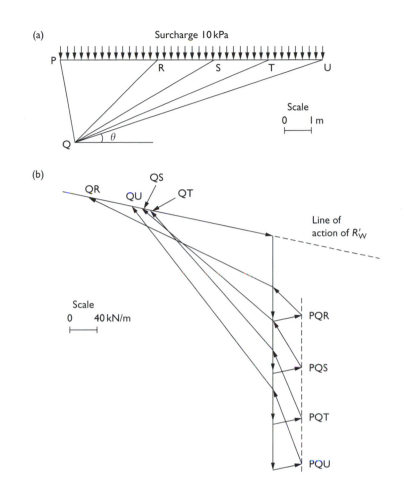

Figure 7.35 Graphical construction for 'least optimistic' sliding wedge failure mechanism: succession of trial wedges and superposition of force polygons (passive side, long-term analysis).

Table 7.5 Numerical values used in and obtained from Figure 7.35(b) (passive side of wall, long-term analysis)

Block	Weight W (kN/m)	Force due to surcharge (kN/m)	$\tan\theta$	θ (degrees)	$90° - \theta$ (degrees)	$U_R = 45/\sin\theta$ (kN/m)	R'_W (kN/m) from Figure 7.35(b)
PQRP	105	35	$3\div 3$	45	45	64	305
PQSP	165	55	$3\div 5$	31	59	87	215
PQTP	225	75	$3\div 7$	23	67	114	210
PQUP	285	95	$3\div 9$	19	71	135	230

Example 7.7 Investigating the long-term stability of the Cricklewood wall

We can again investigate the stability of the wall in terms of failure by toppling and sliding, with reference to the free body diagram shown in Figure 7.36.

It is immediately obvious that the wall is much less stable in the long term, because the active (destabilizing) thrust calculated is larger than in the short-term analysis, while the passive (resisting) thrust is much smaller.

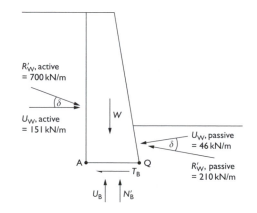

Figure 7.36 Free body diagram for the wall.

(a)   Sliding

The safety of the wall against sliding can again be investigated by comparing the magnitude of the lateral active thrust with the total available resistance, which comes from the passive pressure of the soil in front of the wall and the shear force T_B along the base. In an effective stress analysis, the maximum base shear force $T_{B\,max}$ depends on the effective normal reaction on the base of the wall, N'_B. N'_B may be calculated from the condition of vertical equilibrium,

$$[R'_{W,active} \times \sin \delta] + W_1 + W_2 - [R'_{W,passive} \times \sin(\delta - 9.8°)]$$

$$+ [U_{W,passive} \times \sin 9.8°] = N'_B + [0.5 \times (u_A + u_Q) \times 4.4 \, \text{m}]$$

where $0.5 \times (u_A + u_Q) \times 4.4 \, \text{m}$ is the pore water reaction on the base U_B, calculated according to the idealized pore water pressure distribution shown in Figure 7.37.

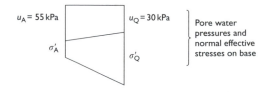

Figure 7.37 Assumed distributions of pore water pressure and effective stress on the base of the retaining wall, long-term analysis.

Substituting the numerical values already calculated (with $\delta = \phi'_{crit} = 20°$),

$$(700\,\text{kN/m} \times 0.342) + 675\,\text{kN/m} + 322.5\,\text{kN/m} - (210\,\text{kN/m} \times 0.177)$$

$$+ (46\,\text{kN/m} \times 0.17) = 1207.6\,\text{kN/m} = N'_B + 2.2\,\text{m} \times (55 + 30)$$

or $N'_B = 1021\,\text{kN/m}$. Thus the maximum base shear force $T_{B\,max} = N'_B \times \tan\delta = 371.6\,\text{kN/m}$ (with $\delta = 20°$).

The total horizontal thrust exerted by the retained soil and groundwater is

$$U_{W,active} + R'_{W,active} \times \cos\delta = 151\,\text{kN/m} + 700\,\text{kN/} \times \text{m} \times \cos 20°$$

$$= 809\,\text{kN/m}$$

The lateral component of the passive side forces $U_{W,passive}$ and $R'_{W,passive}$ is

$$(U_{W,passive} \times \cos 9.8°) + (R'_{W,passive} \times \cos(\delta - 9.8°))$$

$$= 46\,\text{kN/m} \times \cos 9.8° + 210\,\text{kN/m} \times \cos 10.2° = 252\,\text{kN/m}$$

The available resisting force of $(252 + 372)\,\text{kN/m} = 624\,\text{kN/m}$ is now significantly less than the destabilizing active thrust of 809 kN/m, with a ratio of $627/809 = 0.77$. This means that even the maximum possible resistance is insufficient to prevent sliding: equilibrium of the wall cannot be maintained under the action of the active and passive forces calculated.

(b) Toppling

The safety of the wall against toppling may again be investigated by comparing the resisting and disturbing moments about Q, ignoring the effective stress reaction on the base, N'_B, which would be zero if the wall were on the verge of overturning. The pore water pressure reaction might also reasonably be ignored in a clay soil, but should probably be included in the calculation in the case of a sand: it is a question of whether water would be able to fill the opening gap between the base of the wall and the underlying soil quickly enough to maintain the equilibrium pore water pressures.

The destabilizing moment about Q (ignoring U_B) is

$$(U_{W,active} \times 5.5\,\text{m}/3) + (R'_{W,active} \times \cos\delta \times [12.5\,\text{m}/3])$$

$$= (151\,\text{kN/m} \times 5.5\,\text{m}/3) + (700\,\text{kN/m} \times \cos 20° \times [12.5\,\text{m}/3])$$

$$= 3018\,\text{kNm/m}$$

(12.5 m/3 is a slight underestimate of the lever arm, as the lateral effective stress distribution will be approximately bilinear, with the rate of increase of lateral stress with depth greater above the groundwater level than below it.)

The resisting moment about Q is

$$(W_1 \times 3.275\,\text{m}) + (W_2 \times 1.43\,\text{m}) + (R'_{W,active} \times \sin\delta \times 4.4\,\text{m})$$

$$\times (U_{W,passive} \times 3\,\text{m}/3 \cos 9.8°) + (R'_{W,passive} \times \cos\delta \times 0.375 \times 3\,\text{m}/\cos 9.8°)$$

$$= (675 \, \text{kN/m} \times 3.275 \, \text{m}) + (322.5 \, \text{kN/m} \times 1.43 \, \text{m}) + (700 \, \text{kN/m} \times 0.342$$
$$\times 4.4 \, \text{m}) + (46 \, \text{kN/m} \times 1.015 \, \text{m}) + (146 \, \text{kN/m} \times 0.94 \times 1.142 \, \text{m})$$
$$= 3928 \, \text{kNm/m}$$

assuming that the resultant passive thrust again acts at a height of $0.375 \times d$ above the toe of the wall, where $d = 3 \, \text{m}$ is the depth of soil in front of the wall. The calculation shows that the available resisting moment is greater than the overturning moment, so that the wall would not be expected to fail by toppling. However, this is immaterial because the horizontal equilibrium calculation indicates that the wall should have been expected to fail by sliding some time after construction, probably before steady state pore water pressures had been established.

It is in principle also necessary to check that the vertical effective stresses along the base AQ will not be large enough to cause a bearing capacity failure. In the present case, however, there is no point, because the wall cannot be in equilibrium under the combined action of the limiting soil forces calculated.

In a true upper bound solution, the mechanism analysed must be **kinematically admissible** – that is, able physically to occur. In addition to the requirements of normality, other restraints may be imposed by the form of the structure under consideration. Kinematic admissibility should be investigated by drawing a vector diagram showing the velocities of all of the components of the mechanism, in terms of a reference velocity assigned to one of the components. Such a diagram is known as a **velocity diagram** or a **hodograph**.

In examples 7.4 and 7.6, the assumed sliding blocks all remain undistorted. Sometimes it is necessary for a block to distort by shearing as it moves in order to maintain kinematic admissibility. If this occurs, additional energy will be dissipated within the block, as illustrated in section 9.9.

Figure 7.38 shows the most critical failure mechanisms identified for the Cricklewood retaining wall, in both the short-term and the long-term conditions. To construct the hodograph, it is necessary to imagine that the velocity of the sliding block behind the wall has a certain magnitude v_0, in terms of which the magnitudes of all other velocities can then be calculated. Hodographs for the two mechanisms are shown in Figure 7.39.

We could have analysed the assumed failure mechanisms by equating the rate at which work is done by the external forces to the rate at which internal energy is dissipated during a small movement of the mechanism, with the relative rates of displacement determined from the hodographs (as in section 9.9). Repetition of the hodograph calculation to find the most critical mechanism is, however, rather more tedious than the superposition of force polygons used in the equilibrium approach (Figures 7.26, 7.28, 7.33(b) and 7.35(b)). Also, provided that the mechanism is real (i.e. is kinematically admissible), the answer obtained using the limit equilibrium approach should be the same.

Apart from its inherent optimism (i.e. solutions tend to err on the unsafe side), the main shortcoming of the mechanism analysis is that it gives a resultant thrust rather

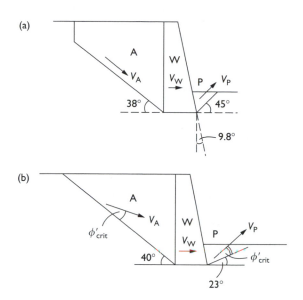

Figure 7.38 Critical failure mechanisms for Cricklewood wall. (a) Short- and (b) long-term analyses.

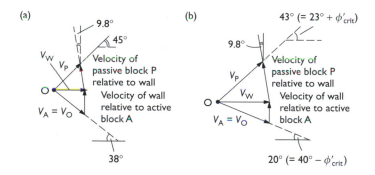

Figure 7.39 Hodographs for Cricklewood retaining wall. (a) Short-term mechanism; (b) long-term mechanism, assuming zero dilation at soil/wall interfaces.

than a distribution of pressure on the retaining wall. This can be overcome to some extent by assuming that the calculated thrust results from a stress distribution which increases linearly with depth down the wall. Indeed, having obtained the resultant thrust, it was necessary to make this assumption (together with a suitable modification for the effect of the surcharge in front of the wall) in order to proceed with the investigation of the safety of the wall against toppling.

The mechanism or equilibrium of blocks approach is particularly useful for dealing with features such as line loads and strip surcharges running parallel to the wall, which cannot easily be taken into consideration in a stress field analysis.

In both the short-term and the long-term cases, we have only investigated failure mechanisms comprising plane slip surfaces. If in reality the shape of the slip surface is different, our solution will err on the unsafe side. This is generally more likely to be significant on the passive side (i.e. in front) of a retaining wall than on the active side (i.e. behind).

7.8 Limit equilibrium stress distributions for embedded retaining walls

7.8.1 Long-term pore water pressures

The long-term stability of permanent structures must be investigated by means of an effective stress analysis. It is necessary therefore at the design stage to estimate the long-term equilibrium pore water pressure distribution around a retaining wall. In the case of an embedded wall in a uniform soil, where the wall is effectively impermeable and weepholes or other drainage measures behind the wall are not installed, the long-term equilibrium pore water pressure distribution will correspond to steady state seepage from a high water table behind the wall to a reduced groundwater level in front. The pore water pressure distribution may be obtained by sketching a flownet, but an approximation in which the fall in hydraulic head is assumed to be distributed linearly around the wall is often close enough for design purposes (Figure 7.40; Symons, 1983).

The idealized pore water pressure distribution shown in Figure 7.40 is known as the **linear seepage approximation**. It is only applicable to wide excavations in fairly uniform soils. If the excavation is narrow (cf. Figures 3.20 and 3.21), strong upward seepage between the two retaining walls will lead to much higher pore water pressures than indicated by the linear seepage approximation. In these conditions, there is a danger of boiling or piping, which is best investigated by means of a flownet.

The presence of sand partings or other more permeable layers could lead to higher pore water pressures on the back of the wall than indicated by Figure 7.40. If the clay is underlain by an aquifer in which pore water pressures remain high following excavation, this could also adversely affect the pore water pressure distribution. On the other hand, the provision of weepholes and an effective drainage system behind the wall could reduce pore water pressures in the retained ground, provided that any settlements that this might cause can be tolerated. However, the installation of effective wall drainage may only be a realistic option in the case of backfilled walls or those retaining relatively free-draining soils.

In summary, it cannot be emphasized too strongly that although the linear seepage approximation can be useful in the analysis of wide excavations in uniform ground, it is not necessarily a substitute for sketching a flownet.

7.8.2 Unpropped embedded walls: fixed earth support conditions

Unpropped embedded walls rely entirely for their stability on an adequate depth of embedment: they are not supported in any other way. They will tend to fail by rotation about a **pivot point** near the toe. The idealized stress distribution at failure is shown in Figure 7.41. Active and passive zones develop where the wall moves away from and into the soil, respectively.

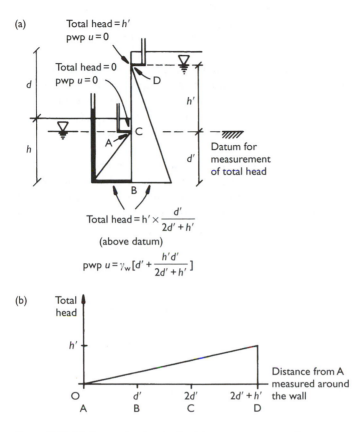

(a) Total head = h'
pwp $u = 0$

Total head = 0
pwp $u = 0$

D

h'

Datum for
measurement
of total head

A

C

d'

B

Total head = $h' \times \dfrac{d'}{2d' + h'}$

(above datum)

pwp $u = \gamma_w [d' + \dfrac{h'd'}{2d' + h'}]$

(b)

Total
head

h'

Distance from A
measured around
the wall

O d' $2d'$ $2d' + h'$
A   B   C   D

Figure 7.40 (a) Approximate steady state seepage pore water pressure distribution, corresponding to (b) fall in hydraulic head distributed linearly around the wall.

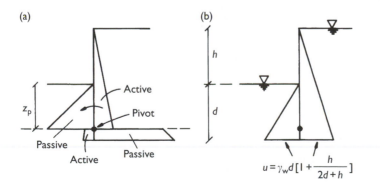

(a)

z_p

Active

Pivot

Passive

Active   Passive

(b)

h

d

$u = \gamma_w d [1 + \dfrac{h}{2d + h}]$

Figure 7.41 Idealized stress distribution for an unpropped embedded cantilever wall at failure: (a) effective stresses; (b) pore water pressures.

The conditions represented by the stress distribution in Figure 7.41(a) are sometimes known as **fixed earth support**, because the depth of embedment must be large enough to prevent free translational movement of the toe. The wall acts as an unpropped cantilever, built into the ground.

Given the retained height h and the soil strength ϕ', there are two unknowns that must be calculated in an investigation of conditions at collapse. These are the depth of embedment, d, required just to prevent collapse, and the depth of the pivot point (about which the wall can be imagined to rotate) below formation level, z_p. The equations of horizontal and moment equilibrium can be used to find these two unknowns, so the system is statically determinate.

If the linear approximation to the steady state pore water pressure distribution is used, the two equilibrium equations are simultaneous and quartic in the two unknowns, and it is necessary to adopt an iterative solution technique such as that outlined by Bolton and Powrie (1987). The inconvenience of the iterative solution in the days before personal computers and programmable calculators led to the development of an approximation to the exact calculation, in which the resultant of the stresses below the pivot point is replaced by a single point force Q acting at the pivot (Figure 7.42).

The portion of the wall below the pivot does not feature in the approximate analysis. The two unknowns are now the depth to the pivot z_p and the equivalent point force Q. Solution is simpler in this case, since moments can be taken about the pivot, eliminating Q from the moment equilibrium equation. The value obtained for z_p is increased by an empirical factor (usually 1.2) to arrive at the overall depth of embedment d. This factor of 1.2 is nothing to do with distancing the wall from collapse (i.e. it is *not* a factor of safety), but is necessary because the calculation is incorrect. If the simplified

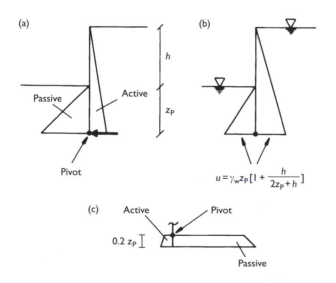

Figure 7.42 Approximate stress analysis for unpropped walls: (a) effective stresses; (b) pore water pressures; (c) check that added depth can mobilize at least the required point load Q.

procedure is used, then a check should be carried out to ensure that the added depth is sufficient to mobilize at least the calculated value of Q (Figure 7.42(c)).

7.8.3 Embedded walls propped at the crest: free and fixed earth support conditions

If the possibility of a structural failure of the wall of the props is neglected, an embedded wall propped at the crest will tend to fail by rigid-body rotation about the position of the prop, as discussed in section 7.6. The idealized effective stress distribution at failure is reproduced in Figure 7.43(a). Pore water pressures according to the linear seepage model are shown in Figure 7.43(b).

The conditions giving rise to the effective stress distribution shown in Figure 7.43(a) are sometimes known as **free earth support**, because the toe of the wall is relatively free to move laterally. In other words no fixity is developed at the toe. In this case, the two unknowns are the prop force F and the depth of embedment d required just to prevent failure. As in section 7.6, the depth of embedment d can be calculated by taking moments about the prop, and F then follows from the condition of horizontal force equilibrium. The introduction of a factor of safety or a strength mobilization factor to calculate the design value of the depth of embedment has already been mentioned, and is described in more detail in section 10.2.

Some authors (e.g. Scott, 1969; Williams and Waite, 1993) describe the use of a 'fixed earth support' calculation for a propped cantilever wall. The idealized and simplified effective stress distributions are shown in Figure 7.44.

Although in a rigid wall, this stress distribution might correspond to a mechanism of failure involving the formation of a plastic hinge at the point of maximum bending moment, this is not really the intention. The approach is usually adopted in an attempt to take account of the effects of wall bending, which may result in a reversal in the sign of the bending moment if the wall is sufficiently flexible. For stiff walls in clay soils, the use of the fixed earth support analysis is not generally considered appropriate (Gaba *et al.*, 2003).

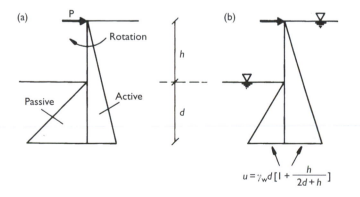

$$u = \gamma_w d \left[1 + \frac{h}{2d+h} \right]$$

Figure 7.43 Idealized stress distribution at failure for a stiff wall propped rigidly at the crest; (a) effective stresses; (b) pore water pressures.

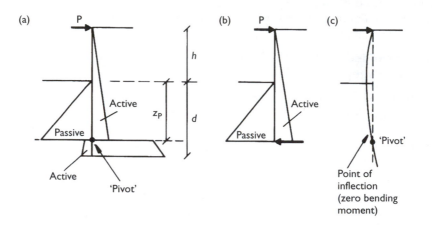

Figure 7.44 'Fixed earth support' effective stress distributions and deformations for a propped embedded wall: (a) idealized stresses; (b) simplified stresses; (c) deformed shape.

In the absence of a true plastic hinge (which would define the wall bending moment at this point), both the idealized and the simplified stress distributions shown in Figure 7.44 are statically indeterminate. To calculate the prop force and the depth of embedment, it is necessary to assume, for example, that the point of contraflexure, at which the bending moment is zero, occurs at the level where the net pressure acting on the wall is zero (Williams and Waite, 1993).

The stress distribution shown in Figure 7.44 would correspond to the 'correct' failure mechanism for a propped wall where the prop yields at a constant load. Such a system is statically determinate, provided that the prop yield load is known.

7.9 Soil/wall friction

The simple lower bound solutions developed in section 7.6 are too conservative for use in design, because the effects of soil/wall friction or adhesion were neglected. The generation of shear stresses at the soil/wall interface is usually of considerable benefit to wall stability.

According to Rowe (1963), the maximum soil/wall friction angle that can possibly be developed, δ_{max}, has two components,

$$\delta_{max} = \phi_w + r \tag{7.12}$$

where ϕ_w is the true friction angle between the soil grains and the material of the wall, and r is the wall roughness angle. Rowe notes that ϕ_w is generally close to ϕ_μ, the true friction angle between soil grains.[1] The angle of soil/wall friction δ is unaffected by the component of soil strength due to dilatancy, except in so far as the peak angle of shearing resistance ϕ'_{peak} represents an extreme upper limit. For a relatively smooth[2] wall, r could be of order 0.5°: real walls made from steel sheet piles, concrete or timber would be expected to be significantly rougher. In general, the roughness of the wall will compensate approximately for the difference between the critical state angle of shearing resistance ϕ'_{crit} and the true friction angle between the soil grains ϕ'_μ ($\phi'_{crit} - \phi'_\mu \approx 5°$ for quartz sand), so that for practical purposes $\delta_{max} \approx \phi'_{crit}$.

Alternatively, it might be argued that a surface on which there is insignificant opportunity for dilation, but whose roughness is comparable with the typical particle size of the soil (D_{50}), would be expected to mobilize $\delta_{max} = \phi'_{crit}$. If the wall is very rough, so that a rupture surface is forced to develop within the body of the soil rather than exactly along the interface, some or all of the dilatant strength of the soil might be mobilized, giving an upper limit of $\delta_{max} = \phi'_{peak}$. In laboratory tests in modified shearboxes, this is only apparent when the soil is compacted against the wall material, (Subba Rao *et al.*, 1998). If the soil adjacent to the wall is loosened or disturbed by wall installation, the upper limit to the soil/wall friction angle is $\delta_{max} = \phi'_{crit}$. In general, it is appropriate to take $\delta_{max} = \phi'_{crit}$.

Rowe and Peaker (1965) show that the wall friction actually developed depends on the direction and degree of wall movement. They suggest that in carrying out an analysis of conditions at collapse, the wall friction angle δ_{mob} mobilized at a wall movement of 5% of the retained height (which is actually quite large) should be used. For sheet pile walls freely embedded in sand, where the relative movement between the wall and the soil in front of it is likely to be such that the wall moves downward relative to the soil, they recommend $\delta_{mob} = \delta_{max}$ on the passive side, where δ_{max} is given by equation (7.12). For sheet pile walls bearing on rock, however, where the likely magnitude of downward wall movement is small, significant friction is less likely to be developed and the recommended values of δ_{mob} are much reduced.

The same arguments regarding the roughness of the wall relative to the particle size of the soil would apply in principle to a total stress analysis. In design, the soil/wall adhesion τ_w is often assumed to be a factor of approximately 2 smaller than the undrained shear strength of the soil τ_u (i.e. $\tau_w = \alpha \times \tau_u$, where $\alpha \approx 0.5$). This is primarily to account for softening of the soil at the soil/wall interface during wall construction or installation, and to allow for the fact that the relative motion between the wall and the soil may be insufficient to mobilize the full shear strength of the interface.

7.10   Earth pressure coefficients taking account of shear stresses at the soil/wall interface

7.10.1   Effects of soil/wall friction

The earth pressure coefficients derived in section 7.3 for a frictionless wall will lead to uneconomical designs when the wall is rough. The stress field analyses are extended in section 9.6 to account for wall friction (in the case of the $(\tau/\sigma')_{max} = \tan \phi'_{crit}$ failure criterion) or adhesion (in the case of the $\tau_{max} = \tau_u$ failure criterion). It is common, however – particularly in an effective stress analysis – to use tabulated values of earth pressure coefficients, such as those given by Caquot and Kerisel (1948), or Kerisel and Absi (1990).

It was suggested in section 7.9 that the surface of the wall will generally be rough in comparison with the typical soil particle size, so that full wall friction $\delta = \phi'_{crit}$ might in many cases be expected to be mobilized following sufficient displacement. For embedded retaining walls that are either unpropped or propped near the crest, the onset of large deformations may often be reasonably well-predicted by the limit equilibrium stress distributions shown in Figures 7.41 and 7.43, using the earth pressure coefficients given by Caquot and Kerisel (1948) based on critical state angles of soil friction, with wall friction angle $\delta = \phi' = \phi'_{crit}$ on both sides of the wall (Bolton and Powrie, 1987; Powrie, 1996).

Codes of practice have traditionally advocated the use of values of soil/wall friction angles δ which are somewhat less than the soil strength ϕ'. This is partly because such advice dates back to the days when the differences between peak and critical state strengths were not fully appreciated: if ϕ'_{peak} were used as a design parameter, an assumed soil/wall friction angle of $\delta = \phi'_{peak}$ would be unrealistically high in most circumstances. The assumption $\delta < \phi'$ also takes account of the fact that the relative movement between the soil and the wall may be insufficient (or in the wrong direction) to generate full soil/wall friction. For thin walls (e.g. sheet piles), the generation of full friction on both sides of the wall may well be incompatible with vertical equilibrium of the wall.

The advice given by successive codes of practice concerning the degree of wall friction allowable in design is not consistent. Terzaghi (1954) recommends $\delta = \phi'/2$ behind the wall and $\delta = 2\phi'/3$ in front. Gaba et al. (2003) and Eurocode 7 (BSI, 1995) give limiting values of $\delta_{max} = \phi'_{crit}$ for rough (cast in situ) concrete, and $\delta_{max} = 0.67\phi'_{crit}$ for smooth (precast) concrete or sheet piling supporting sand and gravel. Gaba et al. (2003) state that design values of δ should be selected on the basis of factors including the wall roughness, the magnitude and direction of relative soil/wall movement, and the requirement for vertical equilibrium of the wall. BS 8002 (BSI, 2001) recommends $\tan \delta = 0.75 \times \tan \phi'_{mob}$ (which equates to δ/ϕ'_{mob} in the range 0.76 to 0.8 for $15° \leq \phi'_{mob} \leq 40°$), both behind and in front of the wall. ϕ'_{mob} is the soil strength that must be mobilized to maintain equilibrium of the wall.

Caquot and Kerisel's (1948) tables suggest that the calculation of earth pressure coefficients using $\delta = \phi'_{mob}/2$ (rather than $\delta = \phi'_{mob}$) is equivalent to an additional strength mobilization factor (i.e. an additional factor of safety on soil strength) of approximately 1.11 over the range $18° < \phi'_{mob} < 36°$. Using $\delta = 2\phi'_{mob}/3$, the additional factor of safety is approximately 1.05. Owing to the non-linearity of the relationship, however (Rowe and Peaker, 1965), the additional factor of safety with $\tan \delta = 0.75 \cdot \tan \phi'_{mob}$ on the passive side is very small, decreasing from 1.02 to less than 1.01 as ϕ'_{mob} increases from 15° to 35°. On the active side with $\tan \delta = 0.75 \cdot \tan \phi'_{mob}$, the additional factor of safety is about 1.03.

In carrying out a design using a code of practice, the specified procedure must be followed in its entirety, because otherwise there is a danger that some component of the overall factor of safety intended (e.g. due to the use of a reduced soil/wall friction angle) may be omitted.

For vertical walls and level backfills, active and passive earth pressure coefficients $\sigma'_h/(\gamma z - u)$ with $\delta = \phi', \tan \delta = 0.75 \times \tan \phi', \delta = 0.67\phi', \delta = 0.5\phi'$ and $\delta = 0$ (equations (7.3) and (7.4)) are reproduced in Tables 7.6 and 7.7. It is assumed in these tables that on the active side, the soil moves downward relative to the wall, while on the passive side, the soil moves upward. If these directions of relative soil/wall movement are reversed, the active earth pressure coefficients are increased, and the passive pressure coefficients are reduced very considerably. This emphasizes the need to consider

Table 7.6 Active earth pressure coefficients $\sigma_h'/(\gamma z - u)$ calculated using the method given by Sokolovskii (1960) for various angles of soil/wall friction angle δ (vertical wall and level backfill, component of reaction normal to the wall)

ϕ', degrees	K_a with $\delta = 0$ (eqn (7.3))	K_a with $\delta = \phi'/2$	K_a with $\delta = 2\phi'/3$	K_a with $\tan \delta = 0.75 \times \tan \phi'$ (BS8002)	K_a with $\delta = \phi'$
12	0.6558	0.6112	0.6003	0.5952	0.5842
13	0.6327	0.5870	0.5758	0.5706	0.5593
14	0.6104	0.5638	0.5524	0.5470	0.5355
15	0.5888	0.5416	0.5300	0.5244	0.5128
16	0.5678	0.5202	0.5085	0.5028	0.4910
17	0.5475	0.4996	0.4879	0.4821	0.4702
18	0.5279	0.4799	0.4681	0.4622	0.4503
19	0.5088	0.4609	0.4491	0.4431	0.4312
20	0.4903	0.4426	0.4308	0.4248	0.4129
21	0.4724	0.4250	0.4133	0.4072	0.3954
22	0.4550	0.4081	0.3964	0.3903	0.3786
23	0.4381	0.3918	0.3802	0.3740	0.3624
24	0.4217	0.3760	0.3647	0.3584	0.3470
25	0.4059	0.3609	0.3497	0.3434	0.3321
26	0.3905	0.3463	0.3352	0.3289	0.3178
27	0.3755	0.3322	0.3213	0.3150	0.3041
28	0.3610	0.3187	0.3080	0.3016	0.2909
29	0.3470	0.3056	0.2951	0.2887	0.2783
30	0.3333	0.2930	0.2827	0.2763	0.2661
31	0.3201	0.2808	0.2707	0.2644	0.2543
32	0.3073	0.2691	0.2592	0.2528	0.2431
33	0.2948	0.2577	0.2481	0.2417	0.2322
34	0.2827	0.2468	0.2374	0.2311	0.2218
35	0.2710	0.2362	0.2270	0.2207	0.2117
36	0.2596	0.2260	0.2171	0.2108	0.2021
37	0.2486	0.2161	0.2074	0.2012	0.1927
38	0.2379	0.2066	0.1982	0.1920	0.1838
39	0.2275	0.1974	0.1892	0.1831	0.1751
40	0.2174	0.1885	0.1806	0.1745	0.1668

the likely magnitude and direction of relative soil/wall movement, before invoking interface friction to help support the wall, as pointed out by Rowe and Peaker (1965).

The earth pressure coefficients given in Tables 7.6 and 7.7 were calculated by Richard and John Harkness using the method proposed by Sokolovskii (1960). They are generally consistent with the earth pressure charts given in Figures A1 and A2 in Annex A of BS8002 (BSI, 2001) and with the tables by Kerisel and Absi (1990). Discrepancies between these values and those given by Caquot and Kerisel (1948) are generally less than 5% for the passive case and 0.5% for the active case when ϕ' is less than 30°.

Annex G of Eurocode 7 (BSI, 1995), and Appendix A6 of Gaba *et al.* (2003) give charts of active and passive earth pressure coefficients calculated using a form of equations (9.23) and (9.27). Numerical values of these earth pressure coefficients, which are slightly more conservative than those given in Tables 7.6 and 7.7 (i.e. the

Table 7.7 Passive earth pressure coefficients $\sigma_h'/(\gamma z - u)$ calculated using the method given by Sokolovskii (1960) for various angles of soil/wall friction angle δ (vertical wall and level backfill, component of reaction normal to the wall)

ϕ', degrees	K_p with $\delta = 0$ (eqn (7.4))	K_p with $\delta = \phi'/2$	K_p with $\delta = 2\phi'/3$	K_p with $\tan \delta = 0.75 \times \tan \phi'$ (BS8002)	K_p with $\delta = \phi'$
12	1.5250	1.6993	1.7458	1.7674	1.8128
13	1.5805	1.7811	1.8351	1.8605	1.9130
14	1.6383	1.8680	1.9303	1.9600	2.0204
15	1.6984	1.9603	2.0320	2.0665	2.1357
16	1.7610	2.0585	2.1406	2.1807	2.2596
17	1.8263	2.1630	2.2568	2.3033	2.3931
18	1.8944	2.2743	2.3814	2.4351	2.5370
19	1.9655	2.3932	2.5150	2.5770	2.6925
20	2.0396	2.5203	2.6586	2.7302	2.8608
21	2.1171	2.6562	2.8132	2.8958	3.0433
22	2.1980	2.8019	2.9799	3.0751	3.2414
23	2.2826	2.9583	3.1599	3.2696	3.4571
24	2.3712	3.1264	3.3546	3.4812	3.6923
25	2.4639	3.3073	3.5657	3.7116	3.9493
26	2.5611	3.5024	3.7949	3.9633	4.2309
27	2.6629	3.7131	4.0443	4.2388	4.5399
28	2.7698	3.9411	4.3163	4.5411	4.8801
29	2.8821	4.1883	4.6135	4.8736	5.2553
30	3.0000	4.4568	4.9390	5.2404	5.6704
31	3.1240	4.7491	5.2963	5.6460	6.1309
32	3.2546	5.0679	5.6896	6.0960	6.6432
33	3.3921	5.4164	6.1237	6.5967	7.2148
34	3.5371	5.7983	6.6039	7.1557	7.8547
35	3.6902	6.2178	7.1369	7.7818	8.5734
36	3.8518	6.6798	7.7302	8.4855	9.3836
37	4.0228	7.1901	8.3927	9.2796	10.300
38	4.2037	7.7551	9.1350	10.179	11.341
39	4.3955	8.3838	9.9687	11.202	12.529
40	4.5989	9.0823	10.912	12.371	13.889

active earth pressure coefficients are larger and the passive earth pressure coefficients are smaller) may be found in Tables 9.3 and 9.4 (section 9.6) of this book.

The influence of soil/wall friction on the stability of embedded walls is illustrated in Figure 7.45. Figure 7.45 gives the results of the limit equilibrium calculation shown in Figure 7.41 for unpropped cantilever walls retaining a soil of unit weight $\gamma = 2\gamma_w$, and for wall friction angles $\delta = 0$ and $\delta = \phi'_{mob}$. The results are shown in terms of the mobilized soil strength required for stability ϕ'_{mob}, and the normalized pivot depth $(h + z_p)/(h + d)$, as functions of the embedment to retained height ratio d/h. Two different depths a to the water table on the retained side are considered: $a = 0$ and $a = h$. With a full height groundwater level ($a/h = 0$), pore water pressures were calculated using the linear seepage model. With the groundwater level at formation level ($a/h = 1$), pore water pressures were assumed to be hydrostatic below the water table, and zero above it.

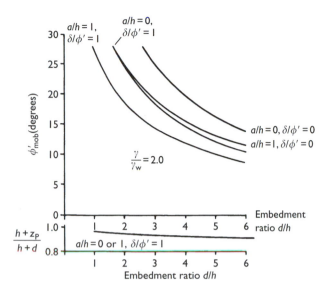

Figure 7.45 (a) Mobilized soil strength ϕ'_{mob} and (b) normalized pivot depth $(h + z_p)/(h + d)$, as functions of the embedment to retained height ratio d/h for unpropped embedded walls in uniform soil. (Redrawn with permission from Bolton et al., 1990.)

For an embedded wall propped at the crest, Figure 7.46 shows the mobilized soil strength ϕ'_{mob} required for equilibrium and the normalized prop load $F/0.5\gamma h^2$ (F has units of kN/m, i.e. kN per metre of wall) as functions of the embedment to retained height ratio, d/h, according to the stress analysis shown in Figure 7.43. Again, curves are given for $\delta = 0$ and $\delta = \phi'_{mob}$, and groundwater levels on the retained side at depths $a/h = 0$ (i.e. at the retained soil surface) and $a/h = 1$ (i.e. at formation level). The soil has unit weight $\gamma = 2\gamma_w$.

7.10.2 Effects of soil/wall adhesion

For a total stress analysis with wall adhesion τ_w, modified expressions for the relationship between the vertical and horizontal total stresses at the active and passive limits are derived in section 9.6.2. Alternatively, expressions for the active and passive limits to the horizontal total stress may be derived from the analysis of sliding wedges of unit depth, following the method described in section 7.7. For a level backfill and a vertical wall:

$$\sigma_{h,min} = (\gamma z + q) - [2\sqrt{(1 + \tau_w/\tau_u)}] \times \tau_u \quad \text{(Active)} \tag{7.13a}$$

$$\sigma_{h,max} = (\gamma z + q) + [2\sqrt{(1 + \tau_w/\tau_u)}] \times \tau_u \quad \text{(Passive)} \tag{7.13b}$$

where γ is the unit weight of the soil, z is the depth and q is the surface surcharge.

Equations (7.13) are less cumbersome than those derived from the stress field analysis of section 9.6.2, and are generally quoted in codes of practice for use in design. In BS 8002 (BSI, 2001), the mobilized wall adhesion τ_w is in general taken as $0.75 \times$ the mobilized undrained shear strength τ_{mob}, where $\tau_{mob} = \tau_u/M$, and the strength

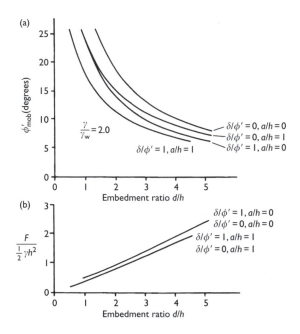

Figure 7.46 (a) Mobilized soil strength ϕ'_{mob} and (b) normalized prop load $F/0.5\gamma h^2$, as functions of the embedment to retained height ratio d/h for embedded walls in uniform soil, propped at the crest. (Redrawn with permission from Bolton *et al.*, 1990.)

mobilization factor $M = 1.5$. More conservatively, Gaba *et al.* (2003) suggest that the limiting value of τ_w should be $\tau_{w,max} = 0.5\,\tau_{u,design}$ with $\tau_{u,design} = \tau_u/1.5$.

The expression $[2\sqrt{(1 + \tau_w/\tau_u)}]$ is sometimes called a **total stress earth pressure coefficient,** and given the symbol K_{ac} (for active conditions) or K_{pc} (for passive conditions). On the active side of the wall, the depth to which a tension crack may remain open is usually taken as $2K_{ac}/\gamma$ (dry) or $2K_{ac}/[\gamma - \gamma_w]$ (flooded): this is not entirely satisfactory, because it requires the soil close to the wall to be able to carry a tensile stress in some direction other than the horizontal.

Key points

- The concepts of **engineering plasticity** can be used to analyse a structure such as a retaining wall on the verge of collapse. Solutions may be either **lower bounds,** based on a system of stresses which is in **equilibrium** and does not violate the **failure criterion** for the soil, which err on the safe side, or **upper bounds,** based on an assumed **mechanism** of collapse, which err on the unsafe side.
- Soils fail according to a frictional failure criterion expressed in terms of effective stresses:

$$(\tau/\sigma')_{max} = \tan\phi' \tag{7.1}$$

where ϕ' is the **effective angle of friction** of the soil. For retaining walls, the appropriate strength to use in collapse calculations is the critical state strength, ϕ'_{crit}.

- As a special case, the undrained failure of clay soils which are sheared at constant volume may be analysed in terms of a total stress failure criterion

$$\tau_{max} = \tau_u \qquad (7.5)$$

where τ_u is the **undrained shear strength**.

- By considering Mohr circles of effective stress at failure, it is possible to calculate the minimum and maximum possible ratios of horizontal to vertical effective stress. The minimum ratio is quantified by the **active earth pressure coefficient** K_a, and applies when the vertical stress is being increased or the horizontal stress is being reduced, such as in the zone of soil behind a retaining wall. The maximum ratio is quantified by the **passive earth pressure coefficient** K_p, and applies when the vertical stress is being reduced or the horizontal stress is being increased, such as in the zone of soil in front of an embedded retaining wall. For a frictionless retaining wall, it is easy to calculate the values of K_a and K_p:

$$K_a = 1/K_p = ([1 - \sin\phi']/[1 + \sin\phi']) \qquad (7.3; 7.4)$$

When soil/wall friction is present, the active and passive earth pressure coefficients K_a and K_p are not as easy to calculate, and tabulated values are generally used.

- For a soil obeying the maximum shear stress failure criterion $\tau_{max} = \tau_u$, the minimum and maximum total stresses are given by

$$\sigma_{h,min} = \sigma_v - 2\tau_u \quad \text{(Minimum, Active)} \qquad (7.6a)$$

$$\sigma_{h,max} = \sigma_v + 2\tau_u \quad \text{(Maximum, Passive)} \qquad (7.6b)$$

for a frictionless wall.

- Stress fields based on active and passive zones can be used to investigate failure conditions for retaining walls. However, it may not be easy to extend the stress field to infinity, and it is common in practice to focus on the stresses acting on the wall. This leads to a **limit equilibrium** solution, rather than a true lower bound.
- Retaining walls on the verge of failure can also be analysed by considering a potential mechanism of failure, which involves blocks of soil sliding along plane rupture surfaces. Analysis of the sliding blocks by statical equilibrium should give the same answer as an analysis based on an energy or work balance, provided that the mechanism is kinematically admissible. If the system of sliding blocks does not form a kinematically admissible mechanism, analysis using the condition of equilibrium will again lead to a limit equilibrium solution rather than a true upper bound.
- Because the mechanism-based approach is inherently unsafe, it is necessary to repeat the calculation until the least unsafe solution – for example, the largest calculated active thrust, and the smallest calculated passive thrust – is found.
- In design, the wall must be sufficiently remote from collapse not to move excessively under working conditions. One of the ways in which this is achieved is by carrying out the collapse calculation with the soil strength reduced by a **mobilization factor** M. This issue is addressed more fully in section 10.2.

Questions

Calculation of lateral earth pressures and prop loads

7.1 (a) Explain the terms 'active' and 'passive' in the context of a soil retaining wall.

(b) Figure 7.47 shows a cross-section through a trench support system, which is formed of a rigid reinforced concrete U-section. Assuming that the retained soil is in the active state, and that the interface friction between the soil and the wall is zero, calculate and sketch the short-term distributions of horizontal total and effective stress and pore water pressure acting on the vertical member AB.

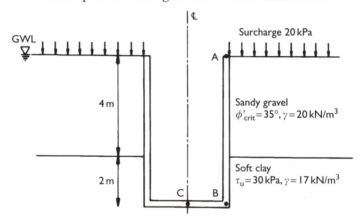

Figure 7.47 Cross-section through trench support system.

(c) Hence calculate the axial load (in kN per metre length of the trench) in the horizontal member BC, and the bending moment (in kNm/m) at B.

(d) Would you expect the axial load in BC and the bending moment at B to increase or decrease in the long term, and why?

[*University of London 2nd year BEng (Civil Engineering) examination, Queen Mary and Westfield College*]

((c) Axial load in BC = 237.4 kN/m; bending moment at B = 528 kN/m based on fully-active stresses in the retained soil.)

Stress field limit equilibrium analysis of an embedded retaining wall

7.2 (a) Figure 7.48 shows a cross-section through a frictionless embedded retaining wall, propped at the crest. Show that the wall would be on the verge of failure if the strength (effective angle of friction) of the soil were 18°. (Take the unit weight of water $\gamma_w = 10\,\text{kN/m}^3$.)

(b) Sketch the distributions of lateral stress on both sides of the wall, and calculate the bending moment at formation level and the prop force.

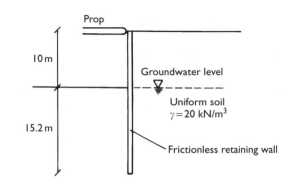

Figure 7.48 Cross-section through embedded retaining wall.

(c) If in fact the critical state strength of the soil is 24°, calculate the mobilization factor $M = \tan \phi'_{crit} / \tan \phi'_{mob}$.

[*University of London 3rd year BEng (Civil Engineering) examination, Queen Mary and Westfield College (part question)*]

((b) Prop load = 555 kN/m; bending moment at formation level = 3791 kNm/m; (c) $M = 1.37$.)

7.3 (a) Figure 7.49 shows a cross-section through a rough embedded retaining wall, propped at the crest. Stating any assumptions you make, estimate the long-term pore water pressure distribution around the wall.

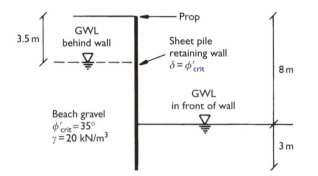

Figure 7.49 Cross-section through embedded retaining wall.

(b) Assuming that the critical state angle of soil friction of 35° is fully mobilized in the *retained* soil, calculate the earth pressure coefficient (based on effective stresses) in the soil in front of the wall required for moment equilibrium about the prop. Using Table 7.7, estimate the corresponding mobilized friction angle in the soil in front of the wall.

(c)   Is the wall safe? Explain briefly your reasoning.
(Take the unit weight of water $\gamma_w = 10\,kN/m^3$.)

[*University of London 3rd year BEng (Civil Engineering) examination,
Queen Mary and Westfield College*]

((a) Pore water pressure at base of wall $= 42.9\,kPa$, using the linear seepage approximation. (b) For equilibrium with fully-active stresses behind the wall, $K_p = 9.01$ which requires ϕ'_{mob} (with $\delta = \phi'_{mob}$) $\approx 35.5°$. (c) The wall is therefore not safe.)

7.4   Figure 7.50 shows a cross-section through a long excavation whose sides are supported by propped cantilever retaining walls. Calculate the depth of embedment needed just to prevent undrained failure by rotation about the prop if the groundwater level behind the wall is

(i)   below formation level; and
(ii)   at original ground level.

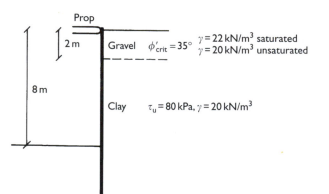

Figure 7.50 Cross-section through embedded retaining wall.

Neglect the effects of friction/adhesion at the soil/wall interface, and take the unit weight of water as $10\,kN/m^3$.
What is the strut load in each case?

((i)   embedment 0.01 m; strut load 9 kN/m (clay is self-supporting with a dry tension crack);
(ii)   embedment 2.09 m; strut load 137.5 kN/m (assuming flooded tension crack).)

Mechanism-based limit equilibrium analysis of gravity retaining walls

7.5 (a) Figure 7.51 shows a cross-section through a mass retaining wall. By means of a graphical construction, estimate the minimum lateral thrust that the wall must be able to resist. (Assume that the angle of friction between the soil and the concrete is equal to $0.67 \times \phi'_{crit}$.)

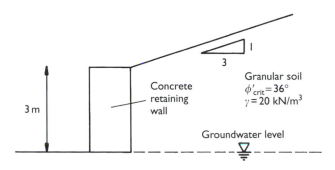

3 m

Concrete retaining wall

Granular soil
$\phi'_{crit} = 36°$
$\gamma = 20$ kN/m^3

Groundwater level

Figure 7.51 Cross-section through mass retaining wall.

(b) If the available frictional resistance to sliding on the base of the wall must be twice the active lateral thrust, calculate the necessary mass and width of the wall. (Take the unit weight of concrete as 24 kN/m^3.)

(c) What other checks would you need to carry out before the design of the wall could be considered to be acceptable?

[*University of London 3rd year BEng (Civil Engineering) examination, Queen Mary and Westfield College*]

((a) Horizontal component of thrust is approximately 24 kN/m, with the slip plane at an angle of 53° to the horizontal. (b) Required width of wall is 1.36 m.)

7.6 (a) Figure 7.52 shows a cross-section through a masonry retaining wall, with a partly sloping backfill which is subjected to a line load of 100 kN/m as indicated. Use a graphical construction to estimate the lateral thrust which must be resisted by friction on the base of the wall in order to prevent failure by the formation of a slip plane extending upward from the base of the wall, such as OA.

(b) Is your answer likely to be greater or less than the true value, and why?

(c) Suggest one way in which the ability of the wall to resist the thrust from the backfill could be improved.

[*University of London 2nd year BEng (Civil Engineering) examination, Queen Mary and Westfield College*]

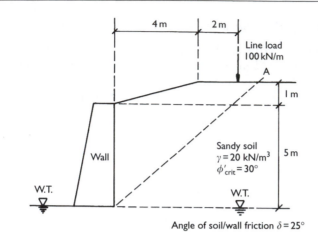

Figure 7.52 Cross-section through masonry retaining wall.

((a) Lateral component of active thrust is approximately 98 kN/m when slip plane OA is at 45° to the horizontal)

7.7   Figure 7.53 shows a cross-section through a mass concrete retaining wall. Estimate the minimum lateral thrust that the wall must be able to resist to maintain the stability of the retained soil. Hence investigate the safety of the wall against sliding.

[*University of London 2nd year BEng (Civil Engineering) examination, King's College*]

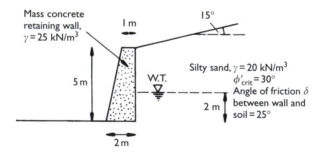

Figure 7.53 Cross-section through mass concrete retaining wall.

(Horizontal thrust (including pore water pressure component) is approximately 98 kN/m, with the slip plane at an angle of approximately 50° to the horizontal. Maximum available resistance to sliding is about 107.5 kN/m (depending on assumed pore water pressures on base and whether downward force from backfill is taken into account), so the wall is unacceptably close to sliding failure.)

7.8 (a) Figure 7.54(a) shows a cross-section through a gravity retaining wall retaining a partly sloping backfill of soft clay. By means of a graphical construction, estimate the minimum (active) lateral thrust that the wall must be able to resist in the short term. How does this compare with the maximum available sliding resistance on the base?

(Assume that the limiting adhesion between the wall and the clay is equal to 0.4 × the undrained shear strength τ_u, and that the angle of soil/wall friction between the wall and the underlying sand is equal to $0.67 \times \phi'$.)

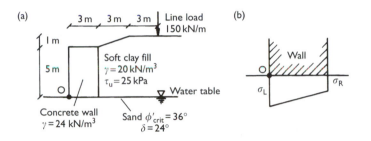

Figure 7.54 Cross-section through gravity retaining wall.

(b) If the thrust from the backfill acts on the back of the wall at a distance of one-third of the height of the wall above the base, and the normal total stress distribution on the base is as shown in Figure 7.54(b), calculate the values of σ_L and σ_R.

(c) What further investigations would you need to carry out, before the design of the wall could be considered acceptable?

((a) Horizontal thrust is approximately 134 kN/m, with the slip plane at an angle of approximately 45° to the horizontal. Maximum available resistance to sliding is about 182.5 kN/m (including the effect of the downward shear force on the back of the wall); (b) $\sigma_L = 80.7$ kPa; $\sigma_R = 192.6$ kPa.)

Notes

1 ϕ_μ is the angle of friction that would be measured between two solid flat blocks, made of the same material as the soil grains. It is different from ϕ', which is why ϕ' is **effective** or known as the **apparent** angle of soil friction.

2 A *smooth* surface is not necessarily *frictionless*: the two words are not synonymous.

References

Atkinson, J.H. (1981) *Foundations and Slopes*. McGraw-Hill, Maidenhead.

Baker, J.F. and Heyman, J. (1969) *Plastic Design of Frames 1: Fundamentals*, Cambridge University Press, Cambridge.

Bolton, M.D. (1991) *A Guide to Soil Mechanics*, M.D. & K. Bolton, Cambridge.

Bolton, M.D. and Powrie, W. (1987) Collapse of diaphragm walls retaining clay. *Géotechnique*, 37(3), 335–53.

Bolton, M.D., Powrie, W. and Symons, I.F. (1989–1990) The design of stiff, *in situ* walls retaining overconsolidated clay, Part I (short-term behaviour) & Part II (long-term behaviour). *Ground Engineering*, 22(8), 44–8; 22(9), 34–40; 23(2), 22–8.

British Standards Institution (1995) Eurocode 7: *Geotechnical design*. Part 1: General rules. DDENV 1997–1. British Standards Institution, London.

British Standards Institution (2001) *Code of Practice for Earth Retaining Structures BS8002* incorporating Amendments Nos 1 and 2 and Corrigendum No. 1. British Standards Institution, London.

Caquot, A. and Kerisel, J. (1948) *Tables for the Calculation of Passive Pressure, Active Pressure and Bearing Capacity of Foundations*, Gauthier Villars, Paris.

Coulomb, C.A. (1776) Essai sur une application des régeles des maximus et minimus a quelque problémes de statique rélatif à l'architecture. *Memoirs Divers Savants* 7, Académie Sciences, Paris.

Drescher, A. and Detournay, E. (1993) Limit load in translational failure mechanisms for associative and non-associative materials. *Géotechnique*, 43(3), 443–56.

Fellenius, W. (1916) Kaj-och jordrasen i Goteborg (The quay and earth slides in Gothenburg). *Tekn. Tidskrift*, 46, 133–8.

Gaba, A.R., Simpson, B., Powrie, W. and Beadman, D.R. (2003) Embedded retaining Walls – guidance for economic design. *Report C580, Construction Industry Research and Information Association*, London.

Heyman, J. (1972) *Coulomb's Memoir on Statics: an Essay in the History of Civil Engineering*, Cambridge University Press, Cambridge.

Institution of Structural Engineers (1951) *Civil Engineering Code of Practice No. 2: Earth Retaining Structures (CP2)*, Institution of Structural Engineers, London.

Kerisel, J. and Absi, E. (1990) *Active and Passive Pressure Tables*, A.A. Balkema, Rotterdam.

Mawditt, J.M. (1989) Discussion of Symons and Murray (1988). *Proceedings of the Institution of Civil Engineers Pt 1*, 86, 980–6.

Petterson, K.E. (1916) Kajraseti Goteburg des 5 Mars. *Tek. Tidskrift*, 46, 289.

Powrie, W. (1996) Limit equilibrium analysis of embedded retaining walls. *Géotechnique*, 46(4), 709–23.

Puller, M. (1996) *Deep excavations – a practical manual*. Thomas Telford, London.

Rankine, W.J.M. (1857) On the stability of loose earth. *Phil. Trans. Royal Society*, 147.

Rowe, P.W. (1963) Stress-dilatancy, earth pressure and slopes. *Proceedings of ASCE. Journal of Soil Mechanics and Foundations Division*, 89(SM3), 37–61.

Rowe, P.W. and Peaker, K. (1965) Passive earth pressure measurements. *Géotechnique*, 15(1), 57–78.

Scott, C.R. (1969) *An Introduction to Soil Mechanics and Foundations*, Applied Science Publishers, London.

Skempton, A.W. (1946) *Principles and Applications of Soil Mechanics. Lecture II: Earth Pressure and the Stability of Slopes*, Institution of Civil Engineers, London.

Sokolovskii, V.V. (1960) *Statics of soil media* (tr D. H. Jones and A. N. Schofield), Butterworths, London.

Subba Rao, K.S., Allam, M.M., and Robinson, R.G. (1998) Interfacial friction between sands and solid surfaces. *Proceedings of the Institution of Civil Engineers, (Geotechnical Engineering)* 131, April, 75–82.

Symons, I.F. (1983) Assessing the stability of a propped *in situ* retaining wall in overconsolidated clay. *Proceedings of the Institution of Civil Engineers*, Pt 2, 75, 617–33.

Terzaghi, K. (1954) Anchored bulkheads. *Transactions of the American Society of Civil Engineers*, **119**, 1243–80.

Tsagareli, Z.V. (1967) *New methods of lightweight wall construction* (in Russian), Stroiizdat, Moscow.

Watson, J.D. (1956) Earth movement affecting LTE railway in deep cutting east of Uxbridge. *Proceedings of the Institution of Civil Engineers*, Pt 2, 5, 302–31.

Whitlow, R. (1995) *Basic Soil Mechanics*, 3rd edn, Longman Scientific & Technical, Harlow.

Williams, B.P. and Waite, D. (1993) *The Design and Construction of Sheet-piled Cofferdams*, Special publication 95, Construction Industry Research and Information Association, London.

Chapter 8

Foundations and slopes

8.1 Introduction and objectives

In this chapter, the concepts of engineering plasticity introduced in Chapter 7 are used to calculate the ultimate (i.e. collapse) loads of different types of foundation, and also to assess the stability of slopes.

This chapter could be divided into three sub-areas: shallow foundations, deep foundations and slopes. The fundamental aspects of these topics are covered in sections 8.2–7.7 and 8.10. The material in sections 8.8 (pile groups and piled rafts), 8.9 (lateral loading of piles) and 8.11 (general slope analysis) is more advanced, and in many first degree courses will probably be taught at a later stage. Sections 8.8, 8.9 and 8.11 could therefore justifiably be omitted on first reading. For completeness, the objectives given below relate to the entire chapter.

Objectives

After having worked through this chapter, you should be able to:

- apply the concepts of engineering plasticity to calculate upper and lower bounds to the collapse loads of idealized shallow strip foundations or **footings** (sections 8.2 and 8.3)
- use empirical enhancement factors which take account of the effects of foundation shape and depth to estimate the ultimate (i.e. collapse) and working loads of more realistic foundations (section 8.4)
- calculate the ultimate and working loads of single piles and isolated deep foundations (section 8.6)
- assess the stability of long uniform slopes, using the infinite slope analysis (section 8.10).

You should have an appreciation of:

- the effects of horizontal and moment loading on shallow foundations (section 8.5)
- the potential interaction effects between closely spaced piles in pile groups, and the potential benefits in terms of improved economy and reduced settlements of using a piled raft (section 8.8)

- the methods used to estimate the lateral load capacity of an individual pile (section 8.9)
- the overwhelming importance of groundwater and pore water pressures on slope stability (section 8.10)
- the methods used to investigate the stability of slopes which cannot reasonably be idealized as being long and uniform, and the approximations that these entail (section 8.11).

8.2   Shallow strip foundations (footings): simple lower bound (safe) solutions

The walls of domestic buildings and other low-rise structures on reasonably firm soil are often founded on concrete strip foundations or **footings**, extending to a depth of 1–2 m below ground level. Design loads for foundations of this type are usually based on collapse loads calculated using a reduced soil strength – that is, the actual soil strength $\tan\phi'$ or τ divided by a suitable factor of safety F_s or strength mobilization factor M. In this section, we shall use the concepts of engineering plasticity to calculate collapse loads for simple shallow foundations.

Figure 8.1(a) shows a schematic cross-section through a typical shallow foundation. For the purpose of analysis, the foundation is idealized as shown in Figure 8.1(b): the soil above the bearing level or **founding plane** is modelled as a surcharge of $\sigma_0 = \gamma D$ on either side of the foundation. In an effective stress analysis, the effective surcharge on either side of the foundation σ_0' is obtained by subtracting the pore water pressure at bearing level, $\sigma_0' = \gamma D - u$.

8.2.1   *Effective stress analysis:* $(\tau/\sigma')_{\mathrm{max}} = \tan\phi'$ *failure criterion*

The collapse load of a long strip footing in a soil obeying the failure criterion $(\tau/\sigma')_{\mathrm{max}} = \tan\phi'$ may be investigated using an idealized stress field made up of

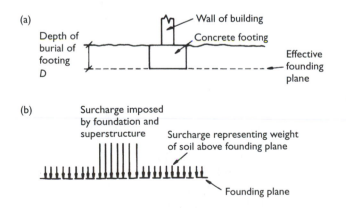

Figure 8.1 (a) Schematic cross-section and (b) idealization of a typical shallow footing.

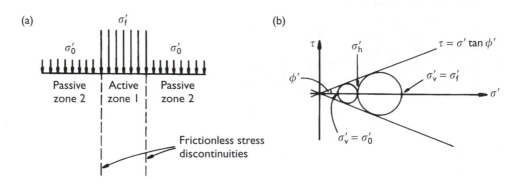

Figure 8.2 Simple stress field for a long strip footing in a soil obeying the failure criterion $(\tau/\sigma')_{max} = \tan\phi'$. (a) Division of soil into active and passive zones; (b) Mohr circles of effective stress just below the founding plane.

active and passive zones, separated by frictionless stress discontinuities, as shown in Figure 8.2. (Stress discontinuities were used in the analysis of an embedded retaining wall in section 7.6, and are discussed in more detail in sections 9.1–9.8.)

Immediately below the founding plane in zone 1, the vertical effective stress is equal to σ_f'. At failure, this will be an active zone (cf. Figure 7.3(a)), and the horizontal effective stress will be equal to $K_a \times \sigma_f'$. In zone 2 at the same level, the vertical effective stress is equal to σ_0'. At failure, zone 2 will be a passive zone (cf. Figure 7.4(a)), and the horizontal effective stress will be equal to $K_p \times \sigma_0'$. For equilibrium across the frictionless stress discontinuity, these two horizontal stresses must be the same, $\sigma_h' = K_a \times \sigma_f' = K_p \times \sigma_0'$, giving a ratio of surface stresses at failure of

$$(\sigma_f'/\sigma_0') = (K_p/K_a) = K_p^2 = N_q \tag{8.1}$$

where $N_q = (\sigma_f'/\sigma_0')$ is termed the **bearing capacity factor**.

The solution shown in Figure 8.2 is over-conservative (because of the introduction of the frictionless stress discontinuities) and may easily be improved. However, the form of the solution, $\sigma_f' = \sigma_0' N_q$, remains the same. It is shown in section 9.5.1 that the least conservative lower bound solution gives

$$N_q = K_p \, e^{\pi \tan\phi'} \tag{8.2}$$

where K_p is the passive earth pressure coefficient, $K_p = (1 + \sin\phi')/(1 - \sin\phi')$.

Various empirical adjustments may then be applied to the basic bearing capacity factor N_q, to take account of features not considered in the analysis. These include the self-weight of the soil (which causes σ_0' and hence σ_f' to increase with depth below the founding plane), the strength of the soil above foundation level, and the finite length of many real foundations. Some of these are discussed in section 8.4: a more comprehensive account is given by Bowles (1996).

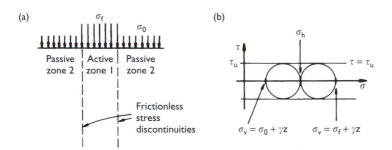

Figure 8.3 Simple stress field for a long strip footing on a clay soil of undrained shear strength τ_u. (a) Division of soil into active and passive zones; (b) Mohr circles of total stress at a depth z below the founding plane.

8.2.2 Short-term total stress analysis: $\tau_{max} = \tau_u$ failure criterion

The corresponding stress field for the rapid (undrained) failure of a clay having undrained shear strength τ_u is shown in Figure 8.3. At a depth z below the founding plane in zone 1, the vertical total stress is equal to $\sigma_f + \gamma z$. At failure, zone 1 is an active zone, and the horizontal total stress at depth z is equal to $(\sigma_f + \gamma z) - 2\tau_u$. In zone 2 at the same level, the vertical total stress is equal to $\sigma_0 + \gamma z$. At failure, zone 2 is a passive zone, and the horizontal total stress is equal to $(\sigma_0 + \gamma z) + 2\tau_u$. For equilibrium, these two horizontal stresses must be the same, $\sigma_h = (\sigma_f + \gamma z) - 2\tau_u = (\sigma_0 + \gamma z) + 2\tau_u$, giving a difference between the stresses below the founding plane at failure of

$$(\sigma_f - \sigma_0) = 4\tau_u \tag{8.3a}$$

In this case, the solution is of the form

$$\sigma_f - \sigma_0 = N_c \tau_u \tag{8.3b}$$

where N_c is the **bearing capacity factor**.

As before, the stress field shown in Figure 8.3 can easily be improved, and further empirical modifications may be made to the value of N_c to account for the shape of the foundation and the strength of the soil above the founding plane. In contrast to the effective stress analysis, the increase in vertical stress with depth below the founding plane due to the self-weight of the soil is of no benefit, because the bearing capacity is expressed in terms of the difference between the vertical total stresses in zones 1 and 2.

8.3 Simple upper bound (unsafe) solutions for shallow strip footings

8.3.1 Short-term total stress analysis: $\tau_{max} = \tau_u$ failure criterion

In section 8.2.2, the bearing capacity $\sigma_f - \sigma_0 = N_c \tau_u$ of a long strip footing on a clay soil of undrained shear strength τ_u was calculated using a lower bound plasticity

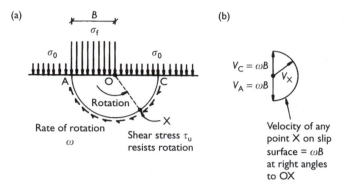

Figure 8.4 Assumed slip circle failure mechanism and hodograph for a long strip footing on a clay soil of undrained shear strength τ_u.

approach. This involved a stress field in equilibrium with the applied loads, which did not violate the failure criterion $\tau_{max} = \tau_u$. The bearing capacity may also be calculated using an upper bound plasticity approach, by considering an assumed mechanism of collapse.

Kinematically admissible mechanisms for plastic materials characterized by the failure criterion $\tau_{max} = \tau_u$ are made up of straight lines or circles, as outlined in section 7.7. In the case of a footing, the simplest class of mechanism is a slip circle as shown in Figure 8.4. The mechanism shown in Figure 8.4 may be analysed by considering the equilibrium of the semicircle of soil AOCA or by equating the rate at which work is done by the external forces to the rate at which energy is dissipated along the slip surface as the mechanism rotates. The same answer is obtained in each case.

(i) *By equilibrium, taking moments about the centre of the slip circle at O.* For a unit length of the footing perpendicular to the plane of the paper, the destabilizing moment is

$[\sigma_f \times B] \times [B/2]$

that is,

[load on foundation AO] × [lever arm of the centre of pressure about O]

The resisting moment is

$\{[\sigma_0 \times B] \times [B/2]\} + \{[\tau_u \times \pi B] \times [B]\}$

that is,

{[load on OC] × [lever arm]}

   +{[shear stress on slip surface× length of slip surface]

      ×[lever arm, which is the radius of the slip circle, B]}

Equating these

$$[\sigma_f \times B^2/2] = [\sigma_0 \times B^2/2] + [\tau_u \times \pi B^2]$$

or

$$(\sigma_f - \sigma_0) = 2\pi \tau_u$$

so that the bearing capacity factor

$$N_c = [(\sigma_f - \sigma_0)/\tau_u] = 2\pi \tag{8.4}$$

(ii) *Alternatively, by considering the rates of work and energy dissipation as the slip mechanism rotates about O at an angular velocity ω, and recalling that work = force × distance moved, so that work rate (= power) = force × velocity.*

The rate of work done by foundation load AO as it falls with a centroidal velocity of $\omega B/2$ downward is $[\sigma_f \times B] \times [\omega B/2]$.

The rate of work done against the surcharge σ_0 on OC as it is raised at a centroidal velocity of $\omega B/2$ is $[\sigma_0 B] \times [\omega B/2]$. Thus the net rate at which work is done by the external loads is

$$[(\sigma_f - \sigma_0) \times B] \times [\omega B/2]$$

The rate at which energy is dissipated at the slip surface is $[\tau_u \times \pi B] \times [\omega B]$ (i.e. the total force on slip surface × the relative velocity of slip).

Equating the net rate at which work is done by the external forces to the rate at which energy is dissipated at the slip surface:

$$[(\sigma_f - \sigma_0) \times \omega B^2/2] = [\tau_u \times \pi B^2]$$

or

$$(\sigma_f - \sigma_0) = 2\pi \tau_u$$

exactly as from the equilibrium analysis.

The problem with the upper bound calculation is that, if we have identified the wrong (i.e. not the most critical) failure mechanism, the solution obtained will err on the unsafe side. Some indication of the error in the solutions we have so far derived for a long strip footing on a clay soil may be obtained by comparison of the two results:

Lower bound (stress field):

$$N_c = (\sigma_f - \sigma_0)/\tau_u = 4 \quad \text{(from Equation 8.3)}$$

Upper bound (mechanism):

$$N_c = (\sigma_f - \sigma_0)/\tau_u = 2\pi = 6.28 \quad \text{(from Equation 8.4)}$$

The upper bound solution may be improved by searching for a more critical mechanism as shown in Figure 8.5, in which the centre of the circular slip is located at some distance vertically above the edge of the foundation. The slip surface is no longer a complete semicircle, but subtends an angle 2α at the centre of the arc.

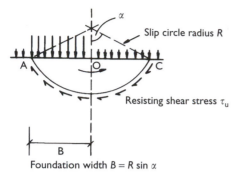

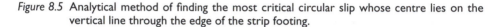

Foundation width $B = R \sin \alpha$

Figure 8.5 Analytical method of finding the most critical circular slip whose centre lies on the vertical line through the edge of the strip footing.

Again considering a unit length of the footing perpendicular to the plane of the paper, the destabilizing moment about the centre of the slip circle is

$$[\sigma_f \times B] \times [B/2]$$

that is,

[load on foundation AO]

   × [perpendicular distance from O to the line of action of the resultant force]

The resisting moment is

$$\{[\sigma_0 \times B] \times [B/2]\} + \{[\tau_u \times 2\alpha R] \times [R]\}$$

that is,

{[load on OC]

   × [perpendicular distance from O to the line of action of the resultant force]}

   + {[shear stress on slip surface × length of slip surface]

   × [lever arm, which is the radius R of the slip circle]}

In this case, the length of the slip surface is equal to the radius $R \times$ the angle 2α. From the geometry of the mechanism, $R = B/\sin \alpha$.

Substituting for R in terms of B and α, and equating the destabilizing and resisting moments,

$$[(\sigma_f - \sigma_0) \times B^2/2] = [\tau_u \times (2\alpha/\sin^2\alpha) \times B^2]$$

or

$$(\sigma_f - \sigma_0) = 4\tau_u \times (\alpha/\sin^2 \alpha)$$

The most critical mechanism within the range currently under investigation is the one giving the smallest value of $(\sigma_f - \sigma_0)$, which in turn requires that the expression $(\alpha/\sin^2\alpha) = (\alpha\mathrm{cosec}^2\alpha)$ is a minimum.

This occurs when

$$d(\alpha \, \text{cosec}^2\alpha)/d\alpha = 0 \quad \text{and} \quad d^2(\alpha \, \text{cosec}^2\alpha)/d\alpha^2 > 0$$

$$d(\alpha \, \text{cosec}^2\alpha)/d\alpha = (\text{cosec}^2\alpha) - (2\alpha \, \text{cosec}^2\alpha \, \cot\alpha)$$

which is zero when $2\alpha \cot\alpha = 1$, or $\tan\alpha = 2\alpha$ (α in radians). This occurs at $\alpha \approx$ 1.17 radian $\approx 67°$, giving a bearing capacity factor $N_c = [(\sigma_f - \sigma_0)/\tau_u] = 5.52$.

This narrows the range for N_c to $4 < N_c < 5.52$. The correct solution (which may be obtained by means of either the lower or upper bound approach: see sections 9.5.2 and 9.9.1) is

$$N_c = 2 + \pi = 5.14$$

8.3.2 Effective stress analysis: $(\tau/\sigma')_{max} = \tan\phi'$ failure criterion

If the condition of normality is assumed to apply to a soil obeying the effective stress failure criterion $(\tau/\sigma')_{max} = \tan\phi'$, the implied angle of dilation at failure is $\psi = \phi'$ (section 7.7). This means that a slip circle is not a kinematically admissible mechanism, because it does not allow for dilation.

The consequence of the application of the normality condition to the frictional failure criterion is that, as relative sliding takes place along an assumed slip surface, the direction of motion is not parallel to the slip surface (as it is with the $\tau_{max} = \tau_u$ failure criterion), but is at an angle $\psi = \phi'$ to it (Figure 8.6(a)). To accommodate this

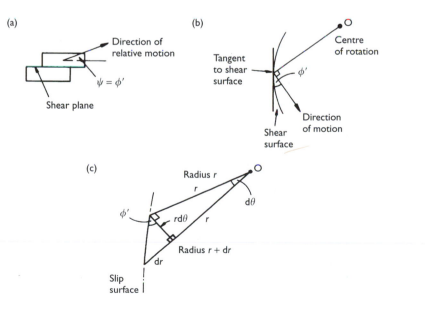

Figure 8.6 (a) Effect of normality applied to the $(\tau/\sigma')_{max} = \tan\phi'$ failure criterion; (b) relative orientation of curved failure surface and radius of rotation; (c) geometry of curved failure surface.

movement, a slip surface which is curved (i.e. not a straight line) must always be at an angle of ϕ' to the direction of motion, and hence at an angle of $(90° + \phi')$ to the radius of rotation, as indicated in Figure 8.6(b). As we move through an angle $d\theta$ at the centre of rotation, the radius increases by an amount dr as shown in Figure 8.6(c). From the geometry of Figure 8.6(c) with $d\theta \to 0$,

$$\tan \phi' = dr/(r\,d\theta)$$

or

$$(dr/r) = \tan \phi'\,d\theta$$

Integrating this between limits of $r = r_0$ at the start of the curve $\theta = 0$, and a general radius $r = r$ following an angle of rotation θ,

$$\ln(r/r_0) = \theta \tan \phi'$$

or

$$r = r_0\,e^{\theta \tan \phi'} \tag{8.5}$$

The curve defined by equation (8.5) is known as a **logarithmic spiral**.

The logarithmic spiral failure surface, in combination with the $(\tau/\sigma')_{max} = \tan \phi'$ failure criterion, has two particular consequences for analysis:

- The resultant of the stresses (τ and σ') on the slip surface is always directed towards the centre of rotation (Figure 8.7). This means that the moment of the stresses on the slip surface about the centre of rotation is zero.
- The resultant stress at any point on the slip surface is perpendicular to the direction of movement. This means that no energy is dissipated along the slip line as relative movement takes place, because the component of movement in the direction of the resultant stress is zero (Figure 8.7).

Figure 8.8 shows a mechanism of failure, comprising a single logarithmic spiral slip surface, for a shallow foundation on a soil obeying the $(\tau/\sigma')_{max} = \tan \phi'$ failure

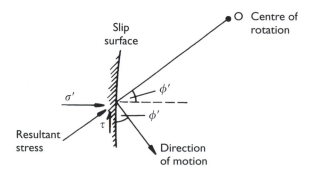

Figure 8.7 Resultant stress and direction of movement on slip surface, $(\tau/\sigma')_{max} = \tan \phi'$ failure criterion.

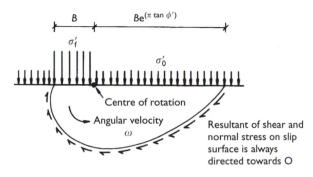

Figure 8.8 Logarithmic spiral failure mechanism for a shallow foundation, $(\tau/\sigma')_{\mathrm{max}} = \tan\phi'$ failure criterion.

criterion. This failure mechanism is analogous to the simple slip circle investigated for the $\tau_{\mathrm{max}} = \tau_{\mathrm{u}}$ failure criterion in section 8.3.1.

Ignoring the self-weight of the soil, and making use of the fact that the moment about O of the stresses on the slip surface is zero, the relationship between σ'_{f} and σ'_0 may be determined from the condition of moment equilibrium about O:

$$(B\sigma'_{\mathrm{f}}) \times (0.5B) = (Be^{\pi \tan\phi'}\sigma'_0) \times (0.5Be^{\pi \tan\phi'})$$

or

$$N_{\mathrm{q}} = \sigma'_{\mathrm{f}}/\sigma'_0 = e^{2\pi \tan\phi'} \tag{8.6}$$

Alternatively, equating the rate at which potential energy is lost by the foundation load σ'_{f} to the rate at which potential energy is gained by the surcharge on the surrounding soil σ'_0 as the soil mass within the slip surface rotates at an angular velocity ω:

$$(B\sigma'_{\mathrm{f}}) \times (0.5B\omega) = (Be^{\pi \tan\phi'})\sigma'_0 \times (0.5B\omega e^{\pi \tan\phi'})$$

or

$$N_{\mathrm{q}} = \sigma'_{\mathrm{f}}/\sigma'_0 = e^{2\pi \tan\phi'}$$

which is exactly the same as equation (8.6). In the work balance calculation, we have again neglected the self-weight of the soil, and also made use of the fact that the energy dissipated along the slip surface is zero.

8.4   Bearing capacity enhancement factors to account for foundation shape and depth, and soil weight

8.4.1   *Effective stress analysis:* $(\tau/\sigma')_{\mathrm{max}} = \tan\phi'$ *failure criterion*

Equation (8.1), which may be written in the form $\sigma'_{\mathrm{f}} = N_{\mathrm{q}} \times \sigma'_0$, gives the bearing capacity of a shallow foundation calculated using a simplified but rigorous theoretical analysis. It is usually used in practice in the form

$$\sigma'_f = \{N_q \times s_q \times d_q\} \times \sigma'_0 + \{N_\gamma \times s_\gamma \times d_\gamma \times r_\gamma \times [0.5\gamma B - \Delta u]\} \tag{8.7}$$

where

- N_q is the **bearing capacity factor** (generally taken as $K_p e^{\pi \tan \phi'}$).
- S_q (the **shape factor**) is an enhancement factor to allow for the fact that the footing is in reality not infinitely long.
- d_q (the **depth factor**) is an enhancement factor which takes account of the fact that, in reality, the soil above the founding plane has some strength and therefore does more than just act as a surcharge.

The purpose of the second term on the right-hand side of equation (8.7) is as follows. In the derivation of equation (8.1), we calculated the collapse load on the basis of the stress state immediately below the founding plane. This is equivalent to assuming that the entire footing would collapse when the failure condition for the soil is reached at the level of the founding plane. In reality, this is not the case, because a kinematically admissible failure mechanism will occupy a depth approximately equal to the foundation width, B. As σ'_0 increases with depth below the founding plane (due to the self-weight of the soil) so does σ'_f. The bearing capacity factor N_q is significantly greater than one, so that σ'_f increases much more rapidly with depth than σ'_0.

The consequence of this is that the first term on the right-hand side of equation (8.7), which is based on the vertical effective stress σ'_0 on either side of the footing at the level of the founding plane, will significantly underestimate the actual bearing capacity of the foundation. Assuming that the soil must be brought to failure over a depth of approximately one footing width B for a mechanism of collapse to form, the average vertical effective stress in the failure zone is $\sigma'_0 + (0.5\gamma B - \Delta u)$, where γ is the unit weight of the soil and Δu is the increase in pore water pressure between the founding plane and a depth of $B/2$ below it. The purpose of the second term on the right-hand side of equation (8.7) is therefore to enhance the bearing capacity to take account of the difference between the vertical effective stress at the level of the founding plane and the average vertical effective stress in the failure zone $(0.5\gamma B - \Delta u)$. In the second term on the right-hand side of equation 8.7,

- N_γ is analogous to the bearing capacity factor N_q
- s_γ is a shape factor, which is analogous to s_q but applied to N_γ
- d_γ is a depth factor, which is analogous to d_q but applied to N_γ
- r_γ is a reduction factor, to account for the fact that the N_γ effect does not increase indefinitely with the width of the footing B, as the second part of the right-hand side of equation (8.7) would otherwise imply.

An alternative approach to the quantification of the increase in bearing capacity due to the increase in vertical effective stress below the founding plane is given by Bolton (1991). He suggests that the second term on the right-hand side of equation (8.7) should be ignored, and the values of σ'_0 and σ'_f in the first term on the right-hand side should be taken as those on an equivalent founding plane, at a depth

Table 8.1 Bearing capacity enhancement factors, $(\tau/\sigma')_{max} = \tan\phi'$ failure criterion

Parameter	Meyerhof (1963)	Brinch Hansen (1970)
Shape factor, s_q	$1 + 0.1 K_p(B/L)$	$1 + [(B/L)\tan\phi']$
Depth factor, d_q	$1 + 0.1\sqrt{K_p}(D/B)$	$1 + 2\tan\phi'(1 - \sin\phi')k$
N_γ	$(N_q - 1)\tan(1.4\phi')$	$1.5 \times (N_q - 1)\tan\phi'$
Shape factor, s_γ (for N_γ)	$= s_q$	$1 - 0.4(B/L)$
Depth factor, d_γ (for N_γ)	$= d_q$	1

Notes
Meyerhof's expressions apply for $\phi' > 10°$.
$k = D/B$ if $D/B \leq 1$; $k = \tan^{-1}(D/B)$ (in radian) if $D/B > 1$.
$K_p = (1 + \sin\phi')/(1 - \sin\phi')$.
$r_\gamma = 1 - 0.25\log_{10}(B/2)$ for $B \geq 2\,\mathrm{m}$ (Bowles, 1996).
Foundation length L, breadth B and depth D.

of $B/2$ below the base of the actual foundation. This leads to the expression

$$\sigma'_f = \{(N_q \times s_q \times d_q) \times \sigma'_e\} - (\sigma'_e - \sigma'_0) \tag{8.8}$$

where $\sigma'_e = \sigma'_0 + [\gamma B/2 - \Delta u]$, and Δu is the increase in pore water pressure between the base of the foundation and a depth $B/2$ below it. Bolton's approach is arguably more elegant and conceptually transparent, but could lead to the overestimation of the bearing capacity of a wide foundation, because the reduction factor r_γ of equation (8.7) is omitted.

In some texts and codes of practice (including Annex B of Eurocode 7 Part 1 (BSI, 1995)), you will see the term $(0.5\gamma B - \Delta u)$ written as $0.5\gamma'B$ where $\gamma' = \gamma - \gamma_w$. This assumes that the increase in pore water pressure below the founding plane is hydrostatic with depth and that the water table is at or above the base of the foundation so that $\Delta u = 0.5B\gamma_w$. Neither of these assumptions is necessarily correct.

Numerical values of $s_q, d_q, N_\gamma, s_\gamma$ and d_γ given by Meyerhof (1963) and Brinch Hansen (1970), which are commonly used in the calculation of the ultimate bearing capacity of shallow foundations, are given in Table 8.1. Bowles (1996) details several others. These enhancement factors are generally empirical, being based on model test results together with a degree of field experience. The most significant discrepancies between the various suggested numerical values occur with the parameters associated with N_γ, which are intended to account for the increase in bearing capacity with depth below the founding plane due to the self-weight of the soil.

8.4.2 Short-term total stress analysis: $\tau_{max} = \tau_u$

For the total stress analysis based on the undrained shear strength τ_u, equation (8.3b) is conventionally modified and used in the form

$$(\sigma_f - \sigma_0) = \{N_c \times s_c \times d_c\} \times \tau_u \tag{8.9}$$

where $N_c = 5.14$ is the bearing capacity factor, and s_c and d_c are enhancement factors to take account of the shape and depth of the footing respectively. In this case, there is no N_γ effect because the difference between the total stresses σ_f and σ_0 is unaffected by the self-weight of the soil. Brinch Hansen (1970) suggests an equation of the form

$$(\sigma_f - \sigma_0) = N_c \times \{1 + s_c^* + d_c^*\} \times \tau_u \tag{8.10}$$

Values of s_c, d_c, s_c^* and d_c^* suggested by Skempton (1951), Meyerhof (1963) and Brinch Hansen (1970) are given in Table 8.2.

Brinch Hansen (1970) also gives adjustment factors which take account of non-vertical loads, non-horizontal footing bases and inclined founding planes. These are detailed by Bowles (1996). Annex B of Eurocode 7 Part 1 (BSI, 1995) gives equations of a similar form to equations (8.7) and (8.9), with factors i_q and i_c intended to account for the effect of inclined loads but with the depth factors d_q and d_c omitted. However, a completely general bearing capacity equation, in which parameter values are inserted as required to suit individual circumstances, can lead to confusion. Furthermore, it will be seen in Chapter 9 that inclined loads, non-horizontal footing bases and sloping natural ground will alter the numerical value of the bearing capacity factor N_q or N_c in a way that can be quantified analytically, without needing to resort to empirical adjustment factors.

Moments and horizontal loads, which may either be applied to a foundation directly (Figure 8.9) or result from the application of an inclined point load that does not act through the centre of the footing (Figure 8.10), can have an important destabilizing

Table 8.2 Bearing capacity enhancement factors, $\tau_{max} = \tau_u$ failure criterion

Parameter	Skempton (1951)	Meyerhof (1963)	Brinch Hansen (1970)
Shape factor, s_c	$1 + 0.2(B/L)$	$1 + 0.2(B/L)$	$s_c^* = 0.2B/L$
Depth factor, d_c	$1 + 0.23\sqrt{(D/B)}$, up to a maximum of 1.46 ($[D/B] = 4$)	$1 + 0.2(D/B)$	$d_c^* = 0.4k$ $k = D/B$ if $D/B \le 1$; $k = \tan^{-1}(D/B)$ (in radian) if $D/B > 1$

Note
Foundation length L, breadth B and depth D.

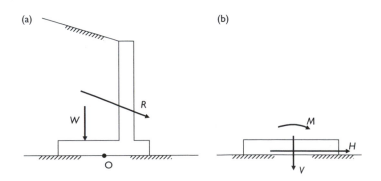

Figure 8.9 Combined vertical, horizontal and moment loads on a shallow foundation: (a) forces on a gravity retaining wall; (b) schematic representation of loads on the foundation. (Redrawn from Butterfield, 1993.)

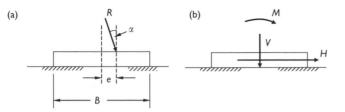

Figure 8.10 Statical equivalence of (a) an eccentric, inclined point load and (b) combined vertical, horizontal, and moment loading through the centroid of a shallow foundation.
(Redrawn from Gottardi and Butterfield, 1993.)

effect and must be considered in design. Shallow foundations subjected to horizontal
and moment loads are discussed briefly in section 8.5.

8.5   Shallow foundations subjected to horizontal and moment loads

In sections 8.1–8.4, we have investigated the failure of shallow foundations subjected
to a vertical load acting through the centroid, which might reasonably be represented
as a uniform vertical stress. In reality, many foundations may be required to carry
horizontal and moment loads as well. For example, the pressures exerted by the wind
on a wall, by waves and currents on the legs of an offshore wind turbine, or by the soil
behind a gravity retaining wall will all result in the application of both a horizontal
force and a moment to the foundation (Figure 8.9).

   An inclined point load that does not act through the centre of the foundation is statically equivalent to a combined horizontal, vertical and moment loads acting through
or about the centroid (Figure 8.10). Example 8.1 illustrates this with reference to the
stresses on the base of the gravity retaining wall analysed in section 7.7.

Example 8.1 Equivalent loads and pressure distributions on the base of a gravity retaining wall

Figure 8.11 shows the distribution of normal and shear stresses needed for
short-term equilibrium of the Cricklewood retaining wall shown in Figure 7.23,
as calculated in Example 7.5 with the fully passive normal force $N_{W,passive}$,
the maximum available base friction T_{Bmax}, and the soil/wall interface forces
on both sides of the wall $T_{W,passive}$ and $T_{W,active}$ all reduced by the factor
1.362, and the fully active normal force $N_{W,active}$ increased by the same amount.
(This gives values of normal contact stress $\sigma_A = 209\,\text{kPa}, \sigma_Q = 319$ kPa, and
$\tau_b = 30\,\text{kPa} \div 1.362 = 22$ kPa). Calculate (a) the equivalent vertical force, horizontal force and moment acting through the mid-point of the base of the wall,
and (b) the eccentricity e and the angle of inclination to the vertical α of the
equivalent single point load.

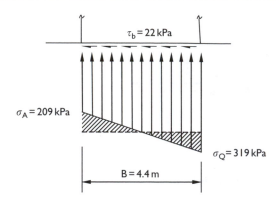

Figure 8.11 Equilibrium short-term total stress distribution on the base of the Cricklewood retaining wall (from Example 7.5).

Solution

(a)   The equivalent horizontal force H is given by the mobilized interface shear stress τ_b multiplied by the width of the wall base:

$$H = 22 \, \text{kPa} \times 4.4 \, \text{m} = \underline{96.8 \, \text{kN/m}}$$

The equivalent vertical force V is given by the average vertical pressure $1/2(\sigma_A + \sigma_Q)$ multiplied by the width of the wall base:

$$V = 1/2(209 + 319)\text{kPa} \times 4.4 \, \text{m} = \underline{1161.6 \, \text{kN/m}}$$

The equivalent moment M is given by the turning effect of the difference between the actual pressure distribution and the average line, indicated by the shaded triangles in Figure 8.11:

$$M = 2 \times 1/2 \times 1/2(319 - 209) \, \text{kPa} \times 2.2 \, \text{m} \times (2.2 \, \text{m} \times 2/3)$$
$$= \underline{177.5 \, \text{kNm/m}}$$

(b)   From Figure 8.10, if the resultant force is R acting at an angle α to the vertical at an eccentricity e to the right of the centre of the base of the wall:

$$H = R \sin \alpha = 97 \, \text{kN/m}$$

$$V = R \cos \alpha = 1162 \, \text{kN/m}$$

$$\Rightarrow \alpha = \tan^{-1}(H/V) = \tan^{-1}(97/1162) \Rightarrow \underline{\alpha = 4.8°}$$

$$R = V/\cos \alpha \Rightarrow \underline{R = 1166 \, \text{kN/m}}$$

$$M = R \cos \alpha \times e = V \cdot e$$

$$\Rightarrow e = M/V = 177.5 \, \text{kN m} \div 1162 \, \text{kN} \Rightarrow \underline{e = 0.153 \, \text{m}}$$

The adequacy of the foundation of the wall under this loading system is assessed in Example 9.3.

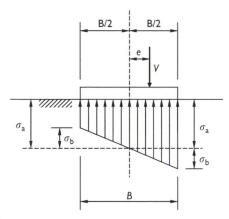

Figure 8.12 Contact stresses below a rigid footing subject to a point vertical load V
acting at a distance e from the centre of the footing.

In the general case of an eccentric point load V acting vertically at a distance e from the centre of a foundation of width B, the contact stress between the foundation and the soil may be thought of as varying from $(\sigma_a - \sigma_b)$ at one side to $(\sigma_a + \sigma_b)$ at the other (Figure 8.12). σ_a is the average vertical stress, and σ_b is the component associated with the eccentricity e which balances the moment $V \cdot e$ about the centroid.

Vertical equilibrium of the footing requires that

$$V = \sigma_a \cdot B \tag{8.11}$$

while the condition of moment equilibrium gives

$$M = 2 \times \sigma_b/2 \times B/2 \times 2/3 \cdot (B/2) = 1/6 \cdot \sigma_b \cdot B^2 = V \cdot e \tag{8.12}$$

If the contact stress is to be greater than zero everywhere below the footing, σ_a must be greater than σ_b (so that $(\sigma_a - \sigma_b)$ remains positive). Hence

$$\frac{V}{B} > \frac{6V \cdot e}{B^2} \quad \text{or} \quad e < B/6 \tag{8.13}$$

Thus there is an overall range of $B/3$ ($B/6$ on either side of the centre) within which a vertical point load must act if it is not to cause a tendency towards tensile stresses and/or separation at one end of the footing. This is traditionally known as the **middle third rule**.

Having determined the components of load (either H, M and V or R, e and α) to which a foundation will be subjected, it is necessary to check that the footing will have a sufficient margin of safety against collapse and will also not deform excessively under working conditions. As already mentioned in section 8.4, Annex B of Eurocode 7 Part 1 (BSI, 1995) gives bearing capacity equations similar to equations (8.7) and (8.9), with an additional factor to account for the inclination of the load. Eurocode 7 Part 1 requires that "special precautions shall be taken where the eccentricity of loading exceeds 1/3 of the width of a rectangular footing or 0.6 of the radius of a circular footing".

A possible approach to calculating the bearing capacity of a surface footing subjected to a simultaneous vertical (V), horizontal (H) and moment (M) loading is given by Butterfield and Gottardi (1994), who carried out a large number of small-scale (model) tests to investigate combinations of V, H and M that would cause failure. They showed that combinations of H, M and V at failure lay on a unique surface or three-dimensional failure envelope when plotted on a three-dimensional graph with axes V, M/B and H where B is the breadth (i.e. the width) of the footing. Division of the moment M (in kNm) by the footing width B (in m) is necessary to make the units of the three axes the same (kN). The failure surface may be described as cigar-shaped, and is parabolic when viewed in either the V, H or the V, M/B plane (Figure 8.13).

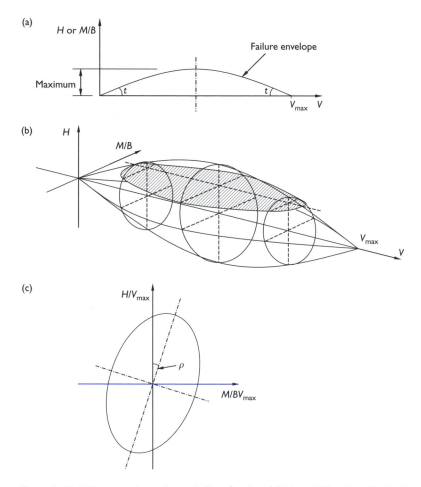

Figure 8.13 Failure envelopes for a shallow footing: (a) H or M/B against V; (b) three-dimensional view in $(V, M/B, H)$ space; (c) normalized cross-section perpendicular to the V axis at $V = V_{max}/2$. (Redrawn from Butterfield and Gottardi, 1994.)

Butterfield and Gottardi (1994) normalized the experimental values of V, H and M/B with respect to the vertical load V_{max} that would cause failure of the footing when acting on its own. For a foundation placed directly on the soil surface, the bearing capacity in pure vertical loading σ'_f may be calculated using equation (8.7) with the surcharge on either side due to the weight of soil above the founding plane $\sigma'_0 = $ zero. The depth factor d_γ and the associated reduction factor r_γ are also omitted, because the depth of burial of the foundation D is zero, giving

$$\sigma'_f \text{ (kPa)} = N_\gamma \times s_\gamma \times (0.5\gamma B - \Delta u) \tag{8.14}$$

If the foundation is of width B and length L, the vertical load at failure, acting through the centroid, is

$$V_{max}\text{(kN)} = \sigma'_f \cdot B \cdot L = N_\gamma \cdot s_\gamma \cdot (0.5\gamma B - \Delta u) \cdot B \cdot L \tag{8.15}$$

The parabolas representing the views of the three-dimensional failure surface in the V, H and $V, M/B$ planes have equations

$$H/t_h = V \cdot (V_{max} - V)/V_{max} \tag{8.16a}$$

and

$$(M/B)/t_m = V \cdot (V_{max} - V)/V_{max} \tag{8.16b}$$

where t_h and t_m are the slopes of the parabolas at the origin in the V, H and $V, M/B$ planes respectively (Figure 8.13(a)). Failure at low loads with $M = 0$ will be governed by sliding of the footing, $H = V.\tan\delta$ where δ is the angle of friction between the footing and the underlying soil. This line represents the tangent to the failure surface at $V = 0$, hence $t_h = \tan\delta$. For Butterfield and Gottardi's experiments, $H_{max} \approx V_{max}/8$ and $(M/B)_{max} \approx V_{max}/11$, corresponding to $t_h = 0.5$ and $t_m = 0.36$. H_{max} is defined at $M/B = 0$, $(M/B)_{max}$ is defined at $H = 0$, and both occur at $V = V_{max}/2$. Sections through the failure surface in the $M/B, H$ plane (i.e. when viewed along the V axis) are ellipses, rotated through an angle ρ from the H axis towards the positive M/B axis (Figure 8.13(c)). The reason for this rotation is that the H and M/B loads can act either together or in opposition: if H and M/B act in the directions shown in Figure 8.10(b), the horizontal load at failure will be much smaller than if either H or M/B is reversed.

The three-dimensional failure surface may be represented by an equation of the form

$$\left[\frac{H/V_{max}}{t_h}\right]^2 + \left[\frac{M/(B \cdot V_{max})}{t_m}\right]^2 - \left[\frac{2 \cdot C \cdot \{M/(B \cdot V_{max})\} \cdot \{H/V_{max}\}}{t_h \cdot t_m}\right]$$

$$= \left[\frac{V}{V_{max}} \cdot \left(1 - \frac{V}{V_{max}}\right)\right]^2 \tag{8.17}$$

where the constant C is a function of t_h, t_m and the inclination ρ to the H axis of the major axis of the cross-sectional ellipse. Data from three series of plane strain tests on foundations of different widths on different types of sand and/or with different density indices have a best fit failure surface represented by equation (8.17) with $t_h = 0.52$, $t_m = 0.35$, $\rho = 14°$ and $C = 0.22$.

The main practical implications of this are:

- Surface footings are particularly vulnerable to horizontal and moment loading, with $H \approx V_{max}/8$ or $M/B \approx V_{max}/11$ being sufficient to cause failure. For this reason, foundations that are required to carry significant horizontal or moment loads are generally piled (see sections 8.6–8.9), or at least buried to some extent.
- The remoteness from the failure surface of a point on the $(V, M/B, H)$ diagram representing a general load will depend on the load path followed to failure. This makes it difficult to define a meaningful factor of safety.
- If a surface footing is to be as safe as possible when subjected to a general load increment, the design vertical load should be about $V_{max}/2$.

It is difficult, if not impossible, to achieve a high factor of safety on load for a shallow foundation subjected to a load path that involves an increase in H or M coupled with a reduction in V. This is illustrated in Example 8.2.

Example 8.2 Assessment of a shallow foundation subject to vertical, horizontal and moment loading

Figure 8.14 shows a schematic cross-section through an offshore platform, supported by four legs each bearing onto the centre of a square pad foundation on the seabed. In plan view, the pad foundations are located with their centres at the corners of a square of side length 56 m. The dead load acting vertically is 52 MN, distributed equally between the four foundations. Wave and wind loading give rise to a design horizontal load of 8 MN, acting at a height of 98 m above the seabed through the centroid of the platform, as indicated. (In designing against failure in the ground according to Eurocode 7 case C, the design value for unfavourable, variable actions would include a load factor of 1.3. This means that our design value for the wind and wave loading of 8 MN corresponds to an expected value of 8 MN ÷ 1.3 ≈ 6 MN). The seabed soil has an angle of shearing resistance $\phi' = 36°$ and saturated unit weight $\gamma = 20\,\text{kN/m}^3$. Determine the size of foundations if a factor of safety on soil strength (strength mobilization factor) $M = 1.25$ applied to $\tan \phi'$ (consistent with Eurocode 7 case C) is required in the most adverse design loading condition. Construct the failure envelope for the foundation you propose, and sketch on it the load paths followed by both the left-hand and the right-hand footings as the wind and wave load increases from zero to its design value. Assume that the failure envelope for combined horizontal and vertical loading is given by equation (8.16a), with $t_h = 0.5$ when the full strength of the soil is used (i.e. the strength mobilization factor $M = 1$).

Solution

The steady component of vertical load that must be carried by each foundation pad is given by

$$V = 52\,\text{MN} \div 4 = 13\,\text{MN}$$

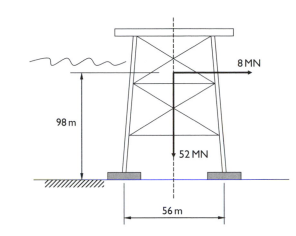

Figure 8.14 Schematic cross-section through an offshore platform with pad foundations below each leg.

The horizontal 8 MN wind and wave loading is shared equally between the four pads, giving a horizontal load on each pad of

$$H = 8\,\text{MN} \div 4 = 2\,\text{MN}$$

The 8 MN wind and wave load also exerts a moment of $8\,\text{MN} \times 98\,\text{m} = 784\,\text{MNm}$ about the seabed. This causes an increase in the vertical load of ΔV on the two right-hand pads, and a corresponding decrease on the left-hand pads, where

$$2 \times \Delta V \times 56\,\text{m} = 784\,\text{MNm} \Rightarrow \Delta V = 7\,\text{MN (per pad)}$$

Hence there are two possible loading cases we must consider:

(i) for the left-hand pads, $(V - \Delta V) = 6\,\text{MN}$ and $H = 2\,\text{MN}$
(ii) for the right-hand pads, $(V + \Delta V) = 20\,\text{MN}$ and $H = 2\,\text{MN}$

All four pads must be designed to withstand the more onerous of these, as the wind and wave loading could act in either direction (actually, the most critical case is when the wind and wave loading acts along the diagonal – real life is three dimensional!).

We can use the equation of the factored failure surface in the $V-H$ plane to calculate the required bearing capacity of the footing in pure vertical loading, V_{max}, based on the allowable mobilized soil strength, in each loading case:

$$H/t_{\text{h}} = V \cdot (V_{\text{max}} - V)/V_{\text{max}} \tag{8.16a}$$

As $t_{\text{h}} = \tan\delta$, this too must be divided by the strength mobilization factor $M = 1.25$ giving

$$t_{\text{h,design}} = 0.5 \div 1.25 = 0.4$$

Case (i): $H = 2\,\text{MN}$ and $V = 6\,\text{MN}$

$$2/0.4 = 6 \cdot (V_{\text{max}} - 6)/V_{\text{max}}$$

$$\Rightarrow 5 \cdot V_{\text{max}} = 6 \cdot V_{\text{max}} - 36 \qquad \Rightarrow V_{\text{max}} = \underline{36\,\text{MN}}$$

Case (ii): $H = 2\,\text{MN}$ and $V = 20\,\text{MN}$

$$2/0.4 = 20 \cdot (V_{\text{max}} - 20)/V_{\text{max}}$$

$$\Rightarrow 5 \cdot V_{\text{max}} = 20 \cdot V_{\text{max}} - 400 \qquad \Rightarrow V_{\text{max}} = \underline{400 \div 15 \approx 27\,\text{MN}}$$

hence the foundation pads that are being unloaded (i.e. on the left-hand side of the platform as shown in Figure 8.14) are the more critical, and all four pads must be designed for $V_{\text{max}} = 36\,\text{MN}$.

V_{max} may be calculated using equation (8.15) with the foundation length $L = B$:

$$V_{\text{max}}(\text{kN}) = N_\gamma \cdot s_\gamma \cdot (0.5\gamma B - \Delta u) \cdot B^2 \tag{8.18}$$

where the various factors including N_γ are based on the allowable mobilized strength ϕ'_{mob} which is given by

$$\phi'_{\text{mob}} = \tan^{-1}\{(\tan\phi')/1.25\}$$

Taking $\phi' = 36°, \phi'_{\text{mob}} = \tan^{-1}(0.5812) \approx 30°$.

Using the expressions for N_γ and s_γ given by Meyerhof (1963: table 8.1) with $\phi' = \phi'_{\text{mob}} = 30°, B = L, K_p = (1 + \sin\phi')/(1 - \sin\phi') = 3$, and $N_q = K_p \cdot \exp\{\pi \cdot \tan\phi'\} = 18.4$,

$$N_\gamma = (N_q - 1) \cdot \tan(1.4\phi') = 17.4 \times \tan 42° = 15.67$$

$$s_\gamma = s_q = 1 + 0.1 \cdot K_p \cdot (B/L) = 1.3$$

Assuming that the increase in pore water pressure below the seabed is hydrostatic, Δu in equation (8.18) $= \gamma_w \times 0.5B = 5B$ (kPa). Substituting the numerical values calculated above into equation (8.18) with $\gamma = 20\,\text{kN/m}^3$ gives

$$V_{\text{max}}(\text{kN}) = N_\gamma \cdot s_\gamma \cdot (0.5\gamma B - \Delta u) \cdot B^2 = 15.67 \times 1.3 \times 5B^3$$

$$= 101.86B^3 \text{ with } B \text{ in metres}$$

Hence $V_{\text{max}} = 36\,\text{MN} = 0.102B^3$.

$$\Rightarrow B^3 = 353\,\text{m}^3 \text{ or } \underline{B = 7.07\,\text{m}}$$

The actual failure load of this foundation in pure vertical loading may be calculated using equation (8.18) with values of N_γ and s_γ based on the full soil strength of 36°. With $\phi' = 36°$; $K_p = 3.852$, $N_q = 37.75$, $N_\gamma = 44.42$ and $s_\gamma = 1.385$ giving

$$V_{\text{max}}(kN) = N_\gamma \cdot s_\gamma \cdot (0.5\gamma B - \Delta u) \cdot B^2 = 44.42 \times 1.385 \times 5B^3 = 307.6B^3$$

or, with $B = 7.07\,\text{m}$,

$$V_{\text{max}} \approx 109\,\text{MN}$$

Note how with $\phi' = 36°$, the strength mobilization factor of 1.25 on $\tan\phi'$ is equivalent to a load factor on V_{max} of $109/36 \approx 3$, not including the load factor of 1.3 already applied to unfavourable variable actions. A load factor of 3 is consistent with normal practice; but while it might seem comforting, overestimating ϕ' by as little as a degree or two would result in a foundation being much closer to failure than the designer intended.

Failure envelopes in the (V, H) plane according to equation (8.16a) are plotted in Figure 8.15 for (i) $V_{max} = 36$ MN, $t_h = 0.4$ and (ii) $V_{max} = 109$ MN, $t_h = 0.5$. Loading paths for both the left-hand (L) and right-hand (R) foundation pads from the point $(V = 13$ MN, $H = 0)$, representing the dead load only, to the points representing the full design wind and wave loading $(V = 6$ MN, $H = 2$ MN) and $(V = 20$ MN, $H = 2$ MN) are also shown. The load path followed by the left-hand footing ends on the limiting envelope based on the factored soil strength $(\phi' = 30°)$ corresponding to $V_{max} = 36$ MN and $t_h = 0.4$, while the load path for the right-hand footing ends well within it. The remoteness of the footing from failure under the design loading is indicated by the distance between the ends of the load paths and the true failure envelope based on the full strength of the soil $(\phi' = 36°)$ and $V_{max} = 109$ MN and $t_h = 0.5$.

It may be seen that, in terms of a factor on the horizontal load, the left-hand footing is very much closer to failure than that on the right-hand side. The value of H at failure in each case may be calculated from the intersection of the failure surface with the relevant loading path. The equation of the real failure surface (with $t_h = 0.5$ and $V_{max} = 109$ MN) is

$$2 \cdot V_{max} \cdot H = V \cdot (V_{max} - V)$$

or in this case, setting $V_{max} = 109$ MN

$$218 \cdot H = V \cdot (109 - V)$$

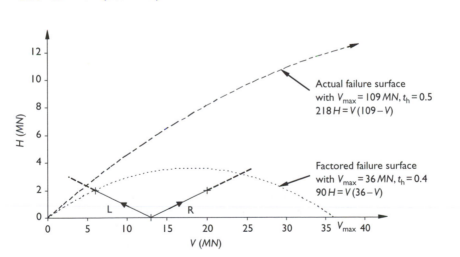

Figure 8.15 Failure envelopes and load paths for the offshore platform foundations under the loading conditions indicated in Figure 8.14.

Considering the geometry of Figure 8.14 for a general wind and wave loading force of $4H$, the loading path for each footing may be expressed as

$$V = 13\,\text{MN} \pm \frac{98 \times 4H}{112} = 13\,\text{MN} \pm 3.5H$$

Solving these equations gives values at failure of ($H = 2.3\,\text{MN}$, $V = 4.9\,\text{MN}$) for the left-hand footing, and ($H = 13.5\,\text{MN}$, $V = 60.2\,\text{MN}$) for the right-hand footing. The corresponding load factors on H (defined as $H_{\text{failure}} \div H_{\text{design}}$) are 1.16 and 6.74, respectively. It can be seen from the relative positions of the actual and factored failure surfaces in Figure 8.15 that a foundation of this type will always be vulnerable to failure in a loading path that involves an increase in H and a decrease in V; and that the load factor in such a loading path will always be small whatever the value of the load factor based on V_{max} or the factor of safety (strength mobilization factor) on $\tan \phi'$. We must also remember that we have not analysed the most critical case, which is when the wind and wave load acts along a diagonal.

Note: this example is based on one originally given by Georgiadis (1985).

8.6 Simple piled foundations: ultimate axial loads of single piles

In masonry buildings, significant line loads arise due to the self-weight of the walls. Live loads are also transmitted to the foundations through the walls, and strip footings represent an appropriate method of load transfer into the underlying soil. In steel or concrete framed buildings, however, the structural loads are transferred downward through the columns or **stanchions**, resulting in the application of a series of high-intensity point loads to the underlying soil. These loads are most effectively transferred to the soil by means of in-ground concrete columns or **piles**. The depth to which the pile penetrates below ground or basement level will depend primarily on the magnitude and nature of the imposed loads and on the stiffness and strength of the soil.

Piles were traditionally designed so that the net load under working conditions was a factor of 2.5 to 3 times smaller than the net load at failure.[1] A more modern approach, following for example Eurocode 7 (BSI, 1995), is to apply the factor of safety to the soil strength ($\tan \phi'$ or τ_u) rather than to the applied load directly. The design load for a foundation in a soil of strength ϕ' (or τ_u) is then taken as the collapse load for the same foundation in a soil of strength ϕ'_d (or $\tau_{u,d}$) where $\tan \phi'_d = (\tan \phi')/M$ (or $\tau_{u,d} = \tau_u/M$) and M is a strength mobilization factor. This approach was adopted for the pad foundation in Example 8.2.

Figure 8.16 illustrates how the foundation load is taken partly by the **skin friction** between the pile and the surrounding soil, and partly by the **bearing pressure** on the pile base. In sandy soils, an effective stress analysis must be carried out. In clay soils, it will usually be necessary to calculate the capacity of the pile in both the undrained condition (using a total stress analysis and the undrained shear strength failure criterion), and in the long term (using effective stresses and the frictional failure criterion).

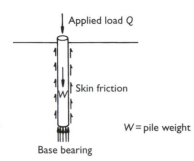

Figure 8.16 Skin friction and base bearing for a single pile.

The bearing capacity at the base of the pile can be calculated using the bearing capacity factors N_c (undrained) and N_q (in terms of effective stresses), modified to take account of the depth of the pile D and its plan dimensions $B \times L$ (where L is greater than or equal to B) using the depth and shape factors given by Skempton (1951) or Brinch Hansen (1970), as indicated in Table 8.1 or Table 8.2. (The Meyerhof depth factors given in Tables 8.1 and 8.2 are not suitable for piles. They were intended for shallow foundations, and have no apparent limit as the D/B ratio is increased).

For piles, the N_γ term is generally ignored unless the base is **under-reamed** (i.e. opened out into a bell-shape), because the foundation width B is small in comparison with the depth D.

Meyerhof (1976) gives an empirical correlation between the blowcount in a **Standard Penetration Test** and the base bearing resistance of piles. (The Standard Penetration Test or SPT is a crude but simple *in situ* soil test, which involves driving a standard sampler into the soil at the bottom of a borehole using a standard hammer. The number of hammerblows required to drive the sampler through a distance of 300 mm is known as the **SPT blowcount**. Empirical correlations between the SPT blowcount and various soil properties are discussed in section 11.3.1.) Similar correlations, based on the more sophisticated *in situ* **cone penetrometer test** (section 11.3.2), are also available (Whitlow, 1995).

Skin friction between the pile and the soil must also be taken into account. Usually, the pile will tend to settle relative to the surrounding soil, so that friction will generally help to support the imposed load. This may not always be the case. For example, if the soil around the pile is subjected to a surcharge loading or groundwater lowering, it may tend to settle relative to the pile: the effect of friction could then be to reduce the capacity of the pile, or at least increase the pile settlement at which it is reached.

In a total stress analysis, a reduction factor α is applied to the undrained shear strength of the soil to take account of disturbance and softening due to the effects of pile installation. In an effective stress analysis, it is necessary to estimate the post-installation lateral stresses in order to determine the contribution of pile/soil friction. The expression $K = \sigma'_h/\sigma'_v = 1 - \sin\phi'$ (originally proposed by Jaky (1944)) is generally accepted to give a reasonable estimate of the *in situ* lateral effective stresses in normally consolidated soils. In overconsolidated soils, the *in situ* earth pressure coefficient may be estimated using $K = \sigma'_h/\sigma'_v = (1 - \sin\phi')$. $OCR^{\sin\phi'}$ (Mayne and

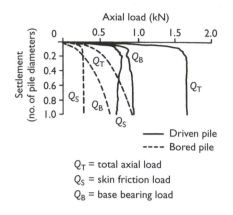

Figure 8.17 Mobilization of skin friction and base bearing pressure with pile displacement. (Redrawn with permission from Fioravante _et al._, 1994.)

Kulhawy, 1982), up to the passive limit $K_p = (1 + \sin \phi')/(1 - \sin \phi')$ (section 4.10.3). However, the expression $K = (1 - \sin \phi')$ is often used to account for possible stress relief during installation.

For bored piles, which are made from reinforced concrete cast in a pre-drilled hole in the ground, the skin friction may be fully mobilized at rather smaller displacements than the base bearing pressure. For driven piles, which are hammered into the ground and may be made of timber, steel or precast reinforced concrete, the rates of mobilization of base and bearing resistance are likely to be similar and rapid. These points are illustrated by Figure 8.17, which gives data from load tests on model piles, carried out in a **geotechnical centrifuge** (section 11.2.2).

The calculation of the ultimate and design loads for a deep foundation, in both the short term and the long term, is illustrated in Example 8.3.

8.7 ϕ'_{crit} or ϕ'_{peak}?

In the design of a pile foundation following Eurocode 7 (BSI, 1995), the design load must not exceed the ultimate load calculated when the soil strengths are reduced by a mobilization factor (factor of safety) of 1.25 (on $\tan \phi'$) or 1.4 (on τ_u). Eurocode 7 does not state explicitly whether it is the critical state or the peak strength to which the mobilization factor should be applied, but says that the strength used should be 'a cautious estimate of the value affecting the occurrence of the [relevant] limit state'. It also requires that 'the stress level of the problem imposed' should be considered when selecting the value of ϕ'.

The arguments in favour of using ϕ'_{crit} rather than ϕ'_{peak} in calculating the ultimate loads of foundations are the same as those given in section 7.3.2 for retaining walls. One of the reasons for using ϕ'_{crit} rather than ϕ'_{peak} is to guard against the possibility of progressive failure in a strain-softening soil. In practice, however, progressive failure does not generally appear to be a significant problem with foundations. Also, in a design based on the collapse load calculated using the critical state strength, a higher load factor (factor of safety) would be needed for a foundation on a loose soil than for an identical foundation on a dense soil, if the settlements under working conditions were to be the same.

Examination of the literature suggests that peak strengths have generally been used in the design of foundations using traditional methods involving the application of a load factor of 2.5–3 to the calculated ultimate capacity. This might indicate a degree of inherent conservatism in the choice of the numerical value of the load factor, and/or the way in which the shape and depth of the foundation, and the effect of the increase in vertical effective stress below the foundation are taken into account. Alternatively, it might be that suitably selected peak strengths can reliably be mobilized all around the foundation at the instant of failure.

If peak strengths are used in design, they must be selected with care. Owing to the exponential term in equation 8.2, the over-estimation of ϕ' by only a few degrees could be catastrophic, especially when $\phi' > 30°$ or so. Substituting $\phi' = 30°$ into equation (8.2) ($N_q = K_p \cdot e^{\pi \tan \phi'}$) gives $N_q = 18.4$; with $\phi' = 33°, N_q = 26.1$. In other words, a 10% increase in ϕ' leads to a 42% increase in the calculated value of N_q. When based on laboratory test data, the peak strength must be measured at the maximum effective stress experienced by the soil in the field, on samples having a representative void ratio and/or stress history. With piles and deep foundations, the effective stresses below the base could be high enough to suppress dilation completely, so that the peak strength may be the critical state strength.

Although the use of a carefully selected value of ϕ'_{peak} in the calculation of the ultimate bearing capacity of a foundation might be appropriate, this is not the case for other forms of geotechnical construction. As mentioned in Chapter 7, the use of critical state strengths appears to lead to a reasonably reliable prediction of the onset of large movements of embedded retaining walls. In the analysis of slopes, where stability rather than the need to limit deformation is the main issue, the critical state strength should be used for intact materials. However, even this is too high where there are pre-existing shear surfaces on which only the residual strength can be mobilized (Section 5.14).

Finally, it must be noted that the application of a given factor of safety to the soil strength results in much smaller load factors for foundations in soils with low values of $\phi'(\sim20°)$ than when ϕ' is high ($\sim35°$). This is illustrated by a comparison of the results of the calculations presented in Examples 8.2 and 8.3. In a design according to Eurocode 7 (BSI, 1995), the possibility of a **serviceability limit state** – for example, structural damage resulting from excessive movement of the foundation – would also need to be considered. In cases where the load factor obtained from the collapse calculation is relatively small, the requirement to avoid a serviceability limit state may well govern the design.

Example 8.3 Calculating the ultimate and design loads for a deep foundation

(a) Explain briefly the two principal mechanisms by which a deep or piled foundation transmits loads to the soil. Are they both always reliable?

Figure 8.18 shows a deep foundation, which consists of a single diaphragm wall panel or **barrette**. Using the data given in Tables 8.2 (Skempton, 1951) and 8.1 (Brinch Hansen, 1970), estimate (b) the short-term and (c) long-term ultimate vertical loads. Also calculate and comment on (d) the short-term and (e) long-term design loads, if strength mobilization factors (factors of safety) of 1.4 (on τ_u) and 1.25 (on $\tan \phi'$) are required.

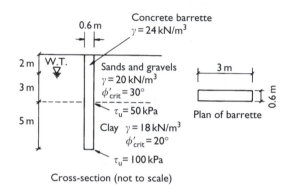

Figure 8.18 Dimensions of barrette foundation and soil profile.

In the short term, assume that the soil/pile adhesion τ_w is equal to $0.5 \times$ the undrained shear strength of the clay τ_u, and that the undrained shear strength of the clay increases linearly from 50 kPa at a depth of 5 m to 100 kPa at the base of the pile. In the long term, assume that the horizontal effective stress at any depth is equal to $(1 - \sin \phi'_{crit})$ times the vertical effective stress at the same depth, that the angle of friction δ between the concrete and the soil is equal to ϕ'_{crit}, and that the long-term pore water pressures are hydrostatic below the indicated water table. Take the unit weight of water as 10 kN/m^3, $N_c = 5.14$ and $N_q = K_p e^{\pi \tan \phi'_{crit}}$, where $K_p = (1 + \sin \phi'_{crit})/(1 - \sin \phi'_{crit})$.

(f) Comment briefly on the assumptions $\sigma'_h = (1 - \sin \phi'_{crit}) \sigma'_v$, $\delta = \phi'_{crit}$ and $\tau_w = 0.5 \tau_u$.

[3rd year BEng (civil engineering) examination, University of London (Queen Mary and Westfield College), modified and extended]

Solution

(a) Load transfer is by (i) skin friction and (ii) base bearing. Skin friction can only be relied upon to support the applied vertical load if the pile tends to settle with respect to the surrounding soil. This may not always be the case, for example if the surrounding ground is surcharged or dewatered so that it compresses or consolidates. Assume in this case that skin friction helps to support the applied vertical load, that is, acts upwards on the pile. The ultimate loads are calculated using the full strengths of the soils.

(b) *Short-term analysis using* $\tau_{max} = \tau_u$ *failure criterion for the clay.* In the sands and gravels, the skin friction is given by $\tau_w = \sigma'_h \tan \delta$. The pore water pressures are hydrostatic below groundwater table, and τ_w is calculated as follows.

The vertical total stress σ_v is calculated from the weight of overburden, using $\gamma = 20\,kN/m^3$ in the sands and gravels, and $\gamma = 18\,kN/m^3$ in the clay.

The vertical effective stress $\sigma'_v = \sigma_v - u$
The horizontal effective stress $\sigma'_h = (1 - \sin \phi'_{crit}) \times \sigma'_v$
The angle of soil/pile friction $\delta = \phi'_{crit}$
The soil/pile shear stress $\tau_w = \sigma'_h \times \tan \phi'_{crit}$

In the clay in the short term, $\tau_w = 0.5\tau_u$, so that over the depth of 5–10 m τ_w increases linearly from $0.5 \times 50 = 25\,kPa$ (at 5 m) to $0.5 \times 100 = 50\,kPa$ (at 10 m).

Values of vertical and horizontal total and effective stress, δ and τ_w for the sands and gravels, and τ_w for the clay in the short term, are given in Table 8.3. The skin friction increases linearly in the range of 0–2 m, 2–5 m and 5–10 m depth.

Table 8.3 Short-term soil/pile shear stresses at key depths

Stratum and depth (m)	σ_v (kPa)	u (kPa)	σ'_v (kPa)	σ'_h (kPa)	δ (degrees)	τ_w (kPa)
Sand and gravel, 0	0	0	0	0	30	0
Sand and gravel, 2	40	0	40	20	30	11.55
Sand and gravel, 5	100	30	70	35	30	20.21
Clay, 5	100	—	—	—	—	25
Clay, 10	190	—	—	—	—	50

The force due to skin friction is equal to the surface (perimeter) area of the barrette × the average shear stress $\tau_{w,av}$ in each zone. The linear distance around the perimeter of the barrette is $2 \times (3\,m + 0.6\,m) = 7.2\,m$.

Thus the total load which may be taken in skin friction is

$$[2\,m \times 7.2\,m \times 0.5 \times (11.5\,kPa)] + [3\,m \times 7.2\,m \times 0.5 \times (11.55\,kPa$$
$$+ 20.21\,kPa)] + [5\,m \times 7.2\,m \times 0.5 \times (25\,kPa + 50\,kPa)]$$
$$= 83\,kN + 343\,kN + 1350\,kN = 1776\,kN$$

The base bearing pressure is calculated using equation (8.9):

$$(\sigma_f - \sigma_0) = \{N_c \times s_c \times d_c\} \times \tau_u$$

where σ_f and σ_0 are the vertical total stresses below and to one side of the pile base at failure (Figure 8.19(a)).

For the barrette, $D = 10\,m$, $B = 0.6\,m$ and $L = 3\,m$, so that the shape factor s_c from Table 8.2 (Skempton, 1951) $= (1 + 0.2\,B/L) = (1 + 0.2 \times 0.6 \div 3) = 1.04$. The depth : breadth ratio $D/B = 16.7$, so that the limiting value of depth factor $d_c = 1.46$ applies. τ_u at the base of the pile is 100 kPa, $\sigma_0 = 190\,kPa$ (from Table 8.3), and $N_c = 5.14$.

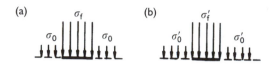

Figure 8.19 Schematic diagram of vertical stresses below and to one side of foundation base: (a) short-term analysis; (b) long-term analysis.

Thus

$$(\sigma_f - 190\,\text{kPa}) = 5.14 \times 1.04 \times 1.46 \times 100\,\text{kPa}$$

or

$$\sigma_f = 970\,\text{kPa}$$

This corresponds to a total base load of $970\,\text{kPa} \times 0.6\,\text{m} \times 3\,\text{m} = 1746\,\text{kN}$.
   The ultimate total load is therefore $1776\,\text{kN} + 1746\,\text{kN} = 3522\,\text{kN}$. This includes the weight of the concrete used to form the foundation, which must be subtracted from the ultimate total load to obtain the ultimate applied load. The weight of the foundation is $[10\,\text{m} \times 0.6\,\text{m} \times 3\,\text{m} \times 24\,\text{kN/m}^3] = 432\,\text{kN}$.
   Thus the ultimate applied load is $3522\,\text{kN} - 432\,\text{kN} = \underline{3090\,\text{kN}}$ in the short term.

(c) *Long-term analysis, using* $(\tau/\sigma')_{\max} = \tan\phi'$ *failure criterion.* In both strata, the long-term skin friction is given by $\tau_w = \sigma'_h \tan\delta$. The long-term values of vertical and horizontal total and effective stress, δ and τ_w for both the sands and gravels and the clay are given in Table 8.4. These were calculated using exactly the same principles as for the sands and gravels in the short-term analysis.

Table 8.4 Long-term soil/pile shear stress at key depths

Stratum and depth (m)	σ_v (kPa)	u (kPa)	σ'_v (kPa)	σ'_h (kPa)	δ (degrees)	τ_w (kPa)
Sand and gravel, 0	0	0	0	0	30	0
Sand and gravel, 2	40	0	40	20	30	11.55
Sand and gravel, 5	100	30	70	35	30	20.21
Clay, 5	100	30	70	46.1	20	16.78
Clay, 10	190	80	110	72.4	20	26.35

Notes
Vertical total stress, σ_v calculated from weight of overburden.
Pore water pressures are hydrostatic below groundwater table.
$\sigma'_v = \sigma_v - u$; $\sigma'_h = (1 - \sin\phi'_{\text{crit}}) \times \sigma'_v$; $\delta = \phi'_{\text{crit}}$; $\tau_w = \sigma'_h \times \tan\phi'_{\text{crit}}$.

As before, the total skin friction force in the long term is given by the surface area of the barrette × the average shear stress $\tau_{w,av}$ in each soil zone:

$$\text{Skin friction force} = [2\,\text{m} \times 7.2\,\text{m} \times 0.5 \times (11.55\,\text{kPa})]$$

$$+ [3\,\text{m} \times 7.2\,\text{m} \times 0.5 \times (11.55\,\text{kPa} + 20.21\,\text{kPa})]$$

$$+ [5\,\text{m} \times 7.2\,\text{m} \times 0.5 \times (16.78\,\text{kPa} + 26.35\,\text{kPa})]$$

$$= 83\,\text{kN} + 343\,\text{kN} + 776\,\text{kN} = 1202\,\text{kN}$$

The base bearing pressure is calculated using equation (8.7), neglecting the N_γ component:

$$\sigma'_f = \{N_q \times s_q \times d_q\} \times \sigma'_0$$

where σ'_f and σ'_0 are the vertical effective stresses below and to one side of the pile base at failure (Figure 8.19(b)).

With $D = 10\,\text{m}$, $B = 0.6\,\text{m}$ and $L = 3\,\text{m}$, the shape factor s_q from Table 8.1 (Brinch Hansen, 1970) = $[1 + (B/L)\tan\phi'_{\text{crit}}] = (1 + 0.2 \times \tan 20°) = 1.073$.

The depth : breadth ratio $D/B = 16.7$, so that the depth factor $d_q = [1 + 2\tan\phi'_{\text{crit}}(1 - \sin\phi'_{\text{crit}})k]$ with $k = \tan^1(D/B) = 1.51$. Thus $d_q = [1 + 2 \times \tan 20°(1 - \sin 20°) \times 1.51] = 1.723$.

σ'_0 at the base of the pile = 110 kPa, from Table 8.4, and $N_q = K_p e^{\pi \tan\phi'_{\text{crit}}} = 6.4$. Hence

$$\sigma'_f = 6.4 \times 1.073 \times 1.723 \times 110\,\text{kPa} = 1301\,\text{kPa}$$

giving a base load due to the effective stress of

$$1301\,\text{kPa} \times 0.6\,\text{m} \times 3\,\text{m} = 2343\,\text{kN}$$

In addition to this there is an upthrust due to the pore water pressure of $3\,\text{m} \times 0.6\,\text{m} \times 80\,\text{kPa} = 144\,\text{kN}$.

The ultimate total load is therefore $1202\,\text{kN} + 2343\,\text{kN} + 144\,\text{kN} = 3689\,\text{kN}$.

As before, the weight of the concrete used to form the foundation must be subtracted from the ultimate total load to obtain the ultimate applied load, giving

$$3689\,\text{kN} - 432\,\text{kN} = \underline{3257\,\text{kN}} \text{ in the long term}$$

The design loads are calculated in the same way, but using factored soil and interface strengths. We will use the strength mobilization factors (factors of safety on soil strength) stipulated in Eurocode 7 (BSI, 1995) for design against failure in the ground (case C) of 1.25 for $\tan\phi'$ and 1.4 for τ_u. We will take interface strengths of $\delta = \phi'_{\text{mob}}$ and, in undrained conditions, $\tau_{w,\text{mob}} = \alpha.\tau_{u,\text{mob}}$ with $\alpha = 0.5$ to match the ultimate load calculations.

(d) *Short-term analysis using* $\tau_{\max} = \tau_u$ *failure criterion for the clay.* In the sands and gravels, the mobilized design strength ϕ'_{mob} is calculated as $\tan\phi'_{\text{mob}} = (\tan\phi'_{\text{crit}})/1.25$ giving $\phi'_{\text{mob}} = 24.8°$. The soil/pile friction angle δ is taken as ϕ'_{mob}, and as before the shear stress due to skin friction is given by $\tau_w = \sigma'_h \tan\delta$. Note that we do NOT factor the soil strength in the calculation of horizontal effective stresses using the expression $\sigma'_h = (1 - \sin\phi'_{\text{crit}}) \times \sigma'_v$. The pore water pressures and the horizontal effective stresses are exactly the same as in Table 8.3. The effect of the reduced soil strength is simply to reduce the skin friction shear stresses in the sands and gravels by the factor $\tan \phi'_{mob} / \tan \phi'_{crit} = 1.25$.

In the clay, the skin friction shear stresses are reduced by a factor of 1.4, so that the design skin friction load in undrained conditions becomes

$$[2\,m \times 7.2\,m \times 0.5 \times (11.55\,kPa) \div 1.25]$$

$$+ [3\,m \times 7.2\,m \times 0.5 \times (11.55\,kPa + 20.21\,kPa) \div 1.25]$$

$$+ [5\,m \times 7.2\,m \times 0.5 \times (25\,kPa + 50\,kPa) \div 1.4]$$

$$= 67\,kN + 274\,kN + 964\,kN = \underline{1305\,kN}$$

The base bearing pressure calculated using equation (8.9) is also simply reduced by the strength mobilization factor for τ_u of 1.4, giving

$$(\sigma_f - 190\,kPa) = 5.14 \times 1.04 \times 1.46 \times 100\,kPa \div 1.4 \quad \text{or}$$

$$\sigma_f = 747\,kPa$$

This corresponds to a total base load of $747\,kPa \times 0.6\,m \times 3\,m = \underline{1345\,kN}$.

The ultimate total load is therefore $1305\,kN + 1345\,kN = 2650\,kN$, from which we must subtract the weight of the foundation (432 kN) to give a design applied load Q_{des} of $\underline{2218\,kN}$. Not surprisingly given the linear relationship between τ_u and the applied load in this case, this is equivalent to a load factor Q_{ult}/Q_{des} of about 1.4.

(e) *Long-term analysis, using* $(\tau/\sigma')_{max} = \tan \phi'$ *failure criterion.* The long-term pore water pressures and effective stresses are the same as in the collapse calculation, so the total long-term skin friction is reduced by the ratio $\tan \phi'_{mob} / \tan \phi'_{crit} = 1.25$. Hence the total design skin friction force is $1202\,kN \div 1.25 = \underline{962\,kN}$.

In the clay, the mobilized effective angle of friction $\phi'_{mob} = \tan^{-1}\{(\tan 20°)/1.25\} = 16.2°$.

The base bearing pressure is again calculated using equation (8.7) neglecting the N_γ component,

$$\sigma'_f = \{N_q \times s_q \times d_q\} \times \sigma'_0$$

$\sigma'_0 = 110\,kPa$ as before, but we must now use $\phi'_{mob} = 16.2°$ to calculate N_q, d_q and s_q (Table 8.2; Brinch Hansen, 1979):

$$s_q = [1 + (B/L) \tan \phi'_{mob}] = (1 + 0.2 \times \tan 16.2°) = 1.058$$

$d_q = [1 + 2 \tan \phi'_{mob} \cdot (1 - \sin \phi'_{mob}) \cdot k]$, with $k = \tan^{-1}(D/B) = 1.51$. Thus $d_q = [1 + 2 \times \tan 16.2° \times (1 - \sin 16.2°) \times 1.51] = 1.633$

$$N_q = K_p e^{\{\pi \tan \phi'_{mob}\}} = 4.42$$

Hence

$$\sigma'_f = 4.42 \times 1.058 \times 1.633 \times 110\,kPa = 840\,kPa$$

giving a base load due to the effective stress of

$$840 \, \text{kPa} \times 0.6 \, \text{m} \times 3 \, \text{m} = \underline{1512 \, \text{kN}}$$

In addition to this, there is an upthrust due to the pore water pressure of $3 \, \text{m} \times 0.6 \, \text{m} \times 80 \, \text{kPa} = 144 \, \text{kN}$.

The design total load is therefore $962 \, \text{kN} + 1512 \, \text{kN} + 144 \, \text{kN} = \underline{2618 \, \text{kN}}$

As before, the weight of the concrete used to form the foundation must be subtracted from the design total load to obtain the design applied load, giving a design applied load of

$$2618 \, \text{kN} - 432 \, \text{kN} = \underline{2186 \, \text{kN}} \text{ in the long term}$$

The load factor in the long-term case is $3257 \div 2186 \approx 1.5$.

The load factors in both the short term and the long term are rather less than the values of 2.5 to 3 that might be applied in a traditional design. This is partly a result of the foundation being mainly in clay: it was shown in Example 8.2 that the application of a strength mobilization factor of 1.25 when $\phi' = 36°$ gave a load factor of about 3 as a result of the exponential increase in N_q with $\tan \phi'$. The calculations we have carried out in (d) and (e) ensure that the foundation will not fail under the expected loading, that is, that a collapse or **ultimate limit state** will not occur. Eurocode 7 (BSI, 1995) also requires a designer to ensure that a **serviceability** failure will not occur. A serviceability failure might involve, for example, a building becoming unserviceable as a result of cracking or other structural damage because the foundations settle too much. This **serviceability limit state** calculation would involve assessing the settlement of the pile, and in cases like the current example where the load factor determined from the ultimate limit state calculation with a factor of safety on soil strength is relatively low, it is likely that the requirement to avoid excessive settlement of the foundation (i.e. the serviceability limit state calculation) would govern the design. Thus for design purposes, we cannot assume that the loads calculated in (d) and (e) represent allowable working loads without also making sure that they will not cause excessive settlement and hence a serviceability limit state. The terms 'ultimate limit state' and 'serviceability limit state' are defined more formally in sections 10.2 and 10.3.

Note that there is no real point in calculating the actual ultimate loads (as in (b) and (c)) if the ultimate limit state design loads (e) and (f) are calculated using the strength mobilization factor or factor of safety on soil strength approach.

(f) The *in situ* horizontal effective stresses in an overconsolidated clay may well be greater than $(1 - \sin \phi'_{\text{crit}})$ times σ'_v. In general, the assumption $\sigma'_h = (1 - \sin \phi'_{\text{crit}})\sigma'_v$ should give a conservative estimate of the lateral stresses, which allows for stress reduction during pile installation.

As the interface between the soils and the concrete is likely to be comparatively rough, the assumption of full soil/pile friction $\delta = \phi'_{\text{mob}}$ is probably reasonable: (for retaining walls, Eurocode 7 (BSI, 1995) and Gaba *et al.* (2003) suggest $\delta \leq \phi_{\text{crit}}$ for rough concrete and $\delta \leq 0.67\phi'_{\text{crit}}$ for smooth (e.g. precast concrete). As the barette is cast in place, the interface between the concrete and the soil is likely to be rough). Also, critical state soil strengths (which are more conservative than peak strengths) have been used in the calculation.

The assumption $\tau_{w,mob} = 0.5\tau_{u,mob}$ is to take account of local softening at the pile/soil interface, during pile installation.

8.8   Pile groups and piled rafts

Piles are often installed in groups. In this case, in investigating the most critical failure mechanism, it is necessary to consider the possibility that some or all of the piles will fail together as a block. Examples of block failure mechanisms are shown in Figure 8.20.

In principle, the analysis of the failing block is exactly the same as the analysis of an individual pile described in section 8.6 and Example 8.3, with the following alterations to the detail.

1. The total 'skin friction' is calculated on the basis of the perimeter area of the entire block.
2. Similarly, the total base bearing force is calculated using the total base area of the block.
3. In an undrained shear strength (total stress) analysis, the full shear strength of the soil may usually be taken to act on the perimeter boundary where there is shearing of soil against soil. (At the corners, and other places where the sliding boundary follows the soil/pile interface, the reduced value of soil/pile shear stress τ_w should still be used.)
4. In an effective stress analysis where the sliding boundary passes through the soil, there is probably no need to reduce the lateral earth pressure to below the *in situ* value to take account of disturbance during pile installation. (Again, where the sliding boundary follows the soil/pile interface, a reduced value of lateral effective stress will probably still be appropriate.)
5. In an effective stress analysis, the N_γ effect will no longer be negligible if the equivalent foundation is now wide in comparison with its depth. Fleming *et al.* (1994) suggest that the base bearing stress at failure will be increased by an amount

$$0.8 \times [0.5\gamma B - \Delta u] \times N_\gamma \tag{8.19}$$

which is equivalent to taking $(s_\gamma \times d_\gamma \times r_\gamma) = 0.8$ in equation (8.7).

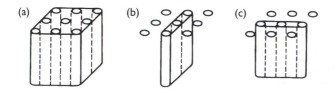

Figure 8.20 Failure of groups of piles en bloc.

Fleming *et al.* (1994) point out that for piles in clay soils, the end bearing pressure q_b is generally of order 10–20 times the average skin friction shear stress τ_w. Thus block failure is only likely to occur when the increase in base bearing area A_b is offset by a much larger (10–20 times) decrease in the surface area of the sliding block A_s. This means that block failure is more likely to be a problem with a large number of long piles, than with a smaller number of shorter piles at the same spacing. In sands, the ratio q_b/τ_w may be of order 50–200, and block failure is even less likely to occur.

Current trends in foundation design tend to favour the use of fewer, more widely spaced piles. In these circumstances, it is unlikely that the failure load calculated assuming failure as a block will be lower than the sum of the failure loads calculated for the piles acting individually. It may therefore be advantageous to prevent the failure of single piles, by ensuring that failure can only occur *en bloc*. For example, a combined pile and raft foundation might be installed, in which the main aim of the piles would be to limit differential settlements to acceptable levels. Randolph and Clancy (1993) suggest that, for this purpose, a small number of piles should be concentrated in the central 25% of the raft area. Horikoshi and Randolph (1996) demonstrate that just nine piles so placed were sufficient to reduce substantially the differential settlement of a raft foundation that, according to a conventional calculation in which the bearing capacity of the raft is ignored, it would require 69 piles to support. Unfortunately, the complex interactions that take place between the raft, the piles and the soil in a piled raft foundation are difficult to analyse, but some simplified design tools are described and assessed by Clancy and Randolph (1996).

For piles of irregular shape, consideration may need to be given to alternative mechanisms of failure, even in the case of an isolated pile. An example of this is a steel pile formed of a universal column, which is H-shaped in cross-section. Failure might take place either by sliding between the steel and the soil (Figure 8.21(a)), or by the formation of slip surfaces between the tips of the flanges (Figure 8.21(b)). In the second case, the area of the shear surface is reduced, but the strength mobilized on the shear surface (taken as τ_u, rather than the reduced value at the soil/pile interface τ_w) is larger. It is not, therefore, immediately obvious which of the failure mechanisms is the more critical.

In the design of piled foundations, it is often necessary to estimate settlements. This is particularly important in the case of large buildings, which may have a combination of piled and raft foundations. The estimation of pile movements under various applied loads is covered in detail by Fleming *et al.* (1994).

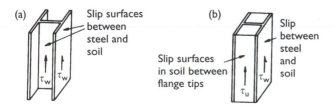

Figure 8.21 Alternative failure mechanisms for a steel H-section pile.

8.9 Lateral loads on piles

Nearly all foundations are subject to some degree of horizontal loading. In a building, this might be induced by the wind; in an offshore oil platform, by wave action; and in a bridge abutment, by vehicle braking or acceleration loads transmitted through the deck. Where the horizontal loads are significant, it has in the past been common practice to install angled or **raking** piles in an attempt to ensure that the resultant load acts along the axis of the pile. This is expensive and in many cases unnecessary, because the capacity of piles to carry lateral loads (i.e. loads perpendicular to their axis) is quite high. It is now more usual to rely on the lateral capacity of vertical piles to carry the horizontal component of the imposed load, unless the horizontal loads are particularly large.

In addition to cases where the load is applied to the top of the pile from the super-structure, piles may be subjected to lateral loading because there is a tendency for the soil to move while the pile remains stationary. One example of this is a piled foundation behind a retaining wall.

For a single pile which is unrestrained at the surface, failure in lateral loading may take place in one of two modes. Short piles will tend to fail by rotation about some point near the toe, as indicated in Figure 8.22(a). Long piles will tend to form a plastic hinge at some depth below the surface, as indicated in Figure 8.22(b). In general, the lateral load H may act at a height or eccentricity e above the soil surface, as shown in Figure 8.22.

If the pile is restrained at the surface by a slab (known as a **pile cap**) which remains horizontal, there are three possible modes of failure in lateral loading. A short, stiff pile could either translate, as indicated in Figure 8.23(a), or it could fail by the formation of a single plastic hinge just below the pile cap (Figure 8.23(b)). A long, slender pile will tend to fail by the formation of two plastic hinges, one just below the pile cap and the other at some depth, as shown in Figure 8.23(c).

The magnitude of the lateral load at failure may be calculated by assuming that the soil is in the passive condition in zones where the pile is being pushed into the soil, and in the active condition in zones where the pile is moving away from the soil. In each of the situations shown in Figures 8.22 and 8.23(b) and (c), there are two unknowns.

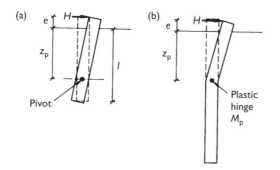

Figure 8.22 Failure modes of single piles in lateral loading. (Redrawn with permission from Fleming *et al.*, 1994.)

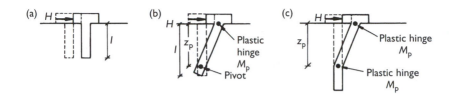

Figure 8.23 Failure modes of single piles in lateral loading: piles restrained at the surface by a pile cap. (Redrawn with permission from Fleming *et al.*, 1994.)

These are the value of H at failure, and the depth z_p to the pivot or the plastic hinge. The values of the two unknowns may be calculated using the equations of horizontal and moment equilibrium for the pile, together with the assumed stress distributions. In Figure 8.23(a), there is apparently only one unknown, H, which is calculated from the condition of horizontal equilibrium. The moment exerted at the top of the pile by the pile cap maintains the pile in moment equilibrium. This moment, which arises from the distribution of vertical stresses on the underside of the pile cap, must be taken into account in the design of the pile cap and the pile–pile cap connection.

In principle, the calculation of the limiting lateral load H is similar to the procedures described in sections 7.6, 7.8.2 and 7.8.3 for embedded cantilever retaining walls. However, the three-dimensional nature of the pile problem results in a tendency for failure at depth to occur by the plastic flow of the soil around the pile. This means that the limiting lateral stresses calculated for the plane strain conditions of the retaining wall analysis will tend to underestimate the net passive resistance at depth.

8.9.1 *Effective stress analysis:* $(\tau/\sigma')_{\mathrm{max}} = \tan\phi'$ *failure criterion*

In an effective stress analysis, the limiting lateral effective force per metre length of pile p'_u may be taken as

$$p'_u = K_p d\sigma'_v \quad \text{at depths } z \leq 1.5d \tag{8.19a}$$

and

$$p'_u = K_p^2 d\sigma'_v \quad \text{at depths } z \geq 1.5d \tag{8.19b}$$

where d is the pile diameter, K_p is the passive earth pressure coefficient $(1 + \sin\phi')/(1 - \sin\phi')$, and σ'_v is the vertical effective stress at depth z. The limiting lateral load at shallow depths corresponds to the formation of a passive type wedge, as in a retaining wall analysis. At greater depths, the mechanism of deformation is horizontal plastic flow, and the limiting lateral load is considerably higher.

Equations (8.19) are to some extent empirical. Fleming *et al.* (1994) quote centrifuge model test data from Barton (1982) which are in agreement with equations (8.19): it is probable that the peak strength, rather than the critical state strength, was used in the computation of K_p. On the basis of equations (8.19), Fleming *et al.* (1994) give charts

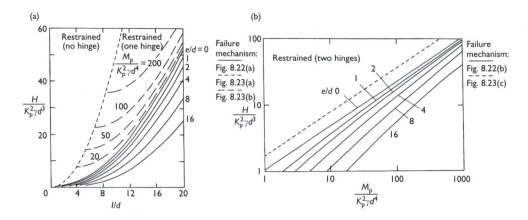

Figure 8.24 Charts giving the ultimate lateral load for piles in non-plastic soils (ϕ' analysis), both unrestrained and restrained at the surface by a pile cap. (a) Short (stiff) piles; (b) long (flexible) piles. (Redrawn with permision from Fleming *et al.*, 1994.)

from which the ultimate lateral load H may be determined for each of the situations shown in Figures 8.22 and 8.23. These charts are reproduced here as Figure 8.24.

For short, stiff piles with no pile cap, which fail as shown in Figure 8.22(a) by rotation about a point near the toe without the formation of a plastic hinge, the solid lines in Figure 8.24(a) give values of the non-dimensionalized lateral load at failure $H/(K_p^2 \gamma d^3)$ as a function of the pile length to diameter ratio l/d, for various values of non-dimensionalized height above ground level at which the lateral load acts, e/d. For short, stiff piles restrained by a pile cap, which fail either in translation as shown in Figure 8.23(a) or by the formation of a plastic hinge at the joint between the pile and the cap as shown in Figure 8.23(b), the dotted lines in Figure 8.24(a) give values of the non-dimensionalized lateral load at failure $H/(K_p^2 \gamma d^3)$ as a function of the pile length to diameter ratio l/d, for various values of the dimensionless parameter $M_p/(K_p^2 \gamma d^4)$. (M_p is the ultimate or **fully plastic** bending moment of the pile at which the entire cross-section is at failure.) It may be seen that for short piles with caps, the failure mechanism (i.e. lateral translation with no hinges, or the formation of one hinge just below the pile cap) depends on the value of $M_p/(K_p^2 \gamma d^4)$ in combination with the length : diameter ratio l/d.

Longer piles without caps fail by the formation of a single plastic hinge as shown in Figure 8.22(b). For these piles, the solid lines in Figure 8.24(b) give values of the non-dimensionalized lateral load at failure $H/(K_p^2 \gamma d^3)$, as a function of the non-dimensionalized pile plastic moment $M_p/(K_p^2 \gamma d^4)$, for various values of the non-dimensionalized height at which the lateral load H acts, e/d. Longer piles with pile caps fail by the formation of two plastic hinges as shown in Figure 8.23(c). In this case, the dotted line in Figure 8.24(b) gives the non-dimensionalized lateral load at failure $H/(K_p^2 \gamma d^3)$, as a function of the non-dimensionalized pile plastic moment $M_p/(K_p^2 \gamma d^4)$.

In using the charts given in Figure 8.24, the value of γ must take account of the effect of pore water pressures in reducing the effective stress. If the soil is dry or the pore water pressures are zero, then the appropriate bulk unit weight of the soil γ_s is used. If the pore water pressures are hydrostatic, then the value of γ must be taken as $\gamma_s - \gamma_w$. If there is vertical seepage, then $\gamma = (\gamma_s - \mathrm{d}u/\mathrm{d}z)$, where $\mathrm{d}u/\mathrm{d}z$ is the rate of increase of pore water pressure with depth.

8.9.2   Total stress analysis: $\tau_{\max} = \tau_u$ failure criterion

Randolph and Houlsby (1984) investigated the lateral load required to push a long circular cylinder through a clay soil of undrained shear strength τ_u, using a lower bound plasticity (stress field) analysis. This solution, which is applicable to the undrained failure of a laterally loaded circular pile in a clay soil, gives a value for the total net lateral load per metre length at failure p_u of between $9.14\,d\tau_u$ (if the pile is frictionless) and $11.94\,d\tau_u$ (if the pile is rough with pile/soil adhesion $\tau_w = \tau_u$), where d is the diameter of the pile. On the basis of this and previous work by Broms (1964), Fleming *et al.* (1994) suggest that the net lateral force at failure per metre length (p_u), acting on a pile moving through a clay of undrained shear strength τ_u, is given by

$$p_u = 2\tau_u d \qquad \text{at the soil surface, } z = 0 \tag{8.20a}$$

$$p_u = [2 + (7z/3d)]\tau_u d \quad \text{for depths } z \le 3d \tag{8.20b}$$

$$p_u = 9\tau_u d \qquad \text{for depths } z \ge 3d \tag{8.20c}$$

As with equation (8.19), the net lateral resisting load at shallow depths corresponds to the formation of a passive wedge, as in a retaining wall analysis. At depths below three times the pile diameter d, the failure mechanism analysed by Randolph and Houlsby (1984) will be developed.

Charts, based on equation (8.20), showing the lateral load which will cause the short-term failure of piles in clay soils of undrained shear strength τ_u, are given by Fleming *et al.* (1994). These charts are reproduced in Figure 8.25.

For short, stiff piles without pile caps, the solid lines shown in Figure 8.25(a) give values of the non-dimensionalized lateral load at failure $H/(\tau_u d^2)$ as a function of the pile length to diameter ratio l/d, for various values of the non-dimensionalized height at which the lateral load H acts, e/d. Short stiff piles with caps may fail either by translation or by the formation of a single plastic hinge. For these piles, the dotted lines in Figure 8.25(a) give non-dimensionalized lateral loads at failure $H/(\tau_u d^2)$ as a function of the pile length to diameter ratio l/d, for various values of non-dimensionalized pile plastic moment $M_p/(\tau_u d^3)$. As before, the failure mechanism for a short pile with a pile cap depends on both the non-dimensionalized plastic moment (in this case, $M_p/[\tau_u d^3]$) and the length : diameter ratio l/d.

For longer piles without caps, the solid lines shown in Figure 8.25(b) give values of the non-dimensionalized lateral load at failure $H/(\tau_u d^2)$ as a function of the non-dimensionalized pile plastic moment $M_p/(\tau_u d^3)$, for various values of the non-dimensionalized height at which the lateral load H acts, e/d. Longer piles with pile caps fail by the formation of two plastic hinges: in this case, the dotted line shown in Figure 8.25(b) gives the non-dimensionalized lateral load at failure $H/(\tau_u d^2)$ as a function of the non-dimensionalized pile plastic moment $M_p/(\tau_u d^3)$.

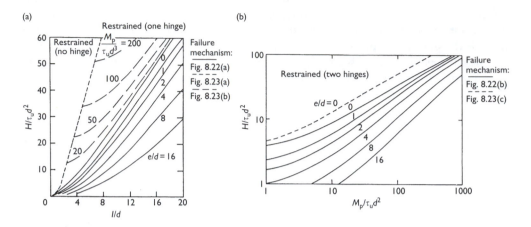

Figure 8.25 Charts for the ultimate undrained lateral load of piles in clay soils (τ_u analysis), both unrestrained and restrained at the surface by a pile cap: (a) short (stiff) piles; (b) long (flexible) piles. (Redrawn with permission from Fleming *et al.*, 1994.)

Figures 8.24 and 8.25 give lateral loads at failure. The design load, as with the axial capacity investigated in section 8.6 and Example 8.3, would be based on the failure load calculated using a design value of soil strength, obtained by reducing the actual soil strength by a mobilization factor M or factor of safety F_s.

8.10   Introductory slope stability: the infinite slope

The stability of slopes in sands, and the long-term stability of slopes in clays, may be investigated by idealizing the slope as infinite and uniform, and considering the effective stresses acting on a plane parallel to the surface of the slope and a depth z below it (Figure 8.26).

For a unit length down a slope at an angle β to the horizontal, the weight W of the block of soil ABCD is $W = \gamma z \cos \beta$. The side forces X are by symmetry equal and opposite, so that the shear stress τ and the normal total stress σ acting on the plane at a depth z below the soil surface may be found by resolution parallel and perpendicular to the slope:

$$\tau = W \sin \beta = \gamma z \cos \beta \sin \beta$$

$$\sigma = W \cos \beta = \gamma z \cos^2 \beta$$

The normal effective stress σ' is given by $\sigma - u$ where u is the pore water pressure. The mobilized soil strength ϕ'_{mob} required to maintain the stability of the slope is given by

$$\tan \phi'_{mob} = \tau / \sigma' = \gamma z \cos \beta \sin \beta / (\gamma z \cos^2 \beta - u)$$

$$= \tan \beta / [1 - (u / \gamma z \cos^2 \beta)] \tag{8.21}$$

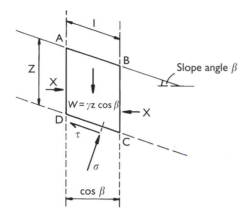

Figure 8.26 Infinite slope analysis.

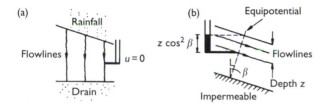

Figure 8.27 Seepage regimes for waterlogged slopes. (a) Downward seepage; (b) flow parallel to slope.

Defining $F_s = \tan\phi'_{crit}/\tan\phi'_{mob}$ as a factor of safety on soil strength,[2] equation (8.21) may be written as

$$F_s = \tan\phi'_{crit}/\tan\phi'_{mob}$$

$$= [1 - (u/\gamma z \cos^2\beta)] \times [\tan\phi'_{crit}/\tan\beta] \tag{8.22}$$

The importance of pore water pressures cannot be overstated. If $u = 0$ (i.e. the slope is dry or there is vertical percolation of water to a drain or an underlying more permeable stratum, Figure 8.27(a)), then ϕ'_{mob} is equal to the slope angle β, and the maximum slope angle is equal to the full (critical state) strength of the soil.

If, at the other extreme, the slope is waterlogged with seepage parallel to the slope (Figure 8.27(b)), $u = \gamma_w z \cos^2\beta$. (This is explained in Example 8.4 and Figure 8.30.) If $\gamma = 2\gamma_w$, $\tan\phi'_{mob} = 2\tan\beta$, that is, the maximum slope angle β_{max} is only approximately half the strength of the soil ϕ'_{crit}. A slope which is placed dry, but without adequate drainage so that it may later become waterlogged, is a disaster waiting to happen.

Example 8.4 Stability of a long slope

Figure 8.28 shows a cross-section through part of a long slope at an angle of inclination to the horizontal β, through which water is flowing at an inclination θ.

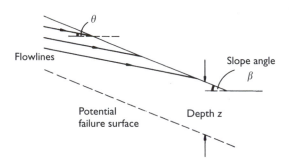

Figure 8.28 Generalized seepage through an infinite slope.

(a) By considering the equipotentials associated with the flow lines shown in Figure 8.28, show that the pore water pressure on a potential failure plane at a depth z vertically below the surface of the slope is given by

$$u = \gamma_w z / (1 + \tan\theta \tan\beta)$$

(b) By considering the limiting equilibrium of an element of the slope, derive an expression relating θ and β to the angle of friction of the soil ϕ'_{crit} when the slope is on the verge of failure. The saturated unit weight of the soil is γ.

(c) What does your expression reduce to in the particular cases

$$\theta = 90° \quad \text{(downward percolation, } u = 0\text{)}$$
$$\theta = \beta \quad \text{(flow parallel to the slope)}$$

and what is the engineering significance of this result?

(d) Figure 8.29 shows a cross-section through a waste tip, which is initially dry. After a prolonged period of heavy rain, the material becomes saturated with water and a shallow pond develops on the top horizontal surface. Assuming that subsequent rainfall is sufficient to keep the top horizontal surface under

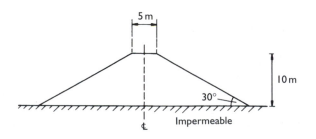

Figure 8.29 Cross-section through waste tip.

a thin layer of water, sketch the steady state seepage flownet corresponding to these new groundwater conditions. Hence show that, if $\phi'_{crit} = 35°$ and $\gamma = 18\,\text{kN/m}^3$, the waste tip will fail.

[*2nd year BEng (civil engineering) examination, University of London (King's College)*]

Solution

(a) Figure 8.30 shows the equipotentials associated with the flowlines indicated in Figure 8.28.

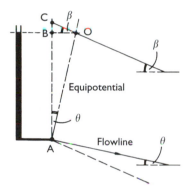

Figure 8.30 Equipotentials and calculation of pore water pressures for infinite slope analysis.

The pore water pressure at the point A, which lies on the potential failure plane at a depth z below the surface of the slope, may be calculated by imagining a standpipe piezometer inserted at A, in which the water level will rise to B. B is at the same level as O. The pore water pressure at A is given by

$$u_A = \gamma_w \times AB$$

Considering the triangle OAB, $OB = AB \times \tan\theta$.
From the triangle OBC, $BC = OB \times \tan\beta$, so that $BC = AB \tan\theta \tan\beta$. But $BC = (z - AB)$ so that

$$z - AB = AB \tan\theta \tan\beta$$

or

$$AB = z/(1 + \tan\theta \tan\beta)$$

and

$$u_A = \gamma_w z/(1 + \tan\theta \tan\beta)$$

Since A is a perfectly general point, this is the pore water pressure at any position along the potential failure plane at a depth z below the surface of the slope.

(b) From Figure 8.26, the weight W of an element of unit width (into the paper), unit length down the slope and depth z is $W = \gamma z \cos \beta$. The side forces cancel, and the shear stress τ and the normal total stress σ acting on the plane at a depth z below the soil surface are found by resolution parallel and perpendicular to the slope:

$$\tau = W \sin \beta = \gamma z \cos \beta \sin \beta$$

$$\sigma = W \cos \beta = \gamma z \cos^2 \beta$$

The normal effective stress σ' is given by $\sigma - u$ where u is the pore water pressure calculated in (a). The soil is in limiting equilibrium when

$$\tau/\sigma' = \tan \phi'_{\text{crit}} = \gamma z \cos \beta \sin \beta / (\gamma z \cos^2 \beta - u)$$

$$= \tan \beta / [1 - (u/\gamma z \cos^2 \beta)]$$

But $u = \gamma_w z / (1 + \tan \theta \tan \beta)$ and $u/\gamma z \cos^2 \beta = [\gamma_w/\gamma][\sec^2 \beta / (1 + \tan \theta \tan \beta)]$, and

$$\sec^2 \beta = 1 + \tan^2 \beta$$

so that

$$u\gamma z \cos^2 \beta = [\gamma_w/\gamma][(1 + \tan^2 \beta)/(1 + \tan \theta \tan \beta)]$$

and failure will occur when

$$\tan \phi'_{\text{crit}} = \tan \beta / \{1 - [\gamma_w(1 + \tan^2 \beta)]/[\gamma(1 + \tan \theta \tan \beta)]\} \quad (8.23)$$

(c) If $\theta = 90°$, $u = 0$ and $\beta = \phi'_{\text{crit}}$ and if $\theta = \beta$,

$$\tan \phi'_{\text{crit}} = \tan \beta / \{1 - [\gamma_w(1 + \tan^2 \beta)]/[\gamma(1 + \tan \beta \tan \beta)]\}$$

$$= \tan \beta / \{1 - [\gamma_w/\gamma]\}$$

or $\tan \beta = 0.5 \tan \phi'_{\text{crit}}$ if $(\gamma_w/\gamma) \approx 0.5$.

The engineering significance of this result is that a waterlogged slope will fail at about half the angle of inclination of a drained slope.

(d) Figure 8.31 shows a flownet for the waste tip, sketched according to the procedures given in sections 3.8 and 3.12.

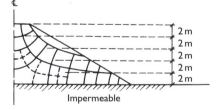

Figure 8.31 Flownet for waterlogged waste tip.

By inspection, the flowlines emerge at between approximately zero and $5°$ to the horizontal. Substituting $\tan\theta = 0$, $\gamma/\gamma_w = 1.8$ and $\phi'_{crit} = 35°$ into equation (8.23), we have

$$\tan 35° = 0.7 = \tan\beta/\{1 - [(1 + \tan^2\beta)/1.8]\}$$

$$\Rightarrow 0.388\tan^2\beta + \tan\beta - 0.311 = 0$$

$$\Rightarrow \tan\beta = 0.28$$

$$\Rightarrow \beta \approx 16°$$

This is the maximum stable slope under the seepage regime shown in Figure 8.31. As it is considerably less than the actual slope of $30°$, the waste tip will certainly fail before steady state seepage has been established.

8.11 Analysis of a more general slope

The infinite slope analysis presented in section 8.10 is appropriate for long, uniform slopes. However, many slopes cannot reasonably be idealized as either long or uniform. Also, it is often important to be able to investigate the essentially localized destabilizing effect of activities such as the construction of a building at the top of a slope, or the formation of an excavation near the bottom. In these cases, the infinite slope analysis is not sufficiently detailed, and it is necessary to use alternative methods which enable variations in slope and/or ground conditions, and the effects of external loads and surcharges to be taken into account.

8.11.1 Total stress analysis for the undrained stability of slopes in clays

The short-term stability of slopes in clay soils is usually investigated by assuming that if failure occurs, the slip surface will be circular in cross-section (Figure 8.32).

If the slipping mass of soil defined by the wedge OAB is in equilibrium, the sum of the disturbing moments about the centre of the circle O must be equal to the sum of the resisting moments. For the particular slope and slip circle shown in Figure 8.32, taking moments about O gives

$$X_W W + R_P P - R_Q Q = R\tau_{mob}\{AB\} \tag{8.24}$$

where τ_{mob} is the shear stress needed to prevent the slope from failing in this particular mechanism and $\{AB\}$ is the arc length along the circumference of the slip circle from A to B.

If the slope is on the verge of failure, the mobilized soil strength τ_{mob} is equal to τ_u, the undrained shear strength of the soil. If the slope is not on the verge of failure τ_{mob} may be expressed as τ_u/F_s, where F_s is the factor of safety of the slope against failure by the particular mechanism shown in Figure 8.32.

The slip circle analysis is essentially an upper bound approach. A number of different slip circles must therefore be investigated in order to find the most critical mechanism, that is the one with the lowest factor of safety F_s.

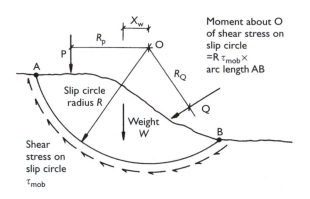

Figure 8.32 Generalized slip circle failure mechanism.

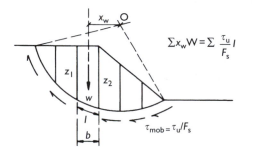

Figure 8.33 Subdivision of slipping soil mass into slices.

Irregular slopes, perhaps with an undrained shear strength which varies with depth, may be investigated by subdividing the mass of soil above the assumed slip surface into a number of slices of width b (Figure 8.33). The weight w of each slice may be calculated according to the trapezium rule, $w = [\gamma b(z_1 + z_2)/2]$. Then, with reference to equation (8.24),

$$X_W W = \sum x_w w \qquad (8.25)$$

where x_w is the lever arm of the weight of the individual slice, and

$$\tau_{mob}\{AB\} = (1/F_s) \int_B^A \tau_u \, d\ell \approx (1/F_s) \sum \tau_u l \qquad (8.26)$$

where the term $(1/F_s) \sum \tau_u \ell$ is the sum of the mobilized shear strength times the length of the base for each slice. In taking F_s outside the summation sign, we have assumed that F_s has the same numerical value for every slice.

In the absence of surface loads such as P and Q, substitution of equations (8.25) and (8.26) into equation (8.24) gives the factor of safety on soil strength $F_s(= \tau_u/\tau_{mob})$:

$$F_s = \left(R \sum \tau_u \ell\right) \Big/ \left(\sum x_w w\right) \qquad (8.27)$$

It must be remembered that the distance x_w for each individual slice may be positive or negative, depending on whether it is to the left or to the right of the centre of the slip circle O. The sign of x_w must be taken into account in evaluating the term $\sum x_w w$.

Repetition of the calculation to find the slip circle for which the factor of safety is lowest is extremely tedious, and is now almost invariably carried out using a computer program. In the days when these calculations were routinely carried out by hand, it was important to work methodically. One common approach was to find F_s for a series of circles of given radius, whose centres lay in a grid pattern as indicated in Figure 8.34. After a few trials had been made, contours of F_s could be plotted and used as a guide to the likely position of the centre of the most critical slip circle. The process had then to be repeated for slip circles of different radii.

Design charts are available for certain idealized slopes. Figure 8.35(a) shows a slope of height H and angle β underlain at a depth Y below the toe by a much stronger and stiffer stratum of soil or rock. This problem was investigated by Taylor (1948),

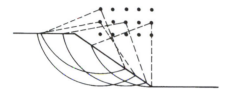

Figure 8.34 Method of searching for the critical slip circle.

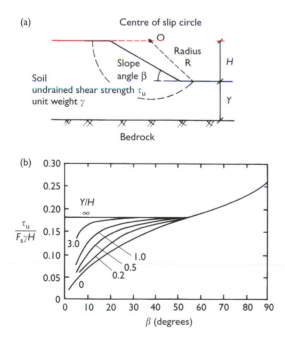

Figure 8.35 (a) Idealized slope geometry studied by Taylor (1948, with permission) and (b) slope stability charts.

who gives curves (reproduced in Figure 8.35(b)) showing the relationship between the dimensionless groups $\tau_u/F_s\gamma H$ and Y/H for various slope angles β, based on the most critical slip circle. The dimensionless group $\tau_u/\gamma H$ is often known as the **stability number**.

Although some slopes can reasonably be represented by a standard case such as that analysed by Taylor (1948), it is now much more common to use one of the many available PC-based programs for slip circle analysis, which can complete the search for the critical slip circle very quickly.

8.11.2   Effective stress analysis of a general slope with a circular slip surface

The generalized slope stability analysis using the total stress failure criterion $\tau_{max} = \tau_u$ is relatively straightforward (the requirement to search for the most critical slip circle being tedious rather than difficult), because there is no need to investigate the stress acting perpendicular to the failure surface. The normal stress has no moment about O, and does not affect the shear stress on the assumed failure surface.

In an effective stress analysis, the shear stress τ on the assumed failure surface is entirely dependent on the normal effective stress σ'. At failure, $(\tau/\sigma') = \tan\phi'_{crit}$ if it is assumed that the stress ratio is greatest on the slip surface. If the slope has a factor of safety on soil strength $F_s = \tan\phi'_{crit}/\tan\phi'_{mob}$, the mobilized shear stress on the slip surface under investigation is

$$\tau_{mob} = (\sigma'\tan\phi'_{crit})/F_s \tag{8.28}$$

Although the normal effective stress σ' has no moment about O, it must be calculated in order to determine the shear stress.

As with the total stress analysis, it is first necessary to divide the sliding mass of soil up into a number of slices. The forces acting on an individual slice are shown in Figure 8.36(a), and the location of the slice within the slipping soil mass in Figure 8.36(b).

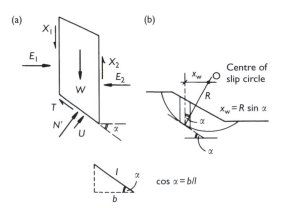

Figure 8.36 (a) Forces acting on a single slice in an effective stress slope stability analysis; (b) location of slice within slipping soil mass.

Assuming that the pore water pressure distribution can be calculated from a flownet, there are at first sight seven unknowns in Figure 8.36(a), even if it is assumed that the side forces act at one-third of the height from the slip surface. These are the horizontal side forces E_1 and E_2, the vertical side forces X_1 and X_2, the effective stress reaction normal to the slip surface $N'(= \sigma' \times l$, where l is the length of the slip surface along the base of the slice), the shear force on the slip surface $T(= \tau_{mob} \times l)$, and the factor of safety on soil strength F_s. (E_1, E_2, X_1 and X_2 are associated with the total stresses acting on the sides of the slice. N' and T are the effective stress resultants, and U is the pore water force, on the base.)

As we have no interest in the side forces, other than to calculate the base reactions N' and T, we can group the resultants $(E_1 - E_2)$ and $(X_1 - X_2)$ together, leaving us with five real unknowns. Unfortunately, we have only four equations relating these unknowns. These are the equations of horizontal and vertical force equilibrium, the equation of moment equilibrium (usually about O), and equation (8.28), which may be used to relate N', T and F_s:

$$(\tau_{mob} \times l) = (\sigma' \times l) \times \tan \phi'_{crit}/F_s$$

or

$$T = N' \tan \phi'_{crit}/F_s \tag{8.29}$$

As we have only four equations in five unknowns, it is necessary to make an additional simplifying assumption. Various different simplifying assumptions have been proposed by different authors: we will now look briefly at the two most common.

(a) Swedish method: Fellenius (1927)

In this method, it is assumed that the inter-slice forces are equal and opposite, so that $(E_1 - E_2) = (X_1 - X_2) = 0$. For Figure 8.36(a) with $(E_1 - E_2) = (X_1 - X_2) = 0$, resolving forces perpendicular to the assumed slip surface,

$$w \cos \alpha = N' + U \tag{8.30}$$

where α is the average angle of inclination of the base of the slice. In this case, α is measured clockwise from the horizontal. Resolving forces parallel to the assumed slip surface,

$$w \sin \alpha = T$$

The disturbing moment about O is

$$\sum x_w w = \sum R \sin \alpha w$$

and the resisting moment is

$$\sum TR$$

which (from equation (8.29)) is equal to

$$\sum (N' \tan \phi'_{crit}/F_s)R \tag{8.31}$$

Substituting for N' from equation (8.30) into equation (8.31), and equating the disturbing and restoring moments about O,

$$\sum R \sin \alpha w = \sum R\{(w \cos \alpha - U) \tan \phi'_{crit}/F_s\}$$

Assuming that R and F_s are the same for all slices,

$$F_s = \frac{\sum[(w \cos \alpha - U) \tan \phi'_{crit}]}{\sum(w \sin \alpha)} \tag{8.32}$$

For a slice of breadth b, average height h and unit weight γ, the weight w is equal to γbh. From Figure 8.36, the length of the base of the slice measured along the slip surface $l = b/\cos \alpha$. In slope stability analysis, the pore water pressure u is usually written as

$$u = r_u \gamma h$$

where r_u is termed the **pore pressure ratio**. r_u is in effect the pore water pressure expressed as a proportion of the average total vertical stress due to the weight of the soil above the slip surface, $r_u = u/\gamma h$. Making these substitutions into equation (8.32), with the pore water force $U = u \times l = ub/\cos \alpha = \gamma h r_u b/\cos \alpha$, we have

$$F_s = \frac{\sum\{[(\gamma bh \cos \alpha) - \gamma h r_u (b/\cos \alpha)] \tan \phi'_{crit}\}}{\sum(\gamma bh \sin \alpha)} \tag{8.33a}$$

If the slice width b is the same for all slices, this can be simplified to:

$$F_s = \frac{\sum\{[(\gamma h \cos \alpha) - (\gamma h r_u/\cos \alpha)] \tan \phi'_{crit}\}}{\sum(\gamma h \sin \alpha)} \tag{8.33b}$$

and if γ is the same for all slices,

$$F_s = \frac{\sum\{[h \cos \alpha - (h r_u/\cos \alpha)] \tan \phi'_{crit}\}}{\sum(h \sin \alpha)} \tag{8.33c}$$

For the infinite slope analysed in section 8.10, h is constant and equal to z. The pore pressure ratio $r_u = u/\gamma h$, and the slip surface angle α is equal to the slope angle β. For any number of slices n, equation (8.33) may then be written as

$$F_s = \frac{n\{[z \cos \beta - (u/\gamma \cos \beta)] \tan \phi'_{crit}\}}{nz \sin \beta}$$

or

$$F_s = \{(z - u/\gamma \cos^2 \beta) \tan \phi'_{crit}\} \div (z \tan \beta)$$
$$= (\tan \phi'_{crit}/\tan \beta) \times (1 - [u/\gamma z \cos^2 \beta])$$

which is identical to equation (8.22). This is not surprising, because the basic assumption made by Fellenius (i.e. that the side forces on each slice cancel each other out exactly) is the same as that used in the infinite slope analysis of section 8.10. We may therefore conclude that Fellenius' method is a reasonable approximation to use when conditions resemble an infinite slope. Essentially, the slope must be reasonably uniform (and therefore probably shallow), with little variation in pore water pressure (i.e. α and r_u are approximately constant).

Equation (8.32) may also be used in the analysis of a non-circular slip, idealized as a series of sliding blocks (Question 8.6).

(b) Bishop's simplified or routine method: Bishop (1955)

Fellenius' assumption that $(E_1 - E_2) = (X_1 - X_2) = 0$ is unnecessarily crude, because it eliminates two unknowns instead of just the one required to make the problem statically determinate. As an alternative, Bishop (1955) suggested that the vertical components of the inter-slice forces might be assumed to be equal and opposite, so that $(X_1 - X_2) = 0$, while the horizontal components need not be assumed to cancel out, i.e. $(E_1 - E_2) \neq 0$. Making this assumption, and (again with reference to Figure 8.36(a)) resolving forces vertically,

$$w = N' \cos\alpha + U \cos\alpha + T \sin\alpha$$

From equation (8.29),

$$T = N' \tan\phi'_{crit}/F_s$$

so that

$$w - U \cos\alpha = N'\{\cos\alpha + (\tan\phi'_{crit}/F_s)\sin\alpha\}$$

or

$$N' = \{w - U\cos\alpha\}/\{\cos\alpha + (\tan\phi'_{crit}/F_s)\sin\alpha\} \tag{8.34}$$

If the effect of the unbalanced side forces on the first and last slices is ignored, the disturbing moment about O for a slipping mass made up of a number of slices is, as in (a) above,

$$\sum x_w w = \sum R \sin\alpha w$$

and the resisting moment is again

$$\sum TR$$

which (from equation 8.29) is equal to

$$\sum (N' \tan\phi'_{crit}/F_s)R \tag{8.31}$$

Substituting for N' from equation (8.34) into equation (8.31), and equating the disturbing and restoring moments about O,

$$\sum wR\sin\alpha = \sum \left[\frac{(w - U\cos\alpha) \times (\tan\phi'_{crit}/F_s) \times R}{[\cos\alpha + (\tan\phi'_{crit}\sin\alpha/F_s)]} \right]$$

Assuming that F_s and R are the same for all slices, and writing the pore water force U in terms of the average pore water pressure u on the base of the slice, $U = ul = ub/\cos\alpha$,

$$F_s = \frac{1}{\sum w \sin\alpha} \times \sum \left[[(w - ub)\tan\phi'_{crit}] \times \left(\frac{1}{[\cos\alpha + (\tan\phi'_{crit}\sin\alpha/F_s)]} \right) \right]$$

(8.35a)

For slices of the same breadth b and unit weight γ, with an average height h and expressing the pore water pressure as $u = r_u \gamma h$, equation (8.35a) may be simplified to

$$F_s = \frac{1}{\sum h \sin\alpha} \times \sum \left\{ [h(1 - r_u)\tan\phi'_{crit}] \times \left(\frac{1}{[\cos\alpha + (\tan\phi'_{crit}\sin\alpha/F_s)]} \right) \right\}$$

(8.35b)

which, since F_s appears on both sides of the equation, must be solved by trial and error, as in Example 8.5. For brevity of notation, the expression

$$\left(\frac{1}{\cos\alpha + (\tan\phi'_{crit}\sin\alpha/F_s)} \right)$$

is often given the symbol n_α.

Both Fellenius' method and Bishop's simplified method are limit equilibrium solutions, based on an assumed circular slip surface. In principle, they can be applied when the slip passes through more than one soil stratum. However, if the soil unit weight γ and the slice width b are not constant, these terms cannot be taken out of the summation. The expressions for the factor of safety F_s cannot then be simplified to the extent of equations (8.33c) and (8.35b). Where γ and/or b vary from slice to slice, equations (8.32), (8.33a) and (8.35a) should be used.

Also, the effects of external forces have not been taken into account in the derivation of equations (8.32), (8.33) and (8.35). If external loads are present, they must be introduced into the calculation at the stage when moments are taken about the centre of the circle O. In evaluating equations (8.32), (8.33) and (8.35), it is important to take account of the sign of $\sin\alpha$, which may be either positive or negative.

In Fellenius' method, it is assumed that the resultant of the inter-slice forces acting on any one slice is zero. In Bishop's simplified method, it is assumed that the resultant of the inter-slice forces is horizontal. Spencer (1967) carried out an analysis of circular slip surfaces, in which it was assumed that the direction of the resultant of the inter-slice forces was the same for all slices, but not necessarily horizontal (i.e. that the inter-slice forces were all parallel). Spencer showed that the calculation was comparatively insensitive to the angle at which the inter-slice forces act, so that the assumption made in Bishop's simplified method (that the inter-slice forces are horizontal) should not generally lead to any significant error.

8.11.3 Non-circular slips

Non-circular slips may be particularly important where there are reasonably well-defined pre-existing planes of weakness, as in the case of the Carsington dam (e.g. Skempton and Coats, 1985; Rowe, 1991). Such planes might correspond to natural layers of weak material, or they may result from the shearing action of compaction plant during the construction of an artificial embankment.

Although derived in section 8.11.2(a) by considering the moment equilibrium of a mass of soil about the centre of a circular slip, equation (8.33a) can also be obtained by resolving forces parallel to the slip surface for each slice, provided it is assumed that the resultant of the inter-slice forces is zero. Equation (8.33a) may therefore be used in the analysis of a non-circular slip divided into a number of conveniently shaped sliding blocks, as in Question 8.6.

Morgernstern and Price (1965) presented a completely general solution for slip surfaces of any shape, in which a functional relationship is introduced between the side shear forces X and the effective stress components of the normal inter-slice forces E'.

A method of slices suitable for general routine use was presented by Janbu (1973). By resolving the forces acting on each slice (Figure 8.36(a)) horizontally and vertically, and assuming (as in the Bishop simplified method) that the resultant of the inter-slice forces is horizontal,

$$F_s = \frac{1}{\sum w \tan \alpha} \times \sum \left[\left[(w - ub) \tan \phi'_{crit} \right] \times \left(\frac{1 + \tan^2 \alpha}{1 + (\tan \phi'_{crit} \tan \alpha / F_s)} \right) \right] \quad (8.36)$$

The main concerns in a slope stability analysis are:

- the appropriate characterization of the soil layers actually present
- the accurate estimation of pore water pressures.

Practically, it will in most cases not be feasible to carry out the search for the most critical slip surface without the aid of a computer program. There are many such programs available. In deciding which one to use, the ability of the program to take account of pore water pressures in a realistic manner is of paramount importance.

A full discussion of slope stability analysis, and slope engineering in general, is given by Bromhead (1986).

Example 8.5 Analysis of a circular slip using the Bishop routine method

The details of a partly completed stability analysis of the slope shown in Figure 8.37(a), using the Bishop routine method, are given in Table 8.5. The configurations of slices 5 and 6 are given in Figure 8.37(b). Complete the analysis to find the factor of safety F_s, to an accuracy of ± 0.1.

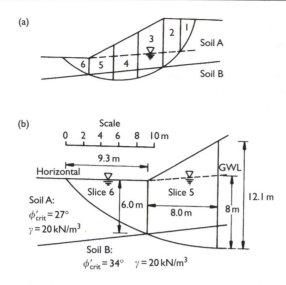

Figure 8.37 (a) Cross-section through slope for stability analysis; (b) detail of slices 5 and 6.

Table 8.5 Details of partly completed Bishop routine analysis of a circular slip: trial $F_s = 2.0$

Slice	Weight w (MN/m)	$\sin \alpha$	$(w - ub) \tan \phi'_{crit}$ (MN/m)	n_α
1	0.61	0.88	0.31	1.44
2	1.42	0.69	0.66	1.11
3	2.59	0.43	1.52	0.95
4	1.57	0.12	0.62	0.97
5		−0.19		
6		−0.53		

Note
$$n_\alpha = \frac{1}{\cos\alpha + (\tan\phi'_{crit} \sin\alpha / F_s)}$$

[University of Southampton, 2nd year BEng (Civil Engineering) examination, slightly modified]

Solution

The first step is to work out the values of w, α, ub and n_α for slices 5 and 6, with the trial factor of safety $F_s = 2$. Idealizing slice 6 as a triangle, the weight of slice 6, w_6, is given by

$$w_6 = 0.5 \times 6\,\text{m} \times 9.3\,\text{m} \times 20\,\text{kN/m}^3 = 0.56\,\text{MN/m}$$

The weight of slice 5, idealized as a trapezium, is

$$w_5 = 0.5 \times (6\,\text{m} + 12.1\,\text{m}) \times 8.0\,\text{m} \times 20\,\text{kN/m}^3 = 1.45\,\text{MN/m}$$

(Although slice 5 is partly in soil A and partly in soil B, both soils have the same unit weight, so there is no need to consider them separately in calculating the weight w_5.)

The pore water pressure acting on the base of slice 6 is zero at the left-hand edge. Scaling from the space diagram (Figure 8.37(b)) and taking the unit weight of water γ_w as $10\,\text{kN/m}^3$, the pore water pressure at the right-hand edge is $6\,\text{m} \times 10\,\text{kN/m}^3 = 60\,\text{kPa}$. The average pore water pressure acting on the base of slice 6 is therefore $30\,\text{kPa}$, giving $(ub)_5 = 30\,\text{kPa} \times 9.3\,\text{m} = 279\,\text{kN/m} = 0.28\,\text{MN/m}$.

Similarly, the average pore water pressure acting on the base of slice 6 is $0.5 \times (60\,\text{kPa} + 80\,\text{kPa}) = 70\,\text{kPa}$, giving $(ub)_6 = 70\,\text{kPa} \times 8\,\text{m} = 0.56\,\text{MN/m}$.

The assumed slip surface at the base of slice 6 passes through soil A, with $\phi'_{crit} = 27°$. For slice 5, $\phi'_{crit} = 34°$ on the base, which is in soil B.

α, the inclination of the base of each slice to the horizontal, is measured from Figure 8.37(b), giving $\alpha_6 = -32°$ and $\alpha_5 = -11°$.

The values of n_α for each slice with $F_s = 2.0$ are calculated using the expression given at the foot of Table 8.5.

We now have sufficient information to calculate both the numerator (i.e. the expression on the top) and the denominator (i.e. the expression on the bottom) of equation (8.35a). If our assumed value of $F_s = 2.0$ is correct, we should find that the numerator, $\sum\{(w-ub)\tan\phi'_{crit}n_\alpha\}$, divided by the denominator, $\sum\{w\sin\alpha\}$, is equal to the assumed factor of safety F_s, in accordance with equation (8.35a). If this is not the case, we must repeat the procedure with new estimates of F_s, until we find the value of F_s that satisfies equation (8.35a). Iteration is relatively straightforward, and convergence should be achieved quite rapidly if the calculated value of $\sum\{(w - ub)\tan\phi'_{crit}n_\alpha\} \div \sum\{w\sin\alpha\}$ is used as the next estimate of F_s.

The calculation is detailed in Table 8.6. In this case, the initial guess of $F_s = 2.0$ is slightly too high. An acceptable value of $F_s = 1.80$ is obtained on the second iteration.

Table 8.6 Completed Bishop routine analysis of a circular slip

Slice	w (MN/m)	$\alpha°$	w sin α (MN/m)	ub (MN/m)	(w − ub) tan ϕ'_{crit} (MN/m)	n_α ($F_s = 2.0$)	n_α ($F_s = 1.8$)	$n_\alpha \times$ (w − ub) tan ϕ'_{crit} ($F_s = 2.0$) (MN/m)	$n_\alpha \times$ (w − ub) tan ϕ'_{crit} ($F_s = 1.8$) (MN/m)
1	0.61	62	0.54		0.31	1.44	1.39	0.45	0.44
2	1.42	43.5	0.98		0.66	1.11	1.09	0.73	0.72
3	2.59	25.5	1.12		1.52	0.95	0.94	1.44	1.43
4	1.57	7	0.19		0.62	0.97	0.96	0.60	0.60
5	1.46	−11	−0.28	0.56	0.60	1.09	1.10	0.65	0.66
6	0.56	−32	−0.30	0.28	0.16	1.40	1.43	0.20	0.20
			$\sum = 2.25$					$\sum = 4.09$	$\sum = 4.05$

Note
The entries in bold type have been calculated; the data in normal type were given in Table 8.5 or Figure 8.37.

With $F_s = 2.0$,

$$\sum\{(w - ub)\tan\phi'_{crit}n_\alpha\} \div \sum\{w\sin\alpha\} = 4.09\,\text{MN/m} \div 2.25\,\text{MN/m} = 1.82$$

Repeating the calculation with $F_s = 1.8$ gives

$$\sum\{(w - ub)\tan\phi'_{crit}n_\alpha\} \div \sum\{w\sin\alpha\} = 4.05\,\text{MN/m} \div 2.25\,\text{MN/m} = 1.80$$

Thus equation (8.35a) is satisfied, and

$$F_s = 1.8$$

In reality, we would have to repeat the entire calculation for a number of different slip circles in order to find the most critical potential failure mechanism (i.e. the one with the lowest factor of safety).

Key points

- The concepts of engineering plasticity can be used to calculate rigorous upper and lower bounds to the vertical loads that will cause the collapse of idealized shallow strip foundations or **footings**.

- Real foundations are not infinitely long. Also, the strength of the soil above the foundation plane is neglected in the idealized plasticity calculations. In an effective stress analysis, the increase in vertical stress with depth below the foundation may add considerably to the bearing capacity. These effects are taken into account by means of empirical factors, whose numerical values depend on the shape and depth of the foundation.

- In general, a foundation may be subjected to a combination of vertical, horizontal and moment loads V, H and M. For a shallow footing of width B, the interaction between them may be investigated by means of a failure surface on a three-dimensional graph with axes V, H and M/B.

- Deep foundations such as **piles** carry some load by **skin friction**, and some in **base bearing**. They may also have a significant capacity to carry a lateral load. The ultimate axial load of a single pile can be estimated by considering the effects of skin friction and base bearing acting together, using a suitably modified form of the bearing capacity equation to calculate the pressure on the base. The ultimate lateral load of a single pile, with or without a pile cap, can be calculated using modified limiting (active and passive) stress distributions.

- The behaviour of a pile group will be influenced by the interaction between neighbouring piles and by the pile cap or, in the case of a piled raft foundation, the raft.

- The stability of a long uniform slope can be investigated using the infinite slope analysis. More generally, the stability of a slope of any shape and finite extent can be analysed by means of a limit equilibrium analysis, in which the soil is divided into a number of slices. The slices are statically indeterminate, so that a simplifying assumption must usually be made. The requirement to search for the most critical slip surface makes the calculation rather tedious: it is best carried out using a computer program.

- The stability of any slope is heavily dependent on the groundwater regime and the associated pore water pressures. The maximum stable angle of a saturated slope with seepage may be only half that of a dry slope. If a steep, dry slope becomes waterlogged, the effects are likely to be catastrophic.

Questions

Shallow foundations

8.1 Figure 8.38 shows a cross-section through a shallow strip footing. Estimate lower and upper bounds to the vertical load Q (per metre length) that will result in the rapid (undrained) failure of the footing.

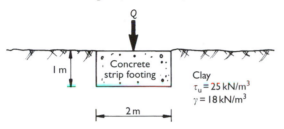

Figure 8.38 Cross-section through shallow strip footing.

[*University of London 2nd year BEng (Civil Engineering) examination, King's College*]

(Lower bound $Q = 236\,\text{kN/m}$; upper bound $Q = 312\,\text{kN/m}$.)

8.2 (a) Explain briefly the essential features of upper and lower bound plasticity analyses as applied to problems in geotechnical engineering.

(b) A long foundation of depth D and width B is built on a clay soil of saturated unit weight γ, undrained shear strength τ_u and effective angle of friction ϕ'. The water table is at a depth D below the soil surface. Show that the vertical load Q, uniformly distributed across the foundation, that will cause failure is given by

$$(Q/B) \geq (\gamma D + 4\tau_u)$$

in the short term, and by

$$(Q/B) \geq (K_p^2 \gamma D)$$

in the long term, where K_p is the passive earth pressure coefficient,

$$K_p = \frac{1 + \sin \phi'}{1 - \sin \phi'}$$

(c) If $\gamma = 20\,\text{kN/m}^3$, $\tau_u = 25\,\text{kPa}$, $\phi' = 22°$ and $D = 1.5\,\text{m}$, is the foundation safer in the short term or in the long term?

[*University of London 2nd year BEng (Civil Engineering) examination, Queen Mary and Westfield College*]

((c) Lower bounds to loads at collapse are 130 kPa (short term) and 145 kPa (long term).)

8.3   A long concrete strip footing founded at a depth of 1 m below ground level is to carry an applied load (not including its own weight) of 300 kN/m. The soil is a clay, with undrained shear strength $\tau_u = 42$ kPa, effective angle of friction $\phi' = 24°$, and unit weight $\gamma = 20$ kN/m^3. Calculate the width of the foundation required to give factors of safety on soil strength of 1.25 (tan ϕ') and 1.4 (τ_u). Both short-term (undrained) and long-term (drained) conditions should be considered. The water table is 1 m below ground level.

Use equations (8.9), with $N_c = (2 + \pi)$, and a depth factor d_c as given by Skempton (Table 8.2), and equation (8.7), with $N_q = K_p e^{\pi \tan \phi'}$ where $K_p = (1 + \sin \phi')/(1 - \sin \phi')$, with d_q, N_γ, d_γ and r_γ as given by Meyerhof and Bowles (Table 8.1). Take the unit weight of concrete as 24 kN/m^3.

[*University of Southampton 2nd year BEng (Civil Engineering) examination, slightly modified*]

(Undrained analysis: required width ≈ 1.93 m. Drained analysis: required width ≈ 2.18 m. As depth factors, etc. depend on footing width, the solutions must be obtained by iteration. Also, it is unusual that the drained analysis gives a more critical result than the undrained analysis.)

Deep foundations

8.4   Figure 8.39 shows a soil profile in which it is proposed to install a foundation made up of a number of circular concrete piles of 1.5 m diameter and 10 m depth. Using the data given below, estimate the long-term design vertical load for a single pile, if a factor of safety of 1.25 on the soil strength tan ϕ' is required.

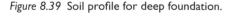

Depth (m)

0 ———————————————————

Groundwater level

2   ▽   Sands and gravels
            $\gamma = 20$ kN/m^3
            $\phi' = 30°$

5 ———————————————————

        Clay
        $\gamma = 18$ kN/m^3
        $\phi' = 20°$

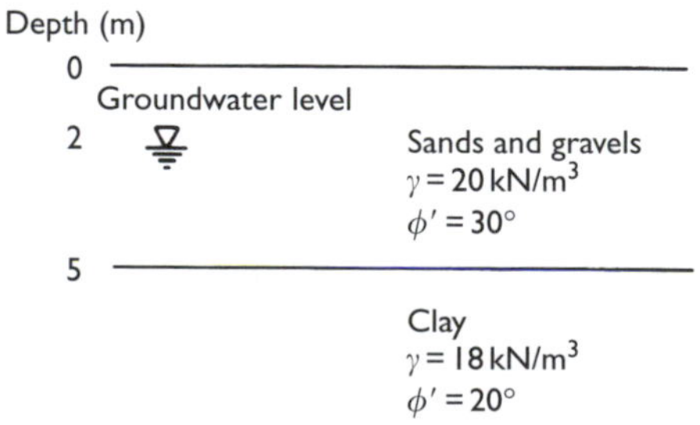

Figure 8.39 Soil profile for deep foundation.

(Assume that the horizontal effective stress at any depth is equal to $(1 - \sin \phi')$ times the vertical effective stress at the same depth, that the angle of friction δ between the concrete and the soil is equal to $0.67\phi'$, and that the long-term pore water pressures are hydrostatic below the indicated water table. Take the unit weight of water as $10\,kN/m^3$, and the unit weight of concrete as $24\,kN/m^3$.)

Data:
Bearing capacity factor $= K_p e^{\pi \tan \phi'} \times$ depth factor $\times$ shape factor, where:

   Depth factor $= (1 + 0.2[D/B])$ up to a limit of 1.5

   Shape factor $= (1 + 0.2[B/L])$

$$K_p = (1 + \sin \phi')/(1 - \sin \phi')$$

and the foundation has width B, length L and depth D.

   Comment briefly on the assumptions $\sigma'_h = (1 - \sin \phi')\sigma'_v$ and $\delta = 0.67\phi'$. Why in reality might it be necessary to reduce the allowable load per pile?

   [*University of London 2nd year BEng (Civil Engineering) examination, Queen Mary and Westfield College*]

(Design applied load = 1675 kN interaction between piles and for the need to limit settlements could reduce this value).

Slopes

8.5   A partly complete stability analysis using the Bishop routine method is given in the table below. The configuration of the remaining slice (slice 4) and other relevant data are given in Figure 8.40. Abstract the necessary additional data from Figure 8.40, and determine the factor of safety of the slope for this slip circle.

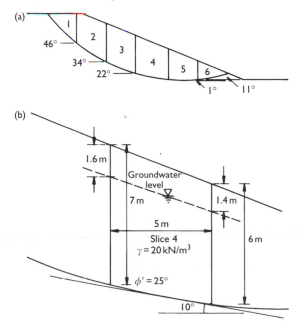

Figure 8.40 (a) Cross-section through slope for stability analysis; (b) detail of slice 4.

Slice	Weight w (kN/m)	ub (kN/m)	$\phi'^{\circ}_{\text{crit}}$	$n_\alpha \times (w - ub)$ $\tan \phi'_{\text{crit}}$ for $F_s = 1.45$ (kN/m)
1	390	0	25	196.5
2	635	90	25	251.8
3	691	163	25	235.1
4	?	?	30	?
5	472	130	30	198.9
6	236	20	30	137.7

[*University of Southampton 2nd year BEng (Civil Engineering)*
examination, slightly modified]

($F_s = 1.3$)

8.6 A slope failure can be represented by the four-slice system shown in
Figure 8.41. By considering the equilibrium of a typical slice (resolving
forces parallel and perpendicular to the local slip surface), and assuming
that the resultant of the inter-slice forces is zero, show that the overall
factor of safety of the slope $F_s = \tan \phi'_{\text{crit}} / \tan \phi'_{\text{mob}}$ may be calculated as

$$F_s = \frac{\sum[(w \cos \alpha - ub) \tan \phi'_{\text{crit}}]}{\sum(w \sin \alpha)}$$

where the symbols have their usual meanings.

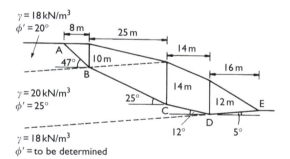

Figure 8.41 Cross-section through non-circular slip.

If the pore pressure conditions which caused failure of the slope shown in
Figure 8.41 can be represented by average pore water pressures of 20 kPa, 30 kPa,
90 kPa and 40 kPa on AB, BC, CD and DE respectively, estimate the value of
ϕ'_{crit} along the failure surface DE.

[*University of Southampton 2nd year BEng (Civil Engineering) examination,*
slightly modified]

(ϕ'_{crit} on DE = 23°)

Notes

1 The net load is the applied load plus the weight of the pile, minus the weight of soil displaced. The latter two items may be disregarded if they approximately cancel each other out.
2 For first-time slides, the long-term stability of slopes is governed by the critical state strength of the soil, ϕ'_{crit}, so that this is the appropriate strength parameter to use in a collapse calculation. If there is a pre-existing slip surface, the long-term stability is likely to be governed by the residual strength of the soil, ϕ'_r, as described in section 5.14. This issue is discussed by Skempton (1964) with reference to a number of case studies in London Clay.

References

Barton, Y.O. (1982) *Laterally Loaded Model Piles in Sand*, PhD dissertation, University of Cambridge.

Bishop, A.W. (1955) The use of the slip circle in the stability analysis of slopes. *Géotechnique*, 5(1), 7–17.

Bolton, M.D. (1991) *A Guide to Soil Mechanics*. MD & K. Bolton, Cambridge.

Bowles, J.E. (1996) *Foundation Analysis and Design*, 5th edn., McGraw-Hill, New York.

Brinch Hansen, J. (1970) *A Revised and Extended Formula for Bearing Capacity*, Danish Geotechnical Institute Bulletin No. 28, DGI, Copenhagen.

British Standards Institution (1995) *Eurocode 7: Geotechnical design, Part 1: general rules*. DDENV 1997-1: 1995. British Standards Institution, London.

Bromhead, E.N. (1986) *The Stability of Slopes*, Surrey University Press, London.

Broms, B.B. (1964) Lateral resistance of piles in cohesive soils. *Journal of the Soil Mechanics and Foundations Division, American Society of Civil Engineers*, 90(SM2), 27–63.

Butterfield, R. (1993) The load displacement response of rigid retaining wall foundations. In *Retaining Structures* (ed. C.R.I. Clayton), pp. 721–730. Thomas Telford, London.

Butterfield, R. and Gottardi, G. (1994) A complete three-dimensional failure envelope for shallow footings on sand. *Géotechnique*, 44(1), 181–184.

Clancy, P. and Randolph, M.F. (1996) Simple design tools for piled raft foundations. *Géotechnique*, 46(2), 313–328.

Fellenius, W. (1927) *Erdstatische Berechnungen mit Reibung und Kohasion (Adhasion) und unter Annahme Kreiszylindrischer Gleitflachen (Earth stability calculations assuming friction and cohesion on circular slip surfaces)*, W. Ernst, Berlin.

Fioravante, V., Jamiolkowski, M. and Pedroni, S. (1994) Modelling the behaviour of piles in sand subjected to axial load, in *Centrifuge 94* (eds C.F. Leung, F.H. Lee and T.S. Tan), pp. 455–460, A.A. Balkema, Rotterdam.

Fleming, W.G.K., Weltman, A.J., Randolph, M.F. and Elson, W.K. (1994) *Piling Engineering*, Blackie and Son, Bishopbriggs.

Gaba, A., Simpson, B., Powrie, W. and Beadman, D.R. (2003) Embedded returning walls – guidance for economic design. Report C580. Construction Industry Research and Information Association, London.

Georgiadis, M. (1985) Load path dependent stability of shallow footings. *Soils and Foundations*, 25(1), 84–88.

Gottardi, G. and Butterfield, R. (1993) On the bearing capacity of surface footings of sand under general planar loads. *Soils and Foundations*, 33(3), 68–79.

Horikoshi, K. and Randolph, M.F. (1996) Centrifuge modelling of piled raft foundations on clay. *Géotechnique*, 46(4), 741–752.

Jaky, J. (1944) The coefficient of earth pressure at rest. *Journal of the Union of Hungarian Engineers and Architects*, 355–358 (in Hungarian).

Janbu, N. (1973) Slope stability computations, in *Embankment Dam Engineering: Casagrande Memorial Volume* (eds R.C. Hirschfeld and S.J. Poulos). John Wiley, New York.

Mayne, P.W. and Kulhawy, F.H. (1982) K_0-OCR relationships in soil. *ASCE Journal of the Geotechnical Engineering Division*, **108**(6), 851–872.

Meyerhof, G.G. (1963) Some recent research on the bearing capacity of foundations. *Canadian Geotechnical Journal*, **1**(1), 16–26.

Meyerhof, G.G. (1976) Bearing capacity and settlement of pile foundations (Terzaghi lecture). *ASCE Journal of the Geotechnical Engineering Division*, **102** GT3, 195–228.

Morgenstern, N.R. and Price, V.E. (1965) The analysis of the stability of general slip circles. *Géotechnique*, **15**(1), 79–93.

Randolph, M.F. and Clancy, P. (1993). Efficient design of piled rafts. *Proceedings of the 2nd International Geotech Seminar on Deep Foundation Bored Auger Piles*, Ghent, Belgium, 119–130.

Randolph, M.F. and Houlsby, G.T. (1984) The limiting pressure on a circular pile loaded laterally in cohesive soil. *Géotechnique*, **34**(4), 613–623.

Rowe, P.W. (1991) A reassessment of the causes of the Carsington embankment failure. *Géotechnique*, **43**(3), 395–421.

Skempton, A.W. (1951) The bearing capacity of clays. *Proceedings of the Building Research Congress*, **1**, 180–189.

Skempton, A.W. (1964) Long-term stability of clay slopes. 4th Rankine Lecture. *Géotechnique*, **14**(2), 75–102.

Skempton, A.W. and Coats, D.J. (1985) Carsington dam failure. *Failures in Earthworks*, 203–220. Thomas Telford, London.

Spencer, E. (1967) A method of analysis of the stability of embankments using parallel interslice forces. *Géotechnique*, **17**(1), 11–26.

Taylor, D.W. (1948) *Fundamentals of Soil Mechanics*, John Wiley, New York.

Whitlow, R. (1995) *Basic Soil Mechanics*, 3rd edn., Longman Scientific & Technical, Harlow.

Chapter 9

Calculation of bearing capacity factors and earth pressure coefficients for more difficult cases, using plasticity methods

9.1   Introduction and objectives

In the zone of soil below a foundation, the major principal stress σ_1' (or σ_1 in a total stress analysis) is vertical. In the zones of soil to each side, the major principal stress is horizontal. There is therefore a 90° rotation in the direction of the principal stresses between the active zone beneath the footing and the passive zones on each side. In the stress field analyses of shallow foundations presented in section 8.2, this 90° rotation in the major principal stress direction was achieved in one 'jump', across a single frictionless **stress discontinuity** (Figure 9.1 (a)).

These frictionless stress discontinuities are extremely unlikely to be present in reality, and an improved (i.e. less conservative) lower bound to the collapse load can be obtained if they are not used in the analysis. There must still be a 90° rotation of the principal stress directions between the active and passive zones, but this can be achieved by means of a series of stress jumps across stronger discontinuities, along which a significant proportion of the soil strength is mobilized. Figure 9.1(b), (c) and (d) show how the 90° rotation can be achieved with two stress jumps of 45°, three jumps of 30°, and an infinite number of infinitesimally small jumps contained within a 'fan zone'.

In some situations – for example, foundations subjected to inclined loads and rough retaining walls – the rotation of the principal stresses through angles other than 90° using non-frictionless stress discontinuities is unavoidable. The aim of this chapter is to show how stress discontinuities and fan zones can be used to calculate bearing capacity factors for foundations with inclined loads, and earth pressure coefficients for rough retaining walls, and retaining walls with non-vertical backs or non-horizontal backfill. In the last section (section 9.9), improved upper bound calculations for the failure of shallow foundations subjected to vertical loads are presented.

The stress analyses presented in this chapter might initially seem somewhat daunting and mathematical. It is true that you will need a thorough understanding of the use of Mohr circles of stress. Having acquired this, and mastered the geometrical relationships introduced in sections 9.2 and 9.3, the calculation of bearing capacity factors for foundations subjected to inclined loads, and earth pressure coefficients for retaining walls with rough or sloping backs and non-horizontal backfills, is simply a matter of working calmly and logically through the procedure given in section 9.4.

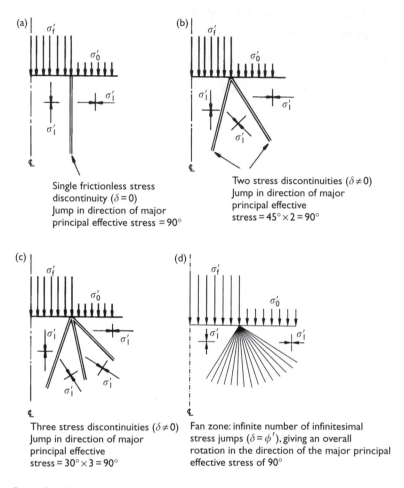

Figure 9.1 Use of stress discontinuities to rotate principal stresses.

Objectives

After having worked through this chapter, you should understand

- how **stress discontinuities,** along which the mobilized soil strength may be less than the full strength of the soil, can be used in lower bound (stress field) solutions to separate zones of stress in which the average total or effective stresses (s or s'), and the principal stress directions, are different (sections 9.1–9.3)
- that in general, better (i.e. less conservative) lower bounds are obtained as the soil strength mobilized along the discontinuities approaches the full strength of the soil (sections 9.1–9.3)
- that the least conservative lower bounds are obtained by using **fan zones,** made up of an infinite number of infinitesimal stress discontinuities along which the

full strength of the soil is mobilized, to rotate the principal stresses as we move between two **uniform stress zones** (sections 9.2–9.4).

You should be able to:

- visualize stress fields in which the principal stress directions are rotated as we move from a **uniform zone** in which the stress state is known in both magnitude and orientation (e.g. a conventional active or passive zone, below a free soil surface), through a **fan zone**, to a second uniform zone in which the stress state is only partly defined (e.g. the zone of soil adjacent to a rough retaining wall, or below a strip foundation) (sections 9.1–9.8)
- calculate the change in average total or effective stress between the two uniform stress zones as a function of the angle through which the principal stress directions are rotated (sections 9.2 and 9.3)
- use the stress field approach to calculate bearing capacity factors N_q and N_c for foundations subjected to inclined loads (section 9.5), and to calculate earth pressure coefficients for rough retaining walls (section 9.6), walls with a non-horizontal backfill (section 9.7), and walls with a sloping or **battered** back (section 9.8).

You should have an appreciation of:

- the significance in a stress field of the α- and β-**characteristics**, along which the full strength of the soil is mobilized (section 9.4)
- the use of mechanisms involving zones which shear as they deform (rather than remaining rigid) to calculate improved upper bounds for the collapse loads of foundations (section 9.9)
- the correlation between the **slip lines** or **velocity discontinuities** in the theoretically correct upper bound mechanism, and the **characteristic directions** in the theoretically correct lower bound stress field (section 9.9).

9.2   Stress discontinuities: ϕ' analysis

In the analysis of section 8.2, the frictionless stress discontinuity separated two zones in which the stress states were different; however, the condition of equilibrium required that the normal effective stress on the plane of the discontinuity was the same in each zone. Imagine now a discontinuity on which the ratio $\tau/\sigma' = \tan\delta$ (where δ is the strength mobilized on the discontinuity, and $\delta \leq \phi'$). For equilibrium both τ and σ' must be the same on both sides of the discontinuity, so that the Mohr circles of effective stress for the two zones intersect, as shown in Figure 9.2(a).

Analysis of the Mohr circles of stress (Figure 9.2) shows that, if the average effective stress s' (which locates the centre of the circle, $s' = 0.5(\sigma'_1 + \sigma'_3)$) increases as we move from zone 1 to zone 2, then

1   The discontinuity is at an angle of $0.5\langle Bs'_1D\rangle = 0.5(\Delta + \delta)$ anticlockwise from the plane on which the major principal effective stress acts in zone 1.
2   The discontinuity is at an angle of $0.5\langle As'_2D\rangle = \pi/2 - 0.5(\Delta - \delta)$ anticlockwise from the plane on which the major principal effective stress acts in zone 2.

3 The jump in the direction of the major principal stress θ is therefore $[\pi/2 - 0.5(\Delta - \delta)] - [0.5(\Delta + \delta)] = (\pi/2 - 0.5\Delta + 0.5\delta - 0.5\Delta - 0.5\delta) = (\pi/2 - \Delta)$.

4 Considering the triangle s_1'DC, DC $= t_1 \sin(\Delta + \delta)$. Considering the triangle s_2'DC, DC $= t_2 \sin(\Delta - \delta)$. Therefore, $(t_2/t_1) = (s_2'/s_1') = \sin(\Delta + \delta)/\sin(\Delta - \delta)$. Substituting for $\Delta(= \pi/2 - \theta)$, $s_2'/s_1' = \sin(\Delta + \delta)/\sin(\Delta - \delta) = \sin(\pi/2 - \theta + \delta)/\sin(\pi/2 - \theta - \delta)$. As $\sin(\pi/2 \pm A) = \cos A$, $\sin(\pi/2 - \theta + \delta)/\sin(\pi/2 - \theta - \delta) = \cos(\delta - \theta)/\cos(\delta + \theta)$.

5 Applying the sine rule to the triangle ODs_1', $s_1'/\sin \Delta = t_1/\sin \delta$, or $t_1/s_1' = \sin \phi' = \sin \delta/\sin \Delta$, so that $\sin \Delta = \sin \delta/\sin \phi'$.

For an infinitesimal discontinuity which is as strong as the soil, $\delta \to \phi'$; the rotation of the major principal effective stress $\theta \to \delta\theta$; and $s_2' - s_1' \to \delta s'$. Setting s_1' equal to a general value, s',

$$s_2'/s_1' = (s' + \delta s')/s' = \cos(\phi' - \delta\theta)/\cos(\phi' + \delta\theta)$$

As $\cos(A \pm B) = \cos A \cos B \mp \sin A \sin B$, and since $\delta\theta$ is small so that $\cos \delta\theta \approx 1$ and $\sin \theta \approx \delta\theta$,

$$1 + (\delta s'/s') \approx (\cos \phi' + \delta\theta \sin \phi')/(\cos \phi' - \delta\theta \sin \phi')$$

Dividing through the top and bottom of the expression on the right-hand side by $\cos \phi'$, and using the binomial expansion $(1 + x)^n \approx 1 + nx$ if n is very small,

$$1 + (\delta s'/s') \approx (1 + \delta\theta \tan \phi') \times (1 - \delta\theta \tan \phi')^{-1}$$
$$\approx (1 + \delta\theta \tan \phi') \times (1 + \delta\theta \tan \phi')$$
$$\approx 1 + 2\delta\theta \tan \phi'$$

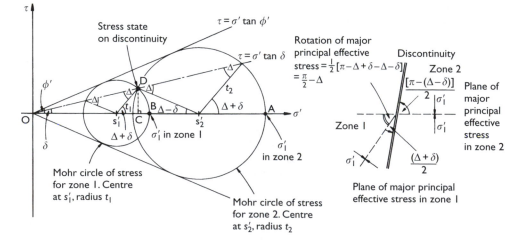

Figure 9.2 (a) Mohr circles of effective stress for (b) zones of soil on either side of a stress discontinuity of strength δ.

or

$$(\delta s'/s') \approx 2\delta\theta \tan\phi' \tag{9.1}$$

For a zone comprising an infinite number of infinitesimal discontinuities (known as a **fan zone** – e.g. Figure 9.1(d)), which results in an overall rotation of the major principal effective stress direction of θ, integration of equation (9.1) between limits of $s' = s'_1$ at $\theta = 0$ and $s' = s'_2$ at $\theta = \theta$ gives

$$s'_2/s'_1 = e^{2\theta \tan\phi'} \tag{9.2}$$

The angle subtended at the centre of the fan zone is equal to the rotation of the principal stress direction, θ.

9.3   Stress discontinuities: τ_u analysis

In the case of an analysis using the undrained shear strength failure criterion, the strong discontinuity has a shear strength $\tau(\leq \tau_u)$. Again, the condition of equilibrium requires that both τ and σ must be the same on both sides of the discontinuity, and the Mohr circles of total stress for the two zones intersect as shown in Figure 9.3.

   If the average total stress s (which locates the centre of the Mohr circle, $s = 0.5(\sigma_1 + \sigma_3)$) increases as we move from zone 1 to zone 2, then

1   The discontinuity is at an angle of $0.5\langle Cs_1D\rangle = 0.5\Delta$ anticlockwise from the plane on which the major principal total stress acts in zone 1.
2   The discontinuity is at an angle of $0.5\langle As_2D\rangle = 0.5(\pi - \Delta)$ anticlockwise from the plane on which the major principal total stress acts in zone 2.
3   The jump in the direction of the major principal total stress θ is therefore $[0.5(\pi - \Delta)] - [0.5\Delta] = (\pi/2 - \Delta)$.
4   From the triangles s_1CD and s_2CD, the difference between the average total stresses $s_2 - s_1$ is $2\tau_u \cos\Delta$ where $\sin\Delta = \tau/\tau_u$.

For an infinitesimal discontinuity which is as strong as the soil, $\tau \to \tau_u$; the rotation of the major principal total stress $\theta \to \delta\theta$; and

$$s_2 - s_1 \to \delta s = 2\tau_u \cos\Delta = 2\tau_u \cos(\pi/2 - \delta\theta) = 2\tau_u\delta\theta \tag{9.3}$$

(because $\cos(\pi/2 - \delta\theta) = \sin\delta\theta$ and $\sin\delta\theta \to \delta\theta$ as $\delta\theta \to 0$).

   For a fan zone comprising an infinite number of infinitesimal discontinuities, which results in an overall rotation of the major principal total stress direction of θ, integration of equation (9.3) between limits of $s = s_1$ at $\theta = 0$ and $s = s_2$ at $\theta = \theta$ gives

$$s_2 - s_1 = 2\tau_u\theta \tag{9.4}$$

Again, the angle subtended at the centre of the fan zone is θ.

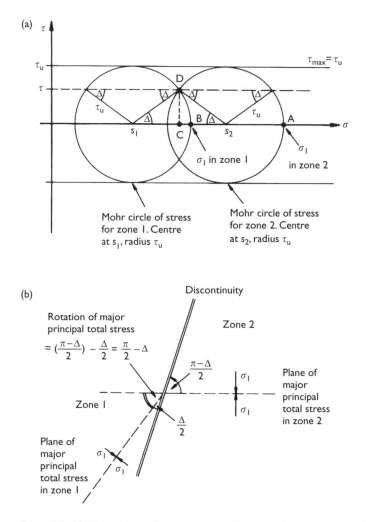

Figure 9.3 (a) Mohr circles of total stress for (b) zones of soil on either side of a stress discontinuity of strength τ.

9.4 Application to stress analysis

9.4.1 General approach

Stress discontinuities may be used to calculate improved lower bounds to the bearing capacity of a foundation, and also to analyse other problems in which there is a rotation in the direction of the major principal stress between two zones of soil. Examples include the bearing capacity of footings subjected to inclined loads; rough retaining walls; and retaining walls with a sloping backfill. The general approach is as follows.

1   Identify the first zone, in which the principal stresses are known in both magnitude and direction. Usually, this zone will be immediately below a free soil surface. The

magnitude of the average stress s'_1 (or s_1 in a total stress analysis) may be a function of depth z. For foundations, it is usual to carry out the analysis at $z = 0$ (i.e. failure at the surface is assumed to correspond to failure of the whole mass, or the weight of the soil is neglected, as in section 8.2). For retaining walls, the analysis is usually carried out for a general depth z (at which $\sigma_v = \gamma z$ in the zone of soil below the free surface), in order to obtain an earth pressure coefficient.

2   Identify the second zone, in which the principal stresses are known in direction but not in magnitude. This might be beneath the load in the case of a footing, or against the wall in the case of a retaining wall analysis.

3   Deduce the angle θ through which the major principal stress rotates between zone 1 and zone 2. Use equation (9.2) (ϕ' analysis) or equation (9.4) (τ_u analysis) to calculate s'_2 (s_2 in a total stress analysis) in terms of s'_1 (or s_1) and θ, both of which are known. Make sure that the relative magnitudes of the average stresses in zones 1 and 2 are correct. In equations (9.2) and (9.4) it is assumed that $s'_2 > s'_1$ and $s_2 > s_1$, but depending on the problem under investigation, the relative magnitudes of s'_2 and s'_1 (or s_2 and s_1) could be reversed.

The application of the general method is illustrated for a number of specific cases in sections 9.5–9.8. In a retaining wall analysis, the invocation of wall friction is usually advantageous. For foundations, it is shown in section 9.5 that the effect of an inclined load on a footing is to reduce quite significantly the bearing capacity factors N_q and N_c.

9.4.2 Visualization of stress fields using characteristic directions

The aim of a lower bound analysis is to demonstrate the existence of a possible equilibrium stress state, in which the combination of shear and normal stresses never moves outside the limits prescribed by the failure criterion for the soil. The stress field in the soil around a retaining wall or a shallow foundation may be built up from zones in which the principal stress directions remain constant (e.g. uniform active and passive zones) and fan zones within which a smooth and continuous rotation of the principal stress directions takes place.

The various stress zones can be represented visually by means of the directions of the planes on which the soil is at failure, that is the planes on which the stress state touches the failure envelope. These are known as the **characteristic directions**. The direction of the plane on which the shear stress is positive (i.e. anticlockwise) is termed the **α-characteristic**, and the direction of the plane on which the shear stress is negative (i.e. clockwise) is termed the **β-characteristic**.

The Mohr circle of effective stress for an analysis using the maximum stress ratio failure criterion $(\tau/\sigma')_{max} = \tan \phi'$ (Figure 9.4(a)) shows that the α-characteristic, on which the shear stress is positive, is at an angle of $(45° + \phi'/2)$ anticlockwise from the plane on which the major principal effective stress acts. The β-characteristic, on which the shear stress is negative, is at an angle of $(45° + \phi'/2)$ clockwise from the plane on which the major principal effective stress acts. In a uniform active zone, the plane on which the major principal effective stress acts is horizontal, so that the characteristic directions are at $\pm(45° + \phi'/2)$ to the horizontal (Figure 9.4(b)). In a uniform passive zone, the plane on which the major principal effective stress acts is vertical, so that

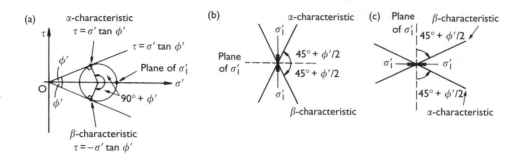

Figure 9.4 Characteristic directions for a $(\tau/\sigma')_{max} = \tan \phi'$ analysis. (a) Mohr circle of effective stress; (b) active zone; (c) passive zone.

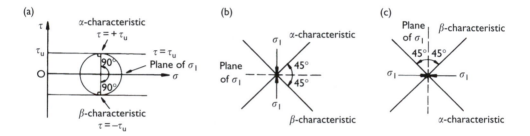

Figure 9.5 Characteristic directions for a $\tau_{max} = \tau_u$ analysis. (a) Mohr circle of total stress; (b) active zone; (c) passive zone.

the characteristic directions are at $\pm(45° + \phi'/2)$ to the vertical, or at $\pm(45° - \phi'/2)$ to the horizontal (Figure 9.4(c)).

In an analysis using the maximum shear stress failure criterion $\tau_{max} = \tau_u$, the characteristic planes are those on which the stress state is $\tau = \pm\tau_u$. The Mohr circle of total stress (Figure 9.5(a)) shows that the α-characteristic, on which the shear stress is positive, is at an angle of 45° anticlockwise from the plane on which the major principal total stress acts. The β-characteristic, on which the shear stress is negative, is at an angle of 45° clockwise from the plane on which the major principal total stress acts. In uniform active and passive zones, the characteristic directions are at $\pm45°$ to the horizontal (Figures 9.5(b) and (c)). The α-characteristic is at 45° anticlockwise from the horizontal in an active zone, and at 45° clockwise from the horizontal in a passive zone.

9.5 Shallow foundations

In sections 9.5.1 and 9.5.2, the introduction of fan zones between the active zone beneath a shallow foundation and the passive zones to either side is used to calculate improved bearing capacity factors N_q and N_c. In Examples 9.1–9.3, the technique is used to calculate the bearing capacities of footings subjected to inclined loads.

9.5.1   Calculation of improved bearing capacity factor for a shallow foundation subjected to a vertical load: effective stress (ϕ') analysis

Figure 9.6 shows the active and passive zones below and beside one half of a shallow strip footing, which is assumed to be symmetrical about the centreline. In the passive zone (zone 1), the vertical effective stress at the surface ($z = 0$) is equal to σ'_0, and the Mohr circle may be drawn as shown in Figure 9.7(a). In the active zone below the foundation (zone 2), the vertical effective stress at $z = 0$ is equal to the vertical stress applied by the foundation σ'_f, which is known in direction (it is vertical) but not in magnitude. As the angular geometry of the Mohr circle is unaffected by the magnitudes of the stresses, the Mohr circle of effective stress for zone 2 may be drawn to an arbitrary scale (Figure 9.7(b)).

From the Mohr circle of effective stress shown in Figure 9.7(a), the average effective stress s'_1 in zone 1 is given by

$$s'_1 = \sigma'_0 + t_1 = \sigma'_0 + s'_1 \sin \phi'$$

or

$$s'_1 = \sigma'_0/(1 - \sin \phi') \tag{9.5}$$

As zone 1 is a passive zone, the plane on which the major principal effective stress acts is vertical.

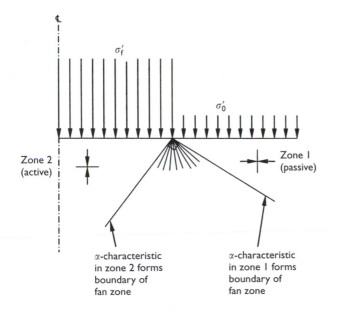

Figure 9.6 Failure of shallow footing with vertical load: fan zone separating active and passive zones.

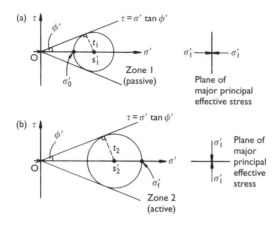

Figure 9.7 Mohr circles of effective stress for (a) passive and (b) active zones adjacent to and below a shallow foundation: ϕ' analysis.

From the Mohr circle of effective stress shown in Figure 9.7(b), the average effective stress s_2' in zone 2 is given by

$$s_2' = \sigma_f' - t_2 = \sigma_f' - s_2' \sin \phi'$$

or

$$\sigma_f' = (1 + \sin \phi')s_2' \tag{9.6}$$

As zone 2 is an active zone, the plane on which the major principal effective stress acts is horizontal. The rotation θ in the direction of the major principal effective stress through the fan zone is therefore 90° or $\pi/2$ radians. As the fan zone is made up of a series of infinitesimal discontinuities on which the full strength of the soil is mobilized (i.e. $\tau = \sigma' \tan \phi'$), the boundaries to the fan zone must be along the α- and β-characteristic directions in the uniform active and passive zones below and to each side of the footing. On the right-hand side of the footing, the fan zone boundaries are the α-characteristics. On the left-hand side of the footing (which is not shown in Figure 9.6), the fan zone boundaries are the β-characteristics. From Figures 9.4(b) and (c), the α-characteristics must intersect the β-characteristics at an angle of $90° \pm \phi'$ in all three zones.

Using equation (9.2) to calculate the ratio s_2'/s_1' for a rotation in the direction of the major principal effective stress $\theta = \pi/2$,

$$s_2'/s_1' = e^{\pi \tan \phi'} \quad \text{or} \quad s_2' = s_1' e^{\pi \tan \phi'} \tag{9.7}$$

Substituting for s_2' from equation (9.7) into equation (9.6),

$$\sigma_f' = (1 + \sin \phi')s_2' = (1 + \sin \phi')s_1' e^{\pi \tan \phi'} \tag{9.8}$$

Substituting for s_1' in terms of σ_0' from equation (9.5) into equation (9.8),

$$\sigma_f' = (1 + \sin \phi')s_1' e^{\pi \tan \phi'} = \sigma_0'[(1 + \sin \phi')/(1 - \sin \phi')]e^{\pi \tan \phi'}$$

or

$$\sigma_f'/\sigma_0' = [(1 + \sin \phi')/(1 - \sin \phi')]e^{\pi \tan \phi'}$$

Thus the bearing capacity factor $N_q = \sigma_f'/\sigma_0'$ (cf. equation (8.1), section 8.2.1) is given by

$$N_q = \sigma_f'/\sigma_0' = K_p e^{\pi \tan \phi'} \tag{9.9}$$

where K_p is the passive earth pressure coefficient, $(1 + \sin \phi')/(1 - \sin \phi')$.

Strictly, the solution is incomplete, as the two fan zones on either side of the centreline will interfere with each other at depth. For a rigorous lower bound solution, it would be necessary to introduce further stress discontinuities to prevent this from happening (see Abbott, 1966).

9.5.2 Calculation of improved bearing capacity factor for a shallow foundation subjected to a vertical load: total stress (τ_u) analysis

Figure 9.8(a) shows the Mohr circle of total stress for the uniform passive zone to one side of the footing (zone 1), in which the vertical total stress at depth $z = 0$ is equal to σ_0. Figure 9.8(b) shows the Mohr circle of total stress for the uniform active zone (zone 2) below the foundation, in which the vertical total stress at depth $z = 0$ is equal to the foundation load σ_f.

From the Mohr circle of total stress shown in Figure 9.8(a), the average total stress s_1 in zone 1 is given by

$$s_1 = \sigma_0 + \tau_u \tag{9.10}$$

From the Mohr circle of total stress shown in Figure 9.8(b), the average total stress s_2 in zone 2 is given by

$$s_2 = \sigma_f - \tau_u \tag{9.11}$$

As before, the rotation θ in the direction of the major principal effective stress through the fan zone is 90° or $\pi/2$ radian. Using equation (9.4) to calculate the difference in

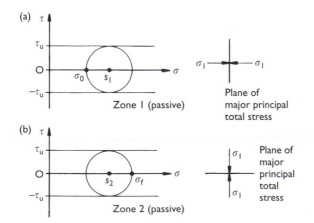

Figure 9.8 Mohr circles of total stress for (a) passive and (b) active zones adjacent to and below a shallow foundation: τ_u analysis.

average total stress $s_2 - s_1$,

$$s_2 - s_1 = 2\tau_u \pi/2 = \tau_u \pi \qquad (9.12)$$

Substituting for s_2 and s_1 from equations (9.11) and (9.10) into equation (9.12),

$$(\sigma_f - \tau_u) - (\sigma_0 + \tau_u) = \tau_u \pi$$

or

$$(\sigma_f - \sigma_0) = (2 + \pi)\tau_u$$

so that the bearing capacity factor N_c is given by

$$N_c = (\sigma_f - \sigma_0)/\tau_u = (2 + \pi) \qquad (9.13)$$

(cf. equation (8.3b), section 8.2.2).

Again, a full lower bound solution would require the introduction of further stress discontinuities to prevent the two fan zones from interfering with each other at depth.

9.5.3 Shallow foundations subjected to inclined loads

The use of stress field analyses with fan zones to calculate the bearing capacity of shallow foundations subjected to inclined loads is illustrated in Examples 9.1–9.3.

Example 9.1 Failure of a strip footing subjected to an inclined load: effective stress (ϕ') analysis

A long wall is to be built on a soil of angle of shearing resistance $\phi'_{crit} = 30°$ and unit weight $20\,kN/m^3$. The wall will be founded above the water table, on a 1 m deep concrete strip footing. The loads transmitted to the soil will be 75 kN/m vertically and 25 kN/m horizontally. Assuming that these loads are uniformly distributed over the area of the foundation, construct Mohr circles of effective stress for the zones of soil beneath and adjacent to the footing, for a footing width such that these zones of soil are on the verge of failure with the mobilization of the critical state soil strength ϕ'_{crit}. Indicate the planes on which the major principal effective stress acts in each zone and calculate the footing width.

(It may be assumed without proof that the ratio of average principal effective stress between two uniform stress zones separated by a fan zone of included angle θ is given by $s'_2/s'_1 = e^{2\theta \tan \phi'}$, and that the rotation of the direction of major principal effective stress is θ).

Is your answer an upper or a lower bound, and why?

[3rd year BEng civil engineering examination, University of London
(Queen Mary and Westfield College), slightly modified]

Solution

The loads given are those which act on the surface of the soil, and therefore include the weight of the foundation itself. If the footing has width b, the normal effective stress applied to the soil is $\sigma'_n = 75/b$ kPa, and the shear stress τ is $25/b$ kPa.

The first step is to investigate the applied stresses which would cause the foundation just to fail.

The footing is shown in cross-section in Figure 9.9(a). Figure 9.9(b) shows the idealized loading applied to the soil, which will be used in the analysis.

Zone 1, to one side of the footing, is a passive zone in which the major principal effective stress is horizontal. The vertical effective stress at founding plane level is $\sigma'_0 = 20$ kPa, due to the weight of the overlying soil. This is the minor principal effective stress, and the Mohr circle of effective stress for zone 1 may be drawn as shown in Figure 9.9(c).

The major principal effective stress in zone 1 is $(1+\sin\phi'_{crit})/(1-\sin\phi'_{crit})\times\sigma'_0$. In the present case, with $\sigma'_0 = 20$ kPa and $\phi'_{crit} = 30°$, the major principal effective stress is equal to 60 kPa, and the average principal effective stress $s'_1 = 40$ kPa.

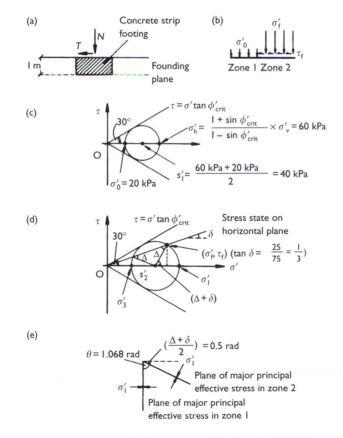

Figure 9.9 Loads on (a) footing and (b) the soil; (c) Mohr circle of effective stress for zone 1; (d) Mohr circle of effective stress for zone 2; (e) rotation of plane of major principal effective stress.

Alternatively, the average effective stress in zone 1 is given by equation (9.5);

$$s_1' = \sigma_0'/(1 - \sin \phi_{crit}') \tag{9.5}$$

In zone 1, the major principal effective stress is horizontal, that is, the plane on which it acts is vertical.

In zone 2 (beneath the footing), the stresses depend on the foundation width, and are therefore unknown in magnitude. It is known, however, that the ratio of $\sigma_f' : \tau_f$ is $75 : 25 = 3 : 1$. On the Mohr circle diagram, the stress state on the horizontal plane in zone 2 must lie on a line drawn at an angle $\delta = \tan^{-1}(1/3) = 18.44°$ from the origin (taking anticlockwise shear stress as positive), as shown in Figure 9.9(d).

From the Mohr circle of effective stress for zone 2 (Figure 9.9(d)), the plane on which the major principal effective stress acts is $(\Delta + \delta)/2$ clockwise from the horizontal (see also Figure 9.2(a)). The rotation of the plane on which the major principal effective stress acts as we go from zone 1 to zone 2 is therefore $\theta = [(\pi/2) - (\Delta + \delta)/2]$ (Figure 9.9(e)).

From the geometry of the Mohr circle (Figure 9.9(d) or Figure 9.2(a)),

$$\sin \Delta = \sin \delta / \sin \phi_{crit}' \tag{9.14}$$

In the present case, $\delta = 18.44°$ and $\phi_{crit}' = 30°$, giving $\Delta = 39.2°$. Thus $(\Delta + \delta)/2 = 28.82° = 0.503$ radians, and $\theta = [(\pi/2) - (\Delta + \delta)/2] = 1.068$ radians.

Applying equation (9.2),

$$s_2' = s_1' \times e^{2\theta \tan \phi_{crit}'} = s_1' \times e^{(\pi - \Delta - \delta) \tan \phi_{crit}'} \tag{9.15}$$

with $\theta = 0.5(\pi - \Delta - \delta) = 1.068$ radians, $\phi_{crit}' = 30°$ and $s_1' = 40\,kPa$,

$$\Rightarrow s_2' = 40\,kPa \times e^{2 \times 1.068 \times \tan 30°} = 137.3\,kPa$$

From the Mohr circle of stress for zone 2 (Figure 9.9(d)), the vertical component of the load on the footing σ_f' is given by

$$\sigma_f' = s_2' + t_2 \cos[\Delta + \delta] = s_2'(1 + \sin \phi_{crit}' \cos[\Delta + \delta]) \tag{9.16}$$

which in the present case is equal to

$$137.3\,kPa \times (1 + \sin 30° \cos 57.64°) = 174\,kPa$$

This is the vertical stress component which will cause failure, which will occur at a footing width b given by

$$75/b = 174$$

or

$$b = 0.43\,m$$

The solution is in principle a lower bound, as it is based on a system of equilibrium stresses which does not violate the failure condition for the soil. However, we have not shown that the stress field can be extended throughout the entire body of soil (i.e. to infinite depth and to infinite lateral extent), so the solution as it stands is strictly incomplete. The extension of the stress field throughout the soil requires the introduction of additional stress discontinuities in order to prevent the fan zones from interfering with each other.

Also, the failure load has been calculated on the basis of the failure of the soil at the surface – in other words, the stabilizing effect of the self-weight of the soil has been neglected. As explained in section 8.4.1, this will lead to a conservative solution in the case of an effective stress analysis. The neglect of the strength of the soil above the founding plane introduces a degree of conservatism in both the effective stress and total stress analyses. These comments apply to all of the analyses presented in section 9.5.

In the general case, equations (9.5), (9.15) and (9.16) may be combined to give σ_f' in terms of σ_0' and a revised bearing capacity factor N_q, which may be used when the load applied to the foundation is not vertical:

$$s_1' = \sigma_0'/(1 - \sin \phi') \tag{9.5}$$

$$s_2' = s_1' \times e^{(\pi - \Delta - \delta) \tan \phi'} = \{\sigma_0'/(1 - \sin \phi')\} \times e^{(\pi - \Delta - \delta) \tan \phi'} \tag{9.15}$$

$$\sigma_f' = s_2'(1 + \sin \phi' \cos[\Delta + \delta]) \tag{9.16}$$

giving

$$\sigma_f' = \sigma_0' \times [(1 + \sin \phi' \cos[\Delta + \delta])/(1 - \sin \phi')] \times e^{(\pi - \Delta - \delta) \tan \phi'}$$

or

$$\sigma_f'/\sigma_0' = N_q = [(1 + \sin \phi' \cos[\Delta + \delta])/(1 - \sin \phi')]$$
$$\times e^{(\pi - \Delta - \delta) \tan \phi'} \tag{9.17}$$

where

$$\sin \Delta = \sin \delta / \sin \phi' \tag{9.14}$$

If the load applied to the foundation is vertical, $\delta = \Delta = 0$ and equation (9.17) reduces to equation (9.9). The value of N_q calculated using equation (9.17) reduces as δ is increased. For $\phi_{crit}' = 30°$ and $\delta = \phi_{crit}'$, $\Delta = 90° = \pi/2$ and $N_q = 2.75$. This may be compared with $N_q = 18.4$ when $\delta = 0$ and $N_q = 8.7$ when $\delta = \tan^{-1}(1/3)$ (i.e. $\delta/\phi' = 0.614$), as in Example 9.1.

Values of $N_q = \sigma_f'/\sigma_0'$, calculated using equations (9.14) and (9.17) for different soil strengths ϕ' and obliquity of load δ/ϕ', are given in Table 9.1. In using Table 9.1, it must be remembered that σ_f' is the vertical component of the load. The resultant stress, applied at an angle δ to the vertical, is equal to $\sigma_f'/\cos \delta$.

Table 9.1 Bearing capacity factors N_q for inclined loads, according to equations (9.14) and (9.17)

ϕ' (degrees)	$\delta/\phi' = 0$	$\delta/\phi' = 0.2$	$\delta/\phi' = 0.4$	$\delta/\phi' = 0.6$	$\delta/\phi' = 0.8$	$\delta/\phi' = 1.0$
15	3.941	3.655	3.335	2.975	2.553	1.788
16	4.335	3.995	3.616	3.195	2.709	1.847
17	4.772	4.368	3.923	3.433	2.874	1.908
18	5.258	4.780	4.257	3.688	3.048	1.969
19	5.798	5.233	4.622	3.962	3.232	2.031
20	6.399	5.734	5.019	4.258	3.426	2.094
21	7.071	6.287	5.454	4.577	3.632	2.157
22	7.821	6.900	5.930	4.920	3.849	2.220
23	8.661	7.580	6.452	5.291	4.079	2.285
24	9.603	8.334	7.024	5.692	4.322	2.349
25	10.66	9.17	7.65	6.13	4.58	2.41
26	11.85	10.11	8.34	6.59	4.85	2.48
27	13.20	11.15	9.10	7.10	5.14	2.55
28	14.72	12.32	9.94	7.65	5.45	2.61
29	16.44	13.62	10.87	8.25	5.78	2.68
30	18.40	15.09	11.89	8.90	6.12	2.75
31	20.63	16.73	13.03	9.61	6.49	2.81
32	23.18	18.59	14.29	10.39	6.88	2.88
33	26.09	20.69	15.70	11.23	7.30	2.95
34	29.44	23.07	17.26	12.15	7.75	3.01
35	33.30	25.77	19.01	13.16	8.22	3.08
36	37.75	28.85	20.97	14.27	8.73	3.15
37	42.92	32.37	23.17	15.49	9.27	3.22
38	48.93	36.39	25.64	16.83	9.85	3.28
39	55.96	41.02	28.43	18.31	10.48	3.35
40	64.20	46.36	31.59	19.95	11.15	3.42

Example 9.2 Failure of a strip footing subjected to an inclined load: total stress (τ_u) analysis

(a) A strip footing is constructed on a stratum of clay of uniform undrained shear strength τ_u. When loaded quickly, the footing fails at an applied vertical stress of 386 kPa. Estimate τ_u.

(b) A second identical footing is built, and a shear stress of 60 kPa is applied. The vertical load is then increased. At what vertical stress will failure occur? (It may be assumed without proof that the ratio of average principal total stress between two uniform stress zones, separated by a fan zone of included angle θ, is given by $s_2 - s_1 = 2\tau_u\theta$, and that the rotation of the direction of major principal total stress is θ).

[*3rd year BEng civil engineering examination, University of London (Queen Mary and Westfield College)*]

Solution

(a) The first step is to estimate τ_u. Figure 9.10(a) shows the zones of soil below and to one side of the footing. Zone 1 is a passive zone, in which the major principal total stress is horizontal and (at the ground surface) $\sigma_v = 0$. From the Mohr circle of total stress shown in Figure 9.10(b),

$$s_1 = \tau_u$$

In the case of the vertical footing load, zone 2 is a conventional uniform active zone, in which the major principal total stress is vertical. From the Mohr circle of total stress shown in Figure 9.10(c), $\sigma_f = s_2 + \tau_u$.

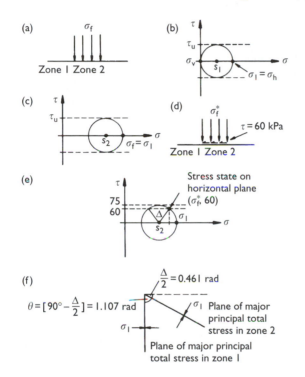

Figure 9.10 (a) Vertical load on footing; (b) Mohr circle of total stress for zone 1; (c) Mohr circle of total stress for zone 2 when the footing load is vertical; (d) inclined load on footing; (e) Mohr circle of total stress for zone 2 when resultant footing load is inclined; (f) rotation of plane of major principal total stress.

The rotation in the direction of principal total stress between zones 1 and 2 is $90° = \pi/2$ radians, so from equation (9.4)

$$s_2 - s_1 = \pi \tau_u; \quad s_2 = s_1 + \pi \tau_u = \tau_u(1 + \pi),$$

and

$$\sigma_f = s_2 + \tau_u = (2 + \pi)\tau_u$$

In the present case, $\sigma_f = 386\,\text{kPa}$:

$$\Rightarrow \tau_u = 386\,\text{kPa}/(2 + \pi)$$

$$\Rightarrow \tau_u = 75\,\text{kPa}$$

With the shear load applied, let the vertical component of the resultant inclined load that causes failure be σ_f^*, as shown in Figure 9.10(d). The Mohr circle of total stress for zone 1 is the same as before, but the Mohr circle of total stress for zone 2 (below the footing) is now as shown in Figure 9.10(e).

From the geometry of the Mohr circle,

$$\sigma_f^* = s_2 + \tau_u \cos \Delta$$

where $\sin \Delta = \tau/\tau_u = 60/75 \Rightarrow \Delta = 53.13°$.

The plane on which the major principal total stress acts is now at an angle of $\Delta/2$ clockwise from the horizontal (Figure 9.10(c); also Figure 9.3(a)), so that the overall rotation of the plane of major principal total stress between zones 1 and 2 is now $\theta = (90° - \Delta/2) = 1.107$ radians (Figure 9.10(f)).

Applying equation (9.4),

$$s_2 - s_1 = 2\tau_u\theta = 2\tau_u(\pi/2 - \Delta/2) = \tau_u(\pi - \Delta)$$

so that

$$\sigma_f^* = s_2 + \tau_u \cos \Delta = s_1 + \tau_u(\pi - \Delta + \cos \Delta) \tag{9.18}$$

In the present case with $s_1 = \tau_u$,

$$\sigma_f^* = \tau_u(1 + \pi - \Delta + \cos \Delta) = 3.814\tau_u$$

or

$$\sigma_f^* = 3.814 \times 75\,\text{kPa} = 286\,\text{kPa}$$

In the general case with the vertical component of the load applied to the foundation denoted by σ_f and $s_1 = \sigma_0 + \tau_u$, equation (9.18) leads to a revised bearing capacity factor N_c:

$$(\sigma_f - \sigma_0)/\tau_u = N_c = (1 + \pi - \Delta + \cos \Delta) \tag{9.19}$$

where $\sin \Delta = \tau/\tau_u$, and τ is the shear stress applied to the foundation. If $\tau = 0$, $\Delta = 0$ and $\cos \Delta = 1$, and equation (9.19) reduces to equation (9.13), giving $N_c = (2 + \pi) = 5.14$. N_c then reduces as τ is increased, until when $\tau = \tau_u$, $N_c = (1 + \pi/2) = 2.57$. This explains why, if a soft mud is just strong enough to support a car when the car is stationary, the car will sink into the mud when the driver tries to move it. (It is sometimes possible to move the car by letting some air out of the tyres, to reduce the normal contact pressure between the tyres and the mud.)

Values of $N_c = (\sigma_f - \sigma_0)/\tau_u$, calculated using equation (9.19) for various values of τ/τ_u, are given in Table 9.2. As with Table 9.1, σ_f is the vertical component of the

Table 9.2 Bearing capacity factors N_c for inclined loads, according to equation (9.19)

τ/τ_u	0	0.1	0.2	0.3	0.4	0.5	0.6	0.7	0.8	0.9	1
N_c	5.142	5.036	4.920	4.791	4.647	4.484	4.298	4.080	3.814	3.458	2.571

load acting on the footing; the horizontal component is τ. The resultant applied stress is inclined at an angle $\tan^{-1}(\tau/\sigma_f)$ to the vertical, and has magnitude $\sqrt{(\sigma_f^2 + \tau^2)}$.

Example 9.3 Bearing capacity of the Cricklewood retaining wall

Assess the adequacy of the Cricklewood retaining wall analysed in Example 7.5 in terms of its short-term bearing capacity, using equation (9.19).

Solution

From Example 8.1, the shear stress on the base of the Cricklewood retaining wall is 22 kPa. The undrained shear strength of the intact soil is 60 kPa. Thus $\sin \Delta = \tau/\tau_u = 22/60$ or $\Delta = 21.5°(= 0.375$ radian). Hence, from equation (9.19), $N_c = (1 + \pi - \Delta + \cos \Delta) \approx 4.7$. Ignoring depth effects (because we have already used the strength of the soil above the founding plane to provide the passive pressure in front of the wall) other than to take $\sigma_0 = (3\,\text{m} \times 20\,\text{kN/m}^3) + 10\,\text{kPa} = 70\,\text{kPa}$, the ultimate (failure) stress is

$$\sigma_f = (N_c \times \tau_u) + \sigma_o = (4.7 \times 60\,\text{kPa}) + 70\,\text{kPa} = 352\,\text{kPa}$$

This is equivalent to an ultimate vertical load of 352 kPa × 4.4 m = 1548 kN, compared with the actual vertical load of 1162 kN. From Example 8.1, the eccentricity of the applied load is 0.153 m or 3.5% of the foundation width: this is well within the middle third and is unlikely to affect significantly the bearing capacity.

The vertical load is about 1.33 times less than that which would cause failure at the same applied horizontal stress, ignoring the slight adverse effect of load eccentricity. It may therefore be concluded that the wall will not suffer a bearing capacity failure in the short term, but on the basis of the assumed undrained shear strength of 60 kPa is unlikely to be sufficiently remote from bearing capacity failure for design purposes.

When a footing is subjected to an inclined load, the problem geometry is not symmetrical. Failure will occur on the side of the footing towards which the shear component of the applied load is directed. This is because the rotation in the principal stress directions (and hence the difference in average stress) between the two zones of soil is smaller in this case. In Examples 9.1 and 9.2, and in Table 9.2, the bearing capacity has been calculated on this basis.

As discussed in section 8.4, the bearing capacity factors N_q and N_c are usually multiplied by empirical enhancement factors to account for effects such as the strength of the soil above the founding plane and – in the case of N_q but not N_c – the self-weight of the soil.

9.6   Calculation of earth pressure coefficients for rough retaining walls

The active and passive earth pressure coefficients derived in section 7.3 on the assumption that the wall was frictionless are true lower bounds, but for design purposes they are unduly over-conservative. This is particularly so in the case of the effective stress analysis (equations (7.3) and (7.4)).

In the absence of wall friction, the stress states behind and in front of a retaining wall may be represented by uniform active and passive zones, as shown in Figure 9.11. Within either of these zones, the orientation of the planes on which the major and minor principal stresses act does not change.

In the effective stress (ϕ') analysis, the characteristic planes are at $45° + (\phi'/2)$ to the horizontal in the active zone behind the wall, and at $45° - (\phi'/2)$ to the horizontal in the passive zone in front of the wall. In the total stress (τ_u) analysis, the characteristic planes are at $45°$ to the horizontal in both the active and the passive zones. In both the ϕ' and the τ_u analyses, the characteristic directions indicate the likely orientation of slip surfaces as shown in Figure 9.11, at least in the absence of external restraints such as props.

Wall friction results in a rotation of the principal stress directions adjacent to the wall, so that the stress state of the soil behind or in front of the wall can no longer be represented by a single active or passive zone. Similarly, there must also be a rotation in the direction of the principal stresses between the free soil surface and a frictionless wall if the soil surface is not horizontal, or if the back of the wall is not vertical. The effect of the rotation in the direction of the principal stresses can be taken into account in all three cases by means of fan zones, as shown in sections 9.6–9.8.

9.6.1   Wall friction: effective stress ϕ' analysis

(a) Active case

Figure 9.12(a) shows the soil behind a rough retaining wall with soil/wall friction angle δ. Zone 1 is a conventional active zone, in which the vertical effective stress at

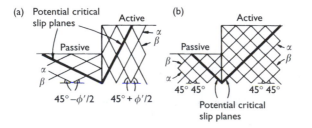

Figure 9.11   Representation of simple active and passive zones behind and in front of a retaining wall using α- and β-characteristics: (a) ϕ' analysis; (b) τ_u analysis.

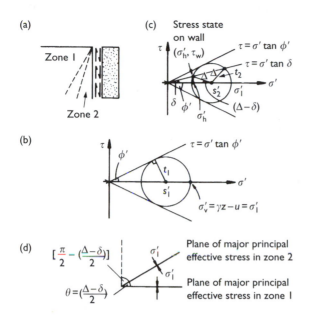

Figure 9.12 Calculation of active earth pressure coefficient for rough retaining wall, ϕ' analysis: (a) division of soil into zones; (b) Mohr circle of effective stress for zone 1; (c) Mohr circle of effective stress for zone 2; (d) rotation in major principal effective stress direction between zones 1 and 2.

a depth z below the free soil surface is $\gamma z - u$, where u is the pore water pressure. In an active zone, the horizontal stress is reduced as the wall moves outward. In zone 1, the major principal effective stress is therefore vertical, and acts on the horizontal plane. The Mohr circle of effective stress for zone 1 is shown in Figure 9.12(b).

Assuming that the soil in zone 2 tends to settle relative to the wall, the soil exerts a downward shear force on the wall. As action and reaction are equal and opposite, the wall exerts an upward shear stress on the soil. With the excavation on the right-hand side of the wall as shown in Figure 9.12(a), the shear stress at the soil/wall interface is anticlockwise as viewed from inside the body of soil, and therefore plots as positive on the Mohr diagram. The Mohr circle of effective stress for zone 2 is shown in Figure 9.12(c).

In zone 1 at a depth z,

$$(\gamma z - u) = s'_1 + t_1 = s'_1(1 + \sin \phi')$$

or

$$s'_1 = (\gamma z - u)/(1 + \sin \phi') \tag{9.20}$$

From the Mohr circle for zone 2 (Figure 9.12(c)), the lateral component of the resultant effective stress acting on the wall is

$$\sigma'_h = s'_2 - t_2 \cos(\Delta - \delta) = s'_2[1 - \sin \phi' \cos(\Delta - \delta)] \tag{9.21}$$

The major principal effective stress in zone 2 acts on a plane which is $[90° - (\Delta - \delta)/2]$ clockwise from the vertical (Figure 9.12(c)), so that the overall rotation in the major principal effective stress direction between zones 1 and 2 is $\theta = (\Delta - \delta)/2$

(Figure 9.12(d)). Substituting this into equation (9.2), and noting that the average principal effective stress decreases as we go from zone 1 to zone 2,

$$s'_1/s'_2 = e^{(\Delta-\delta)\tan\phi'}$$

or

$$s'_2 = s'_1 e^{-\{(\Delta-\delta)\tan\phi'\}} \tag{9.22}$$

Combining equations (9.20), (9.21) and (9.22),

$$\sigma'_h = s'_2[1 - \sin\phi'\cos(\Delta - \delta)]$$

$$= s'_1 e^{-\{(\Delta-\delta)\tan\phi'\}}[1 - \sin\phi'\cos(\Delta - \delta)]$$

$$= (\gamma z - u)e^{-\{(\Delta-\delta)\tan\phi'\}}[1 - \sin\phi'\cos(\Delta - \delta)]/(1 + \sin\phi')$$

so that the active lateral earth pressure coefficient $K_a = \sigma'_h/(\gamma z - u)$ is

$$K_a = \{[1 - \sin\phi'\cos(\Delta - \delta)]/[1 + \sin\phi']\} \times e^{-\{(\Delta-\delta)\tan\phi'\}} \tag{9.23}$$

where $\sin\Delta = \sin\delta/\sin\phi'$.

If the wall is frictionless, $\delta = \Delta = 0$ and equation (9.23) reduces to equation (7.3), with $\sigma'_v = (\gamma z - u)$.

Active earth pressure coefficients calculated using equation (9.23) with various values of ϕ' and wall friction δ are given in Table 9.3. Equation (9.23) forms the basis of the charts of active earth pressure coefficients given in Eurocode 7 (BSI, 1995) and by Gaba *et al.* (2003) for vertical walls with level backfills. The values given in Table 9.3 are generally slightly larger (and therefore more conservative) than those in Table 7.6.

(b) Passive case

Figure 9.13(a) shows the soil in front of a rough retaining wall with soil/wall friction angle δ. Zone 1 is a conventional passive zone, in which the vertical effective stress at a depth z below the free soil surface is $\gamma z - u$, where u is the pore water pressure. In a passive zone, the horizontal stress increases as the wall is pushed into the soil. In zone 1, the major principal effective stress is therefore horizontal, and acts on the vertical plane. The Mohr circle of effective stress for zone 1 is shown in Figure 9.13(b).

Assuming that the soil in zone 2 tends to heave (i.e. move upward) relative to the wall, the soil will exert an upward shear stress on the wall. The wall therefore exerts a downward shear stress on the soil. With the excavation on the right-hand side of the wall as shown in Figure 9.13(a), the shear stress at the soil/wall interface is again anticlockwise as viewed from inside the body of soil, and is plotted as positive on the Mohr diagram. The Mohr circle of effective stress in zone 2 is shown in Figure 9.13(c).

In zone 1 at a depth z,

$$(\gamma z - u) = s'_1 - t_1 = s'_1(1 - \sin\phi')$$

or

$$s'_1 = (\gamma z - u)/(1 - \sin\phi') \tag{9.24}$$

Table 9.3 Active earth pressure coefficients calculated using equation (9.23). These are consistent with the charts given in Eurocode 7 (BSI, 1995) and Gaba *et al.* (2003)

ϕ' (degrees)	K_a with $\delta = 0$	K_a with $\delta = \phi'/2$	K_a with $\delta = 2\phi'/3$	K_a with $\tan \delta = 0.75 \times \tan \phi'$	K_a with $\delta = \phi'$
12	0.6558	0.6133	0.6038	0.5999	0.5931
13	0.6327	0.5892	0.5794	0.5734	0.5683
14	0.6104	0.5660	0.5560	0.5519	0.5446
15	0.5888	0.5437	0.5336	0.5293	0.5219
16	0.5678	0.5223	0.5120	0.5076	0.5002
17	0.5475	0.5017	0.4914	0.4869	0.4793
18	0.5279	0.4894	0.4715	0.4669	0.4593
19	0.5088	0.4629	0.4525	0.4478	0.4402
20	0.4903	0.4446	0.4341	0.4294	0.4218
21	0.4724	0.4269	0.4165	0.4117	0.4041
22	0.4550	0.4100	0.3996	0.3947	0.3872
23	0.4381	0.3956	0.3833	0.3784	0.3709
24	0.4217	0.3778	0.3676	0.3627	0.3552
25	0.4059	0.3626	0.3525	0.3475	0.3402
26	0.3905	0.3480	0.3380	0.3330	0.3257
27	0.3755	0.3339	0.3240	0.3189	0.3118
28	0.3610	0.3202	0.3105	0.3054	0.2984
29	0.3470	0.3071	0.2976	0.2924	0.2855
30	0.3333	0.2944	0.2851	0.2799	0.2731
31	0.3201	0.2822	0.2730	0.2678	0.2612
32	0.3073	0.2704	0.2614	0.2562	0.2497
33	0.2948	0.2590	0.2502	0.2450	0.2387
34	0.2827	0.2479	0.2394	0.2342	0.2280
35	0.2710	0.2373	0.2289	0.2238	0.2177
36	0.2596	0.2270	0.2189	0.2137	0.2078
37	0.2486	0.2171	0.2092	0.2040	0.1983
38	0.2379	0.2075	0.1998	0.1947	0.1891
39	0.2275	0.1983	0.1908	0.1856	0.1803
40	0.2174	0.1893	0.1820	0.1770	0.1718

From the Mohr circle for zone 2 (Figure 9.13(c)), the lateral component of the resultant effective stress acting on the wall is

$$\sigma_h' = s_2' + t_2 \cos(\Delta + \delta) = s_2'[1 + \sin \phi' \cos(\Delta + \delta)] \tag{9.25}$$

The major principal effective stress in zone 2 acts on a plane which is $(\Delta + \delta)/2$ clockwise from the vertical (Figure 9.13(c)). The overall rotation in the major principal effective stress direction between zones 1 and 2 is $\theta = (\Delta + \delta)/2$ (Figure 9.13(d)). Substituting this into equation (9.2), and noting that the average principal effective stress now increases as we go from zone 1 to zone 2,

$$s_2'/s_1' = e^{(\Delta - \delta) \tan \phi'} \tag{9.26}$$

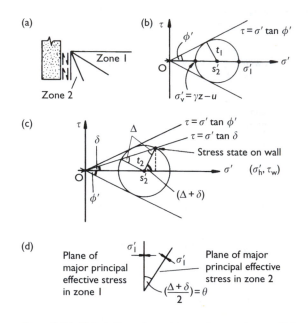

Figure 9.13 Calculation of passive earth pressure coefficient for rough retaining wall, ϕ' analysis: (a) division of soil into zones; (b) Mohr circle of effective stress for zone 1; (c) Mohr circle of effective stress for zone 2; (d) rotation in major principal effective stress direction between zones 1 and 2.

Combining equations (9.24), (9.25) and (9.26),

$$\sigma'_h = s'_2[1 + \sin\phi' \cos(\Delta + \delta)]$$

$$= s'_1 e^{(\Delta+\delta)\tan\phi'}[1 + \sin\phi' \cos(\Delta + \delta)]$$

$$= (\gamma z - u)e^{(\Delta+\delta)\tan\phi'}[1 + \sin\phi' \cos(\Delta + \delta)]/(1 - \sin\phi')$$

so that the passive lateral earth pressure coefficient $K_p = \sigma'_h/(\gamma z - u)$ is

$$K_p = \{[1 + \sin\phi' \cos(\Delta + \delta)]/[1 - \sin\phi']\} \times e^{(\Delta+\delta)\tan\phi'} \qquad (9.27)$$

where $\sin\Delta = \sin\delta/\sin\phi'$.

If the wall is frictionless, $\delta = \Delta = 0$ and equation (9.27) reduces to equation (7.4), with $\sigma'_v = (\gamma z - u)$.

Passive earth pressure coefficients calculated using equation (9.27) with various values of ϕ' and wall friction δ are given in Table 9.4. Equation (9.27) forms the basis of the charts of passive earth pressure coefficients given in Eurocode 7 (BSI, 1995) and by Gaba *et al.* (2003) for vertical walls with level backfills. The values given in Table 9.4 are generally slightly smaller (and therefore more conservative) than those in Table 7.7.

In the analyses presented in section 9.6.1, the stress state on the wall is given by the intersection of the line $\tau = \sigma' \tan\delta$ with the Mohr circle for zone 2. One possible

Table 9.4 Passive earth pressure coefficients calculated using equation (9.27).
These are consistent with the charts given in Eurocode 7 (BSI, 1995)
and Gaba et al. (2003)

ϕ' (degrees)	K_p with $\delta = 0$	K_p with $\delta = \phi'/2$	K_p with $\delta = 2\phi'/3$	K_p with $\tan \delta = 0.75 \times \tan \phi'$	K_p with $\delta = \phi'$
12	1.525	1.6861	1.724	1.739	1.763
13	1.580	1.7657	1.809	1.826	1.855
14	1.638	1.8500	1.900	1.920	1.953
15	1.698	1.9393	1.996	2.020	2.057
16	1.761	2.0341	2.099	2.126	2.168
17	1.826	2.1347	2.209	2.240	2.287
18	1.894	2.2417	2.326	2.361	2.415
19	1.965	2.3556	2.451	2.492	2.552
20	2.040	2.4770	2.584	2.631	2.699
21	2.117	2.6066	2.728	2.782	2.857
22	2.198	2.7449	2.881	2.943	3.028
23	2.283	2.8930	3.047	3.117	3.212
24	2.371	3.0515	3.225	3.305	3.411
25	2.464	3.2215	3.416	3.509	3.627
26	2.561	3.4042	3.623	3.729	3.861
27	2.663	3.6006	3.847	3.969	4.116
28	2.770	3.8123	4.090	4.229	4.393
29	2.882	4.0407	4.353	4.512	4.695
30	3.000	4.2877	4.639	4.822	5.026
31	3.124	4.5550	4.951	5.162	5.389
32	3.255	4.845	5.291	5.534	5.788
33	3.392	5.160	5.664	5.944	6.227
34	3.537	5.504	6.072	6.395	6.712
35	3.690	5.879	6.522	6.895	7.250
36	3.852	6.289	7.017	7.449	7.847
37	4.023	6.738	7.564	8.066	8.512
38	4.204	7.232	8.170	8.754	9.255
39	4.395	7.777	8.844	9.525	10.088
40	4.599	8.378	9.595	10.390	11.026

source of confusion is that, at first sight, this gives two possible stress states. In the active side analysis, the stress state on the retaining wall is the lower of these, while in the passive side analysis, it is the higher.

9.6.2 Wall adhesion: total stress (τ_u) analysis

(a) Active case

Figure 9.14(a) shows the soil behind a rough retaining wall with soil/wall adhesion τ_w. Zone 1 is a conventional active zone, in which the vertical total stress at a depth z below the free soil surface is γz. In a conventional active zone such as zone 1, the major principal total stress is vertical, and acts on the horizontal plane. The Mohr circle of total stress for zone 1 is shown in Figure 9.12(b).

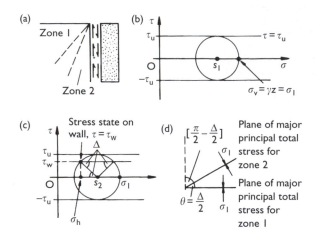

Figure 9.14 Calculation of active earth pressure coefficient for rough retaining wall, τ_u analysis: (a) division of soil into zones; (b) Mohr circle of total stress for zone 1; (c) Mohr circle of total stress for zone 2; (d) rotation in major principal total stress direction between zones 1 and 2.

Assuming that the soil in zone 2 tends to settle relative to the wall, the soil is subjected to an upward shear stress at the soil/wall interface. With the excavation on the right-hand side of the wall as shown in Figure 9.14(a), the shear stress at the soil/wall interface is anticlockwise (as viewed from inside the body of soil), and plots as positive on the Mohr diagram. The Mohr circle of total stress for zone 2 is shown in Figure 9.14(c).

In zone 1 at a depth z,

$$\gamma z = s_1 + \tau_u$$

or

$$s_1 = \gamma z - \tau_u \tag{9.28}$$

From the Mohr circle for zone 2 (Figure 9.14(c)), the lateral component of the resultant total stress acting on the wall is

$$\sigma_h = s_2 - \tau_u \cos \Delta \tag{9.29}$$

As with the ϕ' analysis, the stress state on the active side of the wall is chosen so as to give the lowest possible horizontal stress.

The major principal total stress in zone 2 acts on a plane which is $90° - \Delta/2$ clockwise from the vertical (Figure 9.14(c), so that the overall rotation in the major principal total stress direction between zones 1 and 2 is $\theta = \Delta/2$ (Figure 9.14(d)). Substituting this into equation (9.4), and noting that the average principal total stress decreases as we go from zone 1 to zone 2,

$$s_1 - s_2 = \tau_u \Delta \tag{9.30}$$

Table 9.5 Values of K_{ac} and K_{pc} calculated using equations (9.32) and (9.37) (lower bound approach) and equations (7.13) (upper bound approach)

τ_w/τ_u	0	0.1	0.2	0.3	0.4	0.5	0.6	0.7	0.8	0.9	1
$K_{ac} = K_{pc}$ (equations 9.32 and 9.37)	2	2.095	2.181	2.259	2.328	2.390	2.444	2.490	2.527	2.556	2.571
$K_{ac} = K_{pc}$ (equation 7.13)	2	2.098	2.191	2.280	2.366	2.449	2.530	2.608	2.683	2.757	2.828

Combining equations (9.28), (9.29) and (9.30),

$$\sigma_h = s_2 - \tau_u \cos \Delta = s_1 - \tau_u(\Delta + \cos \Delta) = \gamma z - \tau_u(1 + \Delta + \cos \Delta) \tag{9.31}$$

so that the total lateral earth pressure 'coefficient' K_{ac} (as in the expression $\sigma_h = \gamma z - K_{ac}\tau_u$, section 7.10.2) is given by

$$K_{ac} = (1 + \Delta + \cos \Delta) \tag{9.32}$$

where $\sin \Delta = \tau_w/\tau_u$.

If the wall is frictionless, $\tau_w = \Delta = 0$ and $\cos \Delta = 1$, and equation (9.31) reduces to equation (7.6a), with $\sigma_v = \gamma z$ and $K_{ac} = 2$. If $\tau_w = \tau_u$, $\Delta = 90° = \pi/2$ radians, and $K_{ac} = 2.57$. This may be compared with $K_{ac} = 2.828$ obtained by means of an upper bound approach (equation (7.13a), with $\tau_w = \tau_u$).

Values of K_{ac} calculated using equation (9.32), for various values of τ_w/τ_u are given in Table 9.5.

(b) Passive case

Figure 9.15(a) shows the soil in front of a rough retaining wall with soil/wall adhesion τ_w. Zone 1 is a conventional passive zone, in which the vertical total stress at a depth z below the free soil surface is γz. In a conventional passive zone such as zone 1, the major principal total stress is horizontal, and acts on the vertical plane. The Mohr circle of total stress for zone 1 is shown in Figure 9.15(b).

Assuming that the soil in zone 2 tends to heave (i.e. move upward) relative to the wall, the soil is subjected to a downward shear stress at the soil/wall interface. With the excavation on the right-hand side of the wall, the shear stress at the soil/wall interface is anticlockwise and is plotted as positive. The Mohr circle of total stress in zone 2 is shown in Figure 9.15(c). As this is the passive side, the stress state on the wall is that which has the highest possible lateral stress component.

In zone 1 at a depth z,

$$\gamma z = s_1 - \tau_u$$

or

$$s_1 = \gamma z + \tau_u \tag{9.33}$$

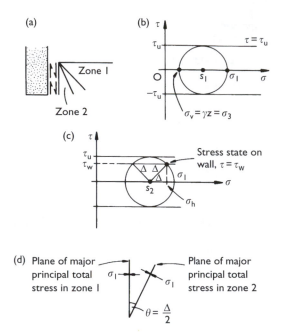

Figure 9.15 Calculation of passive earth pressure coefficient for rough retaining wall, τ_u analysis: (a) division of soil into zones; (b) Mohr circle of total stress for zone 1; (c) Mohr circle of total stress for zone 2; (d) rotation in major principal total stress direction between zones 1 and 2.

From the Mohr circle for zone 2 (Figure 9.15(c)), the lateral component of the resultant total stress acting on the wall is

$$\sigma_h = s_2 + \tau_u \cos \Delta \tag{9.34}$$

The major principal total stress in zone 2 acts on a plane which is $\Delta/2$ clockwise from the vertical (Figure 9.15(c)), so that the overall rotation in the major principal total stress direction between zones 1 and 2 is $\theta = \Delta/2$ (Figure 9.15(d)), as with the active side analysis. Substituting this into equation (9.4), and noting that the average principal total stress now increases as we go from zone 1 to zone 2,

$$s_2 - s_1 = \tau_u \Delta \tag{9.35}$$

Combining equations (9.33), (9.34) and (9.35),

$$\sigma_h = s_2 + \tau_u \cos \Delta = s_1 + \tau_u(\Delta + \cos \Delta) = \gamma z + \tau_u(1 + \Delta + \cos \Delta) \tag{9.36}$$

so that the lateral earth pressure 'coefficient' K_{pc} (as in the expression $\sigma_h = \gamma z + K_{pc} \tau_u$) is given by

$$K_{pc} = (1 + \Delta + \cos \Delta) \tag{9.37}$$

where $\sin \Delta = \tau_w/\tau_u$.

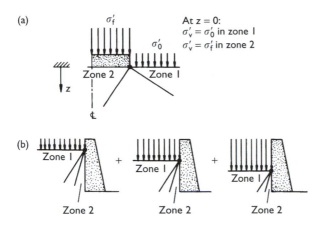

Figure 9.16 Interpretation of stress analyses for (a) footing; (b) retaining wall.

If the wall is frictionless, $\tau_w = \Delta = 0$ so that $\cos \Delta = 1$, and equation (9.36) reduces to equation (7.6b), with $\sigma_v = \gamma z$, and $K_{pc} = 2$. If $\tau_w = \tau_u$, $\Delta = 90° = \pi/2$ radians, and $K_{pc} = 2.57$. Again, this may be compared with $K_{pc} = 2.828$ from an upper bound approach (equation (7.13b), with $\tau_w = \tau_u$). Values of K_{pc} calculated using equation 9.37, for various values of τ_w/τ_u are given in Table 9.5.

In section 9.6, we have calculated earth pressure coefficients which relate the horizontal effective (or total) stress to the nominal vertical effective (or total) stress at depth z, $\sigma'_{v,\text{nominal}} = \gamma z - u$ (or $\sigma_{v,\text{nominal}} = \gamma z$). This is in contrast to the analysis of shallow foundations presented in section 9.5, in which the weight of the soil was completely neglected.

The foundation analyses could be viewed as having been carried out for the soil just below the corner of the footing, where it is possible to progress from zone 1 to zone 2 through the fan zone at depth $z = 0$ (Figure 9.16(a)).

In the retaining wall analyses of section 9.6, we have assumed that the lateral stress σ'_h (or σ_h) acts on the wall in zone 2 at the same depth z at which the nominal vertical stress $\sigma'_{v,\text{nominal}} = \gamma z - u$ (or $\sigma_{v,\text{nominal}} = \gamma z$) was calculated. This is only possible at the upper surface of the soil, adjacent to the wall. In effect, the analysis has been carried out with the soil above each depth z acting as a surcharge, and the resulting stress states have been superimposed (Figure 9.16(b)). For the ϕ' failure criterion, superposition is a perfectly acceptable procedure. It can be shown quite easily that the superposition of two stress states lying on or inside the failure envelope $(\tau/\sigma')_{\text{max}} = \tan \phi'$ will produce a third stress state which lies either inside the failure envelope or on it if the principal effective stress directions in each of the first two stress states are coincident.

9.7   Sloping backfill

If the backfill behind the retaining wall is not level, but slopes upward away from the wall at an angle β to the horizontal,[1] the Mohr circle of stress for the zone of soil immediately below the free surface (zone 1) is altered because:

- The plane on which the stress state is known is the plane parallel to the surface, rather than the horizontal plane.
- This plane is not a **principal plane**, because there is a shear stress acting on it.

The reasoning used in the infinite slope analysis (section 8.10, Figure 8.26) shows that the normal total stress σ and the shear stress τ acting on a plane parallel to the surface and a depth z below it are

$$\sigma = \gamma z \cos^2 \beta$$

and

$$\tau = \gamma z \cos \beta \sin \beta$$

If the pore water pressure at depth z is quantified by means of the pore pressure ratio r_u, so that $u = r_u \gamma z$ (section 8.11.2(a)), the normal effective stress is

$$\sigma' = \gamma z \cos^2 \beta - u = \gamma z \cos^2 \beta - r_u \gamma z = \gamma z (\cos^2 \beta - r_u)$$

This defines the stress state in the zone of soil below the retained surface (zone 1). Construction of the Mohr circle of effective stress for this zone is described in section 9.7.1, and construction of the Mohr circle of total stress in section 9.7.2.

9.7.1 Effective stress (ϕ') analysis

The effective stresses acting on a plane at depth z below a surface which slopes upward at an angle β to the horizontal (Figure 9.17(a)) are

$$\tau = \gamma z \cos \beta \sin \beta,$$
$$\sigma' = \gamma z (\cos^2 \beta - r_u)$$

so that the ratio τ/σ' is given by

$$\tau/\sigma' = \gamma z \cos \beta \sin \beta / \gamma z (\cos^2 \beta - r_u) = (\cos \beta \sin \beta)/(\cos^2 \beta - r_u) = \tan \phi'_{\text{mob}, \beta}$$

where $\phi'_{\text{mob}, \beta}$ is the strength mobilized on planes parallel to the sloping soil surface. With the ground sloping upward from bottom left to top right as shown in Figure 9.17(a), the shear stress is anticlockwise, that is positive. The Mohr circle of effective stress is shown in Figure 9.17(b). The plane of major principal effective stress is at an angle of $[(\Delta + \phi'_{\text{mob}, \beta})/2 - \beta]$ from the horizontal (Figure 9.17(c)), where $\sin \Delta = \sin \phi'_{\text{mob}, \beta} / \sin \phi'$. The maximum possible value of $\phi'_{\text{mob}, \beta}$ is ϕ'.

9.7.2 Total stress (τ_u) analysis

The total stresses acting on a plane at depth z below a surface which slopes upward at an angle β to the horizontal (Figure 9.17(a)) are

$$\tau = \gamma z \cos \beta \sin \beta$$

$$\sigma = \gamma z \cos^2 \beta$$

so that the ratio τ/σ is

$$\tau/\sigma = \gamma z \cos \beta \sin \beta / \gamma z \cos^2 \beta = \tan \beta$$

With the ground sloping upward from bottom left to top right as shown in Figure 9.17(a), the shear stress is anticlockwise, that is positive. The Mohr circle of total stress is shown in Figure 9.18(a). The plane of major principal total stress is at an angle of $[\Delta/2 - \beta]$ clockwise from the horizontal (Figure 9.18(b)), where $\sin \Delta = \gamma z \cos \beta \sin \beta / \tau_u$. The maximum possible value of β occurs when $\gamma z \cos \beta \sin \beta = \tau_u$.

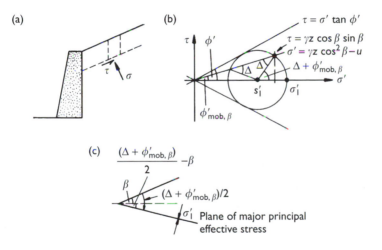

Figure 9.17 (a) Sloping backfill geometry; (b) Mohr circle of effective stress; (c) orientation of plane on which the major principal effective stress acts.

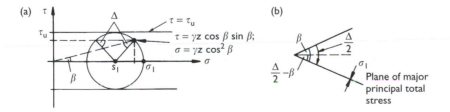

Figure 9.18 (a) Mohr circle of total stress; (b) orientation of plane on which the major principal total stress acts.

To calculate the stresses acting on the retaining wall, using either an effective stress analysis or a total stress analysis, the procedure given in section 9.4.1 is then followed:

- Calculate the principal stress directions in the zone of soil adjacent to the wall (zone 2), on the basis of the known soil/wall friction angle δ or the known soil/wall adhesion τ_w.
- Calculate the angle of rotation θ of the plane on which the major principal stress acts, as we pass through the fan zone that separates zone 2 from zone 1.
- Apply equation (9.2) (effective stress analysis) or equation (9.4) (total stress analysis), to calculate the change in average effective stress s' or average total stress s between zones 1 and 2.

This procedure is illustrated in Example 9.4 (section 9.8).

9.8   A wall with a sloping (battered) back

When, in an active side analysis, the back of a retaining wall is sloping or **battered** at an angle ω to the vertical (measured positive into the soil, as shown in Figure 9.19(a)), the stress state in the zone of soil adjacent to the wall (zone 2) is no longer defined with reference to the vertical plane. This will affect the orientation of the plane on which the major principal stress acts in zone 2. Also, the calculated stress state on the wall (which is in terms of the components normal and parallel to the wall) must be resolved into its horizontal components in order to calculate the horizontal force acting on the wall. In the case of a passive side analysis, the same consideration will apply if the front of the wall is sloping.

9.8.1   Effective stress (ϕ') analysis

The stress state on a wall with soil/wall friction angle δ is (as before) given by (τ_w, σ'_w), where the wall shear stress $\tau_w = \sigma'_w \tan \delta$. Mohr circles of effective stress for the zones of soil adjacent to the wall are shown in Figure 9.19(b) (active case) and (d) (passive case).

In the active case, the effective stress acting normal to the wall is given by

$$\sigma'_n = s'_2 - t_2 \cos(\Delta - \delta) = s'_2[1 - \sin\phi' \cos(\Delta - \delta)]$$

and the plane on which the major principal effective stress acts (Figure 9.19(c)) is at an angle of $[\pi/2 - (\Delta - \delta)/2 + \omega]$ clockwise from the vertical, where $\sin\Delta = \sin\delta/\sin\phi'$.

In the passive case, the effective stress acting normal to the wall is given by

$$\sigma'_n = s'_2 + t_2 \cos(\Delta + \delta) = s'_2[1 + \sin\phi' \cos(\Delta + \delta)]$$

and the plane on which the major principal effective stress acts (Figure 9.19(e)) is at an angle of $[(\Delta + \delta)/2 - \omega]$ clockwise from the vertical, where $\sin\Delta = \sin\delta/\sin\phi'$.

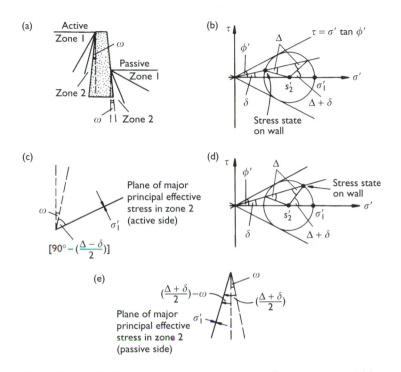

Figure 9.19 (a) Wall geometry; (b) Mohr circle of effective stress and (c) orientation of plane of major principal effective stress for zone of soil adjacent to the wall (zone 2), active case; (d) Mohr circle of effective stress and (e) orientation of plane of major principal effective stress for zone of soil adjacent to the wall (zone 2), passive case.

9.8.2 Total stress (τ_u) analysis

Let the shear stress on a wall with soil/wall adhesion be τ_w. Mohr circles of total stress for the zones of soil adjacent to the wall are shown in Figures 9.20(a) (active case) and 9.20(c) (passive case).

In the active case, the total stress acting normal to the wall is given by

$$\sigma_n = s_2 - \tau_u \cos \Delta$$

and the plane on which the major principal total stress acts (Figure 9.20(b)) is at an angle of $[\pi/2 - \Delta/2 + \omega]$ clockwise from the vertical, where $\sin \Delta = \tau_w/\tau_u$.

In the passive case, the total stress acting normal to the wall is given by

$$\sigma_n = s_2 + \tau_u \cos \Delta$$

and the plane on which the major principal total stress acts (Figure 9.20(d)) is at an angle of $[\Delta/2 - \omega]$ clockwise from the vertical, where $\sin \Delta = \tau_w/\tau_u$.

To calculate the earth pressure coefficients, which relate the lateral stresses on the wall to the nominal vertical stresses at depth z, $\sigma_{v,nominal} = \gamma z$ (total) and $\sigma'_{v,nominal} = \gamma z - u$ (effective) in the zone of soil below the free surface, the procedure given in section 9.4.1 must again be followed, as illustrated in Example 9.4.

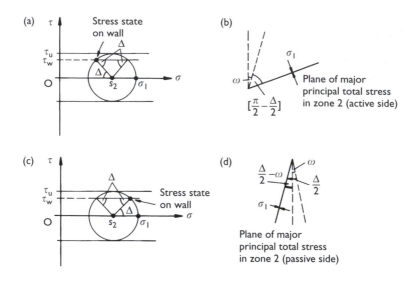

Figure 9.20 (a) Mohr circle of total stress and (b) orientation of plane of major principal total stress for zone of soil adjacent to the wall (zone 2), active case; (c) Mohr circle of total stress and (d) orientation of plane of major principal total stress for zone of soil adjacent to the wall (zone 2), passive case.

Example 9.4 Battered retaining wall with a sloping backfill

Figure 9.21 shows a cross-section through a gravity wall retaining dry soil of unit weight $\gamma = 18\,\mathrm{kN/m^3}$ and critical state angle of internal friction $\phi'_{\mathrm{crit}} = 30°$. The mass of the wall is sufficient to provide a factor of safety (defined as the available sliding resistance ÷ the active horizontal thrust) of 2 against sliding along the base. It is proposed to raise the retained soil surface so that the backfill will slope upward from the wall at an angle of 15°. By what proportion must the mass of the wall be increased to maintain the ratio of available sliding resistance to active horizontal thrust of 2? The angle of soil/wall friction $\delta = 20°$, and the back of the wall has a batter of 5°.

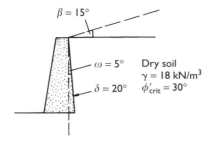

Figure 9.21 Cross-section through gravity retaining wall.

Solution

The problem is to derive a general active earth pressure coefficient which takes account of the effects of wall batter, wall friction and backfill slope. This may then be evaluated for both the original case ($\beta = 0$) and the modified case ($\beta = 15°$), to calculate the change in the lateral thrust.

Figure 9.22(a) shows the Mohr circle of effective stress at a depth z below the free soil surface (zone 1). This is essentially the same as Figure 9.17(b) with $u = 0$ and $\tan \phi'_{mob,\beta} = \beta$.

From Figure 9.22(a),

$$\gamma z \cos^2 \beta = s'_1 + t_1 \cos(\Delta_1 + \beta) = s'_1[1 + \sin \phi'_{crit} \cos(\Delta_1 + \beta)] \tag{9.38}$$

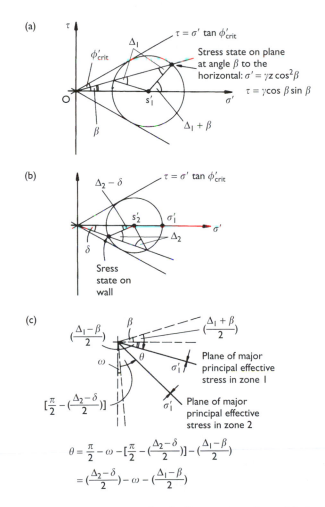

$$\theta = \frac{\pi}{2} - \omega - [\frac{\pi}{2} - (\frac{\Delta_2 - \delta}{2})] - (\frac{\Delta_1 - \beta}{2})$$

$$= (\frac{\Delta_2 - \delta}{2}) - \omega - (\frac{\Delta_1 - \beta}{2})$$

Figure 9.22 (a) Mohr circle of effective stress for soil below the free surface (zone 1); (b) Mohr circle of effective stress for soil adjacent to wall (zone 2); (c) rotation in direction of major principal effective stress between zones 1 and 2.

and the plane on which the major principal effective stress acts is at an angle of $[(\Delta_1 + \beta)/2 - \beta] = [(\Delta_1 - \beta)/2]$ clockwise from the horizontal, where $\sin \Delta_1 = \sin \beta / \sin \phi'_{crit}$.

Figure 9.22(b) shows the Mohr circle of effective stress adjacent to the wall (zone 2) for a general batter angle ω and a general wall friction angle δ. As the wall is the mirror image of that shown in Figure 9.14(a), the upward shear stress on the soil at the soil/wall interface is now clockwise, and therefore plots as negative on the Mohr diagram. (The final results obtained should, of course, be the same whichever way the wall is drawn or built.)

From Figure 9.22(b), the normal component of the effective stress acting on the wall is given by

$$\sigma'_n = s'_2 - t_2 \cos(\Delta_2 - \delta) = s'_2 [1 - \sin \phi'_{crit} \cos(\Delta_2 - \delta)] \tag{9.39}$$

and the plane on which the major principal effective stress acts is at an angle of $[\pi/2 - (\Delta_2 - \delta)/2 + \omega]$ anticlockwise from the vertical, where $\sin \Delta_2 = \sin \delta / \sin \phi'_{crit}$.

The shear stress acting on the wall is $\tau_w = \sigma'_n \tan \delta$.

The rotation in the direction of the major principal effective stress between zones 1 and 2 is $\theta = [(\Delta_2 - \delta)/2 - \omega] - [(\Delta_1 - \beta)/2]$ (Figure 9.22(c)). Using equation (9.2), and noting that in this case $s'_2 < s'_1$,

$$s'_2 = s'_1 e^{-\{(\Delta_2 - \delta - 2\omega - \Delta_1 + \beta) \tan \phi'_{crit}\}} \tag{9.40}$$

Combining equations (9.38), (9.39) and (9.40),

$$\sigma'_n = s'_2 [1 - \sin \phi'_{crit} \cos(\Delta_2 - \delta)]$$
$$= s'_1 e^{-\{(\Delta_2 - \delta - 2\omega - \Delta_1 + \beta) \tan \phi'_{crit}\}} [1 - \sin \phi'_{crit} \cos(\Delta_2 - \delta)]$$
$$= \{\gamma z \cos^2 \beta / [1 + \sin \phi'_{crit} \cos(\Delta_1 - \delta + \beta)]\} e^{-\{(\Delta_2 - \delta - 2\omega - \Delta_1 + \beta) \tan \phi'_{crit}\}}$$
$$\times [1 - \sin \phi'_{crit} \cos(\Delta_2 - \delta)]$$

or

$$\sigma'_n / \gamma z = K_a = \cos^2 \beta \{[1 - \sin \phi'_{crit} \cos(\Delta_2 - \delta)] / [1 + \sin \phi'_{crit} \cos(\Delta_1 + \beta)]\}$$
$$\times e^{-\{(\Delta_2 - \delta - 2\omega - \Delta_1 + \beta) \tan \phi'_{crit}\}} \tag{9.41}$$

If $\sigma'_n = K_a \gamma z$ at depth z, and $\tau_w = \sigma'_n \tan \delta$, then the total force acting normal to the back of the wall due to σ'_n is

$$F_N = \{h / \cos \omega\} \times \{0.5 K_a \gamma h^2\},$$

the total force acting parallel to the back of the wall (due to τ_w) is

$$F_T = \{h / \cos \omega\} \times \{0.5 K_a \gamma h^2 \tan \delta\}$$

and the total horizontal thrust acting on the wall is given by

$$F_H = F_N \cos \omega - F_T \sin \omega = 0.5 K_a \gamma h^3 \times \{1 + \tan \omega \tan \delta\} \tag{9.42}$$

where h is the height of the wall. When the slope of the backfill is raised from $0°$ to $15°$, the only parameter in equation (9.42) which changes is K_a. Thus the increase in lateral thrust is directly proportional to the increase in K_a. If the same ratio of available sliding resistance to active horizontal thrust is to be maintained, the mass of the wall must be increased in the same proportion.

With $\omega = 5°$, $\delta = 20°$, $\phi'_{crit} = 30°$ and $\beta = 0$, $\Delta_1 = 0$ and $\Delta_2 = 43.16°$, giving $\theta = 6.58°$ and $K_{a1} = 0.286$.

With $\omega = 5°$, $\delta = 20°$, $\phi'_{crit} = 30°$ and $\beta = 15°$, $\Delta_1 = 31.17°$ and $\Delta_2 = 43.16°$, giving $\theta = -1.5°$ and $K_{a2} = 0.395$. Thus the proportion by which the mass of the wall must be increased is given by

$$(K_{a2} - K_{a1})/K_{a1} = (0.395 - 0.286) \div 0.286 = 38\%$$

9.9 Improved upper bounds for shallow foundations

9.9.1 Total stress (τ_u) analysis

In section 8.3.1, we analysed the undrained failure of a shallow foundation on a clay soil by considering mechanisms of collapse which consisted of circular arcs. In this section, we shall consider two slightly different classes of collapse mechanism. The first involves a series of blocks which slide relative to one another as rigid bodies, rather like the retaining wall mechanisms shown in Figure 7.38. Although this does not lead to an improved (i.e. reduced) upper bound, it does illustrate a useful principle.

The second class of mechanism involves two blocks which move as rigid bodies, separated by a zone which shears as it deforms, in order to maintain the kinematic admissibility of the mechanism. The shear zone in the second mechanism is directly analogous to the fan zone in the stress field solution given in section 9.5.2. Furthermore, the shear zone mechanism and the fan zone stress field both give the same result $(\sigma_f - \sigma_0 = [2 + \pi]\tau_u)$. As this answer has been obtained using both upper bound and lower bound approaches, it must be the theoretically correct solution.

(a) Failure mechanism consisting of three sliding blocks

Figure 9.23 shows the first of the alternative mechanisms for the undrained failure of a shallow foundation on a clay of undrained shear strength τ_u. This mechanism may be analysed either by work or by statics.

For the analysis by work, it is necessary to construct a velocity diagram or **hodograph**, based on a reference velocity v_0. In this case, v_0 is taken as the downward component of the velocity of the foundation, and the hodograph is shown in Figure 9.24. As the mechanism deforms, energy is dissipated along all of the surfaces along which relative sliding occurs, including those which separate adjacent blocks. In general, the rate at which energy is dissipated is given by the force parallel to the sliding surface multiplied by the relative velocity of sliding. The force parallel to the sliding surface is equal to the undrained shear strength of the clay τ_u multiplied by the length of the sliding surface l. For each of the sliding surfaces involved in the mechanism of Figure 9.23, the lengths, forces and energy dissipation rates are shown in Table 9.6.

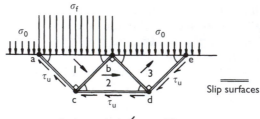

Angles marked △ are 45°

Figure 9.23 Failure mechanism for a shallow foundation on clay, comprising three sliding blocks (τ_u analysis). (Foundation width ab = be = B.)

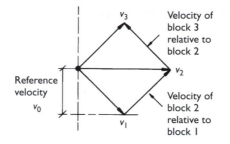

Figure 9.24 Hodograph for the mechanism shown in Figure 9.23.

Table 9.6 Calculation of energy dissipation rate for mechanism shown in Figure 9.23

Sliding surface	Length, l (m)	Force parallel to slip surface $= \tau_u \times l$ (kN/m)	Relative sliding velocity (parallel to slip surface) (m/s)	Rate of energy dissipation = force × velocity (kNm/s per m)
ac	$B/\sqrt{2}$	$\tau_u B/\sqrt{2}$	$v_0\sqrt{2}$	$\tau_u v_0 B$
bc	$B/\sqrt{2}$	$\tau_u B/\sqrt{2}$	$v_0\sqrt{2}$	$\tau_u v_0 B$
cd	B	$\tau_u B$	$2v_0$	$2\tau_u v_0 B$
bd	$B/\sqrt{2}$	$\tau_u B/\sqrt{2}$	$v_0\sqrt{2}$	$\tau_u v_0 B$
de	$B/\sqrt{2}$	$\tau_u B/\sqrt{2}$	$v_0\sqrt{2}$	$\tau_u v_0 B$
Total				$6\tau_u v_0 B$

(These quantities are all expressed per metre length of the foundation, perpendicular to the plane of the paper.)

The rate at which energy is dissipated within the mechanism must be equal to the rate at which potential energy is lost by the foundation load σ_f as it moves downward, minus the rate at which potential energy is gained by the surrounding surcharge σ_0 as it moves upward. From Figure 9.24, the upward velocity of the surcharge is equal to the downward velocity of the foundation, v_0. Hence the net rate at which potential

energy is lost is

$$\sigma_f B v_0 - \sigma_0 B v_0 = (\sigma_f - \sigma_0) B v_0 \text{ kNm/s } (= \text{kW}) \text{ per metre}$$

Equating this to the rate of energy dissipation within the mechanism (from Table 9.6),

$$(\sigma_f - \sigma_0) B v_0 = 6\tau_u v_0 B$$

$$N_c = (\sigma_f - \sigma_0)/\tau_u = 6$$

This is a less unsafe upper bound than that given by the mechanism shown in Figure 8.4, but more unsafe than that given by the mechanism shown in Figure 8.5.

For the mechanism shown in Figure 9.23, the same result may be obtained from a statical equilibrium analysis, provided that all of the forces acting on the mechanism are taken into account. Figure 9.25 shows free body diagrams for each of the three wedges considered separately.

The vertical equilibrium of block 1 requires that

$$\sigma_f B = 2[\tau_u (B/\sqrt{2}) \cos 45°] + 2N_1 \sin 45°$$

Noting that $\cos 45° = \sin 45° = (1/\sqrt{2})$, this gives

$$\sigma_f B = 2[\tau_u (B/2)] + \sqrt{2} N_1$$

or

$$N_1 = (B/\sqrt{2})(\sigma_f - \tau_u) \tag{9.43}$$

For the vertical equilibrium of block 3,

$$\sigma_0 B = -2[\tau_u (B/\sqrt{2}) \cos 45°] + 2N_3 \sin 45°$$

so that

$$\sigma_0 B = -2[\tau_u (B/2)] + \sqrt{2} N_3$$

or

$$N_3 = (B/\sqrt{2}) (\sigma_f + \tau_u) \tag{9.44}$$

For horizontal equilibrium of block 2,

$$N_1 \cos 45° = N_3 \cos 45° + 2[\tau_u (B/\sqrt{2}) \cos 45°] + B\tau_u$$

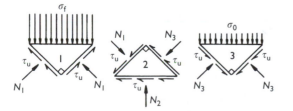

Figure 9.25 Free body diagrams for the three sliding blocks of mechanism shown in Figure 9.23.

or

$$N_1/\sqrt{2} = N_3/\sqrt{2} + 2B\tau_u \qquad (9.45)$$

Substituting for N_1 and N_3 from equations (9.43) and (9.44) into equation (9.45),

$$(B/2)(\sigma_f - \tau_u) = (B/2)(\sigma_f + \tau_u) + 2B\tau_u$$

or

$$(\sigma_f - \sigma_0)/\tau_u = 6$$

exactly as before. Although the self-weight of the soil has been ignored in both the work balance and the equilibrium calculations, in this case its effect is neutral. It is worth noting that the equilibrium analysis relies on the shear stresses on the internal surfaces, bc and bd being known. This would not be the case if bc and bd were not slip planes.

(b) Failure mechanism consisting of two rigid blocks separated by a fan shear zone

Figure 9.26(a) shows the second alternative mechanism for the undrained failure of a shallow foundation on clay. The triangular zones (zones 1 and 3) below and to the side of the footing are similar to those shown in Figure 9.23, but they are now separated by a zone (zone 2) whose third boundary is a circular arc centred on b. Consideration of the kinematics shows that zone 2 cannot simply move as a rigid body. If zone 2 were to rotate about the point b while zone 1 slid down along ac and zone 3 slid upward along de, gaps would open up between zones 1 and 2 at c, and between zones 2 and 3 at b (Figure 9.26(b)): this mechanism is not **kinematically admissible** (i.e. it cannot physically occur).

The movement of the mechanism shown in Figure 9.26(a) can be accommodated by dividing up the middle zone into a number of wedges, each of which is able to slide relative to its neighbours along radial slip lines or **velocity discontinuities** (Figure 9.27(a)).

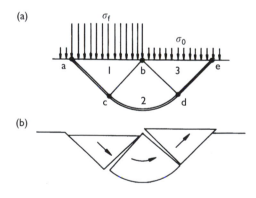

Figure 9.26 (a) Failure mechanism for a shallow foundation on clay, comprising two sliding blocks and a shear zone (τ_u analysis); (b) kinematic inadmissibility if the middle zone moves as a rigid body.

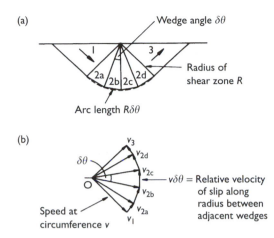

Figure 9.27 (a) Division of zone 2 into wedges which can slip relative to each other; (b) hodograph for calculation of relative sliding velocities.

The hodograph, from which the relative velocities of slip are calculated, is shown in Figure 9.27(b).

For a wedge of radius R which subtends an angle $\delta\theta$ at the centre of the circular arc, forming part of a shear zone moving with a speed at the circumference of v, the rate at which energy is dissipated at the circumference is given by the arc length $\times$ the shear stress on the arc $\times$ the speed of relative displacement, or

$$(R\delta\theta) \times (\tau_u) \times (v)$$

The rate at which energy is dissipated along the radial slip line is radius $\times$ shear stress on radius $\times$ relative velocity of slip, or

$$(R) \times (\tau_u) \times (v\delta\theta)$$

Thus for a shear zone made up of a number of wedges, the overall rate at which energy is dissipated is

$$\sum(2\,R\tau_u v\delta\theta)$$

In the limit as the thickness of each wedge is decreased towards zero, $\delta\theta$ becomes infinitesimally small ($\delta\theta \to d\theta$), and the number of wedges tends towards infinity. The rate of energy dissipation in a shear zone subtending an angle of θ at the centre of the circular arc is then given by integration:

$$\int_0^\theta (2Rv\tau_u)\,d\theta = 2Rv\tau_u\theta \tag{9.46}$$

Figure 9.28 shows the hodograph for the mechanism of undrained failure of the shallow foundation of Figure 9.26(a), with the fan shear zone divided into an infinite number of infinitesimally thin wedges. As before, all velocities are determined in terms of a reference velocity, which is taken as the downward component v_0 of the velocity of the foundation.

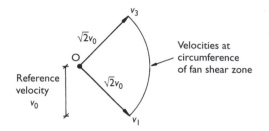

Figure 9.28 Hodograph for the mechanism shown in Figure 9.26(a), with the fan shear zone divided into an infinite number of infinitesimally thin wedges.

The rate at which energy is dissipated is equal to the shear force multiplied by the relative velocity of sliding along the plane slip surfaces ac and de, plus the rate at which energy is dissipated in the fan shear zone, as given by equation (9.46).

The shear force along each of the slip surfaces ac and de is

$$(B/\sqrt{2}) \times \tau_\mathrm{u}$$

The velocity of relative sliding along each of these surfaces is (from Figure 9.28)

$$(\sqrt{2})v_0$$

so that the rate of energy dissipation along ac and de is

$$2 \times (B/\sqrt{2}) \times \tau_\mathrm{u} \times (\sqrt{2})v_0 = 2Bv_0\,\tau_\mathrm{u}$$

For the fan shear zone, the speed at the circumference $v = (\sqrt{2})v_0$, the radius $R = (B/\sqrt{2})$, and the angle subtended at the centre $\theta = 90° = \pi/2$ radians. Thus from equation (9.46), the rate at which energy is dissipated in the fan shear zone is

$$2 \times (B/\sqrt{2}) \times (\sqrt{2})v_0 \times (\pi/2) \times \tau_\mathrm{u} = \pi Bv_0\,\tau_\mathrm{u}$$

Thus the total rate at which energy is dissipated is

$$2Bv_0\,\tau_\mathrm{u}\ (\text{for ac and de}) + \pi Bv_0\,\tau_\mathrm{u}\ (\text{for the fan shear zone}) = (2 + \pi)Bv_0\,\tau_\mathrm{u}$$

The rate at which potential energy is lost by the foundation as it settles with a uniform downward velocity component v_0 is

$$\sigma_\mathrm{f} Bv_0$$

and the rate at which potential energy is gained by the surrounding surcharge as it moves with an upward velocity component v_0 is

$$\sigma_0 Bv_0$$

so that the net rate at which potential energy is lost is

$$(\sigma_\mathrm{f} - \sigma_0)Bv_0$$

Equating this to the rate at which energy is dissipated along the slip lines ac and de and in the fan shear zone,

$$(\sigma_f - \sigma_0)Bv_0 = (2 + \pi)Bv_0\,\tau_u$$

or

$$N_c = (\sigma_f - \sigma_0)/\tau_u = (2 + \pi) \tag{9.47}$$

which is exactly the same as the lower bound solution derived in section 9.5.2 (equation 9.13).

It is no coincidence that the boundaries to the three different zones in the mechanism of Figure 9.26(a) coincide with the stress discontinuities separating the active, passive and fan zones in the analysis of section 9.5.2. The boundaries between the zones in the mechanism of Figure 9.26(a) are lines along which relative slip takes place. In other words, the velocity changes as we move across the boundary from one zone into another.

The boundaries separating different zones in a mechanism analysis are often known as **slip lines** or **velocity discontinuities**. They are analogous to the **stress discontinuities** (across which the state of stress changes as we move from one zone into another) used in the lower bound analysis. In general, if the lower and upper bound solutions which have been obtained for a particular problem are the same, the stress discontinuities in the first will be coincident with the velocity discontinuities (or slip lines) in the second. Furthermore, the stress discontinuities in the lower bound solution will correspond to the α- and β-stress characteristics described in section 9.4. Patterns of stress and velocity characteristics, which are sometimes known as **slip line fields**, are discussed in detail by Atkinson (1981).

9.9.2 *Effective stress (ϕ') analysis*

The upper bound to the collapse load of a shallow foundation derived in section 8.3.2 using the $(\tau/\sigma')_{max} = \tan\phi'$ failure criterion may also be improved by considering a mechanism that includes a shearing zone. Figure 9.29(a) shows such a mechanism, consisting of two rigid zones (zones 1 and 3), separated by a fan shear zone made up of an infinite number of infinitesimally thin wedges (zone 2).

The mechanism shown in Figure 9.29(a) is analogous to the mechanism of Figure 9.26(a) for the undrained shear strength ($\tau_{max} = \tau_u$) analysis. Zone 2 is a fan shear zone, because a rigid block would again be kinematically inadmissible. However, as the relative movement on the shear planes within and at the edges of the fan shear zone is always at 90° to the resultant stress, there is no energy dissipation associated with it. Although this consequence of the normality condition is unrealistic, it is convenient for analysis. Also, as mentioned in section 7.7, it can be shown that collapse loads calculated with an angle of dilation $\psi = \phi'$ will be true upper bounds in that they are greater than or equal to those for real soils with $\psi = 0$ at the critical state (Atkinson, 1981).

The mechanism shown in Figure 9.29(a) may be analysed by means of a work balance. The hodograph for the calculation of relative velocities is shown in Figure 9.29(b). It must be remembered when constructing the hodograph that motion is not parallel to a slip surface, but at an angle $\psi(= \phi')$. As before, the hodograph

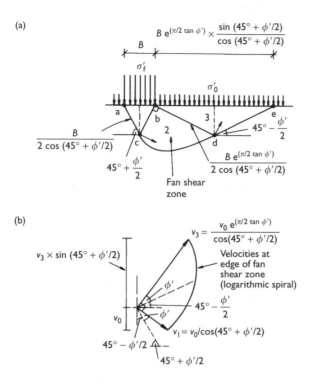

Figure 9.29 (a) Collapse mechanism for a shallow foundation, $(\tau/\sigma')_{max} = \tan\phi'$ failure criterion; (b) hodograph for calculation of relative velocities.

is constructed in terms of a reference velocity v_0, which is taken as the downward component of the velocity of the foundation ab. The locus of velocities at the edge of the fan shear zone is itself a logarithmic spiral, because $v = r\omega$ (where ω is the angular velocity of rotation) and $r = r_0 e^{\theta \tan\phi'}$.

The rate at which potential energy is lost by the foundation load σ'_f is

$$B\sigma'_f v_0$$

From the hodograph shown in Figure 9.29(b), the upward component of the velocity of the surface surcharge to the side of the footing is

$$v_0 \tan(45° + \phi'/2)\, e^{(\pi/2)\tan\phi'}$$

so that the rate at which potential energy is gained is

$$B\, \sigma'_0 \tan(45° + \phi'/2)\, e^{\{(\pi/2)\tan\phi'\}} \times v_0 \tan(45° + \phi'/2) \times e^{(\pi/2)\tan\phi'}$$

Equating these,

$$N_q = \sigma'_f/\sigma'_0 = \tan^2(45° + \phi'/2)\, e^{\pi\tan\phi'} \tag{9.48}$$

which, as $\tan^2(45° + \phi'/2) = (1 + \sin\phi')/(1 - \sin\phi') = K_p$, is exactly the same as equation (9.9). Again, the upper and lower bound solutions are the same, so that equation (9.9) is the theoretically correct answer. The slip lines or velocity discontinuities in the mechanism of Figure 9.29(a) correspond exactly with the stress discontinuities and characteristics associated with the stress field of Figure 9.6.

Key points

- **Stress discontinuities**, along which the mobilized soil strength may be less than the full strength of the soil, can be used in lower bound (stress field) solutions to separate zones of stress in which the average total or effective stress (s or s'), and the principal stress directions, are different.
- In general, better (i.e. less conservative) lower bounds are obtained as the soil strength mobilized along the discontinuities approaches the full strength of the soil.
- The least conservative lower bounds are obtained by using **fan zones**, made up of an infinite number of infinitesimal stress jumps, to rotate the principal stresses as we move between two **uniform stress zones**. The ratio of average effective stresses s_2'/s_1', or the change in average total stress $s_2 - s_1$, is related to the rotation θ of the principal stresses between the two uniform zones, which is the same as the included angle within the fan:

$$s_2'/s_1' = e^{2\theta \tan\phi'} \tag{9.2}$$

or

$$s_2 - s_1 = 2\tau_u \theta \tag{9.4}$$

- Stress fields can be constructed in which the principal stress directions are rotated as we move from a uniform zone, through a fan zone, to a second uniform zone. In the first uniform zone, the stress state is known in both magnitude and orientation (e.g. a conventional active or passive zone below a free soil surface). In the second uniform zone, the stress state is only partly defined (e.g. the zone of soil adjacent to a rough retaining wall, or below a strip foundation). In this way, bearing capacity factors for foundations subjected to inclined loads, and earth pressure coefficients for rough retaining walls, walls with a non-horizontal backfill and walls with a sloping or **battered** back, can be calculated.
- The stress state in each zone can be visualized by means of the α- and β-characteristics, which indicate the directions along which the full strength of the soil is mobilized.
- For strip footings, improved upper bound solutions can be obtained by using mechanisms incorporating **shearing zones** (i.e. zones that shear, rather than remain rigid, as they deform). The **slip lines** or **velocity discontinuities** in the theoretically correct upper bound mechanism coincide with the **characteristic directions** in the theoretically correct lower bound stress field.

Questions

Bearing capacity of foundations

9.1. Figure 9.30 shows a cross-section through the bottom of an under-reamed pile in a clay of undrained shear strength $\tau_u = 75\,\text{kPa}$. By considering the stress state of the soil in the zones immediately below the base of the pile and immediately above the 'bell' of the under-ream, estimate the bearing capacity factor $N_c = (\sigma_f - \sigma_0)/\tau_u$. (Note: the shear stress on the base of the pile must be zero, but what is the effect of shear stress τ_w between the top of the 'bell' of the under-ream and the adjacent soil?). Comment on the shortcomings of your solution.

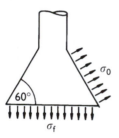

Figure 9.30 Under-reamed piled foundation.

$$N_c = (\sigma_f - \sigma_0)/\tau_u = 9.33 \text{ with } \tau_w = 0; \ N_c = 9.90 \text{ with } \tau_w = \tau_u.$$

Retaining walls and earth pressures

9.2. By considering Mohr circles of stress for zones of soil adjacent to the retaining wall and immediately below the free soil surface, deduce an expression for the active earth pressure coefficient $\sigma_h'/(\gamma z - u)$ behind a retaining wall in a soil having angle of internal friction ϕ', where the angle of soil/wall friction is δ and the retained soil surface is level.

$$\sigma_h'/(\gamma z - u) = [(1 - \sin\phi'\cos(\Delta - \delta)/(1 + \sin\phi')]e^{-\{(\Delta - \delta)\tan\phi'\}}$$

9.3. Figure 9.31 shows a cross-section through a thrust block, which relies on the pressure of the soil behind it to resist an imposed lateral thrust P whose value is a factor of 2.5 times smaller than that which would cause passive failure. Estimate the maximum allowable value of P if the width of the thrust block is 3 m, and its back is smooth. (Ignore the effect of shear stresses on the base of the thrust block and three-dimensional effects.)

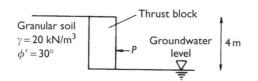

Figure 9.31 Cross-section through thrust block.

If the back of the thrust block is rough, with an angle of interface friction δ, show, by considering Mohr circles of stress at a depth z in zones of soil adjacent to and away from the thrust block, that the earth pressure coefficient $K_p (= \sigma'_h/[\gamma z - u])$ is given by

$$K_p = \{[1 + \sin \phi' \cos(\Delta + \delta)]/[1 - \sin \phi']\} \times e^{(\Delta + \delta) \tan \phi'}$$

where $\sin \Delta = \sin \delta / \sin \phi'$.

Hence calculate the maximum allowable value of P if $\delta = \phi'$.

(It may be assumed without proof that the ratio of average principal effective stress between two uniform stress zones separated by a fan zone of included angle θ is given by $s'_2/s'_1 = e^{2\theta \tan \phi'}$, and that the rotation of the direction of major principal effective stress is θ.)

[3rd year BEng civil engineering examination, University of London (Queen Mary and Westfield College)]

$P_{allowable} = 576\,kN$ if $\delta = 0$ and $965\,kN$ if $\delta = \phi'$.

9.4. (a) A bulldozer has a blade which is 2.5 m wide. Estimate the total force on the bulldozer blade when it is used to move a 1.5 m high bank of dry sand which has $\phi' = 30°$ and unit weight $\gamma = 18\,kN/m^3$, if the angle of friction between the sand and the bulldozer blade is 15°.

(b) If this is 85% of the maximum force the bulldozer can provide, estimate the maximum depth of clay of undrained shear strength $\tau_u = 80\,kPa$ that the bulldozer could move, assuming that the adhesion between the clay and the bulldozer blade is $0.5\tau_u$ and that the depth of the blade is 1.6 m. (Note: conditions in front of the bulldozer blade will be passive in both cases).

((a) Force $= 217\,kN$; maximum force $= 255.4\,kN$; maximum height of clay $= 0.52\,m$.)

Note

1 The slope angle β should not be confused with the β-characteristic defined in section 9.4.2.

References

Abbott, M.B. (1966) *An Introduction to the Method of Characteristics*. Thames and Hudson, London.

Atkinson, J.H. (1981) *Foundations and Slopes*, McGraw-Hill, London.

British Standards Institution (1995) *Eurocode 7: Geotechnical Design*, Part 1: general rules. DDENV 1997–1: 1995. British Standards Institution, London.

Gaba, A., Simpson, B., Powrie, W. and Beadman, D.R. (2003) Embedded retaining walls – guidance for economic design. Report C580, Construction Industry Research and Information Association, London.

Chapter 10

Particular types of earth retaining structure

10.1 Introduction and objectives

This chapter addresses some of the aspects of retaining wall behaviour that have to be taken into account in design. Also, the limit equilibrium concepts developed in Chapter 7 are applied to two rather special forms of earth retaining structure – reinforced soil retaining walls and tunnels.

Objectives

After having worked through this chapter, you should be able to:

- calculate the design length of an embedded retaining wall, following the procedure given in BS8002 (BSI, 2001) (sections 10.2–10.4)
- assess the potential reduction in bending moment and prop load in an embedded retaining wall, due to wall flexibility effects (section 10.6)
- estimate the lateral stresses behind a retaining wall, following compaction of a dry granular backfill (section 10.7.1)
- allow for the effects of a line or strip surcharge in the stress analysis of a retaining wall (section 10.8)
- carry out an approximate stress analysis of a multi-propped retaining wall at any stage during its construction (section 10.9)
- investigate, using limit equilibrium methods, the failure of reinforced soil retaining walls (section 10.10)
- carry out a stress analysis of a cross-section through a circular tunnel at collapse (sections 10.11.1–10.11.3)
- assess the remoteness from collapse of a tunnel heading in a clay soil (section 10.11.4).

You should have an appreciation of:

- the 'limit-state' design philosophy adopted by modern codes of practice such as Eurocode 7 (BSI, 1995), BS8002 (BSI, 2001) and CIRIA report C580 (Gaba *et al.*, 2003) (sections 10.2 and 10.3)
- the shortcomings of various alternative (pre-BS8002) methods of applying a factor of safety to a collapse calculation for an embedded retaining wall (section 10.2)

- the distinction between the actual 'working stress' distributions and the stress distributions used in the design calculation (sections 10.2 and 10.3)
- the particular problems associated with embedded retaining walls in clay soils (section 10.4)
- the linkage between wall movement and mobilized soil strength for an embedded retaining wall (section 10.5)
- the effects of, and the difficulties associated with, the compaction of a clay backfill behind a retaining wall (section 10.7.2)
- the way in which ground movements due to tunnelling can be estimated (section 10.11.5).

10.2   Design calculations for embedded retaining walls: ultimate limit states

The permissible stress states introduced in section 7.3 were used in section 7.6 to develop stress fields or stress distributions that enabled us to calculate the depth of embedment required just to prevent collapse of a propped cantilever retaining wall. In the stress distributions shown in Figures 7.16 and 7.18, the stresses behind the wall are at their minimum possible values (the **active limit**), while the stresses in front of the wall are at their maximum possible values (the **passive limit**). The same is true of the idealized stress distributions for propped and unpropped cantilever walls shown in Figures 7.41–7.44.

Obviously, a real wall must not collapse under working conditions. It is therefore necessary to increase the design depth of embedment beyond that required merely to prevent collapse. As described in Chapter 7, this was traditionally achieved by applying a **factor of safety** – conventionally given the symbol F – to one of the parameters in the collapse calculation. In fact, the aim of the factor of safety in a traditional design is not just to ensure that the wall would not collapse, but also to ensure that it would not deform excessively under working conditions. These two distinct ways in which the wall might fail have come to be termed **limit states**: outright collapse is known as the **ultimate limit state** (**ULS**), while excessive deformation is an example of a possible **serviceability limit state** (**SLS**), beyond which a specific service performance requirement is no longer met.

Modern codes of practice such as BS8002 (BSI, 2001), Eurocode 7 (BSI, 1995) or CIRIA report C580 (Gaba *et al.*, 2003) require a retaining wall (or any other geotechnical structure) to be designed so as to avoid both ultimate and serviceability limit states. The modes of failure described at the end of section 7.4 are all examples of ultimate limit states involving breakage or collapse. Possible serviceability limit states for embedded retaining walls include excessive wall deflection and associated ground movements, and unwanted leakage of groundwater through or beneath the wall.

In all three of the above documents (BS8002, EC7 and C580), the recommended method of design against ultimate limit states involves carrying out a limit equilibrium or other stability calculation with the actual soil strengths ($\tan \phi'$ or τ_{u}) reduced by a factor of safety F_{s} (or strength mobilization factor M). There is also a requirement to use the most onerous pore water pressure regime reasonably likely to occur. Uncertainties in possible future loading conditions are allowed for to some extent by

- increasing the retained height by 10% of the retained height for embedded cantilever walls – normally up to a maximum of 0.5 m – representing an unplanned excavation in front of the wall
- assuming an additional uniform surcharge of 10 kPa to act on the retained soil surface

in all three documents.

As discussed in section 7.3.2, BS8002 (2001) states that the design strength ϕ'_{design} used in the ultimate limit state calculation should be the lesser of $\tan^{-1}\{(\tan \phi'_{peak})/1.2\}$ and ϕ'_{crit}. CIRIA Report C580 (Gaba *et al.*, 2003) recommend $\phi'_{design} = \tan^{-1}\{(\tan \phi')/1.2\}$, and Eurocode 7 (BSI, 1995) case C (failure in the ground) requires $\phi'_{design} = \tan^{-1}\{(\tan \phi')/1.25\}$. In both CIRIA Report C580 and EC7, ϕ' is a moderately conservative estimate of the effective angle of friction relevant to the limit state under consideration, in this case the ultimate limit state or collapse. In this context 'moderately conservative' is a statistical term relating to the likelihood of the strength being less than the chosen value (see e.g. Gaba *et al.*, their section 5.9): it is completely independent of the soil mechanics distinctions that must be made between, (for example) peak, critical state and residual strengths. BS8002 states that the design value of the soil/wall friction angle δ_{design} should not exceed $\tan^{-1}(0.75 \times \tan \phi'_{design})$; Eurocode 7 and Gaba *et al.* (2003) suggest maximum values of $\delta = 0.67\phi'_{crit}$ for smooth concrete and $\delta = \phi'_{crit}$ for rough concrete, which might allow the ultimate limit state calculation to be carried out with $\delta_{design} = \phi'_{design}$.

In a total stress ultimate limit state calculation using the undrained shear strength τ_u, BS8002 and Gaba *et al.* (2003) specify $\tau_{u,design} = \tau_u/M$ with $M = 1.5$. According to BS8002, this value of M may need to be increased in the case of clays that require large strains to mobilize their peak strength, if wall displacements are not to exceed 0.5% of the retained height. In Eurocode 7 for ultimate limit state design against failure in the ground, the required strength mobilization factor M is 1.4, that is, $\tau_{u,design} = \tau_u/1.4$. BS8002 allows a wall adhesion for ultimate limit state design $\tau_{w,design}$ of $0.75 \times \tau_{u,design}$: with the mobilization factor $M = 1.5$, this is equivalent to a maximum value $\tau_{w,max}$ of $0.5 \times \tau_u$. Gaba *et al.* (2003) specify a smaller maximum value of $\tau_{w,design}$ of $0.5 \times \tau_{u,design}$, equivalent to a maximum of $\tau_u/3$. Eurocode 7 is silent on this point.

It is clear from the earlier discussion that the requirements of the various codes of practice, though similar in principle, differ in detail. It is not the purpose of this book to act as a guide to codes of practice and other design documents, and it is important in carrying out a design calculation to follow fully the particular procedure that has been adopted. The brief summaries given earlier are highly simplified, and even with the discussions elsewhere in this book cannot capture the detail and complexity of the various design cases that must be considered. This book is not a substitute for a code of practice or a design guide. If you need to design a retaining wall or any other geotechnical structure, you must obtain a copy of the relevant code and follow completely the procedure it recommends.

In all three of the design guides discussed (BS8002, CIRIA report C580 and Eurocode 7), the factor of safety in the ultimate limit state design is applied mainly to the soil strength. In the past, factors of safety have sometimes been applied to parameters other than the soil strength including, for embedded retaining walls,

(a) the depth of embedment, which is multiplied by a factor F_d
(b) passive earth pressure coefficients, which are reduced by a factor F_p
(c) the moment of the net resisting pressure, which is reduced by a factor F_{np}.

Certain of these definitions of factor of safety are unsafe, while others are inappropriate in some circumstances. Although they are all now out of date, these methods are unlikely to be forgotten for some time to come. To minimize the possibility of their misapplication, some of the more popular – and the disadvantages associated with each one – are summarized later. Further details are given by Burland *et al.* (1981), and extensive comparative calculations are presented by Gaba *et al.* (2003, Appendix A7).

(a) In applying a factor of safety to the depth of embedment, the depth of embedment at collapse (calculated by means of a limit equilibrium analysis using unfactored soil strengths) is simply multiplied by a factor F_d. This method is scientifically unsound and can give misleading results. With the effective stress failure criterion and a water table at the level of the excavated soil surface on both sides of the wall, a value of $F_d = 1.7$ is approximately equivalent to a factor of safety on soil strength $F_s = 1.5$ at $\phi'_{crit} = 15°$, falling to $F_s = 1.4$ at $\phi'_{crit} = 30°$ (Figure 10.1(a) and (b)). With a full-height water table on the retained side, the equivalent values of F_s would be rather smaller. In a total stress analysis, $F_d = 2$ is perilously close to $F_s = 1$ for all values of τ_u (Figure 10.1(c)).

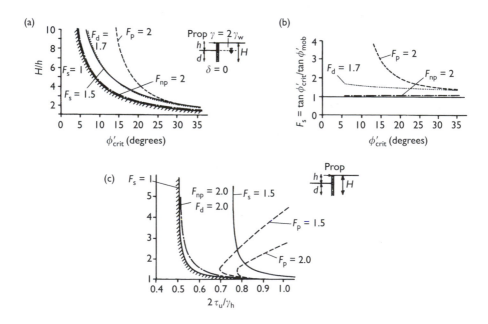

Figure 10.1 (a) Total to retained height ratio H/h and (b) true factor of safety F_s vs. ϕ'_{crit} for different definitions of factor of safety; (c) total to retained height ratio vs. non-dimensionalized undrained shear strength $\tau_u/\gamma h$. (Redrawn with permission from Burland *et al.*, 1981.)

(b) Reducing the passive earth pressure coefficient(s) by the factor F_p was the traditional procedure given in the former UK code of practice for retaining walls, CP2 (IStructE, 1951), and up until the 1970s would probably have been used as a matter of routine. It is likely that the wall bending moments calculated from the equilibrium stress distributions based on earth pressure coefficients K_a behind the wall and K_p/F_p in front would have been used as an estimate of those under working conditions. The justification for this is that in many soils, (e.g. sands and normally consolidated clays), the *in situ* lateral earth pressure coefficient $K_0 = \sigma'_h/\sigma'_v$ is close to the active limit. In these conditions, the stresses in the soil behind the wall fall to their active values after only a small movement of the wall. In front of the wall, rather larger movements are needed for the stresses to rise to the passive limit.

The advent of diaphragm and bored pile wall technology, which enabled embedded walls to be installed in overconsolidated clays having values of ϕ' of the order of 20°, highlighted two problems with this approach. First, the *in situ* lateral stresses in overconsolidated clays are, for the reasons explained in section 10.4.2, likely to be quite high – perhaps approaching the passive limit. It was therefore thought that the wall movement needed to reduce the horizontal stresses behind the wall to the active limit could be rather greater than that needed for the soil in front to reach passive conditions – exactly the opposite from that assumed in the traditional design procedure. This is a cause for concern if the bending moments obtained from the factored limit equilibrium calculation are used for the structural design of the wall. We will return to this point in section 10.4.2, but it is now considered unlikely that the in-service bending moments in an embedded retaining wall in a clay soil could exceed those obtained from a ultimate limit state calculation with the soil strength reduced by the appropriate strength mobilization factor on both sides of the wall.

The second and perhaps more significant problem with the traditional method is that the application of the factor of safety to the earth pressure coefficient, rather than to the soil strength directly, can be misleading especially over a wide range of effective soil strengths ϕ'. For example, in terms of the earth pressure coefficients a value of $F_p = 1.5$ is equivalent to $F_s = 1.2$ when $\phi' = 35°$ (assuming full wall friction $\delta = \phi'$ and using the values of passive pressure coefficient given by Caquot and Kerisel, 1948), but to $F_s = 1.6$ when $\phi' = 20°$. Thus the application of the traditional value of $F_p = 1.5$ to diaphragm walls in clay soils (with $\phi' \approx 20°$) led to the calculation of excessive and uneconomical embedment depths. This point is illustrated by Figure 10.1(b).

(c) In the 'net pressure method', a net pressure diagram based on fully-active and fully-passive pressures is plotted (Figure 10.2a), and the depth of embedment is chosen

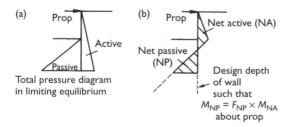

Figure 10.2 'Net pressure' method of factoring collapse calculations for embedded walls.

such that the moment about the prop of the net pressure in front of the wall is equal to a factor F_{np} times the moment of the net pressure behind the wall (Figure 10.2b). This method is potentially dangerous: Figure 10.1 shows that for a wall propped at the top, an apparently satisfactory numerical value of $F_{np} = 2$ may correspond to a factor on soil strength (F_s) of less than 1.1 over the range $10° \leq \phi' \leq 35°$ (Burland et al., 1981). The studies reported by Gaba et al. (2003, Appendix A7) confirm the unreliability of the method for propped walls, but suggest that it may give more satisfactory results if the wall is unpropped. This, together with the selection of conservative soil strength parameters and groundwater conditions, might explain Williams and Waite's (1993) observation that the method has been used successfully in sheet pile cofferdam design for more than 50 years. Nonetheless, it is better to arrive at the right answer for the right reasons, by applying the factor of safety to the soil strength directly as the current codes of practice recommend.

Figure 10.1(a) (from Burland et al., 1981) compares the ratios of the overall wall height H to the retained height h indicated using various factors of safety ($F_p = 2$, $F_s = 1.5$, $F_d = 1.7$, $F_{np} = 2$) with the wall height : retained height ratio at collapse ($F_s = 1$), for various values of ϕ' with a smooth wall ($\delta = 0$) and a water table on both sides of the wall at the level of the excavated soil surface (i.e. there is no seepage around the wall, and the pore water pressures below the water table are hydrostatic). Figures 10.1(a) and (b) illustrate that, for an effective stress analysis, $F_p = 2$ is unduly conservative at values of ϕ' less than about 27°; that $F_{np} = 2$ gives results very close to $F_s = 1.07$; and that $F_d = 1.7$ gives results close to $F_s = 1.5$, becoming less conservative as ϕ' is increased above about 25°.

For a total stress analysis, Figure 10.1(c) shows the ratio of the overall wall length H to the retained height h (H/h) as a function of the non-dimensionalized undrained shear strength $2\tau_u/\gamma_h$, calculated using $F_{np} = 2$, $F_d = 2$, $F_s = 1.5$, $F_s = 1$ (i.e. at limiting equilibrium), $F_p = 1.5$ and $F_p = 2$. The results using $F_{np} = 2$ and $F_d = 2$ are very close to failure ($F_s = 1$). By considering the variation in F_p with H/h at constant $2\tau_u/\gamma h$ (e.g. $2\tau_u/\gamma h = 0.9$), it may be seen that F_p is reduced as H/h is increased. Given that in increase in H/h corresponds to a reduction in the retained height h at constant overall wall length H (i.e. an increase in the depth of embedment d), this result is not sensible.

In summary, the parametric study carried out by Burland, et al. (1981) shows that the most consistent approach for an embedded retaining wall is to apply the factor of safety to the soil strength directly. This is the approach adopted in BS8002 (BSI, 2001), Eurocode 7 (1995) and CIRIA Report C580 (Gaba et al., 2003) once the possibility of an unforeseen surcharge on the retained soil surface and an additional unplanned excavation have been taken into account. Gaba et al. (2003) further recommend the use of numerical soil-structure interaction analysis (with factored soil strength parameters) rather than simple limit equilibrium calculations in ultimate limit state design, particularly for complex structures that are statically indeterminate (e.g. multi-propped walls), where the potential mechanism of collapse is not obvious, of where the construction sequence must be taken into account (e.g. walls with low level props).

Similar considerations apply to the definition of factors of safety for other types of wall, such as the gravity wall analysed in section 7.7. The factor of safety on strength (or strength mobilization factor) approach would require the ultimate limit state design calculation to be carried out with a mobilized soil strength of τ_u/M in the short-term, or $\tan^{-1}\{(\tan \phi')/M\}$ in the long-term analysis. The size of the wall (e.g. its mass and base length) would then be chosen to give an equilibrium-free body diagram, with the thrusts and soil pressures calculated on the basis of the reduced mobilized soil strengths.

10.3 Calculation of bending moments and prop loads: serviceability limit states

In a limit state design, the designer will often need to carry out calculations to demonstrate the avoidance of serviceability limit states, in addition to the collapse or ultimate limit state calculations discussed in section 10.2. Possible serviceability limit states include excessive wall and ground movements, and excessive stresses in walls or props that could lead to bending failure, cracking or buckling. Ideally, serviceability limit states should be assessed with reference to the expected working or service conditions of the wall and the soil, but this is not generally that easy. For example, an ultimate limit state calculation following the procedures given in BS8002 (BSI, 2001) for an embedded retaining wall will give a stress distribution that is in equilibrium and (because of the factor applied to the soil strength, the 10 kPa surcharge behind the wall, and the increased embedment depth) remote from collapse. However, it is unlikely to represent the actual stress distribution under working conditions except for a stiff, unpropped wall. This is because the soil strengths actually mobilized may be different on each side of the wall, depending on the post-installation stress state and the stress–strain response of the soil in the different stress paths followed; and because wall flexibility and other soil–structure interaction effects will lead to a variation in soil strain (and hence mobilized soil strength) with depth. Ways in which the basic limit equilibrium calculation might be modified to take account of different rates of strength mobilization with wall movement and wall flexibility are discussed in sections 10.4–10.6.

The approach to serviceability limit state calculations recommended by Gaba *et al.* (2003) is to carry out some form of numerical analysis that takes account of soil–structure interaction effects. Such an analysis should be based on the actual wall geometry (i.e. without the unplanned depth of excavation) and the actual (unfactored) soil strength parameters: it will also be necessary to specify wall and soil stiffnesses and (depending on the complexity of the analysis) other parameters as well. Detailed guidance is given in Gaba *et al.* (2003). For embedded walls with a single level of props near the top, Gaba *et al.* (2003) point out that a limit equilibrium approach will tend to overestimate wall bending moments and underestimate prop or anchor loads in comparison with a soil–structure interaction analysis. If the limit equilibrium approach is adopted, they suggest that the serviceability limit state bending moment diagram for an embedded wall propped at the top may be obtained using the procedure illustrated in Figure 10.3. First, the bending moment distribution at true limiting equilibrium is calculated, that is, with the wall having the embedment needed just to prevent collapse with unfactored soil strengths, an increased excavation depth and additional external loads as required by the code. The serviceability limit state bending moment diagram is then simply sketched in between the maximum of the limit equilibrium distribution with unfactored soil strengths and the actual toe of the wall. The reasoning behind this procedure, and its consequent limitations, are discussed in section 10.4.2.

In the limit-state design of a structural element (e.g. to BS8110 Part 1; BSI, 1997 or EC2 Part 1, DD ENV 1992-1-1; BSI, 1992 for reinforced concrete and BS5950 Part 1; BSI, 2000 or EC3 Part 5 ENV 1993-5; BSI, 1998 for steel) the same distinction between ultimate limit state and serviceability limit state is made. Gaba *et al.* (2003)

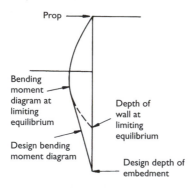

Figure 10.3 Determination of serviceability limit state bending moments for an embedded wall propped near the crest from limit equilibrium calculations, according to CIRIA Report C580 (adapted from Gaba *et al.* 2003).

recommend the use of wall bending moments and shear forces determined from the geotechnical serviceability limit state calculations (e.g. using soil–structure interaction analysis or the approach indicated in Figure 10.3) in the serviceability limit state structural design. For ultimate limit state structural design, they recommend that wall bending moments and shear forces are taken as the greater of

- the values obtained from the ultimate limit state calculation (either limit equilibrium or soil–structure interaction with factored soil strengths and geometry changes and additional loads as required)
- 1.35 times the serviceability limit state values calculated using soil–structure interaction analysis (with actual soil strengths) or the approach indicated in Figure 10.3
- values consistent with increased prop loads calculated independently to guard against accidental loading and progressive failure (e.g. if a temporary prop is accidentally knocked out, the neighbouring props must be able to carry the load it has shed without themselves failing), or using the empirical method for the design of temporary supports suggested by Twine and Roscoe (1999).

Serviceability limit state structural design of the props should be based either on the prop load obtained from an serviceability limit state soil–structure interaction analysis (with actual soil strengths and the expected in-service wall geometry), or 1.85 times the prop load calculated using the modified limit equilibrium approach indicated in Figure 10.3. The designer is also required to check the ability of the props to withstand an increase in load due to a rise in temperature (see Gaba *et al.*, 2003; their section 7.1.3).

Because limit equilibrium methods tend to underestimate prop loads in comparison with soil–structure interaction analyses, the prop load for ultimate limit state structural design should be taken as the greater of

- 1.35 times the serviceability limit state value described above
- the value obtained from a ultimate limit state soil–structure interaction analysis (with factored soil strengths, altered geometry and additional loads as appropriate)

or 1.85 times the value obtained from a limit equilibrium ultimate limit state calculation (depending on which method of calculation has been used)

- the value obtained from an serviceability limit state soil–structure interaction analysis in which accidental loading and progressive failure have been taken into account, or 1.85 times the value obtained from a corresponding limit equilibrium calculation (again, depending on which method has been used).

Gaba *et al.* (2003) emphasize particularly the need for the designer to ensure that the accidental loss of a prop will not result in the sequential overloading and failure of the props all the way along the wall.

It is usually necessary to stiffen the wall at the levels at which the props act, by means of longitudinal beams known as walings. In this way, the effects of minor local variations in ground or loading conditions can to a limited extent be spread along the wall. The detailed design of the system of props and walings is very important: information on this is given by Williams and Waite (1993).

10.4 Embedded walls retaining clay soils

10.4.1 Time-scale over which undrained conditions apply

There are two main problems facing the designer of an embedded wall in a clay soil which are unlikely to arise in the design of walls in sands. The first of these concerns the length of time for which the initial undrained shear strength can be relied upon: in other words, for how long after excavation is an analysis based on total stresses and the 'maximum shear stress' failure criterion $\tau_{max} = \tau_u$ valid?

The answer to this question is extremely complex, and will depend on factors which even in the same soil may vary from site to site, such as the proximity of sources of recharge, the extent of local fissuring and the presence of high-permeability layers. Excavation is an unloading process that will result eventually in the swelling and softening of the clay and hence a reduction in undrained shear strength τ_u. In terms of effective stresses, the negative excess pore water pressures induced on excavation maintain the average effective stress s' ($=0.5(\sigma_1' + \sigma_3')$) temporarily artificially high, so that the Mohr circle of effective stress is initially well within the failure envelope (Figure 10.4). As the clay swells, the negative excess pore water pressures dissipate and the average effective stress s' is reduced. The shear stress required to maintain the stability of the excavation remains constant, and when the Mohr circle of effective stress touches the effective stress failure envelope, collapse will occur.

Many of the factors required to assess the time-scale over which the undrained analysis might be valid are practically unquantifiable. Clearly the approach should never be used for permanent works, and even for temporary works it is advisable to err on the side of caution. Softening is likely to occur first near the retained and excavated soil surfaces, and may be taken into account by using effective stress analysis or reduced undrained shear strengths in these zones. The recommendations of Gaba *et al.* (2003) in this respect are summarized at the end of section 7.6 (p. 389) of this book.

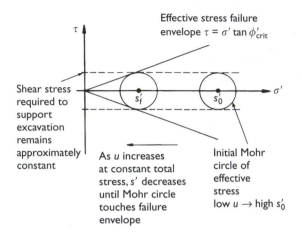

Figure 10.4 Failure after dissipation of negative excess pore water pressures following excavation.

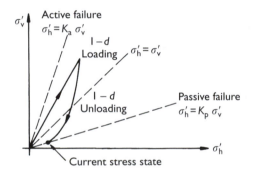

Figure 10.5 Schematic stress history for an overconsolidated clay deposit.

10.4.2   Effect of high in situ lateral stresses

The second question concerns the significance of the high *in situ* lateral earth pressures that are usually present in overconsolidated clay deposits. The *in situ* horizontal stresses in overconsolidated clays are generally high because of the stress history of the deposit. Deposition corresponds approximately to one-dimensional consolidation, during which the horizontal and vertical effective stresses increase in proportion to each other (Figure 10.5). On unloading, which might occur due to the melting of a glacier or the erosion of overlying soil, the horizontal effective stress tends to remain 'locked-in', decreasing proportionately less quickly than the vertical effective stress. In extreme cases (of which London Clay is one example), a zone of undisturbed soil extending to a depth of several metres from the surface may be brought to the verge of passive failure simply as a result of the vertical unloading process.

The process of installing a diaphragm-type retaining wall in an overconsolidated clay will almost certainly result in a reduction in the horizontal effective stresses close to the wall to below their *in situ* values. However, the pre-excavation lateral earth pressure coefficient is still likely to be greater than or equal to one (Powrie, 1985 and Gaba *et al.* 2003, Appendix A5). Furthermore, the soil remaining in front of the wall below formation level might on excavation be brought to passive failure simply as a result of the further unloading without the need for an increase in the horizontal stress. In these circumstances, the wall movement required to mobilize fully passive pressures in front might be thought to be rather less than that needed to reduce the stresses behind to the active limit. This could lead to the development of larger bending movements under in-service or working conditions than might have been allowed for using conventional design methods, in particular if fully active conditions are assumed to have been reached in the soil behind the wall (the F_p method described in section 10.2).

Finite element analyses carried out by Potts and Fourie (1984) appeared to confirm that in-service bending moments in embedded walls in overconsolidated clays could be very much greater than those calculated on the basis of fully active conditions behind the wall. However, the finite element analyses took no account of the stress relief likely to be caused by wall installation, and the potentially very different stress–strain responses of the soil on either side of the wall.

Powrie *et al.* (1998) argue that, when an embedded wall moves as the soil is removed from within the main excavation, the rates of mobilization of soil strength are likely to be different on either side of the wall. This is because the stress paths followed are different in each case, particularly when considered in relation to the stress path imposed on the soil during wall installation. A diaphragm wall is usually excavated panel-by-panel under a bentonite support slurry; the reinforcing cage is lowered in; and concrete is tremied into the bottom of the panel excavation, displacing the bentonite slurry upward. Thus, during wall installation, the soil on either side of a diaphragm wall is subjected to a reduction in horizontal stress (to the hydrostatic pressure of the bentonite slurry) as the panel is excavated, followed by an increase in horizontal stress as the concrete is poured. When the main excavation is made, the soil behind the wall is subjected to a reversal of this stress path (i.e. a reduction in horizontal stress), while the soil remaining in front experiences a continuing increase in horizontal stress. The response of the soil behind the wall is therefore likely to be stiffer than that of the soil in front (see section 6.2), leading to the more rapid mobilization of soil strength with shear strain in the retained soil (Figure 10.6). In other words, as the wall moves, the horizontal stresses behind will probably fall more rapidly than the horizontal stresses in front of it increase – particularly for an unpropped wall, where the relationship between wall movement and shear strain is the same for the soil behind and in front of the wall. For a rigid wall propped at the crest, the shear strain in the soil in front of the wall is $(1 + h/d)$ times that in the retained soil, for a given angle of wall rotation, where h is the retained height and d is the embedded depth (section 10.5).

Powrie *et al.* (1998) suggest that, when the combined effects of the high *in situ* lateral earth pressure coefficient, stress relief due to wall installation, the different rates of mobilization of soil strength with shear strain on each side of the wall, and the enhanced shear strains (for a given wall rotation) in front of a wall propped at the

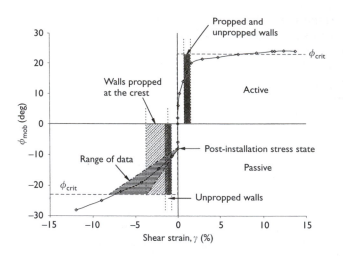

Figure 10.6 Different rates of mobilization of strength with shear strain in stress paths simulating those experienced by soil elements behind and in front of an embedded retaining wall (data from Powrie *et al.* 1998).

top are taken into account, a ultimate limit state calculation based on factored soil strengths, following BS8002, EC7 or Gaba *et al.* (2003), is likely to

- overestimate the in-service lateral stresses in the soil behind an unpropped embedded wall relative to those in front, and
- provide a reasonable estimate of the relative magnitudes of the in-service lateral stresses behind and in front of a rigid embedded wall propped at the crest, although in reality the lateral stresses behind such a wall are likely to be reduced by wall bending effects (section 10.6).

The empirical procedure suggested by Gaba *et al.* (2003) to estimate serviceability limit state bending moments from a limit equilibrium analysis for a wall propped at the top (Figure 10.3) reflects this, in that the serviceability limit state bending moments are based on fully active conditions behind the wall. The validity of this approach depends on a combination of

- a relatively high rate of mobilization of strength with strain in the soil behind the wall, causing horizontal stresses in the retained soil to fall quickly towards the active limit, and/or
- a degree of wall flexibility, so that bending moments are reduced by stress redistribution.

Experience suggests that in most practical situations, one or both of these conditions are generally satisfied. We might note, however, that the approach indicated in Figure 10.3 does not lead to the calculation of greater bending moments as the depth of embedment is increased, as might occur with some other methods of analysis or in reality.

Example 10.1 Calculating the design depth of embedment of an embedded retaining wall

Figure 10.7 shows a cross-section through the retaining wall on one side of a proposed cut-and-cover tunnel. The side walls are embedded cantilevers, propped apart at the crest by a roof slab. Stating carefully the assumptions you make concerning wall friction, pore water pressures, tension cracks etc., and giving reasons for your choice of factor of safety, calculate the depth of embedment required for a design based on the ultimate limit state.

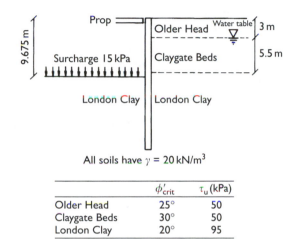

All soils have $\gamma = 20\,\text{kN/m}^3$

	ϕ'_{crit}	τ_u (kPa)
Older Head	25°	50
Claygate Beds	30°	50
London Clay	20°	95

Figure 10.7 Cross-section through proposed retaining wall.

Figure 10.7 is based on the cut-and-cover tunnel on the M25 London orbital motorway at Bell Common, which is described by Hubbard *et al.* (1984) and Tedd *et al.* (1984).

(a) Short-term (total stress) analysis

The ultimate limit state design depth of embedment will be calculated broadly in accordance with BS8002 (BSI, 2001). The recommended mobilization factor M on undrained shear strength τ_u is 1.5. The wall adhesion τ_w is taken as $0.75 \times \tau_{mob}$, giving $\tau_w = 0.75 \times \tau_u/1.5 = 0.5 \times \tau_u$.

The minimum (active) and maximum (passive) lateral total stresses are given by equations (7.13) using the factored values of soil strength τ_{mob}.

On the retained side of the wall, the surface surcharge q is taken as 10 kPa, in accordance with BS8002 (2001). On the excavated side of the wall, the 15 kPa surcharge indicated in Figure 10.7 represents the carriageway. As the carriageway has a stabilizing effect, the most critical design condition in the short term is when the excavation is open to the full depth, before the carriageway has been placed. The 15 kPa surcharge is therefore not included in the short-term analysis.

$$\sigma_{h,min} = q + \gamma z - [2\sqrt{(1 + \tau_w/\tau_{mob})}] \times \tau_{mob} = q + \gamma z - K_{ac} \times \tau_{mob}$$
(Active, 7.13(a))

$$\sigma_{h,max} = \gamma z + [2\sqrt{(1 + \tau_w/\tau_{mob})}] \times \tau_{mob} = \gamma z + K_{pc} \times \tau_{mob}$$
(Passive, 7.13(b))

For design purposes, the retained height is increased by 0.5 m (this being less than 10% of the retained height) to 10.175 m (BS8002, 2001). The calculated wall embedment x is then the depth of embedment required below the design excavation level, that is, the total wall length H is 10.175 m $+ x$.

The first step is to calculate the values of $\tau_{mob}, \tau_w, K_{pc}, \sigma_v$ and σ_h at key depths (i.e. wherever there is a change in the linearity of the stress distribution), behind and in front of the wall. These are tabulated in Table 10.1. The pressure u_{tc} that would be exerted by water in a flooded tension crack behind the wall is given, assuming that the water level in the tension crack is maintained at the natural groundwater level, 3 m below the retained soil surface. Table 10.1 also includes the stresses on both sides of the wall at a general depth x below the revised excavation level, that is, at $(10.175 + x)$ m below the retained soil surface.

Table 10.1 shows that in the Claygate Beds (CB) (neglecting the small negative value at 3 m depth), the active lateral total stress is greater than the hydrostatic pressure of water in a tension crack that floods to the water table level, 3 m below the retained soil surface. This means that a tension crack, either flooded or dry, cannot remain open in the Claygate Beds.

In the London Clay (LC), the hydrostatic pressure exerted by water in a flooded tension crack is greater than the active lateral total stress until a depth of 12.76 m below the retained soil surface is reached. This means that a flooded tension crack could remain open in the London Clay to this depth. In the Older Head (OH), which is entirely above the water table, a tension crack could form but will be assumed to remain dry.

The lateral stress distributions indicated in Table 10.1 are shown in Figure 10.8(a). Figure 10.8(b) shows the division of the lateral stress distributions into rectangular and triangular stress blocks, together with the lever arm (about the prop) of the centre of pressure in each case.

The value x required for short-term (undrained) equilibrium may be calculated by taking moments about the prop. Assuming that $x \leq 2.585$ m, the moment due to the active pressures is:

$[0.5 \times 54.4 \text{ kPa} \times 5.5 \text{ m}]$

$\qquad \times [3 \text{ m} + (5.5 \text{ m} \times 2/3)]$ ⠀⠀⠀⠀(stress block 1 in Figure 10.8(b))

$\qquad + [0.5 \times (71.75 + 10x) \text{ kPa}$

$\qquad \times (7.175 + x) \text{ m}]$

$\qquad \times [3 \text{ m} + \{(2/3) \times (7.175 + x) \text{ m}\}]$ ⠀⠀(stress block 2 in Figure 10.8(b))

Table 10.1 Mobilized undrained soil strengths, vertical and horizontal total stresses etc. at key depths behind and in front of the wall shown in Figure 10.7

Depth (m) behind wall	Soil stratum	τ_u (kPa)	τ_{mob} (kPa)	τ_w (kPa)	K_{ac}	$q + \gamma z$ (kPa)	$\sigma_{h,min}$ (active) = $q + \gamma z - (K_{ac} \times \tau_{mod})$ (kPa)	u_{tc} (kPa)
0	O H	50	33.33	25	2.646	10	[−78.2]	0
3	O H	50	33.33	25	2.646	70	[−18.2]	0
3	C B	40	26.67	20	2.646	70	[−0.6] ≈0	0
8.5	C B	40	26.67	20	2.646	180	109.4	55
8.5	L C	95	63.33	47.5	2.646	180	12.43	55
10.175 + x	L C	95	63.33	47.5	2.646	213.5 + 20x	45.93 + 20x	71.75 + 10x
12.76	L C	95	63.33	47.5	2.646	265.2	97.6	97.6

Depth (m) in front of wall	Soil stratum	τ_u (kPa)	τ_{mob} (kPa)	τ_w (kPa)	K_{pc}	γz (kPa)	$\sigma_{h,max}$ (passive) = $\gamma z - (K_{pc} \times \tau_{mod})$ (kPa)	u_{tc} (kPa)
0	L C	95	63.33	47.5	2.646	0	167.6	n/a
x	L C	95	63.33	47.5	2.646	20x	167.6 + 20x	n/a

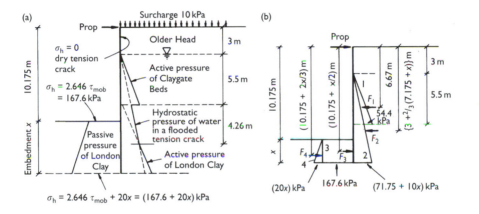

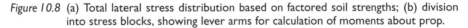

Figure 10.8 (a) Total lateral stress distribution based on factored soil strengths; (b) division into stress blocks, showing lever arms for calculation of moments about prop.

which for equilibrium must be equal to the moment of the passive pressures:

$$[167.6 \, \text{kPa} \times x \, \text{m}] \times [(10.175 + x/2) \, \text{m}] \quad \text{(stress block 3 in Figure 10.8(b))}$$

$$+ [0.5 \times 20x \, \text{kPa} \times x \, \text{m}]$$

$$\times [(10.175 + 2x/3) \, \text{m}] \quad \text{(stress block 4 in Figure 10.8(b))}$$

Solution by trial and error gives

$$x \approx 2.4 \, \text{m}$$

The initial assumption $x \leq 2.585 \, \text{m}$ is therefore justified. The overall wall length H is

$$H = (10.175 + 2.4) \, \text{m} = 12.575 \, \text{m}$$

The wall embedment d at the expected excavation depth of 9.675 m is

$$d = (12.575 - 9.675)\,\text{m}$$

$$\Rightarrow d = 2.9\,\text{m}$$

The short-term prop load F is calculated from the condition of horizontal equilibrium:

$$
\begin{aligned}
F = &[0.5 \times 54.4\,\text{kPa} \times 5.5\,\text{m}] && \text{(stress block 1 in Figure 10.8(b))} \\
&+ [0.5 \times (71.75 + 10x)\,\text{kPa} \\
&\times (7.175 + x)\,\text{m}] && \text{(stress block 2 in Figure 10.8(b))} \\
&- [167.6\,\text{kPa} \times x\,\text{m}] && \text{(stress block 3 in Figure 10.8(b))} \\
&- [0.5 \times 20x\,\text{kPa} \times x\,\text{m}] && \text{(stress block 4 in Figure 10.8(b))}
\end{aligned}
$$

Substituting $x = 2.4\,\text{m}$,

$$F \approx 148\,\text{kN/m}$$

In the case of a temporary structure, it would be prudent to allow for a degree of softening of the clay near the excavated soil surface, for example as described in section 10.4.1 and at the end of section 7.6. This would lead to an increase in the calculated depth of embedment.

In the case of a permanent structure, however, the long-term condition is almost certain to be more critical from a design point of view. Provided that an effective stress analysis which represents fully softened conditions is carried out and found to be more critical, there is no real need to allow for partial softening in the total stress analysis, unless the support conditions change during construction so that the stability of the wall is more critical in the short term than in the long term.

(b)   Long-term (effective stress) analysis

Again following BS8002 (BSI, 2001), the overall wall length H is calculated on the basis of an excavation depth of $9.675 + 0.5 = 10.175$ m. A 10 kPa surcharge is assumed to act on the retained soil surface. The design strengths are taken as the lesser of $\tan \phi'_{\text{crit}}$ and $(\tan \phi'_{\text{peak}})/M$, where the strength mobilization factor $M = 1.2$. The wall friction angle δ is given by $\tan \delta = 0.75 \times \tan \phi'_{\text{mob}}$. In this case, we will assume that the peak and critical state strengths are the same, so that the design soil strengths ϕ'_{mob} are given by $\tan \phi'_{\text{mob}} = (\tan \phi'_{\text{peak}})/1.2 = (\tan \phi'_{\text{crit}})/1.2$.

Values of active and passive earth pressure coefficients corresponding to the design soil strengths, according to Tables 7.6 and 7.7, are given in Table 10.2.

The next step is to calculate the long-term pore water pressures. In this case, the excavation is fairly wide. In the absence of actual data, we will assume that the permeability of the Claygate Beds is similar to that of the London Clay, so that the linear seepage model may be used.

Table 10.2 Earth pressure coefficients based on factored soil strengths for effective stress analysis of embedded retaining wall

Stratum	ϕ'_{crit} $(= \phi'_{max})$ (degrees)	$\tan \phi'_{crit}$	$\tan \phi'_{mob}$ with $M = 1.2$	ϕ'_{mob} (degrees) (rounded to nearest whole degree)	K_a based on ϕ'_{mob} and $\tan \delta$ $= 0.75 \times \tan \phi'_{mob}$ (Table 7.6)	K_p based on ϕ'_{mob} and $\tan \delta$ $= 0.75 \times \tan \phi'_{mob}$ (Table 7.7)
Older Head	25	0.466	0.389	21	0.407	(2.90)
Claygate Beds	30	0.577	0.481	26	0.329	(3.96)
London Clay	20	0.364	0.303	17	0.482	2.30

Figure 10.9(a) shows the pore water pressures calculated using the linear seepage assumption, at key depths on both sides of the wall, for a general wall embedment x. Writing $y = (2x + 14.35)/(2x + 7.175)$, the pore water pressure at the base of the wall is $\gamma_w xy$ (kPa), and the pore water pressure at the interface between the Claygate Beds and the London Clay is $5.5\gamma_w xy/(x + 7.175)$ (kPa). The effective vertical and horizontal stresses at key depths behind and in front of the wall are given in Table 10.3 in terms of x and y; the active earth pressure coefficients in each stratum K_{aOH}, K_{aCB} and K_{aLC}; and the passive pressure coefficient in the London Clay K_{pLC}.

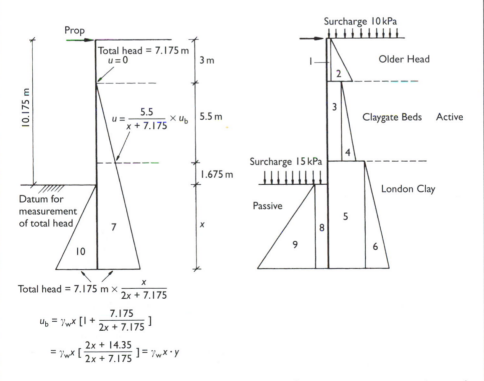

Figure 10.9 (a) Pore water pressure distribution using linear seepage approximation, and (b) effective stress distribution based on factored soil strengths.

Table 10.3 Lateral effective stresses at key depths behind and in front of the wall

Depth (m) behind wall	$\gamma z + 10$ (kPa)	Pore water pressure u (kPa)	$\sigma_h' = K_a(\gamma z + 10 - u)$ (kPa)
0 Older Head	10	0	$10K_{aOH}$
3 Older Head	70	0	$70K_{aOH}$
3 Claygate Beds	70	0	$70K_{aCB}$
8.5 Claygate Beds	180	$55xy/(x + 7.175)$	$[180 - 55xy/(x + 7.175)] \times K_{aCB}$
8.5 London Clay	180	$55xy/(x + 7.175)$	$[180 - 55xy/(x + 7.175)] \times K_{aLC}$
$10.175 + x$ London Clay	$213.5 + 20x$	$10xy$	$[213.5 + 20x - 10xy] \times K_{aLC}$

Depth (m) in front of wall	$\gamma z + 15$ (kPa)	Pore water pressure u (kPa)	$\sigma_h' = K_p(\gamma z + 15 - u)$ (kPa)
0	15	0	$15K_{pLC}$
x	$15 + 20x$	$10xy$	$[20x + 15 - 10xy] \times K_{pLC}$

Note

The unit weight of water has been taken as $10\,\text{kN/m}^3$.

Table 10.4 Resultant forces and lever arms for stress blocks shown in Figure 10.9

Stress block (see Figure 10.9)	Average stress within stress block (kPa)	Depth of stress block (m)	Resultant force (kN/m)	Lever arm about prop
\multicolumn{5}{l}{Stresses and pore water pressure behind the wall}				
1	$10K_{aOH}$	3	$30K_{aOH}$	$3\,\text{m} \div 2 = 1.5\,\text{m}$
2	$0.5 \times 60K_{aOH}$	3	$90K_{aOH}$	$3\,\text{m} \times (2/3) = 2\,\text{m}$
3	$70 \times K_{aCB}$	5.5	$385K_{aCB}$	$3\,\text{m} + (5.5\,\text{m} \div 2)$ $= 5.75\,\text{m}$
4	$0.5 \times [110 - \{55xy/ (x + 7.175)\}] \times K_{aCB}$	5.5	$2.75 \times [110 - \{55xy/ (x + 7.175)\}] \times K_{aCB}$	$3\,\text{m} + (5.5\,\text{m} \times 2/3)$ $= 6.67\,\text{m}$
5	$[180 - \{55xy/ (x + 7.175)\}] \times K_{aLC}$	$x + 1.675$	$(x + 1.675) \times [180 - \{55xy/(x + 7.175)\}] \times K_{aLC}$	$8.5\,\text{m} + [(1.675 + x)\,\text{m}/2]$ $= (9.338 + x/2)\,\text{m}$
6	$0.5 \times [33.5 + 20x - \{10xy(x + 1.675)/ (x + 7.175)\}] \times K_{aLC}$	$x + 1.675$	$(x + 1.675) \times 0.5 \times [33.5 + 20x - \{10xy(x + 1.675)/ (x + 7.175)\}] \times K_{aLC}$	$8.5\,\text{m} + 2(x +1.675)\,\text{m}/3$ $= (9.62 + 2x/3)\,\text{m}$
7 (pwp)	$0.5 \times 10xy$	$x + 7.175$	$(0.5 \times 10xy) \times (x + 7.175)$	$3\,\text{m} + (2/3) \times (7.175 + x)\,\text{m}$ $= (7.783 + 2x/3)\,\text{m}$
\multicolumn{5}{l}{Stresses and pore water pressure in front of wall}				
8	$15K_{pLC}$	x	$15x \times K_{pLC}$	$10.175 + x/2\,\text{m}$
9	$0.5 \times K_{pLC} \times (20x - 10xy)$	x	$x \times K_{pLC} \times (10x - 5xy)$	$10.175 + 2x/3\,\text{m}$
10 (pwp)	$0.5 \times 10xy$	x	$0.5x \times 10xy$	$10.175 + 2x/3\,\text{m}$

The effective stress blocks which must be considered in the analysis are shown in Figure 10.9(b). The ultimate limit state design depth of embedment for long-term conditions may be found by taking moments about the prop.

With reference to Figure 10.9, the resultant forces and the lever arms for each stress block are given in Table 10.4. In each case, the resultant force (in kN/m) is calculated as the average stress (in kPa or kN/m^2), multiplied by the depth over which the average stress acts (in m). The moment of each stress block is given by the resultant force multiplied by the lever arm.

For moment equilibrium about the prop, the sum of the clockwise moments (in this case, the forces acting behind the wall multiplied by the appropriate lever arms) must be equal to the anticlockwise moments (in this case, the sum of the forces acting in front of the wall multiplied by the appropriate lever arms). By trial and error, the condition of moment equilibrium is satisfied with

$$x \approx 12.56\,\text{m}$$

This gives an overall wall length $H = 10.175\,\text{m} + 12.56\,\text{m} = 22.735\,\text{m}$. For the anticipated retained height $h = 9.675\,\text{m}$, the depth of embedment d is

$$d = 22.735\,\text{m} - 9.675\,\text{m}$$

$$\Rightarrow d \approx 13\,\text{m}$$

The prop load will be greater in the long term than in the short term. The ultimate limit state design prop load may be calculated from the condition of horizontal force equilibrium for the wall, with the ultimate limit state design excavation depth of 10.175 m. Subtracting the sum of the forces in front of the wall from the sum of the forces behind the wall (Table 10.4, column 4 with $x = 12.56\,\text{m}$ and $y = 1.2222$), we obtain

$$F = 488\,\text{kN/m}$$

The easiest way of carrying out analyses such as this is to assemble the equations in a spreadsheet. An iterative solution can then be obtained very rapidly.

10.5 Geostructural mechanism to estimate wall movements

One of the main shortcomings of a traditional limit equilibrium calculation is that it gives the designer no explicit information on ground movements that may be relevant to a serviceability limit state. For stiff walls, where the effective stress distributions on either side may be assumed to be approximately linear, an idealized displacement mechanism or **geostructural mechanism** can be used to relate the rigid body rotation of the wall to the maximum shear strain in the adjacent soil, and hence to the ground movements. (Ground movements due to wall bending are neglected.) The shear strain in the adjacent soil may then be related to the mobilized strength required for equilibrium by means of an appropriate stress–strain curve, enabling soil and wall deformations under working conditions to be estimated from the equilibrium calculation. In this approach, the condition of **equilibrium** of the wall is satisfied in order to find the mobilized soil strength ϕ'_{mob}; the **stress–strain** relationship for the soil is used to determine the soil strain corresponding to the mobilized strength ϕ'_{mob}; and the displacements are **compatible** with these strains according to the assumed displacement mechanism.

The stress–strain relationship is used in the form of a graph of mobilized soil strength ϕ'_{mob} as a function of shear strain γ. Since $\phi'_{\text{mob}} = \sin^{-1}[t/s']$ where $t = 0.5[\sigma'_1 - \sigma'_3]$ and $s' = 0.5[\sigma'_1 + \sigma'_3]$ (Figure 5.6), the use of a single ϕ'_{mob}–γ curve for the soil on one side of the wall is equivalent to the assumption of an increasing shear modulus with average effective stress s', and hence with depth. In principle, both the pre-excavation earth pressure coefficient and the probably different stress–strain responses of the soil on either side of the wall could be taken into account in the ϕ'_{mob}–γ relationships used. (e.g. Figure 10.6).

In centrifuge model tests on unpropped embedded walls retaining clay (Bolton and Powrie, 1988), significant soil movements during excavation occurred mainly in the zones defined approximately by lines drawn at 45°, extending upward from the toe of the wall on both sides. For a rigid wall OV rotating about its toe O, this pattern of deformations would be consistent with a shearing triangle AOV, beyond which the soil is effectively rigid (Figure 10.10(a)). From the Mohr circle of strain increment shown in Figure 10.10(b) for the triangle AOV, the shear strain increment $\delta\gamma$ within this zone is uniform and equal to twice the increment of wall rotation $\delta\theta$. For a rigid wall rotating into the soil the pattern of deformation is reversed, but the relation $\delta\gamma = 2\delta\theta$ still holds.

For a rigid unpropped embedded wall VW rotating about a point O near its toe, six of these deforming triangles may be used to construct an idealized displacement mechanism, as shown in Figure 10.10(c). As before, the shear strain increment $\delta\gamma$ in each of the six deforming triangles is uniform and equal to twice the incremental wall rotation $\delta\theta$.

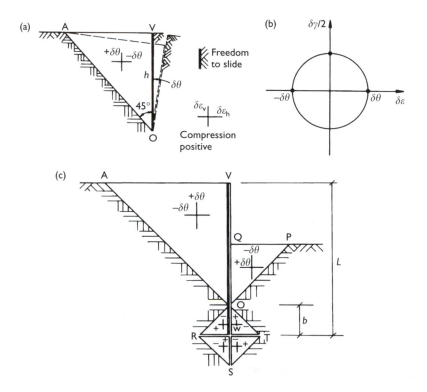

Figure 10.10 (a) Idealized deformation pattern for a rotating bulkhead and (b) corresponding Mohr circle of strain. (c) Synthesis of idealized displacement mechanism for a stiff unpropped wall, rotating about a point near its toe. (Redrawn with permission from Bolton and Powrie, 1988.)

The stress analysis shown in Figure 7.41 can be used for an unpropped wall of any geometry to calculate the mobilized soil strength ϕ'_{mob} required for equilibrium. (The mobilized soil strength ϕ'_{mob} is assumed to be uniform with depth, and in the simplest case is the same on both sides of the wall.) The shear strain corresponding to a given mobilized soil strength can be determined from an appropriate laboratory test on a representative soil element. The idealized strain field shown in Figure 10.10(c) can then be used to estimate the magnitude of the wall rotation and soil movements.

Bolton and Powrie (1988) applied this approach to a centrifuge model test on an unpropped wall of retained height 10 m and embedment depth 20 m. The calculation was based on the undeformed wall geometry, and the pore water pressures measured immediately after excavation in the centrifuge test. In these conditions, a mobilized soil strength ϕ'_{mob} of 17.5° was required for equilibrium, with wall friction $\delta = \phi'_{mob}$. The corresponding shear strain according to plane strain laboratory test data (starting from a mobilized strength of zero, which corresponds to a pre-excavation lateral earth pressure coefficient of 1) is 1.1% (Figure 10.11(a)). This implies an incremental wall rotation $\delta\theta$ of 0.55%. Figure 10.12 shows that the soil movements calculated using the idealized displacement pattern shown in Figure 10.10, with $\delta\theta = 0.55\%$, are very similar to those measured in the centrifuge test.

Alternatively, the idealized displacement pattern may be used to limit the permissible mobilized soil strength ϕ'_{mob}, on the basis of an allowable wall movement. Suppose that the maximum allowable movement at the top of an embedded retaining wall of overall length 20 m was 100 mm. This would imply a maximum wall rotation $\delta\theta$ of approximately (0.1 m ÷ 20 m) or 0.5%. The corresponding increment of shear strain in the retained soil is 1%, which according to the ϕ'_{mob}–γ curve of Figure 10.11(b) places a limit on the mobilized soil strength of 16.7°.

Taking the critical state strength ϕ'_{crit} as 22°, the corresponding soil strength mobilization factor M is $\tan\phi'_{crit}/\tan\phi'_{mob} = 1.35$. This is greater than the strength mobilization factor $M = 1.2$ (applied to the peak soil strength, $\tan\phi'_{peak}$), required by BS8002 (BSI, 2001). This is partly because most natural clays tend to be stiffer than the reconstituted kaolin sample to which Figure 10.11 relates. Also, in a design according to BS8002, the 10 kPa surcharge on the retained soil surface and the increased excavation depth both serve to distance the wall from failure.

A similar analysis for walls propped at the crest is described by Bolton and Powrie (1988). This suggests that for a stiff wall and a given wall rotation $\delta\theta$, the shear strain on the excavated side of the wall is $(1 + h/d)$ times that on the retained side, where h is the retained height and d is the depth of embedment. If this type of calculation were used as an aid to design, the permissible mobilized soil strength would depend on a number of factors, including the acceptable ground movement, the initial stress state of the soil, and its stiffness measured in appropriate stress and strain paths. For a propped wall, the different shear strain/wall rotation relationships on either side would also need to be taken into account. Further details, including a worked example, are given by Bolton et al. (1989, 1990).

The calculation can easily be generalized to allow for a pre-excavation stress state that does not correspond to a mobilized soil strength of zero (i.e. a pre-excavation earth pressure coefficient that is not equal to 1), and different rates of soil strength mobilization with shear strain on each side of the wall. Both of these features are apparent in the data shown in Figure 10.6. The different strengths mobilized on each

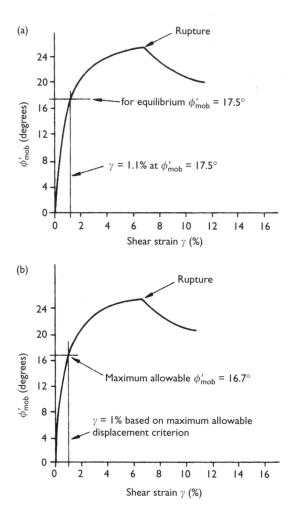

Figure 10.11 (a) Estimation of shear strain from calculated mobilized strength; (b) selection of permissible mobilized strength from allowable strain criterion. (Redrawn with permission from Bolton and Powrie, 1988.)

side of the wall would then have to

- satisfy the conditions of equilibrium, through their associated earth pressure coefficients and the appropriate idealized stress distributions, and
- correspond to the same shear strain increment in the case of an unpropped wall, or a shear strain increment in the soil in front of the wall that is $(1 + h/d)$ times that in the retained soil in the case of a wall propped at the top.

10.6   Effect of relative soil: wall stiffness

The stiffness of an embedded retaining wall may affect both deformations and bending moments under working conditions. However, most methods of analysis based on

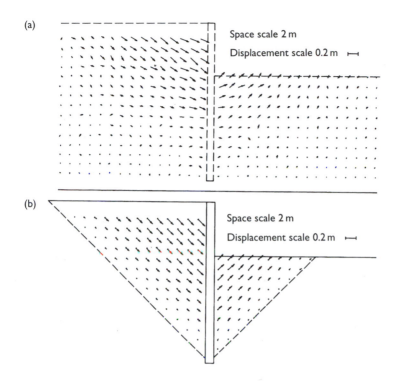

Figure 10.12 Comparison of (a) measured and (b) calculated soil movements for an unpropped wall of retained height 10 m and embedment 20 m. (Redrawn with permission from Bolton and Powrie, 1988.)

limiting linear stress distributions take no account of the flexibility of the wall. It is useful at least to identify a threshold wall stiffness above which the effects of wall bending may reasonably be neglected. For a wall of given overall length H and flexural rigidity EI, bending effects are most significant when the wall is propped at the crest.

In this section, some of the ways in which the effect of wall flexibility on the behaviour of embedded retaining walls has been quantified are described.

10.6.1 Model tests on walls in dry sand

Rowe (1952) reported the results of a series of model tests on anchored sheet pile walls of various stiffness, retaining dry sand. Rowe quantified the stiffness of a wall by means of a flexibility $\rho = H^4/EI$, where H is the overall height of the wall $(= d + h)$, and EI is its flexural rigidity or bending stiffness.

Rowe found that for truly rigid props (or, in the case of his model tests, unyielding tie-back anchors), the horizontal stress distribution on the retained side of the wall was non-linear (Figure 10.13). This is because a reduction in lateral stress at the mid-section of the wall must be accompanied by an increase in lateral stress at the unyielding section near the prop.[1] If the wall is propped just below the crest, there may even be a small backward movement of the wall above the prop, into the retained ground.

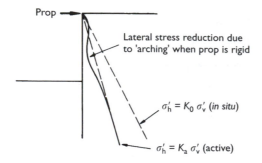

Figure 10.13 Reduction of lateral stress in the retained soil due to arching.

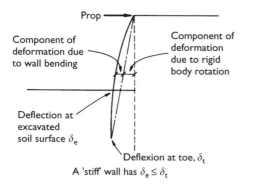

Figure 10.14 Components of wall displacement and definition of a 'stiff' wall.

However, an outward movement at the anchor point of just $H/1000$ was sufficient to generate fully active conditions, and a linear variation in lateral stress with depth behind the wall. In practice, although modern support systems may be rather stiffer than those used by Rowe (1952), movements at the prop or anchor point of this order would probably occur unless the supports were pre-stressed.

Wall deformation occurs partly due to rigid body rotation (in the case of a propped wall, about the position of the prop), and partly due to bending (Figure 10.14). Rowe (1952) found that the lateral stress distribution in front of the wall depended on the relative importance of the bending component of wall deformation, and hence on the bending stiffness of the wall. If the wall was stiff, so that the deflection at the level of the excavated soil surface (also known as the **dredge level**) was of the same order as the deflection at the toe, the stress distribution in front of the wall was approximately triangular. Measured bending moments were in agreement with those from a 'free earth support' calculation based on a fully active triangular stress distribution behind the wall and a smaller-than-passive (i.e. factored) triangular stress distribution in front (Figure 10.15(a)). (This factored free earth support calculation corresponds to the application of a factor of safety F_p to the passive pressure coefficient, method (b) in section 10.2.)

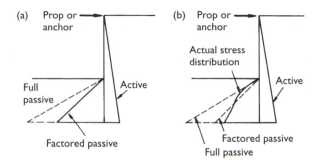

Figure 10.15 Stress distributions behind and in front of stiff and flexible embedded walls (after Rowe, 1952): (a) stiff wall; (b) flexible wall.

If the wall was more flexible, so that the deflection at dredge level was significantly greater than that at the toe, the centroid of the stress distribution in front of the wall was raised (Figure 10.15(b)). This led to smaller anchor loads and bending moments than those given by the (factored) free earth support calculation.

Rowe eventually presented his results as curves of bending moment and tie rod (anchor) load as a function of wall flexibility ρ. The bending moments and anchor loads were expressed as a proportion of the values calculated using a free earth support stress analysis (Figure 7.43), with the passive pressures reduced by a factor F_p. However, the validity of this as a 'benchmark' analysis probably depends on the initial lateral stresses in the soil being low.

In Rowe's tests, the difference in ground level between the two sides of the wall was achieved by excavation in front rather than by backfilling behind. During preparation of the model, the sand was placed loosely on either side of the sheet-pile wall, so that the pre-excavation lateral earth pressure coefficients would have been expected to be low. In these conditions, the measured bending moments were always less than or equal to those given by the factored free earth support calculation, and Rowe normalized his results with respect to this apparent limit. If the pre-excavation lateral earth pressure coefficients are high, as is likely in an overconsolidated clay deposit, the actual bending moments may be higher than those calculated using the free earth support method based on fully active pressures behind the wall (Potts and Fourie, 1985; Bolton and Powrie, 1988; see also section 10.4.2).

The observation that the stress distribution in front of the wall remains approximately linear until the movement of the wall at dredge level begins to exceed the movement at the toe (as indicated in Figure 10.14) might reasonably be expected to apply irrespective of the initial lateral stresses. A 'stiff' wall could then be defined as a wall in which bending deflections are small enough in comparison with displacements due to rigid body rotation not to affect the linearity (but without prejudging the magnitude) of the lateral stress distribution. It would then be possible to apply the proportional reductions in bending moment and anchor load given in Figure 10.16 to bending moments calculated using equilibrium stress distributions based on, for example, a uniform mobilized soil strength, $\phi'_{mob} = \tan^{-1}\{(\tan \phi')/1.2\}$ as in a ultimate limit state calculation according to BS8002 (2001), as well as on fully active stresses behind the wall and lower-than-passive stresses in front, which is the traditional approach for walls retaining sands.

Rowe recognized that the critical wall flexibility ρ_c at which the deflection at dredge level is equal to the deflection at the toe and the stress distribution in front of the wall begins to change significantly from the linear idealization should be related in some way to the stiffness of the soil. He suggested that, within the range of his experiments, ρ_c was approximately inversely proportional to the one-dimensional modulus of compression of the soil, E_0':

$$\rho_c \approx 630 \text{ metres} \div E_0' \tag{10.1}$$

where ρ_c is in m^3/kN and E_0' is in kN/m^2 (kPa).

Rowe's data also show that the critical wall flexibility ρ_c depends on the retained height ratio, $\alpha = h/H$, and the depth to the prop or anchor, βH.

10.6.2 Analytical methods

Rowe (1955) presented an analysis of anchored sheet-pile walls in which it was assumed that the lateral effective stresses behind the wall had fallen to the active limit, and the lateral effective stress p_b in front of the wall at a depth x below formation level was given by the expression

$$p_b = mxy/d \tag{10.2}$$

where d is the embedment of the wall, y is the deflection and m is a soil stiffness parameter. Although Rowe describes how m may be measured or deduced, this definition of soil stiffness is unusual. If y/d is taken as an indication of the magnitude of the linear strain, then mx would be the Young's modulus of the soil, and m a measure of the rate of increase of Young's modulus with depth E^*.

Rowe carried out analyses of walls of various retained height ratio $\alpha = h/H$ and depth βH to the anchor point, with and without surcharges at the retained soil surface, and with different degrees of anchor yield. He concluded that, within the ranges of these variables likely to be encountered in practice, the results of the analyses could be presented as a single moment reduction curve for design office use. This curve (Figure 12 in Rowe's 1955 paper, reproduced here in consistent units as Figure 10.16) shows the bending moment as percentage of the free earth support value (assuming fully active conditions behind the wall) as a function of the logarithm of $m\rho$. (The parameter $m\rho$ is sometimes known as the **flexibility number**, and is given the symbol R.) In consistent units, $m\rho$ is dimensionless: however Rowe's values have been multiplied by 144 to achieve this, because Rowe has m in lb/ft^3 and ρ in ft^5/lb in^2. (On the logarithmic plot, multiplication by 144 is achieved by adding $\log_{10}[144] = 2.16$.)

Rowe's design chart (Figure 10.16) represents his analytical solution to within $\pm10\%$ for anchored walls with retained height ratios h/H (where $H = h + d$) in the range 0.65–0.75; anchor depths βH in the range $0 < \beta < 0.2$; surcharges acting at the retained soil surface of up to $0.2\gamma H$ in magnitude; and a movement at the anchor point of up to $0.008H$. The curves given are for the case of a rigid prop or anchor, which would be expected to give slightly larger prop or anchor loads and bending moments than a yielding support.

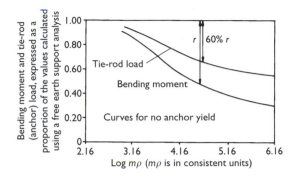

Figure 10.16 Moment reduction as a function of wall flexibility. (Redrawn with permission from Rowe, 1955; $m\rho$ in consistent units, i.e. dimensionless – see text.)

Rowe (1955) notes in his paper that the analysis gives results which are generally in close agreement with those of his earlier (1952) experiments. The theoretical critical flexibility number $m\rho$ (in consistent units), below which the reduction in bending moment due to flexure is small, is generally approaching 1000. For stiffer walls in loose sand, however, the theoretical results and the experimental data show some divergence, and the critical flexibility number determined experimentally in this case was approximately 2500.

Rowe's design chart may not be suitable for walls where the groundwater level in the retained soil is high, because retained height ratios h/H of less than 0.65 would probably be required. If the chart is used in such circumstances, it must be remembered that only the component of the bending moment due to effective stresses should be reduced. The chart may be unsuitable for walls in clay soils where the *in situ* earth pressure coefficient is high, because of the assumption in Rowe's analysis that the retained soil is in the active state (see section 10.4 for a fuller discussion of this point). Also, Figure 10.16 relates to the case of a rigid prop or anchor: the pre-stressing of props or anchors would be expected to lead to higher wall bending moments.

Potts and Fourie (1985) investigated the effect of the flexural rigidity EI on wall movements and bending moments for values of pre-excavation earth pressure coefficient $K_i = 2.0$ and 0.5. (The pre-excavation earth pressure coefficient K_i is the ratio σ'_h/σ'_v after wall installation, immediately prior to excavation of the soil in front of the wall. It is different from the *in situ* earth pressure coefficient K_0, due to the lateral stress relief that occurs during wall installation.) Threshold flexibility numbers were not identified: differences were apparent between the behaviour of a rigid wall ($\rho = 7 \times 10^{-5}\,\mathrm{m^3/kN}$; $E^*\rho = 0.42$) and the next stiffest wall investigated ($\rho = 7 \times 10^{-2}\,\mathrm{m^3/kN}$; $E^*\rho = 420$, where $E^* = 6000\,\mathrm{kN/m^3}$ is the rate of increase of soil Young's modulus E with depth). However, the movement at the toe (which is indicative of the component of deflection due to rigid-body rotation) was similar for all walls with a given K_i.

In Potts and Fourie's finite element analyses, the wall with $E^*\rho = 420$ is not quite stiff according to Rowe's criterion that the deflection at dredge level should not exceed the deflection at the toe. This implies that the critical flexibility number $E^*\rho$ is less than 420.

In reality, however, soil is not a linear elastic material, and the operational values of E (and hence E^*) will decrease with increasing shear strain. The initial Young's modulus may not be the appropriate value to use, and the variation in Young's modulus with shear strain may have been accounted for differently by Rowe than by Potts and Fourie (1985).

In addition, the decrease in soil stiffness with increasing strain will tend to reduce the extent to which high local bending strains affect the linearity of the stress distribution. This may explain why Bolton and Powrie (1988) were apparently able to neglect the effects of flexure in the analysis of a wall with $E^*\rho = 1350$. Also, in Potts and Fourie's (1985) analysis the wall was propped rigidly at the crest, so that redistribution of lateral stresses in the retained soil due to arching may well have occurred.

Rowe's experiments were carried out using dry sand, and in Potts and Fourie's analyses the pore water pressures were set to zero. In reality in a saturated soil, the pore water pressures will contribute significantly to the total stresses acting on both sides of the wall, and in the case of steady state seepage round an impermeable wall will increase approximately linearly with depth (Figure 7.40). Thus, the absence of pore water pressures in Rowe's tests and in Potts and Fourie's analyses will tend to emphasize non-linearities in the lateral effective stress distributions, leading perhaps to the overestimation of the threshold wall stiffness under more realistic conditions.

10.6.3 Characterization of soil/wall stiffness

Figure 10.15 shows that the deflection of an embedded wall at any depth is due partly to rotation as a rigid body about the prop, and partly to the effects of bending. Using the geostructural mechanism approach described in section 10.5, it may be shown that on excavation in front of an *in situ* wall in a soil of unit weight γ, rigid body rotation is governed by γ/G^*, while bending deformation depends on $\gamma H^4/EI$ for undrained conditions. (H^4 is the overall wall length and EI is the bending stiffness per metre run, in kNm^2/m. G^* is the rate of increase of shear modulus G with depth. If Poisson's ratio ν' is in the range 0.2–0.4, $0.36\,E^* \leq G^* \leq 0.42E^*$.)

A flexibility number quantifying the relative importance of wall deflections due to bending and rigid body rotation may then be identified as $(\gamma/G^*) \div (\gamma H^4/EI) = G^*H^4/EI$. If the wall is to be categorized as stiff, the deformation at dredge level should not be greater than that at the toe. This definition may be used to identify critical flexibility numbers for walls of different embedment ratio and different pre-excavation earth pressure coefficient, and different support conditions (e.g. propped at the crest or unpropped). Critical flexibility numbers are generally higher for unpropped walls or walls propped at formation level than for walls propped at the crest, because for a wall of given overall length H and bending stiffness per metre run EI bending effects are most significant when the wall is propped at the crest.

For multi-propped walls, in which the opportunity for rigid body rotation may be limited, a **system stiffness** $EI/\gamma_w h_{av}^4$ has been defined by Clough *et al.* (1989). (γ_w is the unit weight of water and h_{av} is the average distance between the supports.) In undrained conditions, the unit weight of the soil could have been used instead of the unit weight of water, because it is the soil that loads the wall and causes it to deform in bending. This would give a system stiffness of $EI/\gamma h_{av}^4$.

Addenbrooke *et al.* (2000) suggest that the displacement of a multi-propped wall in given ground conditions may be investigated with reference to the parameter $\Delta = EI/h_{av}^5$. Although this is not dimensionless, Addenbrooke *et al.* found that wall displacements were similar (in given soil conditions) for different combinations of EI and h_{av} giving the same value of Δ.

Example 10.2 An anchored embedded retaining wall

Figure 10.17 shows a cross-section through an anchored sheet pile retaining wall.

(a) Assuming that fully active conditions have been reached in the soil behind the wall, calculate the factor of safety on soil strength F_{sp} on the passive side of the wall.

(b) Calculate the maximum bending moment and the anchor load associated with this equilibrium condition.

(c) Explain why the in-service maximum bending moment could be less than this value, and estimate the in-service maximum bending moment and anchor load using Rowe's moment reduction chart (Figure 10.16).

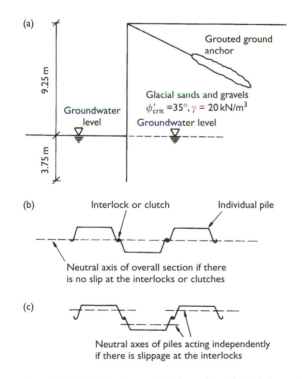

Figure 10.17 (a) Cross-section through anchored sheet pile retaining wall; cross-sections through Larssen 4A steel sheet piles, showing neutral axis if (b) there is no slip at the clutches, and (c) if the piles act independently.

The retaining wall is made from interlocking Larssen 4A steel sheet piles having a Young's modulus 210×10^6 kPa, and a relevant second moment of cross-sectional area of 4.4916×10^{-4} m^4 per metre run if there is no slip at the clutches between adjacent piles (Figure 10.17(b)), or 1.4125×10^{-4} m^4 if friction cannot be relied upon to prevent slip from taking place, so that the piles act independently (Figure 10.17(c)). The soil stiffness parameter m is estimated to be 6000 kPa/m.

(Figure 10.17(a) is based on the sheet-pile wall used in the construction of the road tunnel on the A1(M) at Hatfield in Hertfordshire, approximately 40 km north of London. Details are given by Symons et al. (1987).)

Solution

(a) Assuming fully active conditions, the lateral earth pressure coefficient in the soil behind the wall is $K_a = 0.2207$ (from Table 7.6, for $\phi'_{crit} = 35°$ and wall friction angle δ given by $\tan \delta = 0.75 \times \tan \phi'_{crit}$). The (as yet unknown) earth pressure coefficient in the soil in front of the wall below formation level is K_p. The effective stresses and pore water pressures at key depths (i.e. at the soil surface, at the toe of the wall and at the level of the water table) behind and in front of the wall are given in Table 10.5.

Table 10.5 Stress at key depths behind and in front of the wall

Depth behind wall (m)	γz (kPa)	Pore water pressure u (kPa)	$(\gamma z - u)$ (kPa)	σ'_h (kPa) $= K_a \times (\gamma z - u)$
0	0	0	0	0
9.25	185	0	185	40.8
13.0	260	37.5	222.5	49.1

Depth in front of wall (m)	γz (kPa)	Pore water pressure u (kPa)	$(\gamma z - u)$ (kPa)	σ'_h (kPa) $= K_a \times (\gamma z - u)$
0	0	0	0	0
3.75	75	37.5	37.5	$37.5 K_p$

The unit weight of water has been taken as 10 kN/m^3. The lateral stress distribution is shown in Figure 10.18(a).

Taking moments about the anchor (ignoring the pore water pressures, which balance each other exactly because they are hydrostatic below the same groundwater level on each side of the wall),

$$[0.5 \times 40.8 \text{ kPa} \times 9.25 \text{ m}] \times [9.25 \text{ m} \times 2/3]$$

$$+ [40.8 \text{ kPa} \times 3.75 \text{ m}] \times [9.25 \text{ m} + (3.75 \text{ m}/2)]$$

$$+ [0.5 \times 8.3 \text{ kPa} \times 3.75 \text{ m}] \times [9.25 \text{ m} + (2 \times 3.75 \text{ m}/3)]$$

$$= [0.5 \times 37.5 \ K_p (\text{kPa}) \times 3.75 \text{ m}] \times [9.25 \text{ m} + (2 \times 3.75 \text{ m}/3)]$$

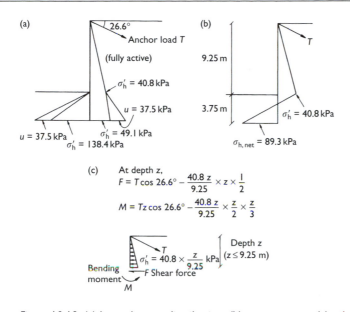

Figure 10.18 (a) Lateral stress distribution; (b) net pressures; (c) calculation of shear force and bending moment.

or

$$3048.6 = 826.2\,K_p$$

$$K_p = 3.69$$

From Table 7.7, this corresponds to a mobilized soil strength of just under $25°$, assuming wall friction such that $\tan \delta = 0.75 \times \tan \phi'_{mob}$. Thus the factor of safety on soil strength on the passive side of the wall is

$$F_{sp} = \tan 35° / \tan 25° = 1.5$$

(b) The net pressure diagram is shown in Figure 10.18(b).

The horizontal component of the anchor load $T \cos 26.6°$ is calculated from the condition of horizontal force equilibrium,

$$[T \cos 26.6°] + [0.5 \times (3.69 \times 37.5)\,\text{kPa} \times 3.75\,\text{m}]$$

$$= [0.5 \times 40.8\,\text{kPa} \times 9.25\,\text{m}] + [40.8\,\text{kPa} \times 3.75\,\text{m}]$$

$$+ [0.5 \times 8.3\,\text{kPa} \times 3.75\,\text{m}]$$

giving

$$[T \cos 26.6°] = 97.8\,\text{kN/m},$$

or

$$T = 109\,\text{kN/m}$$

The maximum bending moment occurs where the shear force is zero. For depths z less than 9.25 m, the calculation of shear force and bending moment is illustrated in Figure 10.18(c). At a depth z (≤ 9.25 m), the lateral effective stress is

$$\sigma'_h = (z/9.25) \times 40.8 \text{ kPa}$$

and the pore water pressure is zero.

The shear force is zero when

$$[T \cos 26.6°] = [(z/9.25) \times 40.8 \times (z/2)]$$

giving $z^2 = 44.35$ m^2 or $z = 6.66$ m.

The bending moment M is given by

$$M = [T \cos 26.6°] \text{ kN/m} \times [z] \text{ m}$$

$$- [(z/9.25) \times 40.8 \text{ kPa} \times (z/2) \text{ m}] \times [z/3] \text{ m}$$

$$= [(T \cos 26.6°) \times z - 0.735z^3] \text{ kNm/m}$$

At $z = 6.66$ m,

$$M_{max} = 434 \text{ kNm/m}$$

(c) The bending moment and anchor load in (b) have been calculated on the basis of a linear stress distribution. If the wall is flexible, the stress distribution in front of it will be non-linear, with comparatively higher lateral stresses just below the excavated soil surface (as in Figure 10.15(b)). The effect of this will be to reduce the wall bending moments and the anchor load.

The moment reduction factor depends on the relative soil–wall flexibility $m\rho = mH^4/EI$. For this example, $m = 6000$ kPa/m, $H = 13$ m, $E = 210 \times 10^6$ kPa, and I is between 4.4916×10^{-4} m^4/m and 1.4125×10^{-4} m^4/m. Thus

$$(mH^4/EI)_{min} = (6000 \times 13^4) \div (210 \times 10^6 \times 4.4916 \times 10^{-4})$$

$$(\text{kPa/m} \times \text{m}^4) \div (\text{kPa} \times \text{m}^4/\text{m})$$

and

$$(mH^4/EI)_{max} = (6000 \times 13^4) \div (210 \times 10^6 \times 1.4125 \times 10^{-4})$$

$$(\text{kPa/m} \times \text{m}^4) \div (\text{kPa} \times \text{m}^4/\text{m})$$

or

$$1816 < mH^4/EI < 5777$$

$$\Rightarrow 3.26 < \log_{10}(mH^4/EI) < 3.76$$

From Figure 10.16, the moment reduction factor is therefore between 0.7 (for $mH^4/EI = 5777$) and 0.85 (for $mH^4/EI = 1816$). The reduction in anchor load is 0.6 times the reduction in moment, giving an anchor load reduction factor of between 0.82 and 0.91. Thus the maximum in-service bending moment could be reduced to

$$304 \, \text{kNm/m} \leq M_{\text{max,des}} \leq 369 \, \text{kNm/m}$$

and the anchor load to

$$89 \, \text{kN/m} \leq T_{\text{des}} \leq 99 \, \text{kN/m}$$

assuming that the anchors are not pre-stressed.

In reality, the anchors were pre-stressed, and the maximum measured bending moment was still only 110 kNm/m.

10.7 Compaction stresses behind backfilled walls

The soil behind a backfilled retaining wall is often placed and compacted in layers. This can lead to higher-than-active lateral stresses in the retained soil – a possibility which should be considered in design. Assuming that compaction is properly carried out with appropriate compaction plant and in layers which are sufficiently thin, the magnitudes of the 'locked-in' lateral stresses will depend on several factors. These include the flexibility of the structural system, and whether the backfill is free-draining or behaves as a clay with respect to the generation of excess pore water pressures during placement and compaction. In this section, the simpler theories that have been proposed for the estimation of compaction stresses for design purposes are described.

10.7.1 Free-draining soils

The following analysis is for free-draining soils, in which excess pore water pressures will not be generated during compaction, and follows Broms (1971) and Ingold (1979).

Consider an element of soil of unit weight γ at a depth z below the surface. The pore water pressure is zero, so that the initial vertical effective stress is γz. It is assumed that sufficient lateral wall movement has occured to mobilize the full strength of the soil, so that the soil is in the active condition with $\sigma'_{\text{h}} = K_a \sigma'_{\text{v}}$.

As a result of the application during compaction of a line load of intensity q (kN/m) at the soil surface vertically above it, the soil element experiences an increase in vertical effective stress of $\Delta \sigma'_{\text{v}}$. It is assumed that during loading, the lateral movement of the wall is sufficient to maintain the soil in the active condition; and that on unloading (when the line load q is removed), the horizontal stress remains constant until the passive failure condition is reached (Figure 10.19(a)). For a given initial stress state γz, there will be a value of $\Delta \sigma'_{\text{v}}$ which is just large enough to result in passive failure on unloading (Figure 10.19(b)). Since the required value of $\Delta \sigma'_{\text{v}}$ increases with γz, it follows that a cycle of vertical effective stress of given magnitude will lead to the development of passive conditions only in the soil above a certain critical depth, z_{c}.

It may be seen from Figure 10.19(b) that

$$K_p \gamma z_c = K_a (\gamma z_c + \Delta \sigma'_v)$$

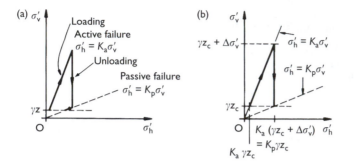

Figure 10.19 (a) Assumed stress path for a soil element at a depth z on loading and unloading, and (b) relationship between $\Delta\sigma_v'$ and z_c.

so that the relationship between the increase in vertical effective stress and the critical depth is

$$\Delta\sigma_v' = \gamma z_c (K_p^2 - 1) \qquad (10.3)$$

which reduces to

$$\Delta\sigma_v' = \gamma z_c K_p^2 \qquad (10.4)$$

if γz_c is small in relation to $\Delta\sigma_v'$.

According to an elastic analysis (Holl, 1941), the increase in vertical effective stress (assuming zero pore water pressure, $u = 0$) at a depth z below an infinitely long line load of intensity q kN/m is

$$\Delta\sigma_v' = 2q/\pi z \qquad (10.5)$$

Assuming this to be applicable here, equations (10.4) and (10.5) may be combined to give the critical depth as a function of the surface line load q:

$$z_c = K_a \sqrt{2q/\pi\gamma} \qquad (10.6)$$

Noting from Figure 10.19(b) that at the critical depth the residual increase in horizontal effective stress $\Delta\sigma_h'$ due to compaction is $K_a \Delta\sigma_v'$, and substituting for $\Delta\sigma_v'$ from equation (10.5), with $z = z_c$ as given by equation (10.6), we have

$$\Delta\sigma_h' (\text{at } z_c) = \sqrt{2q\gamma/\pi} \qquad (10.7)$$

which is independent of the soil strength parameter ϕ'.

The increase in horizontal stress which results from the compaction of a granular backfill by means of a rolling line load of intensity q at the surface is as shown in Figure 10.20(a). The maximum increase in horizontal effective stress occurs at $z = z_c$. Above this depth, passive conditions are reached and the lateral stress increases with depth z, with $\sigma_h' = K_p \times \sigma_v'$. Below z_c, the increase in vertical effective stress is insufficient for the residual lateral stress to reach the passive limit, and $\Delta\sigma_h'$ reduces with depth.

In reality, the fill is likely to be placed and compacted in layers thin enough to generate the maximum increase in horizontal stress all the way down the wall, as

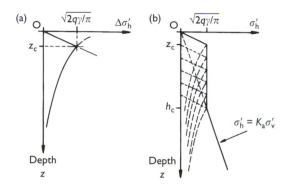

Figure 10.20 (a) Idealized distribution of increase in horizontal effective stresses due to compaction of a single layer; (b) distribution of horizontal stresses in backfill compacted in a number of layers.

shown in Figure 10.20(b). As more and more fill is placed, a point will eventually be reached where the active stresses below a certain depth h_c are greater than those that were induced by compaction when the layer was placed. This depth is given by the condition

$$\sigma_h' = \sqrt{2q\gamma/\pi} = K_a \gamma h_c$$

or

$$h_c = (1/K_a)\sqrt{2q/\pi\gamma} \tag{10.8}$$

In the above analysis, it has been assumed that the lateral movement of the wall is sufficient to mobilize the full strength of the soil, so that active and passive conditions are developed at various stages of placement and compaction of the fill. If the wall is effectively rigid, K_a would be replaced throughout by the earth pressure coefficient following one-dimensional compression K_{nc} ($= 1 - \sin \phi'$), and K_p by $1/K_{nc}$.

Ingold (1979) presents three case histories where the measured stresses conform reasonably well to those calculated using the K_a/K_p procedure for yielding walls, but suggests that the rotation or translation required to reduce stresses to the active limit is of the order of only $h/500$, where h is the height of the wall.

Symons and Murray (1988) present data from inverted T-cantilever walls forming bridge abutments at Guildford and Loudwater. At Guildford, deformations were apparently sufficient to destroy the increased lateral stresses due to compaction except near the root. At Loudwater, the fill was lightweight pulverized fuel ash (pfa) and the structural system was much more rigid: although lateral stresses considerably in excess of the design values were recorded, this probably reflects primarily on the design parameters used. The lateral stresses at both Guildford and Loudwater varied significantly with changes in ambient temperature.

Example 10.3 Calculating the lateral stresses behind a retaining wall, due to compaction of a granular backfill

Figure 10.21 shows a cross-section through a bridge abutment, formed of an inverted T-cantilever wall with a piled foundation. The fill material is to be compacted by a vibrating roller which applies a maximum line load of 25 kN/m.

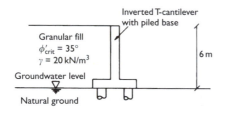

Figure 10.21 Cross-section through inverted T-cantilever retaining wall.

(a) Calculate the maximum thickness of the layers in which the fill may be compacted if fully passive conditions are to be achieved.

(b) Assuming that this procedure is followed on site, estimate the final stress distribution which acts on the back of the wall at the end of placement and compaction of the fill.

(c) Comment briefly on the applicability of this stress distribution, particularly near the top and the bottom of the retaining wall.

   This example is based on the bridge abutment at Guildford described by Symons and Murray (1988).

Solution

(a) Assuming that the wall moves sufficiently during compaction of the fill for the idealized stress path shown in Figure 10.19 to apply, the maximum thickness of the layers in which the fill may be compacted if fully passive conditions are to be achieved is equal to the critical depth given by equation (10.6):

$$z_c = K_a\sqrt{2q/\pi\gamma} \tag{10.6}$$

In the present case, $K_a = (1 - \sin 35°)/(1 + \sin 35°) = 0.271$; $q = 25\,\text{kN/m}$ and $\gamma = 20\,\text{kN/m}^3$. Substituting these values into equation (10.6), the maximum layer thickness is

$$z_c = 0.271 \times \sqrt{[(2 \times 25)/(\pi \times 20)]} = 0.24\,\text{m}$$

(b) The horizontal effective stress at the critical depth z_c (ignoring the effects of self-weight during placement of the layer) is given by equation (10.7):

$$\Delta\sigma'_h \text{ (at } z_c) = \sqrt{2q\gamma/\pi} \tag{10.7}$$

In this case,

$$\Delta\sigma_h' \text{ (at } z_c) = \sqrt{(2 \times 25 \times 20/\pi)} = 17.8 \text{ kPa}$$

This is the lateral effective stress until the depth h_c (given by equation (10.8)) is reached, at which the active lateral stress due to the weight of overburden will begin to exceed the lateral stress due to compaction:

$$h_c = (1/K_a)\sqrt{2q/\pi\gamma} \tag{10.8}$$

or $h_c = (1/K_a)^2 \times z_c$. In the present case,

$$h_c = (1/0.271)^2 \times 0.24 \text{ m} = 3.3 \text{ m}$$

Below h_c, the lateral stress increases linearly with depth until the base of the wall is reached, where

$$\sigma_h' = K_a \gamma z = 0.271 \times 20 \text{ kN/m}^3 \times 6 \text{ m} = 32.5 \text{ kPa}$$

The resulting lateral stress distribution is shown in Figure 10.22.

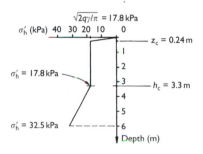

Figure 10.22 Lateral stress distribution after compaction of backfill in layers for the wall shown in Figure 10.21.

(c) Near the bottom of the retaining wall, the freedom of the wall to rotate or move laterally is quite restricted. Although at 6 m depth the effects of compaction are less than the increase in lateral stress due to the placement of overburden, it may be that the lateral stresses near the base of the wall approach $K_0 \times \sigma_v'$ (where $K_0 = 1 - \sin\phi'$), rather than $K_a \times \sigma_v'$ as assumed in the analysis, as there may be insufficient movement to generate fully active lateral stresses.

At mid-height, the lateral stresses established during compaction might be eliminated by further movement of the wall during the placement of the upper layers of fill. At the top of the wall, conditions might be expected to be reasonably close to those assumed in the analysis. However, the lateral stresses would probably be reduced to the active limit by a comparatively small movement of the wall in service (6 m ÷ 500 = 12 mm). Also, the theory is highly idealized, and the in-service lateral stresses will be very strongly influenced by other effects, such as the loads transmitted by a bridge deck as it tends to expand and contract with varying temperature.

10.7.2   Clays

Clayton *et al.* (1991) presented an analysis of the compaction process of clay soils, in which changes in volume are slow so that the 'undrained shear strength' failure criterion $\tau_{max} = \tau_u$ is appropriate.

In general terms, the analysis is similar in principle to that for free-draining soils, given in section 10.7.1. It is assumed that the application and removal of a surface load results in residual passive conditions down to a critical depth z_c below the surface, and that at depths $z > z_c$ the increase in horizontal stress may be derived from an elastic analysis (Figure 10.23(a)). The higher the value of τ_u, the smaller the critical depth z_c. Clayton *et al.* (1991) show that z_c is likely to vary between 100 mm for $\tau_u = 50$ kPa and 30 mm for $\tau_u = 400$ kPa. At these small depths, σ_v may be neglected and the passive total stress may be taken as $2\tau_u$.

If it is assumed that the clay fill is placed and compacted in layers of thickness z_c or less, the maximum horizontal stress will be generated all down the wall to a depth h_c, given by

$$2\tau_u = K_T \gamma h_c$$

or

$$h_c = 2\tau_u / K_T \gamma \tag{10.9}$$

where K_T is the ratio of total stresses ($= \sigma_h/\sigma_v$) in the absence of compaction. This leads to the overall distribution of horizontal total stress shown in Figure 10.23(b).

The tendency of excavated clay to form lumps or **clods** may make it difficult to place for compaction in thin layers, since the clods could well be more than 100 mm in diameter. The bulk undrained shear strength of the assemblage of clods will be smaller than the undrained shear strength of the intact clay: it is the former which should be used for the calculation of lateral stresses due to compaction.

Perhaps the most significant difference between the results of the two analyses is that the maximum lateral stress generated by compaction in granular soils ($= \sqrt{[2q\gamma/\pi]}$)

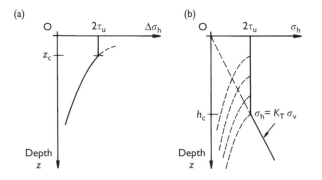

Figure 10.23 (a) Increase in horizontal total stress with depth following placement and compaction of a single layer of clay; (b) distribution of horizontal total stress in backfill compacted in layers.

is independent of the strength of the soil, whereas in the case of clays it is proportional to the bulk undrained shear strength.

Carder *et al.* (1980) describe an experiment carried out at the UK Transport Research Laboratory (TRL) on a retaining wall backfilled with silty clay, in which a deflection at the top of the wall of only $h/1000$ was sufficient to reduce the lateral stresses in the fill to their active limit.

A perhaps more important factor that should be allowed for in the design of walls with clay backfill is the possible development of high stresses due to swelling. If the clay fill is overconsolidated and/or has been excavated from some depth, the pore water pressures may be strongly negative relative to their long-term equilibrium values. With time, the negative excess pore water pressures may dissipate and the clay may swell. Although this may imply a reduction in the horizontal effective stress, the increase in pore water pressure is far more significant and the net result is an increase in the total horizontal stress acting on the wall.

Mawditt (1989) presents a case history of swelling pressures developed in clay fill retained between 9 m-high reinforced concrete inverted T-cantilever walls with piled bases on the M3 motorway at Sunbury Cross, near London (Figure 7.8(c)). The clay fill had been obtained from 14 different sites (mostly deep excavations) across London, including the 30 m deep tunnels for the Victoria Line on the Underground. In this case, the dissipation of negative excess pore water pressures led to lateral stresses significantly greater than those assumed in the initial design.

10.8   Strip loads

Non-uniform surcharges are difficult to take into account in a stress field analysis, because their influence on the horizontal stress in the vicinity of the wall is not easy to quantify. A strip load running parallel to the wall may be modelled using the procedure suggested by Pappin *et al.* (1986), which is illustrated in Figure 10.24.

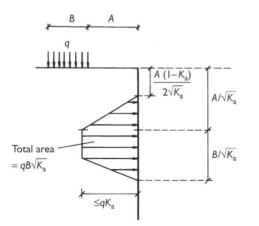

Figure 10.24 Additional lateral effective stress acting on the back of the wall due to a strip load running parallel to the wall. (Redrawn with permission from Pappin *et al.*, 1986.)

Other methods of representing strip, patch and line loads in limit equilibrium analyses are given by Gaba *et al.* (2003, their section 4.1.7).

10.9 Multi-propped embedded walls

In the permanent condition, embedded retaining walls are often propped at more than one level. Examples of this include underground car parks and building basements, in which each floor slab will probably act as a prop; and cut-and-cover tunnels, which may be propped by reinforced concrete slabs at both roof and carriageway level.

A multi-propped wall is a statically indeterminate structure (i.e. the prop forces and wall bending moments cannot be calculated using the condition of statical equilibrium alone). For the purpose of estimating bending moments and prop loads, the wall may be analysed as a series of simply supported beams, spanning between props at adjacent levels (Figure 10.25). The lateral effective stress in the retained soil is often assumed to be at the active limit, provided that this is consistent with the ground conditions. The bottom part of the wall is analysed as a propped embedded wall, at the stage of construction just before the formation level slab is installed. The uppermost prop may be at or below the level of the top of the wall.

A multi-propped wall is likely to act in different ways (e.g. as an unpropped cantilever, an embedded wall propped at the crest, or as an embedded wall with more than one prop) at different stages during its construction. In investigating the bending moments and prop loads, it is necessary to consider each phase of wall construction separately, including stages during which the wall is supported by temporary props, in order to determine the largest load in each part of the structure.

Gaba *et al.* (2003) recommend that, for design purposes, multi-propped walls are analysed using numerical methods so as to take account of soil–structure interaction and construction sequence effects.

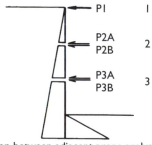

Wall span between adjacent props analysed as simply-supported beam; lowest section analysed as propped cantilever.
Total load in prop 2 = P2A + P2B.
Total load in prop 3 = P3A + P3B etc.

Figure 10.25 Simplified analysis of multi-propped walls.

10.10   Reinforced soil walls

Soil reinforcement is a technique in which a reinforcing material (usually either metal or plastic strips) is placed within the soil, in order to improve its strength and stiffness. A **reinforced soil retaining wall** is made up of a number of vertical or near-vertical facing panels, connected to strips of reinforcement embedded within the soil backfill (Figure 7.10). The vertical stress acting on the reinforcement (due to the weight of soil above it) allows the reinforcement to develop a tensile load, which resists the lateral pressure exerted by the soil on the facing panel. The reinforced soil technique can offer an economical method of constructing a backfilled retaining wall, provided that slightly larger than usual wall movements can be tolerated.

It was mentioned in section 7.4 that the designer of any type of retaining wall must investigate a number of modes of collapse, including monolithic rotation or sliding, the triggering of a landslide, and the failure of the materials used to construct the wall itself. In the case of a reinforced soil wall, the last of these includes the possibility of the failure of the reinforcement, either in **tension** or by **pull-out** (i.e. by slippage between the reinforcement and the surrounding soil).

The internal stability of a reinforced soil wall may be investigated by means of the stress analysis shown in Figure 10.26. This involves the limiting equilibrium of a block of soil ABCD, which extends a distance L (where L is the length of the reinforcement strips) behind the face of the retaining wall.

The block of soil ABCD has a general depth z. It is assumed that the vertical boundaries BC and AD are frictionless, and that the horizontal effective stress σ_h' on BC increases linearly with depth z, with $\sigma_h' = K_a \sigma_v'$. In the absence of pore

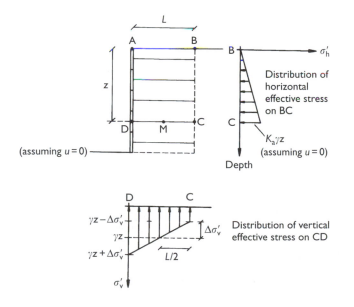

Figure 10.26 Investigation of possible failure of a reinforced soil retaining wall by pull-out or breakage of the reinforcement strips. (Redrawn with permission from Bolton and Pang, 1982.)

water pressures, $\sigma'_v = \gamma z$ along BC, where γ is the unit weight of the retained soil, $K_a = (1 - \sin\phi')/(1 + \sin\phi')$ is the active earth pressure coefficient and ϕ' is the angle of shearing resistance (effective angle of friction) of the soil.

As the lateral stress acting on the front face of the retaining wall AD is zero, the vertical effective stress distribution along the base of the block of soil (CD) must provide a moment to balance the effect of the triangular lateral stress distribution on BC. It is assumed that the vertical effective stress on CD decreases linearly from $(\gamma z + \Delta\sigma'_v)$ at the back of the facing panels to $(\gamma z - \Delta\sigma'_v)$ at the end of the reinforcement, with a mean value of γz. The value of $\Delta\sigma'_v$ may be obtained from the condition of moment equilibrium about M, the mid-point of CD. From Figure 10.26,

$$[0.5 \times K_a \times \gamma z] \times [z] \times [z/3] = 2 \times [0.5 \times \Delta\sigma'_v \times (L/2)]$$
$$\times [(L/2) \times (2/3)]$$

[Average lateral stress on BC] $\times$ [Depth BC] $\times$ [Lever arm]

$$= [\text{moment due to } \Delta\sigma'_v]$$

(Remember that the lateral stress on the front face of the retaining wall, AD, is zero.) Hence

$$K_a \gamma z^3/6 = \Delta\sigma'_v \times (L^2/6)$$

or

$$\Delta\sigma'_v = K_a \gamma z (z/L)^2 \tag{10.10}$$

and the vertical effective stress adjacent to the wall at a depth z is given by

$$\sigma'_v = \gamma z + \Delta\sigma'_v = \gamma z (1 + K_a z^2/L^2) \tag{10.11}$$

so that the horizontal effective stress acting on the facing panel at depth z, at conditions of limiting equilibrium, is

$$\sigma'_h = K_a \sigma'_v = K_a \gamma z (1 + K_a z^2/L^2)$$

The horizontal load on a facing panel of width Δy and depth Δz at a depth z is therefore

$$(K_a \gamma z) \times (1 + K_a z^2/L^2) \times (\Delta y \times \Delta z) \tag{10.12}$$

At the joint between the facing panel and the reinforcement strip, all of this load must be taken by tension in the reinforcement. If P is the load that will cause the reinforcement strip to break in tension, the **factor of safety** or **load factor against tensile failure** F_T for a reinforcement strip at depth z may be defined as

$$F_T = \text{Ultimate tensile load} \div \text{Actual tensile load}$$

or

$$F_T = P \div \{(K_a \gamma z \Delta y \Delta z)(1 + K_a z^2/L^2)\} \tag{10.13}$$

The vertical effective stress, averaged over the entire length of a reinforcement strip at depth z, is γz. If b is the width of a reinforcement strip and μ is the coefficient of friction between a reinforcement strip and the soil, the horizontal load at the joint with the panel that will be required to pull the reinforcement strip out of the soil is

$$\mu \times [2bL\gamma z]$$

The **factor of safety** or **load factor against pull-out** (or **frictional failure**) is defined as

$$F_F = \text{Horizontal load that would cause pull-out} \div \text{Actual horizontal load}$$

which is given by

$$F_F = \{2\mu bL\gamma z\} \div \{(K_a\gamma z\Delta y\Delta z)(1 + K_a z^2/L^2)\} \tag{10.14}$$

On the basis of a series of centrifuge model tests on reinforced soil retaining walls in dry sand, Bolton and Pang (1982) found that:

- Pull-out failure was generally well-predicted using equation (10.14). However, the use of the peak (rather than the critical state) soil strength in equation (10.14) resulted in a tendency to err on the unsafe side. The use of the critical state soil strength would eliminate the tendency of the calculation sometimes to err on the unsafe side, and is probably therefore to be recommended in design.
- Tensile failure was only well-predicted by equation (10.13) for walls with narrow reinforced zones (i.e. with $L/H \leq 0.5$, where H is the overall height of the wall), which were already close to pull-out failure. Otherwise, equation (10.13) was somewhat conservative, even if peak strengths were used.

To investigate the apparent overconservatism of equation (10.13), Bolton and Pang (1982) carried out further tests, in which the model wall was built on a rigid foundation layer rather than on sand. In these tests, equation (10.12) was found to overpredict the tension in the layer of reinforcement nearest the base of the wall by a factor of about 1.4. The largest tensions were measured in the reinforcement strips at a depth of $0.75H$ below the retained soil surface. The reason for this was probably a reduction in vertical effective stress adjacent to the facing panels, due to shear between the facing panels and the soil. The frictional resistance of the foundation may also have contributed to the reduction in load in the reinforcement strips nearest the base of the wall.

The reduction in vertical effective stress at the base of the wall is one of the reasons why equation (10.13) is overconservative in its prediction of tensile failure. However, the vertical stress reduction would probably not occur if the facing panel were smooth or flexible, or if the foundation layer were comparatively compressible, so that significant friction between the facing panels and the retained soil was not developed.

The second reason for the over-conservatism of equation (10.13) is that, as the most heavily loaded reinforcement strips reach their ultimate tensile strength, additional loads may be taken by the less heavily loaded reinforcement strips in a process known as **plastic stress redistribution**. This will delay the overall collapse of the wall. However, if the reinforcement is effectively brittle (e.g. due to local corrosion or a weak joint with the facing panel), or if the load factor against pull-out F_F is close to unity, plastic stress redistribution will not be able to take place.

A further point is that walls with a higher load factor against pull-out F_F are likely to be subjected to higher lateral earth pressures under working conditions, perhaps with $\sigma'_h = K_{nc}\sigma'_v$ acting along BC (where $K_{nc} = 1 - \sin\phi'$), rather than $\sigma'_h = K_a\sigma'_v$.

In summary, the collapse of conventional reinforced soil retaining walls in granular materials can be conservatively predicted using the stress analysis shown in Figure 10.26. The degree of conservatism varies from case to case, and cannot be relied on as it is due to factors such as the exact details of wall construction, which may be difficult to quantify or even control.

Furthermore, a degree of conservatism in design may be desirable with this type of construction. In the centrifuge model tests carried out by Bolton and Pang (1982), there was no warning of incipient collapse from an increase in the rate of wall movement with decreasing factor of safety. Empirical modifications to the stress analysis of Figure 10.26, involving for example the assumption of a 'mechanism' of failure, have been proposed by several authors. In many cases, these may be inappropriate, and should therefore be used (if at all) with considerable caution.

Lee *et al.* (1994) highlight the importance of the quality of construction on the performance of a reinforced soil wall in practice, with reference to a number of wall failures in eastern Tennessee, USA.

10.11   Tunnels

Tunnels are used to carry railways, roads, sewers and other services below ground – usually beneath built-up areas or under rivers. They are constructed in a variety of ways, depending on their size and the prevailing ground conditions. Small diameter tunnels for pipes and service ducts may be constructed using a small tunnel boring machine (known as a **mole**), operated by remote control. Large tunnels must be excavated more conventionally, and although tunnel boring machines are now generally used, some tunnels are still dug out by hand. Tunnels in rock may be substantially self-supporting, but in most soil conditions it will be necessary to construct a permanent lining of steel, masonry or reinforced concrete to support the surrounding ground as soon as possible after excavation.

The construction of large tunnels was revolutionized by Marc Brunel's invention in or before 1825 of the **tunnelling shield**, which offered almost complete support to the advancing face of the tunnel during excavation. The shield was used to construct the world's first tunnel underneath a navigable waterway. The tunnel now carries the East London underground railway line below the River Thames in London, between Wapping and Rotherhithe.

A description of the ground engineering problems encountered during the construction of this tunnel is given by Skempton and Chrimes (1994). The following description of Brunel's shield is an edited version of that given by Beaver (1973) in his book, *A History of Tunnels*. One of the twelve frames which comprised Marc Brunel's 1825 tunnelling shield is illustrated in Figure 10.27.

'The tunnel was formed of two parallel horseshoe-shaped arches, each 14 ft (4.2 m) wide and 17 ft (5.1 m) high, within a rectangular mass of brickwork 37.5 ft (11.25 m) wide and 22 ft (6.6 m) high (Figure 10.28). The shield consisted of twelve cast- and wrought-iron frames, each nearly 22 ft (6.6 m) high and a little over 3 ft (0.9 m) wide. When placed side-by-side against the face of the excavation, like books on a shelf, they formed a shield whose top, bottom and sides supported the earth 9 ft (2.7 m) in advance

Figure 10.27 One of the twelve frames which made up Marc Brunel's 1825 tunnelling shield. (From Beamish, 1862; with permission from Skempton and Chrimes, 1994.)

of the brickwork. Each frame was divided into three storeys or compartments, about 7 ft (2.1 m) high and 3 ft (0.9 m) wide, and occupied by a single miner. The assembled shield accommodated 36 men, all occupying separate working chambers. At the front of the assembly, the face of the tunnel was supported by a total of 504 **poling-boards**, each of which was 3 ft (0.9 m) wide, 6 in (0.15 m) deep and 3 in (0.075 m) thick. The poling-boards were in turn supported against the front of the frame, by means of screw-jacks.

The feet of the frames were broad iron shoes, while the roof of the uppermost cell was formed by a pivoted plate called a **stave**. The earth at the sides was supported by staves fixed to the outermost frames. The tunnel was progressed by each man removing one of his poling-boards and excavating the earth behind it to a depth of 4.5 in (0.113 m). The board was then replaced and driven forward by means of the screw-jacks. This process was repeated until all of the poling-boards had been moved forward. The

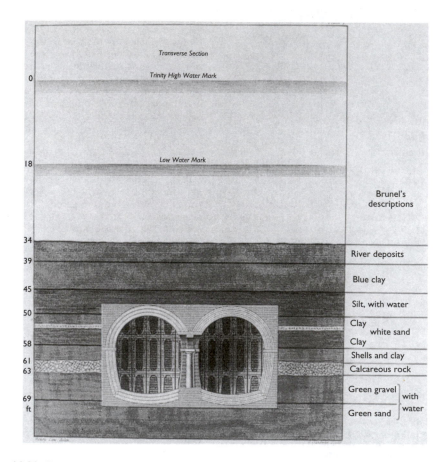

Figure 10.28 Cross-section through the first Thames tunnel. (Law, 1846; with permission from Skempton and Chrimes, 1994.)

frames were then pushed forward, one at a time, by means of large screw-jacks butted against the brickwork just behind the shield. The bricklayers worked so close to the advancing frames that the brickwork was carried right up to the shield. In this way, the only exposed areas of ground were those left by the removal of an individual poling-board.' (Reproduced from P. Beaver, *A History of Tunnels*; published by The Citadel Press, 1973.)

Marc Brunel's first tunnelling shield, which was patented by him in 1818 but never built, was in many ways more similar to a modern tunnelling machine. The 1818 patent was for a 12 ft (3.6 m) diameter auger (i.e. a drill) encased in an iron cylinder. The idea was that the blade could be rotated manually by the miners, while the cylinder was pushed forward (like the 1825 shield) along the line of the tunnel, by means of jacks reacting against the brickwork of the permanent lining. The practical problem with the proposal at the time was that it would not have been feasible to rotate the cutting blade manually. With the increasing availability of relatively compact portable power plant in the twentieth century, this difficulty was overcome.

As an alternative to using a shield or a tunnel boring machine in low-permeability soils, temporary support during construction can be provided by raising the air pressure within the tunnel. This is now avoided unless there is no alternative, because of the health hazards associated with working in compressed air.

Shallow tunnels on land are often constructed in an open excavation and then re-buried, in what is known as the **cut-and-cover** technique. Short, shallow tunnels of up to three or four metres in height – for example a tunnel constructed beneath an existing railway embankment – may be jacked into place using a technique known as **pipe-jacking**, which was originally developed for use with small diameter pipelines. The road tunnels across the River Conwy in North Wales and the River Medway in Kent (south-east England) are **immersed tubes**, installed section-by-section in a shallow trench excavated on the river bed.

Shallow tunnels may be rectangular, but many tunnels are circular in cross-section, or at least have an arched roof. This is because a circular cross-section enables the tunnel to resist the vertical and lateral stresses exerted by the ground by means of compressive stresses rather than in bending, giving a generally more efficient form of construction.

Much of the remainder of this section is concerned with tunnels which may be analysed in plane strain as long cavities of essentially circular cross-section. In the construction of such a tunnel, there are two main issues that must be addressed. These are:

- What support must be provided to the surrounding ground, both in the short term and in the long term, in order for the tunnel not to collapse?
- What are the ground movements associated with the tunnel, and how will these affect existing buildings, pipelines and buried structures?

10.11.1 Stress analysis of a tunnel of circular cross-section

The top of a circular tunnel is known as the **crown**, and the bottom is termed the **invert**. There are also **springings** and **haunches**, as defined in Figure 10.29(a). The diameter of the tunnel is D, and the depth of the crown below ground level is known

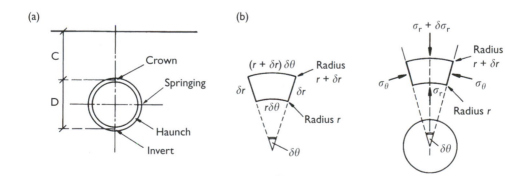

Figure 10.29 (a) Definition of tunnel geometry; (b) equilibrium of soil element above crown.

as the **cover**, C. The internal support pressure required to prevent collapse, σ_{TC}, may be determined by considering the equilibrium of an element of soil above the crown of the tunnel, as shown in Figure 10.29(b).

The total stress in the radial direction is σ_r, and the total stress in the circumferential direction, also known as the **hoop stress**, is σ_θ. The unit weight of the soil is γ, so that the weight of the soil element shown in Figure 10.29(b) is (per metre depth into the paper)

$$w = \gamma \times [r + (\delta r/2)]\delta\theta \times \delta r \approx \gamma r \delta r \delta\theta \tag{10.15}$$

The condition of horizontal equilibrium is satisfied by symmetry. Resolving forces vertically,

$$[(\sigma_r + \delta\sigma_r) \times (r + \delta r)\delta\theta] + [\gamma r \delta r \delta\theta] = [\sigma_r r \delta\theta] + [(2\sigma_\theta \delta r) \times \sin(\delta\theta/2)]$$

Because $\delta\theta$ is small, $\sin(\delta\theta/2) \approx \delta\theta/2$, so that

$$[(\sigma_r + \delta\sigma_r) \times (r + \delta r)\delta\theta] + [\gamma r \delta r \delta\theta] = [\sigma_r r \delta\theta] + [(2\sigma_\theta \delta r) \times (\delta\theta/2)] \tag{10.16}$$

Dividing through by $\delta\theta$, multiplying out and ignoring terms involving the squares or small quantities (e.g. $\delta\sigma_r \times \delta r$),

$$\sigma_r \delta r + r\delta\sigma_r + \gamma r \delta r = \sigma_\theta \delta r$$

Dividing through by $r\delta r$,

$$\frac{\sigma_r}{r} + \frac{\delta\sigma_r}{\delta r} + \gamma = \frac{\sigma_\theta}{r}$$

$$\Rightarrow \frac{\delta\sigma_r}{\delta r} = \frac{\sigma_\theta - \sigma_r}{r} - \gamma$$

In the limit as δr and $\delta\sigma_r \rightarrow 0$, the equation of equilibrium becomes

$$\frac{d\sigma_r}{dr} = \frac{\sigma_\theta - \sigma_r}{r} - \gamma \tag{10.17}$$

10.11.2   Collapse of tunnels in clay: short-term total stress analysis

In a clay soil of undrained shear strength τ_u which is on the verge of undrained failure,

$$(\sigma_\theta - \sigma_r) = 2\tau_u$$

(σ_θ is greater than σ_r because σ_r has been reduced at the boundary of the tunnel until the tunnel is about to collapse). Substituting this into equation (10.17),

$$d\sigma_r/dr = 2\tau_u/r - \gamma$$

or

$$d\sigma_r = [(2\tau_u/r) - \gamma]dr$$

which may be integrated between limits of $\sigma_r = \sigma_{TC}$ inside the tunnel at radius $r = D/2$, and general values $\sigma_r = \sigma_R$ at a radius $r = R$ (where R is the distance from the ground surface to the centreline or **axis** of the tunnel):

$$\int_{\sigma_{TC}}^{\sigma_R} d\sigma_r = \int_{D/2}^{R} \left(\frac{2\tau_u}{r} - \gamma \right) dr$$

giving

$$(\sigma_R - \sigma_{TC}) = 2\tau_u \ln(2R/D) - \gamma(R - [D/2])$$

or

$$\sigma_{TC} = \sigma_R - 2\tau_u \ln(2R/D) + \gamma(R - [D/2]) \tag{10.18}$$

where σ_{TC} is the support pressure when the tunnel is on the verge of collapse. For a tunnel with a **depth of cover** C ($= R - [D/2]$), and a surface surcharge $\sigma_R = q$, the support pressure required just to prevent collapse is

$$\sigma_{TC} = q + \gamma C - 2\tau_u \ln([2C/D] + 1) \tag{10.19}$$

The support pressure required just to prevent the tunnel from collapsing (σ_{TC}) is sometimes expressed in terms of a parameter called the **stability number at collapse**, which is given the symbol T_C and is defined as

$$T_c = \frac{q + \gamma[C + (D/2)] - \sigma_{TC}}{\tau_u} \tag{10.20}$$

10.11.3 Collapse of tunnels: effective stress analysis

In terms of effective stresses, the relationship between σ_r' and σ_θ' at failure is

$$\sigma_\theta' = K_p \sigma_r'$$

where $K_p = (1 + \sin \phi')/(1 - \sin \phi')$ and ϕ' is the critical state strength (effective angle of friction) of the soil. If the pore water pressures are zero, equation (10.17) is valid in terms of effective stresses, giving

$$(d\sigma_r'/dr) = \sigma_r'(K_p - 1)/r - \gamma \tag{10.21}$$

Equation (10.21) may be integrated by making the substitution $y = \sigma_r'/r$. Differentiating the expression $\sigma_r' = yr$ gives

$$d\sigma_r'/dr = r dy/dr + y \tag{10.22}$$

Substituting this into equation (10.21),

$$r dy/dr + y = y(K_p - 1) - \gamma$$

or

$$r dy/dr = y(K_p - 2) - \gamma \tag{10.23}$$

The limits of $\sigma_r' = \sigma_{TC}'$ at the edge of the tunnel, $r = D/2$, and $\sigma_r' = \sigma_R'$ at a radius $r = R$ become $y = 2\sigma_{TC}'/D$ at $r = D/2$, and $y = \sigma_R'/R$ at $r = R$. Thus

$$\int_{2\sigma_{TC}'/D}^{\sigma_R'/R} \frac{dy}{(K_p - 2)y - \gamma} = \int_{R/2}^{R} \frac{dr}{r}$$

giving

$$\ln\left[\frac{\{(K_p - 2)(\sigma_R'/\gamma R)\} - 1}{\{(K_p - 2)(2\sigma_{TC}'/\gamma D)\} - 1}\right] = (K_p - 2)\ln\left[\frac{2R}{D}\right]$$

$$= \ln\left[\left(\frac{2R}{D}\right)^{(K_p - 2)}\right]$$

or

$$[\{(K_p - 2)(\sigma_R'/\gamma R)\} - 1]/[\{(K_p - 2)(2\sigma_{TC}'/\gamma D)\} - 1]$$
$$= (2R/D)^{(K_p - 2)} \tag{10.24}$$

For a tunnel having a depth to the crown (i.e. a cover) of C (= $R - [D/2]$, where R is depth of the tunnel axis below the ground surface), with no surcharge applied to the soil surface so that $\sigma_R' = 0$, equation (10.24) may be rearranged to give the internal support pressure σ_{TC}' required just to prevent collapse:

$$[\{(K_p - 2)(\sigma_{TC}'/\gamma D)\} - 1] = -1/[([2C + D]/D)^{(K_p - 2)}]$$

$$\Rightarrow (K_p - 2)(2\sigma_{TC}'/\gamma D) = 1 - [([2C + D]/D)^{-(K_p - 2)}]$$

$$\Rightarrow \sigma_{TC}' = [\gamma D/\{2(K_p - 2)\}] \times [1 - (D/[2C + D])^{(K_p - 2)}] \tag{10.25}$$

For a deep tunnel with $D \ll C$, σ_{TC}' is approximately equal to $\gamma D/[2(K_p - 2)]$. For a tunnel of diameter $D = 4\,\text{m}$ with a depth of cover $C = 10\,\text{m}$ in a soil with unit weight $\gamma = 20\,\text{kN/m}^3$ and $\phi' = 35°$ ($K_p = 3.69$), the support pressure is (according to equation (10.25))

$$\sigma_{TC}' = [20\,\text{kN/m}^3 \times 2\text{m} \div (3.69 - 2)] \times [1 - (4/24)^{(1.69)}] = 22.5\,\text{kPa}$$

which is very small in comparison with the overburden pressure above the crown of 200 kPa. The error that would be introduced by neglecting the term in $D/(2C + D)$ would in this example have been less than 5%.

It must be reiterated that in the above analysis it has been assumed that the tunnel is on the verge of collapse, and that the pore water pressures are zero. If the pore water pressures are non-zero, the entire component of the total stress due to the pore water pressure must be carried by the tunnel lining, assuming that the lining is impermeable. This is because the arching effect, which reduces the effective stress component of the load on the tunnel lining, cannot occur through the pore water, as water is unable to carry shear stresses.

Similarly, tunnels in coarse sands cannot be supported by increasing the air pressure inside the tunnel, because this would have no effect except to increase the pore water pressure in the surrounding ground. Tunnels in sands must be supported by a structural lining. As these soils cannot sustain negative pore water pressures (and therefore have no undrained shear strength), tunnelling through them is much more difficult than tunnelling through clays.

One final point is that the integrity of the tunnel invert is just as important as the integrity of the tunnel crown. The major UK tunnel collapses at Penmanshiel (on the East Coast railway line between Berwick-upon-Tweed and Edinburgh in Scotland) in 1979, and Heathrow Airport (London) in 1994 (Muir Wood, 2000) both followed disturbance to the invert which left the tunnel linings unable to carry the compressive loads required to support the soil on each side of and below the tunnel cavity.

10.11.4 Collapse of tunnel headings

We have so far considered the collapse of the roof of a long tunnel, which is essentially a plane strain event. The collapse of an advancing tunnel face, and/or the section of tunnel close to it (known as the **heading**) in which the lining has not yet been constructed, involves a complex three-dimensional failure mechanism (Figure 10.30(a)). This was investigated experimentally for tunnels in soft clay by Mair (1979).

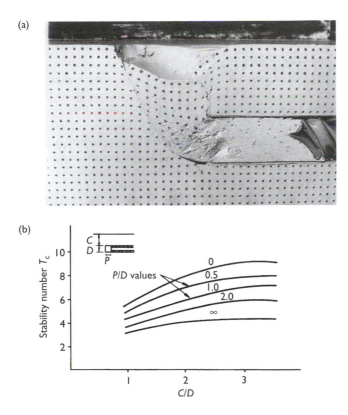

Figure 10.30 (a) Undrained collapse of a tunnel heading in clay ($C/D = 1.5$, $P/D = 0$), and (b) associated stability numbers. (Reproduced with permission from Mair, 1979.)

For tunnels in clay, Figure 10.30(b) gives values of the stability number

$$T_c = \frac{q + \gamma[C + (D/2)] - \sigma_{TC}}{\tau_u} \tag{10.20}$$

determined experimentally by Mair (1979), as a function of the cover to diameter ratio C/D for various heading lengths P/D, where P is the length of the unsupported heading. (In Mair's tests, the surface surcharge q was zero.)

10.11.5 Ground movements due to tunnelling

Experience and observation (e.g. O'Reilly and New, 1982) have shown that in clays, the construction of a tunnel will usually cause a settlement trough at the ground surface, of the shape shown in Figure 10.31. The equation of the settlement trough is

$$S = S_{max}e^{-x^2/2i^2} \tag{10.26}$$

which is sometimes termed a **Gaussian distribution**. S is the settlement at a general horizontal distance x from the centreline of the tunnel, S_{max} is the maximum settlement (at $x = 0$, above the centreline of the tunnel), and i is a parameter which defines the width of the settlement zone.

Equation (10.26) may be integrated to give the volume of the settlement trough:

$$V_s = \sqrt{2\pi} \times i\,S_{max} \tag{10.27}$$

For undrained excavations in clays, where there is no change in the specific volume of the soil, the settlement trough represents an additional volume of material, over and above the volume of the tunnel itself, which must have been excavated in order to form the tunnel. The volume of the settlement trough is normally divided by the nominal volume of the tunnel to give the proportion of additional material excavated. This quantity is known as the **volume loss**, V_L. In a reasonably well-controlled tunnelling operation in London Clay, a volume loss of about 1.4% is achievable.

$$V_L = V_s/V_{tunnel} = \left[\sqrt{(2\pi)}i\,S_{max}\right]/\left[\pi D^2/4\right]$$

$$= \left[(4\sqrt{2})i\,S_{max}\right]/\left[D^2\sqrt{\pi}\right]$$

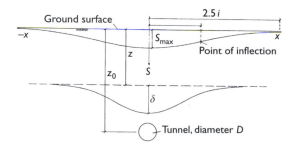

Figure 10.31 Settlement troughs above a tunnel. (Redrawn with permission from Mair et al., 1993.)

giving

$$V_L = 3.192 i\, S_{max}/D^2 \qquad (10.28)$$

or

$$S_{max} = 0.313 V_L D^2/i \qquad (10.29)$$

At horizons between the crown of the tunnel and the ground surface, the shape of the settlement pattern is similar but the values of S_{max} and i will vary with depth z (Figure 10.31). Mair *et al.* (1993) show that a reasonable fit to data from a number of full-scale tunnels in clay, and centrifuge model tests, is given by equations (10.26) and (10.29) with $V_L = 1.4\%$ and

$$i/z_0 = 0.175 + 0.325(1 - [z/z_0]) \qquad (10.30)$$

where is z_0 the depth to the centreline of the tunnel, $z_0 = C + (D/2)$. At the surface $(z = 0)$, $i = 0.5 \times z_0$.

In using equations (10.26), (10.29) and (10.30) to predict the surface and sub-surface settlements associated with tunnels in clays, it is necessary to estimate the volume loss V_L. V_L will depend on the ground conditions, the method of construction and the quality of workmanship. The tunnel support pressure σ_T must be sufficiently in excess of the collapse pressure σ_{TC} (given by equation (10.19) or Figure 10.30) to limit the ground loss to a reasonable value. The remoteness of a tunnel from collapse may be quantified by means of the load factor, LF. With a surface surcharge q,

$$LF = [\gamma z_0 + q - \sigma_T]/[\gamma z_0 + q - \sigma_{TC}] \qquad (10.31)$$

where z_0 is the depth to the tunnel centreline, $z_0 = C + (D/2)$. In this context, LF may be viewed as the inverse of the factor of safety. When $\sigma_T = \gamma z_0 + q$, $LF = 0$. As

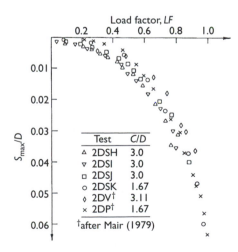

Figure 10.32 Normalized maximum surface settlement as a function of load factor for circular tunnels in soft clay. (Redrawn with permission from Taylor, 1984.)

σ_T is reduced LF increases until, when the tunnel is on the verge of collapse, $\sigma_T = \sigma_{TC}$ and $LF = 1$.

Figure 10.32 shows the experimental relationship given by Taylor (1984), between the maximum settlement at the surface (normalized with respect to the tunnel diameter D) and the load factor LF, for tunnels in speswhite kaolin clay. Figure 10.32 may be used to estimate the load factor – and hence the tunnel support pressure – required in order to limit the maximum surface settlement S_{max} – and hence, using equations (10.29) and (10.30), the volume loss V_L – to an acceptable value.

Key points

- Limit state design involves carrying out calculations to investigate potential ultimate (ULS) and serviceability (SLS) limit states. Ultimate limit states are concerned with outright collapse; serviceability limit states may be concerned with excessive stresses or deformations.
- In carrying out ultimate limit state calculations for an embedded retaining wall, a procedure of the type given in BS8002 (2001) or Gaba *et al.* (2003) should be followed. Although other methods have been used in the past, these can give misleading results in some circumstances, and should therefore be avoided.
- The actual stress distributions acting on an embedded retaining wall under in-service or working conditions will in general be different from those used in the ultimate limit state design calculation. This is because the soil strength mobilized under working conditions may be different on each side of the wall. Also, the effective stress distribution against a flexible wall will be non-linear, as a result of wall bending. The relative soil/wall stiffness is quantified by means of a dimensionless flexibility number R:

$$R = mH^4/EI$$

  where m is a soil stiffness parameter, EI is the flexural rigidity of the wall and H is the total wall length. Walls propped at the crest are generally regarded as 'stiff' (in that the effect of bending on the lateral stress distribution is small) if $R \lesssim 1000$. Bending effects are less significant if the wall is unpropped or propped at formation level, than if the wall is propped at the crest.
- The compaction of the soil behind a retaining wall can lead to increased lateral stresses in the backfill. In theory, the maximum lateral stress induced in a granular backfill, placed in thin layers and compacted by means of a roller that applies a line load of intensity q (kN/m), is

$$\sqrt{2q\gamma/\pi} \tag{10.7}$$

  In practice, the lateral stresses may be reduced to the active limit by a small outward movement of the wall, after compaction.
- A reinforced soil retaining wall may fail by either pull-out or breakage of the reinforcing strips. These failure modes can be assessed by means of a simple limit equilibrium analysis (equations (10.13) and (10.14)). Pull-out failure is generally well-predicted using equation (10.14). Tensile failure is only well-predicted by

equation (10.13) for walls which are already close to pull-out failure. Otherwise, equation (10.13) is somewhat conservative, due to the plastic redistribution of stresses between the reinforcement strips and other factors.

- The collapse of a tunnel of circular cross-section can be investigated by means of a stress analysis of the surrounding soil. This enables the minimum internal support pressure (σ_{TC} in a total stress analysis, or σ'_{TC} in an effective stress analysis) that must be provided by the tunnel lining just to prevent failure, to be calculated. The collapse of a tunnel heading may be assessed by means of the experimental data given in Figure 10.30.
- Soil settlements due to tunnelling are usually estimated semi-empirically, on the basis of the **Gaussian** curve shown in Figure 10.31.

Questions

Retaining Walls

10.1   Figure 10.33 shows the idealized geometry of a proposed retaining wall, together with soil and groundwater conditions.

(a)   Estimate the long-term pore water pressure distribution due to steady state seepage around the wall.

(b)   Using the idealized soil profile, calculate the depth of embedment required for the wall to be just stable in the long term and the corresponding prop load.

(c)   Explaining your method, estimate the ultimate limit state design depth of embedment of the wall and the associated prop load.

All soils have unit weight $20\,\text{kN/m}^3$. Take the unit weight of water as $10\,\text{kN/m}^3$. The upper sands and gravels have $\phi'_{crit} = 30°$ and $\phi'_{peak} = 34°$, the clays have $\phi'_{crit} = 20°$ and $\phi'_{peak} = 22°$, and the lower sands have $\phi'_{crit} = 32°$ and $\phi'_{peak} = 36°$. Use the earth pressure coefficients given in Tables 7.6 and 7.7.

10.2   Discuss the differences, from a design point of view, between *in situ* and backfilled retaining walls. Your answer should address issues such as the form of the retaining structure, the nature of the retained soil, and groundwater effects.

[*(University of London 3rd year BEng (Civil Engineering) examination, Queen Mary and Westfield College)*]

10.3   Explain briefly the terms **upper bound** and **lower bound**, in the context of engineering plasticity, and explain how these concepts can be used as a basis for retaining wall design.

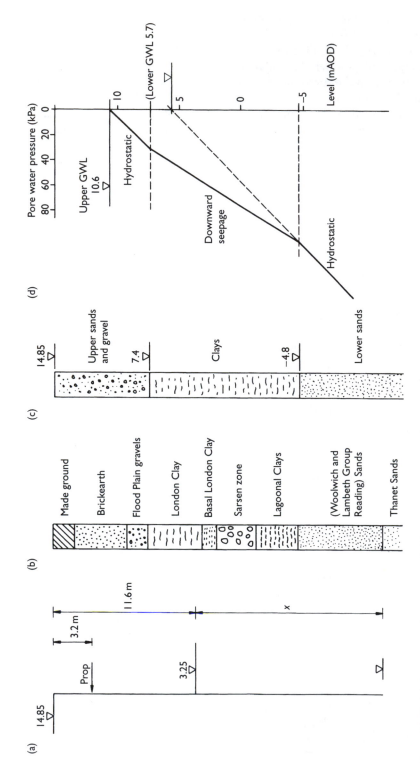

Figure 10.33 (a) Geometry of retaining wall; (b) ground conditions; (c) idealized soil profile; (d) *in situ* pore water pressures.

Discuss the modes of failure which would need to be taken into account in the design of

(a)   a mass retaining wall
(b)   an embedded retaining wall.

Illustrate your answer with sketches and examples where appropriate, and indicate suitable ways of applying a factor of safety to the limit state calculation in each case.

Discuss briefly the main shortcomings of limit-based design, and suggest alternative methods which may be used.

[(University of London 3rd year BEng (Civil Engineering) examination, Queen Mary and Westfield College)]

10.4   Stating carefully the assumptions you make, develop the argument that granular soil of unit weight γ at a depth z behind a yielding retaining wall may be brought to passive failure by the application of a cycle of vertical effective stress of magnitude $\Delta\sigma_v' = \gamma z K_p^2$, where K_p is the passive earth pressure coefficient $(1 + \sin\phi')/(1 - \sin\phi')$.

Hence derive the lateral stress distribution behind a yielding retaining wall where the backfill has been compacted in thin layers by the application of a surface line load of magnitude q kN/m.

(It may be assumed that the increase in vertical effective stress at a depth z below the line load is $2q/\pi z$.)

Sketch the lateral stress distribution in the case of an L-cantilever wall having a retained height $H = 5$ m, with a backfill of unit weight $\gamma = 18$ kN/m^3 and $\phi' = 30°$, compacted using a vibrating roller capable of applying an equivalent line load q of 20 kN/m. Calculate the shear force and bending moment at the corner of the 'L'.

Reinforced soil retaining walls

10.5   A 4 m-high reinforced soil retaining wall is constructed, using galvanized strip reinforcements 3.5 m long, with a cross-section of 3 mm $\times$ 80 mm. The ultimate tensile strength of the strips is 190 N/mm^2, and the design must allow for an overall loss of thickness of 0.75 mm due to corrosion, during the anticipated working life of the structure. The strips are spaced at 1 m vertically and 0.3 m horizontally. The uppermost reinforcement is 0.5 m below the top of the wall. The surface of the backfill is horizontal. The backfill material has unit weight $\gamma = 18$ kN/m^3 and friction angle $\phi' = 30°$. The angle of friction δ between the backfill and the reinforcement is 20° (i.e. the coefficient of friction $\mu = \tan 20°$.) Calculate:

(a)   The ultimate tensile load of a single reinforcement strip, before corrosion.

(b)   The ultimate tensile load of a single reinforcement strip, after corrosion.

(c)   The factor of safety or load factor F_T against tensile failure.

(d)   The factor of safety or load factor F_F against pull-out.

Comment on the adequacy of the design.

[*(University of London 3rd year BEng (Civil Engineering) examination, Queen Mary and Westfield College)*]

((a) 45.6 kN; (b) 34.2 kN; (c) $F_T = 3.31$ on corroded strength; (d) $F_F = 1.42$)

Tunnels

10.6   A circular tunnel, 10 m² in diameter, is to be driven through a soft deposit of marine clay. The axis of the tunnel is at a depth of 20 m below the soil surface. The soft clay has unit weight $\gamma = 16\,\text{kN/m}^3$, and undrained shear strength $\tau_u = 40\,\text{kPa}$. A surcharge of $q = 10\,\text{kPa}$ acts at the surface of the clay.

(a)   calculate the tunnel support pressure required just to prevent collapse, σ_{TC}

If the maximum allowable surface settlement is 50 mm, estimate

(b)   the maximum allowable volume loss, V_L

(c)   the load factor, *LF*; and

(d)   the required tunnel support pressure, σ_T.

((a) 139 kPa; (b) 1.6%; (c) 0.45; (d) 200 kPa.)

Note

1   When a load is applied to a soil surface, the settlement zone will extend beyond the loaded area. (This can be demonstrated using the Newmark chart described in section 6.4.) Similarly, when an initially uniformly-loaded soil surface is unloaded over part of its area, the zone of soil movement will extend beyond the unloaded region. If the soil outside the unloaded area is prevented from moving by an unyielding support, the contact stress between the soil and the support must increase. This process is conventionally termed **arching**.

References

Addenbrooke, T.I., Potts, D.M. and Dabee, B. (2000) Displacement flexibility number for multipropped retaining wall design. *ASCE Journal of Geotechnical and Geoenvironmental Engineering*, **126**(8), 718–26.

Beamish, R. (1862) *Memoir of the Life of Sir Marc Isambard Brunel*, Longman, London.

Beaver, P. (1973) *A History of Tunnels*, The Citadel Press, Secaucus, New Jersey.

Bolton, M.D. and Pang, P.L.R. (1982) Collapse limit states of reinforced earth retaining walls. *Géotechnique*, **32**(4), 349–67.

Bolton, M.D. and Powrie, W. (1988) Behaviour of diaphragm walls in clay prior to collapse. *Géotechnique*, **38**(2), 167–89.

Bolton, M.D., Powrie, W. and Symons, I.F. (1989, 1990) The design of stiff, *in situ* walls retaining overconsolidated clay, Part I & Part II. *Ground Engineering*, **22**(8), 44–8; **22**(9), 34–40; and **23**(2), 22–8.

British Standards Institution (1992) *Eurocode 2: design of concrete structures, Part 1-1: General Rules and Rules for Buildings*. DD ENV 1992-1-1. BSI, London.

British Standards Institution (1995), *Eurocode 7: Geotechnical design. Part 1: General rules.* DD ENV 1997–1: BSI, London.

British Standards Institution (1997) *BS8110 Part 1: Structural use of concrete*, BSI, London.

British Standards Institution (1998) *Eurocode 3: Design of Steel Structures* Part 5: Piling DDENV 1993–5 BSI, London.

British Standards Institution (2000) *BS5950-1 Structural use of steelwork in building. Part 1: Code of Practice for Design-Ruled and Welded Sections*. BSI, London.

British Standards Institution (2001) *BS8002: Code of practice for earth retaining structures*, (Incorporating Amendments Nos 1 and 2 and Corrigendum No. 1) BSI, London.

Broms, B.B. (1971) Lateral earth pressures due to compaction of cohesionless soils. *Proceedings of the 4th Budapest Conference on Soil Mechanics and Foundation Engineering (3rd Danube-European Conference)* (ed. A. Kezdi), pp. 373–84, Akademiai Kiado, Budaapest.

Burland, J.B., Potts, D.M. and Walsh, N.M. (1981) The overall stability of free and propped embedded cantilever walls. *Ground Engineering*, **14**(5), 28–38.

Caquot, A. and Kerisel, J. (1948) *Tables for the Calculation of Passive Pressure etc.* Gauthier Villars, Paris.

Carder, D.R., Murray, R.T. and Krawczyk, J.V. (1980) *Earth pressures against an experimental retaining wall backfilled with silty clay*. TRRL Laboratory Report 946, TRL, Crowthorne.

Clayton, C.R.I., Symons, I.F. and Hiedra-Cobo, J.C. (1991) Pressure of clay backfill against retaining structures. *Canadian Geotechnical Journal*, **28**(2), 309–15.

Clough, G.W., Smith, E.M. and Sweeney, B.P. (1989) Movement control of excavation support systems by iterative design. *Foundation Engineering: Current Principles and Practices*, **2**, 869–84, ASCE, New York.

Gaba, A., Simpson, B., Powrie, W. and Beadman, D.R. (2003) *Embedded retaining walls – guidance for economic design*. CIRIA Report C580. Construction Industry Research and Information Association, London.

Holl, D.L. (1941) Plane strain distribution of stress in elastic media. *Iowa Engng. Exp. Stn. Bull.*, 148–63.

Hubbard, H .W., Potts, D.M., Miller, D. and Burland, J.B. (1984) Design of the retaining walls for the M25 cut and cover tunnel at Bell Common. *Géotechnique*, **34**(4), 495–512.

Ingold, T.S. (1979) The effects of compaction on retaining walls. *Géotechnique*, **29**(3), 265–83.

Institution of Structural Engineers (1951) *Civil Engineering Code of Practice No. 2: Earth Retaining Structures (CP2)*, Institution of Structural Engineers, London.

Law, H. (1846) *Memoir of the Thames Tunnel (to 1828)*. John Weale, London.

Lee, K., Jones, C.J.F.P., Sullivan, W.R. and Trolinger, W. (1994) Failure and deformation of four reinforced soil walls in eastern Tennessee. *Géotechnique*, **44**(3), 397–426.

Mair, R.J. (1979) *Centrifugal modelling of tunnel construction in soft clay.* PhD dissertation, University of Cambridge.

Mair, R.J., Taylor, R.N. and Bracegirdle, A. (1993) Subsurface settlement profiles above tunnels in clays. *Géotechnique*, **43**(2), 315–20.

Mawditt, J.M. (1989) Discussion on Symons and Murray (1988). *Proceedings of the Institution of Civil Engineers, Pt. 1*, **86**, 980–6.

Muir Wood, A. (2000) *Tunnelling–Management by design.* E&F.N. Spon, London.

O'Reilly, M.P. and New, B.M. (1982) Settlements above tunnels in the United Kingdom – their magnitude and prediction, in *Tunnelling 82*, pp. 173–81, IMM, London.

Pappin, J.W., Simpson, B., Felton, P.J. and Raison, C. (1986) Numerical analysis of flexible retaining walls. *Proceedings of a Symposium on Computer Applications in Geotechnical Engineering.* Midland Geotechnical Society, Birmingham, and Arup Geotechnics program FREW.

Potts, D.M. and Fourie, A.B. (1984) The behaviour of a propped retaining wall: results of a numerical experiment. *Géotechnique*, **34**(3), 383–404.

Potts, D.M. and Fourie, A.B. (1985) The effect of wall stiffness on the behaviour of a propped retaining wall. *Géotechnique*, **35**(3), 347–52.

Powrie, W. (1985) Discussion on 5th Géotechnique symposium in print, The performance of propped and cantilevered rigid walls. *Géotechnique*, **35**(4), 546–8.

Powrie, W., Pantelidou, H. and Stallebrass, S.E. (1998) Soil stiffness in stress paths relevant to diaphragm walls in clay. *Géotechnique* **48**(4), 483–94.

Rowe, P.W. (1952) Anchored sheet pile walls. *Proceedings of the Institution of Civil Engineers, Pt 1*, **1**, 27–70.

Rowe, P.W. (1955) A theoretical and experimental analysis of sheet pile walls. *Proceedings of the Institution of Civil Engineers, Pt 1*, **4**, 32–69.

Skempton, A. W. and Chrimes, M. M. (1994) Thames Tunnel: geology, site investigation and geotechnical problems. *Géotechnique*, **44**(2), 191–216.

Symons, I. F. (1991) Discussion on limit equilibrium methods for free embedded cantilever walls in granular materials, by A.V.D. Bica and C.R.I. Clayton. *Proceedings of the Institution of Civil Engineers, Pt. 1*, **90**, 213–16.

Symons, I. F. and Murray, R. T. (1988) Conventional retaining walls: pilot and full-scale studies. *Proceedings of the Institution of Civil Engineers, Pt 1*, **84**, 519–38.

Symons, I.F., Little, J.A., McNulty, T.A., Carder, D.R. and Williams, S.G.O. (1987) Behaviour of a temporary anchored sheet pile wall on the A1(M) at Hatfield. Report RR99, Transport Research Laboratory, Crowthorne.

Taylor, R.N. (1984) *Ground movements associated with tunnels and trenches.* PhD dissertation, University of Cambridge.

Tedd, P., Chard, B.M., Charles, J.A. and Symons, I.F. (1984) Behaviour of a propped embedded retaining wall in stiff clay at Bell Common tunnel. *Géotechnique*, **34**, 513–32.

Twine, D. and Roscoe, H. (1999) *Temporary propping of deep excavations – guidance on design.* CIRIA Report C517, Construction Industry Research and Information Association, London.

Williams, B.P. and Waite, D. (1993) *The design and construction of sheet-piled coffer-dams.* Special publication 95, Construction Industry Research and Information Association, London.

Chapter 11

Modelling, *in situ* testing and ground improvement

11.1 Introduction and objectives

This chapter is divided into three main sections. In section 11.2, the principles of numerical and physical modelling are outlined, and the application of these techniques in geotechnical engineering is discussed. In section 11.3, the estimation or measurement of soil parameters from *in situ* test data is described. In section 11.4, a number of methods of ground improvement are outlined.

Objectives

After having worked through this chapter, you should have an appreciation of:

- the application of numerical and physical models in geotechnical engineering, including the limitations and approximations involved (section 11.2)
- the use of *in situ* tests to measure or estimate soil parameters, including the assumptions and uncertainties associated with the interpretation of field test data (section 11.3)
- some of the methods that can be used to improve the strength and/or stiffness of the natural ground, including the circumstances and soil types in which each is applicable (section 11.4).

You should be able to:

- estimate the soil strength and stiffness, on the basis of the blowcount in a standard penetration test (section 11.3.1)
- carry out a preliminary interpretation of cone penetrometer test data, to determine the soil types present (section 11.3.2)
- estimate the shear modulus G and the undrained shear strength τ_u of a clay soil, on the basis of the results of a pressuremeter test (section 11.3.3).

The stress analysis of the pressuremeter test given in section 11.3.3 (b) is rather complicated, and probably only for the enthusiast.

11.2   Modelling

Modelling is an everyday activity for almost all engineers. Many of the traditional procedures in engineering design are in effect models or idealizations of real behaviour, without which most problems would be too complex to analyse. In some circumstances, direct physical modelling of a real construction at a smaller scale may assist in the development of an understanding of the important behavioural mechanisms involved. Numerical modelling techniques such as **finite element analysis** (section 11.2.1) are becoming increasingly used as aids to geotechnical design.

The essential feature of a model is that a decision is taken at an early stage as to the important aspects of the overall behaviour of the construction under consideration. These are separated out and analysed, while the aspects that are judged not to be important are neglected. If this initial decision is wrong, the model may be inappropriate to the situation to which it is applied. The consequences of this could be disastrous.

In any theoretical analysis, the stress/strain/strength behaviour of the soil must be described in a mathematical form. Soil strength is generally easier to quantify than the complete stress–strain relationship. This has led to the development of analytical models based on the theorems of plasticity and conditions at collapse, which involve only the soil strength. These were described in Chapters 7–9. In some circumstances, it is necessary to attempt to predict soil movements associated with foundations and other geotechnical structures. There is then no alternative but to attempt to describe the stress–strain behaviour in some idealized manner, and carry out an appropriate calculation. The simplest approach – although it is somewhat unrealistic – is to model the soil as an elastic material, as described in Chapter 6.

Models based on the concepts of elasticity, plasticity and limiting equilibrium have been extensively described in Chapters 6–9. These models still form the basis of most routine geotechnical design: the results of a physical model test or a finite element analysis which were obviously incompatible would generally be scrutinized very carefully indeed. In large projects, perhaps involving novel designs, the use of numerical or physical modelling techniques is not uncommon. As more and more powerful personal computers become available, it is likely that numerical modelling will play an increasingly significant role, even in routine geotechnical analysis. The aim of this section is to provide a brief overview of the principles and potential shortcomings of numerical and physical modelling techniques.

11.2.1   Numerical modelling

An example of the use of finite element analysis to investigate a problem of steady state groundwater flow was given in section 3.18. The finite element method may also be used to solve numerically the equilibrium equations for a soil mass, while at the same time satisfying the condition of compatibility and the stress–strain relationship for the soil, for given boundary conditions.

In **finite element analysis**, the region of interest is divided into discrete (finite) elements with common nodes. A typical finite element mesh for the analysis of an embedded retaining wall is shown in Figure 11.1. Smaller elements are used in regions where the changes in stress and strain are expected to be most significant – in this

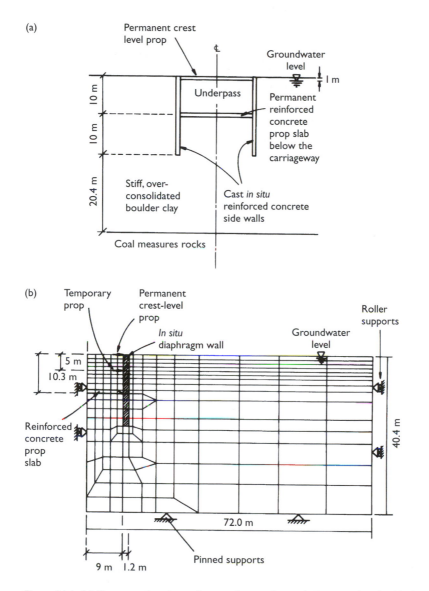

Figure 11.1 (a) Cross section through an underpass formed of propped embedded retaining walls, and (b) typical finite element mesh used in analysis. (Redrawn with permission from Richards and Powrie, 1994.)

case, in the vicinity of the wall and the excavation. Some programs will generate the mesh automatically. Others include an adaptive mesh refinement procedure (AMR), which automatically re-configures the mesh as the analysis progresses, to concentrate elements in zones of extreme deformation (Hicks and Mar, 1994).

The mesh should extend far enough laterally from the feature under investigation for changes in stress to be negligible. The lower boundary to the mesh will often (as in

Figure 11.1) correspond to some underlying stiff stratum. If no such stratum is present, the boundary should ideally be deep enough for changes in stress at this depth to be negligible. For a plane strain problem, this would typically be 6–10 times the width of the loaded or unloaded zone. It is usual to take full advantage of symmetry: for example, in the mesh shown in Figure 11.1(b), only one half of the total excavation is modelled. The centreline and the far vertical boundary are usually free to slide in the vertical direction (i.e. they are supported on rollers), while the bottom boundary is usually taken as pinned (Figure 11.1(b)). Modelling the far vertical boundary in this way means that, strictly, it is a line of symmetry.

Most geotechnical finite element programs use the **displacement method**, in which the displacement within each element is expressed as a polynomial function (known as the **interpolating polynomial**) of the nodal displacements and the position of the element (Zienkiewicz, 1967). The displacements are converted into strains, using the condition of **compatibility**. The element stresses are then determined by means of the material **stress–strain** relationship. Finally, the principle of virtual work is used to calculate the equivalent nodal forces that are in **equilibrium** with the internal stresses. The nodal forces should balance the loads due to the self-weight of the soil and the stresses at the boundaries.

The variation of strain within an individual element is determined by the interpolating polynomial used to calculate the displacements, and defines the **order** of the element. The order of the elements used will determine to a large extent the number of elements required in the finite element mesh for the analysis of a given problem. Elements in which the strain is assumed to be constant – such as **constant strain triangles** – are somewhat crude. Although higher order elements are available, a linear variation in strain within triangular or rectangular elements – known as **linear strain triangles** and **linear strain quadrilaterals** – is normally found to give sufficient accuracy for many geotechnical problems. Full details of the implementation of the finite element method are given in textbooks such as Irons and Ahmad (1980) and Smith and Griffiths (1996).

Finite element methods are used routinely in structural analysis, and many programs are available commercially. However, a commercial finite element package designed for structural problems is unlikely to be suitable for geotechnical applications, for which the following features are required:

- the ability to carry out the analysis in terms of effective stresses, which means that the program must be capable of taking due account of pore water pressures
- the ability to model consolidation of the soil, in which changes in specific volume are coupled to changes in effective stress as non-equilibrium pore water pressures dissipate
- the facility to add elements, to simulate the construction of (for example) embankments, and the facility to remove elements, to simulate excavation processes
- the facility to model structural elements, such as props and retaining walls, using bar or beam elements if required.

With recent advances in computer technology, numerical modelling using finite element analysis is becoming an increasingly popular and powerful analytical tool. Where construction methods are complex, a numerical model may be the only way in which the effects of different construction sequences can be investigated. A potentially major limitation is that most analyses are carried out either in plane strain or in conditions of axial symmetry. Three-dimensional effects are at present rarely modelled fully.

The range of constitutive models required in geotechnical finite element analysis is likely to be rather more extensive than that needed for the analysis of many structures, for which a simple linear elastic model may often be sufficient. For soils, a transition point – dependent on effective stresses – from elastic to plastic behaviour is essential, for example, using a frictional failure criterion, such as $(\tau/\sigma')_{max} = \tan \phi'_{crit}$ (which is known as the **Mohr–Coulomb failure criterion**), or $q = Mp'$ (Cam clay).

A model which treats the soil as an elastic material until a stress state on the failure envelope is reached, after which the soil behaves as an ideal plastic material, is described as **elastic–perfectly plastic**, and represents a bare minimum requirement for geotechnical analysis. Further features may be incorporated, such as anisotropic behaviour, with different stiffness moduli in the vertical (E'_v) and horizontal (E'_h) directions ($E'_v \neq E'_h$); a soil modulus that increases with depth ($E' = E'_{z=0} + mz$); or a modulus that changes with strain in some empirically defined way (Jardine *et al.*, 1986).

Models such as Cam clay (Figures 5.16, 5.21, 5.23, 5.26 and 5.27); modified Cam clay (Figure 5.29); or a finite element formulation of the behavioural regime proposed by Schofield (1980: Figure 5.42), in which the shape of the yield surface on the dry side of the critical state is altered to mimic the Hvorslev surface and the tensile cut-off, may also be used.

Cam clay models will give an elastic stiffness that varies with average effective stress p'. The equation of one of the elastic unload–reload lines (on a graph of v against $\ln p'$) shown in Figure 5.16(b) may be written as

$$v = v_\kappa - \kappa \ln p' \tag{11.1}$$

where v_κ is the intersection of the elastic line with the specific volume (v) axis. (v_κ is not a soil constant: it is different for each elastic line.) Differentiating equation (11.1) with respect to p',

$$dv/dp' = -\kappa/p'$$

By definition, the elastic bulk modulus $K = dp'/d\varepsilon_{vol}$. The volumetric strain increment $d\varepsilon_{vol} = -dv/v$,[1] so that

$$K = -vdp'/dv = vp'/\kappa \tag{11.2}$$

Cam clay does not model the reduction in soil stiffness with strain following a change in the direction of the stress path (the recent stress history effect: Figure 6.2). There are, however, models available which do: these include the three-surface model described in section 5.21 (Stallebrass, 1990), and the Brick model developed by Simpson (1992).

A further potential problem with the use of Cam clay is that the critical state is expressed not in terms of the angle of friction ϕ'_{crit} ($[\tau/\sigma'_{max}] = \tan \phi'_{crit}$), but in terms of the parameter M ($q = Mp'$: equation (5.27)). The relationship between ϕ'_{crit} and M depends on the intermediate principal effective stress σ'_2, as indicated by equation (5.36):

$$\sin \phi'_{crit} = 3M / \left\{ \left[6\sqrt{(1 - b + b^2)} \right] - \left[(2b - 1)M \right] \right\} \tag{5.36}$$

where

$$b = (\sigma'_2 - \sigma'_3) / (\sigma'_1 - \sigma'_3) \tag{5.35}$$

The adoption of a value of M measured in triaxial compression tests will lead to an excessively high value of ϕ'_{crit} in plane strain. This is because in triaxial compression $\sigma'_2 = \sigma'_3$, giving $b = 0$ and

$$M_{triaxial\,compression} = 6 \sin \phi'_{crit} / (3 - \sin \phi'_{crit}) \tag{11.3a}$$

In plane strain, $\sigma'_2 \approx 0.5 \times (\sigma'_1 + \sigma'_3)$, giving $b \approx 0.5$ and

$$M_{plane\,strain} = \sqrt{3} \times \sin \phi'_{crit} \tag{11.3b}$$

Assuming a constant value of ϕ'_{crit}, and eliminating $\sin \phi'_{crit}$ from equations (11.3a) and (11.3b),

$$M_{plane\,strain} = (3\sqrt{3} M_{triaxial\,compression}) / (6 + M_{triaxial\,compression}) \tag{11.3c}$$

For example, with $\phi'_{crit} = 22°$ $M_{triaxial\,compression} = 0.856$ and $M_{plane\,strain} = 0.649$.

It is generally accepted that a finite element analysis in which the high stiffness of soils at small strains is not modelled will tend to overestimate soil movements at the edges of the finite element mesh, where the changes in stress are very small (Jardine *et al.*, 1986). Although the high stiffness of soils at small strains can be modelled empirically by means of a non-linear stress–strain curve, the realistic representation of soil stiffness in a numerical analysis is rather more complex than this. First, the use of the same stress–strain relationship for all soil elements is unlikely to be appropriate in the field. Real stress paths will differ widely (e.g. between elements on different sides of a retaining wall). Second, it is necessary to alter the stiffness of the soil whenever the direction of the stress path changes. These problems can only really be overcome by the use of a soil model that takes into account the recent stress history of the soil, both at the start of and during the analysis.

Although a simple linear elastic/perfectly plastic model is unlikely to represent real soil behaviour it may, with the judicious selection of stiffness parameters, be possible to obtain realistic results. On the basis of the observed movements of a number of retaining walls in London Clay, Burland and Kalra (1986) backcalculated a soil stiffness profile for use in elastic–perfectly plastic finite element analyses in which the soil stiffness increases linearly with depth. Having been calibrated against field data, this profile may be used with some confidence in predictive analyses of retaining wall displacements in similar conditions.

Chandler (1995; see also Powrie *et al.*, 1999) carried out finite element analyses of an embedded retaining wall on the A406 North Circular Road at Waterworks Corner, South Woodford, London, using three different soil models (Cam clay, an elastic–perfectly plastic model, and the Brick model – described by Simpson (1992)). Chandler's results showed that the calculated wall movements depended primarily on the effective soil stiffness. The calculated lateral stresses depended more on the soil model than the soil stiffness, while the calculated soil surface settlement profiles depended on both the soil model and the soil stiffness.

Higgins *et al.* (1993) back-analysed the M25 Bell Common retaining wall, using (i) a linear elastic/perfectly plastic soil model, with parameters derived from the back-analysis of previous excavations in London Clay; and (ii) logarithmic relationships between G/p' and axial strain, and between K/p' and volumetric strain, based on triaxial test data (following Jardine *et al.*, 1986). They concluded that while it was possible to obtain reasonable results in terms of wall movements and bending moments using either method, it was necessary to model the construction sequence in some detail. On the other hand, calculated ground movements (as distinct from retaining wall movements) can be very sensitive to the soil model used in the analysis (Gunn, 1993; Burd *et al.*, 1994).

To summarize, stress-based variables (such as structural bending moments) and perhaps the movements of a propped retaining wall can probably be calculated reasonably closely using simple models, provided that the parameters which control the soil stiffness are selected with care. However, the satisfactory calculation of strain-based variables (such as the soil surface settlements behind a retaining wall or above a tunnel) depends on the use of a realistic soil model as well.

The facility to take account of construction processes is one of the strengths of finite element analysis. However, some degree of approximation will in many cases be necessary, especially where there are localized or three-dimensional effects in an otherwise plane strain analysis. In the case of the *in situ* retaining wall shown in Figure 11.1, the sequence of analysis (starting with the wall already in place) was as follows:

1. removal of elements (over a period representing 22 days), simulating excavation between the walls to 5.5 m below original ground level
2. addition of a bar element, simulating the installation of a temporary prop at a depth of 5 m below original ground level
3. removal of elements (over a period representing 28 days), simulating excavation to 10.3 m below original ground level
4. addition of concrete elements, simulating the placement of a permanent reinforced concrete prop slab at formation level
5. addition of a bar element at the top of the wall, simulating the installation of a permanent prop at crest level
6. removal of the bar element at 5 m below ground level, simulating the removal of the temporary prop
7. 120 years' excess pore water pressure dissipation, modelling the long-term behaviour of the wall.

It is quite common in finite element analyses of *in situ* retaining walls to start with the wall already in place. This is primarily because although it is possible to model wall installation, the complexity of the analysis is increased considerably. Also, the effects of wall installation are in reality three-dimensional (particularly if the wall is made up of individual bored piles), and can only be represented approximately in a two-dimensional analysis.

Wall installation effects may be significant in three different ways:

1. The ground movements that occur during wall installation may be of the same order as those during the main excavation stage.
2. In an overconsolidated clay, the process of wall installation will tend to reduce the horizontal stresses to below their *in situ* values.
3. The wall installation process defines the recent stress history of the soil, thereby influencing the soil stiffness at the start of excavation.

If wall installation is not modelled explicitly, the second of these effects is sometimes approximated by specifying horizontal effective stresses at the start of the analysis which are less than the actual *in situ* values. This is not ideal, because the assumed reduction in lateral stress extends across the entire mesh, whereas in reality it would be confined to the soil in the vicinity of the wall. Nonetheless, provided that the reduced lateral stresses are chosen with regard to the intended construction process (e.g. diaphragm wall panels or bored piles), it is arguably better than doing nothing.

In an analysis in which the soil consolidates as non-equilibrium pore water pressures dissipate, the time increment over which a loading or unloading process (e.g. excavation) takes place must be specified. The program will then automatically calculate the degree of drainage which takes place, without the need for the user to decide whether a process is effectively drained or undrained. Alternatively, if it is obvious that a process is either fully drained or truly undrained, an appropriate non-consolidation analysis can be carried out.

In formulating the stress–strain relationship, some programmers use a **tangent stiffness procedure,** in which the stiffness matrix of the soil at the start of an increment of loading (or unloading) is assumed to apply throughout that load increment. If the stress–strain curve changes from linear elastic to plastic during the load increment, this will lead to errors. If the stress–strain curve is non-linear, the error will build up as the analysis progresses (Figure 11.2(a)). The size of the error can be reduced by applying the load in a number of small increments, as shown in Figure 11.2(b). A more satisfactory approach is to use an iterative procedure, in which the stress–strain state after the application of the load is repeatedly checked and corrected until it lies on the stress–strain curve (Figure 11.2(c)). Although the iteration process adds considerably to the time taken for each load increment, the overall time taken for an analysis to run can be reduced because the load can be applied in far fewer increments. A further substantial benefit is that there is considerably less uncertainty that the correct solution has in fact been obtained.

In many programs, it is assumed that the strain increment vector at failure is perpendicular to the failure surface when the stress and strain increment axes are superimposed (the normality condition: Figure 7.21, section 7.7). For frictional failure criteria such as $q = Mp'$ and $(\tau/\sigma')_{max} = \tan \phi'$, this implies a non-zero rate of dilation at failure: thus the calculated volumetric strain rates (or negative pore water pressures in undrained conditions) are likely to be unrealistically high.

Finite element analysis has become much more accessible in recent years, and is no longer the preserve of specialists. Analyses that were formerly left to run overnight on a mainframe computer can now be carried out in half an hour or so on a personal

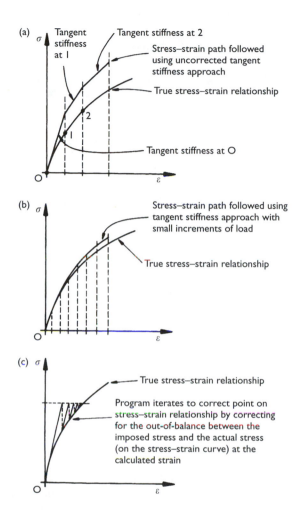

Figure 11.2 (a) Cumulative error due to tangent stiffness solution routine; (b) reduction in error by use of small load increments; (c) elimination of error by use of iterative solution technique.

computer. The use of finite element programs, however, requires considerable care if misleading results are to be avoided – or at least identified, investigated and discounted.

Woods and Clayton (1993) catalogue some of the problems experienced by users of a finite element program (CRISP) based on the concepts of critical state soil mechanics (Britto and Gunn, 1987). CRISP was originally developed as a research tool at Cambridge University. The code is widely available, and various modifications and special purpose elements have been added by research workers elsewhere. The issues discussed by Woods and Clayton (1993) include:

- the use of the normality condition (also termed an **associated flow rule**), which can lead to excessive dilation, strength and stiffness

- the need for special interface elements in soil–structure interaction problems (e.g. a retaining wall which tends to separate from the soil)
- the use of a tangent stiffness solution routine, which may require a large number of increments for numerical accuracy
- the fact that bending moments in retaining walls should be calculated from the stresses within the wall, rather than from the stresses in the adjacent soil.

It must also be recognized that there is no point in carrying out a sophisticated analysis unless it is justified by the quality of the data. A finite element analysis will not compensate for an inadequate site investigation and laboratory testing programme. Finally, in some soil models, the effect on the results of an analysis of the parameter values used can be disproportionately large. One possible example of this is the slope of the Hvorslev surface in the Schofield model (Figure 5.42), which governs the peak stress ratios q/p' sustainable by the soil (Richards, 1995).

11.2.2   Physical modelling: geotechnical centrifuge testing

Physical modelling tends to be used for the investigation of basic mechanisms of collapse or deformation, parametric studies and the validation or calibration of numerical codes. In principle, physical modelling techniques could be used directly in design. In Britain this seems to be a less common application, but in countries such as Japan it is quite routine.

It has already been mentioned that in numerical modelling it can be difficult to describe the stress–strain relationship of the soil in a way which is appropriate to the stress–strain paths being followed by all soil elements at all times. Small-scale physical modelling at normal gravity (1g) does not solve this problem, because the stresses due to the soil self weight are too low. As stress–strain relationships depend on stress history, stress state and stress path, the stress–strain behaviour in the field will not be replicated in the model. Even if it is only conditions at failure that are being investigated, the low stresses in 1g tests in sands can give misleading results, because the effects of dilation may be far more significant than they would be at higher stresses in a field construction. In clays, the stresses which drive failure may be much too small in relation to the undrained shear strength. These problems can be overcome by testing a $1/n$ scale model in a geotechnical centrifuge in a radial acceleration field of n times normal gravity, so that self weight stresses are the same at corresponding points in the model and in the field (Figure 11.3).

Most of the centrifuges used to test geotechnical models consist in essence of a rotating arm, balanced by means of a counterweight. The model itself is contained within a strongbox, which is placed on the end of the centrifuge arm. Centrifuges used in geotechnical model testing range from a few centimetres in radius to perhaps 10 m. The very small machines are used primarily to investigate mechanisms of collapse in small-scale models at a high centrifugal acceleration: there is no room to install instrumentation to measure soil movements, stresses or pore water pressures. The very large machines can be unwieldy and expensive to operate. Many geotechnical centrifuges therefore have radii in the range 1.5–4 m, with 2 m being a suitable radius for convenient and economical operation. A review of the historical development of

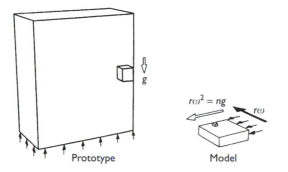

$$r\omega^2 = ng$$

Prototype   Model

Figure 11.3 Inertial (centrifugal acceleration) effects in a centrifuge model replace gravitational effects in a field construction. (Redrawn with permission from Schofield, 1980.)

centrifuge testing is given by Craig (1995), while a useful description of the principles and some applications of centrifuge modelling is given by Schofield (1980).

In centrifuge modelling, considerable thought must be given to the scaling relationships. In a $1:n$ scale centrifuge model, linear dimensions are reduced by a factor of n, while stresses and pore water pressures are the same at corresponding depths in the model (at ng) and in the full-scale $1g$ prototype. (The term prototype is used to describe the full-scale construction to which the model would correspond, without meaning to imply that such a structure would ever actually be built.) From these, it follows that areas and forces are reduced in the model by a factor of n^2, and volumes and bending moments by a factor of n^3. Consolidation times scale in proportion to the square of the drainage path length, so that they are reduced in the centrifuge by a factor of n^2. This is a consequence of the reduced scale, not of testing the model in a centrifugal acceleration field. It is also one of the major advantages of small-scale testing: at a scale of 1:100, a year's consolidation can be observed in less than an hour.

For true similarity in dynamic tests (in which accelerations and inertial forces are important), the frequency in the model must be n times greater than in the prototype. This means that for dynamic events, time in the model is reduced by the scale factor n. In centrifuge tests in which an earthquake is simulated, the time scale factor associated with consolidation events (such as non-equilibrium pore pressure dissipation) is therefore different from that for dynamic events. This problem can be overcome by using silicone oil, with a viscosity n times greater than the viscosity of water, as a pore fluid. The increased viscosity of the silicone oil slows down its flow velocity, so that the timescales for consolidation processes and dynamic events are the same. In quasi-static tests, in which loading and unloading take place relatively slowly (and which cover most types of geotechnical engineering construction in the majority of situations), the discrepancy in the dynamic timescale is irrelevant.

It is usual to use the same soil in the centrifuge model as in the prototype construction. This means that the ratio of the dimension of the structure L (e.g. the depth of an excavation or the width of a foundation) to the typical particle size D_{50} is n times smaller in the model than in the prototype. Usually, this will not have a significant effect, provided that the relative structure to particle size ratio L/D_{50} is still large (Davis and Auger, 1979).

Sometimes, however, the particle size may be large in comparison with the feature being tested. One example of this is a 1 m thick resistive layer within a landfill cap, which when tested at a scale of 1 : 40 would be only 25 mm thick. It is not yet clear whether the onset of rupture following subsidence of the landfill depends on the differential settlement relative to the thickness of the capping layer, or on the differential settlement relative to the individual particle size. If the second of these is in fact the case, it might be necessary to carry out centrifuge tests using a soil in which the particle size had been reduced by a factor of n. This could change soil parameters such as the consolidation coefficient quite considerably, with potentially significant implications as far as other aspects of the test were concerned. A number of tests at different scales could be carried out – a procedure known as **modelling of models** – in order to establish the maximum scale factor at which similar results will be obtained using the same material in the model as in the prototype.

Some of the main scaling relationships are given in Table 11.1. If there is ever any doubt as to the validity of the modelling procedure in any particular case, models should be tested at two or more different scales. If the scaled results are not the same, the implication is that something has been missed in the modelling process. Scaling considerations in general are discussed by Taylor (1995: pp. 19–33).

Centrifuge model tests have been carried out to investigate the behaviour of almost every conceivable geotechnical construction and process, including trench excavations, most kinds of retaining wall, reinforced soil, soil nails, tunnels, embankments, piled foundations, shallow foundations, landfill caps and liners, and pollution migration. Apparatus has been developed to construct embankments in stages, to drive piles into the soil, to load foundations to failure, to simulate or even carry out excavation processes, and to measure the properties of the soil while the model is spinning round in the centrifuge. Some idea of the scope of centrifuge modelling activity worldwide is given by the proceedings of the last few international conferences on the topic (Ko and McLean, 1991; Leung *et al.*, 1994; Kimura *et al.*, 1998; Phillips *et al.*, 2002), in the book edited by Taylor (1995), and in the collection of papers by Davies *et al.* (1998). A useful review of modelling contaminant transport is given by Culligan and Barry (1998).

Table 11.1 Scale factors for centrifuge modelling, assuming that the soil in the model has the same properties as the soil in the prototype

Quantity	Value in prototype	Value in model at a scale of 1 : n
Linear dimension	l_p	l_p/n
Area	A_p	A_p/n^2
Volume	V_p	V_p/n^3
Stress	σ_p	σ_p
Force	F_p	F_p/n^2
Moment	M_p	M_p/n^3
Displacement	δ_p	δ_p/n
Strain	ε_p	ε_p
Consolidation time	t_{cp}	t_{cp}/n^2
Frequency (dynamic tests)	f_p	nf_p
Time for dynamic events	t_{dp}	t_{dp}/n
Speed (dynamic tests)	v_p	v_p
Acceleration (dynamic tests)	a_p	a_p/n

Standard transducers are available to measure soil surface movements and pore water pressures. Structural loads, such as axial forces in props and bending moments in piles or retaining walls, are usually measured by means of specially designed load cells, or strain gauges arranged in an appropriate bridge network. Miniature video cameras can be used to record the progress of a centrifuge test, and the images can be analysed after the test to determine patterns of soil displacement from black marker beads embedded in the visible face of the model. More recently, techniques have been developed to measure soil movements directly (i.e. without the need for markers), using digital image analysis and particle image velocimetry (PIV) techniques (White *et al.*, 2003). The measurement of total stress within the soil mass, or total stress against a retaining wall, is difficult but has been attempted by some researchers (e.g. Page, 1996). A review of centrifuge model instrumentation is given by Phillips (1995).

Figure 11.4(a) shows a schematic cross-section of a centrifuge model of a propped embedded retaining wall in clay. The model is contained within a strongbox suitable for testing on a 1.8 m radius centrifuge, as shown in Figure 11.4(b). The strongbox is constructed from aluminium alloy, and incorporates a perspex viewing window in

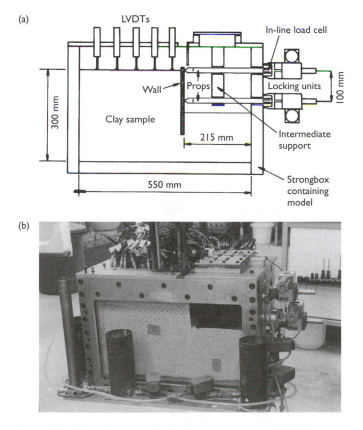

Figure 11.4 Centrifuge model of a twin-propped embedded retaining wall: (a) schematic cross-section. (Redrawn with permission from Powrie *et al.*, 1994.); (b) strongbox assembly. (From Richards, 1995, with permission.)

the front face so that the cross-sectional plane of the model may be observed during the test.

Clay models are usually consolidated to a known equilibrium state before the changes in stress comprising the actual test are imposed. The one possible exception to this is a study carried out to investigate mechanisms of undrained collapse, in which the centrifugal acceleration might be increased steadily to initiate the rapid collapse of a model in a clay sample having a uniform profile of undrained shear strength τ_u with depth. In such a case, a low permeability clay should be used to ensure that any changes in volume – and hence in undrained shear strength – which occur during the gradual increase in centrifugal acceleration are insignificant.

In the case of the model shown in Figure 11.4, the actual test is the excavation of the soil from in front of the retaining wall and the installation of props at two levels. This must be carried out or simulated without stopping the centrifuge, otherwise the known equilibrium stress state will be lost. The clay used to make the model shown in Figure 11.4 was prepared by one-dimensional consolidation in a large press. On removal from the press, a slot was cut in the clay sample to receive the model wall. The soil within the main excavation was also removed at this stage, and replaced by a rubber bag filled with a heavy fluid having the same unit weight as the soil. During reconsolidation, the vertical stress exerted by the fluid in the rubber bag on the soil below excavation level in front of the wall is the same as would have been exerted by the soil removed.

At the end of reconsolidation, the clay is in a state corresponding to an idealized field condition, in which the vertical effective stress increases with depth, and the pore water pressures are hydrostatic below the groundwater level set by the modeller. A valve-controlled wastepipe is used to drain the heavy fluid from the rubber bag, simulating the excavation of the soil from in front of the wall, at an appropriate stage following reconsolidation of the clay sample in the centrifuge. This is the most common technique for simulating excavation processes in the centrifuge.

As with the finite element analysis of Figure 11.1, the centrifuge model test shown in Figure 11.4 started with the wall in place. It has already been mentioned that the installation of a diaphragm wall may change the lateral stresses significantly from their initial *in situ* values. The exact effect of the installation of the wall depends on a number of factors, but a lateral earth pressure coefficient prior to excavation in front of the wall (K_i) of unity in the soil close to the wall perhaps represents a reasonable estimate (Tedd *et al.*, 1984; Powrie, 1985). If, as in the example shown in Figure 11.4, the heavy fluid is mixed to the same unit weight as the soil it replaced, the vertical and horizontal stresses in front of the wall above formation level are the same at any depth. This is consistent with a pre-excavation lateral earth pressure coefficient $K_i = 1$, representing the effect of diaphragm wall installation. Unfortunately, the pre-excavation lateral stresses on the wall below formation level are unknown.

The draining of fluid from a rubber bag to simulate the stress changes which result from excavation can be used to simulate pre-excavation earth pressure coefficients other than unity. For example, Lade *et al.* (1981) used paraffin oil with a density of $7.65\,kg/m^3$ to give a pre-excavation earth pressure coefficient of approximately $(1 - \sin\phi')$ in a granular material. At the other extreme, Powrie and Kantartzi (1993, 1996) describe the use of sodium chloride solution of density $1162\,kg/m^3$, contained in a rubber bag filled to a height h above the level of the soil surface, to impose a profile of lateral earth pressure coefficient with depth similar to the *in situ* conditions in an overconsolidated clay deposit in the field. (The fluid height h and the unit weight of the salt solution were chosen so that at the base of the excavation the vertical stresses inside and outside the rubber bag were equal.) Following reconsolidation, the salt solution was drained to the level of the soil surface and diluted, to simulate the stress changes caused by excavation of a diaphragm wall trench under bentonite slurry.

Other methods of simulating excavation in the centrifuge include the removal of rigid supports (Craig and Yildirim, 1976). Alternatively, the soil within the excavation could be contained within a flexible porous fabric bag, which could be winched clear at the appropriate stage of the centrifuge test using an electric motor, as proposed by Ko *et al.* (1982). The extent to which the latter method has actually been used successfully in practice is, however, unclear. Kimura *et al.* (1994) describe the development and operation of a mechanical scraper physically to excavate the soil from in front of a pre-installed retaining wall.

The evolution of centrifuge modelling techniques to investigate the behaviour of geotechnical structures such as embedded retaining walls has reflected the development of methods of construction in geotechnical engineering practice. For example, centrifuge model tests on embedded retaining walls carried out by Powrie (1986) demonstrated the efficiency of props at formation level in terms of minimizing wall movements for a given retained height to depth of embedment ratio. The main shortcoming of these tests (some of whose results were summarized by Powrie and Li (1991)) was that the propping system had to be installed prior to reconsolidation and excavation in the centrifuge.

The tests carried out by Richards (1995: Figure 11.4) to investigate the behaviour of *in situ* walls propped at both crest and formation level incorporated props with locking devices so that the props would not begin to take load or resist movement until required to do so by the modeller. (See also Richards and Powrie, 1998.) Walls propped at two levels are often used for underpasses at major interchanges on roads in urban areas.

Open excavations of all kinds are usually modelled by draining a heavy fluid from a rubber bag. The technique is also used for closed excavations such as tunnels, although in some cases the tunnelling process is replicated by reducing the air pressure (starting from the average *in situ* soil stress) within a sealed membrane, although it is now possible to excavate a tunnel using a miniature tunnelling machine as described by Nomoto *et al.* (1994).

It may be difficult to justify the costs involved in the development of sophisticated construction and excavation techniques unless some important aspect of soil behaviour is neglected or obliterated by the use of a simple method. Sometimes this will be the case. It is now recognized that, if realistic results are to be obtained from centrifuge model tests on driven piles, the piles must be installed in the centrifuge at the test acceleration (Craig, 1984). This has necessitated the development of in-flight pile driving hammers (e.g. de Nicola and Randolph, 1994), which are now used as a matter of routine.

The modelling method adopted in any given situation will depend on a number of factors. These include the nature of the problem being investigated, the purpose of the model, and the constraints of time and cost. The various modelling techniques complement rather than compete against each other. Each has a contribution to make, and it falls to the geotechnical engineer to draw on the totality of these contributions both to reach an understanding of the problem in hand and to develop an appropriate method of geotechnical engineering design.

11.3 *In situ* testing

The main problems associated with the laboratory testing of soils are disturbance during sampling and the difficulty of testing samples large enough to be representative of the soil in the field, where the effects of structure and fabric can be highly significant. In an attempt to overcome these problems, several methods of testing the soil *in situ* have been developed. Some of these are described in this section: a more detailed critical appraisal is given by Wroth (1984).

It is pointed out by Wood and Wroth (1977) that the mode of deformation imposed on the soil may differ widely between different forms of soil test. This will lead to discrepancies between the values of soil parameters (such as the undrained shear strength) measured using different techniques (e.g. a **pressuremeter** – section 11.3.3 – or a **shear vane** – section 11.3.4 – in the field, and a shearbox or a triaxial cell in the laboratory). These discrepancies are in addition to the effects of sample size and sampling disturbance, and must be borne in mind when attempting to compare data from different soil tests.

11.3.1 The Standard Penetration Test

The **Standard Penetration Test** (or **SPT** for short) is probably the oldest and simplest form of *in situ* soil test. It is carried out in boreholes during site investigation. A split barrel sampler attached to the end of a series of rods is driven into the soil below the bottom of a borehole to a depth of 150 mm, by means of a falling weight arrangement known as a **drop-hammer**. The number of hammer blows required to drive the sampler a further 300 mm is then recorded: this is the **SPT blowcount**, which is conventionally given the symbol N.

Although the test is described as 'standard', the energy actually delivered to the split barrel sampler depends on the design of the hammer arrangement, which varies from country to country according to local tradition and equipment. The energy delivered to the sampler may be as much as 85% of the potential energy of the hammer (i.e. the falling weight) at the top of its travel, or as little as 45%.

Skempton (1986) shows that many apparent inconsistencies in SPT data from around the world may be attributed to differences in this energy ratio. Following Seed *et al.* (1984), he recommends that an equivalent energy ratio of 60% should be adopted as an internationally recognized standard. This is consistent with current UK practice for rod lengths (which are approximately the same as borehole depths) in excess of 10 m and borehole diameters of up to 115 mm. Correction factors for other types of equipment are given by Skempton (1986). Correction factors for rod lengths less than 10 m and borehole diameters greater than 115 mm are given in Table 11.2.

The resistance of the soil to penetration would be expected to increase with relative density, soil strength and stiffness, and empirical correlations between the SPT blowcount and each of these parameters are widely used in practice. The SPT

Table 11.2 Correction factors to SPT blowcount for rod lengths <10 m and borehole diameters >115 mm (after Skempton, 1986, with permission)

		Correction factor
Rod length (m)	>10	1.0
	6–10	0.95
	4–6	0.85
	3–4	0.75
Borehole diameter (mm)	65–115	1.0
	150	1.05
	200	1.15

blowcount would also be expected to increase with the average effective stress p', which itself increases with depth and, because of the effect of the lateral stress, with overconsolidation ratio. Skempton (1986) presents data which illustrate that:

- the SPT blowcount N increases almost linearly with overburden pressure σ_v' at a constant density index I_D,
- at a constant overburden pressure, N increases roughly in proportion to I_D^2, so that, as pointed out by Meyerhof (1957),

$$N/(I_D^2) = a + b\sigma_v'$$

- at a given density index and overburden pressure, N is higher for sands with a larger mean grain size (D_{50}).

Stroud (1989) points out that it is reasonable to correlate the soil stiffness with the SPT blowcount directly, because both depend on the average effective stress p', and therefore increase with depth. However, in the case of the frictional soil strength ϕ' and the density index I_D of a granular soil, which would not generally be expected to change very significantly with depth, it is necessary to normalize the SPT blowcount to a reference vertical effective stress, which is conventionally taken as 100 kPa. This is to separate out increases in SPT blowcount due to increases in over-burden pressure alone. The normalized SPT blowcount is denoted by the symbol N_1, or $(N_1)_{60}$ when the energy ratio is 60%.

Normalization of SPT blowcounts is carried out by means of a correction factor C_N:

$$N_1 = C_N N \tag{11.4}$$

Skempton (1986) shows that, for normally consolidated sands, C_N varies between

$$C_N = 200/[100 + \sigma_v'(\text{kPa})] \tag{11.5a}$$

for fine sands of medium relative density, to

$$C_N = 300/[200 + \sigma_v'(\text{kPa})] \tag{11.5b}$$

for dense, coarse sands. For overconsolidated fine sands,

$$C_N = 170/[70 + \sigma_v'(\text{kPa})] \tag{11.5c}$$

Equation 11.5(b) gives numerically similar results to the correction factor suggested by Peck *et al.* (1974),

$$C_N = 0.77 \log_{10}[2000/\sigma_v'(\text{kPa})] \tag{11.5d}$$

The density index of sands may be estimated on the basis of the normalized SPT blowcount, $(N_1)_{60}$, according to Table 11.3.

Table 11.3 shows that, for $I_D > 0.35$,

$$(N_1)_{60}/I_D^2 \approx 60 \tag{11.6}$$

Stroud (1989) argues that the correlation between $(N_1)_{60}$ and I_D depends on ϕ_{crit}'. In this context, it should be noted that Table 11.3 and equation (11.6) are based on data from angular to sub-angular quartz sands, with $\phi_{crit}' \approx 33°$ to $34°$.

It is now well known that the stiffness of a soil is not constant, but reduces with increasing strain following the last significant change in direction of the stress path. Stroud (1989) takes this into acount, by presenting correlations between the effective stress secant Young's modulus E' divided by the SPT blowcount N_{60} (not normalized, because both stiffness and SPT blowcount increase with increasing overburden pressure), and the parameter q_{net}/q_{ult}. q_{net} is the net load applied to a foundation, and q_{ult} is the ultimate load or the bearing capacity (given by, e.g. σ_f' in equation (8.7) or σ_f in equation (8.9)). The expectation is that the values of Young's modulus derived from these correlations will be used primarily to estimate the settlement of foundations. The settlement of the foundation – and the deformation of the soil – increases with q_{net}/q_{ult}, which may therefore be regarded as an indicator of the shear strain.

Table 11.3 Correlation of density index with normalized SPT blowcount (Terzaghi and Peck, 1948, as modified by Skempton (1986, with permission))

I_D	Classification	$(N_1)_{60}$	$(N_1)_{60}/I_D^2$
	Very loose		
0.15		3	–
	Loose		
0.35		8	65
0.5	Medium	15	60
0.65		25	59
	Dense		
0.85		42	58
	Very dense		
1.00		58	58

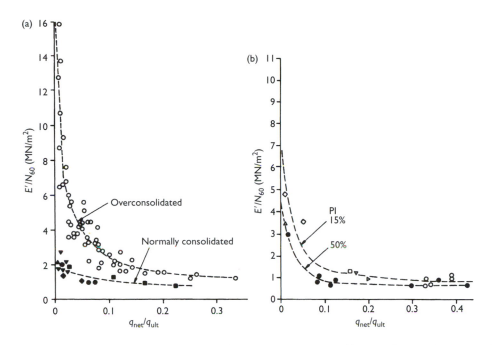

Figure 11.5 Empirical correlations between E'/N_{60} and q_{net}/q_{ult} for (a) normally consolidated and overconsolidated sands and gravels and (b) overconsolidated clays. (Redrawn with permission from Stroud, 1989.)

The correlations given by Stroud (1989) between E'/N_{60} and q_{net}/q_{ult} for normally consolidated and overconsolidated sands and gravels are reproduced in Figure 11.5(a). The corresponding relationships for overconsolidated clays with plasticity indices (PI) of 15% and 50% are shown in Figure 11.5(b). In each case, the actual data points on which the correlations are based are included, as an indication of the potential error.

Correlations between ϕ'_{peak} and $(N_1)_{60}$ at different overconsolidation ratios are also given by Stroud (1989). These are reproduced in Figure 11.6. Peck *et al.* (1974) take this a step further by giving correlations between the SPT blowcount and the bearing capacity factors N_q and N_γ (N_q and N_γ are defined in sections 8.2–8.4). These bearing capacity factors are calculated on the basis of the peak strength ϕ'_{peak}, and the Peck *et al.* correlation takes no account of differences in overconsolidation ratio. Terzaghi and Peck (1967) give relationships between the SPT blowcount and the allowable bearing pressure on a strip footing, in order to limit the maximum likely settlement to less than 25 mm.

The undrained shear strength τ_u of an overconsolidated clay may be related to the SPT blowcount using the simple expression

$$\tau_u = f_1 N_{60} \tag{11.7}$$

(Stroud, 1974). As τ_u would generally be expected to increase with depth, it is the corrected blowcount N_{60}, rather than the normalized corrected blowcount, $(N_1)_{60}$, which is used in the correlation. The parameter f_1 (which has units of kPa) depends

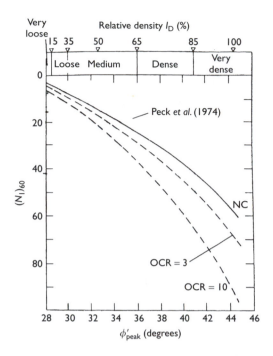

Figure 11.6 Empirical correlations between $(N_1)_{60}$, I_D and ϕ'_{peak} for quartz sands and gravels at overconsolidation ratios of 1 (*NC*), 3 and 10. (Redrawn with permission from Stroud, 1989.)

on the plasticity index (PI) of the clay. Data presented by Stroud and Butler (1975) suggest that $f_1 = 4.5$ kPa for plasticity indices in excess of 30%. As the plasticity index decreases from 30% to 15%, the parameter f_1 increases from 4.5 kPa to 6 kPa.

For weak rocks such as mudstone and marl, the relationship between E'/N_{60} and $q_{net}/q_{ult,u}$ (where $q_{ult,u}$ is the ultimate load in undrained conditions)[2] is similar in form to that shown in Figure 11.5 for sands and overconsolidated clays. For moderate degrees of loading ($0.2 \le [q_{net}/q_{ult,u}] \le 0.4$), $E'/N_{60} \approx 1$ MPa. The ultimate shear strength τ_u of a weak rock[3] may be estimated approximately using equation (11.7) with $f_1 = 5$ kPa, giving $\tau_u/N_{60} = 5$ kPa. For chalk, $E'/N_{60} \approx 5$ MPa at $0.2 \le (q_{net}/q_{ult,u}) \le 0.4$, and $\tau_u/N_{60} \approx 25$ kPa (Stroud, 1989).

11.3.2 The cone penetration test

In a **cone penetration test** (CPT), a cone at the end of a series of rods is pushed at a steady rate of 15–25 mm/s into the soil, and the resistance to penetration Q_c (a force in kN) is measured by means of a load cell just behind the cone. The force due to side friction immediately above the cone is also measured, using a sleeve (known as a **friction sleeve**) mounted on strain-gauged supports. As both loads are measured electrically, continuous readings are usually produced. Most cones have a diameter of 37.5 mm, an apex angle of 60°, and a projected area (perpendicular to the direction of penetration) of 1000 mm².

Other cone geometries are sometimes used, and the geometry of the cone will affect the results obtained. Some versions of the apparatus incorporate a pore water pressure transducer with its filter either on or just behind the cone: such a device is known as a **piezocone**. The pore water pressures measured during piezocone penetration can help to identify the soil type: generally, excess pore water pressures (i.e. pressures over and above the *in situ* equilibrium values) will be developed in clays, but not in sands. A typical piezocone tip, with the filter for pore water pressure measurement in the generally preferred position just behind the cone (known as the u_2 position), is shown schematically in Figure 11.7(a).

Considerable attention must be paid to the design and specification of the load cells and the seals and ensuring that the pore pressure measurement system is fully saturated if reliable piezocone test data are to be obtained. It is also necessary to correct the measured cone resistance and sleeve friction load to account for the fact that the pore water pressure u_2 acts on the shoulder at the back of the cone and on

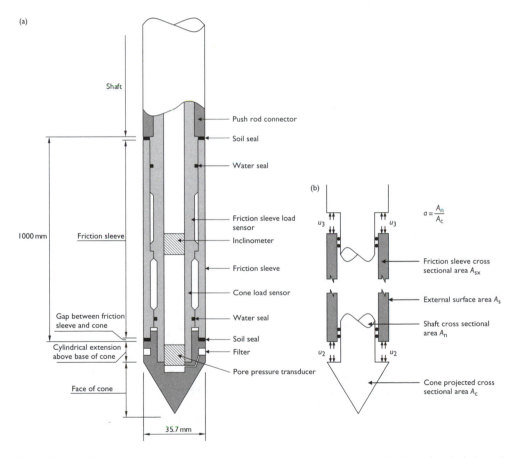

Figure 11.7 (a) Schematic cross section through a piezocone tip showing load cells and seals (adapted from Lunne *et al.*, 1997 and Zuidberg, 1988); (b) correction of measured loads for u_2 and u_3 pore water pressure effects (adapted from Lunne *et al.*, 1997).

the end of the friction sleeve, as illustrated in Figure 11.7(b). This is important mainly in soft, fine grained soils where the porewater pressures generated during driving can be large in comparison with the cone resistance q_c and the sleeve friction f_s. Further details of the quite complex preparation, calibration and correction procedures needed to ensure a successful cone penetration or piezocone test are given by Meigh (1987) and Lunne *et al.* (1997).

Raw cone penetration test data are normally expressed as the **cone resistance** q_c (calculated as the force, Q_c, divided by the projected area of the cone, A_c), and the **sleeve friction** f_s (calculated as the total force acting on the friction sleeve, F_s, divided by the surface area of the friction sleeve, A_s). Typical piezocone test data plotted as pore water pressure, f_s, q_c and **friction ratio** f_s/q_c against depth are shown in Figure 11.8(a).

Wroth (1984) argued that, to compensate for the effects of increasing overburden stresses with depth, cone penetration test data should be presented in terms of the **normalized cone resistance** Q_t, the **normalized friction ratio** F_r and the **pore pressure ratio** B_q, where

$$Q_t = \frac{q_t - \sigma_{vo}}{\sigma'_{vo}} \tag{11.8a}$$

$$F_r = \frac{f_s}{q_t - \sigma_{vo}} \tag{11.8b}$$

and

$$B_q = \frac{\Delta u}{q_t - \sigma_{vo}} \tag{11.8c}$$

q_t is calculated from the measured cone resistance, q_c, by correcting for the fact that the pore water pressure behind the cone, u_2, acts on the back of the cone between the cone and the friction sleeve: $q_t = q_c + u_2 \cdot (1-a)$, where a = the area of the shaft, A_n, divided by the projected area of the cone, A_c (Figure 11.8b). σ_{vo} is the vertical total stress, u_o the *in situ* pore water pressure and σ'_{vo} the effective stress (= $\sigma_{vo} - u_o$) at the level of the cone. f_s is the measured sleeve friction: ideally the sleeve friction corrected for pore pressure effects, $f_t = f_s - (u_2 - u_3) \cdot (A_{sx}/A_s)$, should be used in place of f_s, but u_3 is rarely measured (A_{sx} is the cross sectional area of the friction sleeve, which is here assumed to be constant – Figure 11.7(b)). Δu is the excess pore water pressure during driving, $\Delta u = u_2 - u_o$.

Different soil types may be identified by means of empirical correlations between the three normalized parameters Q_t, F_r and B_q suggested by Robertson (1990) and shown in Figures 11.8(b) and (c). For data from a basic cone penetration test in which pore water pressures have not been measured, Figure 11.8(b) alone may be used.

The main advantages of the cone penetration or piezocone test are that the disturbance to the soil is minimal, and that a continuous record of the soil profile can be produced. However, the need to use an empirical correlation to determine the soil type is a potential weakness. Also, thin layers of sand (less than 100 mm thick) in a clay stratum, and thin layers of clay (less than 150–200 mm thick) in a sand stratum, might well remain undetected.

For sands, a correlation between q_c and ϕ'_{peak} may be obtained by means of bearing capacity theory. This requires certain assumptions to be made, for example concerning

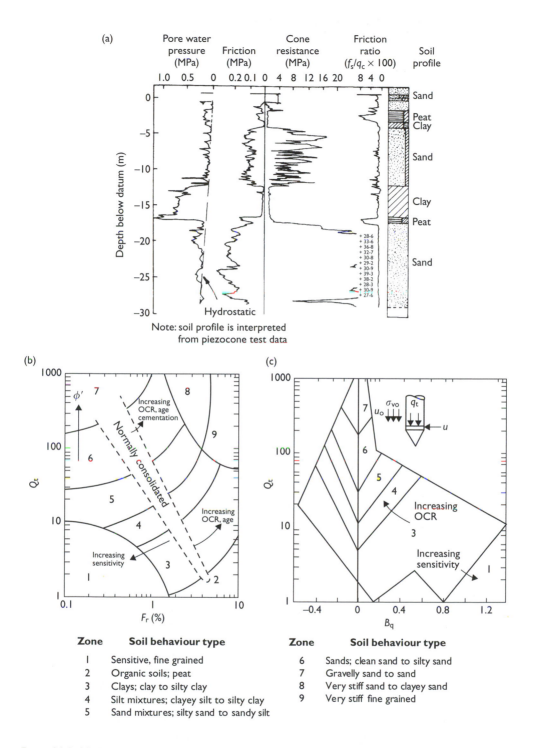

Figure 11.8 (a) Typical piezocone test data (Redrawn with permission from Zuidberg *et al.*, 1982, reprinted from Verruijt *et al.*, 1982.); (b) and (c) identification of soil types from piezocone test data. (Redrawn with permission from Lunne *et al.*, 1997.)

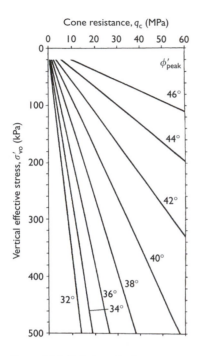

Figure 11.9 Relationship between q_c and ϕ'_{peak} for uncemented, normally consolidated quartz sand. (Durgunoglu and Mitchell, 1975, redrawn with permission from Robertson and Campanella, 1983.)

the degree of friction between the cone and the soil, and the mode of soil deformation. A correlation between q_c and ϕ'_{peak} for clean, relatively uniform, uncemented, unaged sands presented by Durgunoglu and Mitchell (1975) is reproduced in Figure 11.9.

A correlation – again for sands and based on somewhat limited experimental data – between the cone resistance q_c and the effective stress secant Young's modulus E' at 25% and 50% of the deviator stress at failure (corresponding to equivalent degrees of loading $\{q/q_{ult}\} = 0.25$ and 0.5: see section 11.3.1) is given in Figure 11.10.

In clays, cone penetration is likely to be undrained and the cone resistance q_c may be expressed in the form

$$q_c = \sigma_v + N_k \cdot \tau_u \tag{11.9a}$$

where σ_v is the vertical total stress at the depth of the cone, τ_u is the undrained shear strength of the soil, and N_k is termed the **cone factor**. N_k is analogous to the bearing capacity factor N_c, introduced in section 8.2. Rearranging equation (11.9a),

$$\tau_u = (q_c - \sigma_v)/N_k \tag{11.9b}$$

N_k depends on both the geometry of the cone and the rate of penetration. A further problem of interpretation arises, because the undrained shear strength corresponding to the mode of deformation imposed in the cone penetration test may not be the same as the undrained shear strength measured in some other test (e.g. shear box, triaxial or

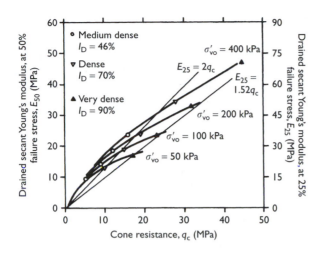

Figure 11.10 Effective stress secant Young's modulus as a function of cone resistance q_c, at 25% and 50% of the deviator stress at failure ($q_{net}/q_{ult} = 0.25$ and $q_{net}/q_{ult} = 0.5$), at different vertical effective stresses, for uncemented, normally consolidated quartz sands. (Redrawn with permission from Robertson and Campanella, 1983; based on data from Baldi *et al.*, 1981.)

Table 11.4 Typical values of cone factor N_k for marine and glacial clays (Reproduced with permission from Meigh, 1987)

	Average N_k	Range of N_k
Stiff, fissured marine clays (e.g. London Clay)	27	24–30
Glacial clays	18	14–22

shear vane), which might be carried out as a calibration. If the calibration test is carried out in the laboratory, the effects of sample disturbance and the limited sample size may also be significant. Bearing these difficulties in mind, Meigh (1987) gives the ranges of values for N_k quoted in Table 11.4, for a 60° cone of diameter 37.5 mm, tested at a rate of penetration of 15–25 mm/s, with τ_u determined from plate loading tests.

Other correlations that may be used to quantify engineering parameters such as soil strength and stiffness or to estimate the bearing capacity and settlement of foundations directly from cone penetration test data are given by Lunne *et al.* (1997).

11.3.3 Pressuremeter tests

(a) General description

The pressuremeter is a cylindrical device designed to apply a uniform radial pressure to the sides of a borehole in which it is placed. There are two different basic types:

- the **Menard pressuremeter** (MPM), which is lowered into a pre-formed borehole

- the **self-boring pressuremeter** (SBP), which forms its own borehole and thus causes much less disturbance to the soil prior to testing.

In both cases, the pressuremeter test involves the application of known stresses to the soil and the measurement of the resulting soil deformation. The interpretation of pressuremeter test data in engineering terms does not, therefore, rely on empirical correlations. In this respect, the pressuremeter test is in a different class from the standard penetration and cone penetration tests.

A diagrammatic representation of the essential features of a pressuremeter is given in Figure 11.11.

The sides of the borehole are loaded by pressurizing a fluid contained within a flexible rubber membrane. The outside of the rubber membrane is usually protected by steel strips. The expansion of the cavity is determined either by measuring the volume of fluid needed to pressurize the membrane, and/or by measuring the movement of the soil at the cavity wall directly using either feeler arms or displacement transducers (lvdts). Three feeler arms or lvdts, located in the same horizontal plane at a spacing of 120°, are usually used.

Generally, pressuremeters are designed for maximum inflation pressures in the ranges 2.5–10 MPa in soils and 10–20 MPa in very stiff soils and weak rocks. The interpretation of pressuremeter test data is based on the analysis of an expanding cylindrical cavity, with deformation in the horizontal plane only. The length of the expanding portion of a pressuremeter should therefore be at least six times the diameter to avoid significant end effects.

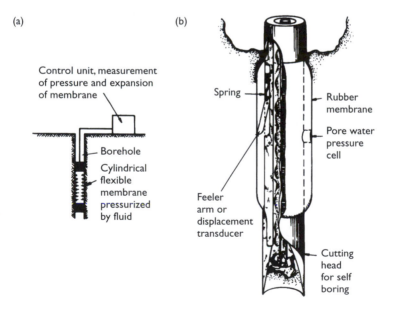

Figure 11.11 Schematic diagram of pressuremeter apparatus: (a) general arrangement. (Redrawn with permission from Mair and Wood, 1987.); (b) self-boring pressuremeter in test mode. (Redrawn with permission from Hughes *et al.*, 1977.)

Corrections must be made to the measured pressure, volume change and cavity deformation to account for the compressibility of the fluid and pipework, differences in elevation between the instrument and the pressure transducer, and the stiffness of the membrane. Spurious volume changes, due to the expansion of pipework and the compression of the fluid, are likely to be most significant in stiff soils. Pressure loss (where the pressure actually applied to the soil is less than the pressure inside the membrane, due to the stiffness of the membrane), is likely to be most significant in soft soils. The calibration procedures that must be carried out to quantify these effects are described by Mair and Wood (1987).

The corrected data from a pressuremeter test are plotted as a graph of cavity pressure p against the increase in cavity volume ΔV or the cavity strain ε_c, as shown in Figure 11.12. The cavity strain ε_c is defined as the outward movement of the cavity wall y_c divided by the original cavity radius (i.e. the radius at the start of the pressuremeter test) ρ_0. It is shown in section 11.3.3(b) that for small strains this is approximately equal to half the proportional increase in cavity volume, that is, $\varepsilon_c = y_c/\rho_0 \approx 0.5 \Delta V/V_0$, where V_0 is the initial volume of the cavity.

The main difference between a Menard pressuremeter and a self-boring pressuremeter is that the Menard pressuremeter is inserted into a pre-formed borehole. Some disturbance is therefore inevitable, and at the start of a Menard pressuremeter test the instrument will probably not be in contact with the sides of the borehole. This difference is reflected in the data from the early stages of the test, before the Menard pressuremeter makes contact with the borehole wall.

In a Menard pressuremeter test (Figure 11.12(a)), the stress–strain curve should steepen sharply at a point such as A, as the device comes into contact with the

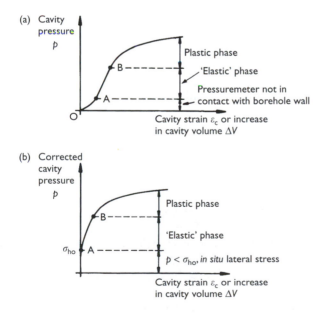

Figure 11.12 Corrected pressuremeter test data plotted as cavity pressure vs. cavity strain or volume change: (a) Menard pressuremeter; (b) self-boring pressuremeter.

pre-formed borehole wall and compresses any softened soil. Unless the soil is a normally consolidated clay, the relationship between the increase in pressure and the increase in cavity strain will then be approximately linear until a point B is reached. At B, the soil at the borehole wall starts to deform plastically. As the cavity pressure is increased further, the plastic zone extends further into the surrounding soil, until eventually some limiting pressure p_L is attained.

In a high-quality self-boring pressuremeter test, the corrected pressure–volume change curve should start with the pressuremeter in contact with the sides of the borehole. No softening or disturbance of the surrounding ground should be apparent. Ideally, there should be no increase in cavity volume until the cavity pressure exceeds the *in situ* lateral total stress in the ground, σ_{ho}. The *in situ* horizontal total stress may therefore be identified from the point A in Figure 11.12(b), at which the cavity volume starts to increase significantly with increasing pressure. In a self-boring pressuremeter which detects increases in cavity volume by means of displacement transducers contained within the device, each transducer may give a different 'lift-off' pressure. (The 'lift-off' pressure is the cavity pressure at which the displacement transducer or feeler arm starts to indicate significant movement, and would be expected in a self-boring pressuremeter test to be equal to the *in situ* lateral stress.) The estimation of the *in situ* horizontal total stress from the test data therefore requires some care and experience.

After the *in situ* lateral effective stress has been exceeded, the cavity volume should increase approximately linearly with the applied pressure, until the soil at the borehole wall starts to deform plastically at B. In reality, the 'linear' portion AB may be very short or almost non-existent (Figure 11.15. The same is true for the Menard pressuremeter test.). As the cavity pressure is increased further, the radius of the plastic zone increases until eventually a limiting pressure p_L is reached.

It is not straightforward to estimate the *in situ* lateral total stress from Menard pressuremeter test data, owing to the disturbance and softening of the soil at the borehole wall. Even if the point A in Figure 11.12(a) corresponds to the device making contact with intact soil at the side of the borehole, it cannot be taken as an indication of the *in situ* lateral stress. This is because the soil at A is being loaded not from its *in situ* condition, but from an unloaded state following pressuremeter installation. Mair and Wood (1987) describe an iterative procedure proposed by Marsland and Randolph (1977), which can be used to estimate the *in situ* lateral total stress from Menard pressuremeter test results in high-plasticity overconsolidated clays such as London Clay, which behave approximately elastically during initial loading.

In addition to the *in situ* horizontal total stress, the graph of cavity pressure against cavity strain may be used to estimate the shear modulus of the soil, the undrained shear strength of a clay, and the peak strength ϕ'_{peak} and the angle of dilation ψ of a sand. By incorporating pore water pressure transducers into the device, it is also possible to measure the horizontal consolidation coefficient c_h (which governs consolidation due to horizontal strain with horizontal drainage) and the angle of friction ϕ' for clays. In the present discussion, we will focus on the most common applications, which are the determination of the *in situ* horizontal total stress and the shear modulus, and the undrained shear strength of clays. Details of the procedures used to determine the other parameters are given by Mair and Wood (1987).

(b) Stress analysis

The interpretation of pressuremeter test data is based on the analysis of an expanding cylindrical cavity in an infinite body of soil. Deformation of the soil is assumed to take place in the horizontal plane (Figure 11.13(a)) while the vertical total stress σ_z remains constant. (The same results are obtained if it is initially assumed that the vertical strain ε_z, rather than the vertical total stress increment $\Delta\sigma_z$, is zero.)

If the cavity pressure is increased by an amount Δp, the cavity radius increases by an amount y_c from its initial value ρ (i.e. its value before the cavity pressure increment Δp is applied). In the soil outside the cavity, a general cylindrical 'shell' of initial radius r is pushed outward to a new radius $r+y$, as shown in cross-section in Figure 11.13(b). Similarly, a cylindrical 'shell' in the surrounding soil slightly further away from the cavity, of initial radius $r + dr$, is pushed out to a new radius $(r + dr) + (y + dy)$.

The radial strain increment $\Delta\varepsilon_r$ at radius r is defined as

$\Delta\varepsilon_r =$ change in radial distance between shells

$\div$ initial radial distance between shells

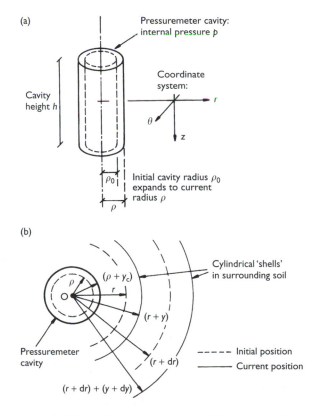

Figure 11.13 (a) Idealized geometry of deformation during pressuremeter test and (b) cross-section showing radial displacement of cylindrical shells of soil of different initial radii.

From Figure 11.13(b),

$$\Delta\varepsilon_r = -[(dr + dy) - dr] \div [(r + dr) - r] = -dy/dr \tag{11.10}$$

where the negative sign indicates that, for the sense of y shown in Figure 11.13(b), the radial strain increment is tensile.

The **circumferential** or **tangential** strain increment $\Delta\varepsilon_\theta$ at radius r is defined as

$$\Delta\varepsilon_\theta = \text{change in circumference} \div \text{initial circumference}$$

From Figure 11.13(b),

$$\Delta\varepsilon_\theta = -[2\pi(r + y) - 2\pi r] \div [2\pi r] = -y/r \tag{11.11}$$

where the negative sign indicates that, for the sense of y shown in Figure 11.13(b), the circumferential strain is tensile.

If the soil can be idealized as a uniform, isotropic, linear elastic material that obeys Hooke's law, equations (6.1) can be written in terms of the principal total stress and strain increments in the radial (r), circumferential (θ) and vertical (z) directions:

$$\Delta\varepsilon_r = (1/E_u)(\Delta\sigma_r - \nu_u\Delta\sigma_\theta - \nu_u\Delta\sigma_z) \tag{11.12a}$$

$$\Delta\varepsilon_\theta = (1/E_u)(\Delta\sigma_\theta - \nu_u\Delta\sigma_r - \nu_u\Delta\sigma_z) \tag{11.12b}$$

$$\Delta\varepsilon_z = (1/E_u)(\Delta\sigma_z - \nu_u\Delta\sigma_r - \nu_u\Delta\sigma_\theta) \tag{11.12c}$$

But $\Delta\sigma_z = 0$ (because σ_z is constant), so that

$$\Delta\varepsilon_r = (1/E_u)(\Delta\sigma_r - \nu_u\Delta\sigma_\theta) \tag{11.13a}$$

$$\Delta\varepsilon_\theta = (1/E_u)(\Delta\sigma_\theta - \nu_u\Delta\sigma_r) \tag{11.13b}$$

$$\Delta\varepsilon_z = (1/E_u)(-\nu_u\Delta\sigma_r - \nu_u\Delta\sigma_\theta) \tag{11.13c}$$

Adding equation (11.13a) to ν_u times equation (11.13b) to eliminate $\Delta\sigma_\theta$,

$$\Delta\varepsilon_r + \nu_u\Delta\varepsilon_\theta = [(1 - \nu_u^2)/E_u]\Delta\sigma_r$$

or

$$\Delta\sigma_r = [E_u/(1 - \nu_u^2)][\Delta\varepsilon_r + \nu_u\Delta\varepsilon_\theta] \tag{11.14a}$$

Similarly, adding equation (11.13b) to ν_u times equation (11.13a) to eliminate $\Delta\sigma_r$,

$$\Delta\sigma_\theta = [E_u/(1 - \nu_u^2)][\Delta\varepsilon_\theta + \nu_u\Delta\varepsilon_r] \tag{11.14b}$$

We have now derived the radial and circumferential stress increments $\Delta\sigma_r$ and $\Delta\sigma_\theta$, in terms of the undrained (total stress) elastic parameters E_u (Young's modulus) and ν_u (Poisson's ratio), and the radial and circumferential strain increments $\Delta\varepsilon_r$ and $\Delta\varepsilon_\theta$.

The equation of equilibrium for an annular segment is the same as it is in the analysis of a tunnel (section 10.11), except that gravity now acts out of (perpendicular to) the plane under consideration rather than in-plane, as in the case of a horizontal tunnel. Deleting the unit weight term from equation (10.17), and writing equation (10.17) in terms of stress increments $\Delta\sigma_r$ and $\Delta\sigma_\theta$ rather than stresses σ_r and σ_θ, the equation of equilibrium becomes

$$r(\mathrm{d}\Delta\sigma_r/\mathrm{d}r) = (\Delta\sigma_\theta - \Delta\sigma_r) \tag{11.15}$$

Substituting from equation (11.10) for $\Delta\varepsilon_r$ and from equation (11.11) for $\Delta\varepsilon_\theta$ into equations (11.14),

$$\Delta\sigma_r = [E_\mathrm{u}/(1-v_\mathrm{u}^2)][-(\mathrm{d}y/\mathrm{d}r) - (v_\mathrm{u}y/r)] \tag{11.16a}$$

and

$$\Delta\sigma_\theta = [E_\mathrm{u}/(1-v_\mathrm{u}^2)][-(y/r) - (v_\mathrm{u}\mathrm{d}y/\mathrm{d}r)] \tag{11.16b}$$

Substituting equations (11.16) into equation (11.15), and noting that the expression $[E_\mathrm{u}/(1-v_\mathrm{u}^2)]$ appears on both sides and therefore cancels out:

$$r\frac{\mathrm{d}[(-\mathrm{d}y/\mathrm{d}r) - v_\mathrm{u}y/r]}{\mathrm{d}r} = \left(\frac{-y}{r} - v_\mathrm{u}\frac{\mathrm{d}y}{\mathrm{d}r}\right) - \left(\frac{-\mathrm{d}y}{\mathrm{d}r} - v_\mathrm{u}\frac{y}{r}\right)$$

$$\Rightarrow \left(-r\frac{\mathrm{d}^2y}{\mathrm{d}r^2} - v_\mathrm{u}\frac{\mathrm{d}y}{\mathrm{d}r} + v_\mathrm{u}\frac{y}{r}\right) = \left(-\frac{y}{r} - v_\mathrm{u}\frac{\mathrm{d}y}{\mathrm{d}r} + \frac{\mathrm{d}y}{\mathrm{d}r} + v_\mathrm{u}\frac{y}{r}\right)$$

$$\Rightarrow \frac{\mathrm{d}^2y}{\mathrm{d}r^2} + \frac{1}{r}\frac{\mathrm{d}y}{\mathrm{d}r} - \frac{y}{r^2} = 0 \tag{11.17}$$

Equation (11.17) is a differential equation in terms of the single variable y. Its general solution, which may be verified by substitution, is

$$y = A/r + Br$$

where A and B are constants whose values depend on the boundary conditions. As y must approach zero with increasing radial distance from the cavity, $B = 0$ and

$$y = A/r \tag{11.18}$$

In the pressuremeter test, the value of the constant A can be determined because the outward radial movement of the cavity wall y_c is measured or can be deduced from the cavity volume change,

$$A = y_\mathrm{c}\rho \tag{11.19}$$

where ρ is the current radius of the pressuremeter cavity.

Alternatively, defining the **cavity strain** as $\varepsilon_\mathrm{c} = y_\mathrm{c}/\rho_0$ (where ρ_0 is the cavity radius at the start of the pressuremeter test),

$$A = \varepsilon_\mathrm{c}\rho\rho_0 \tag{11.20}$$

Substitution of equation (11.20) into equations (11.10) and (11.11) gives the following expressions for the radial and circumferential strains (measured from zero at the start of the pressuremeter test, so that the 'strain increment' notation can be dropped):

$$\varepsilon_r = -\mathrm{d}y/\mathrm{d}r = A/r^2 = \varepsilon_c \rho \rho_0/r^2 \tag{11.21a}$$

$$\varepsilon_\theta = -y/r = -A/r^2 = -\varepsilon_c \rho \rho_0/r^2 = -\varepsilon_r \tag{11.21b}$$

Substitution of equations (11.21) into equations (11.14) gives the following expressions for radial and circumferential total stress increments:

$$\Delta\sigma_r = [E_u/(1 + \nu_u)][\varepsilon_c \rho \rho_0/r^2] \tag{11.22a}$$

$$\Delta\sigma_\theta = -[E_u/(1 + \nu_u)][\varepsilon_c \rho \rho_0/r^2] = -\Delta\sigma_r \tag{11.22b}$$

From equation (6.6), $E_u/(1 + \nu_u) = 2G$, so that

$$\Delta\sigma_r = 2G\varepsilon_c \rho \rho_0/r^2 = 2G\varepsilon_r \tag{11.23a}$$

$$\Delta\sigma_\theta = -2G\varepsilon_c \rho \rho_0/r^2 = 2G\varepsilon_\theta \tag{11.23b}$$

This result is remarkable in two respects.

- Substitution of equations (11.22) into equation (11.12c) with $\Delta\sigma_z = 0$ shows that $\varepsilon_z = 0$, so that all deformation occurs in the horizontal plane. The vertical strain is zero, while the vertical stress remains constant.
- During cavity expansion, the surrounding soil deforms at constant volume. This is because the radial and circumferential strains are equal and opposite and the vertical strain is zero. Furthermore, there is no change in average total stress p, because the vertical total stress remains constant while the increase in radial total stress $\Delta\sigma_r$ is exactly countered by the decrease in circumferential total stress $\Delta\sigma_\theta$. This means that, if the soil behaviour is approximately isotropic and elastic, so that deformation at constant volume takes place at constant average effective stress p', there is no change in pore water pressure. Cavity expansion is entirely a shearing process, rather than the compression process which it might at first sight appear.

At the cavity wall, the increase in radial total stress $\Delta\sigma_r$ is equal to the increase in cavity pressure Δp. Substituting $r = \rho$, the current cavity radius, into equation (11.23a),

$$\Delta p = 2G\varepsilon_c \rho \rho_0/\rho^2 = 2G\varepsilon_c \rho_0/\rho$$

so that the graph of cavity pressure p against cavity strain ε_c will have slope $(\mathrm{d}p/\mathrm{d}\varepsilon_c) = 2G(\rho_0/\rho)$, or

$$G = (1/2)(\rho/\rho_0)(\mathrm{d}p/\mathrm{d}\varepsilon_c) \tag{11.24}$$

At the start of a test, the cavity radius ρ is equal to the initial cavity radius ρ_0, giving $\rho/\rho_0 = 1$ and $G = (1/2)(\mathrm{d}p/\mathrm{d}\varepsilon_c)$. It is usual, however, to measure the shear modulus during an unload/reload cycle (e.g. Figure 11.15), in order to ensure that all of the soil is deforming approximately elastically. In this case, the factor (ρ/ρ_0) will be significant if the cavity strain at the start of the unload/reload cycle is comparatively large. A second reason for determining the shear modulus from an unload/reload cycle is to avoid the effects of the disturbance caused by pressuremeter installation. Some installation disturbance is almost inevitable, even with a self-boring pressuremeter.

As an alternative to the cavity strain ε_c, pressuremeter test data may be presented in terms of the cavity volume change, ΔV. As the cavity increases in radius from ρ to $\rho + \Delta y_c$, the volume increases from $\pi\rho^2 h$ to $\pi(\rho + \Delta y_c)^2 h$, where h is the height of the cavity. Thus the proportional increase in volume (calculated with reference to the current cavity volume) is

$$\Delta V/V = [\pi(\rho + \Delta y_c)^2 h - \pi\rho^2 h]/[\pi\rho^2 h]$$
$$= [2\rho\Delta y_c + \Delta y_c^2]/\rho^2$$

If $\Delta y_c \ll \rho$, the term in Δy_c^2 can be neglected and

$$\Delta V/V \approx 2\Delta y_c/\rho = 2\Delta\varepsilon_c \times \rho_0/\rho$$

where $\Delta\varepsilon_c$ is the increase in cavity strain.

In the limit as $\Delta\varepsilon_c \to d\varepsilon_c$ and $\Delta V \to dV$,

$$d\varepsilon_c = (1/2) \times \rho/\rho_0 \times dV/V \tag{11.25}$$

and, substituting for $d\varepsilon_c$ into equation (11.24),

$$G = V(dp/dV) \tag{11.26}$$

In determining the shear modulus from an unload/reload cycle, the magnitude of the stress cycle must be small enough to ensure that the soil behaviour remains approximately elastic, and that plastic yielding in unloading does not commence. Consideration of the Mohr circle of total stress (Figure 11.14) shows that, starting from an *in situ* state of $\sigma_r = \sigma_\theta = \sigma_{ho}$, the shear stress in the wall of the cavity at a cavity pressure of $\sigma_{ho} + \Delta p$ is given by

$$\tau_c = \Delta\sigma_r = \Delta p \tag{11.27}$$

An ideal clay soil will start to deform plastically when the cavity shear stress τ_c reaches the undrained shear strength of the material τ_u,

$$\tau_c = \tau_u \tag{11.28}$$

Plastic behaviour will therefore start to occur at a cavity pressure of p_p, where

$$p_p = \sigma_{ho} + \tau_u \tag{11.29}$$

After this point, the elastic solution for stresses and strains is not valid. As the cavity pressure is increased, the plastic zone spreads outward into the surrounding soil.

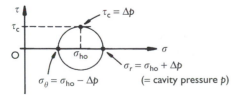

Figure 11.14 Mohr circle of total stress for the soil adjacent to the pressuremeter cavity.

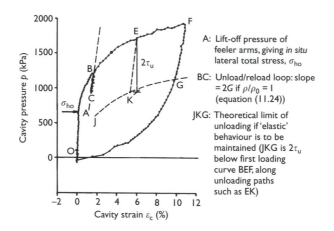

Figure 11.15 Interpretation of self-boring pressuremeter test results in Bartoon Clay, Zeebrugge, Belgium. (Redrawn with permission from Wroth, 1984.)

If the cavity pressure has been increased sufficiently to cause failure of the soil in the borehole wall (i.e. to a value $p = p_p = \sigma_{ho} + \tau_u$), the behaviour of the soil on unloading will be elastic until the cavity pressure has been reduced to such an extent that the shear stress at the borehole wall is equal to τ_u in the opposite sense, that is, $p = \sigma_{ho} - \tau_u$. Thus the reduction in cavity pressure during an unload/reload cycle should not exceed $2\tau_u$, if the behaviour of the soil is to remain approximately elastic. This limit to the magnitude of an unload/reload cycle carried out to measure the shear modulus G is indicated in Figure 11.15, which illustrates the interpretation of the data obtained from a well-executed pressuremeter test.

The undrained shear strength of a clay can be estimated from the relationship between the cavity pressure and the cavity volume after first yield (i.e. at cavity pressures $p > \sigma_{ho} + \tau_u$). At a general stage of the pressuremeter test with $p > \sigma_{ho} + \tau_u$, an annular zone of soil of external radius r_p around the cavity is at failure, with

$$\sigma_r - \sigma_\theta = 2\tau_u \tag{11.30}$$

(Figure 11.16. Note that $\sigma_r > \sigma_\theta$ because the cavity is expanding into the surrounding soil. This is in contrast to the circular tunnel analysed in section 10.11, because in the case of the tunnel the surrounding soil collapses into the cavity.)

Substituting equation (11.30) into equation (11.15), the condition of equilibrium in the plastic zone becomes

$$r(d\sigma_r/dr) = (\sigma_\theta - \sigma_r) = -2\tau_u \tag{11.31}$$

which may be rearranged and integrated between limits of $\sigma_r = p$ (the current cavity pressure) at the cavity wall, $r = \rho$, and $\sigma_r = \sigma_{ho} + \tau_u$ (which is just sufficient to cause plastic behaviour) at the outer limit of the plastic zone, which is at a radius $r = r_p$:

$$\int_p^{(\sigma_{ho}+\tau_u)} d\sigma_r = \int_p^{r_p} \frac{-2\tau_u}{r}\, dr$$

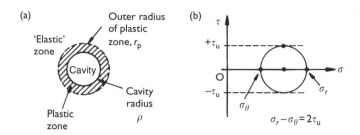

Figure 11.16 (a) Plastic and elastic zones around the pressuremeter at a general stage of the test ($p > \sigma_{ho} + \tau_u$) and (b) Mohr circle of total stress in plastic zone.

giving

$$p = (\sigma_{ho} + \tau_u) + 2\tau_u[\ln(r_p/\rho)] \tag{11.32}$$

We must now find r_p, the outer radius of the plastic region. Inside the plastic zone, the relationship between radius r and displacement y given by equations (11.18) and (11.20) does not apply, because equation (11.18) was derived on the assumption that the material behaved elastically, obeying Hooke's law. If, however, the plastic zone deforms at constant volume, we can say that the volume of soil which now occupies an annular zone of inner radius ρ and outer radius r_p originally occupied an annular zone of inner radius $(\rho - y_c)$ and outer radius $(r_p - y_{rp})$. (r_p is the current radius of the edge of the plastic zone and y_{rp} is its outward radial displacement from its original position. ρ is the current radius of the pressuremeter cavity, and y_c is the increase in radius from the original position.) Equating the two volumes, for a cavity of height h:

$$[\pi r_p^2 - \pi \rho^2]h = [\pi (r_p - y_{rp})^2 - \pi (\rho - y_c)^2]h$$

Dividing through by πh, multiplying out the terms in brackets, and assuming that $y_{rp} \ll r_p$ and $y_c \ll \rho$ gives

$$r_p y_{rp} = \rho y_c = \rho \rho_0 \varepsilon_c \tag{11.33}$$

where $\varepsilon_c = y_c/\rho_0$. Just outside the plastic zone, the elastic relationships still apply. In particular,

$$\varepsilon_r = -\varepsilon_\theta = y_{rp}/r_p \quad \text{(from equations 11.21)}$$

and

$$\Delta \sigma_r = 2G\varepsilon_r \quad \text{(from equation 11.23a)}$$

so that, with $\Delta \sigma_r = \tau_u$ at $r = r_p$,

$$\varepsilon_r = -\varepsilon_\theta = y_{rp}/r_p = \tau_u/2G \tag{11.34}$$

Combining equations (11.33) and (11.34) to eliminate y_{rp},

$$y_{rp} = \tau_u r_p/2G = \rho \rho_0 \varepsilon_c/r_p$$

or

$$r_p = \sqrt{\{(2G\rho\rho_0\varepsilon_c)/\tau_u\}} \tag{11.35}$$

Substituting equation (11.35) into equation (11.32), and recalling that $\ln\{\sqrt{x}\} = 0.5 \times \ln\{x\}$,

$$p = (\sigma_{ho} + \tau_u) + \tau_u\{\ln[(2G\rho_0\varepsilon_c)/(\rho\tau_u)]\} \tag{11.36a}$$

Using equation (11.25), equation (11.36a) may be rewritten in terms of $\Delta V/V$ rather than ε_c,

$$p = (\sigma_{ho} + \tau_u) + \tau_u\{\ln[(G/\tau_u) \times (\Delta V/V)]\}$$

or

$$p = \sigma_{ho} + \tau_u \times \{1 + \ln(G/\tau_u) + \ln(\Delta V/V)\} \tag{11.36b}$$

The significance of equation (11.36b) is that, during the plastic phase of the pressuremeter test, a graph of cavity pressure p against the natural logarithm of the proportional cavity volume change $\Delta V/V$ should have slope τ_u. Also, if the line is projected to the point $\ln(\Delta V/V) = 0$ (which corresponds to $\Delta V/V = 1$), the corresponding value of p is p_L, where

$$p_L = \sigma_{ho} + \tau_u \times \{1 + \ln(G/\tau_u)\} \tag{11.37}$$

which gives an additional check on the values of G and τ_u determined using equations (11.24) (or 11.26) and (11.36b).

The above analysis of the pressuremeter test is basically that given by Gibson and Anderson (1961). They also present a similar analysis of a pressuremeter test in sand, which is in terms of effective stresses and the frictional failure criterion $\tau = \sigma' \tan\phi'$. Starting from an *in situ* horizontal effective stress σ'_{ho}, the increase in radial effective stress $\Delta\sigma'_r$ as the cavity pressure is increased is equal in magnitude to the decrease in circumferential effective stress $\Delta\sigma'_\theta$. It is assumed that there is no change in pore water pressure, so that all changes in cavity pressure are carried by the soil skeleton as changes in effective stress. Provided that the material behaviour is essentially elastic, this is consistent with deformation at constant volume because the average effective stress s' remains the same.

From the Mohr circle of effective stress shown in Figure 11.17, the increase in cavity pressure Δp_p when the soil at the borehole wall starts to fail is given by

$$\Delta p_p = \Delta\sigma'_r = t = s' \sin\phi'$$

where $s' = \sigma'_{ho}$ the initial *in situ* horizontal effective stress. Hence

$$\Delta p_p = \Delta\sigma'_r = \sigma'_{ho} \sin\phi' \tag{11.38}$$

The maximum reduction in cavity pressure $\Delta p_{u/r,max}$ that can then be applied in an unload/reload cycle without causing failure in the opposite sense (i.e. with $\sigma'_\theta > \sigma'_r$) is given by the diameter of the Mohr circle of effective stress shown in Figure 11.17. This is because, as σ'_r is reduced, σ'_θ increases by the same amount. The Mohr circle of effective stress at the onset of plastic behaviour in cavity collapse is therefore the

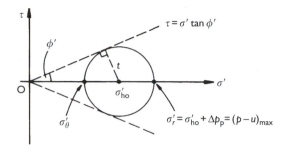

Figure 11.17 Mohr circle of effective stress for pressuremeter test in sand, at onset of plasticity at the cavity wall.

same as that at the onset of plastic behaviour in cavity expansion, with the values of σ'_r and σ'_θ reversed. Hence

$$\Delta p_{u/r,max} = 2[(p - u)_{max} - \sigma'_{ho}] \tag{11.39}$$

where $[(p - u)_{max} - \sigma'_{ho}] = \sigma'_{ho} \sin \phi'$

$$\Rightarrow \sigma'_{ho} = [(p - u)_{max}]/[1 + \sin \phi'] \tag{11.40}$$

where p is the cavity pressure and u is the pore water pressure in the soil at the cavity wall, and $(p - u)_{max}$ is the maximum difference between these quantities. Substituting equation (11.40) into equation (11.39) to eliminate σ'_{ho},

$$\Delta p_{u/r,max} = 2\{(p - u)_{max}\}\{1 - [1/(1 + \sin \phi')]\}$$

$$\Rightarrow \Delta p_{u/r,max} = 2[(p - u)_{max}][\sin \phi'/(1 + \sin \phi')] \tag{11.41}$$

For pressuremeter tests in clays, the assumption that shear takes place at constant volume is generally not unreasonable. For sands, however, this assumption is less realistic, because a sand is likely to dilate strongly, at least following the onset of plastic behaviour. The analysis presented by Hughes *et al.* (1977) takes into account the effects of dilation by means of an idealized relationship between volumetric strain ε_{vol} and shear strain γ:

$$\varepsilon_{vol} = c - \gamma \sin \psi \tag{11.42}$$

(where c is a constant: Figure 11.18), together with a **stress–dilatancy equation**, which relates the peak and critical state angles of friction and the angle of dilation ψ:

$$\frac{1 + \sin \phi'_{peak}}{1 - \sin \phi'_{peak}} = \frac{1 + \sin \phi'_{crit}}{1 - \sin \phi'_{crit}} \times \frac{1 + \sin \psi}{1 - \sin \psi} \tag{11.43}$$

Equation (11.43) was originally proposed by Rowe (1962): an example of a simpler, empirical stress–dilatancy equation relating ϕ'_{peak}, ϕ'_{crit} and ψ was given in equation (2.14) (Bolton, 1986).

In the analysis presented by Hughes *et al.* (1977), it is assumed that the sand will dilate indefinitely with constant peak angle of friction ϕ'_{peak} and constant angle of

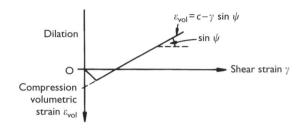

Figure 11.18 Idealized shear–volumetric strain relationship used in the analysis of a pressuremeter test in sand. (Redrawn with permission from Hughes *et al.*, 1977.)

dilation ψ. The failure criterion $\tau = \sigma' \tan \phi'$ is expressed in terms of the peak strength ϕ'_{peak} (as are equations 11.38–11.41). Making these assumptions, Hughes *et al.* show that, after the onset of plasticity at the borehole wall, the cavity pressure p and the cavity strain ε_c are related by:

$$\ln(p - u_0) = s \ln\{\varepsilon_c/(1 + \varepsilon_c) + c/2\} + A \tag{11.44}$$

where A is a constant, u_0 is the initial *in situ* pore water pressure, and

$$s = [(1 + \sin \psi) \sin \phi'_{peak}]/(1 + \sin \phi'_{peak}) \tag{11.45}$$

If ε_c is small, so that $1 + \varepsilon_c \approx 1$, equation (11.44) becomes

$$\ln(p - u_0) \approx s \ln\{\varepsilon_c + c/2\} + A \tag{11.46}$$

and the value of s may be deduced from the slope of a graph of $\ln(p - u_0)$ against $\ln\{\varepsilon_c + c/2\}$. To plot this graph, the value of the constant c (defined in Figure 11.18) must be estimated for the sand under consideration.

The stress–dilatancy relationship, equation (11.43), may be used to eliminate either ϕ'_{peak} or ψ from equation (11.45), giving (after a considerable amount of algebra)

$$\sin \phi'_{peak} = s/[1 + (s - 1) \sin \phi'_{crit}] \tag{11.47}$$

and

$$\sin \psi = s + (s - 1) \sin \phi'_{crit} \tag{11.48}$$

11.3.4 The vane shear test

The **shear vane** apparatus consists of four blades on the end of a shaft, which is pushed into a clay soil and rotated at a constant angular speed of between 6 and 12°/minute. The torque T required to do this is related to the undrained shear strength of the soil, assuming that failure occurs by the rotation of a cylinder of soil of depth D and diameter B. It is conventionally assumed that the undrained shear strength τ_u is mobilized on all shearing surfaces (Figure 11.19).

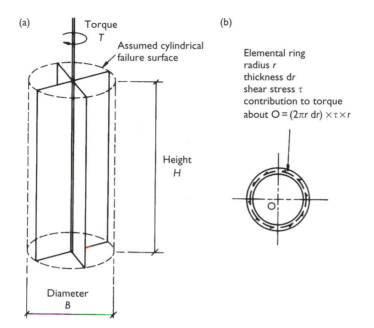

Figure 11.19 Vane shear test (a) schematic arrangement; (b) calculation of torque due to shear stress on cylinder ends.

Taking moments about the axis of the shaft, and with reference to Figure 11.19, the torque T is given by

$$T = \{(\pi BH) \times \tau_u \times (B/2)\} + 2 \int_0^{B/2} (2\pi r \tau_u r)\mathrm{d}r$$

or

$$T = \tau_u \times \left[\frac{\pi B^2 H}{2} + \frac{\pi B^3}{6}\right] \tag{11.49}$$

Usually, the height of the vane is equal to twice the overall diameter, that is, $H = 2B$. Field vanes for use in weaker soils ($\tau_u < 50\,\mathrm{kPa}$) are generally $150\,\mathrm{mm}$ long, while field vanes for use in stronger soils ($50\,\mathrm{kPa} < \tau_u < 100\,\mathrm{kPa}$) are $100\,\mathrm{mm}$ long. Pocket and laboratory versions are also available, for the rapid assessment of the undrained shear strength of smaller samples of soil.

Wroth (1984) quotes theoretical and experimental evidence presented by Donald *et al.* (1977) and Menzies and Merrifield (1980), which shows that, while the shear stress distribution on the vertical surface of the rotating cylinder of soil is approximately uniform, the shear stress distribution on the cylinder ends is very different from that assumed in the conventional analysis (Figure 11.20).

If the shear stress distribution on the top and bottom surfaces is represented by the expression

$$\tau = \tau_u \times [r/(B/2)]^n \tag{11.50}$$

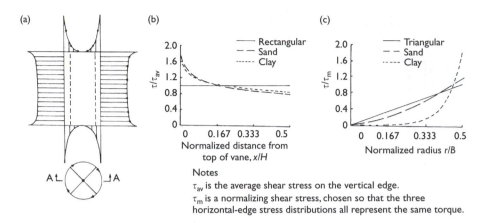

Figure 11.20 Theoretical and experimental distributions of shear stress at the edges of a shear vane. (a) Computed by Donald *et al.*, 1977; (b) vertical edge; (c) horizontal edge, based on measurements made by Menzies and Merrifield (1980). (Redrawn with permission from Wroth, 1984.)

where n is a constant, the torque due to the shear stress on the ends of the cylinder becomes

$$T_{h} = 2 \int_{0}^{B/2} \{2\pi r \tau_{u}[r/(B/2)]^{n} r\}\, \mathrm{d}r = \int_{0}^{B/2} \{2^{n+2}\pi \tau_{u} r^{n+2} B^{-n}\}\, \mathrm{d}r$$

or

$$T_{h} = \tau_{u} \times \left[\frac{\pi B^{3}}{2(n+3)} \right] \tag{11.51}$$

giving an overall torque of

$$T = \tau_{u} \times \left[\frac{\pi B^{2} H}{2} + \frac{\pi B^{3}}{2(n+3)} \right] \tag{11.52}$$

(Wroth, 1984). Equation (11.49) is a special case of equation (11.52), with $n = 0$. The experimental data of Menzies and Merrifield (1980) for London Clay conform approximately to equation (11.50) with $n = 5$. The relationship between the undrained shear strength τ_{u} and the overall torque T according to the revised analysis (equation (11.52) with $n = 5$ and $H = 2B$) is compared in Table 11.5 with that obtained using the conventional assumption (equation (11.49) with $H = 2B$, or equation (11.52) with $n = 0$ and $H = 2B$). Table 11.5 also compares the proportions of the total torque due to shear stresses on the vertical surface (T_{v}/T) in each case.

Table 11.5 has two main implications:

- The shear strength deduced from a given torque using the conventional analysis will err on the conservative side by about 9%.
- The shear strength measured is essentially that which operates on vertical planes. As this accounts for 94% of the resistive torque (compared with 86% in the

Table 11.5 Effect of assumed distribution of shear stress on cylinder ends on τ_u/T and T_v/T. T is the total torque; T_v is the component of torque due to shear on vertical surfaces

	Conventional analysis (equation (11.49) with $H = 2B$)	Revised analysis (equation (11.52) with $n = 5$ and $H = 2B$)
τ_u/T (m^{-3})	$0.273/B^3$	$0.3/B^3$
T_v/T	$0.857(= 6/7)$	$0.941(=16/17)$

conventional analysis), the effects of anisotropy will be comparatively small. The analysis and interpretation of the shear vane test is discussed in more detail by Wroth (1984).

11.3.5 The plate bearing test

The **plate bearing test** involves the application of a gradually increasing vertical load to a miniature foundation, represented by a circular plate of between 300 mm and 1 m in diameter, and the measurement of the resulting settlement. The main problem associated with either using the resulting load–settlement curve directly, or deducing the effective strength of the soil from the plate load at failure, is that of scale. It is apparent from Chapter 6 that the depth to which the soil below a foundation suffers a significant increase in stress depends primarily on the extent of the loaded surface area. It is unlikely, therefore, that all of the soil which will contribute to the settlement of the full-size foundation will feature in the behaviour of the plate loading test. This may not be a problem, provided that there are no undetected soft or weak layers outside the zone of influence of the plate test, and provided that the stiffness of the soil generally increases with depth. These issues must be investigated by means of site investigation boreholes and appropriate *in situ* or laboratory tests: the results of a plate loading test should not be considered in isolation. A plate loading test is not an adequate substitute for a suitably detailed site investigation and soil testing programme.

11.4   Ground improvement techniques

Various methods can be used to improve the strength and the stiffness of the ground by treating it *in situ*. These include densifying treatments such as compaction or pre-loading; pore water pressure reduction techniques such as dewatering (see Chapter 3, and in particular section 3.19) or electro-osmosis; the bonding of soil particles by ground freezing, grouting, and cement and lime stabilization; and the addition of reinforcing elements such as geotextiles and stone columns. The aim of this section is to give a brief overview of some of the more popular techniques of ground improvement, with particular emphasis on the basic physical and chemical mechanisms involved. It is not intended to make you into an expert on ground improvement techniques: you will need to consult specialist texts such as those by Hausmann (1990) when the need arises. Some more recent developments are described in the 11th *Géotechnique* symposium in print (Raison *et al.*, 2000).

11.4.1 *Electro-osmosis*

Groundwater will flow in response to an electrical (as well as a hydraulic) potential difference: this process is known as **electro-osmosis**. In the application of electro-osmosis in soils, metal rods or **electrodes** are inserted into the ground and an electrical potential (voltage) difference is maintained between them. The positive electrode is termed the **anode,** and the negative electrode is termed the **cathode**. In most soils, water flows towards the cathode (i.e. from positive to negative, in the direction of conventional electric current). A typical arrangement is shown schematically in Figure 11.21.

Applications of electro-osmosis include the control of pore water pressures to maintain excavation stability, the consolidation of soil around piles and foundations to improve their load-carrying capacity, and inducing an increase in water content around a pile to facilitate driving or extraction.

Excess water is removed from the cathode well by pumping. If required, a greater average degree of consolidation can be obtained by reversing the polarity of the electrodes after initial equilibrium has been achieved. Although there will be an increase in water content near the new cathode, the reduction in water content at the new anode will more than compensate for this. There is some evidence (Jaecklin, 1968) that in clays, the increase in strength and stiffness following a certain reduction in water content by electro-osmosis may be greater than that which would result from the same reduction in water content by another method. This could be due to physical and chemical changes induced in the soil, aided by the release of positively charged ions such as Na^+, K^+ and Ca^{2+} (termed **cations**, because they are attracted to the cathode), as the anode corrodes. This aspect of the process is sometimes known as **electrochemical hardening,** or **electrogrouting** (Hausmann, 1990).

Electro-osmosis works because a typical soil particle in water has a negatively charged surface, which attracts positively charged ions (cations). The concentration of ions in the pore fluid decreases rapidly with distance from the particle surface. The positive ions immediately adjacent to the particle surface are quite firmly attached. Immediately beyond the surface layer, the ion concentration is still reasonably high, but the ions are relatively mobile. The system of the negatively charged particle surface and the high concentration of positive ions in the pore water immediately adjacent to it is known as the **diffuse double layer** (Figure 11.22(a)).

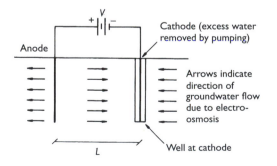

Figure 11.21 Electro-osmosis.

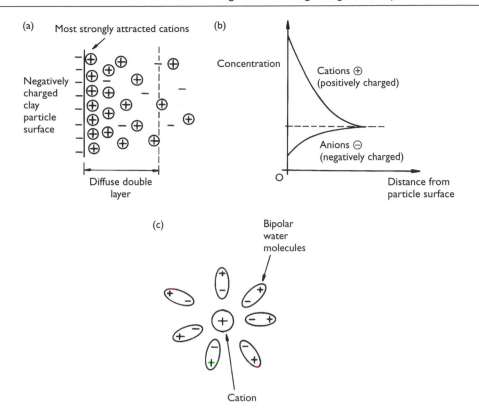

Figure 11.22 (a) Diffuse double layer; (b) variation in ion concentration with distance from the clay particle surface; (c) hydrated cation.

Beyond the diffuse double layer, the ion concentration falls to that of the pore water (Figure 11.22(b)). Water molecules are **bipolar**, with a positive charge at one end and a negative charge at the other. The positive ions in the diffuse layer around a clay particle are surrounded by bipolar water molecules as shown in Figure 11.22(c). When an electrical potential difference is applied to the soil, the positive ions are attracted to and move towards the cathode, carrying the water molecules with them. Ion (and hence water) movement occurs primarily within the diffuse layer surrounding the soil particle.

The rate of movement of pore water depends on the voltage gradient applied (in volts per metre, V/m) and the **electro-osmotic coefficient of permeability** k_e (which has units of m/s per V/m, or m²/Vs). k_e depends in turn on the porosity of the soil, and the viscosity, the **electrical permittivity** (also known as the **dielectric constant**) and the **electro-kinetic potential** (also known as the **zeta potential**, ζ) of the pore water. The zeta potential depends on the **electrolyte** concentration, which does not vary significantly for most natural silicaceous soils (an ectrolyte is a solution that conducts electricity – in this case, the ions dissolved in the pore water). Thus for many natural soils (montmorillonite being a notable exception), k_e is of the order of 0.5×10^{-8} m/s per V/m (Mitchell, 1993). If the concentration of electrolytes in the pore water is high, electro-osmosis will not work because the zeta potential will be too small. In extreme cases, for example in soils composed of calcium carbonates and some industrial wastes, the direction of flow will be reversed so that water flows from the cathode to the anode.

According to the generally accepted Helmholtz–Smoluchowski theory of electro-osmosis (e.g. Mitchell, 1993), the electro-osmotic permeability k_e does not depend on the pore size. This is in contrast to the hydraulic permeability used in Darcy's Law, which was discussed in Chapter 3. The Helmholtz–Smoluchowski theory is based on a large-pore model, which neglects the intrusion of the cation layer into the pores. An alternative model, proposed by Schmid (Mitchell, 1993), which takes account of the intrusion of the cation layer into the pore space, suggests that k_e should be proportional to the square of the pore size. In practice, however, the Helmholtz model gives generally more realistic results for soils. This may be because most clays form aggregations of particles, so that electro-osmotic flow is governed by the larger pores between the aggregations, rather than by the smaller pores between the individual particles (Mitchell, 1993).

By analogy with Darcy's Law (equation 3.2), the electro-osmotic flowrate q_e (m³/s) may be written as

$$q_e = A k_e i_e \tag{11.53}$$

where A (m²) is the gross cross-sectional area of flow, k_e (m²/Vs) is the electro-osmotic permeability and i_e is the electrical potential gradient $-\partial V/\partial x$ (V/m). The flowrate per unit area $v = q/A$ due to the combined effects of an electrical potential gradient and a hydraulic potential gradient is

$$v = -[k_h \partial h/\partial x] - [k_e \partial V/\partial x] \tag{11.54}$$

(assuming horizontal flow; k_h (m/s) is the hydraulic permeability in the horizontal direction).

Following the application of a voltage gradient $\partial V/\partial x$, the pore water pressures will change until an equilibrium condition is reached, in which the pore water flow due to electro-osmosis is exactly balanced by a flow in the opposite direction due to the hydraulic gradient established thereby. When there is no net flow, $v = 0$ in equation (11.54) and

$$[k_h \partial h/\partial x] = -[k_e \partial V/\partial x]$$

or

$$k_h \partial h = -k_e \partial V$$

which may be integrated to give

$$k_h h = -k_e V + C$$

where C is a constant of integration. Assuming that $V = 0$ and $h = 0$ at the cathode, and writing (at constant elevation along a horizontal flowline) $h = u/\gamma_w$,

$$u = -(k_e/k_h)\gamma_w V \tag{11.55}$$

Equation (11.55) shows that the magnitude of the reduction in pore water pressure (and hence, if the total stress remains constant, the increase in effective stress) will increase with k_e/k_h. As for many soils k_e is approximately constant, it follows that the largest increases in effective stress can be achieved in soils where the hydraulic permeability k_h is relatively small, i.e. silts and clays.

The rate of consolidation during electro-osmosis may be investigated by using equation (11.54) instead of Darcy's Law in the derivation of the one-dimensional consolidation equation (section 4.8). A solution for one-dimensional flow (Johnston and Butterfield, 1977) is shown in Figure 11.23. A solution for radial flow is given by Esrig (1968). These solutions are similar in form to, and can be used in the same way as, the solutions to the conventional consolidation problems presented in Chapter 4.

When expressed as a percentage of the eventual settlement or volume change, the rate of proportional consolidation depends only on the consolidation coefficient governing vertical compression due to horizontal flow, $c_{hv} = k_h E_0'/\gamma_w$: it is independent of

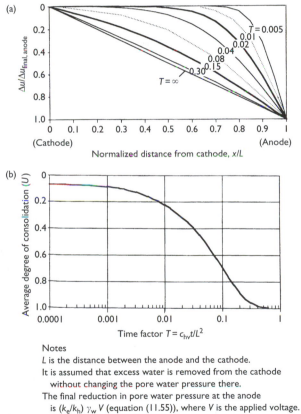

Notes

L is the distance between the anode and the cathode.

It is assumed that excess water is removed from the cathode
without changing the pore water pressure there.

The final reduction in pore water pressure at the anode
is $(k_e/k_h)\,\gamma_w\,V$ (equation (11.55)), where V is the applied voltage.

$c_{hv} = k_h\,E_0'/\gamma_w$ (section 4.7.3).

Figure 11.23 (a) Isochrones of pore water pressure and (b) average degree of consolidation U vs. time factor T for one-dimensional electro-osmotic consolidation with lateral flow. (With permission from Johnson and Butterfield, 1977; redrawn from Hausmann, 1990.)

the electro-osmotic permeability k_e. In many applications, however, it is the absolute change in pore water pressure or the absolute settlement that is important. If electro-osmosis increases the overall amount of consolidation by a certain factor, the absolute rate of settlement or pore pressure change will be increased by the same factor, even though the rate of change expressed as a proportion of that which ultimately results is unaltered. This could be beneficial in pore pressure reduction or pre-loading schemes in low-permeability soils, where consolidation is slow.

Electro-osmosis is only economically viable in soils of low permeability. This is partly because, as indicated by equation (11.55), the change in pore water pressure achievable becomes insignificant when $k_h \gg k_e$. Also, the electrical energy used is proportional to the volume of water removed, which in high-permeability soils is very large. Typical voltage differences used in electro-osmosis are generally in the range 50–100 V, with potential gradients usually less than 50 V/m.

Further details of the mechanisms and effects of electro-osmosis are given by Mitchell (1993) and Hausmann (1990).

11.4.2 Ground freezing

In principle, ground freezing can be used to stabilize almost any excavation regardless of depth and soil permeability, provided that the initial groundwater flow velocities are small. For large, comparatively shallow excavations, however, the technique is generally rather more expensive than conventional dewatering and is therefore not commonly used. Typical applications include deep shafts of small diameter – Powers (1992) quotes 3–5 m diameter and up to 800 m deep – in which a **freezewall** (i.e. a 'wall' of frozen ground) temporarily supports the sides of the excavation in addition to excluding groundwater.

The freezewall is created by pumping a refrigerant through **freezepipes** installed in the ground, typically at a spacing of 1 m or so. Common refrigerants include calcium chloride solution (known as **brine**), which is recirculated through a cooler; and liquid nitrogen, which is disposed of by venting it to the atmosphere after use. As the temperature of liquid nitrogen ($-196°C$) is much lower than that of brine ($-55°C$), freezing occurs much more rapidly (for the same well spacing) with liquid nitrogen as a refrigerant. The liquid nitrogen method is more expensive to operate, even if the liquid nitrogen is passed through more than one freezepipe before being vented to the atmosphere, but it can be economically viable for short-term projects or in an emergency.

Figure 11.24(a) illustrates schematically the development of a freezewall as the pore water in the ground around each freezepipe becomes frozen.

Groundwater flow velocities in excess of about 1 m/day (10^{-5} m/s) can make ground freezing impracticable because of the increased volume of groundwater that must be chilled and the additional cooling capacity this requires (Powers, 1992). Natural groundwater velocities are generally smaller than 1 m/day, but in many circumstances the groundwater regime is altered by the presence of the excavation (e.g. Figure 3.21). Also, dewatering wells may need to be installed to maintain the stability of the base of the excavation, leading to locally high groundwater velocities. In these cases, the additional load on the refrigeration plant must be taken into account at the planning stage.

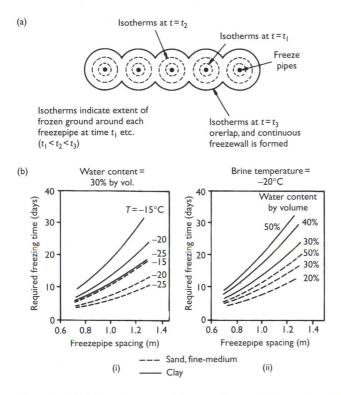

Figure 11.24 a) Development of freezewall around freezepipes (plan view); (b) required freezing time as a function of freezepipe spacing, (i) water content = 30% by volume,
$-25°C < T < -15°C$; and (ii) $T = -20°C$, various water contents by volume.
(Redrawn with permission from Hausmann, 1990, after Jessberger, 1987, quoting
Stoss, 1976.)

The time taken to form a freezewall depends on the soil type, the soil water
content, the temperature of the coolant and the freezepipe spacing, as indicated in
Figure 11.24(b).

11.4.3 Grouting

Soil can be improved, in the sense of making it stronger and stiffer and/or reducing
its permeability, by injecting a fluidized material such as a cement paste suspension
or a chemical solution known as a **grout**. The grout fills the soil pores and then **sets**
or **hardens**, bonding the soil particles together. The pressure required for the grout
to penetrate the voids between the soil particles increases as the pore size decreases,
and increases with the viscosity of the grout. For cement-based grouts made up of fine
particles in suspension, the particle size of the grout in relation to that of the soil being
grouted is also an important consideration.

Cement-based grouts are used for coarse materials such as gravels and fissured rock,
but will not penetrate a material finer than a very coarse sand. For medium sands,
chemical grouts such as sodium silicate in **colloidal suspension** are used. (A colloid

is a solid particle small enough to remain indefinitely in suspension in a fluid, due to random molecular or **Brownian** motion.) The grouting of fine sands and silts requires the use of a **solution grout** such as an acrylic resin. In clay soils, grouts can only be injected into fissures and along rupture surfaces, but this can help to stabilize slopes – at least temporarily.

Generally, groundwater control is easier to achieve using grout injection than ground improvement. This is because a reasonably effective water barrier can be formed by penetrating only the coarser materials (e.g. a gravel lens), whereas for satisfactory ground improvement, virtually all of the particles must be bonded together.

If it cannot permeate between the soil particles, the grout can open and penetrate along a fissure. This process is known as **hydrofracture**. In theory, hydrofracture becomes a possibility when the grout injection pressure exceeds the smaller of the lateral or the vertical stresses in the ground, at the point of injection. In practice, hydrofracture is likely to occur at an injection pressure of 2–6 times the overburden. The creation of long, thin fissures serves no useful purpose. However, the injection of pressurized grout to form short, wedge-shaped fractures might lead to some useful local compaction of the soil adjacent to each fracture. For this purpose, a reasonably viscous grout paste is used.

Other applications of grouting include jacking up buildings which have undergone excessive settlement, underpinning buildings whose foundations are failing, and a technique known as **compensation grouting**. Compensation grouting is a pre-emptive measure, which involves the injection of grout during the construction of an underground excavation such as a tunnel. The aim is to cause a degree of heave at the soil surface which exactly compensates for the surface settlement due to the construction of the tunnel, with the result that there is no net movement at the soil surface (or at the base of the foundation of the building it is sought to protect). In this way, tunnels can be constructed beneath densely built-up areas without causing damage to existing buildings. Figure 11.25 illustrates the use of compensation grouting in connection with tunnelling work for the Waterloo International railway station in London.

In order to control the grouting operation, the parameters influencing the rate of grout penetration into the surrounding soil (i.e. the pumping pressure, the grout flowrate and the grout viscosity) must be monitored continuously. Grout viscosity varies with **gel strength**, which increases as the grout begins to set or **gelate**. The rate of gelation will depend in turn on the nature of the grout, and factors such as the temperature of the ground. In compensation grouting, ground movements must also be monitored in order to provide a continuous check on the effectiveness of the operation.

Further details of grouting processes and applications are given by Hausmann (1990).

11.4.4 Preloading

It is well known that overconsolidated clays have generally higher stiffnesses and undrained shear strengths than normally consolidated clays (Figure 4.6).

The stiffness and undrained shear strength of a normally consolidated soil in the field may be improved by artificial overconsolidation, known as **preloading**. Before construction work commences, the soil is preloaded to a vertical effective stress in excess of the anticipated working load. The preload may be applied by means of

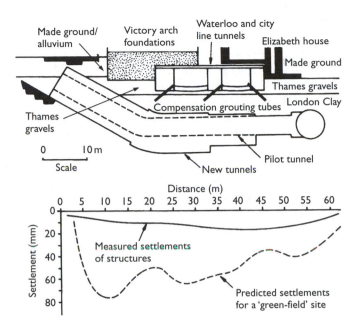

Figure 11.25 Effect of compensation grouting in reducing the settlements of structures during tunnelling near Waterloo railway station, London (Redrawn with permission from Wheeler, 1993.)

a surcharge at the ground surface (e.g. water in tanks or an additional overburden of soil), or by reducing the pore water pressures using vacuum wells (section 3.19).

Since preloading is usually used in connection with clay soils, there will be a delay between the application of the preload and the response of the soil in terms of increased effective stresses and settlements, because of the time taken for the soil to consolidate. Following the application of the preload, soil settlements must be monitored and compared with the expected behaviour, so that the preload can be removed at the optimum time. An example of this is given in Question 6.6. Consolidation times can be reduced by the installation of **vertical drains**, as described in section 4.9.

As a first approximation, rates of consolidation may be estimated using the simple models presented in Chapter 4. Variations in vertical stress increment with distance (both horizontally and vertically) from the loaded area may be estimated using elasticity-based methods, such as the Newmark chart described in section 6.4.

Preloading can also be used with **coralline** or **calcareous** soils, which may exhibit very large strains on first loading due to the crushing of fragile hollow particles.

11.4.5 Surface compaction

Surface compaction of soils may be achieved in the following ways (Van Impe, 1989):

- application of shearing stresses which essentially cause a local failure of the soil (by means of smooth metal rollers, sheepsfoot rollers or pneumatic tyred rollers)
- application of tamping or dynamic energy (using pounders or rammers)

- vibration (using vibrating plates)
- a combination of shear stresses and vibration (using vibrating rollers).

These types of surface compaction are most effective on granular materials compacted in thin layers (0.3–0.5 m deep), such as the soil used to backfill a trench or behind a retaining wall. Specifications for the satisfactory compaction of backfill material used in highway works are issued by the UK Department of Transport (DoT, 1993), and most manufacturers of compaction equipment provide technical literature which enables plant to be selected on the basis of its ability to meet the DoT specification in the circumstances under consideration.

Clay is sometimes used as a backfill, because it is less expensive and/or more readily available than granular material: it is generally, however, more difficult to compact. Excavated clay waste is usually in the form of clods with air voids in between. The overall specific volume of the excavated clods in bulk may be up to 50% higher than that of the intact clay. Compaction of the material is achieved by reshaping the clods so that they become more closely packed, reducing the volume of the air voids to about 5%. This requires sufficient compactive effort to overcome the undrained shear strength, which, if the clay has been excavated from some depth, may be quite high. The satisfactory compaction of clay fill depends on the application of shear stresses large enough to remould the clods, rather than dynamic loads or vibration. Unless there are initially significant air voids, any attempt at compaction of an essentially clayey soil is unlikely to be effective.

11.4.6 Heavy tamping

Heavy tamping involves the repeated dropping of a heavy weight (up to 170 tonnes) from a height of perhaps 22 m, onto a number of places on the surface of the soil, in a grid pattern at a spacing of up to 15 m. The technique was first used to compact layers of loose, granular material from the surface. The vibrations due to heavy tamping comprise both longitudinal waves, which disturb and rearrange the soil skeleton, and transverse waves, which compact the soil at depth.

Heavy tamping may also be used on low-permeability plastic soils, where compaction can result from two mechanisms. First, immediate settlements may occur due to the compression or dissolution into the pore water of small bubbles of gas. These gas bubbles may comprise between 1% and 4% of the total volume, and are found in many alluvial clays because of the presence of organic matter and micro-organisms. Second, the increase in pore water pressure which results from dropping the heavy weight is large enough to reduce the effective stresses to zero and cause fissures to develop in the soil. The fissures act as fast drainage paths, enabling pore water to escape and speeding up the process of consolidation. In low permeability soils, the time intervals between drops of the heavy weight must be long enough to allow consolidation to take place.

The depth D to which heavy tamping is effective may be estimated using equation (11.56) (Hausmann, 1990):

$$D \text{ (metres)} = 0.5\sqrt{WH} \tag{11.56}$$

where W is the mass of the falling weight in tonnes and H is the height of fall in metres. This is an empirical relationship, and is of no help in determining the grid spacing of the impact points required for effective compaction of the entire volume of soil. A more detailed appraisal is given by Gu and Lee (2002).

It is almost always necessary to spread a layer of stone fill 1.5–2 m thick on the surface of a site which is to be compacted using heavy tamping, both to support the plant (which could weigh 200–300 tonnes) and to prevent or limit the formation of craters. Heavy tamping is particularly useful for compacting loose, variable material which perhaps contains large voids (such as waste tips), but should be used with caution near existing buildings because of the possibility of damaging vibrations.

11.4.7 Cement and lime stabilization

Soils can be improved or **stabilized** by mixing in cement or quicklime. In both cases, the principal mechanism of improvement is the formation of chemical bonds between the soil particles. The main benefits are increased stiffness and durability, and better **volume stability** (i.e. the soil becomes less susceptible to shrinkage or swelling). The undrained shear strength (or the unconfined compressive strength, if appropriate) will also increase as a result of the chemical bonding or cementation of the soil particles, but the improved material may be brittle because once the bonds are broken their strength is lost completely. Lime may be added to a clay to improve its workability; the water content at the plastic limit is increased and at a given water content the clay becomes more **friable** (i.e. crumbly). The addition of lime will also make the soil more difficult to compact, so that the dry density (section 1.12) achieved for a given compactive effort will be smaller. However, the compaction curve of dry density against water content (Figure 1.18) will be flatter, so that the dry density achieved by compaction is less dependent on the water content.

Cement stabilization of soils works because cement and water react to form cementitious calcium silicate and aluminate hydrates, which bind the soil particles together. In addition, the hydration reaction releases calcium hydroxide $Ca(OH)_2$, slaked lime, which may in turn react with some components of the soil, in particular clay minerals. Hydration of the cement occurs immediately on contact with water, but the secondary reactions are slower and may continue for many months. Cement stabilization can be used successfully in a wide range of soils, because the primary reaction, hydration, is independent of soil type. Difficulties are normally only encountered with coarse gravels or soils with a high organic content. The effectiveness of the treatment depends on adequate mixing and compaction, which in high-plasticity soils (i.e. clays) can be difficult to achieve.

In **lime stabilization**, no cementitious calcium silicates and aluminates are formed and the technique relies entirely on the reaction between the lime and the soil. **Lime stabilization** therefore only works in soils with a substantial proportion ($>35\%$) of fine particles ($<60\,\mu m$). Lime may be added as **quicklime**, CaO, or as hydrated (**slaked**) lime, $Ca(OH)_2$. Quicklime reacts with water to give hydrated lime plus a considerable amount of heat (65.3 kJ/mol). Quicklime is more cost effective in terms of transport and handling because hydrated lime contains about 25% water.

If quicklime is mixed into the soil it will react immediately with the pore water, resulting in a drying of the soil which is usually beneficial, particularly in terms of improving trafficability on site. On mixing lime with clay, the cations adsorbed onto the surfaces of the clay particles (e.g. sodium) are exchanged for calcium, resulting in a structural change which causes the clay particles to **coagulate** (i.e. form into small clumps). This reduces the plasticity index of the clay, improving its workability and potentially (after compaction) its strength and stiffness. For kaolinite clays, both the plastic limit (w_{PL}) and the liquid limit (w_{LL}) are increased, while for montmorillonite and illite, the plastic limit is increased but the liquid limit is reduced. In both cases, however, there is a reduction in the plasticity index ($PI = w_{LL} - w_{PL}$) (Van Impe, 1989).

The main contribution to the increase in undrained shear strength and stiffness of lime-stabilized soils comes from cementation, as the second stage of the clay/lime reaction removes silica from the clay mineral lattice to form products similar to those resulting from the hydration of cement. The effectiveness of the cementation process increases with the specific surface area of the soil particles: lime stabilization does not work in clean sands and gravels. The degree of cementation is limited by the available silica, and there is no benefit in adding more lime than the amount required to utilize the silica content of the soil. The addition of too much lime can be counterproductive, in contrast to stabilization using cement, where the increase in strength depends on the amount of cement added.

Typically, the ratios of cement or lime to soil used in practice are in the range 2–10%.

11.4.8   Soil reinforcement methods

Ground improvement may be achieved by methods which do not treat the bulk of the volume of soil, but aim to provide sufficient local reinforcement to influence the overall performance. An example of this is the use of horizontal reinforcement strips in a reinforced soil wall, as discussed in section 10.10. Horizontal reinforcement, in the form of a **polymer geotextile** or **geogrid** layer (Figure 11.26), may also be incorporated into the base of an embankment, in order to prevent it from spreading. A comprehensive review of the uses and properties of geotextiles is given by John (1987).

In cases where excessive settlement is a potential problem, the ground may be reinforced vertically by the installation of stone (or gravel or sand) columns. In granular soils, stone columns are often installed using a technique known as **vibroflotation**. This involves the densification and compaction of the natural soil by means of a long, horizontally vibrating lance, typically 0.4 m in diameter, penetrating vertically into the ground. The centre of the lance is hollow, and is used to introduce additional material (the stone column) into the void formed by the compaction of the original ground.

In clay soils, which are not amenable to compaction by vibration, the hole for the stone column may be drilled out (as for a bored concrete pile), or formed by hammering a hollow steel tube into the ground. In the latter case, the stone or gravel is introduced and compacted through the inside of the tube as it is gradually withdrawn. The main benefit is then the reinforcing effect of the stiffer and stronger granular columns. A potential secondary benefit is that the columns may act as vertical drains, reducing drainage path lengths and consolidation times. A case history describing the use of stone columns to stabilize and accelerate consolidation of the foundations for an embankment on alluvial clay is given by Cooper and Rose (1999).

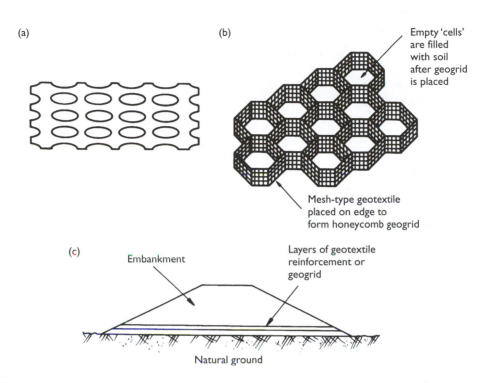

(a)

(b)

Empty 'cells'
are filled
with soil
after geogrid
is placed

Mesh-type geotextile
placed on edge to
form honeycomb geogrid

(c)

Embankment

Layers of geotextile
reinforcement or
geogrid

Natural ground

Figure 11.26 (a) Polymer mesh geotextile; (b) typical geogrid arrangement; (c) reinforcement of embankment base using geotextile or geogrid layer.

Vertical reinforcement in clay soils can also be achieved by means of **lime-stabilized columns**. Lime-stabilized columns are formed by remoulding the soil in place using a vertical mixer, while at the same time introducing an appropriate quantity (usually between 3% and 10% of the dry mass of the soil) of quicklime and/or gypsum. As mentioned in section 11.4.7, some benefit may derive from a reduction in the water content of the immediately surrounding ground as pore water is used in the hydration reaction, in addition to the increased strength of the lime-stabilized column due to the formation of cementitious bonds.

Lime stabilized columns formed by mixing dry lime powder into the soil tend to show an increase in permeability compared with the untreated soil, and so may act to some extent as preferential drainage pathways. On the other hand, lime-stabilized columns installed by injecting a suspension of lime powder in water into the soil seem to have a lower permeability than the untreated ground. Van Impe (1989) suggests that this difference may be due to the fact that the second method results in smaller effective pore spaces.

Applications of lime stabilized columns include the reduction of soil settlements and the stabilization of slopes. However, there may be some uncertainty concerning the long-term durability of the lime-stabilized columns. Further details are given by Van Impe (1989) and Hausmann (1990).

11.4.9 Assessing the success of ground improvement techniques

It is important that the improvement resulting from any form of ground treatment is assessed and quantified, preferably as the work progresses. This must generally be carried out using *in situ* testing techniques. The success or otherwise of any method of soil improvement will depend to a considerable extent on its suitability for the ground conditions to which it are applied. In this respect, it is always worth a search of the literature for previous documented experience before embarking on any programme of ground improvement. While it is true that we learn from mistakes, the best mistakes to learn from are those made by others, so that we do not repeat them ourselves.

Key points

- The behaviour of geotechnical structures can be investigated using numerical modelling techniques. One of the most popular is the finite element method. The features of a commercial finite element package that make it suitable for use in geotechnical analysis include the ability to work in terms of effective stresses, and the facility to add or remove elements to simulate construction and excavation processes. A wide variety of problems can be tackled using numerical modelling techniques. However, a considerable degree of approximation may still be necessary, particularly in terms of the **constitutive relationship** used for the soil, and in the representation of three-dimensional problems by a two-dimensional (either **plane strain** or **axisymmetric**) analysis.
- Small-scale physical model testing can also be used to gain an insight into the behaviour of a geotechnical construction. The main problem is that a small-scale model tested at normal gravity may be unrepresentative of a full-scale structure, because the self weight stresses at corresponding depths in the model and in the full-scale **prototype** are different. This can be overcome by testing a $1/n$ scale model in a geotechnical centrifuge, at a radial acceleration of $n \times$ normal gravity. Approximations are introduced into the modelling procedure by the need to simulate excavation and construction processes while the model is in the centrifuge. Also, the relevant scaling rules between the model and the prototype should be established in each individual case. In necessary, scaling relationships can be confirmed by testing the same model at two or more different scales – a procedure known as **modelling of models**.
- Empirical correlations can be made between the blowcount in a **Standard Penetration Test** (SPT) and a number of soil parameters, including the **density index** I_D, the **stiffness modulus** E', and the **peak strength** ϕ'_{peak}. To account for variations in equipment type and test procedure, the raw SPT blowcount N is conventionally corrected to a value corresponding to the delivery to the sampler of 60% of the potential energy of the hammer at the top of its travel. The corrected SPT blowcount is given the symbol N_{60}. In general, the SPT blowcount N_{60} would be expected to increase with depth. In correlations between the SPT blowcount and soil parameters that also increase significantly with depth (e.g. the stiffness modulus, E'), the corrected SPT blowcount N_{60} should be used. In correlations

between the SPT blowcount and soil parameters that do not generally change very significantly with depth (e.g. the density index I_D and the peak strength ϕ'_{peak}), the corrected SPT blowcount N_{60} should be normalized to a reference stress, which is usually taken as $\sigma'_v = 100\,\text{kPa}$. The normalized, corrected SPT blowcount is given the symbol $(N_1)_{60}$, and is calculated using the expression

$$N_1(\text{ or } (N_1)_{60}) = C_N N(\text{or } C_N N_{60}) \tag{11.4}$$

where the values of the correction factor C_N are as given in equations (11.5).

- The *in situ* horizontal stress σ_{ho}, the soil shear modulus G and the undrained shear strength τ_u of a clay soil can be measured in the field using a **pressuremeter**. Pressuremeter test data are plotted as a graph of cavity pressure p as a function of cavity strain ε_c or increase in cavity volume ΔV. The pressuremeter test is analysed as an expanding cylindrical cavity in an elastic–perfectly plastic material. On this basis, the shear modulus G is determined from the slope of the graph of cavity pressure p against change in cavity volume, dp/dV, during an elastic phase of the test (usually within an unload/reload cycle):

$$G = V(dp/dV) \tag{11.26}$$

where V is the current cavity volume.

The undrained shear strength τ_u should be equal to the slope of a graph of the cavity pressure p against the natural logarithm of the proportional volume change, $\ln(\Delta V/V)$, during a plastic phase of the test.

- Various techniques can be used to improve the strength and the stiffness of the ground, by treating it *in situ*. These include densifying treatments such as compaction or pre-loading; pore water pressure reduction techniques such as dewatering or electro-osmosis; the bonding of soil particles by ground freezing, grouting, and cement and lime stabilization; and the addition of reinforcing elements such as geotextiles and stone columns. The suitability of each technique for any particular application depends on both the soil type and the purpose of the proposed ground treatment.

Questions

Modelling

11.1 Compare and contrast the use of physical and numerical models as aids to design. Your answer should address issues such as the assumptions that have to be made in setting up the model, limitations as to the validity of the results, and other factors which would lead to the use of one in preference to the other.

[University of London 3rd year BEng (Civil Engineering) examination, Queen Mary and Westfield College]

In situ testing

11.2 (a) Describe the principal features of the Menard and self-boring pressuremeters, and compare their advantages and limitations.

 (b) Figure 11.27 shows a graph of corrected cavity pressure p as a function of the cavity strain ε_c for a self-boring pressuremeter test. The test was carried out in a borehole at a depth of 11 m in a stratum of sandy soil of unit weight $20 \, kN/m^3$. The piezometric level was 1 m below the ground surface. Estimate

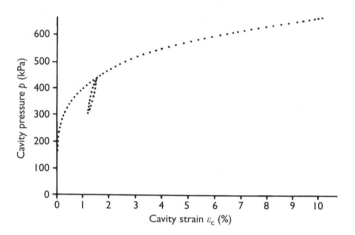

Figure 11.27 Pressuremeter test data.

 (i) the *in situ* horizontal total stress
 (ii) the coefficient of earth pressure at rest, K_0
 (iii) the soil shear modulus, G.

((i) 165 kPa; (ii) 0.54; (iii) 23 MPa based on unload/reload cycle.)

Ground improvement

11.3 Write brief notes on:

 (a) grouting
 (b) surface compaction and heavy tamping
 (c) cement and lime stabilization.

In each case, your answer should include (but not be restricted to) a discussion of the ground conditions and soil types for which the method is suitable.

[University of London 3rd year BEng (Civil Engineering) examination, Queen Mary and Westfield College]

11.4 Give an account of:

(a) the principal applications of grouting in geotechnical engineering
(b) the factors influencing the penetration of grouts into soils
(c) the major differences in properties and performance between cement-based grouts and low-viscosity chemical grouts.

[University of London 3rd year BEng (Civil Engineering) examination,
Queen Mary and Westfield College]

Notes

1 The volumetric strain increment $d\varepsilon_{vol}$ is defined as $-dv/v$ so that a reduction in volume (which is denoted by a negative dv) corresponds to a compressive (i.e. positive) volumetric strain increment.
2 The ultimate undrained load $q_{ult,u}$ is used because, for the case histories on which the correlation is based, it was not possible to calculate q_{ult} in effective stress terms. The value of $q_{ult,u}$ is estimated on the basis of the undrained shear strength τ_u (Stroud, 1989).
3 The strength of rock is more conventionally described by means of the **unconfined compressive strength** σ_{uc}, which is defined as the axial stress σ_a causing failure at zero lateral or confining pressure σ_r. Consideration of the Mohr circle of total stress with $\sigma_a = \sigma_{uc}$ and $\sigma_r = 0$ shows that $\sigma_{uc} = 2\tau_u$.

References

Baldi, G., Bellotti, R., Ghionna, V., Jamiolkowski, M. and Pasqualini, E. (1981) Cone resistance of dry medium sand. *Proceedings of the 10th European Conference on Soil Mechanics & Foundation Engineering*, Stockholm, **2**, 427–32.

Bolton, M.D. (1986) The strength and dilatancy of sands. *Géotechnique*, **36**(1), 65–78.

Britto, A. and Gunn, M.J. (1987) *Critical State Soil Mechanics via Finite Elements*. Ellis Horwood, Chichester.

Burbidge, M.C. (1982) *A case study review of settlements on granular soils*. MSc dissertation, University of London (Imperial College of Science & Technology).

Burd, H.J., Houlsby, G.T., Chow, L., Augarde, C.E. and Liu, G. (1994) Analysis of settlement damage to masonry structures, in *Numerical Methods in Geotechnical Engineering* (ed. I.M. Smith), pp. 203–8, A.A. Balkema, Rotterdam.

Burland, J.B. and Kalra, J.C. (1986) Queen Elizabeth II Conference centre: geotechnical aspects. *Proceedings of the Institution of Civil Engineers, Pt. 1*, **80**, 1479–503.

Chandler, R.J. (1995) *Field studies, analysis and numerical modelling of retaining walls propped at formation level*. PhD dissertation, University of London (Queen Mary and Westfield College).

Cooper, M.R. and Rose, A.N. (1999) Stone column support for an embankment on deep alluvial soils. *Proceedings of the Institution of Civil Engineers (Geotechnical Engineering)*, **137**, 15–25.

Craig, W.H. (1984) Installation studies for model piles, in *Proceedings of a Symposium on the Application of Centrifuge Modelling to Geotechnical Design* (ed. W.H. Craig), pp. 440–55, A.A. Balkema, Rotterdam.

Craig, W.H. (1995) Geotechnical centrifuges: past, present and future, in *Geotechnical Centrifuge Technology* (ed. R.N. Taylor), pp. 1–18, Blackie, London.

Craig, W.H. and Yildirim, S. (1976) Modelling excavations and excavation processes. *Proceedings of the 6th European Conference on Soil Mechanics & Foundation Engineering*, **1**, 33–6.

Culligan, P.J. and Barry, D.A. (1998) Similitude requirements for modelling NAPL movement in a geotechnical centrifuge. *Proceedings of the Institution of Civil Engineers (Geotechnical Engineering)*, **131**, 180–6.

Davies, M.C.R. *et al.* (1998) Applications of centrifuge modelling to geotechnical engineering practice. *Proceedings of the Institution of Civil Engineers (Geotechnical Engineering)*, **131**, 125–86.

Davis, A. and Auger, D. (1979) La butée des sables: essais en vrai grandeur. *Annales de l'institut technique du bâtiment et travaux publics*, 375.

de Nicola, A. and Randolph, M.F. (1994) Development of a miniature pile driving actuator, in *Centrifuge '94* (eds C.F. Leung, F.H. Lee and T.S. Tan), pp. 473–8, A.A. Balkema, Rotterdam.

Department of Transport (1993) *Manual of Contract Documents for Highway Works*, HMSO, London.

Donald, I.B., Jordan, D.O., Parker, R.J. and Toh, C.T. (1977) The vane test – a critical appraisal. *Proceedings of the 9th International Conference on Soil Mechanics*, **1**, 81–8.

Durgunoglu, H.T. and Mitchell, J.K. (1975) Static penetration resistance of soils. *Proceedings of the ASCE Conference on* in situ *Measurement of Soil Properties*, Raleigh, NC, **1**, 151–88.

Esrig, M.I. (1968) Pore pressures, consolidation and electrokinetics. *J. ASCE, Soil Mechanics & Foundations Division*, **94**(SM4), 899–921.

Gibson, R.E. and Anderson, W.F. (1961) *In situ* measurement of soil properties with the pressuremeter. *Civ Engng & Public Works Review*, **56**(658), 615–18.

Gu, Q. and Lee, F.H. (2002) Ground response to dynamic compaction of dry sand. *Géotechnique*, **52**(7), 481–93.

Gunn, M.J. (1993) The prediction of surface settlements due to tunnelling, in *Predictive Soil Mechanics, Proceedings of the Wroth Memorial Symposium*, pp. 304–16, Thomas Telford, London.

Hausmann, M.R. (1990) *Engineering Principles of Ground Modification*. McGraw-Hill, New York.

Hicks, M.A. and Mar, A. (1994) A combined constitutive model-adaptive mesh refinement formulation for the analysis of shear bands in soil, in *Numerical Methods in Geotechnical Engineering* (ed. I.M. Smith), pp. 59–66, A.A. Balkema, Rotterdam.

Higgins, K.G., Potts, D.M. and Symons, I.F. (1993) The use of laboratory derived soil parameters for the prediction of retaining wall behaviour, in *Retaining Structures* (ed. C.R.I. Clayton), pp. 92–101, Thomas Telford, London.

Hughes, J.M.O., Wroth, C.P. and Windle, D. (1977) Pressuremeter tests in sands. *Géotechnique*, **27**(4), 455–77.

Irons, B.M. and Ahmad, S. (1980) *Techniques of Finite Elements*, Ellis Horwood, Chichester.

Jaecklin, F.P. (1968) *Elektrische Bodenstabilisierung*, Swiss Soc. Soil Mechs & Fndn Engng, Publication no. 72.

Jardine, R.J., Potts, D.M., Fourie, A.B. and Burland, J.B. (1986) Studies of the influence of non-linear stress–strain characteristics in soil–structure interaction. *Géotechnique*, **36**(3), 377–96.

Jessberger, H.L. (1987) Artificial freezing of the ground for construction purposes, in *Ground Engineers' Reference Book* (ed. F.G. Bell), Ch. 31, Butterworths, London.

John, N.W.M. (1987) *Geotextiles*, Blackie, London.

Johnston, I.W. and Butterfield, R. (1977) A laboratory investigation of soil consolidation by electro-osmosis. *Australian Geomechanics Journal*, 21–32.

Kimura, T., Takemura, J., Hiro-oka, A., Okamura, M. and Park, J. (1994) Excavation in soft clay using an in-flight excavator, in *Centrifuge '94* (eds C.F. Leung, F.H. Lee and T.S. Tan), pp. 649–54, A.A. Balkema, Rotterdam.

Kimura, T., Kusakabe, O. and Takemura, J. (eds) (1998) *Proceedings of the International Conference Centrifuge '98*, A.A. Balkema, Rotterdam.

Ko, H.-Y. and McLean, F.G. (eds) (1991) *Centrifuge '91*, A.A. Balkema, Rotterdam.

Ko, H.-Y., Azevedo, R. and Sture, S. (1982) Numerical and centrifugal modelling of excavations in sand, in *Deformation and Failure of Granular Materials*, pp. 609–14, A.A. Balkema, Rotterdam.

Lade, P.V., Jessberger, H.L., Makowski, E. and Jordan, P. (1981) Modelling of deep shafts in centrifuge tests. *Proceedings of the 10th International Conference on Soil Mechanics & Foundation Engineering*, **1**, 683–91.

Leung, C.F., Lee, F.H. and Tan, T.S. (eds) (1994) *Centrifuge 94*, A.A. Balkema, Rotterdam.

Lunne, T., Robertson, P.K. and Powell, J.J.M. (1997) *Cone Penetration Testing in Geotechnical Practice*. Blackie Academic and Professional, London.

Mair, R.J. and Wood, D.M. (1987) *Pressuremeter Testing: Methods and Interpretation*. Construction Industry Research & Information Association/Butterworths, London.

Marsland, A. and Randolph, M.F. (1977) Comparisons of the results from pressuremeter tests and large *in situ* plate tests in London Clay. *Géotechnique*, **27**(2), 217–43.

Meigh, A.C. (1987) *Cone Penetrometer Testing*. Construction Industry Research and Information Association/Butterworths, London.

Menzies, B.K. and Merrifield, C.M. (1980) Measurement of shear stress distribution on the edges of a shear vane blade. *Géotechnique*, **30**(3), 314–18.

Meyerhof, G.G. (1957) Discussion on research on determining the density of sands by spoon penetration testing, in *Proceedings of the 4th International Conference on Soil Mechanics & Foundation Engineering*, London, **3**, 110.

Mitchell, J.K. (1993) *Fundamentals of Soil Behavior*, 2nd edn. John Wiley, New York.

Nomoto, T., Mito, K., Imamura, S., Ueno, K. and Kusakabe, O. (1994) A miniature shield tunneling machine for a centrifuge, in *Centrifuge '94* (eds C.F. Leung, F.H. Lee and T.S. Tan), pp. 669–704, A.A. Balkema, Rotterdam.

Page, J.R.T. (1996) *Changes in lateral stress during slurry trench wall installation.* PhD dissertation, University of London (Queen Mary and Westfield College).

Peck, R.B., Hanson, W.B. and Thornburn, T.H. (1974) *Foundation Engineering*, 2nd edn. John Wiley, New York.

Philips, R. (1995) Centrifuge modelling: practical considerations, in *Geotechnical Centrifuge Technology* (ed. R.N. Taylor), pp. 34–60, Blackie, London.

Phillips, R., Guo, P.J. and Popescu, R. (eds) (2002) Physical Modelling in Geotechnics ICPMG 02. A.A. Balkema, Rotterdam.

Powers, J.P. (1992) *Construction Dewatering: New Methods and Applications*, 2nd edn. John Wiley, New York.

Powrie, W. (1985) Discussion on 5th Géotechnique symposium in print, The performance of propped and cantilevered rigid walls. *Géotechnique*, **35**(4), 546–8.

Powrie, W. (1986) *The behaviour of diaphragm walls in clay.* PhD dissertation, University of Cambridge.

Powrie, W., Chandler, R.J., Carder, D.R. and Watson, G.V.R. (1999) Back-analysis of an embedded retaining wall with a stabilizing base slab. *Proceedings of the Institution of Civil Engineers (Geotechnical Engineering)*, **137**, 75–86.

Powrie, W. and Kantartzi, C. (1993) Installation effects of diaphragm walls in clay, in *Retaining Structures* (ed. C.R.I. Clayton), pp. 37–45, Thomas Telford, London.

Powrie, W. and Kantartzi, C. (1996) Ground response during diaphragm wall installation in clay. *Géotechnique*, **46**(4), 725–39.

Powrie, W. and Li, E.S.F. (1991) Analysis of *in situ* retaining walls propped at formation level. *Proceedings of the Institution of Civil Engineers, Pt. 2*, **91**, 853–73.

Powrie, W., Richards, D.J. and Kantartzi, C. (1994) Modelling diaphragm wall installation and excavation processes, in *Centrifuge '94* (eds C.F. Leung, F.H. Lee and T.S. Tam), pp. 655–61, A.A. Balkema, Rotterdam.

Raison, C. *et al.* (2000) Eleventh Geotechnique symposium in print: Ground and soil improvement. *Géotechnique*, **50**(6), 611–748.

Richards, D.J. (1995) *Centrifuge and numerical modelling of twin propped retaining walls*. PhD dissertation, University of London (Queen Mary and Westfield College).

Richards, D.J. and Powrie, W. (1994) Finite element analysis of construction sequences for propped retaining walls. *Proceedings of the Institution of Civil Engineers, Geotechnical Engineering*, **107**, 207–16.

Richards, D.J. and Powrie, W. (1998) Centrifuge model tests on doubly propped embedded retaining walls in overconsolidated Kaolin clay. *Géotechnique*, **48**(6), 833–46.

Robertson, P.K. (1990) Soil classification using the cone penetration test. *Canadian Geotechnical Journal*, **27**(1), 151–8.

Robertson, P.K. and Campanella, R.G. (1983) *Interpretation of cone penetration tests*, parts 1 & 2. *Canadian Geotechnical Journal*, **20**, 718–45.

Rowe, P.W. (1962) The stress–dilatancy relation for static equilibrium of an assembly of particles in contact. *Proceedings of the Royal Society, London*, **A269**, 500–27.

Schofield, A.N. (1980) Cambridge geotechnical centrifuge operations: 20th Rankine lecture. *Géotechnique*, **30**(2), 129–70.

Seed, H.B., Tokimatsu, K., Harder, L.F. and Chung, R.M. (1984) *The influence of SPT procedures in soil liquefaction evaluations*, Report no UCB/EERC-84/15, Berkeley. (Reprinted in *J. ASCE Geotech. Engng Divn*, **111**, 1425–45, but without tabulated data on SPT–liquefaction correlation.)

Simpson, B. (1992) Retaining structures: displacement and design: 32nd Rankine lecture. *Géotechnique*, **42**(4), 541–76.

Skempton, A.W. (1986) Standard penetration test procedures and the effects in sands of overburden pressure, relative density, particle size, ageing and overconsolidation. *Géotechnique*, **36**(3), 425–47.

Smith, I.M. and Griffiths, D.V. (1996) *Programming the Finite Element Method*, 3rd edn. John Wiley, Chichester.

Stallebrass, S.E. (1990) Modelling small strains for analysis in geotechnical engineering. *Ground Engineering*, **22**(9), 26–9.

Stoss, K. (1976) *Die Anwendbarkeit von Bodenvereisung zur Sicherung und Abdichtung von Baugruben*, Vortrag bei der Gesellschaft für Technik und Wirtschaft, Dortmund.

Stroud, M.A. (1974) The standard penetration test in insensitive clays and soft rocks. *Proceedings of the 1st European Symposium on Penetration Testing*, **2.2**, 367–75, Stockholm.

Stroud, M.A. (1989) The standard penetration test: its application and interpretation. *Penetration Testing in the UK*, 29–49, Thomas Telford, London.

Stroud, M.A. and Butler, F.G. (1975) The standard penetration test and the engineering properties of glacial materials. *Proceedings of a Symposium on the Engineering Behaviour of Glacial Materials*, 124–35, Midland Geotechnical Society, Birmingham.

Taylor, R.N. (ed.) (1995) *Geotechnical Centrifuge Technology*, Blackie, London.

Tedd, P., Chard, B.M., Charles, J.A. and Symons, I.F. (1984) Behaviour of a propped embedded retaining wall in stiff clay at Bell Common. *Géotechnique*, **34**(4), 513–32.

Terzaghi, K. and Peck, R.B. (1967) *Soil Mechanics in Engineering Practice*, 2nd edn (1st edn 1948). John Wiley, New York.

Van Impe, W.F. (1989) *Soil Improvement Techniques and Their Evolution*, A.A. Balkema, Rotterdam.

Verruijt, A., Beringen, F.L. and Leeyn, E.H. (eds) Penetration testing. *Proceedings of the Second European Symposium on Penetration Testing*, Amsterdam, 24–27 May 1982. A.A. Balkema, P.O. Box 1675, Rotterdam, Netherlands.

Wheeler, P. (1993) Waterloo compensation. *Ground Engineering*, **26**(7), 14–16.

White, D.J., Take, W.A. and Bolton, M.D. (2003) Soil deformation measurement using particle image velocimetry (PIV) and photogrammetry. *Géotechnique*, **53**(7), 619–31.

Wood, D.M. and Wroth, C.P. (1977) Some laboratory experiments related to the results of pressuremeter tests. *Géotechnique*, **27**, 181–201.

Woods, R.I. and Clayton, C.R.I. (1993) The application of the CRISP finite element program to practical retaining wall problems, in *Retaining Structures* (ed. C.R.I. Clayton), pp. 102–11, Thomas Telford, London.

Wroth, C.P. (1984) The interpretation of *in situ* soil tests. 24th Rankine Lecture. *Géotechnique*, **34**(4), 449–89.

Zienkiewicz, O.C. (1967) *The Finite Element Method in Structural and Continuum Mechanics*. McGraw-Hill, London.

Zuidberg, H.M. (1988) Piezocone penetration testing – probe development. *Proceedings of the 2nd International Symposium on Penetration Testing, ISOPT-1*, Orlando, Specialty Session No. 13, 24 March 1988, A.A. Balkema, Rotterdam.

Zuidberg, H.M., Schaap, L.H.J. and Beringen, F.L. (1982) A penetrometer for simultaneously measuring cone resistance, sleeve friction and dynamic pore pressure. *Proceedings of the 2nd European Symposium on Penetration Testing*, Amsterdam, 963–70.

Index

Page numbers appearing in *italic* refer to tables.